RAY W

Co...ons Manual

Construction Calculations Manual

Sidney M. Levy

AMSTERDAM • BOSTON • HEIDELBERG • LONDON • NEW YORK • OXFORD
PARIS • SAN DIEGO • SAN FRANCISCO • SINGAPORE • SYDNEY • TOKYO
Butterworth-Heinemann is an imprint of Elsevier

Butterworth-Heinemann is an imprint of Elsevier
225 Wyman Street, Waltham, MA 02451, USA
The Boulevard, Langford Lane, Kidlington, Oxford OX5 1GB, UK

Copyright © 2012 Elsevier Inc. All rights reserved

No part of this publication may be reproduced, stored in a retrieval system or transmitted in any form or by any means electronic, mechanical, photocopying, recording or otherwise without the prior written permission of the publisher.

Permissions may be sought directly from Elsevier's Science & Technology Rights Department in Oxford, UK: phone (+44) (0) 1865 843830; fax (+44) (0) 1865 853333; email: permissions@elsevier.com. Alternatively you can submit your request online by visiting the Elsevier web site at http://elsevier.com/locate/permissions, and selecting Obtaining permission to use Elsevier material.

Notice
No responsibility is assumed by the publisher for any injury and/or damage to persons or property as a matter of products liability, negligence or otherwise, or from any use or operation of any methods, products, instructions or ideas contained in the material herein. Because of rapid advances in the medical sciences, in particular, independent verification of diagnoses and drug dosages should be made.

Library of Congress Cataloging-in-Publication Data
Application submitted.

British Library Cataloguing in Publication Data
A catalogue record for this book is available from the British Library

ISBN: 978-0-12-382243-7

For information on all Butterworth-Heinemann publications
visit our website at www.elsevierdirect.com

Transferred to Digital Printing in 2012

Working together to grow
libraries in developing countries

www.elsevier.com | www.bookaid.org | www.sabre.org

ELSEVIER BOOK AID International Sabre Foundation

Contents

1.	The National Institute of Standards and Testing (NIST)	1
2.	Conversion Tables and Conversion Formulas	33
3.	Calculations and Formulas—Geometry, Trigonometry, and Physics in Construction	77
4.	Site Work	155
5.	Calculations Relating to Concrete and Masonry	211
6.	Calculating the Size/Weight of Structural Steel and Miscellaneous Metals	265
7.	Lumber—Calculations to Select Framing and Trim Materials	351
8.	Fasteners for Wood and Steel—Calculations for Selection	441
9.	Calculations to Determine the Effectiveness and Control of Thermal and Sound Transmission	503
10.	Interior Finishes	545
11.	Plumbing and HVAC Calculations	589
12.	Electrical Formulas and Calculations	635

Introduction

Construction Calculations provides the construction, engineering, and project owner community with a single source guide for many of the formulas and conversion factors that are frequently encountered during the design and construction phase of a project.

The geometry and trigonometry lessons learned years ago sometimes need refreshing. *Construction Calculations* provides a refresher course on some of the formulas and concepts that tend to crop up from time to time.

A book divided into sections devoted to most of the common components of construction makes it easier to determine how to achieve a Sound Transmission Coefficient (STC) rating of 50, for example, or how to equate the amperage capacity of copper and aluminum cable of the same wire size.

A detailed index preceding each section makes it easy to locate the answer to one's question or at least points the way to its solution.

This one-source volume can prove invaluable for office- or field-based designers and contractors and will come in handy at project and design development meetings as well as provide assistance in specifying and purchasing materials and equipment.

I have selected material that in my 40-some years in the construction business appears relevant to the many situations where answers to questions are required, and required "yesterday."

I hope you will find *Construction Calculations* a worthy addition to your professional library.

Sidney M. Levy

Section 1

The National Institute of Standards and Testing (NIST)

NIST Handbook 44 - 2007 Edition	2	1. Tables of Metric Units of Measurement	14
1. Introduction	3	2. Tables of U.S. Units of Measurement	16
2. Units and Systems of Measurement	4	3. Notes on British Units of Measurement	18
3. Standards of Length, Mass, and Capacity or Volume	10	4. Tables of Units of Measurement	19
		5. Tables of Equivalents	25
4. Specialized Use of the Terms "Ton" and "Tonnage"	13		

A book on construction calculations that includes references to material dimensions, weight, volume, and conversion factors should introduce the reader to the National Institute of Standards and Testing, generally referred to simply as NIST.

Founded in 1901 under the U.S. Department of Agriculture, NIST is a nonregulatory federal agency whose mission is to "promote U.S. innovation and industrial competitiveness by advancing measurement science, standards and technology in ways that enhance economic security and improve our quality of life."

NIST maintains four cooperative programs to carry out its mission:

- NIST Laboratories, headquartered in Gaithersburg, Maryland, and a campus in Boulder, Colorado, to research and advance U.S. technology infrastructure.
- The Baldrige National Quality Program to promote excellence in the performance of manufacturing, service, educational and health care industries recognizing excellence in those organizations with its highly prized annual Malcolm Baldrige Award.
- The Hollings Manufacturing Extension Partnership consisting of a nationwide network of local centers that offer technical and business assistance to small manufacturers
- The Technology Innovations Program providing cost-shared awards to industry, academia, and key organizations that meet national and societal needs.

The NIST Handbook 44 was first published in 1949 and is issued yearly at the Annual Meeting of the National Conference on Weights and Measures.

The Table of Contents of Handbook 44 reflects the type of information contained in this volume, the contents of which hold much value for the design and construction industry. Copies of the entire handbook can be downloaded from the NIST website.

I have chosen to include only Appendix B: Units and Systems of Measurement—Their Origin, Development, and Present Status, and Appendix C: General Tables of Units of Measurement, which seem to be a fitting start to a book on construction calculations.

NIST Handbook 44 - 2007 Edition

Specifications, Tolerances, and Other Technical Requirements
for Weighing and Measuring Devices

as adopted by the 91st National Conference on Weights and Measures 2006

Table of Contents	W	PDF
Full Document in PDF File Format		•
Cover		•
Title Page	•	•
Foreword	•	•
Committee Members	•	•
Table of Contents	•	•
2006 Amendments		•
2006 Editorial Changes	•	•
Introduction	•	•
Section 1		
1.10	General Code	•	•
Section 2		
2.20	Scales	•	•
2.21	Belt-Conveyor Scale Systems	•	•
2.22	Automatic Bulk Weighing Systems	•	•
2.23	Weights	•	•
2.24	Automatic Weighing Systems	•	•
Section 3		
3.30	Liquid-Measuring Devices	•	•
3.31	Vehicle-Tank Meters	•	•
3.32	Liquefied Petroleum Gas and Anhydrous Ammonia Liquid-Measuring Devices	•	•
3.33	Hydrocarbon Gas Vapor-Measuring Devices	•	•
3.34	Cryogenic Liquid-Measuring Devices	•	•
3.35	Milk Meters	•	•
3.36	Water Meters	•	•
3.37	Mass Flow Meters	•	•
3.38	Carbon Dioxide Liquid-Measuring Devices	•	•
Section 4		
4.40	Vehicle Tanks Used as Measures	•	•
4.41	Liquid Measures	•	•
4.42	Farm Milk Tanks	•	•
4.43	Measure-Containers	•	•
4.44	Graduates	•	•
4.45	Dry Measures	•	•
4.46	Berry Baskets and Boxes	•	•

Section 5				
5.50	Fabric-Measuring Devices		•	•
5.51	Wire- and Cordage-Measuring Devices		•	•
5.52	Linear Measure		•	•
5.53	Odometers		•	•
5.54	Taximeters		•	•
5.55	Timing Devices		•	•
5.56.(a)	Grain Moisture Meters		•	•
5.56.(b)	Grain Moisture Meters		•	•
5.57	Near-Infrared Grain Analyzers		•	•
5.58	Multiple Dimension Measuring Devices		•	•
5.59	Electronic Livestock, Meat, and Poultry Evaluation Systems and/or Devices – Tentative Code		•	•
Appendices				
A	Fundamental Considerations		•	•
B	Units and Systems of Measurement		•	•
C	General Tables of Units of Measurement.		•	•
D	Definitions		•	•
Summary of HB-44 Requirements Becoming Effective in 2007			•	•

APPENDIX B: UNITS AND SYSTEMS OF MEASUREMENT: THEIR ORIGIN, DEVELOPMENT, AND PRESENT STATUS

1. Introduction

The National Institute of Standards and Technology (NIST) (formerly the National Bureau of Standards) was established by Act of Congress in 1901 to serve as a national scientific laboratory in the physical sciences, and to provide fundamental measurement standards for science and industry. In carrying out these related functions, the Institute conducts research and development in many fields of physics, mathematics, chemistry, and engineering. At the time of its founding, the Institute had custody of two primary standards—the meter bar for length and the kilogram cylinder for mass. With the phenomenal growth of science and technology over the past century, the Institute has become a major research institution concerned not only with everyday weights and measures, but also with hundreds of other scientific and engineering standards that are necessary to the industrial progress of the nation. Nevertheless, the country still looks to NIST for information on the units of measurement, particularly their definitions and equivalents.

The subject of measurement systems and units can be treated from several different standpoints. Scientists and engineers are interested in the methods by which precision measurements are made. State weights and measures officials are concerned with laws and regulations that assure equity in the marketplace, protect public health and safety, and with methods for verifying commercial weighing and measuring devices. But a vastly larger group of people is interested in some general knowledge of the origin and development of measurement systems, of the present status of units and standards, and of miscellaneous facts that will be useful in everyday life.

This material has been prepared to supply that information on measurement systems and units that experience has shown to be the common subject of inquiry.

2. Units and Systems of Measurement

The expression "weights and measures" is often used to refer to measurements of length, mass, and capacity or volume, thus excluding such quantities as electrical and time measurements and thermometry. This section on units and measurement systems presents some fundamental information to clarify the concepts of this subject and to eliminate erroneous and misleading use of terms.

It is essential that the distinction between the terms "units" and "standards" be established and kept in mind.

A <u>unit</u> is a special quantity in terms of which other quantities are expressed. In general, a unit is fixed by definition and is independent of such physical conditions as temperature. Examples: the meter, the liter, the gram, the yard, the pound, the gallon.

A <u>standard</u> is a physical realization or representation of a unit. In general, it is not entirely independent of physical conditions, and it is a representation of the unit only under specified conditions. For example, a meter standard has a length of one meter when at some definite temperature and supported in a certain manner. If supported in a different manner, it might have to be at a different temperature to have a length of one meter.

2.1 Origin and Early History of Units and Standards

2.1.1 General Survey of Early History of Measurement Systems

Weights and measures were among the earliest tools invented by man. Primitive societies needed rudimentary measures for many tasks: constructing dwellings of an appropriate size and shape, fashioning clothing, or bartering food or raw materials.

Man understandably turned first to parts of the body and the natural surroundings for measuring instruments. Early Babylonian and Egyptian records and the Bible indicate that length was first measured with the forearm, hand, or finger and that time was measured by the periods of the sun, moon, and other heavenly bodies. When it was necessary to compare the capacities of containers such as gourds or clay or metal vessels, they were filled with plant seeds which were then counted to measure the volumes. When means for weighing were invented, seeds and stones served as standards. For instance, the "carat," still used as a unit for gems, was derived from the carob seed.

Our present knowledge of early weights and measures comes from many sources. Archaeologists have recovered some rather early standards and preserved them in museums. The comparison of the dimensions of buildings with the descriptions of contemporary writers is another source of information. An interesting example of this is the comparison of the dimensions of the Greek Parthenon with the description given by Plutarch from which a fairly accurate idea of the size of the Attic foot is obtained. In some cases, we have only plausible theories and we must sometimes select the interpretation to be given to the evidence.

For example, does the fact that the length of the double-cubit of early Babylonia was equal (within two parts per thousand) to the length of the seconds pendulum at Babylon suggest a scientific knowledge of the pendulum at a very early date, or do we merely have a curious coincidence? By studying the evidence given by all available sources, and by correlating the relevant facts, we obtain some idea of the origin and development of the units. We find that they have changed more or less gradually with the passing of time in a complex manner because of a great variety of modifying influences. We find the units modified and grouped into measurement systems: The Babylonian system, the Egyptian system, the Phileterian system of the Ptolemaic age, the Olympic system of Greece, the Roman system, and the British system, to mention only a few.

2.1.2 Origin and Development of Some Common Customary Units

The origin and development of units of measurement has been investigated in considerable detail and a number of books have been written on the subject. It is only possible to give here, somewhat sketchily, the story about a few units.

Units of length: The cubit was the first recorded unit used by ancient peoples to measure length. There were several cubits of different magnitudes that were used. The common cubit was the length of the forearm from the elbow to the tip of the middle finger. It was divided into the span of the hand (one-half cubit), the palm or width of the hand (one sixth), and the digit or width of a finger (one twenty-fourth). The Royal or Sacred Cubit, which was 7 palms or 28 digits long, was used in constructing buildings and monuments and in surveying. The inch, foot, and yard evolved from these units through a complicated transformation not yet fully understood. Some believe they evolved from cubic measures; others believe they were simple proportions or multiples of the cubit. In any case, the Greeks and Romans inherited the foot from the Egyptians. The Roman foot was divided into both 12 unciae (inches) and 16 digits. The Romans also introduced the mile of 1000 paces or double steps, the pace being equal to five Roman feet. The Roman mile of 5000 feet was introduced into England during the occupation. Queen Elizabeth, who reigned from 1558 to 1603, changed, by statute, the mile to 5280 feet or 8 furlongs, a furlong being 40 rods of 5½ yards each.

The introduction of the yard as a unit of length came later, but its origin is not definitely known. Some believe the origin was the double cubit, others believe that it originated from cubic measure. Whatever its origin, the early yard was divided by the binary method into 2, 4, 8, and 16 parts called the half-yard, span, finger, and nail. The association of the yard with the "gird" or circumference of a person's waist or with the distance from the tip of the nose to the end of the thumb of Henry I are probably standardizing actions, since several yards were in use in Great Britain.

The point, which is a unit for measuring print type, is recent. It originated with Pierre Simon Fournier in 1737. It was modified and developed by the Didot brothers, Francois Ambroise and Pierre Francois, in 1755. The point was first used in the United States in 1878 by a Chicago type foundry (Marder, Luse, and Company). Since 1886, a point has been exactly 0.351 459 8 millimeters, or about 1/72 inch.

Units of mass: The grain was the earliest unit of mass and is the smallest unit in the apothecary, avoirdupois, Tower, and Troy systems. The early unit was a grain of wheat or barleycorn used to weigh the precious metals silver and gold. Larger units preserved in stone standards were developed that were used as both units of mass and of monetary currency. The pound was derived from the mina used by ancient civilizations. A smaller unit was the shekel, and a larger unit was the talent. The magnitude of these units varied from place to place. The Babylonians and Sumerians had a system in which there were 60 shekels in a mina and 60 minas in a talent. The Roman talent consisted of 100 libra (pound) which were smaller in magnitude than the mina. The Troy pound used in England and the United States for monetary purposes, like the Roman pound, was divided into 12 ounces, but the Roman uncia (ounce) was smaller. The carat is a unit for measuring gemstones that had its origin in the carob seed, which later was standardized at 1/144 ounce and then 0.2 gram.

Goods of commerce were originally traded by number or volume. When weighing of goods began, units of mass based on a volume of grain or water were developed. For example, the talent in some places was approximately equal to the mass of one cubic foot of water. Was this a coincidence or by design? The diverse magnitudes of units having the same name, which still appear today in our dry and liquid measures, could have arisen from the various commodities traded. The larger avoirdupois pound for goods of commerce might have been based on volume of water which has a higher bulk density than grain. For example, the Egyptian hon was a volume unit about 11% larger than a cubic palm and corresponded to one mina of water. It was almost identical in volume to the present U.S. pint.

The stone, quarter, hundredweight, and ton were larger units of mass used in Great Britain. Today only the stone continues in customary use for measuring personal body weight. The present stone is 14 pounds, but an earlier unit appears to have been 16 pounds. The other units were multiples of 2, 8, and 160 times the stone, or 28,

112, and 2240 pounds, respectively. The hundredweight was approximately equal to two talents. In the United States the ton of 2240 pounds is called the "long ton." The "short ton" is equal to 2000 pounds.

Units of time and angle: We can trace the division of the circle into 360 degrees and the day into hours, minutes, and seconds to the Babylonians who had a sexagesimal system of numbers. The 360 degrees may have been related to a year of 360 days.

2.2 The Metric System

2.2.1 Definition, Origin, and Development

Metric systems of units have evolved since the adoption of the first well defined system in France in 1791. During this evolution the use of these systems spread throughout the world, first to the non-English speaking countries, and more recently to the English speaking countries. The first metric system was based on the centimeter, gram, and second (cgs), and these units were particularly convenient in science and technology. Later metric systems were based on the meter, kilogram, and second (mks) to improve the value of the units for practical applications. The present metric system is the International System of Units (SI). It is also based on the meter, kilogram, and second as well as additional base units for temperature, electric current, luminous intensity, and amount of substance. The International System of Units is referred to as the modern metric system.

The adoption of the metric system in France was slow, but its desirability as an international system was recognized by geodesists and others. On May 20, 1875, an international treaty known as the International Metric Convention or the Treaty of the Meter was signed by seventeen countries including the United States. This treaty established the following organizations to conduct international activities relating to a uniform system for measurements:

(1) The General Conference on Weights and Measures (French initials: CGPM), an intergovernmental conference of official delegates of member nations and the supreme authority for all actions;
(2) The International Committee of Weights and Measures (French initials: CIPM), consisting of selected scientists and metrologists, which prepares and executes the decisions of the CGPM and is responsible for the supervision of the International Bureau of Weights and Measures;
(3) The International Bureau of Weights and Measures (French initials: BIPM), a permanent laboratory and world center of scientific metrology, the activities of which include the establishment of the basic standards and scales of the principal physical quantities and maintenance of the international prototype standards.

The National Institute of Standards and Technology provides official United States representation in these organizations. The CGPM, the CIPM, and the BIPM have been major factors in the continuing refinement of the metric system on a scientific basis and in the evolution of the International System of Units.

Multiples and submultiples of metric units are related by powers of ten. This relationship is compatible with the decimal system of numbers, and it contributes greatly to the convenience of metric units.

2.2.2 International System of Units

At the end of World War II, a number of different systems of measurement still existed throughout the world. Some of these systems were variations of the metric system, and others were based on the customary inch-pound system of the English-speaking countries. It was recognized that additional steps were needed to promote a worldwide measurement system. As a result the 9th GCPM, in 1948, asked the ICPM to conduct an international study of the measurement needs of the scientific, technical, and educational communities. Based on the findings of this study, the 10th General Conference in 1954 decided that an international system should be derived from six base units to provide for the measurement of temperature and optical radiation in addition to mechanical and electromagnetic quantities. The six base units recommended were the meter, kilogram, second, ampere, Kelvin degree (later renamed the kelvin), and the candela.

In 1960, the 11th General Conference of Weights and Measures named the system based on the six base quantities of the International System of Units, abbreviated SI from the French name: Le Système International d'Unités. The SI metric system is now either obligatory or permissible throughout the world.

2.2.3 Units and Standards of the Metric System

In the early metric system there were two fundamental or base units, the meter and the kilogram, for length and mass. The other units of length and mass, and all units of area, volume, and compound units such as density were derived from these two fundamental units.

The meter was originally intended to be one ten-millionth part of a meridional quadrant of the earth. The Meter of the Archives, the platinum length standard which was the standard for most of the 19th century, at first was supposed to be exactly this fractional part of the quadrant. More refined measurements over the earth's surface showed that this supposition was not correct. In 1889, a new international metric standard of length, the International Prototype Meter, a graduated line standard of platinum-iridium, was selected from a group of bars because precise measurements found it to have the same length as the Meter of the Archives. The meter was then defined as the distance, under specified conditions, between the lines on the International Prototype Meter without reference to any measurements of the earth or to the Meter of the Archives, which it superseded. Advances in science and technology have made it possible to improve the definition of the meter and reduce the uncertainties associated with artifacts. From 1960 to 1983, the meter was defined as the length equal to 1 650 763.73 wavelengths in a vacuum of the radiation corresponding to the transition between the specified energy levels of the krypton 86 atom. Since 1983 the meter has been defined as the length of the path traveled by light in a vacuum during an interval of $1/299792458$ of a second.

The kilogram, originally defined as the mass of one cubic decimeter of water at the temperature of maximum density, was known as the Kilogram of the Archives. It was replaced after the International Metric Convention in 1875 by the International Prototype Kilogram which became the unit of mass without reference to the mass of a cubic decimeter of water or to the Kilogram of the Archives. Each country that subscribed to the International Metric Convention was assigned one or more copies of the international standards; these are known as National Prototype Meters and Kilograms.

The liter is a unit of capacity or volume. In 1964, the 12th GCPM redefined the liter as being one cubic decimeter. By its previous definition—the volume occupied, under standard conditions, by a quantity of pure water having a mass of one kilogram—the liter was larger than the cubic decimeter by 28 parts per 1 000 000. Except for determinations of high precision, this difference is so small as to be of no consequence.

The modern metric system (SI) includes two classes of units:

> base units for length, mass, time, temperature, electric current, luminous intensity, and amount of substance; and
> derived units for all other quantities (e.g., work, force, power) expressed in terms of the seven base units.

For details, see NIST Special Publication 330 (2001), The International System of Units (SI) and NIST Special Publication 811 (1995), Guide for the Use of the International System of Units.

2.2.4 International Bureau of Weights and Measures

The International Bureau of Weights and Measures (BIPM) was established at Sèvres, a suburb of Paris, France, by the International Metric Convention of May 20, 1875. The BIPM maintains the International Prototype Kilogram, many secondary standards, and equipment for comparing standards and making precision measurements. The Bureau, funded by assessment of the signatory governments, is truly international. In recent years the scope of the work at the Bureau has been considerably broadened. It now carries on researches in the fields of electricity, photometry and radiometry, ionizing radiations, and time and frequency besides its work in mass, length, and thermometry.

2.2.5 Status of the Metric System in the United States

The use of the metric system in this country was legalized by Act of Congress in 1866, but was not made obligatory then or since.

Following the signing of the Convention of the Meter in 1875, the United States acquired national prototype standards for the meter and the kilogram. U.S. Prototype Kilogram No. 20 continues to be the primary standard for mass in the United States. It is recalibrated from time to time at the BIPM. The prototype meter has been replaced by modern stabilized lasers following the most recent definition of the meter.

From 1893 until 1959, the yard was defined as equal exactly to $3600/3937$ meter. In 1959, a small change was made in the definition of the yard to resolve discrepancies both in this country and abroad. Since 1959, we define the yard as equal exactly to 0.9144 meter; the new yard is shorter than the old yard by exactly two parts in a million. At the same time, it was decided that any data expressed in feet derived from geodetic surveys within the United States would continue to bear the relationship as defined in 1893 (one foot equals $1200/3937$ meter). We call this foot the U.S. Survey Foot, while the foot defined in 1959 is called the International Foot. Measurements expressed in U.S. statute miles, survey feet, rods, chains, links, or the squares thereof, and acres should be converted to the corresponding metric values by using pre-1959 conversion factors if more than five significant figure accuracy is required.

Since 1970, actions have been taken to encourage the use of metric units of measurement in the United States. A brief summary of actions by Congress is provided below as reported in the Federal Register Notice dated July 28, 1998.

Section 403 of Public Law 93-380, the Education Amendment of 1974, states that it is the policy of the United States to encourage educational agencies and institutions to prepare students to use the metric system of measurement as part of the regular education program. Under both this act and the Metric Conversion Act of 1975, the "metric system of measurement" is defined as the International System of Units as established in 1960 by the General Conference on Weights and Measures and interpreted or modified for the United States by the Secretary of Commerce (Sec. 4(4)—Pub. L. 94-168; Sec. 403(a)(3)—Pub. L. 93-380). The Secretary has delegated authority under these subsections to the Director of the National Institute of Standards and Technology.

Section 5164 of Public Law 100-418, the Omnibus Trade and Competitiveness Act of 1988, amends Public Law 94-168, The Metric Conversion Act of 1975. In particular, Section 3 Metric Conversion Act is amended to read as follows:

"Sec. 3. It is therefore the declared policy of the United States–

(1) to designate the metric system of measurement as the preferred system of weights and measures for United States trade and commerce;
(2) to require that each federal agency, by a date certain and to the extent economically feasible by the end of the fiscal year 1992, use the metric system of measurement in its procurements, grants, and other business-related activities, except to the extent that such use is impractical or is likely to cause significant inefficiencies or loss of markets to U.S. firms, such as when foreign competitors are producing competing products in non-metric units;
(3) to seek ways to increase understanding of the metric system of measurement through educational information and guidance and in government publications; and
(4) to permit the continued use of traditional systems of weights and measures in nonbusiness activities."

The Code of Federal Regulations makes the use of metric units mandatory for agencies of the federal government. (Federal Register, Vol. 56, No. 23, Page 160, January 2, 1991.)

2.3 British and United States Systems of Measurement

In the past, the customary system of weights and measures in the British Commonwealth countries and that in the United States were very similar; however, the SI metric system is now the official system of units in the United Kingdom, while the customary units are still predominantly used in the United States. Because references to the units of the old British customary system are still found, the following discussion describes the differences between the U.S. and British customary systems of units.

After 1959, the U.S. and the British inches were defined identically for scientific work and were identical in commercial usage. A similar situation existed for the U.S. and the British pounds, and many relationships, such as 12 inches = 1 foot, 3 feet = 1 yard, and 1760 yards = 1 international mile, were the same in both countries; but there were some very important differences.

In the first place, the U.S. customary bushel and the U.S. gallon, and their subdivisions differed from the corresponding British Imperial units. Also the British ton is 2240 pounds, whereas the ton generally used in the United States is the short ton of 2000 pounds. The American colonists adopted the English wine gallon of 231 cubic inches. The English of that period used this wine gallon and they also had another gallon, the ale gallon of 282 cubic inches. In 1824, the British abandoned these two gallons when they adopted the British Imperial gallon, which they defined as the volume of 10 pounds of water, at a temperature of 62 °F, which, by calculation, is equivalent to 277.42 cubic inches. At the same time, they redefined the bushel as 8 gallons.

In the customary British system, the units of dry measure are the same as those of liquid measure. In the United States these two are not the same; the gallon and its subdivisions are used in the measurement of liquids and the bushel, with its subdivisions, is used in the measurement of certain dry commodities. The U.S. gallon is divided into four liquid quarts and the U.S. bushel into 32 dry quarts. All the units of capacity or volume mentioned thus far are larger in the customary British system than in the U.S. system. But the British fluid ounce is smaller than the U.S. fluid ounce, because the British quart is divided into 40 fluid ounces whereas the U.S. quart is divided into 32 fluid ounces.

From this we see that in the customary British system an avoirdupois ounce of water at 62 °F has a volume of one fluid ounce, because 10 pounds is equivalent to 160 avoirdupois ounces, and 1 gallon is equivalent to 4 quarts, or 160 fluid ounces. This convenient relation does not exist in the U.S. system because a U.S. gallon of water at 62 °F weighs about $8\frac{1}{3}$ pounds, or $133\frac{1}{3}$ avoirdupois ounces, and the U.S. gallon is equivalent to 4×32, or 128 fluid ounces.

1 U.S. fluid ounce	= 1.041 British fluid ounces
1 British fluid ounce	= 0.961 U.S. fluid ounce
1 U.S. gallon	= 0.833 British Imperial gallon
1 British Imperial gallon	= 1.201 U.S. gallons

Among other differences between the customary British and the United States measurement systems, we should note that they abolished the use of the troy pound in England on January 6, 1879; they retained only the troy ounce and its subdivisions, whereas the troy pound is still legal in the United States, although it is not now greatly used. We can mention again the common use, for body weight, in England of the stone of 14 pounds, this being a unit now unused in the United States, although its influence was shown in the practice until World War II of selling flour by the barrel of 196 pounds (14 stone). In the apothecary system of liquid measure the British add a unit, the fluid scruple, equal to one third of a fluid drachm (spelled dram in the United States) between their minim and their fluid drachm. In the United States, the general practice now is to sell dry commodities, such as fruits and vegetables, by their mass.

2.4 Subdivision of Units

In general, units are subdivided by one of three methods: (a) decimal, into tenths; (b) duodecimal, into twelfths; or (c) binary, into halves (twos). Usually the subdivision is continued by using the same method. Each method has its advantages for certain purposes, and it cannot properly be said that any one method is "best" unless the use to which the unit and its subdivisions are to be put is known.

For example, if we are concerned only with measurements of length to moderate precision, it is convenient to measure and to express these lengths in feet, inches, and binary fractions of an inch, thus 9 feet, 4⅜ inches. However, if these lengths are to be subsequently used to calculate area or volume, that method of subdivision at once becomes extremely inconvenient. For that reason, civil engineers, who are concerned with areas of land, volumes of cuts, fills, excavations, etc., instead of dividing the foot into inches and binary subdivisions of the inch, divide it decimally; that is, into tenths, hundredths, and thousandths of a foot.

The method of subdivision of a unit is thus largely made based on convenience to the user. The fact that units have commonly been subdivided into certain subunits for centuries does not preclude their also having another mode of subdivision in some frequently used cases where convenience indicates the value of such other method. Thus, while we usually subdivide the gallon into quarts and pints, most gasoline-measuring pumps, of the price-computing type, are graduated to show tenths, hundredths, or thousandths of a gallon.

Although the mile has for centuries been divided into rods, yards, feet, and inches, the odometer part of an automobile speedometer shows tenths of a mile. Although we divide our dollar into 100 parts, we habitually use and speak of halves and quarters. An illustration of rather complex subdividing is found on the scales used by draftsmen. These scales are of two types: (a) architects, which are commonly graduated with scales in which $\frac{3}{32}$, $\frac{3}{16}$, $\frac{1}{8}$, $\frac{1}{4}$, $\frac{3}{8}$, $\frac{1}{2}$, $\frac{3}{4}$, 1, 1$\frac{1}{12}$, and 3 inches, respectively, represent 1 foot full scale, and also having a scale graduated in the usual manner to $\frac{1}{16}$ inch; and (b) engineers, which are commonly subdivided to 10, 20, 30, 40, 50, and 60 parts to the inch.

The dictum of convenience applies not only to subdivisions of a unit but also to multiples of a unit. Land elevations above sea level are given in feet although the height may be several miles; the height of aircraft above sea level as given by an altimeter is likewise given in feet, no matter how high it may be.

On the other hand, machinists, toolmakers, gauge makers, scientists, and others who are engaged in precision measurements of relatively small distances, even though concerned with measurements of length only, find it convenient to use the inch, instead of the tenth of a foot, but to divide the inch decimally to tenths, hundredths, thousandths, etc., even down to millionths of an inch. Verniers, micrometers, and other precision measuring instruments are usually graduated in this manner. Machinist scales are commonly graduated decimally along one edge and are also graduated along another edge to binary fractions as small as 1/64 inch. The scales with binary fractions are used only for relatively rough measurements.

It is seldom convenient or advisable to use binary subdivisions of the inch that are smaller than $\frac{1}{64}$. In fact, $\frac{1}{32}$-, $\frac{1}{16}$-, or $\frac{1}{8}$-inch subdivisions are usually preferable for use on a scale to be read with the unaided eye.

2.5 Arithmetical Systems of Numbers

The subdivision of units of measurement is closely associated with arithmetical systems of numbers. The systems of units used in this country for commercial and scientific work, having many origins as has already been shown, naturally show traces of the various number systems associated with their origins and developments. Thus, (a) the binary subdivision has come down to us from the Hindus, (b) the duodecimal system of fractions from the Romans, (c) the decimal system from the Chinese and Egyptians, some developments having been made by the Hindus, and (d) the sexagesimal system (division by 60) now illustrated in the subdivision of units of angle and of time, from the ancient Babylonians. The use of decimal numbers in measurements is becoming the standard practice.

3. Standards of Length, Mass, and Capacity or Volume

3.1 Standards of Length

The meter, which is defined in terms of the speed of light in a vacuum, is the unit on which all length measurements are based.

The yard is defined[1] as follows:

    1 yard = 0.914 4 meter

and the inch is exactly equal to 25.4 millimeters.

3.1.1 Calibration of Length Standards

NIST calibrates standards of length including meter bars, yard bars, miscellaneous precision line standards, steel tapes, invar geodetic tapes, precision gauge blocks, micrometers, and limit gauges. It also measures the linear dimensions of miscellaneous apparatus such as penetration needles, cement sieves, and hemacytometer chambers. In general, NIST accepts for calibration only apparatus of such material, design, and construction as to ensure accuracy and permanence sufficient to justify calibration by the Institute. NIST performs calibrations in accordance with fee schedules, copies of which may be obtained from NIST.

    NIST does not calibrate carpenters' rules, machinist scales, draftsman scales, and the like. Such apparatus, if they require calibration, should be submitted to state or local weights and measures officials.

3.2 Standards of Mass

The primary standard of mass for this country is United States Prototype Kilogram 20, which is a platinum-iridium cylinder kept at NIST. We know the value of this mass standard in terms of the International Prototype Kilogram, a platinum-iridium standard which is kept at the International Bureau of Weights and Measures.

    In Colonial Times the British standards were considered the primary standards of the United States. Later, the U.S. avoirdupois pound was defined in terms of the Troy Pound of the Mint, which is a brass standard kept at the United States Mint in Philadelphia. In 1911, the Troy Pound of the Mint was superseded, for coinage purposes, by the Troy Pound of the Institute.

    The avoirdupois pound is defined in terms of the kilogram by the relation:

    1 avoirdupois pound = 0.453 592 37 kilogram.[2]

    These changes in definition have not made any appreciable change in the value of the pound.

    The grain is $\frac{1}{7000}$ of the avoirdupois pound and is identical in the avoirdupois, troy, and apothecary systems. The troy ounce and the apothecary ounce differ from the avoirdupois ounce but are equal to each other, and equal to 480 grains. The avoirdupois ounce is equal to 437.5 grains.

3.2.1 Mass and Weight

The mass of a body is a measure of its inertial property or how much matter it contains. The weight of a body is a measure of the force exerted on it by gravity or the force needed to support it. Gravity on earth gives a body a downward acceleration of about 9.8 m/s^2. (In common parlance, weight is often used as a synonym for mass as in weights and measures.) The incorrect use of weight in place of mass should be phased out, and the term mass used when mass is meant.

    Standards of mass are ordinarily calibrated by comparison to a reference standard of mass. If two objects are compared on a balance and give the same balance indication, they have the same "mass" (excluding the effect of air buoyancy). The forces of gravity on the two objects are balanced. Even though the value of the acceleration of gravity, g, is different from location to location, because the two objects of equal mass in the same location (where both masses are acted upon by the same g) will be affected in the same manner and by the same amount by any change in the value of g, the two objects will balance each other under any value of g.

---

1. See Federal Register for July 1, 1959. See also next to last paragraph of 2.2.5.
2. See Federal Register for July 1, 1959.

However, on a spring balance the mass of a body is not balanced against the mass of another body. Instead, the gravitational force on the body is balanced by the restoring force of a spring. Therefore, if a very sensitive spring balance is used, the indicated mass of the body would be found to change if the spring balance and the body were moved from one locality to another locality with a different acceleration of gravity. But a spring balance is usually used in one locality and is adjusted or calibrated to indicate mass at that locality.

3.2.2 Effect of Air Buoyancy

Another point that must be taken into account in the calibration and use of standards of mass is the buoyancy or lifting effect of the air. A body immersed in any fluid is buoyed up by a force equal to the force of gravity on the displaced fluid. Two bodies of equal mass, if placed one on each pan of an equal-arm balance, will balance each other in a vacuum. A comparison in a vacuum against a known mass standard gives "true mass." If compared in air, however, they will not balance each other unless they are of equal volume. If of unequal volume, the larger body will displace the greater volume of air and will be buoyed up by a greater force than will the smaller body, and the larger body will appear to be of less mass than the smaller body.

The greater the difference in volume, and the greater the density of the air in which we make the comparison weighing, the greater will be the apparent difference in mass. For that reason, in assigning a precise numerical value of mass to a standard, it is necessary to base this value on definite values for the air density and the density of the mass standard of reference.

The apparent mass of an object is equal to the mass of just enough reference material of a specified density (at 20 °C) that will produce a balance reading equal to that produced by the object if the measurements are done in air with a density of 1.2 mg/cm^3 at 20 °C. The original basis for reporting apparent mass is apparent mass versus brass. The apparent mass versus a density of 8.0 g/cm^3 is the more recent definition, and is used extensively throughout the world. The use of apparent mass versus 8.0 g/cm^3 is encouraged over apparent mass versus brass. The difference in these apparent mass systems is insignificant in most commercial weighing applications.

A full discussion of this topic is given in NIST Monograph 133, Mass and Mass Values, by Paul E. Pontius [for sale by the National Technical Information Service, 5285 Port Royal Road, Springfield, VA 22161 (COM 7450309).]

3.2.3 Calibrations of Standards of Mass

Standards of mass regularly used in ordinary trade should be tested by state or local weights and measures officials. NIST calibrates mass standards submitted, but it does not manufacture or sell them. Information regarding the mass calibration service of NIST and the regulations governing the submission of standards of mass to NIST for calibration are contained in NIST Special Publication 250, Calibration and Related Measurement Services of NIST, latest edition.

3.3 Standards of Capacity

Units of capacity or volume, being derived units, are in this country defined in terms of linear units. Laboratory standards have been constructed and are maintained at NIST. These have validity only by calibration with reference either directly or indirectly to the linear standards. Similarly, NIST has made and distributed standards of capacity to the several states. Other standards of capacity have been verified by calibration for a variety of uses in science, technology, and commerce.

3.3.1 Calibrations of Standards of Capacity

NIST makes calibrations on capacity or volume standards that are in the customary units of trade; that is, the gallon, its multiples, and submultiples, or in metric units. Further, NIST calibrates precision-grade volumetric glassware which is normally in metric units. NIST makes calibrations in accordance with fee schedules, copies of which may be obtained from NIST.

3.4 Maintenance and Preservation of Fundamental Standard of Mass

It is a statutory responsibility of NIST to maintain and preserve the national standard of mass at NIST and to realize all the other base units. The U.S. Prototype Kilogram maintained at NIST is fully protected by an alarm system. All measurements made with this standard are conducted in special air-conditioned laboratories to which the standard is taken a sufficiently long time before the observations to ensure that the standard will be in a state of equilibrium under standard conditions when the measurements or comparisons are made. Hence, it is not necessary to maintain the standard at standard conditions, but care is taken to prevent large changes of temperature. More important is the care to prevent any damage to the standard because of careless handling.

4. Specialized Use of the Terms "Ton" and "Tonnage"

As weighing and measuring are important factors in our everyday lives, it is quite natural that questions arise about the use of various units and terms and about the magnitude of quantities involved. For example, the words "ton" and "tonnage" are used in widely different senses, and a great deal of confusion has arisen regarding the application of these terms.

The ton is used as a unit of measure in two distinct senses: (1) as a unit of mass, and (2) as a unit of capacity or volume. In the first sense, the term has the following meanings:

(a) The short, or net ton of 2000 pounds.
(b) The long, gross, or shipper's ton of 2240 pounds.
(c) The metric ton of 1000 kilograms, or 2204.6 pounds.

In the second sense (capacity), it is usually restricted to uses relating to ships and has the following meaning:

(a) The register ton of 100 cubic feet.
(b) The measurement ton of 40 cubic feet.
(c) The English water ton of 224 British Imperial gallons.

In the United States and Canada the ton (mass) most commonly used is the short ton. In Great Britain, it is the long ton, and in countries using the metric system, it is the metric ton. The register ton and the measurement ton are capacity or volume units used in expressing the tonnage of ships. The English water ton is used, chiefly in Great Britain, in statistics dealing with petroleum products.

There have been many other uses of the term ton such as the timber ton of 40 cubic feet and the wheat ton of 20 bushels, but their uses have been local and the meanings have not been consistent from one place to another.

Properly, the word "tonnage" is used as a noun only in respect to the capacity or volume and dimensions of ships, and to the amount of the ship's cargo. There are two distinct kinds of tonnage; namely, vessel tonnage and cargo tonnage and each of these is used in various meanings. The several kinds of vessel tonnage are as follows:

Gross tonnage, or gross register tonnage, is the total cubical capacity or volume of a ship expressed in register tons of 100 cubic feet, or 2.83 cubic meters, less such space as hatchways, bakeries, galleys, etc., as are exempted from measurement by different governments. There is some lack of uniformity in the gross tonnages as given by different nations due to lack of agreement on the spaces that are to be exempted. Official merchant marine statistics of most countries are published in terms of the gross register tonnage. Press references to ship tonnage are usually to the gross tonnage.

The net tonnage, or net register tonnage, is the gross tonnage less the different spaces specified by maritime nations in their measurement rules and laws. The spaces deducted are those totally unavailable for carrying cargo, such as the engine room, coal bunkers, crew quarters, chart and instrument room, etc. The net tonnage is used in computing how much cargo that can be loaded on a ship. It is used as the basis for wharfage and other similar charges.

The register under-deck tonnage is the cubical capacity of a ship under her tonnage deck expressed in register tons. In a vessel having more than one deck, the tonnage deck is the second from the keel.

There are several variations of displacement tonnage.

The dead weight tonnage is the difference between the "loaded" and "light" displacement tonnages of a vessel. It is expressed in terms of the long ton of 2240 pounds, or the metric ton of 2204.6 pounds, and is the weight of fuel, passengers, and cargo that a vessel can carry when loaded to its maximum draft.

The second variety of tonnage, cargo tonnage, refers to the weight of the particular items making up the cargo. In overseas traffic it is usually expressed in long tons of 2240 pounds or metric tons of 2204.6 pounds. The short ton is only occasionally used. Therefore, the cargo tonnage is very distinct from vessel tonnage.

APPENDIX C: GENERAL TABLES OF UNITS OF MEASUREMENT

These tables have been prepared for the benefit of those requiring tables of units for occasional ready reference. In Section 4 of this Appendix, the tables are carried out to a large number of decimal places and exact values are indicated by underlining. In most of the other tables, only a limited number of decimal places are given, therefore making the tables better adopted to the average user.

1. Tables of Metric Units of Measurement

In the metric system of measurement, designations of multiples and subdivisions of any unit may be arrived at by combining with the name of the unit the prefixes deka, hecto, and kilo meaning, respectively, 10, 100, and 1000, and deci, centi, and milli, meaning, respectively, one-tenth, one-hundredth, and one-thousandth. In some of the following metric tables, some such multiples and subdivisions have not been included for the reason that these have little, if any currency in actual usage.

In certain cases, particularly in scientific usage, it becomes convenient to provide for multiples larger than 1000 and for subdivisions smaller than one-thousandth. Accordingly, the following prefixes have been introduced and these are now generally recognized:

yotta,	(Y),	meaning 10^{24}	deci,	(d),	meaning 10^{-1}
zetta,	(Z),	meaning 10^{21}	centi,	(c),	meaning 10^{-2}
exa,	(E),	meaning 10^{18}	milli,	(m),	meaning 10^{-3}
peta,	(P),	meaning 10^{15}	micro,	(μ),	meaning 10^{-6}
tera,	(T),	meaning 10^{12}	nano,	(n),	meaning 10^{-9}
giga,	(G),	meaning 10^{9}	pico,	(p),	meaning 10^{-12}
mega,	(M),	meaning 10^{6}	femto,	(f),	meaning 10^{-15}
kilo,	(k),	meaning 10^{3}	atto,	(a),	meaning 10^{-18}
hecto,	(h),	meaning 10^{2}	zepto,	(z),	meaning 10^{-21}
deka,	(da),	meaning 10^{1}	yocto,	(y),	meaning 10^{-24}

Thus a kilometer is 1000 meters and a millimeter is 0.001 meter.

Units of Length

10 millimeters (mm)	= 1 centimeter (cm)
10 centimeters	= 1 decimeter (dm) = 100 millimeters
10 decimeters	= 1 meter (m) = 1000 millimeters
10 meters	= 1 dekameter (dam)
10 dekameters	= 1 hectometer (hm) = 100 meters
10 hectometers	= 1 kilometer (km) = 1000 meters

Units of Area

100 square millimeters (mm^2)	= 1 square centimeter (cm^2)
100 square centimeters	= 1 square decimeter (dm^2)
100 square decimeters	= 1 square meter (m^2)
100 square meters	= 1 square dekameter (dam^2) = 1 are
100 square dekameters	= 1 square hectometer (hm^2) = 1 hectare (ha)
100 square hectometers	= 1 square kilometer (km^2)

Units of Liquid Volume

10 milliliters (mL)	= 1 centiliter (cL)
10 centiliters	= 1 deciliter (dL) = 100 milliliters
10 deciliters	= 1 liter[1] = 1000 milliliters
10 liters	= 1 dekaliter (daL)
10 dekaliters	= 1 hectoliter (hL) = 100 liters
10 hectoliters	= 1 kiloliter (kL) = 1000 liters

Units of Volume

1000 cubic millimeters (mm^3)	= 1 cubic centimeter (cm^3)
1000 cubic centimeters	= 1 cubic decimeter (dm^3)
	= 1 000 000 cubic millimeters
1000 cubic decimeters	= 1 cubic meter (m^3)
	= 1 000 000 cubic centimeters
	= 1 000 000 000 cubic millimeters

Units of Mass

10 milligrams (mg)	= 1 centigram (cg)
10 centigrams	= 1 decigram (dg) = 100 milligrams
10 decigrams	= 1 gram (g) = 1000 milligrams
10 grams	= 1 dekagram (dag)
10 dekagrams	= 1 hectogram (hg) = 100 grams
10 hectograms	= 1 kilogram (kg) = 1000 grams
1000 kilograms	= 1 megagram (Mg) or 1 metric ton(t)

---

1. By action of the 12th General Conference on Weights and Measures (1964), the liter is a special name for the cubic decimeter.

2. Tables of U.S. Units of Measurement[2]

In these tables where foot or mile is underlined, it is survey foot or U.S. statute mile rather than international foot or mile that is meant.

Units of Length

12 inches (in)	=1 foot (ft)
3 feet	= 1 yard (yd)
16½ feet	= 1 rod (rd), pole, or perch
40 rods	= 1 furlong (fur) = 660 feet
8 furlongs	= 1 U.S. statute mile (mi) = 5280 feet
1852 meters (m)	= 6076.115 49 feet (approximately)
	= 1 international nautical mile

Units of Area[3]

144 square inches (in^2)	= 1 square foot (ft^2)
9 square feet	= 1 square yard (yd^2)
	= 1296 square inches
272¼ square feet	= 1 square rod (rd^2)
160 square rods	= 1 acre = 43 560 square feet
640 acres	= 1 square mile (mi^2)
1 mile square	= 1 section of land
6 miles square	= 1 township
	= 36 sections = 36 square miles

Units of Volume[3]

1728 cubic inches (in^3)	= 1 cubic foot (ft^3)
27 cubic feet	= 1 cubic yard (yd^3)

Gunter's or Surveyors Chain Units of Measurement

0.66 foot (ft)	= 1 link (li)
100 links	= 1 chain (ch)
	= 4 rods = 66 feet
80 chains	= 1 U.S. statute mile (mi)
	= 320 rods = 5280 feet

---

2. This section lists units of measurement that have traditionally been used in the United States. In keeping with the Omnibus Trade and Competitiveness Act of 1988, the ultimate objective is to make the International System of Units the primary measurement system used in the United States.

3. Squares and cubes of customary but not of metric units are sometimes expressed by the use of abbreviations rather than symbols. For example, sq ft means square foot, and cu ft means cubic foot.

Units of Liquid Volume[4]

4 gills (gi)	= 1 pint (pt) = 28.875 cubic inches (in^3)
2 pints	= 1 quart (qt) = 57.75 cubic inches
4 quarts	= 1 gallon (gal) = 231 cubic inches
	= 8 pints = 32 gills

Apothecaries Units of Liquid Volume

60 minims	= 1 fluid dram (fl dr or f ℨ)
	= 0.225 6 cubic inch (in^3)
8 fluid drams	= 1 fluid ounce (fl oz or f ℥)
	= 1.804 7 cubic inches
16 fluid ounces	= 1 pint (pt)
	= 28.875 cubic inches
	= 128 fluid drams
2 pints	= 1 quart (qt) = 57.75 cubic inches
	= 32 fluid ounces = 256 fluid drams
4 quarts	= 1 gallon (gal) = 231 cubic inches
	= 128 fluid ounces = 1024 fluid drams

Units of Dry Volume[5]

2 pints (pt)	= 1 quart (qt) = 67.200 6 cubic inches (in^3)
8 quarts	= 1 peck (pk) = 537.605 cubic inches
	= 16 pints
4 pecks	= 1 bushel (bu) = 2150.42 cubic inches
	= 32 quarts
	Avoirdupois Units of Mass[6]

[The "grain" is the same in avoirdupois, troy, and apothecaries units of mass.]

27^{11}/$_{32}$ grains (gr)	= 1 dram (dr)
16 drams	= 1 ounce (oz)
	= 437½ grains
16 ounces	= 1 pound (lb)
	= 256 drams
	= 7000 grains
100 pounds	= 1 hundredweight (cwt)[7]
20 hundredweights	= 1 ton (t)
	= 2000 pounds[7]

---

4. When necessary to distinguish the liquid pint or quart from the dry pint or quart, the word "liquid" or the abbreviation "liq" should be used in combination with the name or abbreviation of the liquid unit.
5. When necessary to distinguish dry pint or quart from the liquid pint or quart, the word "dry" should be used in combination with the name or abbreviation of the dry unit.
6. When necessary to distinguish the avoirdupois dram from the apothecaries dram, or to distinguish the avoirdupois dram or ounce from the fluid dram or ounce, or to distinguish the avoirdupois ounce or pound from the troy or apothecaries ounce or pound, the word "avoirdupois" or the abbreviation "avdp" should be used in combination with the name or abbreviation of the avoirdupois unit.
7. When the terms "hundredweight" and "ton" are used unmodified, they are commonly understood to mean the 100-pound hundredweight and the 2000-pound ton, respectively; these units may be designated "net" or "short" when necessary to distinguish them from the corresponding units in gross or long measure.

In "gross" or "long" measure, the following values are recognized:

112 pounds (lb)	= 1 gross or long hundredweight (cwt)[7]
20 gross or long hundredweights	= 1 gross or long ton
	= 2240 pounds[7]

Troy Units of Mass

[The "grain" is the same in avoirdupois, troy, and apothecaries units of mass.]

24 grains (gr)	= 1 pennyweight (dwt)
20 pennyweights	= 1 ounce troy (oz t) = 480 grains
12 ounces troy	= 1 pound troy (lb t)
	= 240 pennyweights = 5760 grains

Apothecaries Units of Mass

[The "grain" is the same in avoirdupois, troy, and apothecaries units of mass.]

20 grains (gr)	= 1 scruple (s ap or ℈)
3 scruples	= 1 dram apothecaries (dr ap or ℨ)
	= 60 grains
8 drams apothecaries	= 1 ounce apothecaries (oz ap or ℥)
	= 24 scruples = 480 grains
12 ounces apothecaries	= 1 pound apothecaries (lb ap)
	= 96 drams apothecaries
	= 288 scruples = 5760 grains

3. Notes on British Units of Measurement

In Great Britain, the yard, the avoirdupois pound, the troy pound, and the apothecaries pound are identical with the units of the same names used in the United States. The tables of British linear measure, troy mass, and apothecaries mass are the same as the corresponding United States tables, except for the British spelling "drachm" in the table of apothecaries mass. The table of British avoirdupois mass is the same as the United States table up to 1 pound; above that point the table reads:

14 pounds	= 1 stone
2 stones	= 1 quarter = 28 pounds
4 quarters	= 1 hundredweight = 112 pounds
20 hundredweight	= 1 ton = 2240 pounds

The present British gallon and bushel–known as the "Imperial gallon" and "Imperial bushel"–are, respectively, about 20 % and 3 % larger than the United States gallon and bushel. The Imperial gallon is defined as the volume of 10 avoirdupois pounds of water under specified conditions, and the Imperial bushel is defined as 8 Imperial gallons. Also, the subdivision of the Imperial gallon as presented in the table of British apothecaries fluid measure differs in two important respects from the corresponding United States subdivision, in that the

Imperial gallon is divided into 160 fluid ounces (whereas the United States gallon is divided into 128 fluid ounces), and a "fluid scruple" is included. The full table of British measures of capacity (which are used alike for liquid and for dry commodities) is as follows:

4 gills	= 1 pint
2 pints	= 1 quart
4 quarts	= 1 gallon
2 gallons	= 1 peck
8 gallons (4 pecks)	= 1 bushel
8 bushels	= 1 quarter

The full table of British apothecaries measure is as follows:

20 minims	= 1 fluid scruple
3 fluid scruples	= 1 fluid drachm
	= 60 minims
8 fluid drachms	= 1 fluid ounce
20 fluid ounces	= 1 pint
8 pints	= 1 gallon (160 fluid ounces)

4. Tables of Units of Measurement (all underlined figures are exact)

Units of Length—International Measure[8]

Units	Inches	Feet	Yards	Miles	Centimeters	Meters
1 inch =	1	0.083 333 33	0.027 777 78	0.000 015 782 83	2.54	0.025 4
1 foot =	12	1	0.333 333 3	0.000 189 393 9	30.48	0.304 8
1 yard =	36	3	1	0.000 568 181 8	91.44	0.914 4
1 mile =	63 360	5 280	1 760	1	160 934.4	1609.344
1 centimeter =	0.393 700 8	0.032 808 40	0.010 936 13	0.000 006 213 712	1	0.01
1 meter =	39.370 08	3.280 840	1.093 613	0.000 621 371 2	100	1

---

8. One international foot = 0.999 998 survey foot (exactly)
   One international mile = 0.999 998 survey mile (exactly)

Units of Length—Survey Measure[8]

Units	Links	Feet	Rods	Chains	Miles	Meters
1 link =	1	0.66	0.04	0.01	0.000 125	0.201 168 4
1 foot =	1.515 152	1	0.060 606 06	0.015 151 52	0.000 189 393 9	0.304 800 6
1 rod =	25	16.5	1	0.25	0.003 125	5.029 210
1 chain =	100	66	4	1	0.0125	20.116 84
1 mile =	8 000	5 280	320	80	1	1609.347
1 meter =	4.970 960	3.280 833	0.198 838 4	0.049 709 60	0.000 621 369 9	1

Units of Area—International Measure[9]

Units	Square Inches	Square Feet	Square Yards
1 square inch =	1	0.006 944 444	0.000 771 604 9
1 square foot =	144	1	0.111 111 1
1 square yard =	1296	9	1
1 square mile =	4 014 489 600	27 878 400	3 097 600
1 square centimeter =	0.155 000 3	0.001 076 391	0.000 119 599 0
1 square meter =	1550.003	10.763 91	1.195 990

Note:
1 survey foot = $\frac{1200}{3937}$ meter (exactly)
1 international foot = 12 × 0.0254 meter (exactly)
1 international foot = 0.0254 × 39.37 survey foot (exactly)

Units	Square Miles	Square Centimeters	Square Meters
1 square inch =	0.000 000 000 249 097 7	6.451 6	0.000 645 16
1 square foot =	0.000 000 035 870 06	929.030 4	0.092 903 04
1 square yard =	0.000 000 322 830 6	8361.273 6	0.836 127 36
1 square mile =	1	25 899 881 103.36	2 589 988.110 336
1 square centimeter =	0.000 000 000 038 610 22	1	0.0001
1 square meter =	0.000 000 386 102 2	10 000	1

---

9. One square survey foot = 1.000 004 square international feet
One square survey mile = 1.000 004 square international miles

Units of Area—Survey Measure[9]

Units	Square Feet	Square Rods	Square Chains	Acres
1 square foot =	1	0.003 673 095	0.000 229 568 4	0.000 022 956 84
1 square rod =	272.25	1	0.062 5	0.006 25
1 square chain =	4356	16	1	0.1
1 acre =	43 560	160	10	1
1 square mile =	27 878 400	102 400	6400	640
1 square meter =	10.763 87	0.039 536 70	0.002 471 044	0.000 247 104 4
1 hectare =	107 638.7	395.367 0	24.710 44	2.471 044

Units	Square Miles	Square Meters	Hectares
1 square foot =	0.000 000 035 870 06	0.092 903 41	0.000 009 290 341
1 square rod =	0.000 009 765 625	25.292 95	0.002 529 295
1 square chain =	0.000 156 25	404.687 3	0.040 468 73
1 acre =	0.001 562 5	4 046.873	0.404 687 3
1 square mile =	1	2 589 998	258.999 8
1 square meter =	0.000 000 386 100 6	1	0.000 1
1 hectare =	0.003 861 006	10 000	1

Units of Volume

Units	Cubic Inches	Cubic Feet	Cubic Yards
1 cubic inch =	1	0.000 578 703 7	0.000 021 433 47
1 cubic foot =	1728	1	0.037 037 04
1 cubic yard =	46 656	27	1
1 cubic centimeter =	0.061 023 74	0.000 035 314 67	0.000 001 307 951
1 cubic decimeter =	61.023 74	0.035 314 67	0.001 307 951
1 cubic meter =	61 023.74	35.314 67	1.307 951

Units	Milliliters (Cubic Centimeters)	Liters (Cubic Decimeters)	Cubic Meters
1 cubic inch =	16.387 064	0.016 387 064	0.000 016 387 064
1 cubic foot =	28 316.846 592	28.316 846 592	0.028 316 846 592
1 cubic yard =	764 554.857 984	764.554 857 984	0.764 554 857 984

(Continued)

---

9. One square survey foot = 1.000 004 square international feet
One square survey mile = 1.000 004 square international miles

Units	Milliliters (Cubic Centimeters)	Liters (Cubic Decimeters)	Cubic Meters
1 cubic centimeter =	1	0.001	0.000 001
1 cubic decimeter =	1000	1	0.001
1 cubic meter =	1 000 000	1000	1

Units of Capacity or Volume – Dry Volume Measure

Units	Dry Pints	Dry Quarts	Pecks	Bushels
1 dry pint =	1	0.5	0.062 5	0.015 625
1 dry quart =	2	1	0.125	0.031 25
1 peck =	16	8	1	0.25
1 bushel =	64	32	4	1
1 cubic inch =	0.029 761 6	0.014 880 8	0.001 860 10	0.000 465 025
1 cubic foot =	51.428 09	25.714 05	3.214 256	0.803 563 95
1 liter =	1.816 166	0.908 083 0	0.113 510 4	0.028 377 59
1 cubic meter =	1 816.166	908.083 0	113.510 4	28.377 59

Units	Cubic Inches	Cubic Feet	Liters	Cubic Meters
1 dry pint =	33.600 312 5	0.019 444 63	0.550 610 5	0.000 550 610 5
1 dry quart =	67.200 625	0.038 889 25	1.101 221	0.001 101 221
1 peck =	537.605	0.311 114	8.809 768	0.008 809 768
1 bushel =	2 150.42	1.244 456	35.239 07	0.035 239 07
1 cubic inch =	1	0.000 578 703 7	0.016 387 06	0.000 016 387 06
1 cubic foot =	1728	1	28.316 85	0.028 316 85
1 liter =	61.023 74	0.035 314 67	1	0.001
1 cubic meter =	61 023.74	35.314 67	1000	1

Units of Capacity or Volume – Liquid Volume Measure

Units	Minims	Fluid Drams	Fluid Ounces	Gills
1 minim =	1	0.016 666 67	0.002 083 333	0.000 520 833 3
1 fluid dram =	60	1	0.125	0.031 25
1 fluid ounce =	480	8	1	0.25
1 gill =	1 920	32	4	1
1 liquid pint =	7 680	128	16	4
1 liquid quart =	15 360	256	32	8
1 gallon =	61 440	1024	128	32

Units	Minims	Fluid Drams	Fluid Ounces	Gills
1 cubic inch =	265.974 0	4.432 900	0.554 112 6	0.138 528 1
1 cubic foot =	459 603.1	7660.052	957.506 5	239.376 6
1 milliliter =	16.230 73	0.270 512 2	0.033 814 02	0.008 453 506
1 liter =	16 230.73	270.512 2	33.814 02	8.453 506

Units	Liquid Pints	Liquid Quarts	Gallons	Cubic Inches
1 minim =	0.000 130 208 3	0.000 065 104 17	0.000 016 276 04	0.003 759 766
1 fluid dram =	0.007 812 5	0.003 906 25	0.000 976 562 5	0.225 585 94
1 fluid ounce =	0.062 5	0.031 25	0.007 812 5	1.804 687 5
1 gill =	0.25	0.125	0.031 25	7.218 75
1 liquid pint =	1	0.5	0.125	28.875
1 liquid quart =	2	1	0.25	57.75
1 gallon =	8	4	1	231
1 cubic inch =	0.034 632 03	0.017 316 02	0.004 329 004	1
1 cubic foot =	59.844 16	29.922 08	7.480 519	1728
1 milliliter =	0.002 113 376	0.001 056 688	0.000 264 172 1	0.061 023 74
1 liter =	2.113 376	1.056 688	0.264 172 1	61.023 74

Units	Cubic Feet	Milliliters	Liters
1 minim =	0.000 002 175 790	0.061 611 52	0.000 061 611 52
1 fluid dram =	0.000 130 547 4	3.696 691	0.003 696 691
1 fluid ounce =	0.001 044 379	29.573 53	0.029 573 53
1 gill =	0.004 177 517	118.294 1	0.118 294 1
1 liquid pint =	0.016 710 07	473.176 5	0.473 176 5
1 liquid quart =	0.033 420 14	946.352 9	0.946 352 9
1 gallon =	0.133 680 6	3785.412	3.785 412
1 cubic inch =	0.000 578 703 7	16.387 06	0.016 387 06
1 cubic foot =	1	28 316.85	28.316 85
1 milliliter =	0.000 035 314 67	1	0.001
1 liter =	0.035 314 67	1000	1

Units of Mass Not Less Than Avoirdupois Ounces

Units	Avoirdupois Ounces	Avoirdupois Pounds	Short Hundred-weights	Short tons
1 avoirdupois ounce =	1	0.0625	0.000 625	0.000 031 25
1 avoirdupois pound =	16	1	0.01	0.000 5
1 short hundredweight =	1 600	100	1	0.05
1 short ton =	32 000	2000	20	1
1 long ton =	35 840	2240	22.4	1.12
1 kilogram =	35.273 96	2.204 623	0.022 046 23	0.001 102 311
1 metric ton =	35 273.96	2204.623	22.046 23	1.102 311

Units	Long Tons	Kilograms	Metric Tons
1 avoirdupois ounce =	0.000 027 901 79	0.028 349 523 125	0.000 028 349 523 125
1 avoirdupois pound =	0.000 446 428 6	0.453 592 37	0.000 453 592 37
1 short hundredweight =	0.044 642 86	45.359 237	0.045 359 237
1 short ton =	0.892 857 1	907.184 74	0.907 184 74
1 long ton =	1	1016.046 908 8	1.016 046 908 8
1 kilogram =	0.000 984 206 5	1	0.001
1 metric ton =	0.984 206 5	1000	1

Units of Mass Not Greater Than Pounds and Kilograms

Units	Grains	Apothecaries Scruples	Pennyweights	Avoirdupois Drams
1 grain =	1	0.05	0.041 666 67	0.036 571 43
1 apoth. scruple =	20	1	0.833 333 3	0.731 428 6
1 penny weight =	24	1.2	1	0.877 714 3
1 avdp. dram =	27.343 75	1.367 187 5	1.139 323	1
1 apoth. dram =	60	3	2.5	2.194 286
1 avdp. ounce =	437.5	21.875	18.229 17	16
1 apoth. or troy oz. =	480	24	20	17.554 29
1 apoth. or troy pound =	5760	288	240	210.651 4
1 avdp. pound =	7000	350	291.666 7	256
1 milligram =	0.015 432 36	0.000 771 617 9	0.000 643 014 9	0.000 564 383 4
1 gram =	15.432 36	0.771 617 9	0.643 014 9	0.564 383 4
1 kilogram =	15432.36	771.617 9	643.014 9	564.383 4

Units	Apothecaries Drams	Avoirdupois Ounces	Apothecaries or Troy Ounces	Apothecaries or Troy Pounds
1 grain =	0.016 666 67	0.002 285 714	0.002 083 333	0.000 173 611 1
1 apoth. scruple =	0.333 333 3	0.045 714 29	0.041 666 67	0.003 472 222
1 pennyweight =	0.4	0.054 857 14	0.05	0.004 166 667
1 avdp. dram =	0.455 729 2	0.062 5	0.56 966 15	0.004 747 179
1 apoth. dram =	1	0.137 142 9	0.125	0.010 416 67
1 avdp. ounce =	7.291 667	1	0.911 458 3	0.075 954 86
1 apoth. or troy ounce =	8	1.097 143	1	0.083 333 333
1 apoth. or troy pound =	96	13.165 71	12	1
1 avdp. pound =	116.666 7	16	14.583 33	1.215 278
1 milligram =	0.000 257 206 0	0.000 035 273 96	0.000 032 150 75	0.000 002 679 229
1 gram =	0.257 206 0	0.035 273 96	0.032 150 75	0.002 679 229
1 kilogram =	257.206 0	35.273 96	32.150 75	2.679 229

Units	Avoirdupois Pounds	Milligrams	Grams	Kilograms
1 grain =	0.000 142 857 1	64.798 91	0.064 798 91	0.000 064 798 91
1 apoth. scruple =	0.002 857 143	1295.978 2	1.295 978 2	0.001 295 978 2
1 penny weight =	0.003 428 571	1555.173 84	1.555 173 84	0.001 555 173 84
1 avdp. dram =	0.003 906 25	1771.845 195 312 5	1.771 845 195 312 5	0.001 771 845 195 312 5
1 apoth. dram =	0.008 571 429	3887.934 6	3.887 934 6	0.003 887 934 6
1 avdp. ounce =	0.062 5	28 349.523 125	28.349 523 125	0.028 349 523 125
1 apoth. or troy ounce =	0.068 571 43	31 103.476 8	31.103 476 8	0.031 103 476 8
1 apoth. or troy pound =	0.822 857 1	373 241.721 6	373.241 721 6	0.373 241 721 6
1 avdp. pound =	1	453 592.37	453.592 37	0.453 592 37
1 milligram =	0.000 002 204 623	1	0.001	0.000 001
1 gram =	0.002 204 623	1000	1	0.001
1 kilogram =	2.204 623	1 000 000	1000	1

5. Tables of Equivalents

In these tables it is necessary to differentiate between the "international foot" and the "survey foot." Therefore, the survey foot is underlined.

When the name of a unit is enclosed in brackets (thus, [1 hand] . . .), this indicates (1) that the unit is not in general current use in the United States, or (2) that the unit is believed to be based on "custom and usage" rather than on formal authoritative definition.

Equivalents involving decimals are, in most instances, rounded off to the third decimal place except where they are exact, in which cases these exact equivalents are so designated. The equivalents of the imprecise units "tablespoon" and "teaspoon" are rounded to the nearest milliliter.

Units of Length	
angstrom (Å)[10]	0.1 nanometer (exactly) 0.000 1 micrometer (exactly) 0.000 000 1 millimeter (exactly)
1 cable's length	0.000 000 004 inch 120 fathoms (exactly) 720 feet (exactly) 219 meters
1 centimeter (cm)	0.393 7 inch
1 chain (ch) (Gunter's or surveyors)	66 feet (exactly) 20.116 8 meters
1 decimeter (dm)	3.937 inches
1 dekameter (dam)	32.808 feet
1 fathom	6 feet (exactly) 1.828 8 meters
1 foot (ft)	0.304 8 meter (exactly)

Units of Length	
1 furlong (fur)	10 chains (surveyors) (exactly) 660 feet (exactly) ⅛ U.S. statute mile (exactly) 201.168 meters
[1 hand]	4 inches
1 inch (in)	2.54 centimeters (exactly)
1 kilometer (km)	0.621 mile
1 league (land)	3 U.S. statute miles (exactly) 4.828 kilometers
1 link (li) (Gunter's or surveyors)	0.66 foot (exactly) 0.201 168 meter
1 meter (m)	39.37 inches 1.094 yards
1 micrometer	0.001 millimeter (exactly) 0.000 039 37 inch
1 mil	0.001 inch (exactly) 0.025 4 millimeter (exactly)

---

10. The angstrom is basically defined as 10^{-10} meter.

1 mile (mi) (U.S. statute)[11]	5280 feet survey (exactly)
	1.609 kilometers
1 mile (mi) (international)	5280 feet international (exactly)
1 mile (mi) (international nautical)[12]	1.852 kilometers (exactly)
	1.151 survey miles
1 millimeter (mm)	0.039 37 inch
	0.001 meter (exactly)
1 nanometer (nm)	0.000 000 039 37 inch
	0.013 837 inch (exactly)
1 point (typography)	1/72 inch (approximately)
	0.351 millimeter
1 rod (rd), pole, or perch	16½ feet (exactly)
	5.029 2 meters
1 yard (yd)	0.914 4 meter (exactly)
Units of Area	
1 acre[13]	43 560 square feet (exactly)
	0.405 hectare
1 are	119.599 square yards
	0.025 acre
1 hectare	2.471 acres
[1 square (building)]	100 square feet
1 square centimeter (cm^2)	0.155 square inch
1 square decimeter (dm^2)	15.500 square inches
1 square foot (ft^2)	929.030 square centimeters
1 square inch (in^2)	6.451 6 square centimeters (exactly)
1 square kilometer (km^2)	247.104 acres
	0.386 square mile
1 square meter (m^2)	1.196 square yards
	10.764 square feet
1 square mile (mi^2)	258.999 hectares
1 square millimeter (mm^2)	0.002 square inch
1 square rod (rd^2), sq pole, or sq perch	25.293 square meters
1 square yard (yd^2)	0.836 square meter

---

11. The term "statute mile" originated with Queen Elizabeth I who changed the definition of the mile from the Roman mile of 5000 feet to the statute mile of 5280 feet. The international mile and the U.S. statute mile differ by about 3 millimeters, although both are defined as being equal to 5280 feet. The international mile is based on the international foot (0.3048 meter) whereas the U.S. statute mile is based on the survey foot (1200/3937 meter).
12. The international nautical mile of 1852 meters (6076.115 49...feet) was adopted effective July 1, 1954, for use in the United States. The value formerly used in the United States was 6080.20 feet = 1 nautical (geographical or sea) mile.
13. The question is often asked as to the length of a side of an acre of ground. An acre is a unit of area containing 43 560 square feet. It is not necessarily square, or even rectangular. But, if it is square, then the length of a side is equal to $\sqrt{43560 \text{ ft}^2} = 208.710$ ft (not exact).

Units of Capacity or Volume

1 barrel (bbl), liquid	31 to 42 gallons[14]
1 barrel (bbl), standard for fruits, vegetables, and other dry commodities, except cranberries	7056 cubic inches 105 dry quarts 3.281 bushels, struck measure
1 barrel (bbl), standard, cranberry	5826 cubic inches 86 $^{45}/_{64}$ dry quarts 2.709 bushels, struck measure
1 bushel (bu) (U.S.) struck measure	2150.42 cubic inches (exactly) 35.238 liters
[1 bushel, heaped (U.S.)]	2747.715 cubic inches 1.278 bushels, struck measure[15]
[1 bushel (bu) (British Imperial) (struck measure)]	1.032 U.S. bushels, struck measure 2219.36 cubic inches
1 cord (cd) (firewood)	128 cubic feet (exactly)
1 cubic centimeter (cm^3)	0.061 cubic inch
1 cubic decimeter (dm^3)	61.024 cubic inches
1 cubic foot (ft^3)	7.481 gallons 28.316 cubic decimeters

Units of Capacity or Volume

1 cubic inch (in^3)	0.554 fluid ounce 4.433 fluid drams 16.387 cubic centimeters
1 cubic meter (m^3)	1.308 cubic yards
1 cubic yard (yd^3)	0.765 cubic meter
1 cup, measuring	8 fluid ounces (exactly) 237 milliliters ½ liquid pint (exactly)
1 dekaliter (daL)	2.642 gallons 1.135 pecks ⅛ fluid ounce (exactly)

14. There are a variety of "barrels" established by law or usage. For example, federal taxes on fermented liquors are based on a barrel of 31 gallons; many state laws fix the "barrel for liquids" as 31½ gallons; one state fixes a 36-gallon barrel for cistern measurement; federal law recognizes a 40-gallon barrel for "proof spirits"; by custom, 42 gallons comprise a barrel of crude oil or petroleum products for statistical purposes, and this equivalent is recognized "for liquids" by four states.

15. Frequently recognized as 1¼ bushels, struck measure.

1 dram, fluid (or liquid) (fl dr) or f℥ (U.S.)	0.226 cubic inch 3.697 milliliters 1.041 British fluid drachms
[1 drachm, fluid (fl dr) (British)]	0.961 U.S. fluid dram 0.217 cubic inch 3.552 milliliters
1 gallon (gal) (U.S.)	231 cubic inches (exactly) 3.785 liters 0.833 British gallon 128 U.S. fluid ounces (exactly)
[1 gallon (gal) (British Imperial)]	277.42 cubic inches 1 201 U.S. gallons 4.546 liters 160 British fluid ounces (exactly)
1 gill (gi)	7.219 cubic inches 4 fluid ounces (exactly) 0.118 liter
1 hectoliter (hL)	26.418 gallons 2.838 bushels
1 liter (1 cubic decimeter exactly)	1.057 liquid quarts 0.908 dry quart 61.025 cubic inches
1 milliliter (mL)	0.271 fluid dram 16.231 minims 0.061 cubic inch
1 ounce, fluid (or liquid) (fl oz) or f℥ (U.S.)	1.805 cubic inches 29.573 milliliters 1.041 British fluid ounces
[1 ounce, fluid (fl oz) (British)]	0.961 U.S. fluid ounce 1.734 cubic inches 28.412 milliliters
1 peck (pk)	8.810 liters
1 pint (pt), dry	33.600 cubic inches 0.551 liter
1 pint (pt), liquid	28.875 cubic inches exactly 0.473 liter
1 quart (qt), dry (U.S.)	67.201 cubic inches 1.101 liters 0.969 British quart
1 quart (qt), liquid (U.S.)	57.75 cubic inches (exactly) 0.946 liter 0.833 British quart

Units of Capacity or Volume

[1 quart (qt) (British)]	69.354 cubic inches 1.032 U.S. dry quarts 1.201 U.S. liquid quarts
1 tablespoon, measuring	3 teaspoons (exactly) 15 milliliters 4 fluid drams ½ fluid ounce (exactly)
1 teaspoon, measuring	⅓ tablespoon (exactly) 5 milliliters 1⅓ fluid drams[16]
1 water ton (English)	270.91 U.S. gallons 224 British Imperial gallons (exactly)

Units of Mass

1 assay ton[17] (AT)	29.167 grams
1 carat (c)	200 milligrams (exactly) 3.086 grains
1 dram apothecaries (dr ap or ʒ)	60 grains (exactly). 3.888 grams
1 dram avoirdupois (dr avdp)	$27^{11}/_{32}$ (= 27.344) grains 1.772 grams
1 gamma (γ)	1 microgram (exactly)
1 grain	64.798 91 milligrams (exactly)
1 gram (g)	15.432 grains 0.035 ounce, avoirdupois
1 hundredweight, gross or long[18] (gross cwt)	112 pounds (exactly) 50.802 kilograms
1 hundredweight, gross or short (cwt or net cwt)	100 pounds (exactly) 45.359 kilograms
1 kilogram (kg)	2.205 pounds
1 microgram (μg)	0.000 001 gram (exactly)
1 milligram (mg)	0.015 grain
1 ounce, avoirdupois (oz avdp)	437.5 grains (exactly) 0.911 troy or apothecaries ounce 28.350 grams
1 ounce, troy, or apothecaries (oz t or oz ap or ʒ)	480 grains (exactly) 1.097 avoirdupois ounces 31.103 grams
1 pennyweight (dwt)	1.555 grams

16. The equivalent "1 teaspoon = 1⅓ fluid drams" has been found by the Bureau to correspond more closely with the actual capacities of "measuring" and silver teaspoons than the equivalent "1 teaspoon = 1 fluid dram," which is given by a number of dictionaries.
17. Used in assaying. The assay ton bears the same relation to the milligram that a ton of 2000 pounds avoirdupois bears to the ounce troy; hence the mass in milligrams of precious metal obtained from one assay ton of ore gives directly the number of troy ounces to the net ton.
18. The gross or long ton and hundredweight are used commercially in the United States to a limited extent only, usually in restricted industrial fields. These units are the same as the British "ton" and "hundredweight."

Units of Mass

1 point	0.01 carat
	2 milligrams
1 pound, avoirdupois (lb avdp)	7000 grains (exactly)
	1.215 troy or apothecaries pounds
	453.592 37 grams (exactly)
1 pound, troy or apothecaries (lb t or lb ap)	5760 grains (exactly)
	0.823 avoirdupois pound
	373.242 grams
1 scruple (s ap or ℈)	20 grains (exactly)
	1.296 grams
1 ton, gross or long[19]	2240 pounds (exactly)
	1.12 net tons (exactly)
	1.016 metric tons
1 ton, metric (t)	2204.623 pounds
	0.984 gross ton
	1.102 net tons
1 ton, net or short	2000 pounds (exactly)
	0.893 gross ton
	0.907 metric ton

---

19. The gross or long ton and hundredweight are used commercially in the United States to a limited extent only, usually in restricted industrial fields. These units are the same as the British "ton" and "hundredweight."

Section 2

Conversion Tables and Conversion Formulas

2.0.0	Executive Order 12770—the Metric Conversion Law	34
2.0.1	Inches to Feet Conversion Table	34
2.0.2	Inches to Centimeter Conversion	35
2.0.3	Centimeter to Inches Conversion	36
2.0.4	Feet to Meters Conversion	37
2.0.5	Meter to Feet Conversion	38
2.0.6	Acres to Hectares Conversion	39
2.0.7	Hectares to Acres Conversion	40
2.0.8	Square Inch to Square Feet Conversion	40
2.0.9	Square Feet to Square Inch Conversion	41
2.0.10	Square Feet to Square Mile Conversion	42
2.0.11	Square Mile to Square Feet Conversion	43
2.0.12	Square Feet to Acres Conversion	43
2.0.13	Acre to Square Feet Conversion	44
2.0.14	Square Yard to Square Meter Conversion Table	45
2.0.15	Square Meter to Square Yard Conversion Table	45
2.0.16	Square Mile to Square Meter Conversion Table	46
2.0.17	Square Meter to Square Mile Conversion Table	47
2.0.18	Square Mile to Hectare Conversion Table	48
2.0.19	Hectare to Square Mile Conversion Table	48
2.0.20	Miles to Kilometers Conversion	49
2.0.21	Kilometers to Miles Conversion	50
2.0.22	Pounds to Kilograms Conversion	51
2.0.23	Kilograms to Pounds Conversion	52
2.0.24	Fahrenheit to Celsius Conversion	53
2.0.25	Celsius to Fahrenheit Temperature Conversion	54
2.0.26	Fahrenheit to Rankine Temperature Conversion	54
2.0.27	Rankine to Fahrenheit Temperature Conversion	55
2.1.0	Converting Water from One Form to Another	56
2.1.1	U.S. and Metric Lumber Length Conversion Table	56
2.2.0	Conversion Factors—Energy, Volume, Length, Weight, Liquid	57
2.3.0	Conversion of Liquids—Specific Gravity to Degrees Baume	58
2.4.0	Volume-to-Weight Conversion Table	62
2.5.0	Convert Old A.I.S.C. Structural Shapes to New Designations	64
2.6.0	USA Mesh Size—Convert to International Particle Size (Microns)	65
2.7.0	Convert Wire Gauge from Standard (SWG) to American (AWG) to Brown & Sharpe (B&S) to Metric	67
2.8.0	Map of United States Showing Four Continental Time Zones	69
2.8.1	Convert Time Zones—UTC to Four Standard U.S. Time Zones	69
2.8.2	Convert Time Zones—UTC to Four Daylight U.S. Time Zones	70
2.9.0	Convert Roman Numerals to Arabic Dates	71
2.10.0	Converting Speed—Knots to MPH to Kilometers per Hour	72
2.11.0	Conversion Factors for Builders and Design Professionals Who Cook	73

2.0.0 Executive Order 12770—The Metric Conversion Law

On July 25, 1991, President George H.W.Bush signed Executive Order 12770, referred to as the Metric Conversion Law, directing U.S. businesses and professionals to use the SI (metric) system as the "preferred" system for weights and measures.

The designation SI is derived from *Systems International d'unites*, the current international designation for the standard metric system of weights and measures.

Although engineering, design, and construction professionals in the United States have not fully adopted conversion to the metric system, the increased globalization of these professions requires familiarization with metric conversion units.

Avoirdupois units of mass (weight) differ from the Troy system, which is now the accepted weight denomination for precious metals. The Troy pound contains 12 ounces, whereas the Avoirdupois system in current use today for almost everything except precious metals contains 16 ounces to the pound.

Imperial units such as inches, feet, yards, and miles are similar to customary U.S. units, but there are some exceptions. The Imperial system uses the term *stone* instead of pounds—a *stone* is the equivalent of 14 U.S. pounds.

The Imperial system also uses *long hundredweight,* which is equal to 112 U.S. pounds, and it also uses the *long ton*, equivalent to 2240 U.S. pounds.

Metric to U.S. and U.S. to metric conversion tables in this section provide easy conversion for:

1. Length, width, and thickness
2. Area—Volume
3. Weight
4. Temperature

2.0.1 Inches to Feet Conversion Table

Inches	Feet	Inches	Feet	Inches	Feet	Inches	Feet
0	0.00	13	1.08	35	2.91	57	4.75
0.1	0.008	14	1.16	36	3.00	58	4.83
0.2	0.016	15	1.25	37	3.08	59	4.91
0.3	0.025	16	1.33	38	3.16	60	5.00
0.4	0.033	17	1.41	39	3.25	61	5.08
0.5	0.041	18	1.50	40	3.33	62	5.16
0.6	0.05	19	1.58	41	3.41	63	5.25
0.7	0.058	20	1.66	42	3.50	64	5.33
0.8	0.066	21	1.75	43	3.58	65	5.41
0.9	0.075	22	1.83	44	3.66	66	5.50
1	0.08	23	1.91	45	3.75	67	5.58
2	0.16	24	2.00	46	3.83	68	5.66
3	0.25	25	2.08	47	3.91	69	5.75
4	0.33	26	2.16	48	4.00	70	5.83
5	0.41	27	2.25	49	4.08	71	5.91
6	0.5	28	2.33	50	4.16	72	6.00
7	0.58	29	2.41	51	4.25	73	6.08
8	0.66	30	2.50	52	4.33	74	6.16
9	0.75	31	2.58	53	4.41	75	6.25
10	0.83	32	2.66	54	4.50	76	6.33
11	0.91	33	2.75	55	4.58	77	6.41
12	1.00	34	2.83	56	4.66	78	6.50

Inches	Feet	Inches	Feet	Inches	Feet	Inches	Feet
79	6.58	87	7.25	95	7.91	400	33.30
80	6.66	88	7.33	96	8.00	500	41.60
81	6.75	89	7.41	97	8.08	600	50.00
82	6.83	90	7.50	98	8.16	700	58.30
83	6.91	91	7.58	99	8.25	800	66.60
84	7.00	92	7.66	100	8.30	900	75.00
85	7.08	93	7.75	200	16.60		
86	7.16	94	7.83	300	25.00		

By permission: www.metric-conversions.org

2.0.2 Inches to Centimeter Conversion

Inches	Centimeters	Inches	Centimeters	Inches	Centimeters	Inches	Centimeters
0	0.00	28	71.12	65	165.1	102	259.08
0.1	0.254	29	73.66	66	167.64	103	261.62
0.2	0.508	30	76.20	67	170.18	104	264.16
0.3	0.762	31	78.74	68	172.72	105	266.7
0.4	1.016	32	81.28	69	175.26	106	269.24
0.5	1.27	33	83.82	70	177.8	107	271.78
0.6	1.524	34	86.36	71	180.34	108	274.32
0.7	1.778	35	88.90	72	182.88	109	276.86
0.8	2.032	36	91.44	73	185.42	110	279.4
0.9	2.286	37	93.98	74	187.96	111	281.94
1	2.54	38	96.52	75	190.5	112	284.48
2	5.08	39	99.06	76	193.04	113	287.02
3	7.62	40	101.6	77	195.58	114	289.56
4	10.16	41	104.14	78	198.12	115	292.1
5	12.70	42	106.68	79	200.66	116	294.64
6	15.24	43	109.22	80	203.2	117	297.18
7	17.78	44	111.76	81	205.74	118	299.72
8	20.32	45	114.3	82	208.28	119	302.26
9	22.86	46	116.84	83	210.82	120	304.8
10	25.40	47	119.38	84	213.36	121	307.34
11	27.94	48	121.92	85	215.9	122	309.88
12	30.48	49	124.46	86	218.44	123	312.42
13	33.02	50	127.0	87	220.98	124	314.96
14	35.56	51	129.54	88	223.52	125	317.5
15	38.10	52	132.08	89	226.06	126	320.04
16	40.64	53	134.62	90	228.6	127	322.58
17	43.18	54	137.16	91	231.14	128	325.12
18	45.72	55	139.7	92	233.68	129	327.66
19	48.26	56	142.24	93	236.22	130	330.2
20	50.80	57	144.78	94	238.76	131	332.74
21	53.34	58	147.32	95	241.3	132	335.28
22	55.88	59	149.86	96	243.84	133	337.82
23	58.42	60	152.4	97	246.38	134	340.36
24	60.96	61	154.94	98	248.92	135	342.9
25	63.50	62	157.48	99	251.46	136	345.44
26	66.04	63	160.02	100	254.0	137	347.98
27	68.58	64	162.56	101	256.54	138	350.52

(Continued)

Inches	Centimeters	Inches	Centimeters	Inches	Centimeters
139	353.06	150	381.0	300	762.0
140	355.6	151	383.54	400	1016
141	358.14	152	386.08	500	1270
142	360.68	153	388.62	600	1524
143	363.22	154	391.16	700	1778
144	365.76	155	393.7	800	2032
145	368.3	156	396.24	900	2286
146	370.84	157	398.78		
147	373.38	158	401.32		
148	375.92	159	403.86		
149	378.46	200	508.0		

By permission: www.metric-conversions.org

2.0.3 Centimeter to Inches Conversion

Centimeters	Inches	Centimeters	Inches	Centimeters	Inches	Centimeters	Inches
0	0.00	24	9.44	57	22.440	90	35.433
0.1	0.039	25	9.84	58	22.834	91	35.826
0.2	0.078	26	10.23	59	23.228	92	36.220
0.3	0.118	27	10.62	60	23.622	93	36.614
0.4	0.157	28	11.02	61	24.015	94	37.007
0.5	0.196	29	11.41	62	24.409	95	37.401
0.6	0.236	30	11.81	63	24.803	96	37.795
0.7	0.275	31	12.20	64	25.196	97	38.188
0.8	0.314	32	12.59	65	25.590	98	38.582
0.9	0.354	33	12.99	66	25.984	99	38.976
1	0.39	34	13.38	67	26.377	100	39.370
2	0.78	35	13.77	68	26.771	101	39.763
3	1.18	36	14.17	69	27.165	102	40.157
4	1.57	37	14.56	70	27.559	103	40.551
5	1.96	38	14.96	71	27.952	104	40.944
6	2.36	39	15.35	72	28.346	105	41.338
7	2.75	40	15.748	73	28.740	106	41.732
8	3.14	41	16.141	74	29.133	107	42.125
9	3.54	42	16.535	75	29.527	108	42.519
10	3.93	43	16.929	76	29.921	109	42.913
11	4.33	44	17.322	77	30.314	110	43.307
12	4.72	45	17.716	78	30.708	111	43.700
13	5.11	46	18.110	79	31.102	112	44.094
14	5.51	47	18.503	80	31.496	113	44.488
15	5.90	48	18.897	81	31.889	114	44.881
16	6.29	49	19.291	82	32.283	115	45.275
17	6.69	50	19.685	83	32.677	116	45.669
18	7.08	51	20.078	84	33.070	117	46.062
19	7.48	52	20.472	85	33.464	118	46.456
20	7.87	53	20.866	86	33.858	119	46.850
21	8.26	54	21.259	87	34.251	120	47.244
22	8.66	55	21.653	88	34.645	121	47.637
23	9.05	56	22.047	89	35.039	122	48.031

Centimeters	Inches	Centimeters	Inches	Centimeters	Inches	Centimeters	Inches
123	48.425	135	53.149	147	57.874	159	62.598
124	48.818	136	53.543	148	58.267	200	78.70
125	49.212	137	53.937	149	58.661	300	118.1
126	49.606	138	54.330	150	59.055	400	157.4
127	50.00	139	54.724	151	59.448	500	196.8
128	50.393	140	55.118	152	59.842	600	236.2
129	50.787	141	55.511	153	60.236	700	275.5
130	51.181	142	55.905	154	60.629	800	314.9
131	51.574	143	56.299	155	61.023	900	354.3
132	51.968	144	56.692	156	61.417		
133	52.362	145	57.086	157	61.811		
134	52.755	146	57.480	158	62.204		

By permission: www.metric-conversions.org

2.0.4 Feet to Meters Conversion

Feet	Meters	Feet	Meters	Feet	Meters	Feet	Meters
0	0.00	25	7.62	59	17.983	93	28.346
0.1	0.030	26	7.92	60	18.288	94	28.651
0.2	0.060	27	8.22	61	18.592	95	28.958
0.3	0.091	28	8.53	62	18.897	96	29.260
0.4	0.121	29	8.83	63	19.202	97	29.565
0.5	0.152	30	9.14	64	19.507	98	29.870
0.6	0.182	31	9.44	65	19.812	99	30.175
0.7	0.213	32	9.75	66	20.116	100	30.48
0.8	0.243	33	10.05	67	20.421	101	30.784
0.9	0.274	34	10.36	68	20.726	102	31.089
1	0.30	35	10.66	69	21.031	103	31.394
2	0.60	36	10.97	70	21.336	104	31.699
3	0.91	37	11.27	71	21.640	105	32.004
4	1.21	38	11.58	72	21.945	106	32.308
5	1.52	39	11.88	73	22.250	107	32.613
6	1.82	40	12.192	74	22.555	108	32.918
7	2.13	41	12.496	75	22.86	109	33.223
8	2.43	42	12.801	76	23.164	110	33.528
9	2.74	43	13.106	77	23.469	111	33.832
10	3.04	44	13.411	78	23.774	112	34.137
11	3.35	45	13.716	79	24.079	113	34.442
12	3.65	46	14.020	80	24.384	114	34.747
13	3.96	47	14.325	81	24.688	115	35.052
14	4.26	48	14.630	82	24.993	116	35.356
15	4.57	49	14.935	83	25.298	117	35.661
16	4.87	50	15.24	84	25.603	118	35.966
17	5.18	51	15.544	85	25.908	119	36.271
18	5.48	52	15.849	86	26.212	120	36.576
19	5.79	53	16.154	87	26.517	121	36.880
20	6.09	54	16.459	88	26.822	122	37.185
21	6.40	55	16.764	89	27.127	123	37.490
22	6.70	56	17.068	90	27.432	124	37.795
23	7.01	57	17.373	91	27.736	125	38.10
24	7.31	58	17.678	92	28.041	126	38.404

(Continued)

Feet	Meters	Feet	Meters	Feet	Meters	Feet	Meters
127	38.709	138		149		200	60.90
128	39.014	139		150		300	91.40
129	39.319	140		151		400	121.9
130		141		152		500	152.4
131		142		153		600	182.8
132		143		154		700	213.3
133		144		155		800	243.8
134		145		156		900	274.3
135		146		157			
136		147		158			
137		148		159			

By permission: www.metric-conversions.org

2.0.5 Meters to Feet Conversion

Meters	Feet	Meters	Feet	Meters	Feet	Meters	Feet
0	0.00	27	88.58	63	206.692	99	324.803
0.1	0.328	28	91.86	64	209.973	100	328.083
0.2	0.656	29	95.14	65	213.254	101	331.364
0.3	0.984	30	98.42	66	216.535	102	334.645
0.4	1.312	31	101.70	67	219.816	103	337.926
0.5	1.640	32	104.98	68	223.097	104	341.207
0.6	1.968	33	108.26	69	226.377	105	344.488
0.7	2.296	34	111.54	70	229.658	106	347.769
0.8	2.624	35	114.82	71	232.939	107	351.049
0.9	2.952	36	118.11	72	236.220	108	354.330
1	3.28	37	121.39	73	239.501	109	357.611
2	6.56	38	124.67	74	242.782	110	360.892
3	9.84	39	127.95	75	246.062	111	364.173
4	13.12	40	131.233	76	249.343	112	367.454
5	16.40	41	134.514	77	252.624	113	370.734
6	19.68	42	137.795	78	255.905	114	374.015
7	22.96	43	141.076	79	259.186	115	377.296
8	26.24	44	144.356	80	262.467	116	380.577
9	29.52	45	147.637	81	265.748	117	383.858
10	32.80	46	150.918	82	269.028	118	387.139
11	36.08	47	154.199	83	272.309	119	390.419
12	39.37	48	157.480	84	275.590	120	393.700
13	42.65	49	160.761	85	278.871	121	396.981
14	45.93	50	164.041	86	282.152	122	400.262
15	49.21	51	167.322	87	285.433	123	403.543
16	52.49	52	170.603	88	288.713	124	406.824
17	55.77	53	173.884	89	291.994	125	410.104
18	59.05	54	177.165	90	295.275	126	413.385
19	62.33	55	180.446	91	298.556	127	416.666
20	65.61	56	183.727	92	301.837	128	419.947
21	68.89	57	187.007	93	305.118	129	423.228
22	72.17	58	190.288	94	308.398	130	426.508
23	75.45	59	193.569	95	311.679	131	429.79
24	78.74	60	196.850	96	314.960	132	433.07
25	82.02	61	200.131	97	318.241	133	436.351
26	85.30	62	203.412	98	321.522	134	439.632

Meters	Feet	Meters	Feet	Meters	Feet	Meters	Feet
135	442.913	144	472.440	153	501.968	400	1312.3
136	444.194	145	475.721	154	505.249	500	1640.4
137	449.475	146	479.002	155	508.530	600	1968.5
138	452.755	147	482.283	156	511.811	700	2296.5
139	456.036	148	485.564	157	515.091	800	2624.6
140	459.317	149	488.845	158	518.372	900	2952.7
141	462.598	150	492.125	159	521.653		
142	465.879	151	495.406	200	656.1		
143	469.160	152	498.687	300	984.2		

By permission: www.metric-conversions.org

2.0.6 Acres to Hectares Conversion

Acres	Hectares	Acres	Hectares	Acres	Hectares	Acres	Hectares
0	0.00	21	8.49	51	20.63	81	32.77
0.1	0.040	22	8.90	52	21.04	82	33.18
0.2	0.080	23	9.30	53	21.44	83	33.58
0.3	0.121	24	9.71	54	21.85	84	33.99
0.4	0.161	25	10.11	55	22.25	85	34.39
0.5	0.202	26	10.52	56	22.66	86	34.80
0.6	0.242	27	10.92	57	23.06	87	35.20
0.7	0.283	28	11.33	58	23.47	88	35.61
0.8	0.323	29	11.73	59	23.87	89	36.01
0.9	0.364	30	12.14	60	24.28	90	36.42
1	0.40	31	12.54	61	24.68	91	36.82
2	0.80	32	12.94	62	25.09	92	37.23
3	1.21	33	13.35	63	25.49	93	37.63
4	1.61	34	13.75	64	25.89	94	38.04
5	2.02	35	14.16	65	26.30	95	38.44
6	2.42	36	14.56	66	26.70	96	38.84
7	2.83	37	14.97	67	27.11	97	39.25
8	3.23	38	15.37	68	27.51	98	39.65
9	3.64	39	15.78	69	27.92	99	40.06
10	4.04	40	16.18	70	28.32	100	40.40
11	4.45	41	16.59	71	28.73	200	80.90
12	4.85	42	16.99	72	29.13	300	121.4
13	5.26	43	17.40	73	29.54	400	161.8
14	5.66	44	17.80	74	29.94	500	202.3
15	6.07	45	18.21	75	30.35	600	242.8
16	6.47	46	18.61	76	30.75	700	283.2
17	6.87	47	19.02	77	31.16	800	323.7
18	7.28	48	19.42	78	31.56	900	364.2
19	7.68	49	19.82	79	31.97		
20	8.09	50	20.23	80	32.37		

By permission: www.metric-conversions.org

2.0.7 Hectares to Acres Conversion

Hectares	Acres	Hectares	Acres	Hectares	Acres	Hectares	Acres
0	0.00	21	51.89	51	126.02	81	200.15
0.1	0.247	22	54.36	52	128.49	82	202.62
0.2	0.494	23	56.83	53	130.96	83	205.09
0.3	0.741	24	59.30	54	133.43	84	207.56
0.4	0.988	25	61.77	55	135.90	85	210.03
0.5	1.235	26	64.24	56	138.37	86	212.51
0.6	1.482	27	66.71	57	140.85	87	214.98
0.7	1.729	28	69.18	58	143.32	88	217.45
0.8	1.976	29	71.66	59	145.79	89	219.92
0.9	2.223	30	74.13	60	148.26	90	222.39
1	2.47	31	76.60	61	150.73	91	224.86
2	4.94	32	79.07	62	153.20	92	227.33
3	7.41	33	81.54	63	155.67	93	229.80
4	9.88	34	84.01	64	158.14	94	232.27
5	12.35	35	86.48	65	160.61	95	234.75
6	14.82	36	88.95	66	163.08	96	237.22
7	17.29	37	91.42	67	165.56	97	239.69
8	19.76	38	93.90	68	168.03	98	242.16
9	22.23	39	96.37	69	170.50	99	244.63
10	24.71	40	98.84	70	172.97	100	247.1
11	27.18	41	101.31	71	175.44	200	494.2
12	29.65	42	103.78	72	177.91	300	741.3
13	32.12	43	106.25	73	180.38	400	988.4
14	34.59	44	108.72	74	182.85	500	1235.5
15	37.06	45	111.19	75	185.32	600	1482.6
16	39.53	46	113.66	76	187.80	700	1729.7
17	42.00	47	116.13	77	190.27	800	1976.8
18	44.47	48	118.61	78	192.74	900	2223.9
19	46.95	49	121.08	79	195.21		
20	49.42	50	123.55	80	197.68		

By permission: www.metric-conversions.org

2.0.8 Square Inch to Square Feet Conversion

Square Inches	Square Feet	Square Inches	Square Feet	Square Inches	Square Feet	Square Inches	Square Feet
0	0.00	11	0.076	22	0.152	33	0.229
1	0.006	12	0.083	23	0.159	34	0.236
2	0.013	13	0.090	24	0.166	35	0.243
3	0.020	14	0.097	25	0.173	36	0.25
4	0.027	15	0.104	26	0.180	37	0.256
5	0.034	16	0.111	27	0.187	38	0.263
6	0.041	17	0.118	28	0.194	39	0.270
7	0.048	18	0.125	29	0.201	40	0.277
8	0.055	19	0.131	30	0.208	41	0.284
9	0.062	20	0.138	31	0.215	42	0.291
10	0.069	21	0.145	32	0.222	43	0.298

Square Inches	Square Feet	Square Inches	Square Feet	Square Inches	Square Feet	Square Inches	Square Feet
44	0.305	63	0.437	82	0.569	101	0.701
45	0.312	64	0.444	83	0.576	102	0.708
46	0.319	65	0.451	84	0.583	103	0.715
47	0.326	66	0.458	85	0.590	104	0.722
48	0.333	67	0.465	86	0.597	105	0.729
49	0.340	68	0.472	87	0.604	106	0.736
50	0.347	69	0.479	88	0.611	107	0.743
51	0.354	70	0.486	89	0.618	108	0.75
52	0.361	71	0.493	90	0.625	109	0.756
53	0.368	72	0.5	91	0.631	110	0.763
54	0.375	73	0.506	92	0.638	111	0.770
55	0.381	74	0.513	93	0.645	112	0.777
56	0.388	75	0.520	94	0.652	113	0.784
57	0.395	76	0.527	95	0.659	114	0.791
58	0.402	77	0.534	96	0.666	115	0.798
59	0.409	78	0.541	97	0.673	116	0.805
60	0.416	79	0.548	98	0.680	117	0.812
61	0.423	80	0.555	99	0.687	118	0.819
62	0.430	81	0.562	100	0.694	119	0.826

By permission: www.metric-conversions.org

2.0.9 Square Feet to Square Inch Conversion

Square Feet	Square Inches	Square Feet	Square Inches	Square Feet	Square Inches	Square Feet	Square Inches
0	0.00	13	1872	35	5040	57	8208
0.1	14.40	14	2016	36	5184	58	8352
0.2	28.80	15	2160	37	5328	59	8496
0.3	43.20	16	2304	38	5472	60	8640
0.4	57.60	17	2448	39	5616	61	8784
0.5	72.00	18	2592	40	5760	62	8928
0.6	86.40	19	2736	41	5904	63	9072
0.7	100.8	20	2880	42	6048	64	9216
0.8	115.2	21	3024	43	6192	65	9360
0.9	129.6	22	3168	44	6336	66	9504
1	144.0	23	3312	45	6480	67	9648
2	288.0	24	3456	46	6624	68	9792
3	432.0	25	3600	47	6768	69	9936
4	576.0	26	3744	48	6912	70	10080
5	720.0	27	3888	49	7056	71	10224
6	864.0	28	4032	50	7200	72	10368
7	1008	29	4176	51	7344	73	10512
8	1152	30	4320	52	7488	74	10656
9	1296	31	4464	53	7632	75	10800
10	1440	32	4608	54	7776	76	10944
11	1584	33	4752	55	7920	77	11088
12	1728	34	4896	56	8064	78	11232

(Continued)

Square Feet	Square Inches	Square Feet	Square Inches	Square Feet	Square Inches	Square Feet	Square Inches
79	11376	87	12528	95	13680	400	57600
80	11520	88	12672	96	13824	500	72000
81	11664	89	12816	97	13968	600	86400
82	11808	90	12960	98	14112	700	100800
83	11952	91	13104	99	14256	800	115200
84	12096	92	13248	100	14400	900	129600
85	12240	93	13392	200	28800		
86	123845	94	13536	300	43200		

By permission: www.metric-conversions.org

2.0.10 Square Feet to Square Mile Conversion

Square Feet	Square Miles	Square Feet	Square Miles	Square Feet	Square Miles	Square Feet	Square Miles
0	0.00	21	7.53	51	1.82	81	2.90
0.1	3.587	22	7.89	52	1.86	82	2.94
0.2	7.174	23	8.25	53	1.90	83	2.97
0.3	1.076	24	8.60	54	1.93	84	3.01
0.4	1.434	25	8.96	55	1.97	85	3.04
0.5	1.793	26	9.32	56	2.00	86	3.08
0.6	2.152	27	9.68	57	2.04	87	3.12
0.7	2.510	28	1.00	58	2.08	88	3.15
0.8	2.669	29	1.04	59	2.11	89	3.19
0.9	3.228	30	1.07	60	2.15	90	3.22
1	3.58	31	1.11	61	2.18	91	3.26
2	7.17	32	1.14	62	2.22	92	3.30
3	1.07	33	1.18	63	2.25	93	3.33
4	1.43	34	1.21	64	2.29	94	3.37
5	1.79	35	1.25	65	2.33	95	3.40
6	2.15	36	1.29	66	2.36	96	3.44
7	2.51	37	1.32	67	2.40	97	3.47
8	2.86	38	1.36	68	2.43	98	3.51
9	3.22	39	1.39	69	2.47	99	3.55
10	3.58	40	1.43	70	2.51	100	3.50
11	3.94	41	1.47	71	2.54	200	7.10
12	4.30	42	1.50	72	2.58	300	1.00
13	4.66	43	1.54	73	2.61	400	1.40
14	5.02	44	1.57	74	2.65	500	1.70
15	5.38	45	1.61	75	2.69	600	2.10
16	5.73	46	1.65	76	2.72	700	2.60
17	6.09	47	1.68	77	2.76	800	2.80
18	5.45	48	1.72	78	2.79	900	3.20
19	6.81	49	1.75	79	2.83		
20	7.17	50	1.79	80	2.86		

By permission: www.metric-conversions.org

2.0.11 Square Mile to Square Feet Conversion

Square Miles	Square Feet	Square Miles	Square Feet	Square Miles	Square Feet	Square Miles	Square Feet
0	0.00	21	585446400	51	1421798400	81	2258150400
0.1	2767840	22	613324800	52	1449676800	82	2286028800
0.2	5575660	23	641203200	53	1477555200	83	2313907200
0.3	8363520	24	669081600	54	1505433600	84	2341785600
0.4	11151360	25	696960000	55	1533312000	85	2369664000
0.5	13939200	26	724838400	56	1561190400	86	2397542400
0.6	16727040	27	752716800	57	1589068800	87	2425420800
0.7	19514880	28	760595200	58	1616947200	88	2453299200
0.8	22302720	29	808473600	59	1644825600	89	2481177600
0.9	25090560	30	836352000	60	1672704000	90	2509056000
1	27878400	31	864230400	61	1700582400	91	2536934400
2	55756800	32	892108800	62	1728460800	92	2584812800
3	83635200	33	919987200	63	1756339200	93	2592691200
4	111513600	34	947865600	64	1784217600	94	2620569600
5	139392000	35	975744000	65	1812096000	95	2648448000
6	167270400	36	1003622400	66	1839974400	96	2676326400
7	195148800	37	1031500800	67	1867852800	97	2704204800
8	223027200	38	1059379200	68	1895731200	98	2732083200
9	250905600	39	1067257600	69	1923609600	99	2759961600
10	278784000	40	1115136000	70	1951488000	100	2787840000
11	306662400	41	1143014400	71	1979366400	200	5575680000
12	334540800	42	1170892800	72	2007244800	300	8363520000
13	362419200	43	1198771200	73	2035123200	400	11151360000
14	390297600	44	1226649600	74	2063001600	500	13939200000
15	418176000	45	1254526000	75	2090880000	600	16727040000
16	446054400	46	1282406400	76	2118758400	700	19514880000
17	473932800	47	1310284800	77	2146636800	800	22302720000
18	501811200	48	1338153200	78	2174515200	900	25090560000
19	529589600	49	1366041600	79	2202393600		
20	557568000	50	1393920000	80	2230272000		

By permission: www.metric-conversions.org

2.0.12 Square Feet to Acres Conversion

Square Feet	Acres	Square Feet	Acres	Square Feet	Acres	Square Feet	Acres
0	0.00	3	6.88	15	0.00	27	0.00
0.1	2.295	4	9.18	16	0.00	28	0.00
0.2	4.591	5	0.00	17	0.00	29	0.00
0.3	6.887	6	0.00	18	0.00	30	0.00
0.4	9.182	7	0.00	19	0.00	31	0.00
0.5	1.147	8	0.00	20	0.00	32	0.00
0.6	1.377	9	0.00	21	0.00	33	0.00
0.7	1.606	10	0.00	22	0.00	34	0.00
0.8	1.836	11	0.00	23	0.00	35	0.00
0.9	2.066	12	0.00	24	0.00	36	0.00
1	2.29	13	0.00	25	0.00	37	0.00
2	4.59	14	0.00	26	0.00	38	0.00

(Continued)

Square Feet	Acres	Square Feet	Acres	Square Feet	Acres	Square Feet	Acres
39	0.00	57	0.00	75	0.00	93	0.00
40	0.00	58	0.00	76	0.00	94	0.00
41	0.00	59	0.00	77	0.00	95	0.00
42	0.00	60	0.00	78	0.00	96	0.00
43	0.00	61	0.00	79	0.00	97	0.00
44	0.00	62	0.00	80	0.00	98	0.00
45	0.00	63	0.00	81	0.00	99	0.00
46	0.00	64	0.00	82	0.00	100	0.00
47	0.00	65	0.00	83	0.00	200	0.00
48	0.00	66	0.00	84	0.00	300	0.00
49	0.00	67	0.00	85	0.00	400	0.00
50	0.00	68	0.00	86	0.00	500	0.00
51	0.00	69	0.00	87	0.00	600	0.00
52	0.00	70	0.00	88	0.00	700	0.00
53	0.00	71	0.00	89	0.00	800	0.00
54	0.00	72	0.00	90	0.00	900	0.00
55	0.00	73	0.00	91	0.00		
56	0.00	74	0.00	92	0.00		

By permission: www.metric-conversions.org

2.0.13 Acres to Square Feet Conversion

Acres	Square Feet	Acres	Square Feet	Acres	Square Feet	Acres	Square Feet
0	0.00	21	914760	51	2221560	81	3528360
0.1	4356	22	958320	52	2265120	82	3571920
0.2	8712	23	1001880	53	2308680	83	3615480
0.3	13068	24	1045440	54	2352240	84	3659040
0.4	17424	25	1089000	55	2395800	85	3702600
0.5	21780	26	1132560	56	2439360	86	3746160
0.6	26136	27	1176120	57	2482920	87	3789720
0.7	30492	28	1219680	58	2526480	88	3833280
0.8	34848	29	1263240	59	2570040	89	3876840
0.9	39204	30	1306800	60	2613600	90	3920400
1	43560	31	1350360	61	2657160	91	3963960
2	87120	32	1393920	62	2700720	92	4007520
3	130680	33	1437480	63	2744280	93	4051080
4	174240	34	1481040	64	2787840	94	4094640
5	217800	35	1524600	65	2831400	95	4138200
6	261360	36	1568160	66	2874960	96	4181760
7	304920	37	1611720	67	2918520	97	4225320
8	348480	38	1655280	68	2962080	98	4268880
9	392040	39	1698840	69	3005640	99	4312440
10	435600	40	1742400	70	3049200	100	4356000
11	479160	41	1785960	71	3092760	200	8712000
12	522720	42	1829520	72	3136320	300	13068000
13	566280	43	1873080	73	3179880	400	17424000
14	609840	44	1916640	74	3223440	500	21780000
15	653400	45	1960200	75	3267000	600	26136000
16	696960	46	2003760	76	3310560	700	30492000
17	740520	47	2047320	77	3354120	800	34848000
18	784080	48	2090880	78	3397680	900	39204000
19	827640	49	2134440	79	3441240		
20	871200	50	2178000	80	3484800		

By permission: www.metric-conversions.org

Section | 2 Conversion Tables and Conversion Formulas

2.0.14 Square Yard to Square Meter Conversion Table

Square Yards	Square Meters	Square Yards	Square Meters	Square Yards	Square Meters	Square Yards	Square Meters
0	0.00	21	17.55	51	42.64	81	67.72
0.1	0.083	22	18.39	52	43.47	82	68.56
0.2	0.167	23	19.23	53	44.31	83	69.39
0.3	0.250	24	20.06	54	45.15	84	70.23
0.4	0.334	25	20.90	55	45.98	85	71.07
0.5	0.418	26	21.73	56	46.82	86	71.90
0.6	0.501	27	22.57	57	47.65	87	72.74
0.7	0.585	28	23.41	58	48.49	88	73.57
0.8	0.668	29	24.24	59	49.33	89	74.41
0.9	0.752	30	25.08	60	50.16	90	75.25
1	0.83	31	25.91	61	51.00	91	76.08
2	1.67	32	26.75	62	51.83	92	76.92
3	2.50	33	27.59	63	52.67	93	77.75
4	3.34	34	28.42	64	53.51	94	78.59
5	4.18	35	29.26	65	54.34	95	79.43
6	5.01	36	30.10	66	55.18	96	80.26
7	5.85	37	30.93	67	56.02	97	81.10
8	6.68	38	31.77	68	56.85	98	81.94
9	7.52	39	32.60	69	57.69	99	82.77
10	8.36	40	33.44	70	58.52	100	83.60
11	9.19	41	34.28	71	59.36	200	167.2
12	10.03	42	35.11	72	60.20	300	250.8
13	10.86	43	35.95	73	61.03	400	334.4
14	11.70	44	36.78	74	61.87	500	418.0
15	12.54	45	37.62	75	62.70	600	501.6
16	13.37	46	38.46	76	63.54	700	585.2
17	14.21	47	39.29	77	64.38	800	668.9
18	15.05	48	40.13	78	65.21	900	752.5
19	15.88	49	40.97	79	66.05		
20	16.72	50	41.80	80	66.89		

By permission: www.metric-conversions.org

2.0.15 Square Meter to Square Yard Conversion Table

Square Meters	Square Yards	Square Meters	Square Yards	Square Meters	Square Yards	Square Meters	Square Yards
0	0.00	1	1.19	11	13.15	21	25.11
0.1	0.119	2	2.39	12	14.35	22	26.31
0.2	0.239	3	3.58	13	15.54	23	27.50
0.3	0.358	4	4.78	14	16.74	24	28.70
0.4	0.478	5	5.97	15	17.93	25	29.89
0.5	0.597	6	7.17	16	19.13	26	31.09
0.6	0.717	7	8.37	17	20.33	27	32.29
0.7	0.837	8	9.56	18	21.52	28	33.48
0.8	0.956	9	10.76	19	22.72	29	34.68
0.9	1.076	10	11.95	20	23.91	30	35.87

(Continued)

Square Meters	Square Yards	Square Meters	Square Yards	Square Meters	Square Yards	Square Meters	Square Yards
31	37.07	51	60.99	71	84.91	91	108.83
32	38.27	52	62.19	72	86.11	92	110.03
33	39.46	53	63.38	73	87.30	93	111.22
34	40.66	54	64.58	74	88.50	94	112.42
35	41.85	55	65.77	75	89.69	95	113.61
36	43.05	56	66.97	76	90.89	96	114.81
37	44.25	57	68.17	77	92.09	97	116.01
38	45.44	58	69.36	78	93.28	98	117.20
39	46.64	59	70.56	79	94.48	99	118.40
40	47.83	60	71.75	80	95.67	100	119.5
41	49.03	61	72.95	81	96.87	200	239.1
42	50.23	62	74.15	82	98.07	300	358.7
43	51.42	63	75.34	83	99.26	400	478.3
44	52.62	64	76.54	84	100.46	500	597.9
45	53.81	65	77.73	85	101.65	600	717.5
46	55.01	66	78.93	86	102.85	700	837.1
47	56.21	67	80.13	87	104.05	800	956.7
48	57.40	68	81.32	88	105.24	900	1076.3
49	58.60	69	82.52	89	106.44		
50	59.79	70	83.71	90	107.63		

By permission: www.metric-conversions.org

2.0.16 Square Mile to Square Meter Conversion Table

Square Miles	Square Meters	Square Miles	Square Meters	Square Miles	Square Meters	Square Miles	Square Meters
0	0.00	13	33669845.43	35	90649583.86	57	147629322.28
0.1	258998.811	14	36259833.54	36	93239571.97	58	150219310.39
0.2	517997.622	15	38849821.65	37	95829560.08	59	152809298.50
0.3	776996.433	16	41439809.76	38	98419548.19	60	155399286.62
0.4	1035995.244	17	44029797.87	39	101009536.30	61	157989274.73
0.5	1294994.055	18	46619785.98	40	103599524.41	62	160579262.84
0.6	1553992.866	19	49209774.09	41	106189512.52	63	163169250.95
0.7	1812991.677	20	51799762.20	42	108779500.63	64	165759239.06
0.8	2071990.488	21	54389750.31	43	111369488.74	65	168349227.17
0.9	2330989.299	22	56979738.42	44	113959476.85	66	170939215.28
1	2589988.11	23	59569726.53	45	116549464.96	67	173529203.39
2	5179976.22	24	62159714.64	46	119139453.07	68	176119191.50
3	7769964.33	25	64749702.75	47	121729441.18	69	178709179.61
4	10359952.44	26	67339690.86	48	124319429.29	70	181299167.72
5	12949940.55	27	69929678.97	49	126909417.40	71	183889155.83
6	15539928.66	28	72519667.08	50	129499405.51	72	186479143.94
7	18129916.77	29	75109655.19	51	132089393.62	73	189069132.05
8	20719904.88	30	77699643.31	52	134679381.73	74	191659120.16
9	23309892.99	31	80289631.42	53	137269369.84	75	194249108.27
10	25899881.10	32	82879619.53	54	139859357.95	76	196839096.38
11	28489869.21	33	85469607.64	55	142449346.06	77	199429084.49
12	31079857.32	34	88059595.75	56	145039334.17	78	202019072.60

Square Miles	Square Meters	Square Miles	Square Meters	Square Miles	Square Meters	Square Miles	Square Meters
79	204609060.71	87	225328965.59	95	246048870.48	400	1035995244.1
80	207199048.82	88	227918953.70	96	248638858.59	500	1294994055.1
81	209789036.93	89	230508941.81	97	251228846.70	600	1553992866.2
82	212379025.04	90	233098929.93	98	253818834.81	700	1812991677.2
83	214969013.15	91	235688918.04	99	256408822.92	800	2071990488.2
84	217559001.26	92	238278906.15	100	258998811.0	900	2330989299.3
85	220148989.37	93	240868894.26	200	517997622.0		
86	222738977.48	94	243458882.37	300	776996433.1		

By permission: www.metric-conversions.org

2.0.17 Square Meter to Square Mile Conversion Table

Square Meters	Square Miles	Square Meters	Square Miles	Square Meters	Square Miles	Square Meters	Square Miles
0	0.00	21	8.10	51	1.96	81	3.12
0.1	3.861	22	8.49	52	2.00	82	3.16
0.2	7.722	23	8.88	53	2.04	83	3.20
0.3	1.158	24	9.26	54	2.08	84	3.24
0.4	1.544	25	9.65	55	2.12	85	3.28
0.5	1.930	26	1.00	56	2.16	86	3.32
0.6	2.316	27	1.04	57	2.20	87	3.35
0.7	2.702	28	1.08	58	2.23	88	3.39
0.8	3.088	29	1.11	59	2.27	89	3.43
0.9	3.474	30	1.15	60	2.31	90	3.47
1	3.86	31	1.19	61	2.35	91	3.51
2	7.72	32	1.23	62	2.39	92	3.55
3	1.15	33	1.27	63	2.43	93	3.59
4	1.54	34	1.31	64	2.47	94	3.62
5	1.93	35	1.35	65	2.50	95	3.66
6	2.31	36	1.38	66	2.54	96	3.70
7	2.70	37	1.42	67	2.58	97	3.74
8	3.08	38	1.46	68	2.62	98	3.78
9	3.47	39	1.50	69	2.66	99	3.82
10	3.86	40	1.54	70	2.70	100	3.80
11	4.24	41	1.58	71	2.74	200	7.70
12	4.63	42	1.62	72	2.77	300	0.0
13	5.01	43	1.66	73	2.81	400	0.0
14	5.40	44	1.69	74	2.85	500	0.0
15	5.79	45	1.73	75	2.89	600	0.0
16	6.17	46	1.77	76	2.93	700	0.0
17	6.56	47	1.81	77	2.97	800	0.0
18	6.94	48	1.85	78	3.01	900	0.0
19	7.33	49	1.89	79	3.05		
20	7.72	50	1.93	80	3.08		

By permission: www.metric-conversions.org

2.0.18 Square Mile to Hectare Conversion Table

Square Miles	Hectares	Square Miles	Hectares	Square Miles	Hectares	Square Miles	Hectares
0	0.00	21	5438.97	51	13208.93	81	20978.90
0.1	25.899	22	5697.97	52	13467.93	82	21237.90
0.2	51.799	23	5956.97	53	13726.93	83	21496.90
0.3	77.699	24	6215.97	54	13985.93	84	21755.90
0.4	103.599	25	6474.97	55	14244.93	85	22014.89
0.5	129.499	26	6733.96	56	14503.93	86	22273.89
0.6	155.399	27	6992.96	57	14762.93	87	22532.89
0.7	181.299	28	7251.96	58	15021.93	88	22791.89
0.8	207.199	29	7510.96	59	15280.92	89	23050.89
0.9	233.098	30	7769.96	60	15539.92	90	23309.89
1	258.99	31	8028.96	61	15798.92	91	23568.89
2	517.99	32	8287.96	62	16057.92	92	23827.89
3	776.99	33	8546.96	63	16316.92	93	24086.88
4	1035.99	34	8805.95	64	16575.92	94	24345.88
5	1294.99	35	9064.95	65	16834.92	95	24604.88
6	1553.99	36	9323.95	66	17093.92	96	24863.88
7	1812.99	37	9582.95	67	17352.92	97	25122.88
8	2071.99	38	9841.95	68	17611.91	98	25381.88
9	2330.98	39	10100.95	69	17870.91	99	25640.88
10	2589.98	40	10359.95	70	18129.91	100	25899.8
11	2848.98	41	10618.95	71	18388.91	200	51799.7
12	3107.98	42	10877.95	72	18647.91	300	77699.6
13	3366.98	43	11136.94	73	18906.91	400	103599.5
14	3625.98	44	11395.94	74	19165.91	500	129499.4
15	3884.98	45	11654.94	75	19424.91	600	155399.2
16	4143.98	46	11913.94	76	19683.90	700	181299.1
17	4402.97	47	12172.94	77	19942.90	800	207199.0
18	4661.97	48	12431.94	78	20201.90	900	233098.9
19	4920.97	49	12690.94	79	20460.90		
20	5179.97	50	12949.94	80	20719.90		

By permission: www.metric-conversions.org

2.0.19 Hectare to Square Mile Conversion Table

Hectares	Square Miles	Hectares	Square Miles	Hectares	Square Miles	Hectares	Square Miles
0	0.00	5	0.01	19	0.07	33	0.12
0.1	0.000	6	0.02	20	0.07	34	0.13
0.2	0.000	7	0.02	21	0.08	35	0.13
0.3	0.001	8	0.03	22	0.08	36	0.13
0.4	0.001	9	0.03	23	0.08	37	0.14
0.5	0.001	10	0.03	24	0.09	38	0.14
0.6	0.002	11	0.04	25	0.09	39	0.15
0.7	0.002	12	0.04	26	0.10	40	0.15
0.8	0.003	13	0.05	27	0.10	41	0.15
0.9	0.003	14	0.05	28	0.10	42	0.16
1	0.00	15	0.05	29	0.11	43	0.16
2	0.00	16	0.06	30	0.11	44	0.16
3	0.01	17	0.06	31	0.11	45	0.17
4	0.01	18	0.06	32	0.12	46	0.17

Hectares	Square Miles	Hectares	Square Miles	Hectares	Square Miles	Hectares	Square Miles
47	0.18	63	0.24	79	0.30	95	0.36
48	0.18	64	0.24	80	0.30	96	0.37
49	0.18	65	0.25	81	0.31	97	0.37
50	0.19	66	0.25	82	0.31	98	0.37
51	0.19	67	0.25	83	0.32	99	0.38
52	0.20	68	0.26	84	0.32	100	0.3
53	0.20	69	0.26	85	0.32	200	0.7
54	0.20	70	0.27	86	0.33	300	1.10
55	0.21	71	0.27	87	0.33	400	1.50
56	0.21	72	0.27	88	0.33	500	1.90
57	0.22	73	0.28	89	0.34	600	2.30
58	0.22	74	0.28	90	0.34	700	2.70
59	0.22	75	0.28	91	0.35	800	3.00
60	0.23	76	0.29	92	0.35	900	3.40
61	0.23	77	0.29	93	0.35		
62	0.23	78	0.30	94	0.36		

By permission: www.metric-conversions.org

2.0.20 Miles to Kilometers Conversion

Miles	Kilometers	Miles	Kilometers	Miles	Kilometers	Miles	Kilometers
0	0.00	22	35.40	53	85.295	84	135.184
0.1	0.160	23	37.01	54	86.904	85	136.794
0.2	0.321	24	38.62	55	88.513	86	138.403
0.3	0.482	25	40.23	56	90.123	87	140.012
0.4	0.643	26	41.84	57	91.732	88	141.622
0.5	0.804	27	43.45	58	93.341	89	143.231
0.6	0.965	28	45.06	59	94.951	90	144.840
0.7	1.126	29	46.67	60	96.560	91	146.450
0.8	1.287	30	48.28	61	98.169	92	148.059
0.9	1.448	31	49.88	62	99.779	93	149.668
1	1.60	32	51.49	63	101.388	94	151.278
2	3.21	33	53.10	64	102.998	95	152.887
3	4.82	34	54.71	65	104.607	96	154.497
4	6.43	35	56.32	66	106.218	97	156.106
5	8.04	36	57.93	67	107.826	98	157.715
6	9.65	37	59.54	68	109.435	99	159.325
7	11.26	38	61.15	69	111.044	100	160.934
8	12.87	39	62.76	70	112.654	101	162.543
9	14.48	40	64.373	71	114.263	102	164.153
10	16.09	41	65.983	72	115.872	103	165.782
11	17.70	42	67.592	73	117.482	104	167.371
12	19.31	43	69.201	74	119.091	105	168.981
13	20.92	44	70.811	75	120.700	106	170.590
14	22.53	45	72.420	76	122.310	107	172.199
15	24.14	46	74.029	77	123.919	108	173.809
16	25.74	47	75.639	78	125.528	109	175.418
17	27.35	48	77.248	79	127.138	110	177.027
18	28.96	49	78.857	80	128.747	111	178.637
19	30.57	50	80.467	81	130.356	112	180.246
20	32.18	51	82.076	82	131.966	113	181.855
21	33.79	52	83.685	83	133.575	114	183.465

(Continued)

Miles	Kilometers	Miles	Kilometers	Miles	Kilometers	Miles	Kilometers
115	185.074	129	207.605	143	230.136	157	252.667
116	186.683	130	209.214	144	231.745	158	254.276
117	188.293	131	210.824	145	233.354	159	255.885
118	189.902	132	212.433	146	234.964	200	321.8
119	191.511	133	214.042	147	236.573	300	482.8
120	193.121	134	215.652	148	238.182	400	643.7
121	194.730	135	217.261	149	239.792	500	804.6
122	196.339	136	218.870	150	241.401	600	965.6
123	197.949	137	220.480	151	243.010	700	1126.5
124	199.558	138	222.089	152	244.620	800	1287.4
125	201.168	139	223.698	153	246.229	900	1448.4
126	202.777	140	225.308	154	247.838		
127	204.386	141	226.917	155	249.448		
128	205.996	142	228.526	156	251.057		

By permission: www.metric-conversions.org

2.0.21 Kilometers to Miles Conversion

Kilometers	Miles	Kilometers	Miles	Kilometers	Miles	Kilometers	Miles
0	0.00	24	14.91	57	35.418	90	55.923
0.1	0.062	25	15.53	58	36.039	91	56.544
0.2	0.124	26	16.15	59	36.660	92	57.166
0.3	0.186	27	16.77	60	37.282	93	57.787
0.4	0.248	28	17.39	61	37.903	94	58.408
0.5	0.310	29	18.01	62	38.525	95	59.030
0.6	0.372	30	18.64	63	39.146	96	59.551
0.7	0.434	31	19.26	64	39.767	97	60.273
0.8	0.497	32	19.88	65	40.389	98	60.894
0.9	0.559	33	20.50	66	41.010	99	61.515
1	0.62	34	21.12	67	41.631	100	62.137
2	1.24	35	21.74	68	42.253	101	62.758
3	1.86	36	22.36	69	42.874	102	63.379
4	2.48	37	22.99	70	43.495	103	64.001
5	3.10	38	23.61	71	44.117	104	64.622
6	3.72	39	24.23	72	44.738	105	65.243
7	4.34	40	24.854	73	45.360	106	65.865
8	4.97	41	25.476	74	45.981	107	66.486
9	5.59	42	26.097	75	46.602	108	67.108
10	6.21	43	26.718	76	47.224	109	67.729
11	6.83	44	27.340	77	47.845	110	68.350
12	7.45	45	27.961	78	48.466	111	68.972
13	8.07	46	28.583	79	49.088	112	69.593
14	8.69	47	29.204	80	49.709	113	70.214
15	9.32	48	29.825	81	50.331	114	70.836
16	9.94	49	30.447	82	50.952	115	71.457
17	10.56	50	31.068	83	51.573	116	72.079
18	11.18	51	31.689	84	52.195	117	72.700
19	11.80	52	32.311	85	52.816	118	73.321
20	12.42	53	32.932	86	53.437	119	73.943
21	13.04	54	33.554	87	54.059	120	74.564
22	13.67	55	34.175	88	54.680	121	75.185
23	14.29	56	34.796	89	55.302	122	75.807

Kilometers	Miles	Kilometers	Miles	Kilometers	Miles	Kilometers	Miles
123	76.428	135	83.885	147	91.341	159	98.798
124	77.050	136	84.506	148	91.962	200	124.2
125	77.671	137	85.127	149	92.584	300	186.4
126	78.292	138	85.749	150	93.205	400	248.5
127	78.914	139	86.370	151	93.827	500	310.6
128	79.535	140	86.991	152	94.448	600	372.8
129	80.156	141	87.613	153	95.069	700	434.9
130	80.778	142	88.234	154	95.691	800	497.0
131	81.399	143	88.856	155	96.312	900	559.2
132	82.020	144	89.477	156	96.933		
133	82.642	145	90.098	157	97.555		
134	83.263	146	90.720	158	98.176		

By permission: www.metric-conversions.org

2.0.22 Pounds to Kilograms Conversion

Pounds	Kilograms	Pounds	Kilograms	Pounds	Kilograms	Pounds	Kilograms
0	0.00	26	11.79	61	27.669	96	43.544
0.1	0.045	27	12.24	62	28.122	97	43.998
0.2	0.090	28	12.70	63	28.576	98	44.452
0.3	0.136	29	13.15	64	29.029	99	44.905
0.4	0.181	30	13.60	65	29.483	100	45.359
0.5	0.226	31	14.06	66	29.937	101	45.812
0.6	0.272	32	14.51	67	30.390	102	46.266
0.7	0.317	33	14.96	68	30.844	103	46.720
0.8	0.362	34	15.42	69	31.297	104	47.173
0.9	0.408	35	15.87	70	31.751	105	47.627
1	0.45	36	16.32	71	32.205	106	48.080
2	0.90	37	16.78	72	32.658	107	48.534
3	1.36	38	17.23	73	33.112	108	48.987
4	1.81	39	17.69	74	33.565	109	49.441
5	2.26	40	18.143	75	34.019	110	49.895
6	2.72	41	18.597	76	34.473	111	50.348
7	3.17	42	19.050	77	34.926	112	50.802
8	3.62	43	19.504	78	35.380	113	51.255
9	4.08	44	19.958	79	35.833	114	51.709
10	4.53	45	20.411	80	36.287	115	52.163
11	4.98	46	20.865	81	36.740	116	52.616
12	5.44	47	21.318	82	37.194	117	53.070
13	5.89	48	21.772	83	37.648	118	53.523
14	6.35	49	22.226	84	38.101	119	53.977
15	6.80	50	22.679	85	38.555	120	54.431
16	7.25	51	23.133	86	39.008	121	54.884
17	7.71	52	23.586	87	39.462	122	55.338
18	8.16	53	24.040	88	39.916	123	55.791
19	8.61	54	24.493	89	40.369	124	56.245
20	9.07	55	24.947	90	40.823	125	56.699
21	9.52	56	25.401	91	41.276	126	57.152
22	9.97	57	25.854	92	41.730	127	57.606
23	10.43	58	26.308	93	42.184	128	58.059
24	10.88	59	26.761	94	42.637	129	58.513
25	11.33	60	27.215	95	43.091	130	58.967

(Continued)

Pounds	Kilograms	Pounds	Kilograms	Pounds	Kilograms	Pounds	Kilograms
131	59.240	143	64.863	155	70.306	900	408.2
132	59.870	144	65.317	156	70.760		
133	60.327	145	65.770	157	71.214		
134	60.781	146	66.224	158	71.667		
135	61.234	147	66.678	159	72.121		
136	61.688	148	67.131	200	90.70		
137	62.142	149	67.585	300	136.0		
138	62.595	150	68.038	400	181.4		
139	63.049	151	68.492	500	226.7		
140	63.502	152	68.946	600	272.1		
141	63.596	153	69.399	700	317.5		
142	64.410	154	69.853	800	362.8		

By permission: www.metric-conversions.org

2.0.23 Kilograms to Pounds Conversion

Kilograms	Pounds	Kilograms	Pounds	Kilograms	Pounds	Kilograms	Pounds
0	0.00	26	57.32	61	134.481	96	211.643
0.1	0.220	27	59.52	62	136.686	97	213.848
0.2	0.440	28	61.72	63	138.891	98	216.053
0.3	0.661	29	63.93	64	141.095	99	218.257
0.4	0.881	30	66.13	65	143.300	100	220.462
0.5	1.102	31	68.34	66	145.605	101	222.666
0.6	1.322	32	70.54	67	147.709	102	224.871
0.7	1.543	33	72.75	68	149.914	103	227.076
0.8	1.763	34	74.95	69	152.118	104	229.280
0.9	1.984	35	77.16	70	154.323	105	231.485
1	2.20	36	79.36	71	156.528	106	233.689
2	4.40	37	81.57	72	158.732	107	235.894
3	6.61	38	83.77	73	160.937	108	238.099
4	8.81	39	85.98	74	163.142	109	240.303
5	11.02	40	88.184	75	165.346	110	242.508
6	13.22	41	90.389	76	167.551	111	244.713
7	15.43	42	92.594	77	169.755	112	246.917
8	17.63	43	94.798	78	171.960	113	249.122
9	19.84	44	97.003	79	174.165	114	251.326
10	22.04	45	99.208	80	176.369	115	253.531
11	24.25	46	101.412	81	178.574	116	255.736
12	26.45	47	103.617	82	180.779	117	257.940
13	28.66	48	105.821	83	182.983	118	260.145
14	30.86	49	108.026	84	185.188	119	262.350
15	33.06	50	110.231	85	187.392	120	264.554
16	35.27	51	112.435	86	189.597	121	266.759
17	37.47	52	114.640	87	191.802	122	268.963
18	39.68	53	116.844	88	194.006	123	271.168
19	41.88	54	119.049	89	196.211	124	273.373
20	44.09	55	121.254	90	198.416	125	275.577
21	46.29	56	123.458	91	200.620	126	277.782
22	48.50	57	125.663	92	202.825	127	279.987
23	50.70	58	127.868	93	205.029	128	282.191
24	52.91	59	130.072	94	207.234	129	284.396
25	55.11	60	132.277	95	209.439	130	286.600

Kilograms	Pounds	Kilograms	Pounds	Kilograms	Pounds	Kilograms	Pounds
131	288.805	141	310.851	151	332.898	300	661.3
132	291.010	142	313.056	152	335.102	400	881.8
133	293.214	143	315.261	153	337.307	500	1102.3
134	295.419	144	317.465	154	339.511	600	1322.7
135	297.624	145	319.670	155	341.716	700	1543.2
136	299.828	146	319.875	156	343.921	800	1763.6
137	302.033	147	324.079	157	346.125	900	1984.1
138	304.237	148	326.284	158	348.330		
139	306.422	149	329.488	159	350.534		
140	308.647	150	330.693	200	440.9		

By permission: www.metric-conversions.org

2.0.24 Fahrenheit to Celsius Conversion

Fahrenheit	Celsius	Fahrenheit	Celsius	Fahrenheit	Celsius	Fahrenheit	Celsius
0	−17.777	21	−6.11	51	10.55	81	27.22
0.1	−17.722	22	−5.55	52	11.11	82	27.77
0.2	−17.666	23	−5.00	53	11.66	83	28.33
0.3	−17.611	24	−4.44	54	12.22	84	28.88
0.4	−17.555	25	−3.88	55	12.77	85	29.44
0.5	−17.5	26	−3.33	56	13.33	86	30.00
0.6	−17.444	27	−2.77	57	13.88	87	30.55
0.7	−17.388	28	−2.22	58	14.44	88	31.11
0.8	−17.333	29	−1.66	59	15.00	89	31.66
0.9	−17.277	30	−1.11	60	15.55	90	32.22
1	−17.22	31	−0.55	61	16.11	91	32.77
2	−16.66	32	0.00	62	16.66	92	33.33
3	−16.11	33	0.55	63	17.22	93	33.88
4	−15.55	34	1.11	64	17.77	94	34.44
5	−15.00	35	1.66	65	18.33	95	35.00
6	−14.44	36	2.22	66	18.88	96	35.55
7	−13.88	37	2.77	67	19.44	97	36.11
8	−13.33	38	3.33	68	20.00	98	36.66
9	−12.77	39	3.88	69	20.55	99	37.22
10	−12.22	40	4.44	70	21.11	100	37.70
11	−11.66	41	5.00	71	21.66	200	93.30
12	−11.11	42	5.55	72	22.22	300	148.8
13	−10.55	43	6.11	73	22.77	400	204.4
14	−10.00	44	6.66	74	23.33	500	260.0
15	−9.44	45	7.22	75	23.88	600	315.5
16	−8.88	46	7.77	76	24.44	700	371.1
17	−8.33	47	8.33	77	25.00	800	426.6
18	−7.77	48	8.88	78	25.55	900	482.2
19	−7.22	49	9.44	79	26.11		
20	−6.66	50	10.00	80	26.66		

By permission: www.metric-conversions.org

2.0.25 Celsius to Fahrenheit Temperature Conversion

Celsius	Fahrenheit	Celsius	Fahrenheit	Celsius	Fahrenheit	Celsius	Fahrenheit
0	32.00	21	69.80	51	123.8	81	177.8
0.1	32.18	22	71.60	52	125.6	82	179.6
0.2	32.36	23	73.40	53	127.4	83	181.4
0.3	32.54	24	75.20	54	129.2	84	183.2
0.4	32.72	25	77.00	55	131.0	85	185.0
0.5	32.90	26	78.80	56	132.8	86	186.8
0.6	33.08	27	80.60	57	134.6	87	188.6
0.7	33.26	28	82.40	58	136.4	88	190.4
0.8	33.44	29	84.20	59	138.2	89	192.2
0.9	33.62	30	86.00	60	140.0	90	194.0
1	33.80	31	87.80	61	141.8	91	195.8
2	35.60	32	89.60	62	143.6	92	197.6
3	37.40	33	91.40	63	145.4	93	199.4
4	39.20	34	93.20	64	147.2	94	201.2
5	41.00	35	95.00	65	149.0	95	203.0
6	42.80	36	96.80	66	150.8	96	204.8
7	44.60	37	98.60	67	152.6	97	206.6
8	45.40	38	100.4	68	154.4	98	208.4
9	48.20	39	102.2	69	156.2	99	210.2
10	50.00	40	104.0	70	158.0	100	212.0
11	51.80	41	105.8	71	159.8	200	392.0
12	53.60	42	107.6	72	161.6	300	572.0
13	55.40	43	109.4	73	163.4	400	752.0
14	57.20	44	111.2	74	165.2	500	932.0
15	59.00	45	113.0	75	167.0	600	1112
16	60.80	46	114.8	76	168.8	700	1292
17	62.60	47	116.6	77	170.6	800	1472
18	64.40	48	118.4	78	172.4	900	1652
19	66.20	49	120.2	79	174.2		
20	68.00	50	122.0	80	176.0		

By permission: www.metric-conversions.org

2.0.26 Fahrenheit to Rankine Temperature Conversion

Fahrenheit	Rankine	Fahrenheit	Rankine	Fahrenheit	Rankine	Fahrenheit	Rankine
0	460.0	7	467.0	23	483.0	39	499.0
0.1	460.1	8	468.0	24	484.0	40	500.0
0.2	460.2	9	469.0	25	485.0	41	501.0
0.3	460.3	10	470.0	26	486.0	42	502.0
0.4	460.4	11	471.0	27	487.0	43	503.0
0.5	460.5	12	472.0	28	488.0	44	504.0
0.6	460.6	13	473.0	29	489.0	45	505.0
0.7	460.7	14	474.0	30	490.0	46	506.0
0.8	460.8	15	475.0	31	491.0	47	507.0
0.9	460.9	16	476.0	32	492.0	48	508.0
1	461.0	17	477.0	33	493.0	49	509.0
2	462.0	18	478.0	34	494.0	50	510.0
3	463.0	19	479.0	35	495.0	51	511.0
4	464.0	20	480.0	36	496.0	52	512.0
5	465.0	21	481.0	37	497.0	53	513.0
6	466.0	22	482.0	38	498.0	54	514.0

Fahrenheit	Rankine	Fahrenheit	Rankine	Fahrenheit	Rankine	Fahrenheit	Rankine
55	515.0	69	529.0	83	543.0	97	557.0
56	516.0	70	530.0	84	544.0	98	558.0
57	517.0	71	531.0	85	545.0	99	559.0
58	518.0	72	532.0	86	546.0	100	560.0
59	519.0	73	533.0	87	547.0	200	660.0
60	520.0	74	534.0	88	548.0	300	760.0
61	521.0	75	535.0	89	549.0	400	860.0
62	522.0	76	536.0	90	550.0	500	960.0
63	523.0	77	537.0	91	551.0	600	1060
64	524.0	78	538.0	92	552.0	700	1160
65	525.0	79	539.0	93	553.0	800	1260
66	526.0	80	540.0	94	554.0	900	1360
67	527.0	81	541.0	95	555.0		
68	528.0	82	542.0	96	556.0		

Note: Rankine is a temperature scale named after Scottish engineer and physicist William Macquorn Rankine, and is based upon zero being Absolute Zero.
By permission: www.metric-conversions.org

2.0.27 Rankine to Fahrenheit Temperature Conversion

Rankine	Fahrenheit	Rankine	Fahrenheit	Rankine	Fahrenheit	Rankine	Fahrenheit
0	−460.00	21	−439.00	51	−409.00	81	−379.00
0.1	−459.9	22	−438.00	52	−408.00	82	−378.00
0.2	−459.8	23	−437.00	53	−407.00	83	−377.00
0.3	−459.7	24	−436.00	54	−406.00	84	−376.00
0.4	−459.6	25	−435.00	55	−405.00	85	−375.00
0.5	−459.5	26	−434.00	56	−404.00	86	−374.00
0.6	−459.4	27	−433.00	57	−403.00	87	−373.00
0.7	−459.3	28	−432.00	58	−402.00	88	−372.00
0.8	−459.2	29	−431.00	59	−401.00	89	−371.00
0.9	−459.1	30	−430.00	60	−400.00	90	−370.00
1	−459.00	31	−429.00	61	−399.00	91	−369.00
2	−458.00	32	−428.00	62	−398.00	92	−368.00
3	−457.00	33	−427.00	63	−397.00	93	−367.00
4	−456.00	34	−426.00	64	−396.00	94	−366.00
5	−455.00	35	−425.00	65	−395.00	95	−365.00
6	−454.00	36	−424.00	66	−394.00	96	−364.00
7	−453.00	37	−423.00	67	−393.00	97	−363.00
8	−452.00	38	−422.00	68	−392.00	98	−362.00
9	−451.00	39	−421.00	69	−391.00	99	−361.00
10	−450.00	40	−420.00	70	−390.00	100	−360.00
11	−449.00	41	−419.00	71	−389.00	200	−260.00
12	−448.00	42	−418.00	72	−388.00	300	−160.00
13	−447.00	43	−417.00	73	−387.00	400	−60.00
14	−446.00	44	−416.00	74	−386.00	500	40.00
15	−445.00	45	−415.00	75	−385.00	600	140.0
16	−444.00	46	−414.00	76	−384.00	700	240.0
17	−443.00	47	−413.00	77	−383.00	800	340.0
18	−442.00	48	−412.00	78	−382.00	900	440.0
19	−441.00	49	−411.00	79	−381.00		
20	−440.00	50	−410.00	80	−380.00		

Note: Rankine is a temperature scale named after Scottish engineer and physicist William Macquorn Rankine, and is based upon zero being Absolute Zero.
By permission: www.metric-conversions.org

2.1.0 Converting Water from One Form to Another

This information is of general interest in converting water information from one form to another. This conversion information can be used in virtually all water measurements.

Acre-feet × 43560 = cubic feet
Acre-feet × 1613.3 = cubic yards
Acre Feet × 325851 = gallons
Acre-feet/day × 0.5 = acre-inches/hour
Acre-feet/day × 226.3 = gallons/minute
Acre-feet/day × 0.3259 = million gallons/day
Cubic feet × 1728 = cubic inches
Cubic feet × 0.03704 = cubic yards
Cubic feet × 7.481 = gallons
Cubic feet/second × 449 = gallons/minute
Cubic feet/second × 38.4 = Colorado miners' inches
Cubic feet/second × 0.02832 = cubic meters/second
Feet of water × .0295 = atmospheres
Feet of water × 62.43 = pounds/square foot
Feet of water × .4335 = pounds/square inch
Gallons × .1337 = cubic feet
Gallons × 3.785 = liters
Gallons of water × 8.33 = pounds of water
Liters × 61.02 = cubic inches
Liters × .001 = cubic meters
Liters × .001308 = cubic yards
Liters × .2642 = gallons

Courtesy of csgnetwork.com/waterconvinformation.html

2.1.1 U.S. and Metric Lumber Length Conversion Table

Actual (ft) to	Metric (m)	Metric (m) to	Actual (ft)
6	1.83	2	6.56
8	2.44	2.50	8.20
10	3.05	3	9.84
12	3.66	3.50	11.48
14	4.27	4	13.12
16	4.88	4.50	14.76
18	5.49	5	16.40
20	6.10	5.50	18.04
22	6.71	6	19.68
24	7.32	6.50	21.32
		7	22.96

Courtesy of csgnetwork.com/lmbrlengthcvttable.html

2.2.0 Conversion Factors—Energy, Volume, Length, Weight, Liquid

Table of Conversion Factors

Multiply	By	To Obtain
Amperes/sq.ft.,	0.108	amperes/sq. dm.
Ampere hours,	3600	coulombs
Amperes/sq. dm.,	9.29	amperes/sq. ft.
Angstrom units,	1×10^{-4}	microns
Centimeters	0.394	inches
Centimeters	393.7	mils
Centimeters	0.0328	feet
Cubic centimeters	3.53×10^{-5}	cubic feet
Cubic centimeters	0.061	cubic inches
Cubic centimeters	2.64×10^{-4}	gallons
Cubic centimeters	0.0338	ounces (fluid)
Cubic feet	28317	cubic centimeters
Cubic feet	1728	cubic inches
Cubic feet	7.48	gallons
Cubic feet of water 60°F	62.37	pounds
Cubic inches	16.39	cubic centimeters
Faradays	9.65×10^{-4}	coulombs
Faraday/second	96500	amperes
Feet	30.48	centimeters
Feet	12	inches
Feet	0.3048	meters
Gallons	4	quarts (liquid)
Gallons	3785.4	cubic centimeters
Gallons (U.S.)	231	cubic inches
Gallons (U.S.)	3.785	liters
Gallons (U.S.)	128	ounces (fluid)
Gallons (U.S.)	8	pints
Gallons (U.S.)	8.34	pounds (av.) of H_2O at 62° F.
Gallons (U.S.)	1.2	gallons (British)
Grains	0.0648	grams
Grains	0.0023	ounces (avoir.)
Grains	0.0021	ounces (troy)
Grains	0.0417	pennyweights (troy)
Grams	15.43	grains
Grams	1000	milligrams
Grams	0.0353	ounces (avoir.)
Grams	0.0321	ounces (troy)
Grams	0.643	pennyweights
Grams/liter	0.122	ounces/gallon (troy)
Grams/liter	0.134	ounces/gallon (avoir.)
Grams/liter	1000	parts per million
Grams/liter	2.44	pennyweights/gallon
Inches	2.54	centimeters
Inches	1000	mils
Kilograms	1000	grams
Kilograms	2.205	pounds (avoir.)
Kilograms	2.679	pounds (troy)
Liters	1000	milliliters
Liters	0.264	gallons
Meters	100	centimeters

(Continued)

Table of Conversion Factors—Cont'd

Multiply	By	To Obtain
Meters	39.37	inches
Microns	3.9×10^{-5}	inches
Milligrams	0.001	grams
Milliliters	1.000027	cubic centimeters
Mils	0.001	inches
Mils	25.4	microns
Ounces (avoir.)	437.5	grains
Ounces (avoir.)	28.35	grams
Ounces (avoir.)	0.911	ounces (troy)
Ounces (avoir.)	18.23	pennyweights
Ounces (avoir.)	0.076	pounds (troy)
Ounces/gallon (avoir.)	7.5	grams/liter
Ounces (troy)	480	grains
Ounces (troy)	31.1	grams
Ounces (troy)	1.097	ounces (avoir.)
Ounces (troy)	20	pennyweights
Ounces/gallon (troy)	8.2	grams/liter
Ounces (fluid)	29.57	cubic centimeters
Ounces/gallon (fluid)	7.7	cc/liter
Pennyweights	24	grains
Pennyweights	1.56	grams
Pennyweights/gallon	0.41	grams/liter
Pints	16	ounces (fluid)
Pounds (avoir.)	453.6	grams
Pounds (avoir.)	16	ounces (avoir.)
Pounds (avoir.)	14.58	ounces (troy)
Pounds (avoir.)	1.215	pounds (troy)
Pounds (troy)	373.24	grams
Pounds (troy)	12	ounces (troy)
Pounds (troy)	0.823	pounds (avoir.)
Quarts (liquid)	946.4	cubic centimeters
Quarts (liquid)	2	pints
Square feet	929.23	square centimeters
Square feet	144	square inches
Square inches	6.45	square centimeters

By permission: Associated Rack Corp., Vero Beach, FL

2.3.0 Conversion of Liquids—Specific Gravity to Degrees Baume

Conversion Table - Specific Gravity, Degrees Baume, Pounds Per Cubic Foot

$°Bé. = 145 - \dfrac{145}{sp.gr.}$ (*heavier than* H_2O); $°Bé. = \dfrac{140}{sp.gr.} - 130$ (*lighter than* H_2O)

Sp. gr. 60°/60°	°Bé	Lb. per gal. at 60°F wt. in air	Lb. per cu. ft. at 60° F wt. in air
1.000	10.00	8.3283	62.300
1.005	0.72	8.3700	62.612
1.010	1.44	8.4117	62.924
1.015	2.14	8.4534	63.236

Conversion Table - Specific Gravity, Degrees Baume, Pounds Per Cubic Foot—Cont'd

Sp. gr. 60°/60°	°Bé	Lb. per gal. at 60°F wt. in air	Lb. per cu. ft. at 60° F wt. in air
1.020	2.84	8.4950	63.547
1.025	3.54	8.5367	63.859
1.030	4.22	8.5784	64.171
1.035	4.90	8.6201	64.483
1.040	5.58	8.6618	64.795
1.045	6.24	8.7035	65.107
1.050	6.91	8.7452	65.419
1.055	7.56	8.7869	65.731
1.060	8.21	8.8286	66.042
1.065	8.85	8.8703	66.354
1.070	9.49	8.9120	66.666
1.075	10.12	8.9537	66.978
1.080	10.74	8.9954	67.290
1.085	11.36	9.0371	67.602
1.090	11.97	9.0787	67.914
1.095	12.58	9.1204	68.226
1.100	13.18	9.1621	68.537
1.105	13.78	9.2038	68.849
1.110	14.37	9.2455	69.161
1.115	14.96	9.2872	69.473
1.120	15.54	9.3289	60.785
1.125	16.11	9.3706	70.097
1.130	16.68	9.4123	70.409
1.135	17.25	9.4540	70.721
1.140	17.81	9.4957	71.032
1.145	18.36	9.5374	71.344
1.150	18.91	9.5790	71.656
1.155	19.46	9.6207	71.968
1.160	20.00	9.6624	72.280
1.165	20.54	9.7041	72.592
1.170	21.07	9.7458	72.904
1.175	21.6	9.7875	73.216
1.180	22.12	9.8292	73.528
1.185	22.64	9.8709	73.840
1.190	23.15	9.9126	74.151
1.195	23.66	9.9543	74.463
1.200	24.17	9.9960	74.775
1.205	24.67	10.0377	75.087
1.210	25.17	10.0793	75.399
1.215	25.66	10.1210	75.711
1.220	26.15	10.1627	76.022
1.225	26.63	10.2044	76.334
1.230	27.11	10.2461	76.646
1.235	27.59	10.2878	76.958
1.240	28.06	10.3295	77.270
1.245	28.53	10.3712	77.582
1.250	29.00	10.4129	77.894
1.255	29.46	10.4546	78.206
1.260	29.92	10.4963	78.518
1.265	30.38	10.5380	78.830
1.270	30.83	10.5797	79.141
1.275	31.27	10.6214	79.453
1.280	31.72	10.6630	79.765
1.285	32.16	10.7047	80.077

(Continued)

Conversion Table - Specific Gravity, Degrees Baume, Pounds Per Cubic Foot—Cont'd

Sp. gr. 60°/60°	°Bé	Lb. per gal. at 60°F wt. in air	Lb. per cu. ft. at 60° F wt. in air
1.290	32.60	10.7464	80.389
1.295	33.03	10.7881	80.701
1.300	33.46	10.8298	81.013
1.305	33.89	10.8715	81.325
1.310	34.31	10.9132	81.636
1.315	34.73	10.9549	81.948
1.320	35.15	10.9966	82.260
1.325	35.57	11.0383	82.572
1.330	35.98	11.0800	82.884
1.335	36.39	11.1217	83.196
1.340	36.79	11.1634	83.508
1.345	37.19	11.2051	83.820
1.350	37.59	11.2467	84.131
1.355	37.99	11.2884	84.443
1.360	38.38	11.3301	84.755
1.365	38.77	11.3718	85.067
1.370	39.16	11.4135	85.379
1.375	39.55	11.4552	85.691
1.380	39.93	11.4969	86.003
1.385	40.31	11.5386	86.315
1.390	40.68	11.5803	86.626
1.395	41.06	11.6220	86.938
1.400	41.43	11.6637	87.250
1.405	41.80	11.7054	87.562
1.410	42.16	11.7471	87.874
1.415	42.53	11.7888	88.186
1.420	42.89	11.8304	88.498
1.425	43.25	11.8721	86.810
1.430	43.60	11.9138	89.121
1.435	43.95	11.9555	89.433
1.440	44.31	11.9972	89.745
1.445	44.65	12.0389	90.057
1.450	45.00	12.0806	90.369
1.455	45.34	12.1223	90.681
1.460	45.68	12.1640	90.993
1.465	46.02	12.2057	91.305
1.470	46.36	12.2473	91.616
1.475	46.69	12.2890	91.928
1.480	47.03	12.3307	92.240
1.485	47.36	12.3724	92.552
1.490	47.68	12.4141	92.864
1.495	48.01	12.4558	93.176
1.500	48.33	12.4975	93.488
1.505	48.65	12.5392	93.800
1.510	48.97	12.5809	94.112
1.515	49.29	12.6226	94.424
1.520	49.61	12.6643	94.735
1.525	49.92	12.7060	95.047
1.530	50.23	12.7477	95.359
1.535	50.54	12.7894	95.671
1.540	50.84	12.8310	95.983
1.545	51.15	12.8727	96.295
1.550	51.45	12.9144	96.606
1.555	51.75	12.9561	96.918
1.560	52.05	12.9978	97.230

Conversion Table - Specific Gravity, Degrees Baume, Pounds Per Cubic Foot—Cont'd

Sp. gr. 60°/60°	°Bé	Lb. per gal. at 60°F wt. in air	Lb. per cu. ft. at 60° F wt. in air
1.565	52.35	13.0395	97.542
1.570	52.64	13.0812	97.854
1.575	52.94	13.1229	98.166
1.580	53.23	13.1646	98.478
1.585	53.52	13.2063	98.790
1.590	53.81	13.2480	99.102
1.595	54.09	13.2897	99.414
1.600	54.38	13.3313	99.725
1.605	54.66	13.3730	100.037
1.610	54.94	13.4147	100.349
1.615	55.22	13.4564	100.661
1.620	55.49	13.4981	100.973
1.625	55.77	13.5398	101.285
1.630	56.04	13.5815	101.597
1.635	56.32	13.6232	101.909
1.640	56.59	13.6649	102.220
1.645	56.85	13.7066	102.532
1.650	57.12	13.7483	102.844
1.655	57.39	13.7900	103.156
1.660	57.65	13.8317	103.468
1.665	57.91	13.8734	103.780
1.670	58.17	13.9150	104.092
1.675	58.43	13.9567	104.404
1.680	58.69	13.9984	104.715
1.685	58.95	14.0401	105.027
1.690	59.20	14.0818	105.339
1.695	59.45	14.1235	105.651
1.700	59.71	14.1652	105.963
1.705	59.96	14.2069	106.275
1.710	60.20	14.2486	106.587
1.715	60.45	14.2903	196.899
1.720	60.70	14.3320	107.210
1.725	60.94	14.3737	107.522
1.730	61.18	14.4153	107.834
1.735	61.34	14.4570	108.146
1.740	61.67	14.4987	108.458
1.745	61.91	14.5404	108.770
1.750	62.14	14.5821	109.082
1.755	62.38	14.6238	109.394
1.760	62.61	14.6655	109.705
1.765	62.85	14.7072	110.017
1.770	63.08	14.7489	110.329
1.775	63.31	14.7906	110.641
1.780	63.54	14.8323	110.953
1.785	63.77	14.8740	111.265
1.790	63.99	14.9157	111.577
1.795	64.22	14.9574	111.889
1.800	64.44	14.9990	112.200
1.805	64.67	15.0407	112.512
1.810	64.89	15.0824	112.824
1.815	65.11	15.1241	113.136
1.820	65.33	15.1658	113.448
1.825	65.55	15.2075	113.760
1.830	65.77	15.2492	114.072

(Continued)

Conversion Table - Specific Gravity, Degrees Baume, Pounds Per Cubic Foot—Cont'd

Sp. gr. 60°/60°	°Bé	Lb. per gal. at 60°F wt. in air	Lb. per cu. ft. at 60° F wt. in air
1.835	65.98	15.2909	114.384
1.840	66.20	15.3326	114.696
1.845	66.41	15.3743	115.007
1.850	66.62	15.4160	115.318
1.855	66.83	15.4577	115.630
1.860	67.04	15.4993	115.943
1.865	67.25	15.5410	116.255
1.870	67.46	15.5827	116.567
1.875	67.67	15.6244	116.879
1.880	67.87	15.6661	117.191
1.885	68.08	15.7078	117.503
1.890	68.28	15.7495	117.814
1.895	68.48	15.7912	118.126
1.900	68.68	15.8329	118.438

By permission: Associated Rack Corp., Vero Beach, FL

2.4.0 Volume-to-Weight Conversion Table

The volume-to-weight conversion table presented on the following pages is a compilation of several sources. Materials converted from volume to weight include paper (high-grade and other), glass, plastic, metals, organics, and other materials (e.g., tires and oil).

It is important to note that although the weight (density) figures presented here are useful for determining rough estimates, they will not be as useful when precise measurements are required. Differences in the way a material is handled, processed, or in the amount of moisture present can make substantial differences in the amount a particular material weighs per specified volume. Because of these differences, it will be important to actually sort and weight materials in your program whenever precise measurements are needed (e.g., recycling contract agreements).

Category	Material (u/c = uncompacted/compacted & baled)	Volume	Estimated Weight (in pounds)
High-Grade Paper	*Computer Paper:*		
	Uncompacted, stacked	1 cu. yd.	655
	Compacted/baled	1 cu. yd.	1,310
	1 case	2800 sheets	42
	White Ledger:		
	(u)stacked/(c)stacked	1 cu. yd.	375465/755-925
	(u)crumpled/(c)crumpled	1 cu. yd.	11 0205/325
	Ream of 20# bond; 8-1/2 × 11	1 ream = 500 sheets	5
	Ream of 20# bond; 8-1/2 × 14	1 ream = 500 sheets	6.4
	White ledger pads	1 case = 72 pads	38
	Tab Cards		
	Uncompacted	1 cu. yd.	605
	Compacted/baled	1 cu. yd.	1,215–1,350

Category	Material (u/c = uncompacted/compacted & baled)	Volume	Estimated Weight (in pounds)
Other Paper	**_Cardboard (Corrugated):_**		
	Uncompacted	1 cu. yd.	50–150
	Compacted	1 cu. yd.	300–500
	Baled	1 cu. yd.	7001,100
	Newspaper:		
	Uncompactad	1 cu. yd.	360–505
	Compacted/baled	1 cu. yd.	7,201,000
	12 stack	—	35
	Miscellaneous Paper:		
	Yellow legal pads	1 case = 72pads	38
	Colored message pads	1 carton = 144 pads	22
	Self-carbon forms; 8-1/2 × 11	1 ream = 500 sheets	50
	Mixed Ledger/Office Paper:		
	Flat (u/c)	1 cu. yd.	380/755
	Crumpled (u/c)	1 cu. yd.	110205/610
Gass	**_Refillable Whole Bottles:_**		
	Refillable beer bottles	1 case = 24 bottles:	14
	Refillable soft drink bottles	1 case = 24 bottles	22
	8 oz. glass container	1 case = 24 bottles	12
	Bottles:		
	Whole	1 cu. yd	500–700
	Semi-crushed	1 cu. yd.	1,0001,800
	Crushed (mechanically)	1 cu. yd.	1,800–2,700
	Uncrushed to manually broken	55 gallon drum	300
Plastic	**_PET (Soda Bottles):_**		
	Whole bottles, uncompacted	1 cu. yd.	30–40
	Whole bottles, compacted	1 cu. yd.	515
	Whole bottles, uncompacted	gaylord	40–53
	Baled	30″ × 62″	500–550
	Granulated	gaylord	700–750
	8 bottles (2-liter size)		1
	HDPE(Dairy):		
	Whole, uncompacted	1 cu. yd.	24
	Whole, compacted	1 cu. yd.	270
	Baled	32″ × 60″	400–500
	HDFE(Mixed):		
	Baled	32″ × 60″	900
	Granulated	semi-load	42,000
	Odd Plastic:		
	Uncompacted	1 cu. yd.	50
	Compacted/baled	1 cu.yd.	400–700
	Mixed PET and HDPE (Dairy):		
	Whole, uncompacted	1 cu. yd.	32

(Continued)

Category	Material (u/c = uncompacted/compacted & baled)	Volume	Estimated Weight (in pounds)
Metals	*Aluminum (Cans):*		
	Whole	1 cu.yd.	50–75
	Compacted (manually)	1 cu. yd.	250–430
	Uncompacted	1 full grocery bag	1.5
		1 case = 24 cans	0.9
	Ferrous (tin-coated steel cans):		
	Whole	1 cu. yd.	150
	Flattened	1 cu. yd.	850
	Whole	1 case = 6 cans	22
Organics	*Yard trmming*:*		
	Leaves (uncompacted)	1 cu. yd.	200–250
	Leaves (compacted)	1 cu. yd.	300–450
	Leaves, vacuumed	1 cu. yd.	350
	Grass clippings (uncompacted)	1 cu. yd.	350–450
	Grass clippings (compacted)	1 cu. yd.	550–1,500
	Finished compost	1 cu. yd.	600
	Scrap wood:		
	Pallets		30–100 (40 avg.)
	Wood chips	1 cu. yd.	500
	Food Waste:		
	Solid/liquid fats	55-gallon drum	400–410
Other Materials	*Tires:*		
	Car	1 tire	12–20
	Truck	1 tire	60–100
	Oil (Used Motor Oil)	1 gallon	7

*Density of yard trimmings is highly variable depending on moisture content.

2.5.0 Convert Old A.I.S.C. Structural Shapes to New Designations

A.I.S.C. Hot-Rolled Structural Steel Shape Designations		
New Designation	Type of Shape	Old Designation
W 24 × 76	W shape	24 WF 76
W 14 × 26	W shape	14 B 26
S 24 × 100	S shape	24 I 100
M 8 × 18.5	M shape	8 M 18.5
M 10 × 9	M shape	10 JR 9.0
M 8 × 34.3	M shape	8 × 8 M 34.3
C 12 × 20.7	American Std. Channel	12 C 20.7
MC 12 × 45	Miscellaneous Channel	12 × 4 C 45.0
MC 12 × 10.6	Miscellaneous Channel	12 JR C 10.6
HP 14 × 73	HP shape	14 BP 73
L6 × 6 × 3/4	Equal Leg Angle	L 6 × 6 × 3/4

Section 2 Conversion Tables and Conversion Formulas

A.I.S.C. Hot-Rolled Structural Steel Shape Designations—Cont'd

New Designation	Type of Shape	Old Designation
L6 × 4 × 5/8	Unequal Leg Angle	L 6 × 4 × 5/8
WT 12 × 38	Structural Tee	ST 12 WF 38
WT 7 × 13	Cut from W shape	ST 7 B 13
ST 12 × 50	Cut from S shape	ST 12 / 50
MT 4 × 9.25	Cut from M shape	ST 4 M 9.25
MT 5 × 4.5	Cut from M shape	ST 5 JR 4.5
MT 4 × 17.15	Cut from M shape	ST 4 M 17.15
PL ½ × 18	Plate	PL 18 × ½
Bar 1□	Square Bar	Bar 1□
Bar 1-1/4 ○	Round Bar	Bar 1-1/4 ○
Bar 2-1/2 × 1/2	Flat Bar	Bar 2-1/2 × 1/2
Pipe 4 Std.	Pipe	Pipe 4 Std.
Pipe 4 × . Strong	Pipe	Pipe 4 ×-Strong
Pipe 4 ×× Strong	Pipe	Pipe 4 ××-Strong
TS 4 × 4 × .375	Structural Tubing: Square	Tube 4 × 4 × .375
TS 5 × 3 × .375	Rectangular	Tube 5 × 3 × .375
TS 3 OD × .250	Circular	Tube 3 OD × .250

Source: Cardinal Metals, Irving, Texas

2.6.0 USA Mesh Size—Convert to International Particle Size (Microns)

International Sieve Chart / Micropowder Grit Chart

ASTM E-11	ANSI B74.12-1992	JIS	BSI	Particle Diameter	AFNOR	DIN	Tyler®	Angstrom Units (Å)
USA	USA	JPN	GBR	USA	FRA	DEU	USA	Global
USS Mesh	d50 (microns) Sedimentation tube method	Microns (μ)	Mesh	Microns (μ)	Microns (μ)	Microns (μ)	Mesh	Angstroms (Å)
3 in.		71		75 mm				
2 in.		50		50				
1.06 in.		26.5		26.5		26.5		1.05 in.
7/8 in.		22.4		22.4	22.4	22.4		0.883 in.
3/4 in.		19		19.0		19		0.742 in.
5/8 in.		16		16.0	16	16		0.624 in.
1/2 in.		12.5		12.5	12.5	12.5		
7/16 in.		11.2		11.2	11.2	11.2		0.441 in.
3/8 in.		9.5		9.5	11.2	9.5		0.371 in.
5/16 in.		8		8.0	8	8	2.5	
0.265 in.		6.7		6.7		6.7	3	
3.5		90		5.6	5.6	5.6	3.5	
5		5		4.00	4	4	5	
8		8		2.36		2.36	8	
12				1.70		1.7	10	
14		1.4	12	1.40	1.4	1.4	12	
16				1.18		1.18	14	
18				1.00	1,0	1.0	16	
20		850	18	850		850	20	

(Continued)

International Sieve Chart / Micropowder Grit Chart—Cont'd

ASTM E-11	ANSI B74.12-1992	JIS	BSI	Particle Diameter	AFNOR	DIN	Tyler®	Angstrom Units (Å)
USA	USA	JPN	GBR	USA	FRA	DEU	USA	Global
USS Mesh	d50 (microns) Sedimentation tube method	Microns (μ)	Mesh	Microns (μ)	Microns (μ)	Microns (μ)	Mesh	Angstroms (Å)
25		710	22	710	710	710	24	
30		600	25	600		600	28	
35		500	30	500	500	500	32	
40		425	36	425		425	35	
45		355	44	355	355	355	42	
50		300	52	300		300	48	
60		250	60	250	250	250	60	
70		212	72	212		212	65	
80		180	85	180	180	180	80	
100		150	100	150		150	100	
120		125	120	125	125	125	115	
140		106		106		106	150	
170		90	170	90	90	90	170	
200		75	200	75		75	200	
230	240 grit = 53.5 – 50	63	240	63	63	63	250	
270	280 grit = 44 – 40.5	53	300	53		53	270	
325	320 grit = 36 – 32.5	45	350	45	45	45	325	
400	360 grit = 28.8 – 25.8	38	400	38		38	400	
450	400 grit = 23.6 – 20.6	32	440	32	32	32	450	
500	500 grit = 19.7 – 16.7			25	25	25	500	
635	600 grit = 16 – 13			20	20	20	635	
	800 grit = 12.3 – 9.8			16		16		
	1000 grit = 9.3 – 6.8			10		10	1,250	
	1200 grit = 6.5 – 4.5			5		5	5,000	
1						1	10,000	
				0.1				1,000
				0.01				100
				0.001				10
				0.0001				1
				0				0

Particle Size Conversion Chart

1 cm	1 mm	100 micrometers	10 micrometers	1 micrometer	100 nanometers	10 nanometers	1 nanometer
0.01 meter	0.001 meter	0.0001 meter	0.00001 meter	0.000001 meter	0.0000001 meter	0.00000001 meter	0.000000001 meter

Source: READE.com

Applicable Standards Applicable standards are ISO 565 (1987), ISO 3310 (1999), ASTM E 11-70 (1995), DIN 4188 (1977), BS 410 (1986) and AFNOR NFX11-501 (1987).

2.7.0 Converting Wire Gauge from

1. Standard Wire Gauge (S.W.G.)
   to
2. Wire Number
   to
3. American Wire Gauge (A.W.G.)
   to
4. A.W.G. and Brown & Sharpe (B&S)-Inches
   to
5. A.W.G. to Metric

S.W.G. (Inches)	Wire Number (Gauge)	A.W.G. or B&S (Inches)	A.W.G. Metric (MM)
0.500	0000000 (7/0)	....	....
0.464	000000 (6/0)	0.580000	....
0.432	00000 (5/0)	0.516500	....
0.400	0000 (4/0)	0.460000	11,684
0.372	000 (3/0)	0.409642	10,404
0.348	00 (2/0)	0.364796	9,266
0.324	0 (1/0)	0.324861	8,252
0.300	1	0.289297	7,348
0.276	2	0.257627	6,543
0.252	3	0.229423	5,827
0.232	4	0.2043	5,189
0.2120	5	0.1819	4,621
0.1920	6	0.1620	4,115
0.1760	7	0.1443	3,665
0.1600	8	0.1285	3,264
0.1440	9	0.1144	2,906
0.1280	10	0.1019	2,588
0.1160	11	0.0907	2,304
0.1040	12	0.0808	2,052
0.0920	13	0.0720	1,829
0.0800	14	0.0641	1,628
0.0720	15	0.0571	1,450
0.0640	16	0.0508	1,291
0.0560	17	0.0453	1,150
0.0480	18	0.0403	1,024
0.0400	19	0.0359	0,9119
0.0360	20	0.0320	0,8128
0.0320	21	0.0285	0,7239
0.0280	22	0.0253	0,6426

(Continued)

S.W.G. (Inches)	Wire Number (Gauge)	A.W.G. or B&S (Inches)	A.W.G. Metric (MM)
0.0240	23	0.0226	0,5740
0.0220	24	0.0201	0,5106
0.0200	25	0.0179	0,4547
0.0180	26	0.0159	0,4038
0.0164	27	0.0142	0,3606
0.0148	28	0.0126	0,3200
0.0136	29	0.0113	0,2870
0.0124	30	0.0100	0,2540
0.0116	31	0.0089	0,2261
0.0108	32	0.0080	0,2032
0.0100	33	0.0071	0,1803
0.0092	34	0.0063	0,1601
0.0084	35	0.0056	0,1422
0.0076	36	0.0050	0,1270
0.0068	37	0.0045	0,1143
0.0060	38	0.0040	0,1016
0.0052	39	0.0035	0,0889
0.0048	40	0.0031	0,0787
0.0044	41	0.0028	0,0711
0.0040	42	0.0025	0,0635
0.0036	43	0.0022	0,0559
0.0032	44	0.0020	0,0508
0.0028	45	0.0018	0,0457
0.0024	46	0.0016	0,0406
0.0020	47	0.0014	0,0350
0.0016	48	0.0012	0,0305
0.0012	49	0.0011	0,0279
0.0010	50	0.0010	0,0254
	51	0.00088	0,0224
	52	0.00078	0,0198
	53	0.00070	0,0178
	54	0.00062	0,0158
	55	0.00055	0,0140
	56	0.00049	0,0124

Source: READE.com

2.8.0 Map of United States Showing Four Continental Times Zones

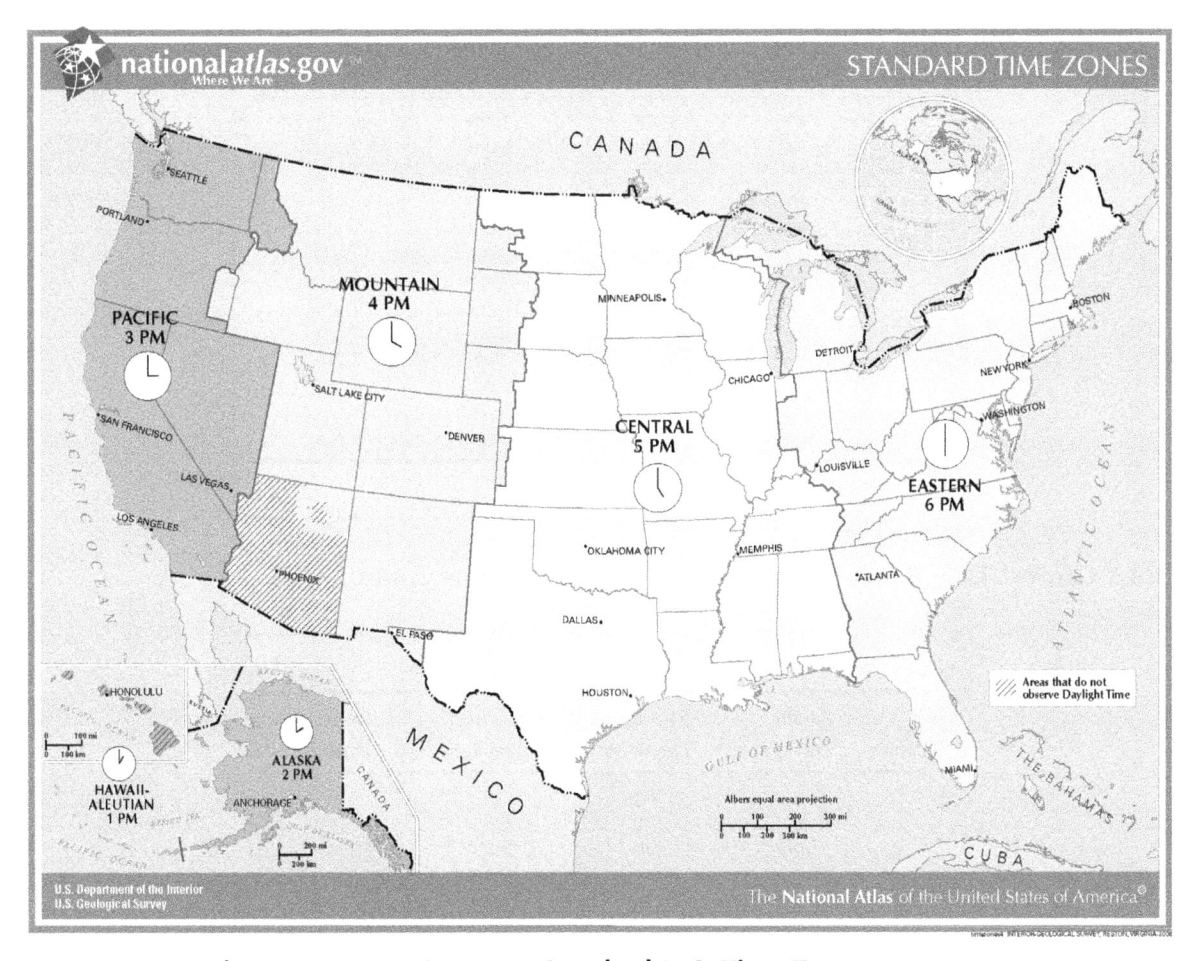

2.8.1 Convert Time Zones—UTC to Four Standard U.S. Time Zones

Conversions from UTC to some U.S. time zones				
UTC (GMT)	Pacific Standard	Mountain Standard	Central Standard	Eastern Standard
0	4 pm *	5 pm *	6 pm *	7 pm *
1	5 pm *	6 pm *	7 pm *	8 pm *
2	6 pm *	7 pm *	8 pm *	9 pm *
3	7 pm *	8 pm *	9 pm *	10 pm *
4	8 pm *	9 pm *	10 pm *	11 pm *
5	9 pm *	10 pm *	11 pm *	12 mid
6	10 pm *	11 pm *	12 mid	1 am
7	11 pm *	12 mid	1 am	2 am
8	12 mid	1 am	2 am	3 am
9	1 am	2 am	3 am	4 am
10	2 am	3 am	4 am	5 am

Conversions from UTC to some U.S. time zones—Cont'd

UTC (GMT)	Pacific Standard	Mountain Standard	Central Standard	Eastern Standard
11	3 am	4 am	5 am	6 am
12	4 am	5 am	6 am	7 am
13	5 am	6 am	7 am	8 am
14	6 am	7 am	8 am	9 am
15	7 am	8 am	9 am	10 am
16	8 am	9 am	10 am	11 am
17	9 am	10 am	11 am	12 noon
18	10 am	11 am	12 noon	1 pm
19	11 am	12 noon	1 pm	2 pm
20	12 noon	1 pm	2 pm	3 pm
21	1 pm	2 pm	3 pm	4 pm
22	'2 pm	3 pm	4 pm	5 pm
23	3 pm	4 pm	5 pm	6 pm

* = previous day

2.8.2 Convert Time Zones-UTC to Four Daylight U.S. Time Zones

Conversions from UTC to some U.S. time zones

UTC (GMT)	Pacific Daylight	Mountain Daylight	Central Daylight	Eastern Daylight
0	5 pm *	6 pm *	7 pm *	8 pm *
1	6 pm *	7 pm *	8 pm *	9 pm *
2	7 pm *	8 pm *	9 pm *	10 pm *
3	8 pm *	9 pm *	10 pm *	11 pm *
4	9 pm *	10 pm *	11 pm *	12 mid
5	10 pm *	11 pm *	12 mid	1 am
6	11 pm *	12 mid	1 am	2 am
7	12 mid	1 am	2 am	3 am
8	1 am	2 am	3 am	4 am
9	2 am	3 am	4 am	5 am
10	3 am	4 am	5 am	6 am
11	4 am	5 am	6 am	7 am
12	5 am	6 am	7 am	8 am
13	6 am	7 am	8 am	9 am
14	7 am	8 am	9 am	10 am
15	8 am	9 am	10 am	11 am
16	9 am	10 am	11 am	12 noon
17	10 am	11 am	12 noon	1 pm
18	11 am	12 noon	1 pm	2 pm
19	12 noon	1 pm	2 pm	3 pm
20	1 pm	2 pm	3 pm	4 pm
21	2 pm	3 pm	4 pm	5 pm
22	3 pm	4 pm	5 pm	6 pm
23	4 pm	5 pm	6 pm	7 pm

*= previous day

2.9.0 Convert Roman Numerals to Arabic Dates

Roman numerals from 1 to 1 million and their Arabic equivalent are listed below. Arabic to Roman dates for the 19th and 20th centuries and a portion of the 21st century are also included in this conversion exercise.

Roman	Arabic
I	1
V	5
X	10
L	50
C	100
D	500
M	1,000
i	1,000
v	5,000
x	10,000
l	50,000
c	100,000
d	500,000
m	1,000,000

19th Century
1801 = MDCCCI
1802 = MDCCCII
1803 = MDCCCIII
1804 = MDCCCIV
1805 = MDCCCV
1806 = MDCCCVI
1807 = MDCCCVII
1808 = MDCCCVIII
1809 = MDCCCIX
1810 = MDCCCX
1811 = MDCCCXI
1812 = MDCCCXII
1813 = MDCCCXIII
1814 = MDCCCXIV
1815 = MDCCCXV
1816 = MDCCCXVI
1817 = MDCCCXVII
1818 = MDCCCXVIII
1819 = MDCCCXIX
1820 = MDCCCXX
1821 = MDCCCXXI
1822 = MDCCCXXII
1823 = MDCCCXXIII
1824 = MDCCCXXIV
1825 = MDCCCXXV
1826 = MDCCCXXVI
1827 = MDCCCXXVII
1828 = MDCCCXXVIII
1829 = MDCCCXXIX
1830 = MDCCCXXX
1831 = MDCCCXXXI
1832 = MDCCCXXXII
1833 = MDCCCXXXIII
1834 = MDCCCXXXIV
1835 = MDCCCXXXV
1836 = MDCCCXXXVI
1837 = MDCCCXXXVII
1838 = MDCCCXXXVIII
1839 = MDCCCXXXIX
1840 = MDCCCXL
1841 = MDCCCXLI
1842 = MDCCCXLII
1843 = MDCCCXLIII
1844 = MDCCCXLIV
1845 = MDCCCXLV
1846 = MDCCCXLVI
1847 = MDCCCXLVII
1848 = MDCCCXLVIII
1849 = MDCCCXLIX
1850 = MDCCCL
1851 = MDCCCLI
1852 = MDCCCLII
1853 = MDCCCLIII
1854 = MDCCCLIV
1855 = MDCCCLV
1856 = MDCCCLVI
1857 = MDCCCLVII
1858 = MDCCCLVIII
1859 = MDCCCLIX
1860 = MDCCCLX
1861 = MDCCCLXI
1862 = MDCCCLXII
1863 = MDCCCLXIII
1864 = MDCCCLXIV
1865 = MDCCCLXV
1866 = MDCCCLXVI
1867 = MDCCCLXVII
1868 = MDCCCLXVIII
1869 = MDCCCLXIX
1870 = MDCCCLXX
1871 = MDCCCLXXI
1872 = MDCCCLXXII
1873 = MDCCCLXXIII
1874 = MDCCCLXXIV
1875 = MDCCCLXXV
1876 = MDCCCLXXVI
1877 = MDCCCLXXVII
1878 = MDCCCLXXVIII
1879 = MDCCCLXXIX
1880 = MDCCCLXXX
1881 = MDCCCLXXXI
1882 = MDCCCLXXXII
1883 = MDCCCLXXXIII
1884 = MDCCCLXXXIV
1885 = MDCCCLXXXV
1886 = MDCCCLXXXVI
1887 = MDCCCLXXXVII
1888 = MDCCCLXXXVIII
1889 = MDCCCLXXXIX
1890 = MDCCCXC
1891 = MDCCCXCI
1892 = MDCCCXCII
1893 = MDCCCXCIII
1894 = MDCCCXCIV
1895 = MDCCCXCV
1896 = MDCCCXCVI
1897 = MDCCCXCVII
1898 = MDCCCXCVIII
1899 = MDCCCXCIX
1900 = MCM or MDCCCC

20th Century
1901 = MCMI
1902 = MCMII
1903 = MCMIII
1904 = MCMIV
1905 = MCMV
1906 = MCMVI
1907 = MCMVII
1908 = MCMVIII
1909 = MCMIX
1910 = MCMX
1911 = MCMXI
1912 = MCMXII
1913 = MCMXIII
1914 = MCMXIV
1915 = MCMXV
1916 = MCMXVI
1917 = MCMXVII
1918 = MCMXVIII
1919 = MCMXIX
1920 = MCMXX
1921 = MCMXXI
1922 = MCMXXII
1923 = MCMXXIII
1924 = MCMXXIV
1925 = MCMXXV
1926 = MCMXXVI
1927 = MCMXXVII
1928 = MCMXXVIII
1929 = MCMXXIX
1930 = MCMXXX
1931 = MCMXXXI
1932 = MCMXXXII
1933 = MCMXXXIII
1934 = MCMXXXIV
1935 = MCMXXXV
1936 = MCMXXXVI
1937 = MCMXXXVII
1938 = MCMXXXVIII
1939 = MCMXXXIX
1940 = MCMXL

1941 = MCMXLI
1942 = MCMXLII
1943 = MCMXLIII
1944 = MCMXLIV
1945 = MCMXLV
1946 = MCMXLVI
1947 = MCMXLVII
1948 = MCMXLVIII
1949 = MCMXLIX
1950 = MCML
1951 = MCMLI
1952 = MCMLII
1953 = MCMLIII
1954 = MCMLIV
1955 = MCMLV
1956 = MCMLVI
1957 = MCMLVII
1958 = MCMLVIII
1959 = MCMLIX
1960 = MCMLX
1961 = MCMLXI
1962 = MCMLXII
1963 = MCMLXIII
1964 = MCMLXIV
1965 = MCMLXV
1966 = MCMLXVI
1967 = MCMLXVII
1968 = MCMLXVIII
1969 = MCMLXIX
1970 = MCMLXX
1971 = MCMLXXI
1972 = MCMLXXII
1973 = MCMLXXIII
1974 = MCMLXXIV
1975 = MCMLXXV
1976 = MCMLXXVI
1977 = MCMLXXVII
1978 = MCMLXXVIII
1979 = MCMLXXIX
1980 = MCMLXXX
1981 = MCMLXXXI

1982 = MCMLXXXII
1983 = MCMLXXXIII
1984 = MCMLXXXIV
1985 = MCMLXXXV
1986 = MCMLXXXVI
1987 = MCMLXXXVII
1988 = MCMLXXXVIII
1989 = MCMLXXXIX
1990 = MCMXC
1991 = MCMXCI
1992 = MCMXCII
1993 = MCMXCIII
1994 = MCMXCIV
1995 = MCMXCV
1996 = MCMXCVI
1997 = MCMXCVII
1998 = MCMXCVIII
1999 = MCMXCIX
2000 = MM

21st Century
2001 = MMI
2002 = MMII
2003 = MMIII
2004 = MMIV
2005 = MMV
2006 = MMVI
2007 = MMVII
2008 = MMVIII
2009 = MMIX
2010 = MMX
2011 = MMXI
2012 = MMXII
2013 = MMXIII
2014 = MMXIV
2015 = MMXV
2016 = MMXVI
2017 = MMXVII
2018 = MMXVIII
2019 = MMXIX
2020 = MMXX

2.10.0 Converting Speed—Knots to MPH to Kilometers per Hour

Speed Conversions - Knots, MPH, KPH		
Knots	Miles per Hour	Kilometers per Hour
1	1.152	1.85
2	2.303	3.70
3	3.445	5.55
4	4.606	7.41
5	5.758	9.26

Speed Conversions - Knots, MPH, KPH—Cont'd

Knots	Miles per Hour	Kilometers per Hour
6	6.909	11.13
7	8.061	12.98
8	9.212	14.83
9	10.364	16.68
10	11.515	18.55

Source: Glen-L Marine Designs-glen-l.com

2.11.0 Conversion Factors for Builders and Design Professionals Who Cook

Common Kitchen Equivalents

Standard	Equivalent	Liquid Equivalent
One pinch or dash	1/16 teaspoon	
1 teaspoon	5 ml	1/6 ounce
3 teaspoons	1 tablespoon	1/2 ounce
4 tablespoons	1/4 cup	2 ounces
1/3 cup	5 tablespoon + 1 teaspoon	3 ounces
1/2 cup	8 tablespoons	4 ounces
1 gill	1/2 cup	4 ounces
1 cup	16 tablespoons	8 ounces
2 cups		1 pint (16 ounces)
4 cups		1 quart (32 ounces)
2 pints		1 quart (32 ounces)
4 quarts		1 gallon (128 ounces)
8 quarts	1 peck	
4 pecks	1 bushel	
16 ounces	1 pound dry measure	

tsp = teaspoon
t = tablespoon
oz = ounce
c = cup
pt = pint
qt = quart
bu = bushel
lb = pound

Oven Temperatures

°F	°C	Gas Mark	Description
225°F	110°C	Gas Mark 1/4	Very slow / Very cool
250	120	Gas Mark 1/2	Very slow / Very cool
275	140	Gas Mark 1	Slow / Cool
300	150	Gas Mark 2	Slow / Cool
325	160	Gas Mark 3	Warm / Very Moderate
350	180	Gas Mark 4	Moderate
375	190	Gas Mark 5	Moderately hot / Fairly hot
400	200	Gas Mark 6	Moderately hot / Fairly hot
425	220	Gas Mark 7	Hot
450	230	Gas Mark 8	Very hot
475	240	Gas Mark 9	Very hot

- Temperature equivalents are not exact; they are only a guide. Consult your oven's manual for its particular settings.
- Settings for convection ovens should be about 50°F (c. 25°C) less than the above.
- Remember to use an oven thermometer to check the actual temperature of your oven, as they do occasionally need to be recalibrated.

By permission: Fante's Kitchen Wares Shop, Philadelphia, PA-www.fantes.com

Liquid (Fluid) Measure Equivalents

0.125 oz	1 fl dram	60 minims			
1 oz	8 fl drams				
4 oz	1 gill				
8 oz	1 c				
16 oz	2 c	1 pt			
32 oz	4 c	2 pt	1 qt		
33.8 oz	4.23 c	2.1134 pt	1.0567 qt	0.264 gal	1 l
128 oz	16 c	8 pt	4 qt	1 gal	3.7853 l
4032 oz	1 bbl	3.94 pt	7.875 qt	31.5 gal	
8064 oz	1 hhd	7.875 pt	15.75 qt	63 gal	

fl = fluid
oz = ounce
c = cup
pt = pint
qt = quart
gal = gallon
bbl = barrel
hhd = hogshead
l = liter

Metric Equivalents, Liquid or Fluid Measure

	Or	Dry	Liquid
1 centiliter		0.6102 cubic inches	0.338 ounces
1 deciliter	10 centiliters	6.102 cubic inches	.0845 gill
1 liter	10 deciliters	0.908 quart	1.0567 quarts
1 decaliter	10 liters	9.08 quarts	2.64 gallons

cl = centiliter
dl = deciliter
l = liter
dal = decaliter
cu in = cubic inch
oz = ounce
qt = quart
gal = gallon

By permission: Fante's Kitchen Wares Shop, Philadelphia, PA-www.fantes.com

Weight Equivalents

Avoirdupois

16 drams	437.5 grains	1 ounce	28.35 grams
16 ounces	7000 grains	1 pound	453.59 grams
1 pound	0.45 kilograms		
1 kilogram	2.2 pounds		
100 pounds	1 central	1 hundredweight	
2000 pounds	1 short ton		
2204.6 pounds	1 metric ton	1000 kilograms	
2240 pounds	1 long ton or gross ton		

Also (in Great Britain)

14 pounds	1 stone	
2 stones	1 quarter	
4 quarters	112 pounds	1 hundredweight
20 hundredweight	1 long ton	

Troy (Precious Metals)

24 grains	1 pennyweight	
20 pennyweights	480 grains	1 ounce
12 ounces	5760 grains	1 pound

Apothecaries' Weight

20 grains	1 scruple	
3 scruples	1 dram	
8 drams	1 ounce	
12 ounces	5760 grains	1 pound

Paper

24 sheets	1 quire	
20 quires	1 short ream	480 sheets
500 sheets	1 ream	
10 reams	1 bale	

Weight of Water

1 cubic inch	.0360 pound
1 cubic foot	62.3 pounds
1 cubic foot	7.48052 U.S. gallons
1 Imperial gallon	10.0 pounds
1 U.S. gallon	8.33 pounds

By permission: Fante's Kitchen Wares Shop, Philadelphia, PA-www.fantes.com

Section 3

Calculations and Formulas—Geometry, Trigonometry, and Physics in Construction

3.0.0	Useful Formulas—Water, Pressure, Heat, Cooling, Horsepower	78
3.1.0	Basic Mathematics—Algebra	81
3.1.1	What Are Roots?	83
3.1.2	Area and Circumference of a Circle by Archimedes	83
3.1.3	Explaining Exponential Functions—Concepts, Solutions	84
3.2.0	General Geometric Formulas	85
3.2.1	Volume of Prisms	90
3.2.2	Volume of Pyramids	91
3.2.3	Area of Prisms and Right Area Prisms	91
3.2.4	Cylinder Volume Theorem	92
3.2.5	Cone Volume Theorem	93
3.2.6	Sphere Volume and Area Theorem	93
3.2.7	The Isosceles and Equilateral Triangles	94
3.2.8	Theorems That Apply to Parallelograms	95
3.2.9	The Small Angle Formula	96
3.2.10	Geometric Surface Area Formulas for Cubes, Spheres, Cones	99
3.2.11	Angles of an N-gon	99
3.3.0	Basic Trigonometric Functions	100
3.3.1	Trigonometry's Sine and Cosine	101
3.3.2	Trigonometric Ratios	101
3.3.3	Basic Trigonometry Formulas	102
3.3.4	What Is the Pythagorean Theorem?	109
3.3.5	Pythagorean and Quotient Identities	110
3.3.6	Properties of Triangles	111
3.3.7	Law of Sines	112
3.3.8	Law of Cosines	112
3.3.9	Reciprocal Ratios	113
3.3.10	Cofunctions	113
3.3.11	A Table of Common Logarithms	116
3.4.0	Physics—Basic Formulas	117
3.4.1	Physics Concepts—Force, Pressure, and Energy	130
3.4.2	Physics—Circular Motion	131
3.4.3	Physics—Gravitation	132
3.4.4	Physics—Work Energy Power	132
3.4.5	Physics—Laws of Motion	133
3.4.6	Physics—One-, Two-, Three-Dimensional Motion	134
3.4.7	Physics—Electricity	135
3.5.0	Financial Formula Calculations—Net Present Value and Compounding	136
3.6.0	Formulas for Calculating the Volume of Cylindrical Tanks	138
3.7.0	Round Tank Volume Tables for One Foot of Depth—1 to 32 Feet in Diameter	142
3.8.0	Capacity of Round Tanks—1 to 30 Feet in Diameter	146
3.9.0	Capacity of Rectangular Tanks—1 to 6 Feet in Depth, 1 to 10 Feet in Length	146
3.10.0	Testing for Hardness in Metal—Mohs, Brinell, Rockwell, Scleroscope, Durometer	147
3.11.0	Comparison of Hardness Scale Shown Graphically	151
3.12.0	Comparison of Hardness Testing in Chart Form	152

3.0.0 Useful Formulas—Water, Pressure, Heat, Cooling, Horsepower

Total Heat (BTU/hr) = $4.5 \times$ cfm $\times \Delta h$ (std air)
Sensible Heat (BTU/hr) = $1.1 \times$ cfm $\times \Delta t$ (std air)
Latent Heat (BTU/hr) = $0.69 \times$ cfm $\times \Delta gr$ (std air)

NOTE: For conditions other than standard air please see this page.

Total Heat (BTU/hr) = $500 \times$ gpm $\times \Delta t$ (water)
TONS = $24 \times$ gpm $\times \Delta t$ (water)
GPM cooler = $(24 \times$ TONS$) / \Delta t$ (water)
Fluid Mixture $T_m = (Xt_1 + Yt_2) / X + Y$ (this works for air or water)
BTU/hr = $3.413 \times$ watts = HP $\times 2546$ = Kg Cal $\times 3.97$
Lb. = 453.6 grams = 7000 grains
psi = ft water/2.31 = in. hg/2.03 = in. water/27.7 = $0.145 \times$ kPa
Ton = $12,000$ BTU/hr = $0.2843 \times$ KW
HP (air) = cfm $\times \Delta p$ (in. H_2O)/$6350 \times$ Eff
HP (water) = gpm $\times \Delta p$ (ft)/$3960 \times$ Eff
Gal. = FT^3/7.48 = 3.785 Liters = 8.33 lb (water) = 231 in.^{3}
gpm = $15.85 \times$ L/S
cfm = $2.119 \times$ L/S
Liter = $3.785 \times$ gal = $0.946 \times$ quart = $28.32 \times ft^3$
Therm = $100,000$ BTU = MJ/105.5
Watt/sq ft = $0.0926 \times W/M^2$
yd = $1.094 \times$ M
ft = $3.281 \times$ M
$ft^2 = 10.76 \times M^2$
$ft^3 = 35.31 \times M^3$
ft/min = $196.9 \times$ M/S
PPM (by mass) = mg/kg

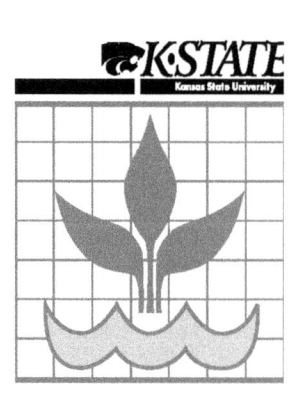

NOTE: Liter/sec is the proper SI term for liquid flow. M^3/sec is the proper SI term for airflow. Due to the awkward nature of using M^3/S at low air flow rates (lots of decimal points), L/S is commonly used to express air flow for HVAC applications.

Source: www.gorhamschaffler.com/formulas.html

1. **Water Measurement**
    1 cubic foot = 7.48 gallons = 62.4 pounds of water
    1 acre-foot = 43,560 cubic feet = 325,851 gallons = 12 acre-inches
    1 acre-inch = 27,154 gallons
    1 acre-foot is the volume of water that would cover 1 acre of land 1 foot deep
    1 acre-inch per hour = 450 gallons per minute (gpm)
    = 1 cubic foot per second (cfs)
    1 cubic meter = 1000 liters = 264 gallons
    1 gallon = 128 ounces = 3785 mililiters
    1 ounce = 29.56 mililiters
    1 liter = 1.06 quarts

2. **Pressure**
    1 pound per square inch (psi) = 2.31 feet of water = 6.9 kpa (kilopascal)
    = 0.0703 kilogram per square centimeter (kg/cm^2)
    = 0.704 meter of water
    A column of water 2.31 feet deep exerts a pressure of 1 psi at the bottom of the column.
    Total dynamic head (TDH) = pumping lift + elevation change + friction loss + irrigation system operating pressure

3. **Area/Length/Weight/Yield**
    1 acre = 0.405 hectare (ha) = 43,560 feet^2
    1 hectare = 2.47 acres
    1 mile = 5280 feet = 1.61 kilometers
    1 foot = 0.305 meter (m)
    1 meter = 3.28 feet
    1 inch = 2.54 centimeters
    1 pound = 454 grams
    1 kilogram per hectare (kg/ha) = 1 metric ton/ha (MT/ha)
    = 0.0149 bushel (60 pounds) per acre

4. **Temperature**
    °F = 1.8 (°C) + 32   °C = (°F − 32)/1.8

°C	−40	−20	0	20	37	60	80	100
°F	−40		0 32		80 98.6		160	212
			(water freezes)		(body temperature)			(water boils)

5. **Horsepower**
    1 horsepower = 0.746 kilowatts (kw) = 33,000 foot-pounds per minute

    Water horsepower (WHP) is the power required to lift a given quantity of water against a given total dynamic head.

    WHP = (Q × H) ÷ 3960, where Q = flow rate in GPM and H = total dynamic head in feet

    Brake horsepower (BHP) is the required power input to the pump.

    BHP = WHP/E, where E = pump efficiency

    Power unit horsepower

    Electric power units: approximate name plate horsepower = BHP ÷ 0.9

Internal combustion units:
Must derate 20% for continuous duty (= 80% efficiency)
        5% for right-angle drive (= 95% efficiency)
        3% for each 1000 feet above sea level (= 91% for 3000 feet)
        1% for each 10° above 60°F (= 96% for 100°F)

Approximate engine horsepower required = BHP ÷ deratings
$$= \text{BHP} \div (0.80 \times 0.95 \times 0.91 \times 0.96)$$

Plastic Pipe Friction Loss (psi loss per 100 feet of pipe) for C = 150

Pipe size (inches)	Flow rate (gpm)					
	10	25	50	75	100	150
	psi loss per 100 feet					
$1\frac{1}{2}$	0.26	1.40	5.50	—	—	—
2	0.09	0.52	1.90	4.10	—	—
$2\frac{1}{2}$	0.03	0.17	0.65	1.35	2.40	5.00
3	—	0.07	0.26	0.38	0.95	2.05
4	—	0.01	0.06	0.14	0.24	0.50
Pipe size (inches)	Flow rate (gpm)					
	200	400	600	800	1000	1200
	psi loss per 100 feet					
4	0.85	3.20	—	—	—	—
6	0.12	0.42	0.93	1.60	2.40	3.40
8	0.03	0.11	0.22	0.38	0.60	0.85
10	—	0.04	0.08	0.16	0.19	0.28
12	—	—	0.03	0.06	0.08	0.11

Economical Pipe Size Selection (flow in gpm)

Size	Aluminum *	Plastic *
(inches)	gpm	
4	200	275
6	450	620
8	800	1100
10	1250	1720
12	1800	2480

*Aluminum pipe velocity limited to 5 ft/sec. Plastic pipe velocity limited to 7 ft/sec.

Maximum Economical Pipe-flow Capacities

A rule of thumb for coupled and gated pipe:

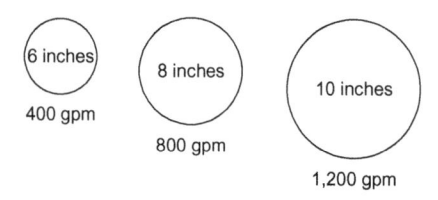

6 inches — 400 gpm

8 inches — 800 gpm

10 inches — 1,200 gpm

This and other publications from Kansas State University are available on the World Wide Web at *http://www.oznet.ksu.edu*.

These materials may be freely reproduced for educational purposes. All other rights reserved. In each case, credit Danny H. Rogers and Mahbub Alam, *Useful Conversions and Formulas*, Kansas State University, December 1997.

Source: Kansas State University

3.1.0 Basic Mathematics—Algebra

Algebra is a type of mathematics that is used to determine unknown variables. This section is not designed to teach Algebra, but to help remind the reader what Algebra looks like (I know for me, the last time I looked at a variety of Algebra problems was well over 10 years ago!).

$$x + y = z$$

The above sample is an example of a simple algebra problem. To solve this equation, we must know at least two of the variables:

$$x = 2$$
$$z = 4$$
$$2 + y = 4$$
$$y = 2$$

Proportions are just another version of ratios and are found in the following manner:

$$\frac{2}{4} = \frac{x}{7}$$
$$4 * x = 2 * 7$$
$$4x = 14$$
$$x = 3.5$$

A monomial is an expression for a single number variable: $a = 5$ for example. Algebra is used to solve a combination of monomials such as binomials and polynomials. The method for solving a binomial or polynomial equation is called FOIL (First, Outer, Inner, Last).

A binomial:

$$(3 + 7x)(6 + 2x)$$
$$F.O.I.L$$
$$18 + 6x + 42x + 14x^2$$
$$18 + 48x + 14x^2$$

A polynomial:

$$x^6 + x^3 y^4 + x^3 y^4 + y^8$$
$$x^6 + 2x^3 y^4 + y^8$$

Other algebraic techniques include a distribution function. This is used to simplify two or more terms in order to make a problem easier to work:

$$4(2 + 3x)$$
$$4 * 2 + 4 * 3x$$
$$8 + 12x$$

Graphing is also a part of Algebra. Two famous types of equations that are used to plot a graph is the linear equation and the quadratic equation.

A linear equation is used to plot a single line based on a slope:
Following is a graph that plots the following linear equation:

$$y = 2x + 1$$

Image Credit

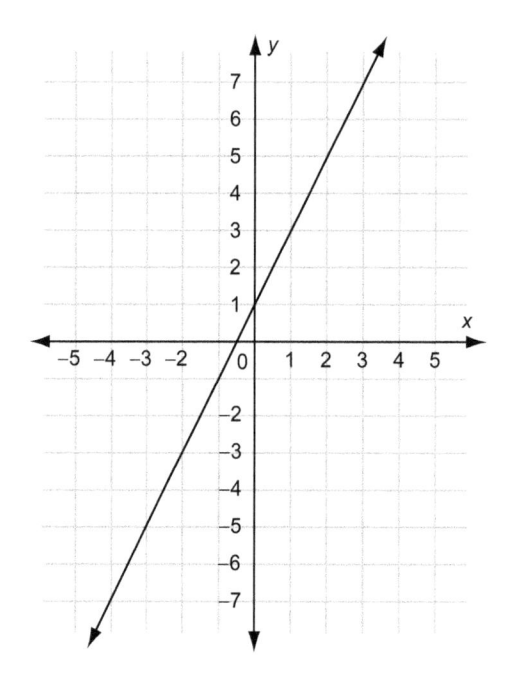

The quadratic equation is used to plot curves and parabolas based on more than one variable:

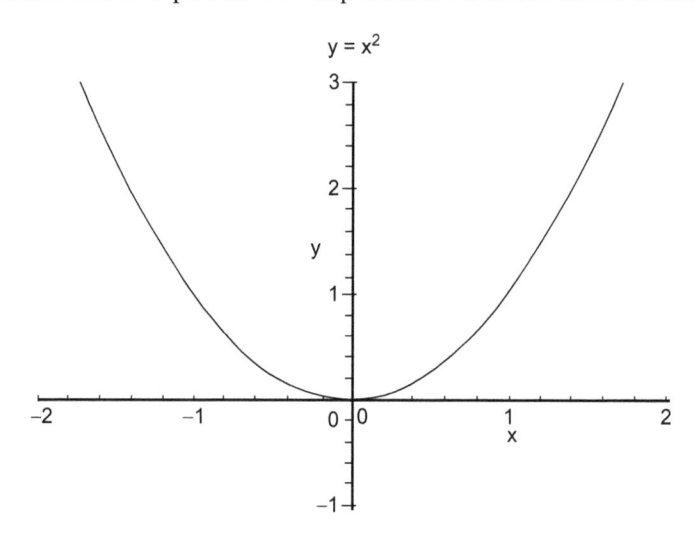

$$y = x^2$$

Above is a graph of the following quadratic equation:

$$y = 2x^2 + 2x - 2$$

3.1.1 What Are Roots?

Roots are numbers that make an expression equal to zero. If it's an equation like $y = mx + b$ or $y = ax^2 + bx + c$ or any other equation involving x and y, the roots are those points where the graph crosses the x-axis.

Example 1: $y = 2x - 3$. The root of this equation is 1.5.

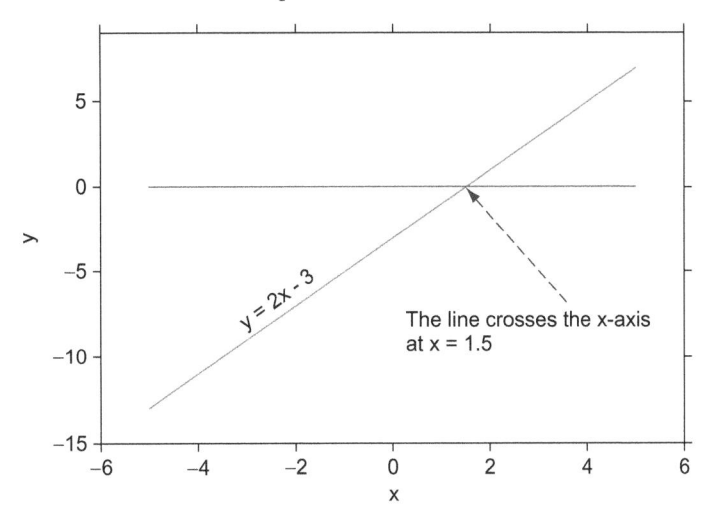

Example 2: $y = 2x^2 - x - 15$. The roots of this equation are -2.5 and 3.

Source: mathwizz.Com

3.1.2 Area and Circumference of a Circle by Archimedes

Goals

1. Understand the relationship among Π, area, and circumference of a circle.
2. Understand how Archimedes used the unit circle (radius = 1) and the concept of mathematical limits to show that $A = \Pi R^2$ and $C = 2\Pi R \cdot b_n 2$.

Basic Principles

1. The area of a polygon with n sides (n-gon) inscribed in a unit circle approaches the special number pi as n increases.
2. The area of an inscribed polygon approaches the area of a circle as the number of sides on the polygon increases.
3. The perimeter of an inscribed polygon approaches the perimeter (circumference) of a circle as the number of sides on the polygon increases.

Givens

- n = number of sides on the inscribed circle
- 1 = radius of unit circle
- h_n = height of an isosceles triangle inscribed in the inner circle
- b_n = base of an isosceles triangle inscribed in the inner circle
- R = radius of the n-gon (Note that this radius is visualized in this applet as being greater than one, but R could be any value greater than zero.)

- Rh_n = height of each inscribed isosceles triangle (based on hn = 1)
- Rb_n = base of each inscribed isosceles triangle (based on hn = 1)
- P_n = the perimeter of the unit circle
- A_n = area of the unit circle
- A = the sum of the areas of the inscribed triangles, the limit of which is the area of the circle
- C = the sum of the "bases" of the inscribed isosceles triangles, the limit of which is the circumference of the circle.

Knowns and Assumptions

- Area of a triangle $= \dfrac{(\text{base} * \text{height})}{2}$
- Other basic rules of algebra and geometry

Method

With the unit circle (red) as the basis, Archimedes used the limiting process on the area and base of polygons (n-gons) inscribed in circles (as n approaches infinity) to determine Π and at the same time verify the formulas for the area and circumference of any circle (blue).

What to Do

1. Notice how R, h_n, b_n, and the unit circle are used in the graphical Definition of Variables.
2. Observe how each of these variables is positioned in the equation in the center bottom of the screen.
3. Follow how the rules of algebra are used to rearrange the variables from the center equation outward to the left and outward to the right. Also, note the use of P_n in the right-hand limit.

   Source: math.psu.edu

4. Increase n. Note that as you increase n, the values of n A_n on the left and the value of $P_n h_n$ on the right side simultaneously come closer and closer to the limiting value Π.
5. Notice: The limit at the left is A, the area of the circle. The limit at the right is $\dfrac{CR}{2}$, the perimeter (or circumference) of the circle.

3.1.3 Explaining Exponential Functions—Concepts, Solutions

Exponential Functions

Exponential functions are functions where $f(x) = a^x + B$ where a is any real constant and B is any expression. For example, $f(x) = e^{-x} - 1$ is an exponential function.

To graph exponential functions, remember that unless they are transformed, the graph will always pass through *(0, 1)* and will approach, but not touch or cross, the *x*-axis. Example:

1. **Problem:** Graph $f(x) = 2^x$.

   Solution: Plug in numbers for *x* and find values for *y*, as we have done with the table below.

x	0	1	2	3
y	1	2	4	8

Now plot the points and draw the graph (shown below).

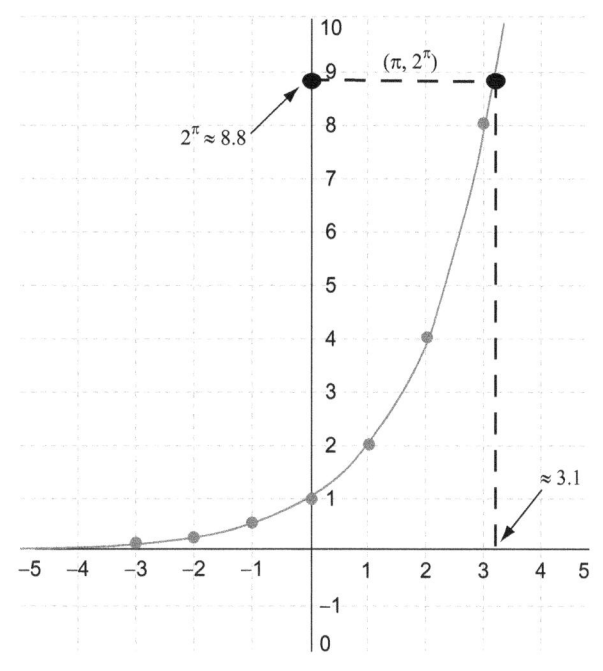

Source: Thinkquest.org

3.2.0 General Geometric Formulas

General Formulas

Here are some common simple geometric formulas useful in estimating sizes, quantities, and amounts:

Volume of a cube in cubic yards

cu yd's = (length' × width' × height') / 27

Why 27? Because 3' × 3' × 3' = 27 cu ft = 1 cu yd

Area of a parallelogram

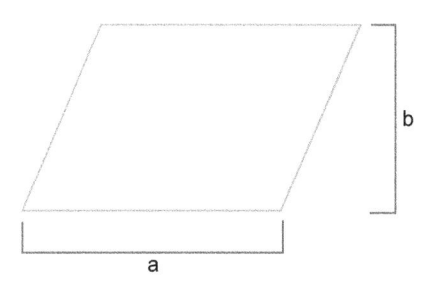

area = a × b

Area of a trapezoid

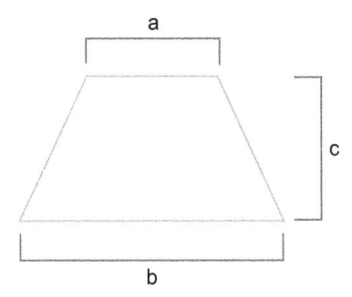

area = $((a + b)/2) \times c$

Hypotenuse of a right triangle

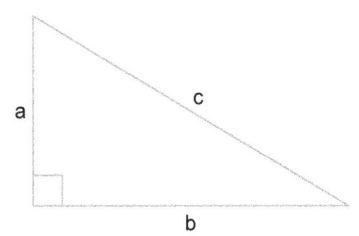

hypotenuse (c) = $\sqrt{a^2 + b^2}$

Area of any triangle

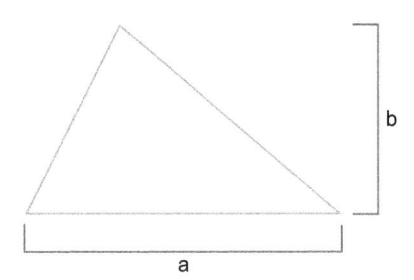

area = $(a \times b)/2$

pi

What exactly is **pi**?

Why is it 3.1416?

If you take any circle, measure its circumference and its diameter, and divide the circle's circumference by its diameter, the answer is always 3.1416. This ratio of the circumference to the diameter is always the same, regardless of the size of the circle or its units of measure.

Volume of a cylinder

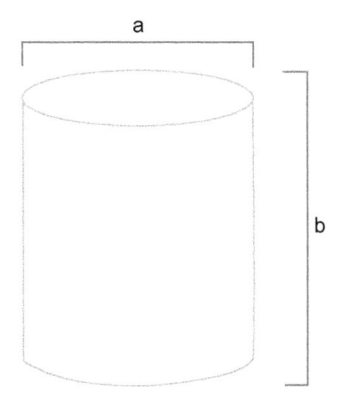

volume = (pi / 4) × a² × b and volume = pi × a/2 × a/2 × b

Volume of a frustum of a pyramid

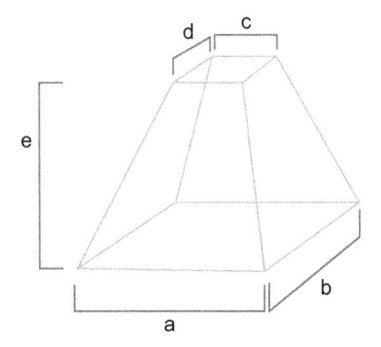

volume = (e/3) × ((a × b) + (c × d) + √((a × b) × (c × d)))

Volume of a cone

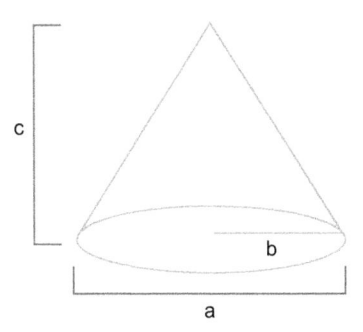

volume = (pi / 3) × b² × c

Volume of a frustum of a cone

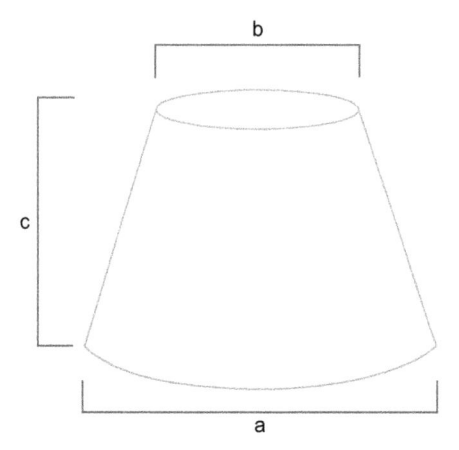

$$\text{volume} = (\pi / 12) \times c \times (a^2 + (a \times b) + b^2)$$

Frustum's surface area

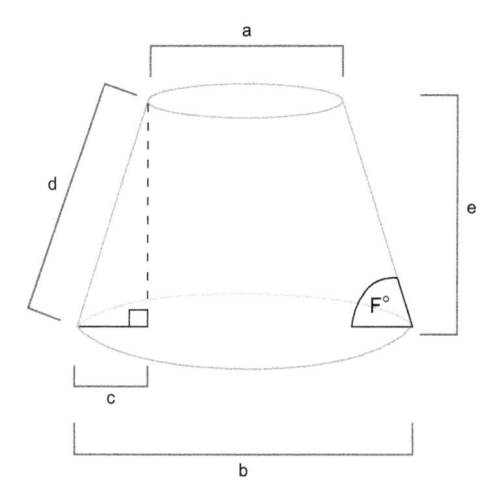

Angle of the cone	$F°$
Frustum's altitude	$e = \text{tangent } F° \times c$
Triangle's base	$c = (b - a)/2$
Length of Frustum's side	$d = \sqrt{c^2 + e^2}$
Frustum's surface area	$\text{area} = (\pi/2) \times d \times (a + b)$

Tangent Examples:
tangent $30° = 0.57735$
tangent $35° = 0.70021$

tangent 40° = 0.83910
tangent 45° = 1.00000
tangent 50° = 1.19175
tangent 55° = 1.42815
tangent 60° = 1.73205
tangent 65° = 2.14451
tangent 70° = 2.74748

Volume of a sphere

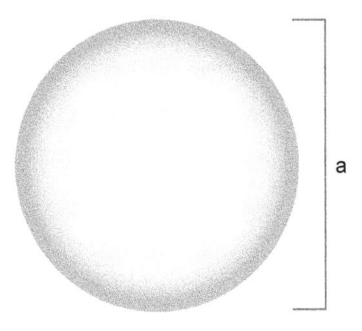

volume = (pi / 6) × a³

Area of a circle

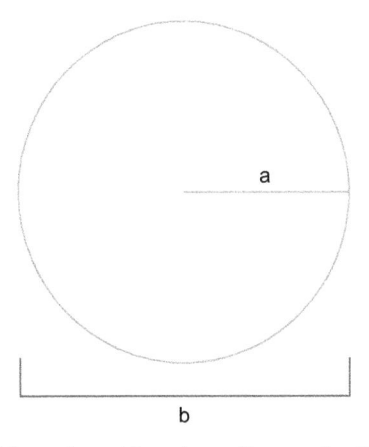

A circle's area can be determined by using either the radius or the diameter

area = pi a²
area = 0.785 b²

Circumference of a circle

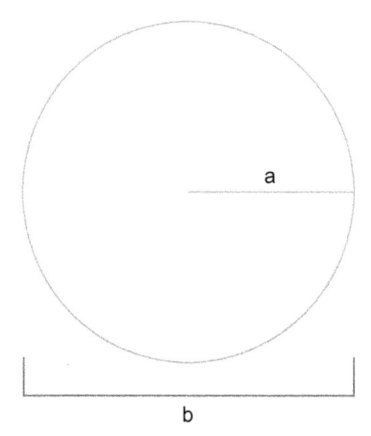

circumference = pi b

Volume of a pyramid

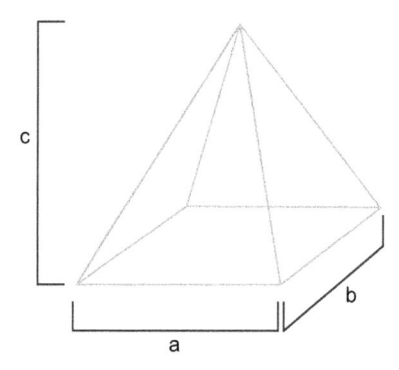

volume = (a × b × c) / 3

3.2.1 Volume of Prisms

Right Prism Volume Postulate

The volume V of any right prism is the product of B, the area of the base, and the height h of the prism.

Formula: $V = Bh$
$B = lw$

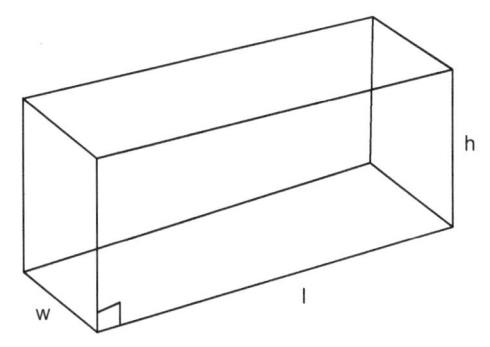

(The base's formula could change depending on the base's shape.)

3.2.2 Volume of Pyramids

Pyramids

A *pyramid* is a polyhedron with a single base and lateral faces that are all triangular. All lateral edges of a pyramid meet at a single point, or *vertex*.

Pyramid Volume Theorem

The volume V of any pyramid with height h and a base with area B is equal to one-third the product of the height and the area of the base.

Formula: $V = (1/3)Bh$

3.2.3 Area of Prisms and Right Area Prisms

There are special formulas that deal with prisms, but they only deal with right prisms. *Right prisms* are prisms that have two special characteristics—all lateral edges are perpendicular to the bases, and lateral faces are rectangular. The figure below depicts a right prism.

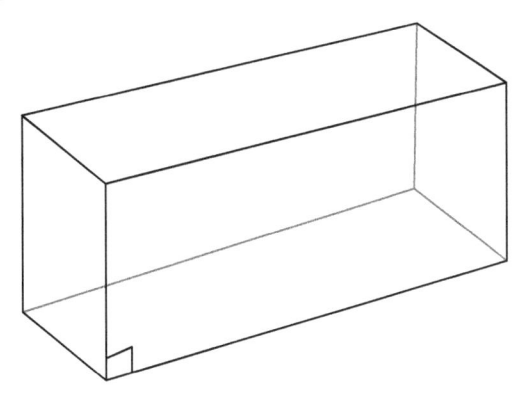

Right Prism Area

The lateral area L (area of the vertical sides only) of any right prism is equal to the perimeter of the base times the height of the prism ($L = Ph$).

The total area T of any right prism is equal to two times the area of the base plus the lateral area.

Formula: T = 2B + Ph

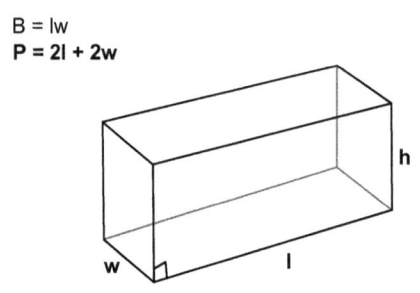

B = lw
P = 2l + 2w

(The base's formula could change depending on the base's shape.)
(The perimeter's formula could change depending on the base's shape.)

3.2.4 Cylinder Volume Theorem

Cylinders

Cylinder Volume Theorem

The volume V of any cylinder with radius r and height h is equal to the product of the area of a base and the height.

Formula: $V = (PI)r^2h$

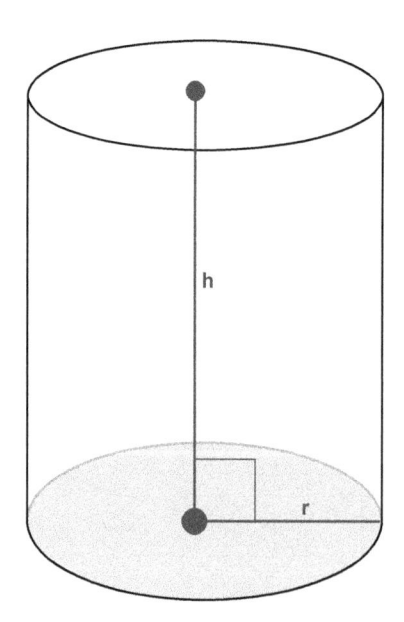

Cylinder Area Theorem

For any right circular cylinder with radius r and height h, the total area T is two times the area of the base plus the lateral area (2(PI)rh).

Formula: $T = 2(PI)rh + 2(PI)r^2$

3.2.5 Cone Volume Theorem

Cones

Cone Volume Theorem

The volume V of any cone with radius r and height h is equal to one-third the product of the height and the area of the base.

Formula: $V = (1/3)(PI)r^2h]$

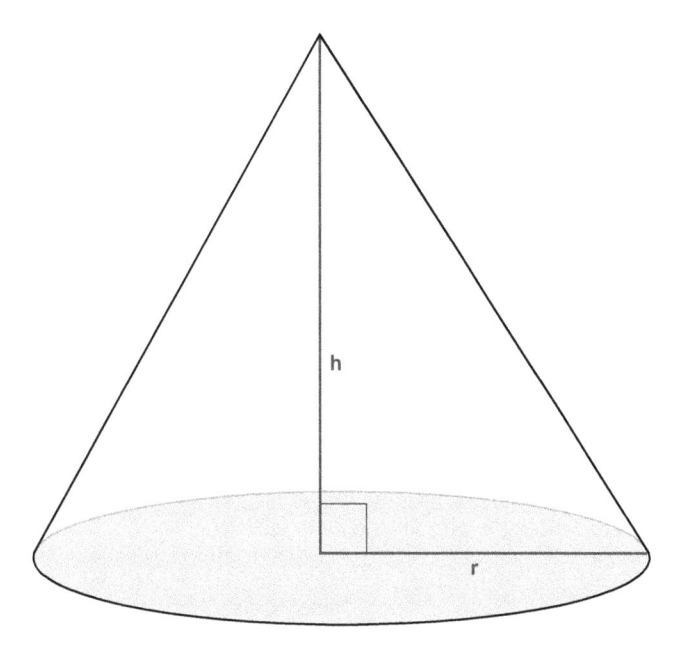

Cone Area Theorem

The total area T of a cone with radius r and slant height l is equal to the area of the base plus PI times the product of the radius and the slant height.

Formula: $T = (PI)rl + (PI)r^2$

3.2.6 Sphere Volume and Area Theorem

Spheres

Sphere Volume and Area Theorem

The volume V for any sphere with radius r is equal to four-thirds times the product of PI and the cube of the radius. The area A of any sphere with radius r is equal to 4(PI) times the square of the radius.

Volume Formula: $V = (4/3)(PI)r^3$

Area Formula: $A = 4(PI)r^2$

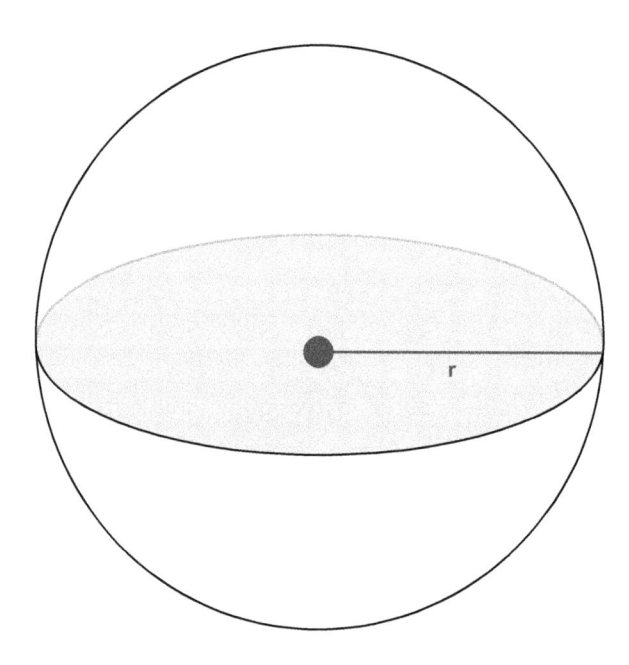

3.2.7 The Isosceles and Equilateral Triangles

An *isosceles triangle* has two congruent sides called *legs* and a third side called the *base*. The *vertex angle* is the angle included by the legs. The other two angles are called *base angles*. The base angles are congruent. The figure below depicts an isosceles triangle with all the parts labeled.

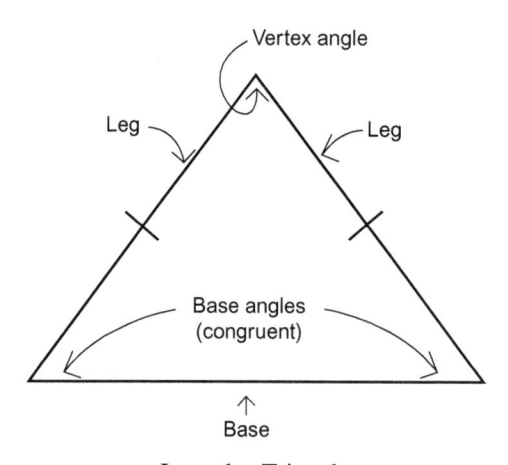

Isosceles Triangle

An *equilateral triangle* is a special isosceles triangle in which all three sides are congruent. Equilateral triangles are also equiangular, which means all three angles are congruent. The measure of each angle is 60 degrees. The figure below depicts an equilateral triangle with all the parts labeled.

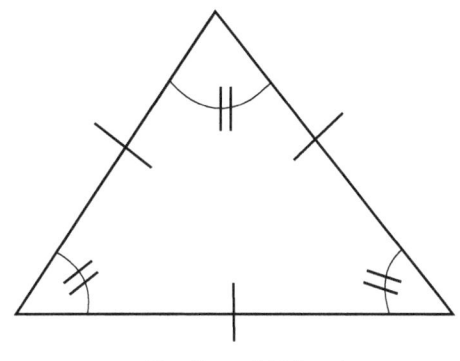

Equilateral Triangle

3.2.8 Theorems That Apply to Parallelograms

A parallelogram is so named because it has two pairs of opposite sides that are parallel. Four theorems apply to parallelograms only. They are as follows.

1. A diagonal of any parallelogram forms two *congruent triangles*. Example:

   Problem: Prove *triangle ABC* is congruent to *triangle CDA*.

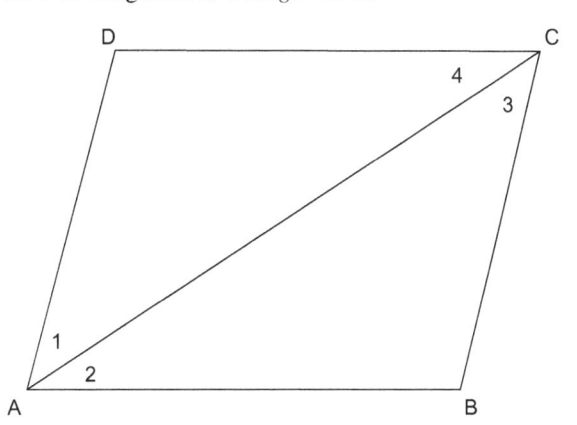

   Solution: Since the figure is a parallelogram, *segment AB* is parallel to *segment DC* and the two segments are also congruent.
   Angle 2 is congruent to *angle 4* and *angle 1* is congruent to *angle 3*. This is true because alternate interior angles are congruent when parallel lines are cut by a transversal.
   Segment AC is congruent to *segment CA* by the **Reflexive Property of Congruence**, which says any figure is congruent to itself.
   Triangle ABC is congruent *triangle CDA* by Angle-Side-Angle.

2. Both pairs of opposite sides of a parallelogram are congruent.
3. Both pairs of opposite angles of a parallelogram are congruent.
4. The diagonals of any parallelogram bisect each other. Example:

   Problem: Prove *segment AE* is congruent to *segment CE* and *segment DE* is congruent to *segment BE*.

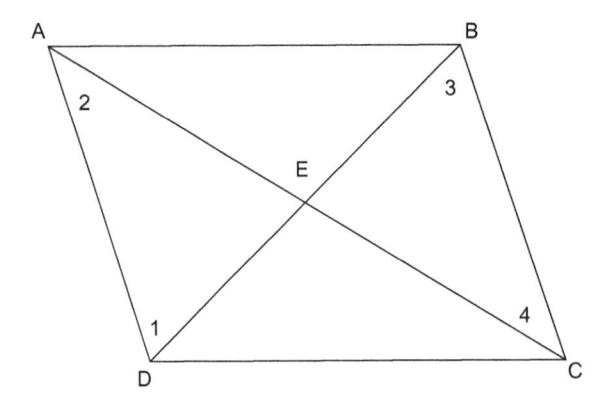

Solution: By the definition of a parallelogram, *segment AD* and *segment BC* are parallel and congruent. *Angle 1* is congruent to *angle 3,* and *angle 2* is congruent to *angle 4*. This is true because alternate interior angles are congruent when parallel lines are cut by a transversal.

Triangle AED and *triangle CEB* are congruent by Angle-Side-Angle.

The segments we were asked to prove as congruent are congruent by CPCTC.

The three theorems that tell us how to find a parallelogram are as follows.

1. If both pairs of opposite sides of a quadrilateral are congruent, the quadrilateral is a parallelogram.
2. If the diagonals of a quadrilateral bisect each other, then the quadrilateral is a parallelogram.
3. If one pair of opposite sides of a quadrilateral is both parallel and congruent, then the quadrilateral is a parallelogram.

3.2.9 The Small Angle Formula

There is a very powerful formula relating the *size* of an object to its *distance* and its *angular size*. This formula, the **small angle formula**, comes from considering a circle of radius **r**. Remember that the circumference **c** is the distance all the way around the circle and $c = 2\pi r$. What if we are not interested in the distance *all* the way around the circle, but instead want to know the distance around *part* of the circle, say the length of the arc marked s?

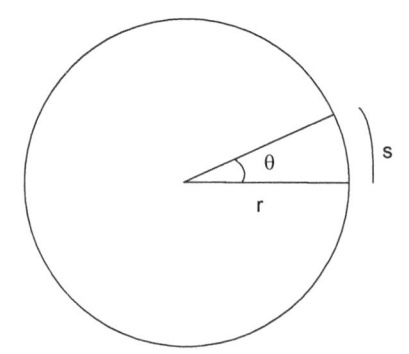

For this we can set up a ratio:

$$\frac{\text{length of s}}{\text{distance around whole circle}} = \frac{\text{angle subtended by s}}{\text{angle around whole circle}}$$

so that

$$\frac{s}{2\pi r} = \frac{\theta}{360°}$$

This is the small angle formula.

Why is this formula so great? Because it can even be used for things that are not part of a circle, as long as the angle is small! For example, when the angle is small (say less than 25), the triangle below looks an awful lot like the wedge from the circle above.

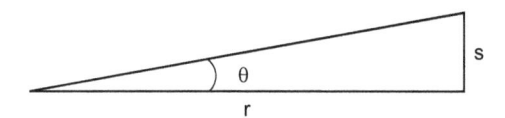

Now we have a very powerful tool indeed because we can turn a lot of astronomy problems into pictures involving skinny triangles—as you will see as you read on!

Angular Sizes

In astronomy, we study the universe while sitting comfortably here on good ol' Terra Firma. This means that we cannot generally measure the sizes of objects using rulers. Let's face it, even if we were to visit Jupiter, it would be awfully hard to find a ruler big enough to measure it...

So from our earthly vantage point, we often describe the size of an object using an angular measure rather than a linear (ruler-like) one. If we are lucky enough to know something about an object's distance, then we can relate its **angular size** to its **linear size** using the small angle formula. This method is *very* frequently used to measure things in astronomy.

As an example, imagine that you are looking at the Green Hall Tower from a distance of 200 meters. You estimate from your point of view that the Tower covers an angle of 10°. We can draw the following picture:

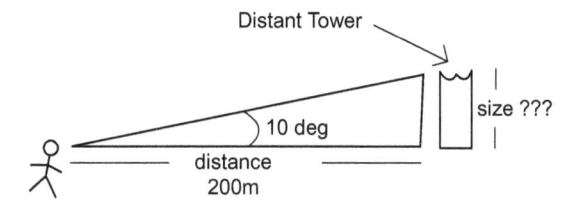

Note that this looks very similar to the skinny triangle picture above. In fact, we can apply the small angle formula to the triangle originating at your eyeball to get the height of the Tower:

$$\text{size} = \frac{\theta}{360°} \times 2\pi \times \text{distance}$$

$$= \frac{\theta}{57.3°} \times \text{distance}$$

$$= \frac{10°}{57.3°} \times 200m = 35m$$

So we were able to measure the height of the Tower without actually going there!

The angle that an object covers when we trace it back to your eyeball is called its **angular size**. Consider the following pictures.

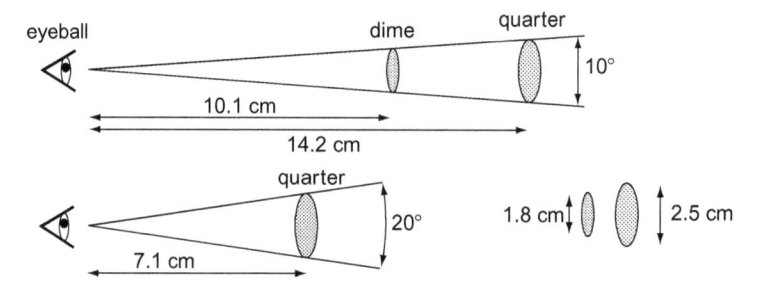

The top drawing demonstrates that two objects having different linear sizes can have the same angular size ( ) if they are viewed from different distances. The object's angular size is determined by the ratio size/distance. The quarter's linear size (2.5 cm) is 1.4 times as big as the dime's (1.8 cm) and so must be placed 1.4 times farther away to subtend the same angle. Now move the quarter two times closer as in the lower drawing, and its angular size is twice as big (20 degrees instead of 10 degrees).

You can practice measuring the angular sizes of things (trees, constellations, friends) using various body parts! As shown in the picture below, the angular size of your fist when you put your arm *straight* out in front of you is approximately 10 degrees. Also with a straight arm, your pinky fingernail subtends about 1 degree.

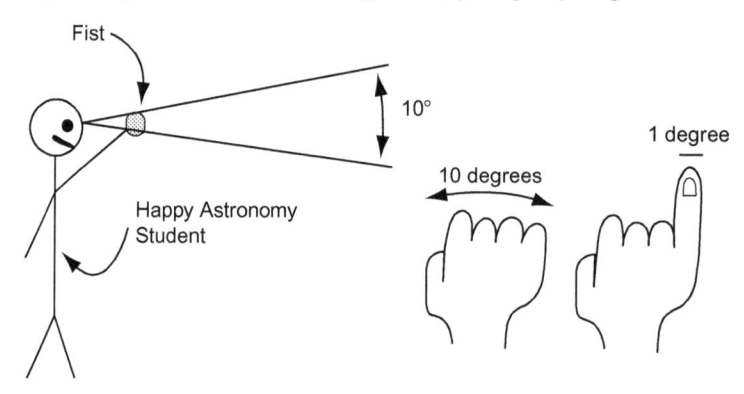

Summary

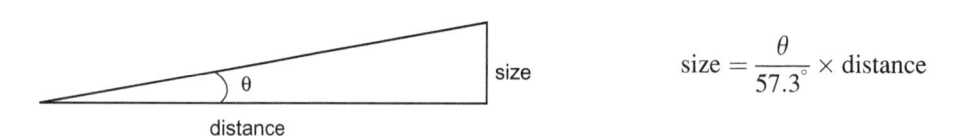

$$\text{size} = \frac{\theta}{57.3°} \times \text{distance}$$

Note that you have to express the angle in degrees in order to get the units to work out properly. In astronomy we are often working with very small angles, measured in <u>arcseconds</u>. We can change the degrees in this equation to arcseconds using our <u>units conversion</u> methods:

$$\text{size} = \frac{\theta}{57.3°} \times \text{distance} \times \frac{1°}{60'} \times \frac{1'}{60''} = \frac{\theta}{206265''} \times \text{distance}$$

You will often see the small angle formula written in the textbooks in this form.

3.2.10 Geometric Surface Area Formulas for Cubes, Spheres, Cones

Geometric Surface Area Formulas—Cube, Sphere, Cone

Surface Area of a Cube

$$SA = 6 x^2$$

where: SA is the surface area and x is the side of the cube

Surface Area of a Cuboid

$$SA = 2(xy + yz + zx)$$

where: x, y, and z are the adjacent three sides of the cuboid

Surface Area of a Sphere

$$SA = 4 \text{ pi } r^2$$

where: r is the radius of the sphere and pi = 3.14

Surface Area of a Cone

$$SA = \text{pi} \times r \times L$$

where: r is the radius of the cone and L is the slant height of the cone and pi = 3.14

Total Surface Area of a Cone (including the area of the base)

$$SA = (\text{pi} \times r \times L) + \text{pi} \times r^2$$

where: r is the radius of the cone and L is the slant height of the cone and pi = 3.14

3.2.11 Angles of an N-gon

Although you won't encounter many odd shapes, such as shapes with t12 sides, it *can* happen. On most instances of this, you will need to find the sum of the measures of the angles. There is a special theorem that says, *if n is the number of sides of any polygon, the sum (S) of the measure of its angles is given by the formula*

$$S = (n - 2)180°$$

The figure and table below will help this theorem make more sense.

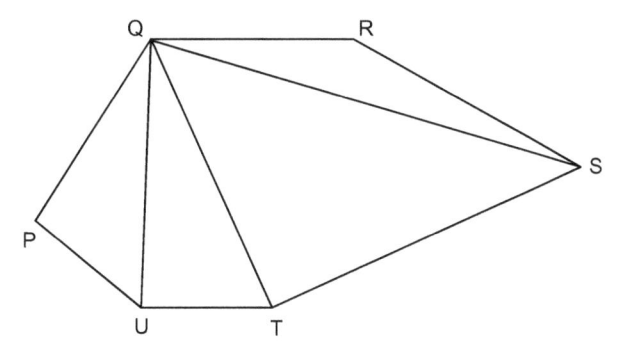

Hexagon

(3 diagonals from one vertex.)
   (4 triangles formed.)

Polygon	No. Sides	Total No. of Diagonals fr. 1 vertex	No. Triangles Formed	Sum of Angle Measures
Triangle	3	0	1	180°
Quad.	4	1	2	360°
Pentagon	5	2	3	540°
Hexagon	6	3	4	720°
.	.	.	.	.
.	.	.	.	.
.	.	.	.	.
n-gon	n	n − 3	n − 2	(n−2)(180°)

3.3.0 Basic Trigonometric Functions

Basic Trigonometric Functions:

Trigonometry is a specialty of Geometry that focuses mostly on angles. If we want the sum of all internal angles or if we want a ratio of angles for example, we will turn to Trigonometry.

Two recurrent terms to know are the Sine and Cosine

- Sine = ratio of the height to the hypotenuse (see image below)
- Cosine = ratio of the base of the hypotenuse (see image below)

A simple example: A straight line has an angle—its 0 degrees.

Sine = 0°
Cosine = 1°

To introduce the working formulas for sine and cosine, we will use a right triangle.

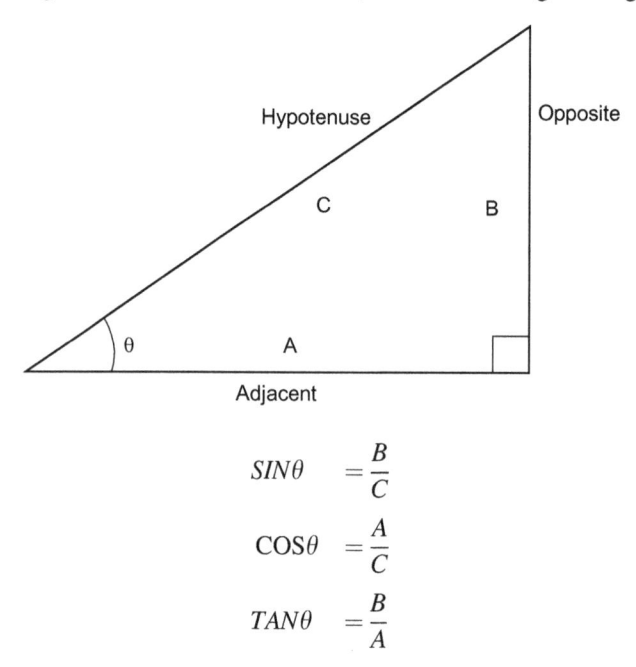

$$SIN\theta = \frac{B}{C}$$

$$COS\theta = \frac{A}{C}$$

$$TAN\theta = \frac{B}{A}$$

Pythagorean Theorem:
$$A^2 + B^2 = C^2$$

Source: astronomyonline.org

3.3.1 Trigonometry's Sine and Cosine

A right-angled triangle has a hypotenuse, base, and perpendicular side. Let angle A be the angle between the base and the hypotenuse.
Then

Sin A = perpendicular / hypotenuse
Cos A = base / hypotenuse
Tan A = (Sin A) / (Cos A) = perpendicular / base
Cot A = 1 / Tan A = (Cos A) / (Sin A)
Sec A = 1 / Cos A
Cosec A = 1/Sin A

Certain basic trigonometric identities applicable to any angle are:

Sin (-A) = - Sin A
Cos (-A) = Cos A
Tan (-A) = - Tan A
$Sin^2 A + Cos^2 A = 1$
$1 + Tan^2 A = Sec^2 A$
$1 + Cot^2 A = Cosec^2 A$

Source: Tutor4Physics.com

3.3.2 Trigonometric Ratios

Ratios

The trigonometric ratios, *sine*, *cosine*, and *tangent* are based on properties of right triangles. The function values depend on the measure of the angle. The functions are outlined below.

sine x = (side opposite x)/hypotenuse
cosine x = (side adjacent x)/hypotenuse
tangent x = (side opposite x)/(side adjacent x)

In the figure below, sin A = a/c, cosine A = b/c, and tangent A = a/b.

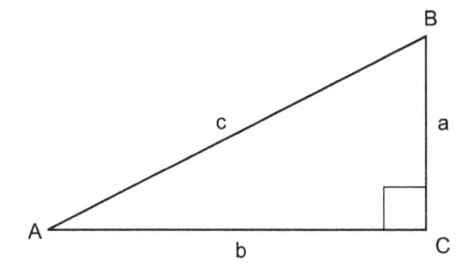

There are two special triangles you need to know, 45-45-90 and 30-60-90 triangles. They are depicted in the figures below.

Isosceles Right Triangle

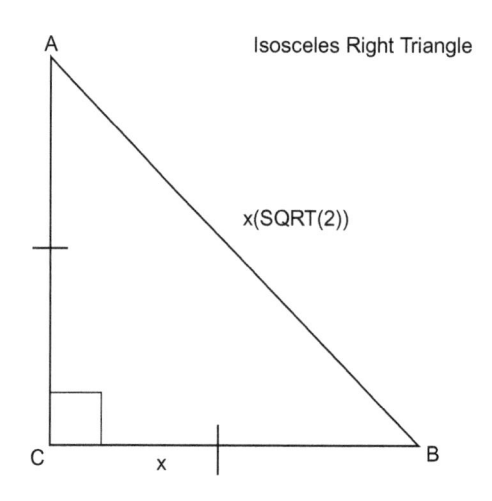

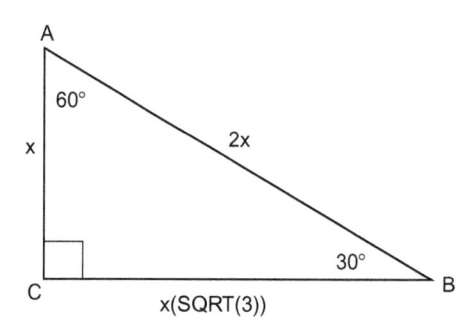

3.3.3 Basic Trigonometry Formulas

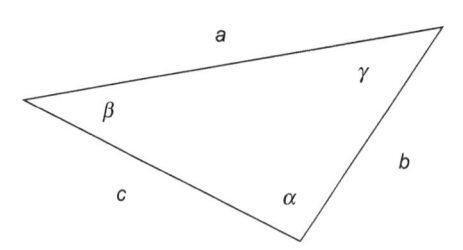

Law of Sines

$$\frac{\sin \alpha}{a} = \frac{\sin \beta}{b} = \frac{\sin \gamma}{c}$$

Law of Cosines

$$a^2 = b^2 + c^2 - 2bc \cos \alpha$$
$$b^2 = a^2 + c^2 - 2ac \cos \beta$$
$$c^2 = a^2 + b^2 - 2ab \cos \gamma$$

Law of Tangents

$$\frac{a+b}{a-b} = \frac{\tan\frac{1}{2}(\alpha+\beta)}{\tan\frac{1}{2}(\alpha-\beta)}$$

$$\frac{a+c}{a-c} = \frac{\tan\frac{1}{2}(\alpha+\gamma)}{\tan\frac{1}{2}(\alpha-\gamma)}$$

$$\frac{b+c}{b-c} = \frac{\tan\frac{1}{2}(\beta+\gamma)}{\tan\frac{1}{2}(\beta-\gamma)}$$

Mollweide's Formulas

$$\frac{b-c}{a} = \frac{\sin\frac{1}{2}(\beta-\gamma)}{\cos\frac{1}{2}\alpha}$$

$$\frac{c-a}{b} = \frac{\sin\frac{1}{2}(\gamma-\alpha)}{\cos\frac{1}{2}\beta}$$

$$\frac{a-b}{c} = \frac{\sin\frac{1}{2}(\alpha-\beta)}{\cos\frac{1}{2}\gamma}$$

Newton's Formulas

$$\frac{b+c}{a} = \frac{\cos\frac{1}{2}(\beta-\gamma)}{\sin\frac{1}{2}\alpha}$$

$$\frac{c+a}{b} = \frac{\cos\frac{1}{2}(\gamma-\alpha)}{\sin\frac{1}{2}\beta}$$

$$\frac{a+b}{c} = \frac{\cos\frac{1}{2}(\alpha-\beta)}{\sin\frac{1}{2}\gamma}$$

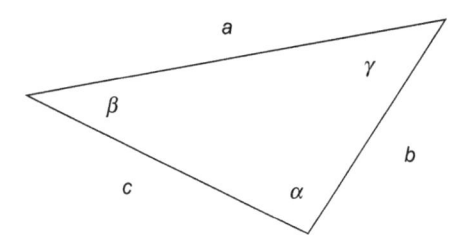

$$\text{Semiperimeter} = s = \frac{1}{2}(a+b+c)$$

Heron's Formula

$$\text{Area} = \sqrt{s(s-a)(s-b)(s-c)}$$

Other Area Formulas

$$\text{Area} = \frac{bc \sin \alpha}{2} = \frac{ac \sin \beta}{2} = \frac{ab \sin \gamma}{2}$$

$$\text{Area} = \frac{c^2 \sin \alpha \sin \beta}{2 \sin \gamma}$$

$$= \frac{b^2 \sin \alpha \sin \gamma}{2 \sin \beta}$$

$$= \frac{a^2 \sin \beta \sin \gamma}{2 \sin \alpha}$$

Triangle Sides

$$a = b \cos \gamma + c \cos \beta$$
$$b = c \cos \beta + a \cos \alpha$$
$$c = a \cos \alpha + b \cos \gamma$$

Radius of Inscribed Circle

$$r = \sqrt{\frac{(s-a)(s-b)(s-c)}{s}} = \frac{\text{Area}}{s}$$

Radius of Circumscribed Circle

$$R = \frac{a}{2 \sin \alpha} = \frac{b}{2 \sin \beta} = \frac{c}{2 \sin \gamma} = \frac{abc}{4(\text{Area})}$$

Angles

$$\sin \alpha = \frac{2}{bc} = (\text{Area})$$

$$\cos \alpha = \frac{c^2 + b^2 - a^2}{2bc}$$

$$\sin \frac{\alpha}{2} = \sqrt{\frac{(s-b)(s-c)}{bc}}$$

$$\cos \frac{\alpha}{2} = \sqrt{\frac{s(s-a)}{bc}}$$

Analogous formulas hold for other angles.

Right Triangle Definitions

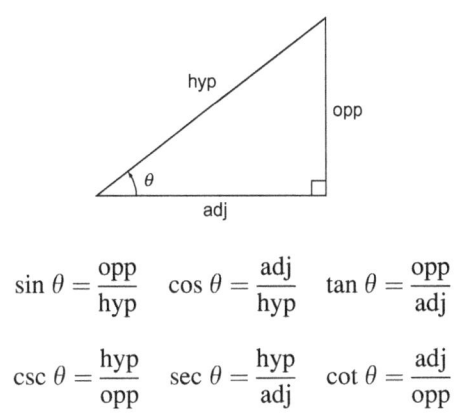

$$\sin \theta = \frac{\text{opp}}{\text{hyp}} \quad \cos \theta = \frac{\text{adj}}{\text{hyp}} \quad \tan \theta = \frac{\text{opp}}{\text{adj}}$$

$$\csc \theta = \frac{\text{hyp}}{\text{opp}} \quad \sec \theta = \frac{\text{hyp}}{\text{adj}} \quad \cot \theta = \frac{\text{adj}}{\text{opp}}$$

Basic Identities

$$\sin x = \frac{1}{\csc x} \quad \cos x = \frac{1}{\sec x} \quad \tan x = \frac{1}{\cot x}$$

$$\csc x = \frac{1}{\sin x} \quad \sec x = \frac{1}{\cos x} \quad \cot x = \frac{1}{\tan x}$$

$$\tan x = \frac{\sin x}{\cos x} \quad \cot x = \frac{\cos x}{\sin x}$$

Pythagorean Identities

$$\sin^2 x + \cos^2 x = 1$$
$$\tan^2 x + 1 = \sec^2 x$$
$$1 + \cot^2 x = \csc^2 x$$

Symmetry Properties

$$\cos(-x) = \cos x \qquad \sec(-x) = \sec x$$
$$\sin(-x) = -\sin x \qquad \csc(-x) = -\csc x$$
$$\tan(-x) = -\tan x \qquad \cot(-x) = -\cot x$$

Sum and Difference Formulas

$$\sin(x \pm y) = \sin x \cos y \pm \cos x \sin y$$
$$\cos(x \pm y) = \cos x \cos y \mp \sin x \sin y$$
$$\tan(x \pm y) = \frac{\tan x \pm \tan y}{1 \mp \tan x \tan y}$$

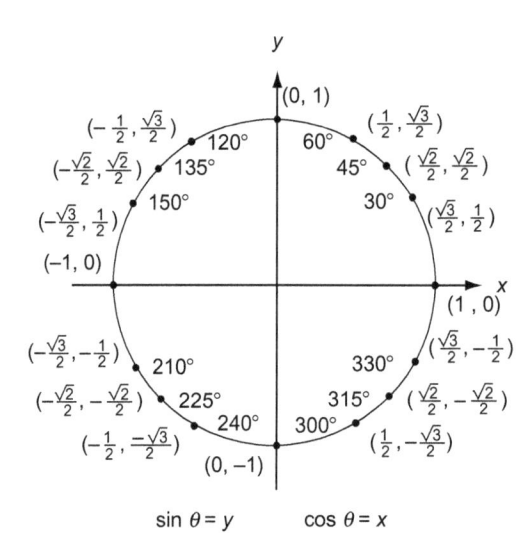

$\sin \theta = y \qquad \cos \theta = x$

Double Angle Formulas

$$\sin 2x = 2 \sin x \cos x$$
$$\cos 2x = \cos^2 x - \sin^2 x = 2\cos^2 x - 1 = 1 - 2\sin^2 x$$
$$\tan 2x = \frac{2 \tan x}{1 - \tan^2 x}$$

Power-Reducing Formulas

$$\sin^2 x = \frac{1 - \cos 2x}{2}$$
$$\cos^2 x = \frac{1 + \cos 2x}{2}$$
$$\tan^2 x = \frac{1 - \cos 2x}{1 + \cos 2x}$$

Product-to-Sum Formulas

$$\sin x \sin y = \frac{1}{2}[\cos(x-y) - \cos(x+y)]$$

$$\sin x \cos y = \frac{1}{2}[\sin(x+y) + \sin(x-y)]$$

$$\cos x \cos y = \frac{1}{2}[\cos(x-y) + \cos(x+y)]$$

Sum-to-Product Formulas

$$\sin x + \sin y = 2 \sin\left(\frac{x+y}{2}\right) \cos\left(\frac{x-y}{2}\right)$$

$$\sin x - \sin y = 2 \cos\left(\frac{x+y}{2}\right) \sin\left(\frac{x-y}{2}\right)$$

$$\cos x + \cos y = 2 \cos\left(\frac{x+y}{2}\right) \cos\left(\frac{x-y}{2}\right)$$

$$\cos x - \cos y = -2 \sin\left(\frac{x+y}{2}\right) \sin\left(\frac{x-y}{2}\right)$$

Pascal's Triangle

```
                    1
                  1   1
                1   2   1
              1   3   3   1
            1   4   6   4   1
          1   5  10  10   5   1
        1   6  15  20  15   6   1
      1   7  21  35  35  21   7   1
    1   8  28  56  70  56  28   8   1
```

The coefficients of $(x+y)^n$ can be obtained from the $(n+1)^{st}$ row of Pascal's Triangle.

For example:

$$(x+y)^4 = x^4 + 4x^3y + 6x^2y^2 + 4xy^3 + y^4$$
$$(x-y)^3 = x^3 - 3x^2y + 3xy^2 - y^3$$

Quadratic Formula

$$ax^2 + bx + c = 0 \iff x = \frac{-b \pm \sqrt{b^2 - 4ac}}{2a}$$

Completing the Square

$$ax^2 + bx + c = a\left(x + \frac{b}{2a}\right)^2 + \left(c - \frac{b^2}{4a}\right)$$

Special Factors

$$A^2 - B^2 = (A - B)(A + B)$$
$$A^3 - B^3 = (A - B)(A^2 + AB + B^2)$$
$$A^3 + B^3 = (A + B)(A^2 - AB + B^2)$$
$$A^4 - B^4 = (A + B)(A - B)(A^2 + B^2)$$

Logarithms

$$\log_a 1 = 0$$
$$\log_a a^x = x$$
$$\log_a xy = \log_a x + \log_a y$$
$$\log_a \frac{x}{y} = \log_a x - \log_a y$$
$$\log_a x^y = y \log_a x$$
$$\log_a x = \frac{\log_b x}{\log_b a}$$

Absolute Value

$$|x| = \begin{cases} x & \text{if } x \geq 0 \\ -x & \text{if } x < 0 \end{cases}$$

$$|-x| = |x|$$
$$|xy| = |x| \cdot |y|$$
$$\left|\frac{x}{y}\right| = \frac{|x|}{|y|}$$
$$|x|^2 = x^2$$
$$\sqrt{x^2} = |x|$$

$|x| = p \quad \Longleftrightarrow \quad x = -p \text{ or } x = p$

$|x| < p \quad \Longleftrightarrow \quad -p < x < p$

$|x| > p \quad \Longleftrightarrow \quad x < -p \text{ or } x > p$

Triangle Inequality

$$|x+y| \leq |x| + |y|$$

Lines

Slope: $m = \dfrac{y_2 - y_1}{x_2 - x_1}$

Point-Slope Form: $y - y_1 = m(x - x_1)$

Slope-Intercept Form: $y = mx + b$

Standard Form: $Ax + By = C$

Vertical Lines: $x = a$

Horizontal Lines: $y = b$

Distance Formula

$$D = \sqrt{(x_2 - x_1)^2 + (y_2 - y_1)^2}$$

Midpoint Formula

$$\left(\frac{x_1 + x_2}{2}, \frac{y_1 + y_2}{2}\right)$$

Inverse Functions

$$y = f(x) \iff x = f^{-1}(y)$$

Compound Interest

n times per year: $A = P\left(1 + \dfrac{i}{n}\right)^{nt}$

Continuously: $A = Pe^{it}$

3.3.4 What Is the Pythagorean Theorem?

What Is the Pythagorean Theorem?

The Pythagorean theorem states that if you have a right-angle triangle, the square of the hypotenuse (that's the side opposite the right angle) equals the sum of the squares of the other two sides.

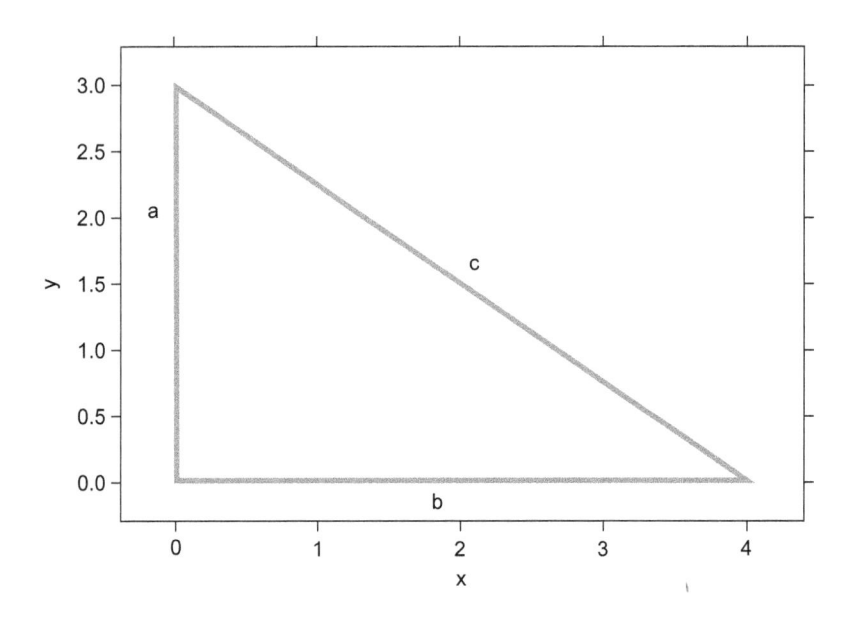

Using this triangle as an example, the hypotenuse is side c. We have a = 3, and b = 4. So:

$$c^2 = 3^2 + 4^2$$
$$c^2 = 9 + 16$$
$$c^2 = 25$$

To find the length of c, all we do is take the square root of both sides. In this case, c = 5.
Source: mathwizz.com

3.3.5 Pythagorean and Quotient Identities

There are two quotient identities. They tell us that the tangent and cotangent functions can be expressed in terms of the sine and cosine functions. They are as follows.

$$\tan x = \frac{\sin x}{\cos x}, \cos x <> 0$$
$$\cot x = \frac{\cos x}{\sin x}, \sin x <> 0$$

Three other identities are very important. They are called the *Pythagorean Identities*. They will come in handy later on when you need to prove more complicated trigonometric identities equal. The *Pythagorean Identities* are:

$$\sin^2 x + \cos^2 x = 1$$
$$1 + \cot^2 x = \csc^2 x$$
$$1 + \tan^2 x = \sec^2 x$$

Remember that $\sin^2 x = (\sin x)^2$.

Special Rules

There are a few special rules you ought to remember when dealing with isosceles and/or equilateral triangles, notably:

1. If a triangle is equilateral, it is equiangular.
2. If two angles of a triangle are congruent, they are the base angles of an isosceles triangle.
3. If a triangle is equiangular, it is equilateral.

3.3.6 Properties of Triangles

Pythagorus's Theorem

$$a^2 + b^2 = c^2$$

where:
c is the hypotenuse of a right angle triangle, and a and b are two sides containing the right angle.

Cosine Law

$$c^2 = a^2 + b^2 - 2ab(\text{Cos C})$$

This formula is applicable to any triangle. Here a, b, and c are the three sides of the triangle, and A, B, and C are the angle opposite to these sides, respectively.
   Similarly

$$a^2 = b^2 + c^2 - 2bc(\text{Cos A})$$

and

$$b^2 = c^2 + a^2 - 2ca(\text{Cos B})$$

Note: This is a general law, and if you make any of the angles equal to 90 degrees it gives the Pythagorean theorem as Cos of 90 degrees = 0.

Sine Law

$$(\text{Sin A})/a = (\text{Sin B})/b = (\text{Sin C})/c$$

This formula is applicable to any triangle. Here a, b and c are the three sides of the triangle, and A, B and C are the angle opposite to these sides, respectively.

3.3.7 Law of Sines

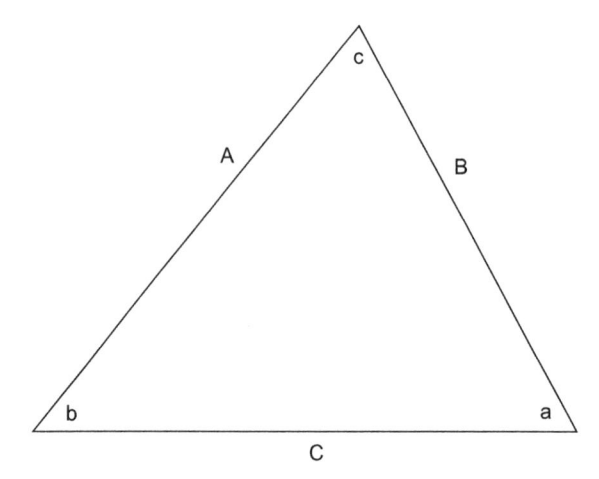

Given a triangle with sides A, B, and C and opposite angles a, b, and c, the Law of Sines states

$$\frac{\sin a}{A} = \frac{\sin b}{B} = \frac{\sin c}{C}$$

Of course, since we can cross-multiply, we can flip the fractions to get

$$\frac{A}{\sin a} = \frac{B}{\sin b} = \frac{C}{\sin c}$$

Source: mathwizz.com

3.3.8 Law of Cosines

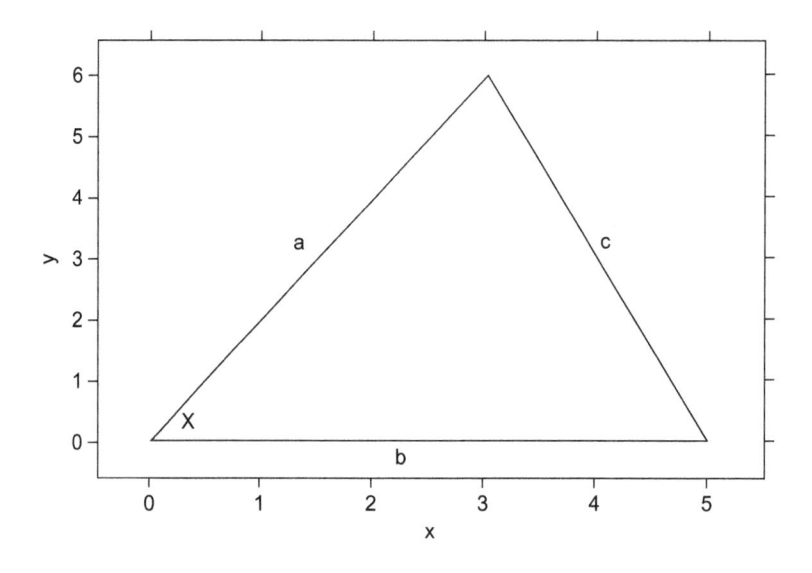

Given a triangle with sides a, b, and c and angle X, the Law of Cosines states

$$c^2 = a^2 + b^2 - (2)(a)(b)(\cos X)$$

Of course, the *Pythagorean theorem* is a special case of the Law of Cosines because we have a right angle triangle and in this case angle X = 90°, and we know that cos 90° = 0 and so we are left with $c^2 = a^2 + b^2$.
Source: mathwizz.com

3.3.9 Reciprocal Ratios

The reciprocal ratios are trigonometric ratios, too. They are as follows:

cotangent x = 1/tan x = (adjacent side)/(opposite side)
secant x = 1/cos x = (hypotenuse)/(adjacent side)
cosecant x = 1/sin x = (hypotenuse)/(opposite side)

Back to Top

3.3.10 Cofunctions

In a right triangle, the two acute angles are complementary. Thus, if one acute angle of a right triangle is x, the other is 90° − x. Therefore, if *sin x* = *(a/c)* then *cos (90° − x)* = *(a/c)*. A table of all the cofunctions is displayed below.

$$\sin x = \cos (90° - x)$$
$$\tan x = \cot (90° - x)$$
$$\sec x = \csc (90° - x)$$
$$\cos x = \sin (90° - x)$$
$$\cot x = \tan (90° - x)$$

A Table of the Common Logarithms

1.026	0.01114736	1.26	0.1003705	2.26	0.3541084	3.26	0.5132176	4.26	0.6294096	5.26	0.7209857	6.26	0.7965743	7.26	0.8609366	8.26	0.9169800	9.26	0.9666110
1.027	0.01157044	1.27	0.1038037	2.27	0.3560259	3.27	0.5145478	4.27	0.6304279	5.27	0.7218106	6.27	0.7972675	7.27	0.8615344	8.27	0.9175055	9.27	0.9670797
1.028	0.01199311	1.28	0.1072100	2.28	0.3579348	3.28	0.5158738	4.28	0.6314438	5.28	0.7226339	6.28	0.7979596	7.28	0.8621314	8.28	0.9180303	9.28	0.9675480
1.029	0.01241537	1.29	0.1105897	2.29	0.3598355	3.29	0.5171959	4.29	0.6324573	5.29	0.7234557	6.29	0.7986506	7.29	0.8627275	8.29	0.9185545	9.29	0.9680157
1.030	0.01283722	1.30	0.1139434	2.30	0.3617278	3.30	0.5185139	4.30	0.6334685	5.30	0.7242759	6.30	0.7993405	7.30	0.8633229	8.30	0.9190781	9.30	0.9684829
1.031	0.01325867	1.31	0.1172713	2.31	0.3636120	3.31	0.5198280	4.31	0.6344773	5.31	0.7250945	6.31	0.8000294	7.31	0.8639174	8.31	0.9196010	9.31	0.9689497
1.032	0.01367970	1.32	0.1205739	2.32	0.3654880	3.32	0.5211381	4.32	0.6354837	5.32	0.7259116	6.32	0.8007171	7.32	0.8645111	8.32	0.9201233	9.32	0.9694159
1.033	0.01410032	1.33	0.1238516	2.33	0.3673559	3.33	0.5224442	4.33	0.6364879	5.33	0.7267272	6.33	0.8014037	7.33	0.8651040	8.33	0.9206450	9.33	0.9698816
1.034	0.01452054	1.34	0.1271048	2.34	0.3692159	3.34	0.5237465	4.34	0.6374897	5.34	0.7275413	6.34	0.8020893	7.34	0.8656961	8.34	0.9211661	9.34	0.9703469
1.035	0.01494035	1.35	0.1303338	2.35	0.3710679	3.35	0.5250450	4.35	0.6384893	5.35	0.7283538	6.35	0.8027737	7.35	0.8662873	8.35	0.9216865	9.35	0.9708116
1.036	0.01535976	1.36	0.1335389	2.36	0.3729120	3.36	0.5263393	4.36	0.6394865	5.36	0.7291648	6.36	0.8034571	7.36	0.8668778	8.36	0.9222063	9.36	0.9712758
1.037	0.01577876	1.37	0.1367206	2.37	0.3747483	3.37	0.5276299	4.37	0.6404814	5.37	0.7299743	6.37	0.8041394	7.37	0.8674675	8.37	0.9227255	9.37	0.9717396
1.038	0.01619735	1.38	0.1398791	2.38	0.3765770	3.38	0.5289167	4.38	0.6414741	5.38	0.7307823	6.38	0.8048207	7.38	0.8680564	8.38	0.9232440	9.38	0.9722028
1.039	0.01661555	1.39	0.1430148	2.39	0.3783979	3.39	0.5301997	4.39	0.6424645	5.39	0.7315888	6.39	0.8055009	7.39	0.8686444	8.39	0.9237620	9.39	0.9726656
1.040	0.01703334	1.40	0.1461280	2.40	0.3802112	3.40	0.5314789	4.40	0.6434527	5.40	0.7323938	6.40	0.8061800	7.40	0.8692317	8.40	0.9242793	9.40	0.9731279
1.041	0.01745073	1.41	0.1492191	2.41	0.3820170	3.41	0.5327544	4.41	0.6444386	5.41	0.7331973	6.41	0.8068580	7.41	0.8698182	8.41	0.9247960	9.41	0.9735896
1.042	0.01786772	1.42	0.1522883	2.42	0.3838154	3.42	0.5340261	4.42	0.6454223	5.42	0.7339993	6.42	0.8075350	7.42	0.8704039	8.42	0.9253121	9.42	0.9740509
1.043	0.01828431	1.43	0.1553360	2.43	0.3856063	3.43	0.5352941	4.43	0.6464037	5.43	0.7347998	6.43	0.8082110	7.43	0.8709888	8.43	0.9258276	9.43	0.9745117
1.044	0.01870050	1.44	0.1583625	2.44	0.3873898	3.44	0.5365584	4.44	0.6473830	5.44	0.7355989	6.44	0.8088859	7.44	0.8715729	8.44	0.9263424	9.44	0.9749720
1.045	0.01911629	1.45	0.1613680	2.45	0.3891661	3.45	0.5378191	4.45	0.6483600	5.45	0.7363965	6.45	0.8095597	7.45	0.8721563	8.45	0.9268567	9.45	0.9754318
1.046	0.01953168	1.46	0.1643529	2.46	0.3909351	3.46	0.5390761	4.46	0.6493349	5.46	0.7371926	6.46	0.8102325	7.46	0.8727388	8.46	0.9273704	9.46	0.9758911
1.047	0.01994668	1.47	0.1673173	2.47	0.3926970	3.47	0.5403295	4.47	0.6503075	5.47	0.7379873	6.47	0.8109043	7.47	0.8733206	8.47	0.9278834	9.47	0.9763500
1.048	0.02036128	1.48	0.1702617	2.48	0.3944517	3.48	0.5415792	4.48	0.6512780	5.48	0.7387806	6.48	0.8115750	7.48	0.8739016	8.48	0.9283959	9.48	0.9768083
1.049	0.02077549	1.49	0.1731863	2.49	0.3961993	3.49	0.5428254	4.49	0.6522463	5.49	0.7395723	6.49	0.8122447	7.49	0.8744818	8.49	0.9289077	9.49	0.9772662
1.050	0.02118930	1.50	0.1760913	2.50	0.3979400	3.50	0.5440680	4.50	0.6532125	5.50	0.7403627	6.50	0.8129134	7.50	0.8750613	8.50	0.9294189	9.50	0.9777236
1.051	0.02160272	1.51	0.1789769	2.51	0.3996737	3.51	0.5453071	4.51	0.6541765	5.51	0.7411516	6.51	0.8135810	7.51	0.8756399	8.51	0.9299296	9.51	0.9781805
1.052	0.02201574	1.52	0.1818436	2.52	0.4014005	3.52	0.5465427	4.52	0.6551384	5.52	0.7419391	6.52	0.8142476	7.52	0.8762178	8.52	0.9304396	9.52	0.9786369
1.053	0.02242837	1.53	0.1846914	2.53	0.4031205	3.53	0.5477747	4.53	0.6560982	5.53	0.7427251	6.53	0.8149132	7.53	0.8767950	8.53	0.9309490	9.53	0.9790929
1.054	0.02284061	1.54	0.1875207	2.54	0.4048337	3.54	0.5490033	4.54	0.6570559	5.54	0.7435098	6.54	0.8155777	7.54	0.8773713	8.54	0.9314579	9.54	0.9795484
1.055	0.02325246	1.55	0.1903317	2.55	0.4065402	3.55	0.5502284	4.55	0.6580114	5.55	0.7442930	6.55	0.8162413	7.55	0.8779470	8.55	0.9319661	9.55	0.9800034
1.056	0.02366392	1.56	0.1931246	2.56	0.4082400	3.56	0.5514500	4.56	0.6589648	5.56	0.7450748	6.56	0.8169038	7.56	0.8785218	8.56	0.9324738	9.56	0.9804579
1.057	0.02407499	1.57	0.1958997	2.57	0.4099331	3.57	0.5526682	4.57	0.6599162	5.57	0.7458552	6.57	0.8175654	7.57	0.8790959	8.57	0.9329808	9.57	0.9809119
1.058	0.02448567	1.58	0.1986571	2.58	0.4116197	3.58	0.5538830	4.58	0.6608655	5.58	0.7466342	6.58	0.8182259	7.58	0.8796692	8.58	0.9334873	9.58	0.9813655
1.059	0.02489596	1.59	0.2013971	2.59	0.4132998	3.59	0.5550944	4.59	0.6618127	5.59	0.7474118	6.59	0.8188854	7.59	0.8802418	8.59	0.9339932	9.59	0.9818186
1.060	0.02530587	1.60	0.2041200	2.60	0.4149733	3.60	0.5563025	4.60	0.6627578	5.60	0.7481880	6.60	0.8195439	7.60	0.8808136	8.60	0.9344985	9.60	0.9822712
1.061	0.02571538	1.61	0.2068259	2.61	0.4166405	3.61	0.5575072	4.61	0.6637009	5.61	0.7489629	6.61	0.8202015	7.61	0.8813847	8.61	0.9350032	9.61	0.9827234
1.062	0.02612452	1.62	0.2095150	2.62	0.4183013	3.62	0.5587086	4.62	0.6646420	5.62	0.7497363	6.62	0.8208580	7.62	0.8819550	8.62	0.9355073	9.62	0.9831751
1.063	0.02653326	1.63	0.2121876	2.63	0.4199557	3.63	0.5599066	4.63	0.6655810	5.63	0.7505084	6.63	0.8215135	7.63	0.8825245	8.63	0.9360108	9.63	0.9836263
1.064	0.02694163	1.64	0.2148438	2.64	0.4216039	3.64	0.5611014	4.64	0.6665180	5.64	0.7512791	6.64	0.8221681	7.64	0.8830934	8.64	0.9365137	9.64	0.9840770
1.065	0.02734961	1.65	0.2174839	2.65	0.4232459	3.65	0.5622929	4.65	0.6674530	5.65	0.7520484	6.65	0.8228216	7.65	0.8836614	8.65	0.9370161	9.65	0.9845273
1.066	0.02775720	1.66	0.2201081	2.66	0.4248816	3.66	0.5634811	4.66	0.6683859	5.66	0.7528164	6.66	0.8234742	7.66	0.8842288	8.66	0.9375179	9.66	0.9849771
1.067	0.02816442	1.67	0.2227165	2.67	0.4265113	3.67	0.5646661	4.67	0.6693169	5.67	0.7535831	6.67	0.8241258	7.67	0.8847954	8.67	0.9380191	9.67	0.9854265
1.068	0.02857125	1.68	0.2253093	2.68	0.4281348	3.68	0.5658478	4.68	0.6702459	5.68	0.7543483	6.68	0.8247765	7.68	0.8853612	8.68	0.9385197	9.68	0.9858754
1.069	0.02897771	1.69	0.2278867	2.69	0.4297523	3.69	0.5670264	4.69	0.6711728	5.69	0.7551123	6.69	0.8254261	7.69	0.8859263	8.69	0.9390198	9.69	0.9863238

1.070	0.02938378	1.70	0.2304489	2.70	0.4313638	3.70	0.5682017	4.70	0.6720979	5.70	0.7558749	6.70	0.8260748	7.70	0.8864907	8.70	0.9395193	9.70	0.9867717
1.071	0.02978947	1.71	0.2329961	2.71	0.4329693	3.71	0.5693739	4.71	0.6730209	5.71	0.7566361	6.71	0.8267225	7.71	0.8870544	8.71	0.9400182	9.71	0.9872192
1.072	0.03019479	1.72	0.2355284	2.72	0.4345689	3.72	0.5705429	4.72	0.6739420	5.72	0.7573960	6.72	0.8273693	7.72	0.8876173	8.72	0.9405165	9.72	0.9876663
1.073	0.03059972	1.73	0.2380461	2.73	0.4361626	3.73	0.5717088	4.73	0.6748611	5.73	0.7581546	6.73	0.8280151	7.73	0.8881795	8.73	0.9410142	9.73	0.9881128
1.074	0.03100428	1.74	0.2405492	2.74	0.4377506	3.74	0.5728716	4.74	0.6757783	5.74	0.7589119	6.74	0.8286599	7.74	0.8887410	8.74	0.9415114	9.74	0.9885590
1.075	0.03140846	1.75	0.2430380	2.75	0.4393327	3.75	0.5740313	4.75	0.6766936	5.75	0.7596678	6.75	0.8293038	7.75	0.8893017	8.75	0.9420081	9.75	0.9890046
1.076	0.03181227	1.76	0.2455127	2.76	0.4409091	3.76	0.5751878	4.76	0.6776070	5.76	0.7604225	6.76	0.8299467	7.76	0.8898617	8.76	0.9425041	9.76	0.9894498
1.077	0.03221570	1.77	0.2479733	2.77	0.4424798	3.77	0.5763414	4.77	0.6785184	5.77	0.7611758	6.77	0.8305887	7.77	0.8904210	8.77	0.9429996	9.77	0.9898946
1.078	0.03261876	1.78	0.2504200	2.78	0.4440448	3.78	0.5774918	4.78	0.6794279	5.78	0.7619278	6.78	0.8312297	7.78	0.8909796	8.78	0.9434945	9.78	0.9903389
1.079	0.03302144	1.79	0.2528530	2.79	0.4456042	3.79	0.5786392	4.79	0.6803355	5.79	0.7626786	6.79	0.8318698	7.79	0.8915375	8.79	0.9439889	9.79	0.9907827
1.080	0.03342376	1.80	0.2552725	2.80	0.4471580	3.80	0.5797836	4.80	0.6812412	5.80	0.7634280	6.80	0.8325089	7.80	0.8920946	8.80	0.9444827	9.80	0.9912261
1.081	0.03382569	1.81	0.2576786	2.81	0.4487063	3.81	0.5809250	4.81	0.6821451	5.81	0.7641761	6.81	0.8331471	7.81	0.8926510	8.81	0.9449759	9.81	0.9916690
1.082	0.03422726	1.82	0.2600714	2.82	0.4502491	3.82	0.5820634	4.82	0.6830470	5.82	0.7649230	6.82	0.8337844	7.82	0.8932068	8.82	0.9454686	9.82	0.9921115
1.083	0.03462846	1.83	0.2624511	2.83	0.4517864	3.83	0.5831988	4.83	0.6839471	5.83	0.7656686	6.83	0.8344207	7.83	0.8937618	8.83	0.9459607	9.83	0.9925535
1.084	0.03502928	1.84	0.2648178	2.84	0.4533183	3.84	0.5843312	4.84	0.6848454	5.84	0.7664128	6.84	0.8350561	7.84	0.8943161	8.84	0.9464523	9.84	0.9929951
1.085	0.03542974	1.85	0.2671717	2.85	0.4548449	3.85	0.5854607	4.85	0.6857417	5.85	0.7671559	6.85	0.8356906	7.85	0.8948697	8.85	0.9469433	9.85	0.9934362
1.086	0.03582983	1.86	0.2695129	2.86	0.4563660	3.86	0.5865873	4.86	0.6866363	5.86	0.7678976	6.86	0.8363241	7.86	0.8954225	8.86	0.9474337	9.86	0.9938769
1.087	0.03622954	1.87	0.2718416	2.87	0.4578819	3.87	0.5877110	4.87	0.6875290	5.87	0.7686381	6.87	0.8369567	7.87	0.8959747	8.87	0.9479236	9.87	0.9943172
1.088	0.03662890	1.88	0.2741578	2.88	0.4593925	3.88	0.5888317	4.88	0.6884198	5.88	0.7693773	6.88	0.8375884	7.88	0.8965262	8.88	0.9484130	9.88	0.9947569
1.089	0.03702788	1.89	0.2764618	2.89	0.4608978	3.89	0.5899496	4.89	0.6893089	5.89	0.7701153	6.89	0.8382192	7.89	0.8970770	8.89	0.9489018	9.89	0.9951963
1.090	0.03742650	1.90	0.2787536	2.90	0.4623980	3.90	0.5910646	4.90	0.6901961	5.90	0.7708520	6.90	0.8388491	7.90	0.8976271	8.90	0.9493900	9.90	0.9956352
1.091	0.03782475	1.91	0.2810334	2.91	0.4638930	3.91	0.5921768	4.91	0.6910815	5.91	0.7715875	6.91	0.8394780	7.91	0.8981765	8.91	0.9498777	9.91	0.9960737
1.092	0.03822264	1.92	0.2833012	2.92	0.4653829	3.92	0.5932861	4.92	0.6919651	5.92	0.7723217	6.92	0.8401061	7.92	0.8987252	8.92	0.9503649	9.92	0.9965117
1.093	0.03862016	1.93	0.2855573	2.93	0.4668676	3.93	0.5943926	4.93	0.6928469	5.93	0.7730547	6.93	0.8407332	7.93	0.8992732	8.93	0.9508515	9.93	0.9969492
1.094	0.03901732	1.94	0.2878017	2.94	0.4683473	3.94	0.5954962	4.94	0.6937269	5.94	0.7737864	6.94	0.8413595	7.94	0.8998205	8.94	0.9513375	9.94	0.9973864
1.095	0.03941412	1.95	0.2900346	2.95	0.4698220	3.95	0.5965971	4.95	0.6946052	5.95	0.7745170	6.95	0.8419848	7.95	0.9003671	8.95	0.9518230	9.95	0.9978231
1.096	0.03981055	1.96	0.2922561	2.96	0.4712917	3.96	0.5976952	4.96	0.6954817	5.96	0.7752463	6.96	0.8426092	7.96	0.9009131	8.96	0.9523080	9.96	0.9982593
1.097	0.04020663	1.97	0.2944662	2.97	0.4727564	3.97	0.5987905	4.97	0.6963564	5.97	0.7759743	6.97	0.8432328	7.97	0.9014583	8.97	0.9527924	9.97	0.9986952
1.098	0.04060234	1.98	0.2966652	2.98	0.4742163	3.98	0.5998831	4.98	0.6972293	5.98	0.7767012	6.98	0.8438554	7.98	0.9020029	8.98	0.9532763	9.98	0.9991305
1.099	0.04099769	1.99	0.2988531	2.99	0.4756712	3.99	0.6009729	4.99	0.6981005	5.99	0.7774268	6.99	0.8444772	7.99	0.9025468	8.99	0.9537597	9.99	0.9995655

Table of Common Logs, p. 3
By permission: SOS Mathematics

3.3.11 A Table of Common Logarithms

The table below lists the common logarithms (with base 10) for numbers between 1 and 10. The logarithm is denoted in bold face. For instance, the first entry in the third column means that the common log of 2.00 is 0.3010300.

1.000 0.00000000		**2.00** 0.3010300	**3.00** 0.4771213	**4.00** 0.6020600	**5.00** 0.6989700	**6.00** 0.7781513	**7.00** 0.8450980	**8.00** 0.9030900	**9.00** 0.9542425
1.001 0.00043408		**2.01** 0.3031961	**3.01** 0.4785665	**4.01** 0.6031444	**5.01** 0.6998377	**6.01** 0.7788745	**7.01** 0.8457180	**8.01** 0.9036325	**9.01** 0.9547248
1.002 0.00086772		**2.02** 0.3053514	**3.02** 0.4800069	**4.02** 0.6042261	**5.02** 0.7007037	**6.02** 0.7795965	**7.02** 0.8463371	**8.02** 0.9041744	**9.02** 0.9552065
1.003 0.00130093		**2.03** 0.3074960	**3.03** 0.4814426	**4.03** 0.6053050	**5.03** 0.7015680	**6.03** 0.7803173	**7.03** 0.8469553	**8.03** 0.9047155	**9.03** 0.9556878
1.004 0.00173371		**2.04** 0.3096302	**3.04** 0.4828736	**4.04** 0.6063814	**5.04** 0.7024305	**6.04** 0.7810369	**7.04** 0.8475727	**8.04** 0.9052560	**9.04** 0.9561684
1.005 0.00216606		**2.05** 0.3117539	**3.05** 0.4842998	**4.05** 0.6074550	**5.05** 0.7032914	**6.05** 0.7817554	**7.05** 0.8481891	**8.05** 0.9057959	**9.05** 0.9566486
1.006 0.00259798		**2.06** 0.3138672	**3.06** 0.4857214	**4.06** 0.6085260	**5.06** 0.7041505	**6.06** 0.7824726	**7.06** 0.8488047	**8.06** 0.9063350	**9.06** 0.9571282
1.007 0.00302947		**2.07** 0.3159703	**3.07** 0.4871384	**4.07** 0.6095944	**5.07** 0.7050080	**6.07** 0.7831887	**7.07** 0.8494194	**8.07** 0.9068735	**9.07** 0.9576073
1.008 0.00346053		**2.08** 0.3180633	**3.08** 0.4885507	**4.08** 0.6106602	**5.08** 0.7058637	**6.08** 0.7839036	**7.08** 0.8500333	**8.08** 0.9074114	**9.08** 0.9580858
1.009 0.00389117		**2.09** 0.3201463	**3.09** 0.4899585	**4.09** 0.6117233	**5.09** 0.7067178	**6.09** 0.7846173	**7.09** 0.8506462	**8.09** 0.9079485	**9.09** 0.9585639
1.010 0.00432137	**1.10** 0.0413927	**2.10** 0.3222193	**3.10** 0.4913617	**4.10** 0.6127839	**5.10** 0.7075702	**6.10** 0.7853298	**7.10** 0.8512583	**8.10** 0.9084850	**9.10** 0.9590414
1.011 0.00475116	**1.11** 0.0453230	**2.11** 0.3242825	**3.11** 0.4927604	**4.11** 0.6138418	**5.11** 0.7084209	**6.11** 0.7860412	**7.11** 0.8518696	**8.11** 0.9090209	**9.11** 0.9595184
1.012 0.00518051	**1.12** 0.0492180	**2.12** 0.3263359	**3.12** 0.4941546	**4.12** 0.6148972	**5.12** 0.7092700	**6.12** 0.7867514	**7.12** 0.8524800	**8.12** 0.9095560	**9.12** 0.9599948
1.013 0.00560945	**1.13** 0.0530784	**2.13** 0.3283796	**3.13** 0.4955443	**4.13** 0.6159501	**5.13** 0.7101174	**6.13** 0.7874605	**7.13** 0.8530895	**8.13** 0.9100905	**9.13** 0.9604708
1.014 0.00603795	**1.14** 0.0569049	**2.14** 0.3304138	**3.14** 0.4969296	**4.14** 0.6170003	**5.14** 0.7109631	**6.14** 0.7881684	**7.14** 0.8536982	**8.14** 0.9106244	**9.14** 0.9609462
1.015 0.00646604	**1.15** 0.0606978	**2.15** 0.3324385	**3.15** 0.4983106	**4.15** 0.6180481	**5.15** 0.7118072	**6.15** 0.7888751	**7.15** 0.8543060	**8.15** 0.9111576	**9.15** 0.9614211
1.016 0.00689371	**1.16** 0.0644580	**2.16** 0.3344538	**3.16** 0.4996871	**4.16** 0.6190933	**5.16** 0.7126497	**6.16** 0.7895807	**7.16** 0.8549130	**8.16** 0.9116902	**9.16** 0.9618955
1.017 0.00732095	**1.17** 0.0681859	**2.17** 0.3364597	**3.17** 0.5010593	**4.17** 0.6201361	**5.17** 0.7134905	**6.17** 0.7902852	**7.17** 0.8555192	**8.17** 0.9122221	**9.17** 0.9623693
1.018 0.00774778	**1.18** 0.0718820	**2.18** 0.3384565	**3.18** 0.5024271	**4.18** 0.6211763	**5.18** 0.7143298	**6.18** 0.7909885	**7.18** 0.8561244	**8.18** 0.9127533	**9.18** 0.9628427
1.019 0.00817418	**1.19** 0.0755470	**2.19** 0.3404441	**3.19** 0.5037907	**4.19** 0.6222140	**5.19** 0.7151674	**6.19** 0.7916906	**7.19** 0.8567289	**8.19** 0.9132839	**9.19** 0.9633155
1.020 0.00860017	**1.20** 0.0791812	**2.20** 0.3424227	**3.20** 0.5051500	**4.20** 0.6232493	**5.20** 0.7160033	**6.20** 0.7923917	**7.20** 0.8573325	**8.20** 0.9138139	**9.20** 0.9637878
1.021 0.00902574	**1.21** 0.0827854	**2.21** 0.3443923	**3.21** 0.5065050	**4.21** 0.6242821	**5.21** 0.7168377	**6.21** 0.7930916	**7.21** 0.8579353	**8.21** 0.9143432	**9.21** 0.9642596
1.022 0.00945090	**1.22** 0.0863598	**2.22** 0.3463530	**3.22** 0.5078559	**4.22** 0.6253125	**5.22** 0.7176705	**6.22** 0.7937904	**7.22** 0.8585372	**8.22** 0.9148718	**9.22** 0.9647309
1.023 0.00987563	**1.23** 0.0899051	**2.23** 0.3483049	**3.23** 0.5092025	**4.23** 0.6263404	**5.23** 0.7185017	**6.23** 0.7944880	**7.23** 0.8591383	**8.23** 0.9153998	**9.23** 0.9652017
1.024 0.01029996	**1.24** 0.0934217	**2.24** 0.3502480	**3.24** 0.5105450	**4.24** 0.6273659	**5.24** 0.7193313	**6.24** 0.7951846	**7.24** 0.8597386	**8.24** 0.9159272	**9.24** 0.9656720
1.025 0.01072387	**1.25** 0.0969100	**2.25** 0.3521825	**3.25** 0.5118834	**4.25** 0.6283889	**5.25** 0.7201593	**6.25** 0.7958800	**7.25** 0.8603380	**8.25** 0.9164539	**9.25** 0.9661417

By permission: SOS Mathematics

3.4.0 Physics—Basic Formulas
Reference Guide & Formula Sheet for Physics
Dr. Hoselton & Mr. Price

#3   **Components of a Vector**
if $V = 34$ m/sec $\angle 48°$
then
$V_i = 34$ m/sec $\cdot (\cos 48°)$; and $V_J = 34$ m/sec $\cdot (\sin 48°)$

#4   **Weight** $= m \cdot g$
$g = 9.81$ m/sec^2 near the surface of the Earth
$ = 9.795$ m/sec^2 in Fort Worth, TX
Density = mass / volume
$\rho = \dfrac{m}{V}$ (unit : kg/m^3)

#7   **Ave speed** = distance / time = $v = d/t$
Ave velocity = displacement / time = $v = d/t$
Ave acceleration = change in velocity / time

#8   **Friction Force**
$F_F = \mu \cdot F_N$
If the object is not moving, you are dealing with static friction and it can have any value from zero up to $\mu_s F_N$.
If the object is sliding, then you are dealing with kinetic friction and it will be constant and equal to $\mu_k F_N$.

#9   **Torque**
$\tau = F \cdot L \cdot \sin \theta$
where θ is the angle between F and L; unit: Nm

#11   **Newton's Second Law**
$F_{net} = \sum F_{Ext} = m \cdot a$

#12   **Work** $= F \cdot D \cdot \cos \theta$
where D is the distance moved and θ is the angle between **F** and the direction of motion, unit: J

#16   **Power** = rate of work done
$Power = \dfrac{Work}{time}$ unit:watt
Efficiency = Work$_{out}$ / Energy$_{in}$
Mechanical Advantage = force out / force in
M.A.= F_{out} / F_{in}

#19   **Constant-Acceleration Linear Motion**

$$v = v_o + a \cdot t \qquad\qquad x$$
$$(x - x_o) = v_o \cdot t + \tfrac{1}{2} a \cdot t^2 \qquad v$$
$$v^2 = v_o^2 + 2 \cdot a \cdot (x - x_o) \qquad t$$
$$(x - x_o) = \tfrac{1}{2} \cdot (v_o + v) \cdot t \qquad a$$
$$(x - x_o) = v \cdot t - \tfrac{1}{2} \cdot a \cdot t^2 \qquad v_o$$

#20 **Heating a Solid, Liquid, or Gas**
$Q = m \cdot c \cdot \Delta T$ (no phase changes!)
Q = the heat added
c = specific heat
ΔT = temperature change, K

#21 **Linear Momentum**
momentum = $\mathbf{p} = m \cdot v$ = mass · velocity
Momentum is conserved in collisions.

#23 **Center of Mass** - point masses on a line
$x_{cm} = \sum (mx)/M_{total}$

#25 **Angular Speed vs. Linear Speed**
Linear speed = $v = r \cdot \omega = r \cdot$ angular speed

#26 **Pressure under Water**
$P = \rho \cdot g \cdot h$
h = depth of water
ρ = density of water

#28 **Universal Gravitation**
$F = G \dfrac{m_1 m_2}{r^2}$
$G = 6.67 \text{ E-}11 \text{ N m}^2/\text{kg}^2$

#29 **Mechanical Energy**
$PE_{Grav} = P = m \cdot g \cdot h$
$KE_{Linear} = K = \frac{1}{2} \cdot m \cdot v^2$

#30 **Impulse = Change in Momentum**
$F \cdot \Delta t = \Delta(m \cdot v)$

#31 **Snell's Law**
$n_1 \cdot \sin\theta_1 = n_2 \cdot \sin\theta_2$
Index of Refraction
$n = c/v$
c = speed of light = $3 \text{ E}+8 \text{ m/s}$

#32 **Ideal Gas Law**
$P \cdot V = n \cdot R \cdot T$
n = # of moles of gas
R = gas law constant
     = 8.31 J / K mole.

#34 **Periodic Waves**
$v = f \cdot \lambda$
$f = 1/T$ T=period of wave

#35 **Constant-Acceleration Circular Motion**

$\omega = \omega_o + \alpha \cdot t$ θ
$\theta - \theta_o = \omega_o \cdot t + \frac{1}{2} \cdot \alpha \cdot t^2$ ω
$\omega^2 = \omega_o + 2 \cdot \alpha \cdot (\theta - \theta_o)$ t
$\theta - \theta_o = \frac{1}{2} \cdot (\omega_o + \omega) \cdot t$ α
$\theta - \theta_o = \omega \cdot t - \frac{1}{2} \cdot \alpha \cdot t^2$ ω_o

Section | 3 Calculations and Formulas—Geometry, Trigonometry, and Physics in Construction

#36 **Buoyant Force - Buoyancy**
$F_B = \rho \cdot V \cdot g = m_{\text{Displaced fluid}} \cdot g = \text{weight}_{\text{Displaced fluid}}$
ρ = density of the fluid
V = volume of fluid displaced

#37 **Ohm's Law**
$V = I \cdot R$
V = voltage applied
I = current
R = resistance
Resistance of a Wire
$R = \rho \cdot L / A_x$
ρ = resistivity of wire material
L = length of the wire
A_x = cross-sectional area of the wire

#39 **Heat of a Phase Change**
$Q = m \cdot L$
L = Latent heat of phase change

#41 **Hooke's Law**
$F = k \cdot x$
Potential Energy of a spring
$W = \frac{1}{2} \cdot k \cdot x^2 =$ work done on spring

#42 **Electric Power**
$p = I^2 \cdot R = V^2 / R = I \cdot V$

#44 **Speed of a Wave on a String**
$T = \dfrac{mv^2}{L}$
T = tension in string
m = mass of string
L = length of string

#45 **Projectile Motion**
Horizontal: $x - x_o = v_o \cdot t + 0$
Vertical: $y - y_o = v_o \cdot t + \frac{1}{2} \cdot a \cdot t^2$

#46 **Centripetal Force**
$F = \dfrac{mv^2}{r} = m\omega^2 r$

#47 **Kirchhoff's Laws**
Loop Rule: $\sum_{\text{Around any loop}} \Delta V_i = 0$
Node Rule: $\sum_{\text{at any node}} I_i = 0$

#51 **Minimum Speed at the top of a Vertical Circular Loop**
$v = \sqrt{rg}$
SERIES
$R_{eq} = R_1 + R_2 + R_3 + \ldots$
PARALLEL
$\dfrac{1}{R_{eq}} = \dfrac{1}{R_1} + \dfrac{1}{R_2} + \ldots + \dfrac{1}{R_n} = \sum_{i=1}^{n} \dfrac{1}{R_i}$

#54 **Newton's Second Law and Rotational Inertia**
τ = torque = $I \cdot \alpha$
I = moment of inertia = $m \cdot r^2$ (for a point mass)
(See table in Lesson 58 for I of 3D shapes.)

#55 **Circular Unbanked Tracks**
$$\frac{mv^2}{r} = \mu mg$$

#56 **Continuity of Fluid Flow**
$A_{in} \cdot v_{in} = A_{out} \cdot v_{out}$
A = Area
v = velocity

#58 **Moment of Inertia**   - I
cylindrical hoop	$m \cdot r^2$
solid cylinder or disk	$\frac{1}{2} m \cdot r^2$
solid sphere	$\frac{2}{5} m \cdot r^2$
hollow sphere	$\frac{2}{3} m \cdot r^2$
thin rod (center)	$\frac{1}{12} m \cdot L^2$
thin rod (end)	$\frac{1}{3} m \cdot L^2$

#59 **Capacitors**
$Q = C \cdot V$
Q = charge on the capacitor
C = capacitance of the capacitor
V = voltage applied to the capacitor

RC Circuits (Discharging)
$V_c = V_o \cdot e^{-t/RC}$
$V_c - I \cdot R = 0$

#60 **Thermal Expansion**
Linear: $\Delta L = L_o \cdot \alpha \cdot \Delta T$
Volume: $\Delta V = V_o \cdot \beta \cdot \Delta T$

#61 **Bernoulli's Equation**
$P + \rho \cdot g \cdot h + \frac{1}{2} \cdot \rho \cdot v^2$ = constant
$Q_{\text{volume Flow Rate}} = A_1 \cdot v_1 = A_2 \cdot v_2$ = constant

#62 **Rotational Kinetic Energy** (See LEM, pg 8)
$KE_{\text{rotational}} = \frac{1}{2} \cdot I \cdot \omega^2 = \frac{1}{2} \cdot I \cdot (v/r)^2$
$KE_{\text{rolling w/o slipping}} = \frac{1}{2} \cdot m \cdot v^2 + \frac{1}{2} \cdot I \cdot \omega^2$
Angular Momentum = $L = I \cdot \omega = m \cdot v \cdot r \cdot \sin\theta$
Angular Impulse equals
CHANGE IN Angular Momentum
$\Delta L = \tau_{\text{orque}} \cdot \Delta t = \Delta(I \cdot \omega)$

#63 Period of Simple Harmonic Motion

$$T = 2\pi\sqrt{\frac{m}{k}}$$ where k = spring constant

$f = 1/T = 1/\text{period}$

#64 Banked Circular Tracks

$v^2 = r \cdot g \cdot \tan\theta$

#66 First Law of Thermodynamics

$\Delta U = Q_{Net} + W_{Net}$

Change in Internal Energy of a system =
+Net Heat added to the system
+Net Work done on the system

Flow of Heat through a Solid

$\Delta Q/\Delta t = k \cdot A \cdot \Delta T/L$

k = thermal conductivity
A = area of solid
L = thickness of solid

#68 Potential Energy stored in a Capacitor

$P = \frac{1}{2} \cdot C \cdot V^2$

RC Circuit formula (Charging)

$V_c = V_{cell} \cdot (1 - e^{-1/RC})$

$R \cdot C = \tau = $ time constant

$V_{cell} - V_{capacitor} - I \cdot R = 0$

#71 Simple Pendulum

$$T = 2\pi\sqrt{\frac{L}{g}} \text{ and } f = 1/T$$

#72 Sinusoidal motion

$x = A \cdot \cos(\omega \cdot t) = A \cdot \cos(2 \cdot \pi \cdot f \cdot t)$

$\omega = $ angular frequency
$f = $ frequency

#73 Doppler Effect

$$f' = f \frac{343 \pm^{Toward}_{Away} v_o}{343 \mp^{Toward}_{Away} v_s}$$

$v_o = $ velocity of observer: $v_s = $ velocity of source

#74 2nd Law of Thermodynamics

The change in internal energy of a system is

$\Delta U = Q_{Added} + W_{Done\ On} - Q_{lost} - W_{Done\ By}$

Maximum Efficiency of a Heat Engine (Carnot Cycle) (Temperatures in Kelvin)

$\%Eff = \left(1 - \frac{T_c}{T_h}\right) \cdot 100\%$

$\frac{1}{f} = \frac{1}{D_o} + \frac{1}{D_i} = \frac{1}{o} + \frac{1}{i}$

$f = $ focal length
$i = $ image distance
$O = $ object distance

Magnification
$$M = -D_i/D_o = -i/o = H_i/H_o$$

Helpful reminders for mirrors and lenses

Focal Length of:	Positive	Negative
Mirror	concave	convex
Lens	converging	diverging
Object distance = 0	all objects	
Object height = H_o	all objects	
Image distance = i	real	virtual
Image height = H_i	virtual, upright	real, inverted
Magnification	virtual, upright	real, inverted

#76 **Coulomb's Law**
$$F = k\frac{q_1 q_2}{r^2}$$

$$k = \frac{1}{4\pi\epsilon_o} = 9E9 \frac{N \cdot m^2}{C^2}$$

#77 **Capacitor Combinations**
PARALLEL
$C_{eq} = C_1 + C_2 + C_3 + \ldots$
SERIES
$$\frac{1}{C_{eq}} = \frac{1}{C_1} + \frac{1}{C_2} + \ldots + \frac{1}{C_n} = \sum_{i=1}^{n} \frac{1}{C_i}$$

#78 **Work Done on a Gas or by a Gas**
$W = P \cdot \Delta V$

#80 **Electric Field around a Point Charge**
$$E = k\frac{q}{r^2}$$

$$k = \frac{1}{4\pi\varepsilon_o} = 9E9 \frac{N \cdot m^2}{C^2}$$

#82 **Magnetic Field around a Wire**
$$B = \frac{\mu_o I}{2\pi r}$$
Magnetic Flux
$\Phi = B \cdot A \cdot \cos\theta$
Force caused by a magnetic field on a moving charge
$F = q \cdot v \cdot B \cdot \sin\theta$

#83 **Entropy Change at Constant T**
$\Delta S = Q/T$
(Phase changes only: melting, boiling, freezing, etc)

#84 **Capacitance of a Capacitor**
$C = \kappa \cdot \epsilon_o \cdot A/d$
κ = dielectric constant
A = area of plates
d = distance between plates
$\varepsilon_o = 8.85\ E(-12)$ F/m

#85 Induced Voltage

$$Emf = N \frac{\Delta \Phi}{\Delta t}$$

N = # of loops

Lenz's Law—induced current flows to create a B-field opposing the change in magnetic flux.

#86 Inductors during an Increase in Current

$V_L = V_{cell} \cdot e^{-t/(L/R)}$

$I = (V_{cell}/R) \cdot [1 - e^{-t/(L/R)}]$

$L/R = \tau$ = time constant

#88 Transformers

$N_1/N_2 = V_1/V_2$

$I_1 \cdot V_1 = I_2 \cdot V_2$

#89 Decibel Scale

B (Decibel level of sound) $= 10 \log (I/I_o)$

I = intensity of sound

I_o = intensity of softest audible sound

#92 Poiseuille's Law

$\Delta P = 8 \cdot \eta \cdot L \cdot Q/(\pi \cdot r^4)$

η = coefficient of viscosity

L = length of pipe

r = radius of pipe

Q = flow rate of fluid

Stress and Strain

Y or S or B = stress / strain

stress = F/A

Three kinds of strain: unit-less ratios

I. Linear: strain = $\Delta L / L$

II. Shear: strain = $\Delta x / L$

III. Volume: strain = $\Delta V / V$

#93 Postulates of Special Relativity

1. Absolute, uniform motion cannot be detected.
2. No energy or mass transfer can occur at speeds faster than the speed of light.

#94 Lorentz Transformation Factor

$\beta = \sqrt{1 - \dfrac{v^2}{c^2}}$

$\Delta t = \Delta t_o / \beta$

#96 Relativistic Length Contraction

$\Delta x = \beta \cdot \Delta x_o$

Relativistic Mass Increase

$m = m_o / \beta$

#97 Energy of a Photon or a Particle

$E = h \cdot f = m \cdot c^2$

h = Planck's constant = 6.63 E(-34) J sec

f = frequency of the photon

#98 Radioactive Decay Rate Law

$A = A_o \cdot e^{-kt} = (1/2^n) \cdot A_0$ (after n half-lives)

Where $k = (\ln 2) /$ half-life

#99   **Blackbody Radiation and the Photoelectric Effect**
$E = n \cdot h \cdot f$ where h = Planck's constant

#100   **Early Quantum Physics Rutherford-Bohr Hydrogen-like Atoms**

$$\frac{1}{\lambda} = R \cdot \left(\frac{1}{n_s^2} - \frac{1}{n^2} \right) meters^{-1}$$

or

$$f = \frac{c}{\lambda} = cR \left(\frac{1}{n_s^2} - \frac{1}{n^2} \right) Hz$$

R = Rydberg's Constant
   = $1.097373143 \text{ E}7 \text{ m}^{-1}$
n_s = series integer (2 = Balmer)
n = an integer $> n_s$

Mass-Energy Equivalence
$m_v = m_o / \beta$
Total Energy = $KE + m_o c^2 = m_o c^2 / \beta$
Usually written simply as $E = m c^2$

de Broglie Matter Waves
For light: $E_p = h \cdot f = h \cdot c / \lambda = p \cdot c$
Therefore, momentum: $p = h / \lambda$
Similarly for particles, $p = m \cdot v = h / \lambda$,
so the matter wave's wavelength must be
$\lambda = h / m v$

Energy Released by Nuclear Fission or Fusion Reaction
$E = \Delta m_o \cdot c^2$

Miscellaneous Formulas

Quadratic Formula

if $a x^2 + b x + c = 0$
then

$$x = \frac{-b \pm \sqrt{b^2 - 4ac}}{2a}$$

Trigonometric Definitions

$\sin \theta$ = opposite/hypotenuse
$\cos \theta$ = adjacent/hypotenuse
$\tan \theta$ = opposite/adjacent
$\sec \theta = 1 / \cos \theta$ = hyp/adj
$\csc \theta = 1 / \sin \theta$ = hyp/opp
$\cot \theta = 1 / \tan \theta$ = adj/opp

Inverse Trigonometric Definitions

$\theta = \sin^{-1} (opp/hyp)$
$\theta = \cos^{-1} (adj/hyp)$
$\theta = \tan^{-1} (opp/hyp)$

Law of Sines

$$a/\sin A = b/\sin B = c/\sin C$$

or

$$\sin A/a = \sin B/b = \sin C/c$$

Law of Cosines

$$a^2 = b^2 + c^2 - 2bc \cos A$$
$$b^2 = c^2 + a^2 - 2ca \cos B$$
$$c^2 = a^2 + b^2 - 2ab \cos C$$

T-Pots

For the functional form

$$\frac{1}{A} = \frac{1}{B} + \frac{1}{C}$$

You may use "The Product over the Sum" rule.

$$A = \frac{B \cdot C}{B + C}$$

For the Alternate Functional form

$$\frac{1}{A} = \frac{1}{B} - \frac{1}{C}$$

You may substitute T-Pot-d

$$A = \frac{B \cdot C}{C - B} = -\frac{B \cdot C}{B - C}$$

Fundamental SI Units

Unit	Base Unit	Symbol
Length	meter	m
Mass	kilogram	kg
Time	second	s
Electric Current	ampere	A
Thermodynamic Temperature	kelvin	K
Luminous Intensity	candela	cd
Quantity of Substance	moles	mol
Plane Angle	radian	rad
Solid Angle	steradian	sr or str

Some Derived SI Units

Symbol/Unit	Quantity	Base Units
C—coulomb	Electric Charge	$A \cdot s$
F—farad	Capacitance	$A^2 \cdot s4/(kg \cdot m^2)$
H—henry	Inductance	$kg \cdot m^2/(A^2 \cdot s^2)$

Hz—hertz	Frequency	s^{-1}	
J—joule	Energy & Work	$kg \cdot m^2/s^2 = N \cdot m$	
N—newton	Force	$kg \cdot m/s^2$	
Ω—ohm	Elec Resistance	$kg \cdot m^2/(A^2 \cdot s^2)$	
Pa—pascal	Pressure	$kg/(m \cdot s^2)$	
T—tesla	Magnetic Field	$kg/(A \cdot s^2)$	
V—volt	Elec Potential	$kg \cdot m^2/(A \cdot s^3)$	
W—watt	Power	$kg \cdot m^2/s^3$	

Non-SI Units

°C	degrees Celsius	Temperature
eV	electron-volt	Energy, Work

Aa Acceleration, Area, A_x = Cross-sectional Area, Amperes, Amplitude of a Wave, Angle

Bb Magnetic Field, Decibel Level of Sound, Angle

Cc Specific Heat, Speed of Light, Capacitance, Angle, Coulombs, °Celsius, Celsius Degrees, Candela

Dd Displacement, Differential Change in a Variable, Distance, Distance Moved, Distance

Ee Base of the Natural Logarithms, Charge on the Electron, Energy

Ff Force, *Frequency of a Wave or Periodic Motion*, Farads

Gg Universal Gravitational Constant, Acceleration due to Gravity, Gauss, Grams, Giga-

Hh Depth of a Fluid, Height, Vertical Distance, Henrys, Hz = Hertz

Ii Current, Moment of Inertia, Image Distance, Intensity of Sound

Jj Joules

Kk K or KE = Kinetic Energy, Force Constant of a Spring, Thermal Conductivity, Coulomb's Law Constant, kg = Kilograms, Kelvins, Kilo-, Rate Constant for Radioactive Decay = $1/\tau$ = ln2 / half-life

Ll Length, Length of a Wire, Latent Heat of Fusion or Vaporization, Angular Momentum, Thickness, Inductance

Mm Mass, Total Mass, Meters, Milli-, Mega-, m_o = Rest Mass, Mol = Moles

Nn index of refraction, Moles of a Gas, Newtons, Number of Loops, Nano-

Pp Power, Pressure of a Gas or Fluid, Potential Energy, Momentum, Power, Pa = Pascal

Qq Heat Gained or Lost, Maximum Charge on a Capacitor, Object Distance, Flow Rate

Rr Radius, Ideal Gas Law Constant, Resistance, Magnitude or Length of a Vector, Rad = Radians

Ss Speed, Seconds, Entropy, Length along an Arc

Tt Time, Temperature, Period of a Wave, Tension, Teslas, $t_{1/2}$ = Half-life

Uu Potential Energy, Internal Energy

Vv velocity, Velocity, Volume of a Gas, Velocity of Wave, Volume of Fluid Displaced, Voltage, Volts

Ww Weight, Work, Watts, Wb = Weber

Xx Distance, Horizontal Distance, x-Coordinate East-and-West Coordinate

Yy Vertical distance, y-Coordinate, North-and-South Coordinate

Zz Z-coordinate, Up-and-Down Coordinate, Linear Expansion

Ββ Beta Coefficient of Volume Expansion, Lorentz Transformation Factor

Χχ Chi

Δδ Delta Δ = Change in a Variable

Εε Epsilon ϵ_o = Permittivity of Free Space

Φφ Phi Magnetic Flux, Angle

Γγ Gamma Surface Tension = F / L, $1/\gamma$ = Lorentz Transformation Factor

Hη **Eta**
Iι **Iota**
ϑ ϕ **Theta and Phi** Lower-case Alternates
Κκ **Kappa** Dielectric Constant
Λλ **Lambda** Wavelength of a Wave, Rate Constant for Radioactive Decay $= 1/\tau = \ln 2/\text{half-life}$
Μμ **Mu** Friction, $\mu_0 =$ Permeability of Free Space, Micro-
Νν **Nu** Alternate Symbol for Frequency
Οο **Omicron**
Ππ **Pi** 3.1425926536...
Θθ **Theta** Angle between Two Vectors
Ρρ **Rho** Density of a Solid or Liquid, Resistivity
Σσ **Sigma** Summation, Standard Deviation
Ττ **Tau** Torque, Time Constant for a Exponential Processes; eg $\tau = RC$ or $\tau = L/R$ or $\tau = 1/k = 1/\lambda$
Υυ **Upsilon**
ς ω **Zeta** and **Omega** Lower-case Alternates
Ωω **Omega** Angular Speed or Angular Velocity, Ohms
Ξξ **Xi**
Ψψ **Psi**
Ζζ **Zeta**

Values of Trigonometric Functions for First Quadrant Angles (simple mostly rational approximations)

θ	sin θ	cos θ	tan θ
0°	**0**	**1**	**0**
10°	1/6	65/66	11/65
15°	1/4	28/29	29/108
20°	1/3	16/17	17/47
29°	$15^{1/2}/8$	7/8	$15^{1/2}/7$
30°	**1/2**	$\mathbf{3^{1/2}/2}$	$\mathbf{1/3^{1/2}}$
37°	3/5	4/5	3/4
42°	2/3	3/4	8/9
45°	$\mathbf{2^{1/2}/2}$	$\mathbf{2^{1/2}/2}$	**1**
49°	3/4	2/3	9/8
53°	4/5	3/5	4/3
60	$\mathbf{3^{1/2}/2}$	**1/2**	$\mathbf{3^{1/2}}$
61°	7/8	$15^{1/2}/8$	$7/15^{1/2}$
70°	16/17	1/3	47/17
75°	28/29	1/4	108/29
80°	65/66	1/6	65/11
90°	**1**	**0**	**∞**

(Memorize the **Bold** rows for future reference.)

Derivatives of Polynomials

For polynomials, with individual terms of the form Ax^n, we define the derivative of each term as

$$\frac{d}{dx}(Ax^n) = nAx^{n-1}$$

To find the derivative of the polynomial, simply add the derivatives for the individual terms:

$$\frac{d}{dx}(3x^2 + 6x - 3) = 6x + 6$$

Integrals of Polynomials

For polynomials, with individual terms of the form Ax^n, we define the indefinite integral of each term as

$$\int (Ax^n)dx = \frac{1}{n+1}Ax^{n+1}$$

To find the indefinite integral of the polynomial, simply add the integrals for the individual terms and the constant of integration, C.

$$\int (6x + 6)dx = \left[3x^2 + 6x + C\right]$$

Prefixes

Factor	Prefix	Symbol	Example
10^{18}	exa-	E	**38 Es** (Age of the Universe in Seconds)
10^{15}	peta-	P	
10^{12}	tera-	T	**0.3 TW** (Peak power of a 1 ps pulse from a typical Nd-glass laser)
10^{9}	giga-	G	**22 G\$** (Size of Bill & Melissa Gates' Trust)
10^{6}	mega-	M	**6.37 Mm** (The radius of the Earth)
10^{3}	kilo-	k	**1 kg** (SI unit of mass)
10^{-1}	deci-	d	**10 cm**
10^{-2}	centi-	c	**2.54 cm** (=1 in)
10^{-3}	milli-	m	**1 mm** (The smallest division on a meter stick)
10^{-6}	micro-	μ	
10^{-9}	nano-	n	**510 nm** (Wavelength of green light)
10^{-12}	pico-	p	**1 pg** (Typical mass of a DNA sample used in genome studies)
10^{-15}	femto-	f	
10^{-18}	atto-	a	**600 as** (Time duration of the shortest laser pulses)

Linear Equivalent Mass

Rotating systems can be handled using the linear forms of the equations of motion. To do so, however, you must use a mass equivalent to the mass of a nonrotating object. We call this the Linear Equivalent Mass (LEM). *(See Example I.)*

For objects that are both rotating and moving linearly, you must include them twice—once as a linearly moving object (using m) and once more as a rotating object (using LEM). *(See Example II.)*

The LEM of a rotating mass is easily defined in terms of its moment of inertia, I.

Section 3 Calculations and Formulas—Geometry, Trigonometry, and Physics in Construction

$$LEM = I/r^2$$

For example, using a standard table of Moments of Inertia, we can calculate the LEM of simple objects rotating on axes through their centers of mass:

	I	LEM
Cylindrical hoop	mr^2	m
Solid disk	$\frac{1}{2}mr^2$	$\frac{1}{2}m$
Hollow sphere	$\frac{2}{5}mr^2$	$\frac{2}{5}m$
Solid sphere	$\frac{2}{3}mr^2$	$\frac{2}{3}m$

Example I

A flywheel, M = 4.80 kg and r = 0.44 m, is wrapped with a string. A hanging mass, m, is attached to the end of the string.

When the hanging mass is released, it accelerates downward at 1.00 m/s². Find the hanging mass.

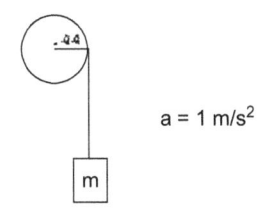

To handle this problem using the linear form of Newton's Second Law of Motion, all we have to do is use the LEM of the flywheel. We will assume, here, that it can be treated as a uniform solid disk.

The only external force on this system is the weight of the hanging mass. The mass of the system consists of the hanging mass plus the linear equivalent mass of the flywheel. From Newton's Second Law we have

F = ma, therefore,
$$mg = [m + (LEM=\tfrac{1}{2}M)]a$$
$$mg = [m + \tfrac{1}{2}M]\,a$$
$$(mg - ma) = \tfrac{1}{2}M\,a$$
$$m(g - a) = \tfrac{1}{2}Ma$$
$$m = \tfrac{1}{2} \cdot M \cdot a / (g - a)$$
$$m = \tfrac{1}{2} \cdot 4.8 \cdot 1.00 / (9.81 - 1)$$
$$M = 0.27 \text{ kg}$$

If a = g/2 = 4.905 m/s²,   m = 2.4 kg
If a = ³⁄₄g = 7.3575 m/s²,   m = 7.2 kg

Note, too, that we do not need to know the radius unless the angular acceleration of the flywheel is requested. If you need α, and you have r, then α = a/r.

Example II

Find the kinetic energy of a disk, m = 6.7 kg, that is moving at 3.2 m/s while rolling without slipping along a flat, horizontal surface. ($I_{DISK} = \tfrac{1}{2}mr^2$; LEM = $\tfrac{1}{2}m$)

The total kinetic energy consists of the linear kinetic energy, $K_L = \tfrac{1}{2}mv^2$, plus the rotational kinetic energy,

$$K_R = \tfrac{1}{2}(I)(\omega)^2 = \tfrac{1}{2}(I)(v/r)^2 = \tfrac{1}{2}(I/r^2)v^2 = \tfrac{1}{2}(LEM)v^2.$$

$$KE = \tfrac{1}{2}mv^2 + \tfrac{1}{2} \cdot (LEM = \tfrac{1}{2}m) \cdot v^2$$

$$KE = \tfrac{1}{2} \cdot 6.7 \cdot 3.2^2 + \tfrac{1}{2} \cdot (\tfrac{1}{2} \cdot 6.7) \cdot 3.2^2$$

$$KE = 34.304 + 17.152 = 51 \text{ J}$$

3.4.1 Physics Concepts—Force, Pressure, and Energy

Physics Concepts—Force, Pressure, and Energy

Force, pressure, and energy are some basic tenets of Physics. The following terms will hopefully serve as a reminder of Physics 101.

- Force
- Pressure
- Energy

Force

Force is described as what is required to change the *velocity* or *acceleration* of an object. Recall that acceleration is any change in vector. The formula for force is rather simple:

$$F = m * a$$

$$F = Force$$

$$m \; massofobject$$

$$a = accleration$$

Source: astronomyonline.org

Here is Newton's famous Force equation:

$F =$ force required, given in Newton's or Dyne's
$m =$ the mass of the object, and
$a =$ <u>acceleration</u>

The Newton (named after Sir Isaac Newton) is the amount required to move 1 kilogram at a distance of 1 meter in 1 second. It is written as:

$$\frac{kg(m)}{s^2}$$

or:

$$(kg)m/s^2$$

or:

$$kg(m)s^{-2}$$

The equation of force is Newton's second Law of Physics. The three laws are:

- A body at rest must remain at rest. and a body in motion remains in motion unless acted upon by an external force.
- A force (F) on a body (m) gives it an *acceleration* (a) in the direction on the force and is inversely proportional to the mass.
- Whenever a body exerts a force from another body, that body exerts a force equal in magnitude and in the opposite direction of the initial mass.

$$F = G\left(\frac{m_1 m_2}{r^2}\right)$$

F = gravitational force between two objects
m_1 = mass of first object in kilograms
m_2 = mass of second object in kilograms
r = distance between objects
G = gravitational constant
Examples of force: Gravity and Friction

Pressure

Pressure is very similar to Force, but applies force over a particular area.

$$Pressure = \frac{Force}{Area}$$

The units of this equation will be:

$$kg/ms^2 = 1\ Pascal = N/m^2$$

3.4.2 Physics—Circular Motion

Circular Motion

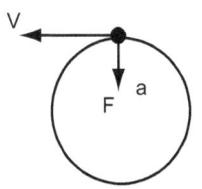

In the diagram **v** is the tangential velocity of the object, **a** is the centripetal (acting towards the center of the circle) acceleration, **F** is the centripetal force, **r** is the radius of the circle, and m is the mass of the object.

$$a = v^2/r$$
$$F = ma = mv^2/r$$

Source: Tutor4Physics.com

3.4.3 Physics—Gravitation

Physics—Gravitation

Kepler's Laws

Toward the end of the sixteenth century, Tycho Brahe collected a huge amount of data giving precise measurements of the position of planets. Johannes Kepler, after a detailed analysis of the measurements, announced three laws in 1619.

1. The orbit of each planet is an ellipse that has the Sun at one of its foci.
2. Each planet moves in such a way that the (imaginary) line joining it to the Sun sweeps out equal areas in equal times.
3. The squares of the periods of revolution of the planets about the Sun are proportional to the cubes of their mean distances from it.

Newton's Law of Universal Gravitation

About 50 years after Kepler announced the laws now named after him, Isaac Newton showed that every particle in the universe attracts every other with a force that is proportional to the products of their masses and inversely proportional to the square of their separation.

Hence:

If **F** is the force due to gravity, **g** the acceleration due to gravity, **G** the Universal Gravitational Constant (6.67×10^{-11} N m²/kg²), **m** the mass, and **r** the distance between two objects, then

$$F = G m_1 m_2 / r^2$$

Acceleration Due to Gravity Outside the Earth

It can be shown that the acceleration due to gravity outside of a spherical shell of uniform density is the same as it would be if the entire mass of the shell were to be concentrated at its center.

Using this principle, we can express the acceleration due to gravity (**g′**) at a radius (**r**) outside the Earth in terms of the Earth's radius (**r_e**) and the acceleration due to gravity at the Earth's surface (**g**).

$$g' = (r_e^2 / r^2) g$$

Acceleration Due to Gravity Inside the Earth

Here let **r** represent the radius of the point inside the Earth. The formula for determining the acceleration due to gravity at this point becomes:

$$g' = (r / r_e) g$$

In both of the above formulas, as expected, g′ becomes equal to g when $r = r_e$.

3.4.4 Physics—Work Energy Power

Work and Energy

As we know from the law of conservation of energy, energy is always conserved.

Work is the product of force and the distance over which it moves. Imagine you are pushing a heavy box across the room. The further you move, the more work you do! If W is work, F the force, and x the distance, then

$$W = Fx$$

Energy comes in many shapes. The energies we see over here are kinetic energy (KE) and potential energy (PE).

$$\text{Transitional KE} = \frac{1}{2} mv^2$$

$$\text{Rotational KE} = \frac{1}{2} I w^2$$

Here **I** is the moment of inertia of the object (a simple manner in which one can understand moment of inertia is to consider it to be similar to mass in transitional KE) and w is angular velocity

$$\text{Gravitational PE} = mgh$$

where **h** is the height of the object

$$\text{Elastic PE} = \frac{1}{2} k L^2$$

where **k** is the spring constant (it gives how much a spring will stretch for a unit force) and L is the length of the spring. Simple isn't it!!

Power

Power (**P**) is work (**W**) done in unit time (t).

$$P = W/t$$

As work and energy (**E**) are the same, it follows that power is also energy consumed or generated per unit time.

$$P = E/t$$

In measuring power, horsepower is a unit that is in common use. However, in physics we use Watt. So the first thing to do in solving any problem related to power is to convert horsepower to Watts. 1 horsepower (hp) = 746 Watts

3.4.5 Physics—Laws of Motion

Newton's Laws of Motion

Through Newton's second law—*The acceleration of a body is directly proportional to the net unbalanced force and inversely proportional to the body's mass*—a relationship is established between Force (**F**), Mass (**m**), and Acceleration (**a**). This is of course a wonderful relation and of immense usefulness.

$$F = m \times a$$

Knowing any two of the quantities automatically gives you the third !!

Momentum

Momentum (**p**) is the quantity of motion in a body. A heavy body moving at a fast velocity is difficult to stop. A light body at a slow speed, on the other hand, can be stopped easily. So momentum has to do with both mass and velocity.

$$p = mv$$

Often physics problems deal with momentum before and after a collision. In such cases the total momentum of the bodies before collision is taken as equal to the total momentum of the bodies after collision. That is to say: momentum is conserved.

Impulse

This is the change in the momentum of a body caused over a very short time. Let **m** be the mass and **v** and **u** be the final and initial velocities of a body.

$$\text{Impulse} = Ft = mv - mu$$

Source: Tutor4Physics.com

3.4.6 Physics—One-, Two-, and Three-Dimensional Motion

One-Dimensional Motion

By one dimension we mean that the body is moving only in one plane and in a straight line. We would be undergoing one-dimensional motion, for example, if we were to roll a marble on a flat table, and if we rolled it in a straight line (not easy!).

Four variables put together in an equation can describe this motion. These are Initial Velocity (**u**); Final Velocity (**v**); Acceleration (**a**); Distance Traveled (**s**) and Time Elapsed (**t**). The equations that tell us the relationship between these variables are as follows:

$$v = u + at$$

$$v^2 = u^2 + 2as$$

$$s = ut + 1/2\, at^2$$

$$\text{average velocity} = (v + u)/2$$

Armed with these equations you can do wonderful things such as calculating a car's acceleration from zero to whatever in 60 seconds !!

Source: Tutor4physics

Two- and Three-Dimensional Motion

Scalar or Vector?

To explain the difference between scalar and vector, we use two words: magnitude and direction. By magnitude we mean how much of the quantity is there. By direction we mean the quantity having a direction that defines it. Physical quantities that are completely specified by just giving out their magnitude are known as scalars. Examples of scalar quantities are distance, mass, speed, volume, density, and temperature. Other physical quantities cannot be defined by just their magnitude. To define them completely, we must also specify their direction. Examples of these quantities are velocity, displacement, acceleration, force, torque, and momentum.

Vector Addition

Parallelogram Law of Vector Addition

Let us assume we were to represent two vectors' magnitude and direction by two adjacent sides of a parallelogram. The resultant could then be represented in magnitude and direction by the diagonal. This diagonal is the one that passes through the point of intersection of these two sides.

Resolution of a Vector

It is often necessary to split a vector into its components. Splitting a vector into its component parts is called resolution of the vector. The original vector is the resultant of these components. When the components of a vector are at right angles to each other, they are called the rectangular components of a vector.

Rectangular Components of a Vector

As the rectangular components of a vector are perpendicular to each other, we can do mathematics on them. This allows us to solve many real-life problems. After all, the best thing about physics is that it can be used to solve real-world problems.

Note: As it is difficult to use vector notations on computer word processors, we will coin our own notation. We will show all **vector quantities in bold**. For example, A will be scalar quantity and **A** will be a vector quantity.

Let $\mathbf{A_x}$ and $\mathbf{A_y}$ be the rectangular components of a vector **A**
then

$$\mathbf{A} = \mathbf{A_x} + \mathbf{A_y}$$

This means that vector **A** is the resultant of vectors $\mathbf{A_x}$ and $\mathbf{A_y}$.

A is the magnitude of vector **A**, and similarly A_x and A_y are the magnitudes of vectors $\mathbf{A_x}$ and $\mathbf{A_y}$.

As we are dealing with rectangular components that are at right angles to each other, we can say that:

$$A = (A_x + A_y)^{1/2}$$

Similarly, the angle Q that the vector **A** makes with the horizontal direction will be

$$Q = \tan^{-1}(A_x/A_y)$$

Physics—One-,Two-,Three-Dimension Motion, p. 3
  *Source:*Tutor4Physics.com

3.4.7 Physics—Electricity

Electricity

According to Ohm's Law, electric potential difference(**V**) is directly proportional to the product of the current(**I**) times the resistance(**R**).

$$V = IR$$

The relationship between power (**P**) and current and voltage is

$$P = IV$$

Using the equations above, we can also write

$$P = V^2/R$$

and

$$P = I^2 R$$

Resistance of Resistors in Series

The equivalent resistance (R_{eq}) of a set of resistors connected in series is

$$R_{eq} = R_1 + R_2 + R_3 + \cdots$$

Resistance of Resistors in Parallel

The equivalent resistance (R_{eq}) of a set of resistors connected in parallel is

$$1/R_{eq} = 1/R_1 + 1/R_2 + 1/R_3$$

3.5.0 Financial Formula Calculations—Net Present Value and Compounding

Present Value Formulas

Calculate the Present Value of a Single Sum

$$PV = \frac{FV}{(1+r)^n}$$

Calculate the Present Value with Compounding

$$PV = \frac{FV}{\left(1+\frac{r}{m}\right)^{(n-m)}}$$

Calculate the Present Value of a Cash-Flow Series

$$PV = \sum_{j=1}^{n} \left[\frac{FV_j}{\left(1+\frac{r}{m}\right)^j}\right]$$

Calculate the Present Value of an Annuity with Continuous Compounding

$$PV_{acp} = \frac{1 - e^{(-rt)}}{r}$$

Calculate the Present Value of a Growing Annuity with Continuous Compounding

$$PV_{ga} = \frac{PMT\left(1 - e^{-(r-g)t}\right)}{e^{(r-g)} - 1}$$

Calculate the Net Present Value of a Cash-Flow Series

$$\text{NPV} = \sum_{j=1}^{n} \frac{\text{CF}_j}{(1+r)^j} = 0$$

Calculate the Future Value with Compounding

$$\text{FV} = \text{PV}\left(1 + \frac{r}{m}\right)^{(n-m)}$$

Calculate the Future Value of a Cash-Flow Series

$$\text{FV} = \sum_{j=1}^{n} \text{CF}_j (1+r)^j$$

Calculate the Future Value of an Annuity

$$\text{FV}_a = \text{PMT}\left[\frac{(1+r)^n - 1}{r}\right]$$

Calculate the Future Value of an Annuity Due

$$\text{FV}_{ad} = \text{PMT}\left[\frac{(1+r)^n - 1}{r}\right](1+r)$$

Calculate the Future Value of an Annuity with Compounding

$$\text{FV}_a = \text{PMT} \cdot \frac{\left(1 + \frac{r}{m}\right)^{(m-n)} - 1}{r/m}$$

Payment Calculations

Calculate Monthly Payment

$$\text{PMT} = P \cdot \frac{r(1+r)^n}{(1+r)^n - 1}$$

Symbols and Variables in Financial Formulas

One important key to understanding formulas for financial math is knowledge of what financial symbols and variables represent.

Where N = Number of Periods
g = Rate of Growth
m = Compounding Frequency
r = Interest Rate
PMT = Periodic Payment

The chart of symbols (located at the right) offers an explanation of the financial variables used in the formulas provided below.

FV = Future Value
PV = Present Value
CF = Cash Flow
j = the jth Period

Calculate the Number of Payments

$$N = \frac{-\log(1 - rFV/PMT)}{\log(1+r)}$$

Convert Interest Rate Compounding Bases

Where r_1 = Original Interest Rate with Compounding Frequency n_1 and r_2 is the stated interest rate with compounding frequency n_2

$$r_2 = \left[\left(1+\frac{r_1}{n_2}\right)^{\frac{n_1}{n_2}} - 1\right] \cdot n_2$$

Calculate Sinking Fund

$$PMT = Goal \cdot \frac{r/m}{\left(1+\frac{r}{m}\right)^{(n-m)} - 1}$$

Future Value Formulas

Calculate the Future Value of a Single Sum

$$FV = PV(1+r)^n$$

Cash-Flow Series Calculations

Calculate the Present Worth Cost of a Cash-Flow Series

$$PWC = \sum_{j-1}^{n} \frac{CF_j}{(1+r)^j} \text{ where } CF_j < 0$$

Calculate the Present Worth Revenue of a Cash-Flow Series

$$PWR = \sum_{j-1}^{n} \frac{CF_j}{(1+r)^j} \text{ where } CF_j > 0$$

3.6.0 Formulas for Calculating the Volume of Cylindrical Tanks

Cylindrical Tank Capacity

We establish capacity in gallons of water in a cylindrical tank by using formulas. The plumbing trade rarely exposes a need for such calculations on a job site. Most state plumbing exams require an individual to be knowledgeable pertaining to a tank's capacity in gallons of water.

Section | 3   Calculations and Formulas—Geometry, Trigonometry, and Physics in Construction

Several steps are required, and you must know how many cubic inches are in 1 gallon of water and how many gallons are in a cubic foot. Dimensions of a tank are in feet and inches or may only be in feet form. You must use the relevant formula for each dimensional situation. When a dimension is in feet and inches, calculations must be used in inch form. Conversion from fraction of an inch to decimal of an inch is also required.

Example: $10'\text{-}2\text{-}1/2'' = 122\text{-}1/2''$, which converts to $122.50''$

The first approach is to find an area of the base of a tank. This is achieved by knowing the diameter of a tank, then multiplying by the diameter by the diameter ($D \times D$), and next by multiplying by 0.7854. At this point in your calculations, you have achieved the square area of a circle.

Area of a cylinder base = Diameter $\times$ Diameter $\times$ 0.7854 ($D \times D \times 0.7854$)

The next step is to know a length or height of a tank. You multiply the square area of a tank's base by the length or height dimension. This gives you the volume of a tank or pipe.

Volume of a cylinder = Height $\times$ Square area of cylindrical base (height also = length or depth)

The next step is to calculate the volume when dimensions are in **inch** form in order to determine the capacity in gallons of a tank. There are 231 cubic inches in 1 gallon of water, so you would **divide** the volume of a tank by **231** to achieve the gallon capacity.

Tank capacity in gallons when measurements are in inch form = Volume $\div$ 231
   Diameter $\times$ Diameter $\times$ 0.7854 $\times$ Length $\div$ 231 ($D \times D \times 0.7854 \times L \div 231$)

The next step is to calculate the volume when dimensions are in **feet** form in order to determine the capacity in gallons of a tank. There are 7.48 gallons in 1 cubic foot, so you would **multiply** the volume of a tank by **7.48**.

Tank capacity in gallons when measurements are in feet form = Volume $\times$ 7.48 Diameter $\times$ Diameter $\times$ 0.7854 $\times$ Length $\times$ 7.48 ($D \times D \times 0.7854 \times L \times 7.48$)

A problem you may have is remembering what formula to use correctly. Because you are rarely faced with a need to make such calculations on a job site, you must understand what the formulas represent. The formula 0.7854 is derived from the circular area of a square or 78-1/2 % of a square. The remaining 21-1/2 % is not relevant because we are calculating a circular area. Imagine placing a circle of equal diameter into a square of equal length (an 8-foot diameter into an 8-foot square).

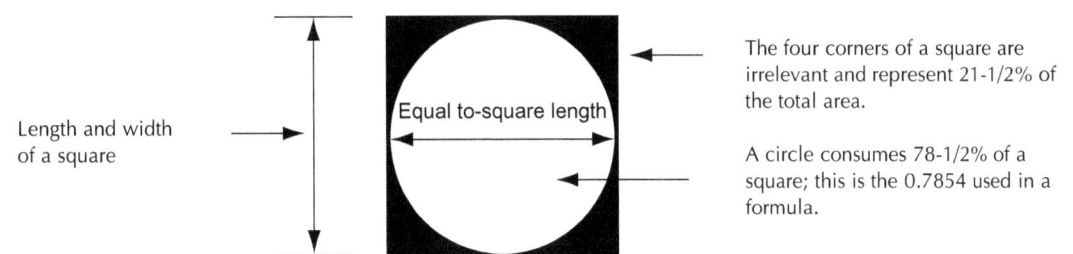

Length and width of a square

Equal to-square length

The four corners of a square are irrelevant and represent 21-1/2% of the total area.

A circle consumes 78-1/2% of a square; this is the 0.7854 used in a formula.

When a circle's diameter is equal size of a square, the area of a circle consumes 0.7854 or 78-1/2% of a square. The remaining 21-1/2% is not used in cylindrical calculations.
   By permission: Joyce Company, Inc., Cary, North Carolina

Cylindrical Tank WorkSheet

Work Sheet

Below is an illustration of a tank that can be viewed as horizontal or vertical. Its dimensions are in feet and inch form. This requires you to calculate all dimensions in inch form and convert to decimal form.

Plan or Side View

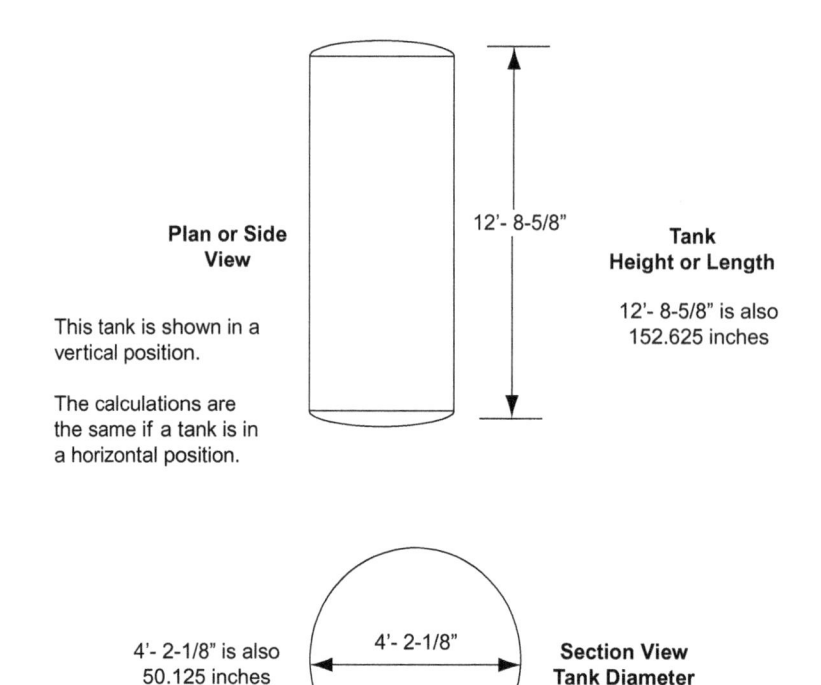

This tank is shown in a vertical position.

The calculations are the same if a tank is in a horizontal position.

Tank Height or Length

12'- 8-5/8"

12'- 8-5/8" is also 152.625 inches

4'- 2-1/8" is also 50.125 inches

4'- 2-1/8"

Section View Tank Diameter

Formulas must relate to cubic inches because the physical dimensions of the tank are in feet and inch form.

- Tank dimensions are: Length = 12' − 8-5/8" × Diameter = 4'-2-1/8"
- Converts to 152.625 inches × 50.125"
- Formula is D × D × 7854 × L ÷ 231 = Gallon Capar

Diameter	×	Diameter	×	Percent	×	Length	=	Cub	÷	Cu. In./1 Gal.	=	Gallon Capacity
↓		↓		↓		↓		↓		↓		↓
50.125	×	50.125	×	.7854	×	152.625	=	301,179.44	÷	231	=	1,303.807

An exam answer in a multiple-choice format may have possible selections that indicate an exact answer remaining in its decimal form. If gallons were rounded to a nearest whole number, the answer would be 1304 gallons.

Use the following dimensions to calculate the gallon capacity of tanks listed below.

Diameter	Length or Height	Gallon Capacity
4'-0"	16'-0"	
6'-3"	12'-8"	
8'-1-5/8"	11' 7-3/8"	
7'-0"	9'-2"	
5'-6-7/8"	5'-6-7/8"	

Section | 3   Calculations and Formulas—Geometry, Trigonometry, and Physics in Construction

The next illustration can help you remember that there are 231 cubic inches in a gallon of water. Imagine a narrow structure that is 19-1/4" (19.25") long, 1" wide, and 12" high. It would take 1 gallon of water to fill such a structure. Now picture freezing 231 ice cubes that are 1" square and 1" high: this is how many 1" cubed-shaped ice cubes you can freeze from a gallon of water. You would have 12 rows high of 1" square and 1" high ice cubes with 19-1/4 cubes per row.

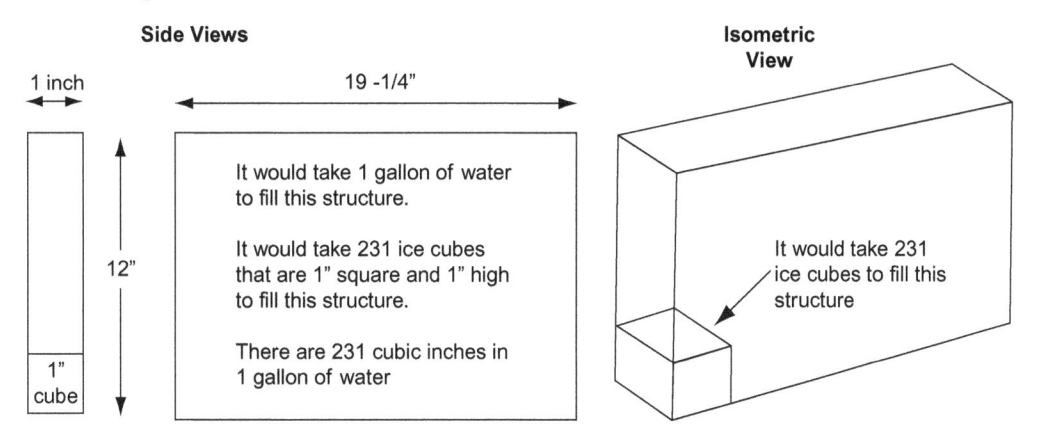

To arrive at the dimensions of this structure, divide 231 cubic inches by the structure height (12). This structure changes dimensionally if you divide by any height. Imagine a structure that is only 1" high. You divide 231 by 1 and have a structure that is dimensioned to be 1" high by 1" wide and 231 inches long. So regardless of the size of a tank, pipe, or structure, there are 231 cubic inches in a gallon of water.

Another portion of a calculating formula that you have to know concerns when measurements are given in feet form. Remember 1 cubic foot holds 7.48 (7-1/2) gallons of water. This next illustration will help you remember how many gallons are in a cubic foot.

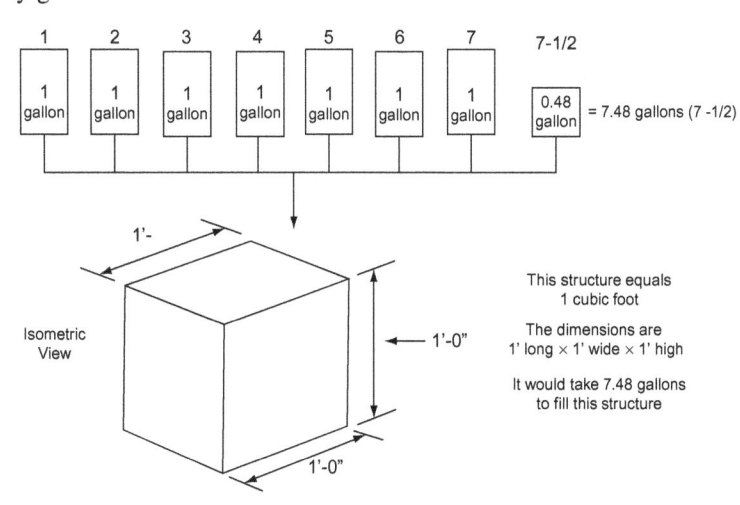

When using this formula for exam purposes and multiple-choice selections are present; you may calculate all gallons in the form of 7.5. If you are faced with answers that are close in numerical sequence or if exact answers are required, use the 7.48 formula, not 7.5.

3.7.0 Round Tank Volume Tables for One Foot of Depth—1 to 32 Feet in Diameter
Calculating the Volume of a Round Tank 1 Foot to 32 Feet in Diameter

Tank Volume
U.S. Gallons in Round Tanks
for One Foot in Depth

Diameter of Tank		No. U.S.Gals.	Cu Ft and Area in Sq Ft
Ft	In.		
1	0	5.87	.785
1	1	6.89	.922
1	2	8.00	1.069
1	3	9.18	1.227
1	4	10.44	1.396
1	5	11.79	1.576
1	6	13.22	1.767
1	7	14.73	1.969
1	8	16.32	2.182
1	9	17.99	2.405
1	10	19.75	2.640
1	11	21.58	2.885
2	0	23.50	3.142
2	1	25.50	3.409
2	2	27.58	3.687
2	3	29.74	3.976
2	4	31.99	4.276
2	5	34.31	4.587
2	6	36.72	4.909
2	7	39.21	5.241
2	8	41.78	5.585
2	9	44.43	5.940
2	10	47.16	6.305
2	11	49.98	6.681
3	0	52.88	7.069
3	1	55.86	7.467
3	2	58.92	7.876
3	3	62.06	8.296
3	4	65.28	8.727
3	5	68.58	9.168
3	6	71.97	9.621
3	7	75.44	10.085
3	8	78.99	10.559
3	9	82.62	11.045
3	10	86.33	11.541
3	11	90.13	12.048
4	0	94.00	12.566
4	1	97.96	13.095
4	2	102.00	13.635
4	3	106.12	14.186
4	4	110.32	14.748
4	5	114.61	15.321

Tank Volume—Cont'd
U.S. Gallons in Round Tanks
for One Foot in Depth

Diameter of Tank		No. U.S.Gals.	Cu Ft and Area in Sq Ft
Ft	In.		
4	6	118.97	15.90
4	7	123.42	16.50
4	8	127.95	17.10
4	9	132.56	17.72
4	10	137.25	18.35
4	11	142.02	18.99
5	0	146.88	19.63
5	1	151.82	20.29
5	2	156.83	20.97
5	3	161.93	21.65
5	4	167.12	22.34
5	5	177.38	23.04
5	6	177.72	23.76
5	7	183.15	24.48
5	8	188.66	25.22
5	9	194.25	25.97
5	10	199.92	26.73
5	11	205.67	27.49
6	0	211.51	28.27
6	3	229.50	30.68
6	6	248.23	33.18
6	9	267.69	35.78
7	0	287.88	38.48
7	3	308.81	41.28
7	6	330.48	44.18
7	9	352.88	47.17
8	0	376.01	50.27
8	3	399.88	53.46
8	6	424.48	56.75
8	9	449.82	60.13
9	0	475.89	63.62
9	3	502.70	67.20
9	6	530.24	70.88
9	9	558.51	74.66
10	0	587.52	78.54
10	3	617.26	82.52
10	6	647.74	86.59
10	9	678.95	90.76
11	0	710.90	95.03
11	3	743.58	99.40
11	6	776.99	103.87
11	9	811.14	108.43

(Continued)

Tank Volume—Cont'd
U.S. Gallons in Round Tanks
for One Foot in Depth

Diameter of Tank		No. U.S.Gals.	Cu Ft and Area in Sq Ft
Ft	In.		
12	0	846.03	113.10
12	3	881.65	117.86
12	6	918.00	122.72
12	9	955.09	127.68
13	0	201.06	132.73
13	3	1031.50	137.89
13	6	1070.80	143.14
13	9	1110.80	148.49
14	0	1151.50	153.94
14	3	1193.00	159.48
14	6	1235.30	165.13
14	9	1278.20	170.87
15	0	1321.90	176.71
15	3	1366.40	182.65
15	6	1411.50	188.69
15	9	1457.40	194.83
16	0	1504.10	201.06
16	3	1551.40	207.39
16	6	1599.50	213.82
16	9	1648.40	220.35
17	0	1697.90	226.98
17	3	1748.20	233.71
17	6	1799.30	240.53
17	9	1851.10	247.45
18	0	1903.60	254.47
18	3	1956.80	261.59
18	6	2010.80	268.80
18	9	2065.50	276.12
19	0	2120.90	283.53
19	3	2177.10	291.04
19	6	2234.00	298.65
19	9	2291.70	306.35
20	0	2350.10	314.16
20	3	2409.20	322.06
20	6	2469.10	330.06
20	9	2529.60	338.16
21	0	2591.00	346.36
21	3	2653.00	354.66
21	6	2715.80	363.05
21	9	2779.30	371.54
22	0	2843.60	380.13
22	3	2908.60	388.82
22	6	2974.30	397.61
22	9	3040.80	406.49

Tank Volume—Cont'd
U.S. Gallons in Round Tanks
for One Foot in Depth

Diameter of Tank			
Ft	In.	No. U.S.Gals.	Cu Ft and Area in Sq Ft
23	0	3108.00	415.48
23	3	3175.90	424.56
23	6	3244.60	433.74
23	9	3314.00	443.01
24	0	3384.10	452.39
24	3	3455.00	461.86
24	6	3526.60	471.44
24	9	3598.90	481.11
25	0	3672.00	490.87
25	3	3745.80	500.74
25	6	3820.30	510.71
25	9	3895.60	520.77
26	0	3971.60	530.93
26	3	4048.40	541.19
26	6	4125.90	551.55
26	9	4204.10	562.00
27	0	4283.00	572.56
27	3	4362.70	583.21
27	6	4443.10	593.96
27	9	4524.30	604.81
28	0	4606.20	615.75
28	3	4688.80	626.80
28	6	4772.10	637.94
28	9	4856.20	649.18
29	0	4941.00	660.52
29	3	5026.60	671.96
29	6	5112.90	683.49
29	9	5199.90	695.13
30	0	5287.70	706.86
30	3	5376.20	718.69
30	6	5465.40	730.62
30	9	5555.40	742.64
31	0	5646.10	754.77
31	3	5737.50	766.99
31	6	5829.70	779.31
31	9	5922.60	791.73
32	0	6016.20	804.25
32	3	6110.60	816.86
32	6	6205.70	829.58
32	9	6301.50	842.39

31-1/2 Gallons equal 1 Barrel

To find the capacity of tanks greater than the largest given in the table, look in the table for a tank of one-half of the given size and multiply its capacity by 4, or one of one-third its size and multiply its capacity by 9, etc.
Source: Cardinal Metal, Inc., Irving, TX

3.8.0 Capacity of Round Tanks—1 to 30 Feet in Diameter

Diameter	gal/ft*
1'-0"	5.88
1'-3"	9.18
1'-6"	13.22
1'-9"	17.99
2'-0"	23.50
2'-3"	29.75
2'-6"	36.72
2'-9"	44.43
3'-0"	52.88
3'-6"	71.98
4'-0"	94.01
4'-6"	118.98
5'	146.89
6'	211.52
7'	287.90
8'	376.04
9'	475.92
10'	587.56
11'	710.94
12'	846.08
15'	1322.00
20'	2350.23
25'	3672.23
30'	5288.01

Capacity of Round Tanks by Diameter & Gallons per Foot
Basis: Inside Measure, 1 cubic foot = 7.481 gallons ● Table Gives Gallons Per Foot of Height

3.9.0 Capacity of Rectangular Tanks—1 to 6 Feet in Depth, 1 to 10 Feet in Length

Capacity of Rectangular Tanks

Length	1'-0"	1'-6"	2'-0"	2'-6"	3'-0"	4'-0"	5'-0"	6'-0"	7'-0"	8'-0"	9'-0"	10'-0"
1'-0"	7.48	11.22	14.96	18.70	22.44	29.92	37.41	44.89	52.37	59.85	67.33	74.81
1'-3"	9.35	14.03	18.70	23.38	28.05	37.41	46.76	56.11	65.46	74.81	84.16	93.51
1'-6"	11.22	16.83	22.44	28.05	33.66	44.89	56.11	67.33	78.55	89.77	100.99	112.22
1'-9"	13.09	19.64	26.18	32.73	39.28	52.37	65.46	78.55	91.64	104.73	117.83	130.92
2'-0"	14.96	22.44	29.92	37.41	44.89	59.85	74.81	89.77	104.73	119.70	134.66	149.62
2'-3"	16.83	25.25	33.66	42.08	50.50	67.33	84.16	100.99	117.83	134.66	151.49	168.32
2'-6"	18.70	28.05	37.41	46.76	56.11	74.81	93.51	112.22	130.92	149.62	168.32	187.03
2'-9"	20.57	30.86	41.15	51.43	61.72	82.29	102.86	123.44	144.01	164.58	185.15	205.73
3'-0"	22.44	33.66	44.89	56.11	67.33	89.77	112.22	134.66	157.10	179.54	201.99	224.43
3'-6"	26.18	39.28	52.37	65.46	78.55	104.73	130.92	157.10	183.28	209.47	235.65	261.84

Capacity of Rectangular Tanks—Cont'd												
Length	1'-0"	1'-6"	2'-0"	2'-6"	3'-0"	4'-0"	5'-0"	6'-0"	7'-0"	8'-0"	9'-0"	10'-0"
4'-0"	29.92	44.89	59.85	74.81	89.77	119.70	149.62	179.54	209.47	239.39	269.32	299.24
4'-6"	33.66	50.50	67.33	84.16	100.99	134.66	168.32	201.99	235.65	269.32	302.98	336.65
5'-0"	37.41	56.11	74.81	93.51	112.22	149.62	187.03	224.43	261.84	299.24	336.65	374.05
6'-0"	44.89	67.33	89.77	112.22	134.66	179.54	224.43	269.32	314.20	359.09	403.97	448.86

Basis: Inside Measure, 1 cubic foot = 7.481 gallons (U.S.) ● Table Gives Gallons Per Foot of Depth

3.10.0 Testing for Hardness in Metal—Mohs, Brinell, Rockwell, Scleroscope, Durometer

Metals

Metals account for about two-thirds of all the elements and about 24% of the mass of the planet. Metals have useful properties, including strength, ductility, high melting points, thermal and electrical conductivity, and toughness. As shown in the periodic table, a large number of the elements are classified as being a metal. A few of the common metals and their typical uses are presented below.

Common Metallic Materials

- Iron/Steel—Steel alloys are used for strength-critical applications.
- Aluminum—Aluminum and its alloys are used because they are easy to form, readily available, inexpensive, and recyclable.
- Copper—Copper and copper alloys have a number of properties that make them useful, including high electrical and thermal conductivity, high ductility, and good corrosion resistance.
- Titanium—Titanium alloys are used for strength in higher temperature (~1000° F) applications, when component weight is a concern, or when good corrosion resistance is required.
- Nickel—Nickel alloys are used for still higher temperatures (~1500–2000° F) applications or when good corrosion resistance is required.
- Refractory materials are used for the highest temperature (> 2000° F) applications.

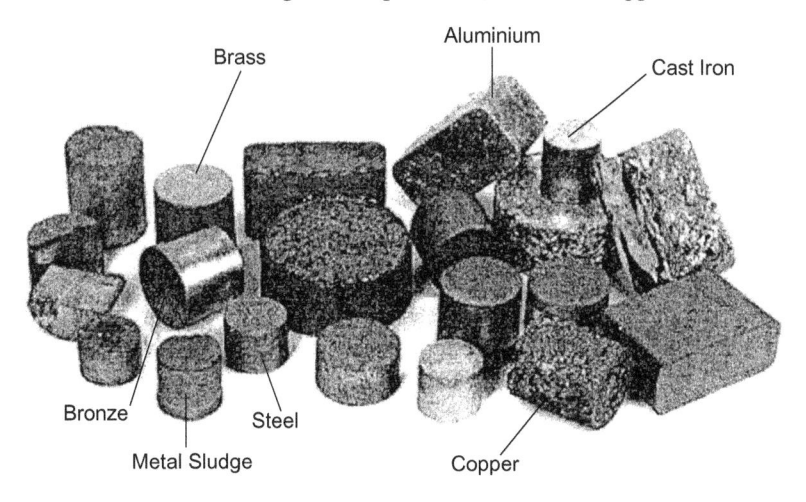

Hardness

Hardness is the resistance of a material to localized deformation. The term can apply to deformation from indentation, scratching, cutting, or bending. In metals, ceramics, and most polymers, the deformation considered is plastic deformation of the surface. For elastomers and some polymers, hardness is defined as the resistance to elastic deformation of the surface. The lack of a fundamental definition indicates that hardness is not a basic property of a material, but rather a composite one with contributions from yield strength, work hardening, true tensile strength, modulus, and others factors. Hardness measurements are widely used for the quality control of materials because they are quick and are considered to be nondestructive tests when the marks or indentations produced by the test are in low-stress areas.

A large variety of methods are used for determining the hardness of a substance. A few of the more common methods are introduced next.

Mohs Hardness Test

One of the oldest ways of measuring hardness was devised by the German mineralogist Friedrich Mohs in 1812. The Mohs hardness test involves observing whether the surface of a material is scratched by a substance of known or defined hardness. To give numerical values to this physical property, minerals are ranked along the Mohs scale, which is composed of 10 minerals that have been given arbitrary hardness values. The Mohs hardness test, while greatly facilitating the identification of minerals in the field, is not suitable for accurately gauging the hardness of industrial materials such as steel or ceramics. For engineering materials, a variety of instruments have been developed over the years to provide a precise measure of hardness. Many apply a load and measure the depth or size of the resulting indentation. Hardness can be measured on the macro-, micro- or nanoscale.

Brinell Hardness Test

The oldest of the hardness test methods in common use on engineering materials today is the Brinell hardness test. Dr. J. A. Brinell invented this test in Sweden in 1900. The Brinell test uses a desktop machine to apply a specified load to a hardened sphere of a specified diameter. The Brinell hardness number, or simply the Brinell number, is obtained by dividing the load used, in kilograms, by the measured surface area of the indentation, in square millimeters, left on the test surface. The Brinell test is frequently used to determine the hardness metal forgings and castings that have large grain structures. The Brinell test provides a measurement over a fairly large area that is less affected by the course grain structure of these materials than are Rockwell or Vickers tests.

A wide range of materials can be tested using a Brinell test simply by varying the test load and indenter ball size. In the United States, Brinell testing is typically done on iron and steel castings using a 3000 kg test force and a 10 mm diameter ball. A 1500 kg load is usually used for aluminum castings. Copper, brass, and thin stock are frequently tested using a 500 kg test force and a 10 or 5 mm ball. In Europe Brinell testing is done using a much wider range of forces and ball sizes; it is common to perform Brinell tests on small parts using a 1 mm carbide ball and a test force as low as 1 kg. These low-load tests are commonly referred to as baby Brinell tests. The test conditions should be reported along with the Brinell hardness number. A value reported as 60 HB 10/1500/30 means that a Brinell hardness of 60 was obtained using a 10 mm diameter ball with a 1500 kg load applied for 30 seconds.

Rockwell Hardness Test

The Rockwell hardness test also uses a machine to apply a specific load and then measure the depth of the resulting impression. The indenter may either be a steel ball of some specified diameter or a spherical diamond-tipped cone of 120° angle and 0.2 mm tip radius, called a brale. A minor load of 10 kg is first applied, which causes a small initial penetration to seat the indenter and remove the effects of any surface irregularities. Then, the dial is set to zero and the major load is applied. Upon removal of the major load, the depth reading is taken while the minor load is still on. The hardness number may then be read directly from the scale. The indenter and the test load used determine the hardness scale that is used (A, B, C, etc).

For soft materials such as copper alloys, soft steel, and aluminum alloys, a 1/16″ diameter steel ball is used with a 100 kg load, and the hardness is read on the B scale. In testing harder materials, hard cast iron, and many steel alloys, a 120 degrees diamond cone is used with up to a 150 kg load, and the hardness is read on the C scale. There are several Rockwell scales other than the B and C scales (which are called the common scales). A properly reported Rockwell value will have the hardness number followed by HR (Hardness Rockwell) and the scale letter. For example, 50 HRB indicates that the material has a hardness reading of 50 on the B scale.

- A -Cemented carbides, thin steel, and shallow case-hardened steel
- B -Copper alloys, soft steels, aluminum alloys, malleable iron, etc.
- C -Steel, hard cast irons, pearlitic malleable iron, titanium, deep case-hardened steel and other materials harder than B 100
- D -Thin steel and medium case-hardened steel and pearlitic malleable iron
- E -Cast iron, aluminum and magnesium alloys, bearing metals
- F -Annealed copper alloys, thin soft sheet metals
- G -Phosphor bronze, beryllium copper, malleable irons
- H -Aluminum, zinc, lead
- K, L, M, P, R, S, V—Bearing metals and other very soft or thin materials, including plastics.

Rockwell Superficial Hardness Test

The Rockwell Superficial Hardness Tester is used to test thin materials, lightly carburized steel surfaces, or parts that might bend or crush under the conditions of the regular test. This tester uses the same indenters as the standard Rockwell tester, but the loads are reduced. A minor load of 3 kg is used, and the major load is either 15 or 45 kg depending on the indenter used. Using the 1/16″ diameter, steel ball indenter, a "T" is added (meaning thin sheet testing) to the superficial hardness designation. An example of a superficial Rockwell hardness is 23 HR15T, which indicates the superficial hardness as 23, with a load of 15 kg using the steel ball.

Vickers and Knoop Microhardness Tests

The Vickers and Knoop hardness tests are a modification of the Brinell test and are used to measure the hardness of thin film coatings or the surface hardness of case-hardened parts. With these tests, a small diamond pyramid is pressed into the sample under loads that are much less than those used in the Brinell test. The difference between the Vickers and the Knoop Tests is simply the shape of the diamond pyramid indenter. The Vickers test uses a square pyramidal indenter, which is prone to crack brittle materials. Consequently, the Knoop test using a rhombic-based (diagonal ratio 7.114:1) pyramidal indenter was developed that produces longer but shallower indentations. For the same load, Knoop indentations are about 2.8 times longer than Vickers indentations.

An applied load ranging from 10 g to 1000 g is used. This low amount of load creates a small indent that must be measured under a microscope. The measurements for hard coatings like TiN must be taken at very high magnification (i.e., 1000X) because the indents are so small. The surface usually needs to be polished. The diagonals of the impression are measured, and these values are used to obtain a hardness number (VHN), usually from a lookup table or chart. The Vickers test can be used to characterize very hard materials, but the hardness is measured over a very small region.

The values are expressed like 2500 HK25 (or HV25), meaning 2500 Hardness Knoop at 25 gram force load. The Knoop and Vickers hardness values differ slightly, but for hard coatings, the values are close enough to be within the measurement error and can be used interchangeably.

Scleroscope and Rebound Hardness Tests

The Scleroscope test is a very old test that involves dropping a diamond-tipped hammer, which falls inside a glass tube under the force of its own weight from a fixed height, onto the test specimen. The height of the rebound travel of the hammer is measured on a graduated scale. The scale of the rebound is arbitrarily chosen and consists of Shore units, divided into 100 parts, which represent the average rebound from pure hardened high-carbon steel. The scale is continued higher than 100 to include metals having greater hardness. The Shore Scleroscope measures hardness in terms of the elasticity of the material, and the hardness number depends on the height to which the hammer rebounds: the harder the material, the higher the rebound.

The Rebound hardness test method is a recent advancement that builds on the Scleroscope test. A variety of electronic instruments are on the market that measure the loss of energy of the impact body. These instruments typically use a spring to accelerate a spherical, tungsten carbide-tipped mass toward the surface of the test object. When the mass contacts the surface, it has a specific kinetic energy, and the impact produces an indentation (plastic deformation) on the surface that takes some of this energy from the impact body. The impact body will lose more energy, and its rebound velocity will be less when a larger indentation is produced on softer material. The velocities of the impact body before and after impact are measured, and the loss of velocity is related to Brinell, Rockwell, or other common hardness values.

Durometer Hardness Test

A Durometer is an instrument that is commonly used for measuring the indentation hardness of rubbers/elastomers, and soft plastics such as polyolefin, fluoropolymer, and vinyl. A Durometer simply uses a calibrated spring to apply a specific pressure to an indenter foot. The indenter foot can be either cone- or sphere-shaped. An indicating device measures the depth of indentation. Durometers are available in a variety of models, and the most popular testers are the Model A used for measuring softer materials and the Model D for harder materials.

Barcol Hardness Test

The Barcol hardness test obtains a hardness value by measuring the penetration of a sharp steel point under a spring load. The specimen is placed under the indenter of the Barcol hardness tester, and uniform pressure is applied until the dial indication reaches a maximum. The Barcol hardness test method is used to determine the hardness of both reinforced and nonreinforced rigid plastics, as well as to determine the degree of cure of resins and plastics.

3.11.0 Comparison of Hardness Scales Shown Graphically
Approximate Comparison of Hardness Scales

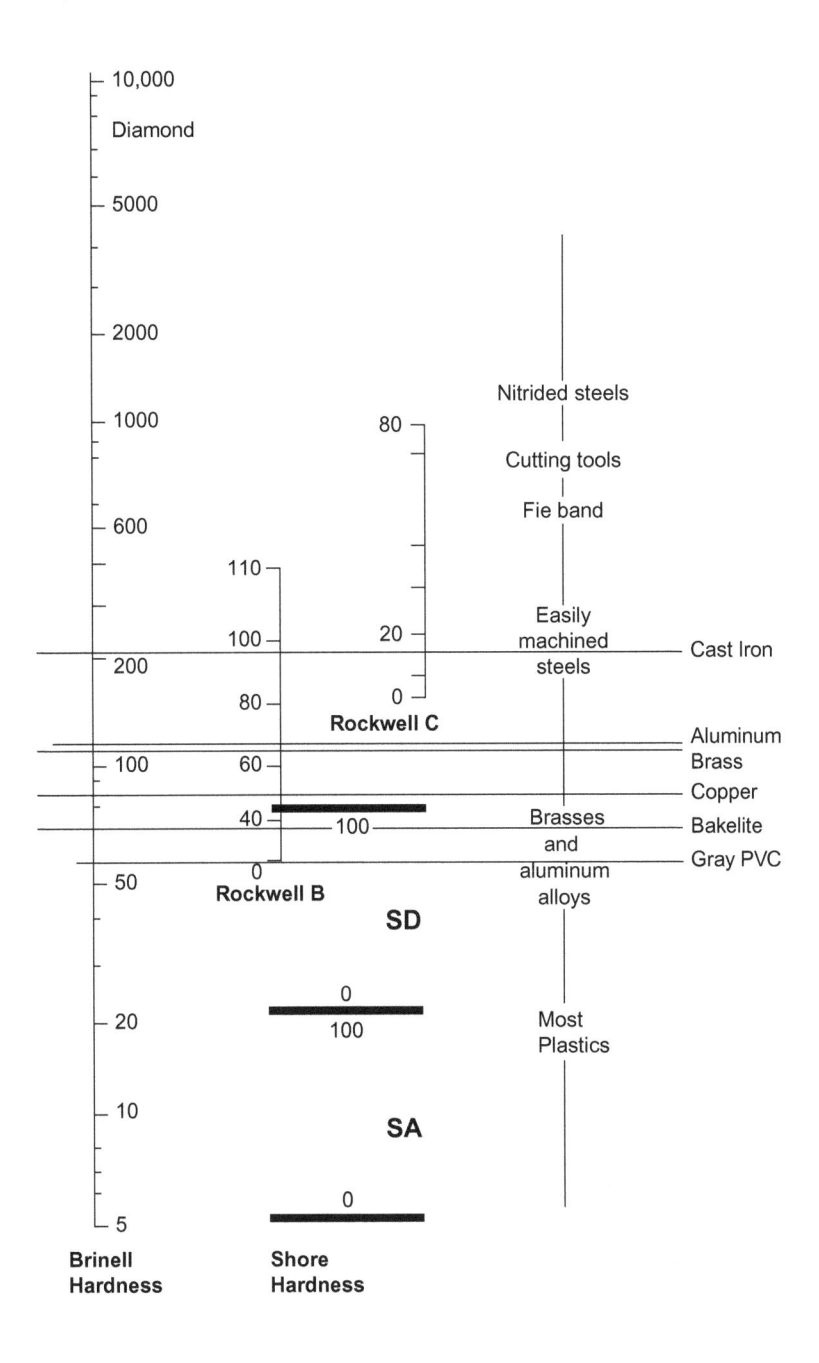

Source: www.calce.umd.edu

3.12.0 Comparison of Hardness Testing in Chart Form

Hardness Comparison Table

Vickers Hardness (DPH)	Brinell Hardness 10mm Diam Ball Load: 3000 kg			Rockwell Hardness (2)				Rockwell Special Hardness Special Brale Equipment				Shore Hardness	Tensile Strength (100sq/m² approx.)	Vickers Hardness Load: 50 kg
	Std Ball	Hultgren Ball	Tungstic Carbide Ball	A Scale Load: 60 kg Brale Equip.	B Scale Load: 100 kg 1/16in. Diam Ball	C Scale Load: 150 kg Brale Equip.	D Scale Load 100 kg Brale Equip.	15-N Scale Load: 15 kg	30-N Scale Load: 30 kg	45-N Scale Load: 45 kg				
410	388	388	388	71.4	-	41.8	56.8	81.4	61.1	45.3	-	195	410
400	379	379	379	70.8	-	40.8	56.0	81.0	60.2	44.1	55	190	400
390	369	369	369	70.3	-	39.8	55.2	80.3	59.3	42.9	-	185	390
380	360	360	360	69.8	(110.0)	38.8	54.4	79.8	58.4	41.7	52	180	380
370	350	350	350	69.2	-	37.7	53.6	79.2	57.4	40.4	-	175	370
360	341	341	341	68.7	(109.0)	36.6	52.8	78.6	56.4	39.1	50	170	360
350	331	331	331	68.1	-	35.5	51.9	78.0	55.4	37.8	-	166	350
340	322	322	322	67.6	(108.0)	34.4	51.1	77.4	54.4	36.5	47	161	340
330	313	313	313	67.0	-	33.3	50.2	76.8	53.6	35.2	-	156	330
320	303	303	303	66.4	(107.0)	32.2	49.4	76.2	52.3	33.9	45	151	320
310	294	294	294	65.8	-	31.0	48.4	75.6	51.3	32.5	-	146	310
300	284	284	284	65.2	(105.5)	29.8	47.5	74.9	50.2	31.1	42	141	300
295	280	280	280	64.8	-	29.2	47.1	74.6	49.7	30.4	-	139	295
290	275	275	275	64.5	(104.5)	28.5	46.5	74.2	49.0	29.5	41	136	290
285	270	270	270	64.2	-	27.8	46.0	73.8	48.4	28.7	-	134	285
280	265	265	265	63.8	(103.5)	27.1	45.3	73.4	47.8	27.9	40	131	280
275	261	261	261	63.5	-	26.4	44.9	73.0	47.2	27.1	-	129	275
270	256	256	256	63.1	(102.0)	25.4	44.3	72.6	46.4	26.2	38	126	270
265	252	252	252	62.7	-	24.8	43.7	72.1	45.7	25.2	-	124	265

260	247	247	247	62.4	(101.0)	24.0	43.1	71.6	45.0	24.3	37	121	260
255	243	243	243	62.0	-	23.1	42.2	71.1	44.2	23.2	-	119	255
250	238	238	238	61.6	99.5	22.2	41.7	70.6	43.4	22.2	36	116	250
245	233	233	233	61.2	-	21.3	41.1	70.1	42.5	21.1	-	114	245
240	228	228	228	60.7	98.1	20.3	40.3	69.6	41.7	19.9	34	111	240
230	219	219	219	-	96.7	(18.0)	-	-	-	-	33	106	230
220	209	209	209	-	95.0	(15.7)	-	-	-	-	32	101	220
210	200	200	200	-	93.4	(13.4)	-	-	-	-	30	97	210
200	190	190	190	-	91.5	(11.0)	-	-	-	-	29	92	200
190	181	181	181	-	89.4	(8.5)	-	-	-	-	28	88	190
180	171	171	171	-	87.1	(6.0)	-	-	-	-	26	84	180
170	162	162	162	-	85.0	(3.0)	-	-	-	-	25	79	170
160	152	152	152	-	81.7	0.0)	-	-	-	-	24	75	160
150	143	143	143	-	78.7	-	-	-	-	-	22	71	150
140	133	133	133	-	75.0	-	-	-	-	-	21	66	140
130	124	124	124	-	71.2	-	-	-	-	-	20	62	130
120	114	114	114	-	66.7	-	-	-	-	-	-	57	120
110	105	105	105	-	62.3	-	-	-	-	-	-	-	110
100	95	95	95	-	56.2	-	-	-	-	-	-	-	100
95	90	90	90	-	52.0	-	-	-	-	-	-	-	95
90	86	86	86	-	48.0	-	-	-	-	-	-	-	90
85	81	81	81	-	41.0	-	-	-	-	-	-	-	85

Source :www.calce.umd.edu

Section 4

Site Work

4.0.0	The Rudiments of Excavation—Classification, Use of Materials, Measurement, and Payment	156
4.1.0	The Unified Soil Classification and Constituents—Explained	159
4.1.0.1	The "Word Picture" of Soil Grain Size	164
4.1.1	ASTM and AASHTO Aggregate and Soil Terminology	165
4.1.2	Field Method of Classification	166
4.1.3	Sediment Classification According to United Soil Classification System	167
4.1.4	USCS Classification Flowline	167
4.1.5	Group Names for Gravelly Soils	168
4.1.6	Group Names for Sandy Soils	168
4.1.7	Calculations to Determine USCS Classification Based upon Percentage Passing Through Sieve	169
4.1.8	Correlation between AASHTO and USCS Systems	169
4.1.9	Classification of Soil and Soil-Aggregate Mixtures for Highway Construction Purposes AASHTO M-145-91 (2000) (Modified)	169
4.1.10	USDA Soil Textural Classification Chart	172
4.2.0	Soil Taxonomy—Formative Elements and Names of Soil Suborders	173
4.3.0	Calculating Soil Compaction Utilizing Various Methods	177
4.3.0.1	Soil Testing—Types—Hand, Proctor, Nuclear Density, Sand Cone	179
4.3.1	Relative Desirability of Soils as Compacted Fill	180
4.3.2	Calculating the Bearing Capacity of Soils	181
4.3.3	Calculating Vibration Control	182
4.3.4	Calculating Earth-Moving Equipment Production	183
4.3.5	Calculating Production of Roller-Type Compaction Equipment	185
4.3.6	Compaction Equipment Type—Applications and Illustrations	186
4.3.6.1	List of Compaction Measuring Devices by Type and Manufacture	189
4.3.6.2	Moisture Density Relation—Compaction Test—Proctor and Modified Proctor	190
4.4.0	Calculating the Maximum Dry Density and Optimum Moisture Content of Soil	190
4.4.1	Calculating the In Situ Dry Density of Soil by the Sand Replacement Method	193
4.5.0	Calculating the Percent of Slope	196
4.5.0.1	Calculating Grade from a Map	197
4.5.0.2	Calculating Grade by Measuring the Road Distance	197
4.5.0.3	Calculating Grade by Using Slope Distance	197
4.5.0.4	Formulas Showing Grade, Ratio, and Angle Relationships	198
4.5.0.5	Chart Showing Slope Angles—0 Degrees to 80 Degrees	199
4.5.1	Illustration of Slope Layback	200
4.5.2	Common Stable Slope Ratios for Varying Soil/Rock Conditions	200
4.5.3	Illustrations of Various Cut/Fill Configurations—Typical Fill, Benched Fill, Reinforced Fill	201
4.5.4	Illustration of Various Cut/Fill Configurations—Balanced Cut-Fill, Full Cut, Through Cut	202

4.5.5	Calculating the Design of Gabion Retaining Walls to 20 Feet (6 Meters) in Height	203	4.7.0	Material Density Chart—Ashes to Wood 204
4.5.6	Calculating the Design of Common Types of Retaining Structures	204	4.8.0	Calculating the Density of Rock, Sand, Till 205

4.0.0 The Rudiments of Excavation—Classification, Use of Materials, Measurement, and Payment

1. SCOPE

The work shall consist of the excavation required by the drawings and specifications and disposal of the excavated materials.

2. CLASSIFICATION

Excavation will be classified as common excavation or rock excavation in accordance with the following definitions or will be designated as unclassified.

Common excavation shall be defined as the excavation of all materials that can be excavated, transported, and unloaded by the use of heavy ripping equipment and wheel tractor-scrapers with pusher tractors or that can be excavated and dumped into place or loaded onto hauling equipment by means of excavators having a rated capacity of one cubic yard or larger and equipped with attachments (such as shovel, bucket, backhoe, dragline or clam shell) appropriate to the material type, character, and nature of the materials.

Rock excavation shall be defined as the excavation of all hard, compacted or cemented materials that require blasting or the use of ripping and excavating equipment larger than defined for common excavation. The excavation and removal of isolated boulders or rock fragments larger than one (1) cubic yard encountered in materials otherwise conforming to the definition of common excavation shall be classified as rock excavation. The presence of isolated boulders or rock fragments larger than one (1) cubic yard will not in itself be sufficient cause to change the classification of the surrounding material.

For the purpose of these classifications, the following definitions shall apply:

Heavy ripping equipment shall be defined as a rear-mounted, heavy duty, single-tooth, ripping attachment mounted on a track type tractor having a power rating of at least 250 flywheel horsepower unless otherwise specified in Section 11.

Wheel tractor-scraper shall be defined as a self-loading (not elevating) and unloading scraper having a struck bowl capacity of at least twelve (12) cubic yards.

Pusher tractor shall be defined as a track type tractor having a power rating of at least 250 flywheel horsepower equipped with appropriate attachments.

3. UNCLASSIFIED EXCAVATION

Excavation designated as "Unclassified Excavation" shall include all materials encountered regardless of their nature or the manner in which they are removed. When excavation is unclassified, none of the definitions or classifications stated in Section 2, CLASSIFICATION, shall apply.

4. BLASTING

The transportation, handling, storage, and use of dynamite and other explosives shall be directed and supervised by person(s) of proven experience and ability who are authorized and qualified to conduct blasting operations.

Source: U.S. Department of Agriculture—Natural Resources Conservation Service.

Blasting shall be done in such a manner as to prevent damage to the work or unnecessary fracturing of the foundation and shall conform to any special requirements in Section 11 of this specification. When specified in Section 11, the Contractor shall furnish the Engineer in writing, a blasting plan prior to blasting operations.

5. USE OF EXCAVATED MATERIALS

Method 1 To the extent they are needed, all suitable materials from the specified excavations shall be used in the construction of required permanent earthfill or rockfill. The suitability of materials for specific purposes will be determined by the Engineer. The Contractor shall not waste or otherwise dispose of suitable excavated materials.

Method 2 Suitable materials from the specified excavations may be used in the construction of required earthfill or rockfill. The suitability of materials for specific purposes will be determined by the Engineer.

6. DISPOSAL OF WASTE MATERIALS

Method 1 All surplus or unsuitable excavated materials will be designated as waste and shall be disposed of at the locations shown on the drawings.

Method 2 All surplus or unsuitable excavated materials will be designated as waste and shall be disposed of by the Contractor at sites of his own choosing away from the site of the work in an environmental acceptable manner and that does not violate local rules and regulations.

7. EXCAVATION LIMITS

Excavations shall comply with OSHA Construction Industry Standards (29CFR Part 1926) Subpart P, Excavations, Trenching, and Shoring. All excavations shall be completed and maintained in a safe and stable condition throughout the total construction phase. Structure and trench excavations shall be completed to the specified elevations and to the length and width required to safely install, adjust, and remove any forms, bracing, or supports necessary for the installation of the work. Excavations outside of the lines and limits shown on the drawings or specified herein required to meet safety requirements shall be the responsibility of the Contractor in constructing and maintaining a safe and stable excavation.

8. BORROW EXCAVATION

When the quantities of suitable materials obtained from specified excavations are insufficient to construct the specified earthfills and earth backfills, additional materials shall be obtained from the designated borrow areas. The extent and depth of borrow pits within the limits of the designated borrow areas shall be as specified in Section 11 or as approved by the Engineer.

Borrow pits shall be excavated and finally dressed to blend with the existing topography and sloped to prevent ponding and to provide drainage.

9. OVER-EXCAVATION

Excavation in rock beyond the specified lines and grades shall be corrected by filling the resulting voids with portland cement concrete made of materials and mix proportions approved by the Engineer. Concrete that will be exposed to the atmosphere when construction is completed shall meet the requirements of concrete selected for use under Construction Specification 31, Concrete for Major Structures, or 32, Structure Concrete, as appropriate.

Concrete that will be permanently covered shall contain not less than five (5) bags of cement per cubic yard. The concrete shall be placed and cured as specified by the Engineer.

Excavation in earth beyond the specified lines and grades shall be corrected by filling the resulting voids with approved compacted earthfill, except that, if the earth is to become the subgrade for riprap, rockfill, sand or gravel bedding, or drainfill, the voids may be filled with material conforming to the specifications for the riprap, rockfill, bedding or drainfill. Prior to correcting an over-excavation condition, the Contractor shall review the planned corrective action with the Engineer and obtain approval of the corrective measures.

10. MEASUREMENT AND PAYMENT

For items of work for which specific unit prices are established in the contract, the volume of each type and class of excavation within the specified pay limits will be measured and computed to the nearest cubic yard by the method of average cross-sectional end areas or by methods outlined in Section 11 of this specification. Regardless of quantities excavated, the measurement for payment will be made to the specified pay limits, except that excavation outside the specified lines and grades directed by the Engineer to remove unsuitable material will be included. Excavation required because unsuitable conditions result from the Contractor's improper construction operations, as determined by the Engineer, will not be included for measurement and payment.

Method 1 The pay limits shall be as designated on the drawings.

Method 2 The pay limits shall be defined as follows:
- **a.** The upper limit shall be the original ground surface as it existed prior to the start of construction operations except that where excavation is performed within areas designated for previous excavation or earthfill the upper limit shall be the modified ground surface resulting from the specified previous excavation or earthfill.
- **b.** The lower and lateral limits shall be the neat lines and grades shown on the drawings.

Method 3 The pay limits shall be defined as follows:
- **a.** The upper limit shall be the original ground surface as it existed prior to the start of construction operations except that where excavation is performed within areas designated for previous excavation or earthfill the upper limit shall be the modified ground surface resulting from the specified previous excavation or earthfill.
- **b.** The lower and lateral limits shall be the true surface of the completed excavation as directed by the Engineer.

Method 4 The pay limits shall be defined as follows:
- **a.** The upper limit shall be the original ground surface as it existed prior to the start of construction operations except that where excavation is performed within areas designated for previous excavation or earthfill the upper limit shall be the modified ground surface resulting from the specified previous excavation or earthfill.
- **b.** The lower limit shall be at the bottom surface of the proposed structure.
- **c.** The lateral limits shall be 18-inches outside of the outside surfaces of the proposed structure or shall be vertical planes 18-inches outside of and parallel to the footings, whichever gives the larger pay quantity, except as provided in d, below.
- **d.** For trapezoidal channel linings or similar structures that are to be supported upon the sides of the excavation without intervening forms, the lateral limits shall be at the under side of the proposed lining or structure.
- **e.** For the purposes of the definitions in b, c, and d, above, any specified bedding or drainfill directly beneath or beside the structure will be considered to be a part of the structure.

All Methods The following provisions apply to all methods of measurement and payment.

Payment for each type and class of excavation will be made at the contract unit price for that type and class of excavation. Such payment will constitute full compensation for all labor, materials, equipment, and all other items necessary and incidental to the performance of the work, except that extra payment for backfilling over-excavation will be made in accordance with the following provisions:

Payment for backfilling over-excavation, as specified in Section 9 of this specification, will be made only if the excavation outside specified lines and grades is directed by the Engineer to remove unsuitable material and if the unsuitable condition is not a result of the Contractor's improper construction operations as determined by the Engineer.

Compensation for any item of work described in the contract but not listed in the bid schedule will be included in the payment for the item of work to which it is made subsidiary. Such items and the items to which they are made subsidiary are identified in Section 11 of this specification.

4.1.0 The Unified Soil Classification and Constituents—Explained

Uniform Field Soil Classification System

(Modified Unified Description)

Introduction April 6, 2009

The purpose of this system is to establish guidelines for the uniform classification of soils by inspection for MDOT Soils Engineers and Technicians. It is the intent of this system to describe only the soil constituents that have a significant influence on the visual appearance and engineering behavior of the soil. This system is intended to provide the best word description of the sample to those involved in the planning, design, construction, and maintenance processes. A method is presented for preparing a "word picture" of a sample for entering on a subsurface exploration log or other appropriate data sheet. The classification procedure involves visually and manually examining soil samples with respect to texture (grain-size), plasticity, color, structure, and moisture. In addition to classification, this system provides guidelines for assessment of soil strength (relative density for granular soils, consistency for cohesive soils), which may be included with the field classification as appropriate for engineering requirements. A glossary of terms is included at the end of this document for convenient reference.

It should be understood that the soil descriptions are based upon the judgment of the individual making the description. Laboratory classification tests are not intended to be used to verify the description, but to further determine the engineering behavior for geotechnical design and analysis, and for construction.

Primary Soil Constituents

The primary soil constituent is defined as the material fraction which has the greatest impact on the engineering behavior of the soil, and which usually represents the soil type found in the largest percentage. To determine the primary constituent, it must first be determined whether the soil is "Fine-Grained" or "Coarse-Grained" or "Organic" as defined below. The field soil classification "word picture" will be built around the primary constituent as defined by the soil types described below.

Coarse-Grained Soils: More than 50% of the soil is *RETAINED* on the (0.075 mm) #200 sieve. A good rule of thumb to determine if particles will be retained or pass the #200 sieve: If individual particles can be distinguished by the naked eye, then they will likely be retained. Also, the finest sand particles often can be identified by their sparkle or glassy quality.

Source: Michigan Department of Transportation

> **Gravel** Identified by particle size, gravel consists of rounded, partially angular, or angular (crushed faces) particles of rock. Gravel size particles usually occur in varying combinations with other particle sizes. Gravel is subdivided into particle size ranges as follows: (Note that particles > (75 mm) 3 inches are cobbles or boulders, as defined in the Glossary of Terms.)
>> *Coarse* - Particles passing the (75 mm) 3 inch sieve, and retained on the (19 mm) 3/4 inch sieve.
>> *Fine* - Gravel particles passing the (19 mm) 3/4 inch sieve, and retained on the (4.76 mm) #4 U.S. standard sieve.

Note: The term "gravel" in this system denotes a particle size range and should not be confused with "gravel" used to describe a type of geologic deposit or a construction material.

Sand Identified by particle size, sand consists of rock particles, usually silicate (quartz) based, ranging between gravel and silt sizes. Sand has no cohesion or plasticity. Its particles are gritty grains that can easily be seen and felt, and may be rounded (natural) or angular (usually manufactured). Sand is subdivided into particle size ranges as follows:

Coarse - Particles that will pass the (4.76 mm) #4 U.S. Standard sieve and be retained on the (2 mm) # 10 U.S. Standard sieve.

Medium - Particles that will pass the (2 mm) #10 U.S. Standard sieve and be retained on the (0.425 mm) # 40 U.S. Standard sieve.

Fine - Particles that will pass the (0.425 mm) #40 U.S. Standards sieve and be retained on the (0.075 mm) # 200 U.S. Standard sieve.

Well-Graded - Indicates relatively equal percentages of Fine, Medium, and Coarse fractions are present.

Note: The particle size of coarse-grained primary soils is important to the Soil Engineer! Always indicate the particle size or size range immediately before the primary soil constituent.

Exception: The use of 'Gravel' alone will indicate both coarse and fine gravel are present.
Examples: **Fine & Medium** Sand; **Coarse** Gravel.

Include the particle shape (angular, partially angular, or rounded) when appropriate, such as for aggregates or manufactured sands.
Example: **Rounded** Gravel.

Fine-Grained Soils: More than 50% of the soil PASSES the (0.075 mm) #200 sieve.

Silt Identified by behavior and particle size, silt consists of material passing the (0.075 mm) #200 sieve that is nonplastic (no cohesion) and exhibits little or no strength when dried. Silt can typically be rolled into a ball or strand, but it will easily crack and crumble. To distinguish silt from clay, place material in one hand and make 10 brisk blows with the other; if water appears on the surface, creating a glossy texture, then the primary constituent is silt.

Clay Identified by behavior and particle size, clay consists of material passing the (0.075 mm) #200 sieve AND exhibits plasticity or cohesion (ability of particles to adhere to each other, like putty) within a wide range of moisture contents. Moist clay can be rolled into a thin (3 mm)1/8 inch thread that will not crumble. Also, clay will exhibit strength increase with decreasing moisture content, retaining considerable strength when dry.

Clay is often encountered in combination with other soil constituents such as silt and sand. If a soil exhibits plasticity, it contains clay. The amount of clay can be related to the degree of plasticity; the higher the clay content, the greater the plasticity.

Note: When applied to laboratory gradation tests, silt size is defined as that portion of the soil finer than the (0.075 mm) # 200 U.S. Standard sieve and coarser than the 0.002 mm. Clay size is that portion of soil finer than 0.002 mm. For field classification, the distinction will be strictly based upon cohesive characteristics.

Organic Soils:
Peat Highly organic soil, peat consists primarily of vegetable tissue in various stages of decomposition, accumulated under excessive moisture conditions, with texture ranging from fibrous to amorphous. Peat is usually black or dark brown in color, and has a distinct organic odor. Peat may have minor amounts of sand, silt, and clay in various proportions.

Fibrous Peat - Slightly or undecomposed organic material having identifiable plant forms. Peat is relatively very lightweight and usually has spongy, compressible consistency.

Amorphous Peat (Muck) - Organic material which has undergone substantial decomposition such that recognition of plant forms is impossible. Its consistency ranges from runny paste to compact rubbery solid.

Marl   Marl consists of fresh water sedimentary deposits of calcium carbonate, often with varying percentages of calcareous fine sand, silt, clay and shell fragments. These deposits are unconsolidated, so marl is usually lightweight. Marl is white or light-gray in color with consistency ranging from soft paste to spongy. It may also contain granular spheres, organic material, or inorganic soils. Note that marl will react (fizz) with weak hydrochloric acid due to the carbonate content.

Secondary Soil Constituents

Secondary soil constituents represent one or more soil types other than the primary constituent which appear in the soil in significant percentages sufficient to readily affect the appearance or engineering behavior of the soil. To correlate the field classification with laboratory classification, this definition corresponds to amounts of secondary soil constituents > 12% for fine-grained and >30% for coarse-grained secondary soil constituents. The secondary soil constituents will be added to the field classification as an adjective preceding the primary constituent. Two or more secondary soil constituents should be listed in ascending order of importance. Examples: **Silty** Fine Sand; **Peaty** Marl; **Gravelly**, **Silty** Medium Sand; **Silty**, **Sandy** Clay.

Tertiary Soil Constituents

Tertiary soil constituents represent one or more soil types that are present in a soil in quantities sufficient to readily identify, but NOT in sufficient quantities to significantly affect the engineering behavior of the soil. The tertiary constituent will be added to the field classification with the phrase "with ____" at the end, following the primary constituent and all other descriptors. This definition corresponds to approximately 5–12% for fine-grained and 15–29% for coarse-grained tertiary soil constituents. Example: Silty Fine to Coarse Sand with **Gravel and Peat**.

    Soil types that appear in the sample in percentages below tertiary levels need not be included in the field classification. However, the slight appearance of a soil type may be characteristic of a transition in soil constituents (more significant deposits nearby), or may be useful in identifying the soil during construction. These slight amounts can be included for descriptive purposes at the end of the field classification as "Trace of ____."

Additional Soil Descriptors

Additional descriptors should be added as needed to adequately describe the soil for the purpose required. These descriptors should *typically* be added to the field classification before the primary and secondary constituents, in ascending order of significance (Exceptions noted below). Definitions for several descriptive terms can be found in the Glossary of Terms below. Other terms may be used as appropriate for descriptive purposes, but not for soil constituents.

Color	Brown, Gray, Yellow, Red, Black, Light-, Dark-, Pale-, etc.
Moisture Content	Dry, Moist, Saturated. Judge by appearance of sample before manipulating.
Structure	Fissured, Friable, Blocky, Varved, Laminated, Lenses, Layers, etc.
Examples:	**Gray-Brown Laminated** Silty Clay; **Light-Brown Saturated** Fine & Medium Sand.
Exceptions:	Certain descriptive terms such as "Fill" may be more appropriate after the primary constituent or at the end of the field classification. Also, the description of distinct soils (inclusions) within a larger stratum should be added after the complete field classification of the predominant soil.
Examples of exceptions:	Stiff Brown Sandy Clay **Fill,** with Coarse Angular Gravel and Asphalt; Gray Silty Clay with Saturated Marl, **Lenses of Saturated Fine Sand.**

Soil Strength Assessment

Soil strength refers to the degree of load-carrying capacity and resistance to deformation which a particular soil may develop. For cohesionless granular soils (sand, gravel, and silt) the relative in-place density is a measure of strength. The in-place consistency for cohesionless soils can be estimated by the Standard Penetration Test (SPT—Blow counts) and by resistance to drilling equipment or "pigtail" augers as described below. For cohesive soils, "consistency" is a measure of cohesion, or shear strength. The shear strength of clay soils can be estimated in the field using the manual methods described below, the SPT, or resistance to drilling equipment. Note that for clay soils, loss of moisture will result in increased strength; therefore, consistency of clay soils should be estimated at the natural moisture content.

The soil consistency, when appropriate and available, should be added to the field classification at the very beginning, using the terminology described below. Examples: **Loose** Brown Rounded Fine Gravel; **Medium Stiff** Gray Moist Sandy Clay.

Cohesionless Soil

Classification	Standard Penetration, N	Relative Density, %	Resistance to Advancement of a (1.2 m) 4 ft. Long, (38 mm) 1.5 inch Diameter Spiral (Pigtail) Auger
Very Loose	< 4	0–15	The auger can be forced several inches into the soil, without turning, under the bodyweight of the technician.
Loose	4–10	15–35	The auger can be turned into the soil for its full length without difficulty. It can be chugged up and down after penetrating about (1/3 m) 1 ft, so that it can be pushed down (25 mm) 1 inch into the soil.
Medium Dense	10–30	35–65	The auger cannot be advanced beyond ±(3/4 m) 2.5 ft without great difficulty. Considerable effort by chugging required to advance further.
Dense	30–50	65–85	The auger turns until tight at ±(1/3 m) 1 ft; cannot be advanced further.
Very Dense	> 50	85–100	The auger can be turned into the soil only to about the length of its spiral section.

Cohesive Soil

Classification	Manual Index for Consistency	Cohesion (psf)	Cohesion (kPa)	Standard Penetration, N
Very Soft	Extrudes between fingers when squeezed	0–250	0–12	< 2
Soft	Molded by light to moderate finger pressure	250–500	12–24	2–4
Medium Stiff	Molded by moderate to firm finger pressure	500–1000	24–48	4–8
Stiff	Readily indented by thumb, difficult to penetrate	1000–2000	48–96	8–15
Very Stiff	Readily indented by thumbnail	2000–4000	96–192	15–30
Hard	Indented with difficulty by thumbnail	4000–8000	192–384	> 30

Glossary of Terms

Blocky — Cohesive soil that can be broken down into small angular lumps that resist further breakdown.

Boulder — A rock fragment, usually rounded by weathering or abrasion, with average dimension of (300 mm) 12″ or more.

Calcareous — Soil containing calcium carbonate, either from limestone deposits or shells. The carbonate will react (fizz) with weak hydrochloric acid.

Cemented — The adherence or bonding of coarse soil grains due to presence of a cementicious material. May be *weak* (readily fragmented), *firm* (appreciable strength), or *indurated* (very hard, water will not soften, rock-like)

Cobble — A rock fragment, usually rounded or partially angular, with an average dimension (75 to 300 mm) 3″ - 12″.

Dry — No appreciable moisture is apparent in the soil.

Fat Clay — Fine-grained soil with very high plasticity and dry strength. Usually has a sticky or greasy texture due to very high affinity for water. Remains plastic at very high water contents (Liquid Limit >50).

Fill — Man-made deposits of natural soils and/or waste materials. Document the components carefully since presence and depth of fill are important engineering considerations.

Fissured — The soil breaks along definite planes of weakness with little resistance to fracturing.

Frequent — Occurring more than one per (300 mm) 1 ft thickness.

Friable — A soil that is easily crumbled or pulverized into smaller, nonuniform fragments or clumps.

Laminated — Alternating horizontal strata of different material or color, usually in increments of (6 mm) 1/4″ or less.

Layer — Horizontal inclusion or stratum of sedimentary soil greater than (100 mm) 4″ thick.

Lens — Inclusion of a small pocket of a sedimentary soil between (10 mm) 3/8″ and (100 mm) 4″ thick, often with tapered edges.

Moist — Describes the condition of a soil with moderate to water content relative to the saturated condition (near optimum). Moisture is readily discernible but not in sufficient content to adversely affect the soil behavior.

Mottled — Irregularly marked soil, usually clay, with spots of different colors.

Muck — See *Amorphous Peat*, under Primary Soil Constituents heading.

Occasional — Occurring once or less per (300 mm) 1 ft thickness.

Organic — Indicates the presence of material that originated from living organisms, usually vegetative, undergoing some stage of decay. May range from microscopic size matter to fibers, stems, leaves, wood pieces, shells, etc. Usually dark brown or black in color, and accompanied by a distinct odor.

Parting — A very thin soil inclusion of up to (10 mm) 3/8″ thickness.

Saturated — All of the soil voids are filled with water (zero air voids). Practically speaking, the condition where the moisture content is sufficient to substantially affect the soil behavior.

Trace — Indicates appearance of a slight amount of a soil type, which may be included in the classification for descriptive or identification purposes only. The Trace soil would have no effect on the soil behavior. Other modifiers such as "Slight" or "Heavy" should not be used with "Trace."

Varved — The paired arrangement of laminations in glacial sediments that reflect seasonal changes during deposition; Fine sand and silt are deposited in the glacial lake during summer, and finer particles are usually deposited in thinner laminations in winter.

4.1.0.1 The "Word Picture" of Soil Grain Size

UNIFIED SOIL CLASSIFICATION AND SYMBOL CHART		
COARSE-GRAINED SOILS (more than 50% of material is larger than No. 200 sieve size.)		
GRAVELS More than 50% of coarse fraction larger than No. 4 sieve size	Clean Gravels (Less than 5% fines)	
	GW	Well-graded gravels, gravel-sand mixtures, little or no fines
	GP	Poorly-graded gravels, gravel-sand mixtures, little or no fines
	Gravels with fines (More than 12% fines)	
	GM	Silty gravels, gravel-sand-silt mixtures
	GC	Clayey gravels, gravel-sand-clay mixtures
SANDS 50% or more of coarse fraction smaller than No. 4 sieve size	Clean Sands (Less than 5% fines)	
	SW	Well-graded sands, gravelly sands, little or no fines
	SP	Poorly graded sands, gravelly sands, little or no fines
	Sands with fines (More than 12% fines)	
	SM	Silty sands, sand-silt mixtures
	SC	Clayey sands, sand-clay mixtures
FINE-GRAINED SOILS (50% or more of material is smaller than No. 200 sieve size.)		
SILTS AND CLAYS Liquid limit less than 50%	ML	Inorganic silts and very fine sands, rock flour, silty of clayey fine sands or clayey silts with slight plasticity
	CL	Inorganic clays of low to medium plasticity, gravelly clays, sandy clays, silty clays, lean clays
	OL	Organic silts and organic silty clays of low plasticity
SILTS AND CLAYS Liquid limit 50% or greater	MH	Inorganic silts, micaceous or diatomaceous fine sandy or silty soils, elastic silts
	CH	Inorganic clays of high plasticity, fat clays
	OH	Organic clays of medium to high plasticity, organic silts
HIGHLY ORGANIC SOILS	PT	Peat and other highly organic soils

PLASTICITY CHART

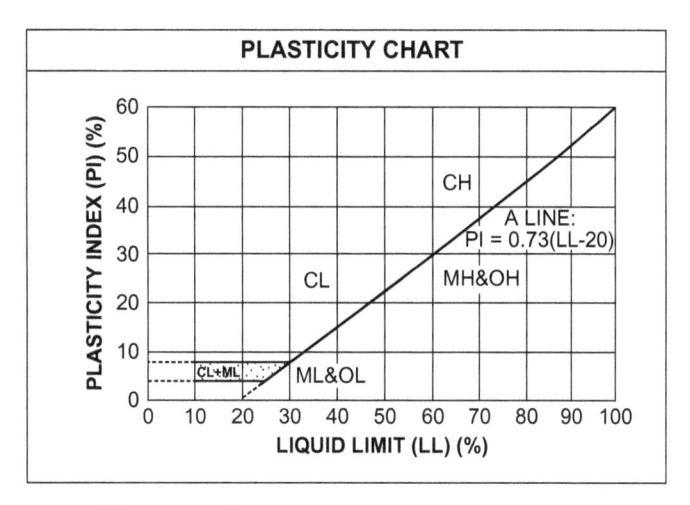

A LINE: PI = 0.73(LL−20)

Source: Virginia Department of Transportation

4.1.1 ASTM and AASHTO Aggregate and Soil Terminology

ASTM Aggregate and Soil Terminology

Terms

The basic reference for the Unified Soil Classification System is ASTM D 2487. Terms include:

- **Coarse-Grained Soils:** More than 50% retained on a 0.075 mm (No. 200) sieve
- **Fine-Grained Soils:** 50% or more passes a 0.075 mm (No. 200) sieve
- **Gravel:** Material passing a 75-mm (3-inch) sieve and retained on a 4.75-mm (No. 4) sieve.
- **Coarse Gravel:** Material passing a 75-mm (3-inch) sieve and retained on a 19.0-mm (3/4-inch) sieve.
- **Fine Gravel:** Material passing a 19.0-mm (3/4-inch) sieve and retained on a 4.75-mm (No. 4) sieve.
- **Sand:** Material passing a 4.75-mm sieve (No. 4) and retained on a 0.075-mm (No. 200) sieve.
- **Coarse Sand:** Material passing a 4.75-mm sieve (No. 4) and retained on a 2.00-mm (No. 10) sieve.
- **Medium Sand:** Material passing a 2.00-mm sieve (No. 10) and retained on a 0.475-mm (No. 40) sieve.
- **Fine Sand:** Material passing a 0.475-mm sieve (No. 40) and retained on a 0.075-mm (No. 200) sieve.
- **Clay:** Material passing a 0.075-mm (No. 200) that exhibits plasticity and strength when dry (PI^3 4).
- **Silt:** Material passing a 0.075-mm (No. 200) that is nonplastic and has little strength when dry ($PI < 4$).
- **Peat:** Soil of vegetable matter.

Note that these definitions are Unified Soil Classification system definitions and are slightly different from those of AASHTO.

Source: pavementinteractive.org

AASHTO soil terminology comes from AASHTO M 145, "Classification of Soils and Soil-Aggregate Mixtures for Highway Construction Purposes." Aggregate terminology comes from AASHTO M 147, "Materials for Aggregate and Soil-Aggregate Subbase, Base and Surface Courses." Basic terms include:

- **Boulders & Cobbles:** Material retained on a 75-mm (3-inch) sieve.
- **Gravel:** Material passing a 75-mm (3-inch) sieve and retained on a 2.00-mm (No. 10) sieve.
- **Coarse Sand:** Material passing a 2.00-mm sieve (No. 10) and retained on a 0.475-mm (No. 40) sieve.

- **Fine Sand:** Material passing a 0.475-mm (No. 40) sieve and retained on a 0.075-mm (No. 200) sieve.
- **Silt-Clay:** Material passing a 0.075-mm (No. 200) sieve.
- **Silt Fraction:** Material passing the 0.075 mm sieve and larger than 0.002 mm.
- **Clay Fraction:** Material smaller than 0.002 mm.
- **Silty:** Material passing a 4.75-mm (No. 4) sieve with a PI £ 10
- **Clayey:** Material passing a 4.75-mm (No. 4) sieve with a PI^3 11
- **Coarse Aggregate:** Aggregate retained on the 2.00 mm sieve and consisting of hard, durable particles or fragments of stone, gravel or slag. A wear requirement (AASHTO T 96) is normally required.
- **Fine Aggregate:** Aggregate passing the 2.00 mm (No. 10) sieve and consisting of natural or crushed sand, and fine material particles passing the 0.075 mm (No. 200) sieve. The fraction passing the 0.075 mm (No. 200) sieve shall not be greater than two-thirds of the fraction passing the 0.425 mm (No. 40) sieve. The portion passing the 0.425 mm (No. 40) sieve shall have a LL $\leq$ 25 and a PI $\leq$ 6. Fine aggregate shall be free from vegetable matter and lumps or balls of clay.

Note that these definitions are AASHTO definitions and are slightly different from those of the Unified Soil Classification system (ASTM).

4.1.2 Field Method of Classification

Procedure

Field Classification Technique for Coarse-Grained Soils

1. Take a representative sample of soil (excluding particles >75 mm) (see Note 1) and classify the soil as coarse-grained or fine-grained by estimating whether 50%, by weight, of the particles can be seen individually by the naked eye. Soils containing >50% of particles that can be seen are coarse-grained soils; soils containing <50% of particles smaller than the eye can see are fine-grained soils. If the soil is predominantly coarse-grained, identify as being a gravel or a sand by estimating whether 50% or more, by weight, of the coarse grains are larger or smaller than 4.75 mm (No. 4 sieve size).
2. If the soil is a gravel, identify as being "clean" (containing little or no fines, <5%) or "dirty" (containing an appreciable amount of fines, >12%). For clean gravels final classification is made by estimating the gradation: the well-graded gravels belong to the GW groups and uniform and gap-graded gravels belong to the GP group. Dirty gravels are of two types: those with nonplastic (silty) fines (GM) and those with plastic (clayey) fines (GC). The determination of whether the fines are silty or clayey is made by the three manual tests for fine-graded soils.
3. If a soil is a sand, the same steps and criteria are used as for gravels in order to determine whether the soil is a well-graded clean sand (SW), poorly graded clean sand (SP), sand with silty fines (SM) or sand with clayey fines (SC).
4. If a material is predominantly (>50% by weight) fine-grained, it is classified into one of six groups (ML, CL, OL, MH, CH, OH) by estimating its dilatancy (reaction to shaking), dry strength (crushing characteristics), and toughness (consistency near the plastic limit) and by identifying it as being organic or inorganic. (See Note 2.)

Source: Florida International University, Miami, FL

4.1.3 Sediment Classification According to United Soil Classification System

GW	Well-graded gravels, gravel/sand mixture, little or no fines
GP	Poorly graded gravels, gravel/sand mixture, little or no fines
GM	Gray calcareous gravel, sand/silt mixture
GC	Gray calcareous gravel/sand/clay mixtures
SW	Well-graded sands, gravelly sands, little or no fines
SP	Poorly graded sands, gravelly sands, little or no fines
SM	Silty sands, sand/silt mixtures
SC	Clayey sands, sand/clay mixtures
ML	Inorganic silts and very fine sands, silty/clayey fine sand
CL	Inorganic clays of low to medium plasticity, sandy silty or lean clays
OL	Organic silts and organic silty clays of low plasticity
MH	Inorganic silts, micaceous or diatomaceous fine sandy or silty soils, elastic silts
CH	Fat clays
OH	Fat organic clays
PT	Peat, humus, and other organic swamp soils
SP-SM	Sandy, gravelly, silty mixtures
L	Limestone
S	Sandstone

Source: Florida International University, Miami, FL

4.1.4 USCS Classification Flowline

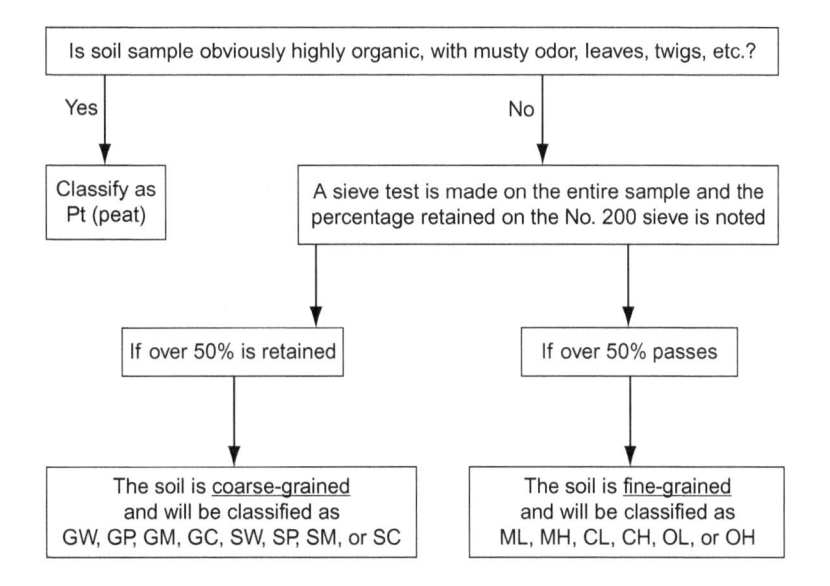

Is soil sample obviously highly organic, with musty odor, leaves, twigs, etc.?

- **Yes** → Classify as Pt (peat)
- **No** → A sieve test is made on the entire sample and the percentage retained on the No. 200 sieve is noted
  - If over 50% is retained → The soil is <u>coarse-grained</u> and will be classified as GW, GP, GM, GC, SW, SP, SM, or SC
  - If over 50% passes → The soil is <u>fine-grained</u> and will be classified as ML, MH, CL, CH, OL, or OH

Source: Florida International University, Miami, FL

4.1.5 Group Names for Gravelly Soils

Group Symbol	Criteria for Group Name: SF[a]	Group Name
GW	< 15	Well-graded gravel
	≥ 15	Well-graded gravel with sand
GP	< 15	Poorly graded gravel
	≥ 15	Poorly graded gravel with sand
GM	< 15	Silty gravel
	≥ 15	Silty gravel with sand
GC	< 15	Clayey gravel
	≥ 15	Clayey gravel with sand
GC-GM	< 15	Silty, clayey gravel
	≥ 15	Silty, clayey gravel with sand
GW-GM	< 15	Well-graded gravel with silt
	≥ 15	Well-graded gravel with silt and sand
GW-GC	< 15	Well-graded gravel with clay
	≥ 15	Well-graded gravel with clay and sand
GP-GM	< 15	Poorly graded gravel with silt
	≥ 15	Poorly graded gravel with silt and sand
GP-GC	< 15	Poorly graded gravel with clay
	≥ 15	Poorly graded gravel with clay and sand

Based on ASTM D-2487
[a] SF = sand fraction = R_{200} - GF, and GF = R_4
Source: Florida International University, Miami, FL

4.1.6 Group Names for Sandy Soils

Group Symbol	Criteria for Group Name: SF[a]	Group Name
GW	< 15	Well-graded sand
	≥ 15	Well-graded sand with gravel
GP	< 15	Poorly graded sand
	≥ 15	Poorly graded sand with gravel
GM	< 15	Silty sand
	≥ 15	Silty sand with gravel
GC	< 15	Clayey sand
	≥ 15	Clayey sand with gravel
GC-GM	< 15	Silty, clayey sand
	≥ 15	Silty, clayey sand with gravel
GW-GM	< 15	Well-graded sand with silt
	≥ 15	Well-graded sand with silt and gravel
GW-GC	< 15	Well-graded sand with clay
	≥ 15	Well-graded sand with clay and gravel
GP-GM	< 15	Poorly graded sand with silt
	≥ 15	Poorly graded sand with silt and gravel
GP-GC	< 15	Poorly graded sand with clay
	≥ 15	Poorly graded sand with clay and gravel

Based on ASTM D-2487
[a] GF = gravel fraction = $R4$
Source: Florida International University, Miami, FL

4.1.7 Calculations to Determine USCS Classification Based upon Percentage Passing through Sieve

Classify the soil by using the **Unified Soil Classification System** if the percentage passing No. 4 sieve = 70, percentage passing No. 200 sieve = 30, the Liquid limit = 33 and the Plasticity index = 12. Give the group symbol and the name.

Solution:

$$F200 = 30$$
$$R200 = 100 - F200 = 100 - 30 = 70$$

Because R200 is greater than 50, the soil is coarse grained.

$$R4 = 100 - F4 = 100 - 70 = 30$$
$$\text{For this soil, } R4/R200 = 30/70 < 0.5$$

This soil is sandy. The group symbol is **SC**. For the group name,

$$GF = \text{gravel fraction} = R4 = 30 \text{ (i.e., } > 15\text{)}$$

Therefore, the group name is <u>clayey sand with gravel</u>.

Source: Florida International University, Miami, FL

4.1.8 Correlation between AASHTO and USCS Systems

AASHTO	Unified
A-2-6	GC, SC
A-2-7	GC, SC
A-3	SP
A-4	ML, OL
A-5	MH
A-6	CL
A-7-5	CL, OL
A-7-6	CH, OH

Source: Florida International University, Miami, FL

4.1.9 Classification of Soil and Soil-Aggregate Mixtures for Highway Construction Purposes AASHTO M-145-91 (2000) (Modified)

This practice describes a procedure for classifying soils into seven groups based on laboratory determination of particle-size distribution, liquid limit, and plasticity index. The group classification should be useful in determining the relative quality of the soil material for use in embankments, subgrades, and backfills. For detailed design of important structures, additional data concerning strength or performance characteristics of the soil under field conditions will usually be required.

Modification: *Determination of Group Index will not be a part of certification, but taught as a useful tool for more accurate determination of soil classification.*

Key Elements

1. **Determine sieve analysis.** Determine sieve analysis using AASHTO T-11 and AASHTO T-27 test procedures (Note 1). The 2.00 mm (No. 10) sieve, 425-μm (No. 40) sieve, and 75-μm (No. 200) sieve must be included to determine the particle size distribution as a basis for classification.
2. **Determine the liquid limit.** Determine the liquid limit of the material using AASHTO T-89 test procedures.
3. **Determine the plastic limit.** Determine the plastic limit and plasticity index of the material using AASHTO T-90 test procedures.
4. **Determine classification of material.** Using the test limits shown in Table 1 of AASHTO M-145, make the classification of the material. If a more detailed classification is desired, a further subdivision of the groups may be made using Table 2 of AASHTO M-145 (**3.1**). With required test data available, proceed from left to right in Table 1 or Table 2 and the correct group will be found by process of elimination (**3.2**). The first group from the left into which the test data will fit is the correct classification (**3.2**).
5. **Report classification.** All limiting test values are shown as whole numbers. If fractional numbers appear on test reports, convert to the nearest whole number for purposes of classification (**3.2**).

Description of Soil Classification Groups:

Soil Fractions: According to the AASHTO system, soils are divided into two major groups as shown in Table 1 of AASHTO M-145. These are the granular materials with 35% or less passing the 75-μm (No. 200) sieve (**5.1, Note 2**) and the silt-clay materials with more than 35% passing the 75-μm (No. 200) sieve (**5.2**). Moreover, five soil fractions are recognized and often used in word descriptions of a material. These five fractions are defined as follows:

Boulders and Cobbles – material retained on the 75 mm (3 in.) sieve. They should be excluded from the portion of a sample to which the classification is applied, but the percentage of such material should be recorded (**4.1.5**).

Gravel – materials passing sieve with 75 mm (3 in.) square openings and retained on the 2.0 mm (No. 10) sieve (**4.1.1**).

Coarse Sand – materials passing the 2.0 mm (No. 10) sieve and retained on the 425-μm (No. 40) sieve (**4.1.2**).

Fine Sand – materials passing the 425-μm (No. 40) sieve and retained on the 75-μm (No. 200) sieve (**4.1.3**).

Combined Silt and Clay – materials passing the 75-μm (No. 200) sieve. The word "silty" is applied to a fine material having a Plasticity Index of **10** or less, and the term "clayey" is applied to fine material having a PI of more than **10** (**4.1.6**).

Granular Materials:

Group A-1: Well-graded mixtures of stone fragments or gravel ranging from coarse to fine with a nonplastic or slightly plastic soil binder (**5.1.1**). However, this group also includes coarse materials without soil binder.
    Subgroup A-1-a: Materials consisting predominantly of stone fragments or gravel, either with or without a well-graded soil binder (**5.1.1.1**).
    Subgroup A-1-b: Materials consisting predominantly of coarse sand either with or without a well-graded soil binder (**5.1.1.2**).
Group A-3: Material consisting of sands deficient in coarse material and soil binder. Typical is fine beach sand or fine desert blow sand, without silt or clay fines or with a very small amount of nonplastic silt. This group also includes stream-deposited mixtures of poorly graded fine sand and limited amounts of coarse sand and gravel (**5.1.2**). These soils make suitable subgrades for all types of pavements when confined and damp. They are subject to erosion and have been known to pump and blow under rigid pavements. (Information: They can be compacted by vibratory, pneumatic-tired, and steel-wheeled rollers but not with a sheepsfoot roller.)

Group A-2: This group includes a wide variety of "granular" materials that are borderline between the materials falling in Groups A-1 and A-3 and silt-clay materials of Groups A-4, A-5, A-6, and A-7. It includes all materials containing 35% or less passing the 75-μm (No. 200) sieve that cannot be classified as A-1 or A-3 (**5.1.3**).

Subgroups A-2-4 and A-2-5: Include various granular materials containing 35% or less passing the 75-μm (No. 200) sieve, and with that portion passing 425-μm (No. 40) sieve having the characteristics of the A-4 and A-5 groups. These groups include such materials as gravel and coarse sand with silt contents or Plasticity Indexes in excess of the limitations of Group A-1, and fine sand with nonplastic silt content in excess of the limitations of Group A-3 (**5.1.3.1**).

Subgroups A-2-6 and A-2-7: Include materials similar to those described under Subgroups A-2-4 and A-2-5, except that the fine portion contains plastic clay having the characteristics of the A-6 or A-7 group (**5.1.3.2**).

A-2 soils are given a poorer rating than A-1 soils because of inferior binder, poor grading, or a combination of the two. Depending on the character and amount of binder, A-2 soils may become soft during wet weather and loose and dusty in dry weather when used as a road surface. If, however, they are protected from these extreme changes in moisture content, they may be quite stable. The A-2-4 and A-2-5 soils are satisfactory as base materials when properly compacted and drained. A-2-6 and A-2-7 soils with low percentages of minus 75-μm (no. 200) sieve material are classified as good bases, whereas these same soils with high percentages of minus 75-μm (No. 200) sieve and PI's of 10 or higher are questionable as a base material. Frequently, the A-2 soils are employed as a cover material for very plastic subgrades.

Silt-Clay Materials

Group A-4: The typical material of this group is a nonplastic or moderately plastic silty soil usually having 75% or more passing the 75 μm (No. 200) sieve. The group includes also mixtures of fine silty soil and up to 64% of sand and gravel retained on the 75-μm (No. 200) sieve (**5.2.1**). These predominantly silty soils are quite common in occurrence. Their texture varies from sandy loams to silty and clayey loams. With the proper amount of moisture present, they may perform well as a pavement component. However, they frequently have an affinity for water and will swell and lose much of their stability unless properly compacted and drained. Moreover, they are subject to frost heave. These soils do not drain readily and may absorb water by capillary action with resulting loss in strength. The silty loams are often difficult to compact properly. Careful field control of moisture content and pneumatic tired rollers are normally required for proper compaction.

Group A-5: The typical material of this group is similar to that described under Group A-4, except that it is usually of diatomaceous or micaceous character and may be highly elastic as indicated by the high liquid limit (**5.2.2**). These soils do not occur as widely as the A-4 soils. They are normally elastic or resilient in both the damp and semi-dry conditions. They are subject to frost heave, erosion, and loss of stability if not properly drained. Since these soils do not drain readily, they may absorb water by capillary action with resulting loss in strength. Careful control of moisture content is normally required for proper compaction.

Group A-6: The typical material of this group is plastic clay soil usually having 75% or more passing the 75-μm (No. 200) sieve. The group also includes mixtures of fine clayey soil and up to 64% of sand and gravel retained on the 75-μm (No. 200) sieve. Materials of this group usually have high-volume change between wet and dry states (**5.2.3**). These soils are quite common in occurrence and are widely used in fills. When moisture content is properly controlled, they compact quite readily with either a sheepsfoot or pneumatic-tired roller. They have high dry strength but lose much of this strength upon absorbing water. The A-6 soils will compress when wet and shrink and swell with changes in moisture content. When placed in the shoulders adjacent to the pavement, they tend to shrink away from the pavement edge upon drying and thereby provide an access route to the underside of the pavement for surface water. The A-6 soils do not drain readily and may absorb water by capillary action with resulting loss in strength.

Group A-7: The typical materials and problems of this group are similar to those described under Group A-6, except that they have the high liquid limits characteristic of the A-5 group and may be elastic as well as subject to high-volume change **(5.2.4)**.

Subgroup A-7-5: Includes those materials with moderate Plasticity Indexes in relation to Liquid Limit and which may be highly elastic as well as subject to considerable volume change **(5.2.4.1)**.

Subgroup A-7-6: Includes those materials with high Plasticity Indexes in relation to Liquid Limit and which are subject to extremely high volume change **(5.2.4.2)**.

Highly organic soils such as peat or muck are not included in this classification. Because of their many undesirable properties, their use should be avoided, if possible, in all types of construction.

4.1.10 USDA Soil Textural Classification Chart

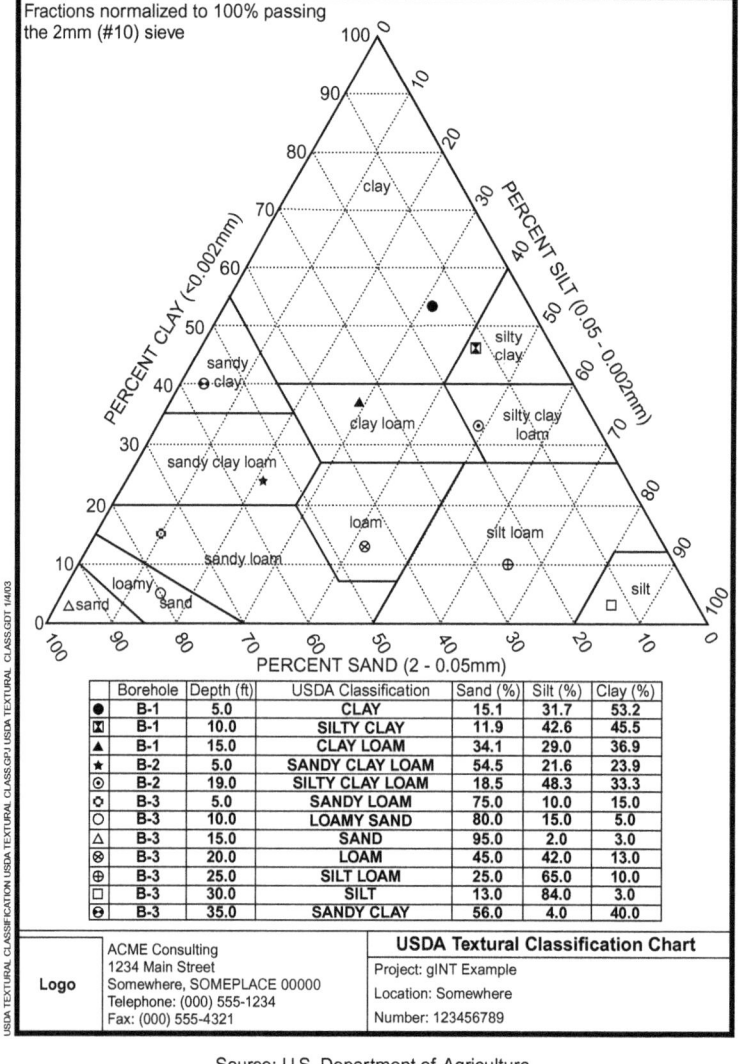

Borehole	Depth (ft)	USDA Classification	Sand (%)	Silt (%)	Clay (%)
B-1	5.0	CLAY	15.1	31.7	53.2
B-1	10.0	SILTY CLAY	11.9	42.6	45.5
B-1	15.0	CLAY LOAM	34.1	29.0	36.9
B-2	5.0	SANDY CLAY LOAM	54.5	21.6	23.9
B-2	19.0	SILTY CLAY LOAM	18.5	48.3	33.3
B-3	5.0	SANDY LOAM	75.0	10.0	15.0
B-3	10.0	LOAMY SAND	80.0	15.0	5.0
B-3	15.0	SAND	95.0	2.0	3.0
B-3	20.0	LOAM	45.0	42.0	13.0
B-3	25.0	SILT LOAM	25.0	65.0	10.0
B-3	30.0	SILT	13.0	84.0	3.0
B-3	35.0	SANDY CLAY	56.0	4.0	40.0

ACME Consulting
1234 Main Street
Somewhere, SOMEPLACE 00000
Telephone: (000) 555-1234
Fax: (000) 555-4321

USDA Textural Classification Chart
Project: gINT Example
Location: Somewhere
Number: 123456789

Source: U.S. Department of Agriculture

4.2.0 Soil Taxonomy—Formative Elements and Names of Soil Suborders

Soil Formation and Classification

The National Cooperative Soil Survey identifies and maps over 20,000 different kinds of soil in the United States. Most soils are given a name, which generally comes from the locale where the soil was first mapped. Named soils are referred to as soil series.

Soil survey reports include the soil survey maps and the names and descriptions of the soils in a report area. These soil survey reports are published by the National Cooperative Soil Survey and are available to everyone.

Soils are named and classified on the basis of physical and chemical properties in their horizons (layers). "Soil Taxonomy" uses color, texture, structure, and other properties of the surface 2 meters deep to key the soil into a classification system to help people use soil information. This system also provides a common language for scientists.

Soils and their horizons differ from one another, depending on how and when they formed. Soil scientists use five soil factors to explain how soils form and to help them predict where different soils may occur. The scientists also allow for additions and removal of soil material and for activities and changes within the soil that continue each day.

Soil-Forming Factors

Parent material. Few soils weather directly from the underlying rocks. These "residual" soils have the same general chemistry as the original rocks. More commonly, soils form in materials that have moved in from elsewhere. Materials may have moved many miles or only a few feet. Windblown "loess" is common in the Midwest. It buries "glacial till" in many areas. Glacial till is material ground up and moved by a glacier. The material in which soils form is called "parent material." In the lower part of the soils, these materials may be relatively unchanged from when they were deposited by moving water, ice, or wind.

Sediments along rivers have different textures, depending on whether the stream moves quickly or slowly. Fast-moving water leaves gravel, rocks, and sand. Slow-moving water and lakes leave fine-textured material (clay and silt) when sediments in the water settle out.

Climate. Soils vary, depending on the climate. Temperature and moisture amounts cause different patterns of weathering and leaching. Wind redistributes sand and other particles especially in arid regions. The amount, intensity, timing, and kind of precipitation influence soil formation. Seasonal and daily changes in temperature affect moisture effectiveness, biological activity, rates of chemical reactions, and kinds of vegetation.

Topography. Slope and aspect affect the moisture and temperature of soil. Steep slopes facing the sun are warmer, just like the south-facing side of a house. Steep soils may be eroded and lose their topsoil as they form. Thus, they may be thinner than the more nearly level soils that receive deposits from areas upslope. Deeper, darker colored soils may be expected on the bottom land.

Biological factors. Plants, animals, microorganisms, and humans affect soil formation. Animals and microorganisms mix soils and form burrows and pores. Plant roots open channels in the soils. Different types of roots have different effects on soils. Grass roots are "fibrous" near the soil surface and easily decompose, adding organic matter. Taproots open pathways through dense layers. Microorganisms affect chemical exchanges between roots and soil. Humans can mix the soil so extensively that the soil material is again considered parent material.

The native vegetation depends on climate, topography, and biological factors plus many soil factors such as soil density, depth, chemistry, temperature, and moisture. Leaves from plants fall to the surface and decompose on the soil. Organisms decompose these leaves and mix them with the upper part of the soil. Trees and shrubs have large roots that may grow to considerable depths.

Department of Agriculture

Time. Time for all these factors to interact with the soil is also a factor. Over time, soils exhibit features that reflect the other forming factors. Soil formation processes are continuous. Recently deposited material, such as the deposition from a flood, exhibits no features from soil development activities. The previous soil surface and underlying horizons become buried. The time clock resets for these soils. Terraces above the active floodplain, while genetically similar to the floodplain, are older land surfaces and exhibit more development features.

These soil-forming factors continue to affect soils even on "stable" landscapes. Materials are deposited on their surface, and materials are blown or washed away from the surface. Additions, removals, and alterations are slow or rapid, depending on climate, landscape position, and biological activity.

When mapping soils, a soil scientist looks for areas with similar soil-forming factors to find similar soils. The colors, texture, structure, and other properties are described. Soils with the same kind of properties are given taxonomic names. A common soil in the Midwest reflects the temperate, humid climate and native prairie vegetation with a thick, nearly black surface layer. This layer is high in organic matter from decomposing grass. It is called a "mollic epipedon." It is one of several types of surface horizons that we call "epipedons." Soils in the desert commonly have an "ochric" epipedon that is light colored and low in organic matter. Subsurface horizons also are used in soil classification. Many forested areas have a subsurface horizon with an accumulation of clay called an "argillic" horizon.

Soil Orders

Soil taxonomy at the highest hierarchical level identifies 12 soil orders. The names for the orders and taxonomic soil properties relate to Greek, Latin, or other root words that reveal something about the soil. Sixty-four suborders are recognized at the next level of classification. There are about 300 great groups and more than 2400 subgroups. Soils within a subgroup that have similar physical and chemical properties that affect their responses to management and manipulation are families. The soil series is the lowest category in the soil classification system.

Soil Order	Formative Terms	Pronunciation
Alfisols	Alf, meaningless syllable	Pedalfer
Andisols	Modified from ando	Ando
Aridisols	Latin, aridies, dry	Arid
Entisols	Ent, meaningless	Recent
Gelisols	Latin gelare, to freeze	Jell
Histosols	Greek, histos, tissue	Histology
Inceptisols	Latin, incepum, beginning	Inception
Mollisols	Latin, mollis, soft	Mollify
Oxisols	French oxide	Oxide
Spodosols	Greek spodos, wood ash	Odd
Ultisols	Latin ultimus, last	Ultimate
Vertisols	Latin verto, turn	Invert

Maps

The distribution of these soil orders in the United States corresponds with the general patterns of the soil-forming factors across the country. A map of soil orders is useful in understanding broad areas of soils.

Detailed soil maps found in soil survey reports, however, should be used for local decision making. Soil maps are like roadmaps; for a very general overview, a small-scale map in an atlas is helpful, but for finding a location of a house in a city, a large-scale detailed map should be used.

Formative Elements in Names of Soil Suborders

Formative Element	Derivation	Sounds Like	Connotation
Alb	L, *albus*, white	Albino	Presence of albic horizon
Anthr	Modified from Gr. anthropes, human	Anthropology	Modified by humans
Aqu	L. *aqua*, water	Aquifer	Aquic conditions
Ar	L. *Arare*, to plow	Arable	Mixed horizons
Arg	Modified from argillic horizon; L. *argilla*, white clay	Argillite	Presence of argillic horizon
Calc	L. calcis, lime	Calcium	Presence of a calcic horizon
Camb	L. cambiare, to exchange	Am	Presence of a cambic horizon
Cry	G. kryos, icy cold	Cry	Cold
Dur	L. durus, hard	Durable	Presence of a duripan
Fibr	L. *fibra*, fiber	Fibrous	Least decomposed stage
Fluv	L. *fluvius*, river	Fluvial	Flood plain
Fol	L. *folia*, leaf	Foliage	Mass of leaves
Gyps	L. gypsum, gypsum	Gypsum	Presence of a gypsic horizon
Hem	Gr hemi, half	Hemisphere	Intermediate stage of decomposition
Hist	Gr. histos, tissue	Histology	Presence of organic materials
Hum	L. *humus*, earth	Humus	Presence of organic matter
Orth	Gr. orthos, true	Orthodox	The common ones
Per	L. Per, throughout in time	Perennial	Perudic moisture regime
Psamm	Gr. psammos, sand	Sam	Sandy texture
Rend	Modified from Rendzina	End	High carbonate content
Sal	L. base of sal, salt	Saline	Presence of a salic horizon
Sapr	Gr. sapros, rotten	Sap	Most decomposed stage
Torr	L. *torridus*, hot and dry	Or	Torric moisture regime
Turb	L. Turbidis, disturbed	Turbulent	Presence of cryoturbation
Ud	L. *udus*, humid	You	Udic moisture regime
Vitr	L. vitrum, glass	It	Presence of glass
Ust	L. *ustus*, burnt	Combustion	Ustic moisture regime
Xer	Gr. *xeros*, dry	Zero	Xeric moisture regime

Formative Elements in Names of Soil Great Groups

Formative Element	Derivation	Sounds Like	Connotation
Acr	Modified from Gr. *Akros*, at the end	Act	Extreme weathering
Al	Modified from aluminum	Algebra	High aluminum, low iron
Alb	L. *Albus*, white	Albino	An albic horizon
Anhy	Gr. *anydros*, waterless	Anhydrous	Very dry
Anthr	Modified from Gr. *anthropos*, human	Anthropology	An anthropic epipedon
Aqu	L. *aqua*, water	Aquifer	Aquic conditions
Argi	Modified from argillic horizon; L. argilla, *white clay*	Argillite	Presence of an argillic horizon
Calci, calc	L. *calcis*, lime	Calcium	A calcic horizon
Cry	Gr. *kryos*, icy cold	Cry	Cold
Dur	L. *durus*, hard	Durable	A duripan
Dystr, dys	Modified from Gr. *dys*, ill; dystrophic infertile	Distant	Low base saturation
Endo	Gr. *endon, endo*, within	Endothermic	Implying a ground water table
Epi	Gr. *epi*, on, above	Epidermis	Implying a perched water table
Eutr	Modified from Gr. *eu*, good; euthrophic, fertile	You	High base saturation
Ferr	L. *ferrum*, iron	Fair	Presence of iron
Fibr	L. *fibra*, fiber	Fibrous	Least decomposed stage
Fluv	L. *fluvius*, river	Fluvial	Flood plain
Fol	L. *folia*, leaf	Foliage	Mass of leaves
Fragi	Modified from L. *fragilis*, brittle	Fragile	Presence of fragipan
Fragloss	Compound of fra (g) and gloss		See the formative elements "frag" and "gloss"
Fulv	L. *fulvus*, dull brownish yellow	Full	Dark brown color, presence of organic carbon
Glac	L. *glacialis*, icy	Glacier	Ice lenses or wedges
Gyps	L. *gypsum*, gypsum	Gypsum	Presence of gypsic horizon
Gloss	Gr. *glossa*, tongue	Glossary	Presence of glossic horizon
Hal	Gr. *hals*, salt	Halibut	Salty
Hapl	Gr. *haplous*, simple	Haploid	Minimum horizon development
Hem	G. *hemi*, half	Hemisphere	Intermediate stage of decomposition
Hist	Gr. *histos*, tissue	History	Presence of organic materials
Hum	L. *humus*, earth	Humus	Presence of organic matter
Hydr	Gr. *hydo*, water	Hydrophobia	Presence of water
Kand, kan	Modified from kandite	Can	1:1 layer silicate clays

Formative Elements in Names of Soil Great Groups—Cont'd

Formative Element	Derivation	Sounds Like	Connotation
Luv	Gr. *louo,* to wash	Ablution	Illuvial
Melan	Gr. *melasanos,* black	Me + Land	Black, presence of organic carbon
Moll	L. *mollis,* soft	Mollusk	Presence of a mollic epipedon
Natr	Modified from *natrium,* sodium	Date	Presence of natric horizon
Pale	Gr. *paleos,* old	Paleontology	Excessive development
Petr	Gr. comb. form of *petra,* rock	Petrified	A cemented horizon
Plac	Gr. base of *plax,* flat stone	Placard	Presence of thin pan
Plagg	Modified from Ger. *plaggen,* sod	Awe	Presence of plaggen epipedon
Plinth	Gr. *plinthos,* brick	In	Presence of plinthite
Psamm	Gr. *psammos,* sand	Sam	Sandy texture
Quartz	Ger. *quarz,* quartz	Quarter	High quartz content
Rhod	Gr. base of *rhodon,* rose	Rhododendron	Dark red color
Sal	L. base of *sal,* salt	Saline	Presence of salic horizon
Sapr	Gr. *saprose,* rotten	Sap	Most decomposed stage
Somb	F. *sombre,* dark	Somber	Presence of sombric horizon
Sphagn	Gr. *sphagnos,* bog	Sphagnum	Presence of Sphagnum
Sulf	L. *sulfur,* sulfur	Sulfur	Presence of sulfides or their oxidation products
Torr	L. *torridus,* hot and dry	Torrid	Torric moisture regime
Ud	L. *udus,* humid	You	Udic moisture regime
Umbr	L. *umbra,* shade	Umbrella	Presence of umbric epipedon
Ust	L. *ustus,* burnt	Combustion	Ustic moisture regime
Verm	L. base of *vermes,* worm	Vermilion	Wormy, or mixed by animals
Vitr	L. *vitrum,* glass	It	Presence of glass
Xer	Gr. *xeros,* dry	Zero	Xeric moisture regime

soils.usda.gov/.../formation.html

4.3.0 Calculating Soil Compaction Utilizing Various Methods

Soil Compaction Tests

1) The Sand-Cone Method

The sand-cone method is used to determine in the field the density of compacted soils in earth embankments, road fill, and structure backfill, as well as the density of natural soil deposits, aggregates, soil mixtures, or other similar materials. It is not suitable, however, for soils that are saturated, soft, or friable (crumble easily).

Characteristics of the soil are computed from

Volume of soil, ft³ (m³) = [weight of sand-filling hole, lb (kg)]/[Density of sand, lb/ft³ (kg/m³)]

% Moisture = 100(weight of moist soil—weight of dry soil)/weight of dry soil

Field density, lb/ft³ (kg/m³) = weight of soil, lb (kg)/volume of soil, ft³ (m³)

Dry density = field density/(1 + % moisture/100)

% Compaction = 100 (dry density)/max dry density

Maximum density is found by plotting a density–moisture curve.

2) Load-Bearing Test

One of the earliest methods for evaluating the in situ deformability of coarse-grained soils is the small-scale load-bearing test. Data developed from these tests have been used to provide a scaling factor to express the settlement r of a full-size footing from the settlement r1 of a 1- ft²(0.0929-m²) plate. This factor r/r1 is given as a function of the width B of the full-size bearing plate as

$$r/r1 = (2B/1 + B)^2$$

From an elastic half-space solution, E's can be expressed from results of a plate load test in terms of the ratio of bearing pressure to plate settlement k_v as

$$E's = \frac{K_v (1 - m^2) p/4}{4B/(1 + B)^2}$$

where m represents Poisson's ratio, which is usually considered to range between 0.30 and 0.40. The E's equation assumes that r1 is derived from a rigid, 1-ft(0.3048-m)-diameter circular plate and that B is the equivalent diameter of the bearing area of a full-scale footing. Empirical formulations, such as the r/r1 equation, may be significantly in error because of the limited footing-size range used and the large scatter of the database. Furthermore, consideration is not given to variations in the characteristics and stress history of the bearing soils.

3) California Bearing Ratio

The California bearing ratio (CBR) is often used as a measure of the quality of strength of a soil that underlies a pavement, for determining the thickness of the pavement, its base, and other layers.

$$CBR = F/F_o$$

where

F = force per unit area required to penetrate a soil mass with a 3-in² (1935.6-mm²) circular piston (about 2 in (50.8 mm) in diameter) at the rate of 0.05 in/min (1.27 mm/min);

F_v = force per unit area required for corresponding penetration of a standard material.

By permission: engineering civil.com

Typically, the ratio is determined at 0.10-in. (2.54-mm) penetration, although other penetrations sometimes are used. An excellent base course has a CBR of 100%. A compacted soil may have a CBR of 50%, whereas a weaker soil may have a CBR of 10.

4) Soil Permeability

The coefficient of permeability k is a measure of the rate of flow of water through saturated soil under a given hydraulic gradient i, cm/cm, and is defined in accordance with Darcy's law as

$$V = kiA$$

where V = rate of flow, cm³/s, and A = cross-sectional area of soil conveying flow, cm².

Coefficient k is dependent on the grain-size distribution, void ratio, and soil fabric and typically may vary from as much as 10 cm/s for gravel to less than 10-7 for clays. For typical soil deposits, k for horizontal flow is greater than k for vertical flow, often by an order of magnitude.

4.3.0.1 Soil Testing—Types—Hand, Proctor, Nuclear Density, Sand Cone

Moisture versus Soil Density

The moisture content of the soil is vital to proper compaction. Moisture acts as a lubricant within soil, sliding the particles together. Too little moisture means inadequate compaction—the particles cannot move past each other to achieve density. Too much moisture leaves water-filled voids and subsequently weakens the load-bearing ability. The highest density for most soils is at a certain water content for a given compaction effort. The drier the soil, the more resistant it is to compaction. In a water-saturated state the voids between particles are partially filled with water, creating an apparent cohesion that binds them together. This cohesion increases as the particle size decreases (as in clay-type soils).

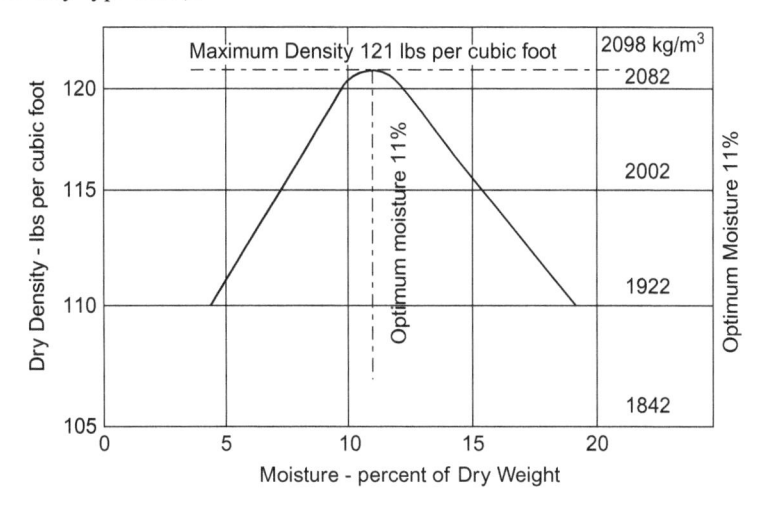

Soil Density Tests

To determine if proper soil compaction is achieved for any specific construction application, several methods were developed. The most prominent by far is soil density.

Why Test?

Soil testing accomplishes the following:

- Measures density of soil for comparing the degree of compaction versus specs
- Measures the effect of moisture on soil density versus specs
- Provides a moisture density curve identifying optimum moisture

Types of Tests

Tests to determine optimum moisture content are done in the laboratory. The most common is the Proctor Test, or Modified Proctor Test. A particular soil needs to have an ideal (or optimum) amount of moisture to achieve maximum density. This is important not only for durability, but will save money because less compaction effort is needed to achieve the desired results.

The Hand Test

A quick method of determining moisture is known as the "Hand Test." Pick up a handful of soil. Squeeze it in your hand. Open your hand. If the soil is powdery and will not retain the shape made by your hand, it is too dry. If it shatters when dropped, it is too wet. It should mold like clay.

Source: concrete-catalog.com

4.3.1 Relative Desirability of Soils as Compacted Fill

RELATIVE DESIRABILITY OF SOILS AS COMPACTED FILL

(NAVFAC DM-7.2, MAY 1982)

Relative Desirability for Various Uses (1=best, 14=least desirable)

* if gravelly
** erosion critical
*** volume change critical
– not appropriate for this type of use

	Group Symbol	Soil Type	Rolled Earth Fill Dams – Homogeneous Embankment	Rolled Earth Fill Dams – Core	Rolled Earth Fill Dams – Shell	Canal Sections – Erosion Resistance	Canal Sections – Compacted Earth Lining	Foundations – Seepage Important	Foundations – Seepage Not Important	Roadways Fills – Frost Heave Not Possible	Roadways Fills – Frost Heave Possible	Roadways – Surfacing
GRAVELS	GW	Well-graded gravels, gravel/sand mixtures, little or no fines	--	--	1	1	--	--	1	1	1	3
GRAVELS	GP	Poorly-graded gravels, gravel/sand mixtures, little or no fines	--	--	2	2	--	--	3	3	3	--
GRAVELS	GM	Silty gravels, poorly-graded gravel/sand/silt mixtures	2	4	--	4	4	1	4	4	9	5
GRAVELS	GC	Clay-like gravels, poorly graded gravel/sand/clay mixtures	1	1	--	3	1	2	6	5	5	1
SANDS	SW	Well-graded sands, gravelly sands, little or no fines	--	--	3*	6	--	--	2	2	2	4
SANDS	SP	Poorly-graded sands, gravelly sands, little or no fines	--	--	4*	7*	--	--	5	6	4	--
SANDS	SM	Silty sands, poorly-graded sand/silt mixtures	4	5	--	8*	5**	3	7	6	10	6
SANDS	SC	Clay-like sands, poorly-graded sand/clay mixtures	3	2	--	5	2	4	8	7	6	2
CLAYS & SILTS LEAN	ML	Inorganic silts and very fine sands, rock flour, silty or clay-like fine sands with slight plasticity	6	6	--	--	6**	6	9	10	11	--
CLAYS & SILTS LEAN	CL	Inorganic clays of low to medium plasticity, gravelly clays, sandy clays, silty clays, lean clays	5	3	--	9	3	5	10	9	7	7
CLAYS & SILTS	OL	Organic silts and organic silt-clays of low plasticity	8	8	--	--	7**	7	11	11	12	--
CLAYS & SILTS FAT	MN	Organic silts, micaceous or diatomaceous fine sandy or silty soils, elastic silts	9	9	--	--	--	8	12	12	13	--
CLAYS & SILTS FAT	CH	Inorganic clays of high plasticity, fat clays	7	7	--	10	8***	9	13	13	8	--
CLAYS & SILTS FAT	OH	Organic clays of medium high plasticity	10	10	--	--	--	10	14	14	14	--

FIGURE 5

4.3.2 Calculating the Bearing Capacity of Soils

Bearing Capacity of Soils

The approximate ultimate bearing capacity under a long footing at the surface of a soil is given by Prandtl's equation:

$$q_n = \left(\frac{c}{\tan \phi}\right) + \frac{1}{2}\gamma_{dry} b \sqrt{K_p}(K_p e^{\pi \tan \phi} - 1)$$

where

q_u = ultimate bearing capacity of soil, lb/ft² (kg/m²)
c = cohesion, lb/ft² (kg/m²)
ϕ = angle of internal friction, degree
γ_{dry} = unit weight of dry soil, lb/ft³ (kg/m³)
b = width of footing, ft (m)
d = depth of footing below surface ft (m)
K_p = coefficient of passive pressure

$$= \left[\tan\left(45 + \frac{\phi}{2}\right)\right]^2$$

$e = 2.718$

For footings below the surface, the ultimate bearing capacity of the soil may be modified by the factor $1 + Cd/b$. The coefficient C is about 2 for cohesionless soils and about 0.3 for cohesive soils. The increase in bearing capacity with depth for cohesive soils is often neglected.

By Permission: lengineeringcivil.com

Typically, the ratio is determined at 0.10-in. (2.54-mm) penetration, although other penetrations sometimes are used. An excellent base course has a CBR of 100%. A compacted soil may have a CBR of 50%, whereas a weaker soil may have a CBR of 10.

4) Soil Permeability

The coefficient of permeability k is a measure of the rate of flow of water through saturated soil under a given hydraulic gradient i, cm/cm, and is defined in accordance with Darcy's law as

$$V = kiA$$

where

V = rate of flow, cm³/s, and A = cross-sectional area of soil conveying flow, cm².

Coefficient k is dependent on the grain-size distribution, void ratio, and soil fabric and typically may vary from as much as 10 cm/s for gravel to less than 10−7 for clays. For typical soil deposits, k for horizontal flow is greater than k for vertical flow, often by an order of magnitude.

Posted in Soil Engineering | 7 Comments

Settlement under Foundations

The approximate relationship between loads on foundations and settlement is

$$q/P = C_1(1 + 2d/b) + C_2/b$$

where

q = load intensity, lb/ft² (kg/m²)
P = settlement, in (mm)
d = depth of foundation below ground surface, ft (m)
b = width of foundation, ft (m)
C_1 = coefficient dependent on internal friction
C_2 = coefficient dependent on cohesion

The coefficients C_1 and C_2 are usually determined by bearing plate loading tests.

Posted in Soil Engineering | 0 Comments

Bearing Capacity of Soils

The approximate ultimate bearing capacity under a long footing at the surface of a soil is given by Prandtl's equation:

$$q_u = \left(\frac{c}{\tan \phi}\right) + \frac{1}{2}\gamma_{dry} \, b\sqrt{K_p}(K_p e^{\pi \tan \phi} - 1)$$

4.3.3 Calculating Vibration Control

Explosive users should take steps to minimize vibration and noise from blasting and protect themselves against damage claims.

Vibrations caused by blasting are propagated with a velocity V, ft/s (m/s), frequency f, Hz, and wavelength L, ft (m), related by

$$L = V/f$$

Velocity v, in/s (mm/s), of the particles disturbed by the vibrations depends on the amplitude of the vibrations A, in (mm):

$$v = 2pfA$$

If the velocity v_1 at a distance D_1 from the explosion is known, the velocity v_2 at a distance D_2 from the explosion may be estimated from

$$v_2 \; ? \; v_1 \, (D_1/D_2)^{1.5}$$

The acceleration a, in/s² (mm/s²), of the particles is given by

$$a = 4 \, p^2 \, f^2 \, A$$

For a charge exploded on the ground surface, the overpressure P, lb/in² (kPa), may be computed from

$$P = 226.62 \, (W^{1/3}/D)^{1.407}$$

where

W = maximum weight of explosives, lb (kg) per delay
D = distance, ft (m), from explosion to exposure.

The sound pressure level, decibels, may be computed from

$$dB = (P/(6.95 \times 10^{-28}))^{0.084}$$

For vibration control, blasting should be controlled with the scaled-distance formula:

$$V = H(D/\ddot{O}W)^{-b}$$

where

b = constant (varies for each site)
H = constant (varies for each site).

Distance to exposure, ft (m), divided by the square root of maximum pounds (kg) per delay, is known as scaled distance.

Most courts have accepted the fact that a particle velocity not exceeding 2 in/s (50.8 mm/s) does not damage any part of any structure. This implies that, for this velocity, vibration damage is unlikely at scaled distances larger than 8

4.3.4 Calculating Earth-Moving Equipment Production

Production is measured in terms of tons or bank cubic yards (cubic meters) of material a machine excavates and discharges, under given job conditions, in 1 h.

Production, bank yd^3/h (m^3/h) = load, yd^3 (m^3) × trips per hour

Trips per hour = working time, min/h/cycle time, min

The load, or amount of material a machine carries, can be determined by weighing or estimating the volume. Payload estimating involves determining the bank cubic yards (cubic meters) being carried, whereas the excavated material expands when loaded into the machine. For determination of bank cubic yards (cubic meters) from loose volume, the amount of swell or the load factor must be known.

Weighing is the most accurate method of determining the actual load. This is normally done by weighing one wheel or axle at a time with portable scales, adding the wheel or axle weights, and subtracting the weight empty. To reduce error, the machine should be relatively level. Enough loads should be weighed to provide a good average:

Bank yd^3 = weight of load, lb(kg)/density of material, lb/bank yd^3 (kg/m^3)

Equipment Required

To determine the number of scrapers needed on a job, required production must first be computed:

Production required, yd^3/h (m^3/h) = quantity, bank yd^3 (m^3)/working time, h
No. of scrapers needed = production required, yd^3/h (m^3/h)/production per unit, yd^3/h (m^3/h)
No. of scrapers a pusher can load = scraper cycle time, min/pusher cycle time, min

Because speeds and distances may vary on haul and return, haul and return times are estimated separately.

Variable time, min = (haul distance, ft/88×speed, mi/h) + (return distance, ft/88×speed, mi/h)

Or

= (haul distance, m/16.7 × speed, km/h) + (return distance, m/16.7 × speed, km/h)

Haul speed may be obtained from the equipment specification sheet when the drawbar pull required is known.

Posted in Soil Engineering | 0 Comments

Earth Quantities Hauled

When soils are excavated, they increase in volume, or swell, because of an increase in voids:

$$V_b = V_b\, L = (100/(100 + \%\ \text{swell}))\, V_L$$

where

V_b = original volume, yd3 (m3), or bank yards
V_L = loaded volume, yd3 (m3), or loose yards
L = load factor

When soils are compacted, they decrease in volume:

$$V_c = V_b\, S$$

where

V_c = compacted volume, yd^3 (m^3)
S = shrinkage factor.

Bank yards moved by a hauling unit equals weight of load, lb (kg), divided by density of the material in place, lb (kg), per bank yard (m^3).

Posted in Soil Engineering | 0 Comments

Formulas for Earth Moving

External forces offer rolling resistance to the motion of wheeled vehicles, such as tractors and scrapers. The engine has to supply power to overcome this resistance; the greater the resistance is, the more power needed to move a load. Rolling resistance depends on the weight on the wheels and the tire penetration into the ground:

$$R = R_f\, W + R_p\, PW$$

where

R = rolling resistance, lb (N)
R_f = rolling-resistance factor, lb/ton (N/tonne)
W = weight on wheels, ton (tonne)
R_p = tire-penetration factor, lb/ton in (N/tonne mm) penetration
p = tire penetration, in (mm)
R_f usually is taken as 40 lb/ton (or 2% lb/lb) (173 N/t) and R_p as 30 lb/ton in (1.5% lb/lb in) (3288 N/t mm).

Hence, the above equation can be written as

$$R = (2\% + 1.5\,\%\ p)\, W' = R'\, W$$

where

W' = weight on wheels, lb(N)
$R' = 2\% + 1.5\%p$.

Additional power is required to overcome rolling resistance on a slope. Grade resistance also is proportional to weight:

$$G = R_g\, s\, W$$

where

G = grade resistance, lb(N)
R_g = grade-resistance factor = 20 lb/ton (86.3 N/t) = 1% lb/lb (N/N)
s = percent grade, positive for uphill motion. Negative for downhill

Thus, the total road resistance is the algebraic sum of the rolling and grade resistances, or the total pull, lb (N), required:

$$T = (R' + R_g s) W' = (2\% + 1.5\%p + 1\%s)W'$$

In addition, an allowance may have to be made for loss of power with altitude. If so, allow 3% pull loss for each 1000 ft (305 m) above 2500 ft (762 m).

Usable pull P depends on the weight W on the drivers:

$$P = fW$$

where

f = coefficient of traction.

4.3.5 Calculating Production of Roller-Type Compaction Equipment

A wide variety of equipment is used to obtain compaction in the field. Sheepsfoot rollers generally are used on soils that contain high percentages of clay. Vibrating rollers are used on more granular soils.

To determine maximum depth of lift, make a test fill. In the process, the most suitable equipment and pressure to be applied, lb/in² (kPa), for ground contact also can be determined. Equipment selected should be able to produce desired compaction with four to eight passes. Desirable speed of rolling also can be determined.

Average speeds, mi/h (km/h), under normal conditions, are given in the table below.

Type	mi/h	km/h
Grid rollers	12	19.3
Sheepsfoot rollers	3	4.8
Tamping rollers	10	16.1
Pneumatic rollers	8	12.8

Compaction production can be computed from

$$\text{yd}^3/\text{h} \ (\text{m}^3/\text{h}) = 16WSLFE/P$$

where

W = width of roller, ft (m)
S = roller speed, mi/h (km/h)
L = lift thickness, in (mm)
F = ratio of pay yd³ (m³) to loose yd³ (m³)
E = efficiency factor (allows for time losses, such as those due to turns); 0.90, excellent; 0.80, average; 0.75, poor
P = number of passes

4.3.6 Compaction Equipment Types—Applications and Illustrations

Equipment Types

Rammers

Rammers deliver a high-impact force (high amplitude), making them an excellent choice for cohesive and semicohesive soils. Frequency range is 500 to 750 blows per minute. Rammers get compaction force from a small gasoline or diesel engine powering a large piston set with two sets of springs. The rammer is inclined at a forward angle to allow forward travel as the machine jumps. Rammers cover three types of compaction: impact, vibration, and kneading.

Vibratory Plates

Vibratory plates are low amplitude and high frequency, designed to compact granular soils and asphalt. Gasoline or diesel engines drive one or two eccentric weights at a high speed to develop compaction force. The resulting vibrations cause forward motion. The engine and handle are vibration-isolated from the vibrating plate. The heavier the plate, the more compaction force it generates. Frequency range is usually 2500 vpm to 6000 vpm. Plates used for asphalt have a water tank and sprinkler system to prevent asphalt from sticking to the bottom of the baseplate. Vibration is the one principal compaction effect.

Reversible Vibratory Plates

In addition to some of the standard vibratory plate features, reversible plates have two eccentric weights that allow smooth transition for forward or reverse travel,

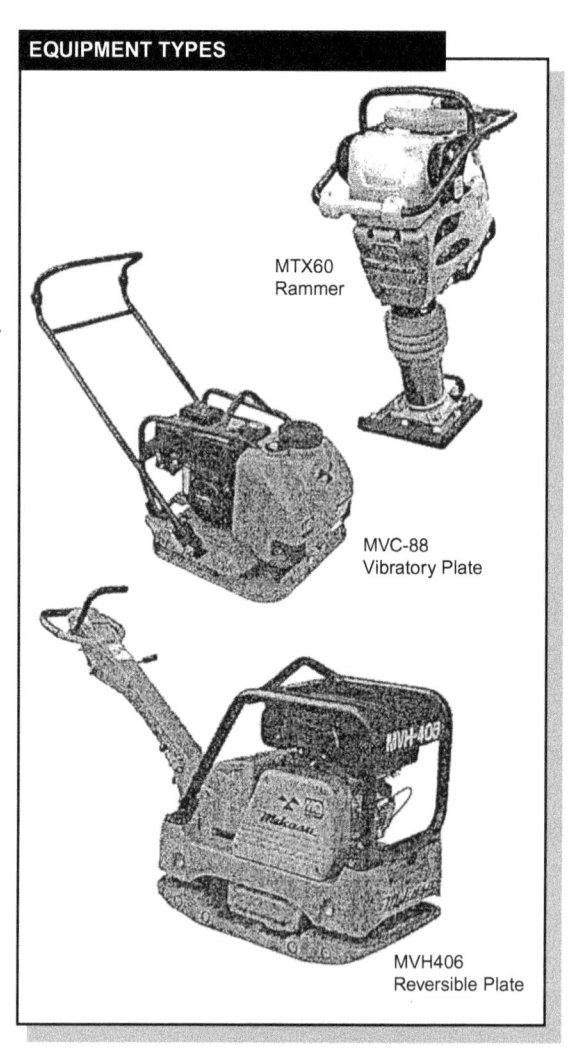

EQUIPMENT TYPES

MTX60 Rammer

MVC-88 Vibratory Plate

MVH406 Reversible Plate

EQUIPMENT APPLICATIONS

	Granular Soils	Sand and Clay	Cohesive Clay	Asphalt
Rammers		B	A	
Vibratory Plates	A	B		A
Reversible Plates	B	A	C	C
Vibratory Rollers	B	A	C	A
Rammax Rollers	C	A	A	

A—Provides optimum performance for most applications.
B—Provides acceptable performance for most applications.
C—Limited performance for most applications. Testing required.

* Chart is provided as a guideline only. Jobsite variables can affect machine performance.

plus increased compaction force as the result of dual weights. Due to their weight and force, reversible plates are ideal for semicohesive soils.

A reversible is possibly the best compaction buy dollar for dollar. Unlike standard plates, the reversible's forward travel may be stopped, and the machine will maintain its force for "spot" compaction.

Rollers

Rollers are available in several categories: walk-behind and ride-on, which are available as smooth drum, padded drum, and rubber-tired models; and are further divided into static and vibratory subcategories.

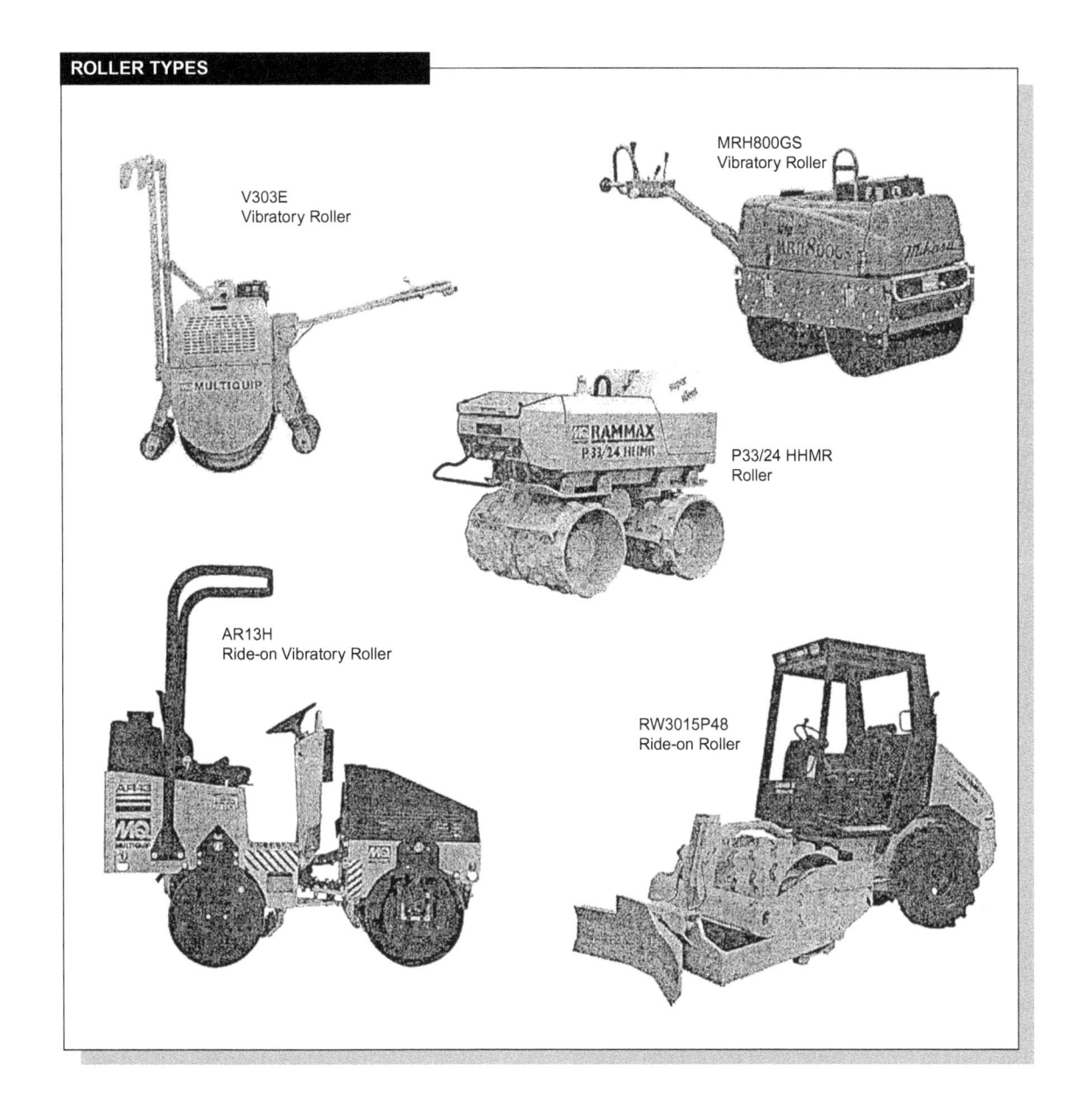

ROLLER TYPES

V303E
Vibratory Roller

MRH800GS
Vibratory Roller

P33/24 HHMR
Roller

AR13H
Ride-on Vibratory Roller

RW3015P48
Ride-on Roller

Walk-behind

Smooth

A popular design for many years, smooth-drum machines are ideal for both soil and asphalt. Dual steel drums are mounted on a rigid frame and powered by gasoline or diesel engines. Steering is done by manually turning the machine handle.

Frequency is around 4000 vpm, and amplitudes range from .018 to .020. Vibration is provided by eccentric shafts placed in the drums or mounted on the frame.

Padded

Padded rollers are also known as trench rollers owing to their effective use in trenches and excavations. These machines feature hydraulic or hydrostatic steering and operation. Powered by diesel engines, trench rollers are built to withstand the rigors of confined compaction. Trench rollers are either skid-steer or equipped with articulated steering. Operation can be by manual or remote control. Large eccentric units provide high-impact force and high amplitude (for rollers) that are appropriate for cohesive soils. The drum pads provide a kneading action on soil. Use these machines for high productivity.

Ride-on

Configured as static steel-wheel rollers, ride-ons are used primarily for asphalt surface sealing and finishing work in the larger (8 to 15 ton) range. Small ride-on units are used for patch jobs with thin lifts.

The trend is toward vibratory rollers. Tandem vibratory rollers are usually found with drum widths of 30″ up to 110″, with the most common being 48″.

Suitable for soil, sub-base, and asphalt compaction, tandem rollers use the dynamic force of eccentric vibrator assemblies for high-production work. Single-drum machines feature a single vibrating drum with pneumatic drive wheels. The drum is available as smooth for sub-base or rock fill, or padded for soil compaction. In addition, a ride-on version of the pad foot trench roller is available for very high productivity in confined areas, with either manual or remote control operation.

Rubber-tire

These rollers are equipped with 7 to 11 pneumatic tires, with the front and rear tires overlapping. A static roller by nature, compaction force is altered by the addition or removal of weight added as ballast in the form of water or sand. Weight ranges vary from 10 to 35 tons. The compaction effort is pressure and kneading, primarily with asphalt finish rolling. Tire pressures on some machines can be decreased while rolling to adjust ground contact pressure for different job conditions.

Safety and General Guidelines

As with all construction equipment, many safety practices should be followed while using compaction equipment. While this handbook is not designed to cover all aspects of job site safety, we wish to mention some of the more obvious items in regard to compaction equipment. Ideally, equipment operators should familiarize themselves with all of their company's safety regulations, as well as any OSHA, state agency, or local agency regulations pertaining to job safety. Basic personal protection, consisting of durable work gloves, eye protection, ear protection, and approved hard hat and work clothes, should be standard issue on any job and available for immediate use.

In the case of walk-behind compaction equipment, additional toe protection devices should be available, depending on applicable regulations. All personnel operating powered compaction equipment should read all operating and safety instructions for each piece of equipment. In addition, training should be provided so that the operator is aware of all aspects of operation.

No minors should be allowed to operate construction equipment. No operator should run construction equipment when under the influence of medication, illegal drugs, or alcohol. Serious injury or death could occur as a result of improper use or neglect of safety practices and attitudes. This applies to both the new worker and the seasoned professional.

Shoring

Trench work brings a new set of safety practices and regulations for the compaction equipment operator. This section does not intend to cover the regulations pertaining to trench safety (OSHA Part 1926, Subpart P). The operator should have knowledge of what is required *before* compacting in a trench or confined area. Be certain a "competent person" (as defined by OSHA in Part 1926.650 revised July 1, 1998) has inspected the trench and follows the OSHA guidelines for inspection during the duration of the job. Besides the obvious danger of a trench cave-in, the worker must also be protected from falling objects. Unshored (or shored) trenches can be compacted with the use of remote control compaction equipment. This allows the operator to stay outside the trench while operating the equipment.

Safety first!

4.3.6.1 List of Compaction Measuring Devices by Type and Manufacturer

	Device	Distributor/Manufacturer
1	Nuclear Density Gauge—[Troxler Model 3440]	Troxler Electronic Laboratories Contact: Michael Dixon 1430 Brook Dr. Downers Grove, IL 60515 Phone: 630-261-9304
2	Sand-Cone Density Apparatus	Humboldt Mfg. Co. 7300 West Agatite Ave. Norridge, IL 60706 Phone: 800-544-7220
3	Soil Compaction Supervisor [SCS]	MBW Incorporated Contact: Frank Multerer P.O. Box 440 Slinger, WI 53086 Phone: 800-678-5237
4	Dynamic Cone Penetrometer [Utility DCP]	SGS Manufacturing Contact: Sandy Golgart 4391 Westgrove Dr. Addison, TX 75001 Phone: 800-526-0747
5	Dynamic Cone Penetrometer [Standard DCP]	Kessler Instruments, Inc. 160 Hicks St. Westbury, NY 11590 Phone: 516-334-4063
6	Geogauge	Humboldt Mfg. Co. 7300 West Agatite Ave. Norridge, IL 60706 Phone: 800-544-7220
7	Clegg Hammer [10-kg & 20-kg Hammers]	Lafayette Instruments Contact: Paul Williams P.O. Box 5729 Lafayette, IN 47903 Phone: 765-423-1505
8	PANDA	Sol Solution Contact: www.sol-solution.com 115 Old Short Hills Rd., Apt. 306 West Orange, NJ 07052 Phone: 973-243-7237

Source: Prof. K. Reddy, University of Illinois, Chicago

4.3.6.2 Moisture Density Relation—Compaction Test—Proctor and Modified Proctor

Purpose

- This laboratory test is performed to determine the relationship between the moisture content and the dry density of a soil for a specified compactive effort. The compactive effort is the amount of mechanical energy that is applied to the soil mass. Several different methods are used to compact soil in the field; some examples include tamping, kneading, vibration, and static load compaction. This laboratory will employ the tamping or impact compaction method using the type of equipment and methodology developed by R. R. Proctor in 1933. Therefore, the test is also known as the Proctor test.
- Two types of compaction tests are routinely performed: (1) the Standard Proctor Test, and (2) the Modified Proctor Test. Each of these tests can be performed by three different methods as outlined in the attached table. In the Standard Proctor Test, the soil is compacted by a 5.5-lb hammer falling a distance of 1 foot into a soil-filled mold. The mold is filled with three equal layers of soil, and each layer is subjected to 25 drops of the hammer. The Modified Proctor Test is identical to the Standard Proctor Test except that it employs a 10-lb hammer falling a distance of 18 inches and uses five equal layers of soil instead of three. Two types of compaction molds are used for testing. The smaller type is 4 in. in diameter and has a volume of about $1/30$ ft^3 (944 cm^3), and the larger type is 6 in. in diameter and has a volume of about $1/13.333$ ft^3 (2123 cm^3). If the larger mold is used, each soil layer must receive 56 blows instead of 25 (see table).

Source: Prof. Krishna Reddy, University of Illinois, Chicago

Alternative Proctor Test Methods

	Standard Proctor ASTM 698			Modified Proctor ASTM 1557		
	Method A	Method B	Method C	Method A	Method B	Method C
Material	≤20% Retained on No.4 Sieve	>20% Retained on No.4 ≤20% Retained on 3/8″ Sieve	>20% Retained on No.3/8″ <30% Retained on 3/4″ Sieve	≤20% Retained on No.4 Sieve	>20% Retained on No.4 ≤20% Retained on 3/8″ Sieve	>20% Retained on No.3/8″ <30% Retained on 3/4″ Sieve
For test sample, use soil passing	Sieve No.4	3/8″ Sieve	¾″ Sieve	Sieve No.4	3/8″ Sieve	¾″ Sieve
Mold	4″ DIA	4″ DIA	6″ DIA	4″ DIA	4″ DIA	6″ DIA
No. of Layers	3	3	3	5	5	5
No. of blows/layer	25	25	56	25	25	56

Note: Volume of 4″ diameter mold = 944 cm^3, Volume of 6″ diameter mold = 2123 cm^3 (verify these values prior to testing)

4.4.0 Calculating the Maximum Dry Density and Optimum Moisture Content of Soil

This test is done to determine the maximum dry density and the optimum moisture content of soil using heavy compaction as per IS: 2720 (Part 8)—983. The apparatus used is

i) Cylindrical metal mould—it should be either of 100 mm dia. and 1000 cc volume or 150 mm dia. and 2250 cc volume and should conform to IS: 10074–1982.
ii) Balances—one of 10 kg capacity, sensitive to 1g and the other of 200 g capacity, sensitive to 0.01g.

iii) Oven—thermostatically controlled with an interior of noncorroding material to maintain temperature between 105 and 110°C.
iv) Steel straightedge—30 cm long.
v) IS Sieves of sizes—4.75 mm, 19 mm, and 37.5 mm.

Preparation of Sample

A representative portion of air-dried soil material, large enough to provide about 6 kg of material passing through a 19 mm IS Sieve (for soils not susceptible to crushing during compaction) or about 15 kg of material passing through a 19 mm IS Sieve (for soils susceptible to crushing during compaction) should be taken. This portion should be sieved through a 19 mm IS Sieve, and the coarse fraction rejected after its proportion of the total sample has been recorded. Aggregations of particles should be broken down so that if the sample was sieved through a 4.75 mm IS Sieve, only separated individual particles would be retained.

Procedure to Determine the Maximum Dry Density and the Optimum Moisture Content of Soil

A) Soil not susceptible to crushing during compaction
   i) A 5 kg sample of air-dried soil passing through the 19 mm IS Sieve should be taken. The sample should be mixed thoroughly with a suitable amount of water depending on the soil type (for sandy and gravelly soil—3 to 5% and for cohesive soil—12 to 16% below the plastic limit). The soil sample should be stored in a sealed container for a minimum period of 16 hrs.
   ii) The mold of 1000 cc capacity with base plate attached should be weighed to the nearest 1g (W_1). The mold should be placed on a solid base, such as a concrete floor or plinth, and the moist soil should be compacted into the mold, with the extension attached, in five layers of approximately equal mass, each layer being given 25 blows from the 4.9 kg rammer dropped from a height of 450 mm above the soil. The blows should be distributed uniformly over the surface of each layer. The amount of soil used should be sufficient to fill the mold, leaving not more than about 6 mm to be struck off when the extension is removed. The extension should be removed, and the compacted soil should be leveled off carefully to the top of the mold by means of the straightedge. The mold and soil should then be weighed to the nearest gram (W_2).
   iii) The compacted soil specimen should be removed from the mold and placed onto the mixing tray. The water content (w) of a representative sample of the specimen should be determined.
   iv) The remaining soil specimen should be broken up, rubbed through 19 mm IS Sieve, and then mixed with the remaining original sample. Suitable increments of water should be added successively and mixed into the sample, and the above operations (i.e., ii to iv) should be repeated for each increment of water added. The total number of determinations made should be at least five, and the moisture content should be such that the optimum moisture content at which the maximum dry density occurs lies within that range.

B) Soil susceptible to crushing during compaction
   Five or more 2.5 kg samples of air-dried soil passing through the 19 mm IS Sieve should be taken. The samples should each be mixed thoroughly with different amounts of water and stored in a sealed container as mentioned in Part A.

C) **Compaction in large size mold**
   For compacting soil containing coarse material up to 37.5 mm size, the 2250 cc mold should be used. A sample weighing about 30 kg and passing through the 37.5 mm IS Sieve is used for the test. Soil is compacted in five layers, each layer being given 55 blows of the 4.9 kg rammer. The rest of the procedure is the same as above.

Reporting of Results

Bulk density Y(gamma) in g/cc of each compacted specimen should be calculated from the equation,

$$Y(gamma) = (W_2 - W_1)/V$$

where

V = volume in cc of the mold.

The dry density Yd in g/cc

$$Yd = 100Y/(100 + w)$$

The dry densities Yd obtained in a series of determinations should be plotted against the corresponding moisture contents, w. A smooth curve should be drawn through the resulting points, and the position of the maximum on the curve should be determined. A sample graph is shown below:

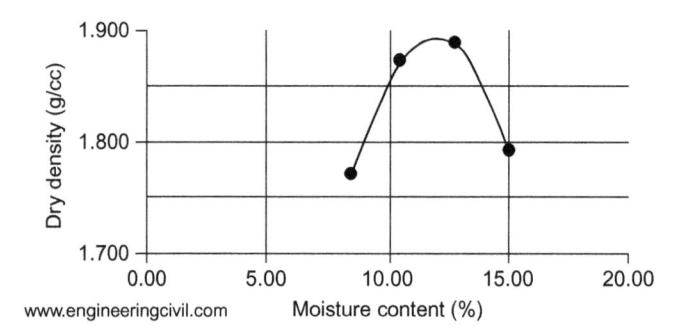

The dry density in g/cc corresponding to the maximum point on the moisture content/dry density curve should be reported as the maximum dry density to the nearest 0.01. The percentage moisture content corresponding to the maximum dry density on the moisture content/dry density curve should be reported as the optimum moisture content and quoted to the nearest 0.2 for values below 5%, to the nearest 0.5 for values from 5 to 10%, and to the nearest whole number for values exceeding 10%.

Water Pressure

The total thrust from water retained behind a wall is

$$P = \tfrac{1}{2} g_o H^2$$

where

H = height of water above bottom of wall, ft (m); and
g_o = unit weight of water, lb/ft^3 (62.4 lb/ft^3 (1001 g/m) for fresh water and 64 lb/ft^3 (1026.7 kg/m^3) for saltwater)

The thrust is applied at a point $H/3$ above the bottom of the wall, and the pressure distribution is triangular, with the maximum pressure of $2P/H$ occurring at the bottom of the wall. Regardless of the slope of the surface behind the wall, the thrust from water is always horizontal.

Section | 4 Site Work

Lateral Pressures in Cohesive Soils

For walls that retain cohesive soils and are free to move a considerable amount over a long period of time, the total thrust from the soil (assuming a level surface) is

$$P = \tfrac{1}{2} g H^2 K_A - 2cH K_A^{1/2}$$

or because highly cohesive soils generally have small angles of internal friction,

$$P = \tfrac{1}{2} g H^2 K_A - 2 c H K_A^{1/2}$$

The thrust is applied at a point somewhat below $H/3$ from the bottom of the wall, and the pressure distribution is approximately triangular.

For walls that retain cohesive soils and are free to move only a small amount or not at all, the total thrust from the soil is

$$P = \tfrac{1}{2} g H^2 K_P$$

because the cohesion would be lost through plastic flow.

Lateral Pressures in Cohesionless Soils

For walls that retain cohesionless soils and are free to move an appreciable amount, the total thrust from the soil is measurable.

Lateral Pressures in Soils

Lateral Pressures in Soils, Forces on Retaining Walls

The Rankine theory of lateral earth pressures, used for estimating approximate values for lateral pressures on retaining walls, assumes that the pressure on the back of a vertical wall is the same as the pressure that would exist on a vertical plane in an infinite soil mass. Friction between the wall and the soil is neglected. The pressure on a wall consists of (1) the lateral pressure of the soil held by the wall, (2) the pressure of the water (if any) behind the wall, and (3) the lateral pressure from any surcharge on the soil behind the wall.

4.4.1 Calculating the In Situ Dry Density of Soil by the Sand Replacement Method

This test is done to determine the in situ dry density of soil by the sand replacement method as per IS: 2720 (Part XXVIII)—1974. The apparatus needed is

i) Sand-pouring cylinder conforming to IS: 2720 (Part XXVIII)—1974
ii) Cylindrical calibrating container conforming to IS: 2720 (Part XXVIII)—1974
iii) Soil-cutting and excavating tools such as a scraper tool, bent spoon
iv) Glass plate—450 mm square and 9 mm thick or larger
v) Metal containers to collect excavated soil
vi) Metal tray—300 mm square and 40 mm deep with a 100 mm hole in the center
vii) Balance, with an accuracy of 1g

Procedure to Determine the In Situ Dry Density of Soil by the Sand Replacement Method

A. Calibration of apparatus
   a) **The method given below should be followed for determining the weight of sand in the cone of the pouring cylinder:**
      i) The pouring cylinder should be filled so that the level of the sand in the cylinder is within about 10 mm of the top. Its total initial weight (W_1) should be maintained constant throughout the tests for which the calibration is used. A volume of sand equivalent to that of the excavated hole in the soil (or equal to that of the calibrating container) should be allowed to run out of the cylinder under gravity. The shutter of the pouring cylinder should then be closed and the cylinder placed on a plain surface, such as a glass plate.
      ii) The shutter of the pouring cylinder should be opened and sand allowed to run out. When no further movement of sand takes place in the cylinder, the shutter should be closed and the cylinder removed carefully.
      iii) The sand that had filled the cone of the pouring cylinder (that is, the sand that is left on the plain surface) should be collected and weighed to the nearest gram.
      iv) These measurements should be repeated at least three times and the mean weight (W_2) taken.
   b) The method described below should be followed for determining the bulk density of the sand (Y_s):
      i) The internal volume (V) in ml of the calibrating container should be determined from the weight of water contained in the container when filled to the brim. The volume may also be calculated from the measured internal dimensions of the container.
      ii) The pouring cylinder should be placed concentrically on the top of the calibrating container after being filled to the constant weight (W_1). The shutter of the pouring cylinder should be closed during the operation. The shutter should be opened and sand allowed to run out. When no further movement of sand takes place in the cylinder, the shutter should be closed. The pouring cylinder should be removed and weighed to the nearest gram.
      iii) These measurements should be repeated at least three times and the mean weight (W_3) taken.
B. **Measurement of soil density**
   The following method should be followed for the measurement of soil density:
   i) A flat area, approximately 450 sq mm of the soil to be tested, should be exposed and trimmed down to a level surface, preferably with the aid of the scraper tool.
   ii) The metal tray with a central hole should be laid on the prepared surface of the soil with the hole over the portion of the soil to be tested. The hole in the soil should then be excavated using the hole in the tray as a pattern, to the depth of the layer to be tested up to a maximum of 150 mm. The excavated soil should be carefully collected, leaving no loose material in the hole and weighed to the nearest gram (W_w). The metal tray should be removed before the pouring cylinder is placed in position over the excavated hole.
   iii) The water content (w) of the excavated soil should be determined as discussed in earlier posts. Alternatively, the whole of the excavated soil should be dried and weighed (W_d).
   iv) The pouring cylinder, filled to the constant weight (W_1), should be so placed that the base of the cylinder covers the hole concentrically. The shutter should then be opened and sand allowed to run out into the hole. The pouring cylinder and the surrounding area should not be vibrated during this period. When no further movement of sand takes place, the shutter should be closed. The cylinder should be removed and weighed to the nearest gram (W_4).

Calculations

i) The weight of sand (W_a) in grams required to fill the calibrating container should be calculated from the formula:

$$W_a = W_1 - W_3 - W_2$$

ii) The bulk density of the sand (γ_s) in kg/m³ should be calculated from the formula:

$$\gamma_s = \frac{W_a}{V} \times 1000$$

iii) The weight of sand (W_b) in grams required to fill the excavated hole should be calculated from the formula:

$$W_b = W_1 - W_4 - W_2$$

iv) The bulk density (γ_b), that is, the weight of the wet soil per cubic meter should be calculated from the formula:

$$\gamma_b = \frac{W_w}{W_b} \times \gamma_a \text{ kg/m}^3$$

v) The dry density (γ_d), that is, the weight of dry soil per cubic meter should be calculated from the formula:

$$\gamma_d = \frac{100\gamma_b}{100 - w} \text{ kg/m}^3$$

$$\gamma_d = \frac{W_d}{W_b} \times \gamma_b \text{ kg/m}^3$$

Reporting of results

The following values should be reported:

i) dry density of soil in kg/m³ to the nearest whole number; also to be calculated and reported in g/cc correct to the second place of decimal
ii) water content of the soil in percent reported to two significant figures

A sample pro forma for the record of the test results is given below.

In Situ Dry Density of Soil by Sand Replacement Method

Calibration of Apparatus

S. No.	Description	Determination
1	Mean weight of sand in cone (of pouring cylinder) (W_2) in g	450
2	Volume of calibrating container (V) in ml	980
3	Weight of sand + Cylinder, before pouring (W_1) in g	11040
4	Mean weight of sand + Cylinder, after pouring (W_3) in g	9120
5	Weight of sand to fill calibrating container ($W_a = W_1 - W_3 - W_2$) in g	1470
6	Bulk density of sand $\gamma_s = \frac{W_a}{V} \times 1000$ kg/m³ =	1500 kg/m³

Source: www.engineeringcivil.com

Measurement of Soil Density

S. No.	Observation and Calculations	Determination No. I	Determination No. II	Determination No. III
1	Weight of wet soil from the hole (W_w) in g	2310	2400	2280
2	Weight of sand + Cylinder, before pouring (W_1) in g	11040	11042	11037
3	Weight of sand + Cylinder, after pouring (W_4) in g	8840	8752	8882
4	Weight of sand in the hole ($W_b = W_1 - W_4 - W_2$) in g	1750	1840	1705
5	Bulk density $\gamma_b = \dfrac{W_w}{W_b} \times \gamma_s$ kg/m^3	1980	1956.5	2005.8
6	Water content (w) in %	18.48	18.81	19.26
7	Dry density $\gamma_d = \dfrac{100\gamma_b}{100 + w}$ kg/m^3	1671.17	1646.75	1681.87
Dry density (Average value)			1667 kg/m^3	

Note: The figures given in the above tables are for illustration purposes only.

4.5.0 Calculating the Percent of Slope

Purpose: To apply measuring and math skills in determining the steepness of a slope.

Season: All

Materials: Yard (meter) stick or measuring tape, a straight stick or board, carpenters level, or flat bottle half filled with colored water

TEKS: 5.1A 5.2A,B,C,D 5.4A,B 5.8B

Procedure: Slope is expressed in percent, meaning the number of units the land falls (or rises) in 100 units of horizontal distance. The higher the percent, the steeper the slope. Example: A slope that drops 10 vertical feet in 100 horizontal feet is a 10% slope (vertical drop/horizontal distance times 100).

A transit is the most accurate instrument for measuring slope.

A transit is a telescopic sighting instrument mounted on a tripod that has adjustable legs and gears for leveling the telescope. Usually a transit is not available for student use. You can measure how steep a slope is using some simple materials.

Place the 50-inch stick horizontally on the ground. Put the level on the 50-inch stick, and move the lower end of the stick up until the bubble shows that the stick is level. Measure the distance from the ground to the end of the level 50-inch stick in inches. To determine percent on slope, you may divide the distance from ground to end of level stick by 50 inches and multiply by 100, or do it the simple way and multiply the distance from ground to end of level stick by two.

Slope is a very important land feature. It often determines whether a piece of land should be used for grass, trees, or cultivated crops. Slope also determines the rate at which water flows downhill. Water flows slowly over a gentle slope and rapidly over a steep one. The steepness of a slope can be evaluated as follows, according to the United States Department of Agriculture's Soil Conservation Service:

Nearly Level (0–2%). Has no limitation on its uses. Any limitations are the result of other factors, such as drainage.

Gently Sloping (3–6%). Desirable for almost any type of development; may have erosion problems; limitations are due mostly to factors other than slope.
Moderately Sloping (6–12%). May have severe erosion problems and has a strong appeal for single-family development.
Strongly Sloping (12–18%). Has severe limitations for all

Source: www.co.bell.tx.us

4.5.0.1 Calculating Grade from a Map

One way to calculate the grade of a hill is with a map that shows the altitudes of locations.
For example, you've measured out a distance of 3 miles (run) with a change in altitude of 396 feet (rise). First, the units must be made consistent, so we convert 3 miles to 15,840 feet.

$$\text{grade} = (\text{rise} \div \text{run}) * 100$$
$$\text{grade} = (396 \div 15{,}840) * 100 = 2.5\%$$

4.5.0.2 Calculating Grade by Measuring the Road Distance

With an altimeter and an odometer, we travel the exact route we measured on the map, and our altimeter indicates a change in altitude of 396 feet which, not surprisingly, is precisely what we had already measured on the map. However, there is a small difference between the 3-mile distance measured on the map and the 3.0009375 miles (15,844.95 feet) we just traveled on the road. The map distance is the true horizontal distance, but the travel distance of 3.0009375 miles is the *slope length* or slope distance. To calculate the true run, we need to use the Pythagorean Theorem.

$$\text{run} = \text{Square Root } (15{,}844.95^2 - 396^2)$$
$$\text{run} = 15{,}840 \text{ feet}$$

Now we can calculate the grade = $(396 \div 15840) * 100 = 2.5\%$

The slope angle *exactly* equals what we previously calculated because instead of using the slope length as the run, we used it to calculate the true horizontal distance.

4.5.0.3 Calculating Grade by Using Slope Distance

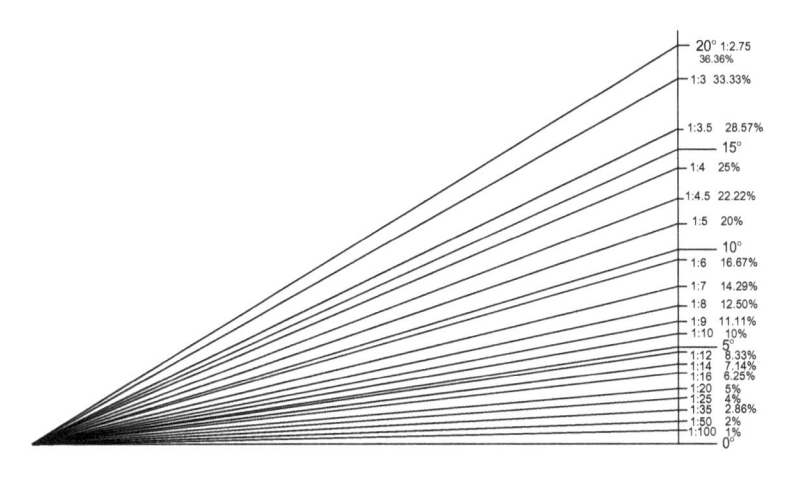

Bicyclists, motorists, carpenters, roofers, and others either need to calculate slope or at least must have some understanding of it.

Slope, tilt, or inclination can be expressed in three ways:

1) As a *ratio* of the rise to the run (for example, 1 in 20)
2) As an *angle* (almost always in degrees)
3) As a *percentage* called the "grade" which is the *(rise ÷ run) * 100*

Of these three ways, slope is expressed as a ratio or a grade much more often than an actual angle, and here's the reason why:

Stating a ratio such as 1 in 20 tells you immediately that for every 20 horizontal units traveled, your altitude increases 1 unit.

Stating this as a percentage, whatever horizontal distance you travel, your altitude increases by 5% of that distance.

Stating this as an angle of 2.8624 degrees doesn't give you much of an idea how the rise compares to the run.

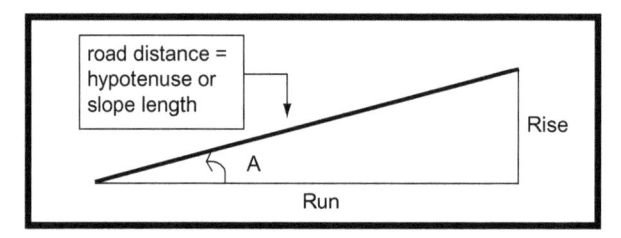

By permisssion: wolf@1728.com:

4.5.0.4 Formulas Showing Grade, Ratio, and Angle Relationships

1) If we know the *ratio* of a road or highway (for example, 1 in 20), then *angle A* = arctangent (rise ÷ run) which equals
   - arctangent (1 ÷ 20) =
   - arctangent (.05) =
   - 2.8624 degrees and the
   - *grade* = (rise ÷ run) * 100 which equals
   - (1 ÷ 20) * 100 =
   - 5%.

2) If we know the *angle* of a road or highway (for example, 3 degrees), then the *ratio* = 1 in (1 ÷ tan (A)), which equals
   - 1 in (1 ÷ tan (3)) =
   - 1 in (1 ÷ .052408) =
   - 1 in 19.081 and the
   - *grade* = (rise ÷ run) * 100 which equals
   - (1 ÷ 19.081) * 100 =
   - 5.2408%

3) If we know the *grade* of a road (for example, 3%), then *angle A* = arctangent (rise ÷ run), which equals
   - arctangent (.03) =
   - 1.7184 degrees and the
   - *ratio* = 1 in (1 ÷ tan(A)) which equals
   - 1 in (1 ÷ tan(1.7184)) =
   - 1 in (1 ÷ .03) =
   - 1 in 33.333

4.5.0.5 Chart Showing Slope Angles—0 Degrees to 80 Degrees

The graph toward the top of the page shows a small range of angles from 0 to 20 degrees. This chart covers a wider range:

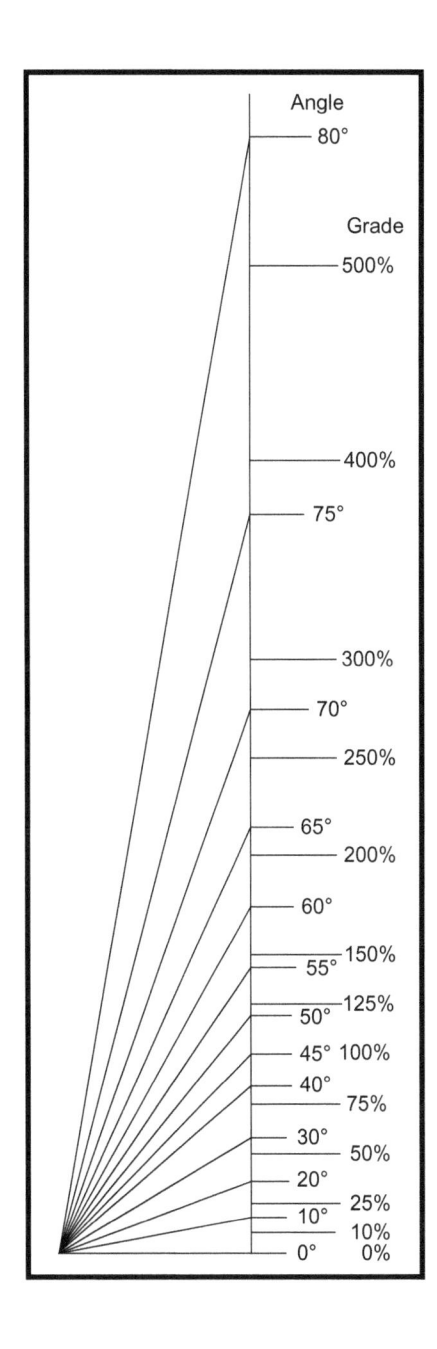

4.5.1 Illustration of Slope Layback

Layback

Layback for excavation is given in units of **run** to **rise.** However, it can be specified in terms of either **variable run** to a **constant rise** of one (1), or it can be given in a **constant run** (1) to **variable rise**.

A layback of zero (0) is simply an excavation with vertical walls (no slope to it).

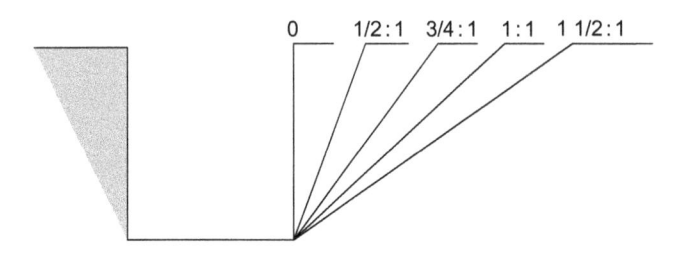

Here are some common laybacks and their equivalents:

1/2 :1 is the same as **1:2**
3/4 :1 is the same as **1:1 1/3**
1:1 is a 45° slope
1-1/2 :1 is the same as **1: 2/3**

By permission: www.mc2-ice.com

4.5.2 Common Stable Slope Ratios for Varying Soil/Rock Conditions

Low-Volume Roads BMPs: 105

TABLE 11.1 Common Stable Slope Ratios for Varying Soil/Rock Conditions

Soil/Rock Condition	Slope Ratio (Hor:Vert)
Most rock	¼:1 to ½:1
Very well cemented soils	¼:1 to ½:1
Most in-place soils	¾:1 to 1:1
Very fractured rock	1:1 to 1 ½:1
Loose coarse granular soils	1 ½:1
Heavy clay soils	2:1 to 3:1
Soft clay rich zones or wet seepage areas	2:1 to 3:1
Fills of most soils	1 ½:1 to 2:1
Fills of hard, angular rock	1 1/3:1
Low cuts and fills (<2-3 m. high)	2:1 or flatter (for revegetation)

4.5.3 Illustrations of Various Cut/Fill Configurations—Typical Fill, Benched Fill, Reinforced Fill

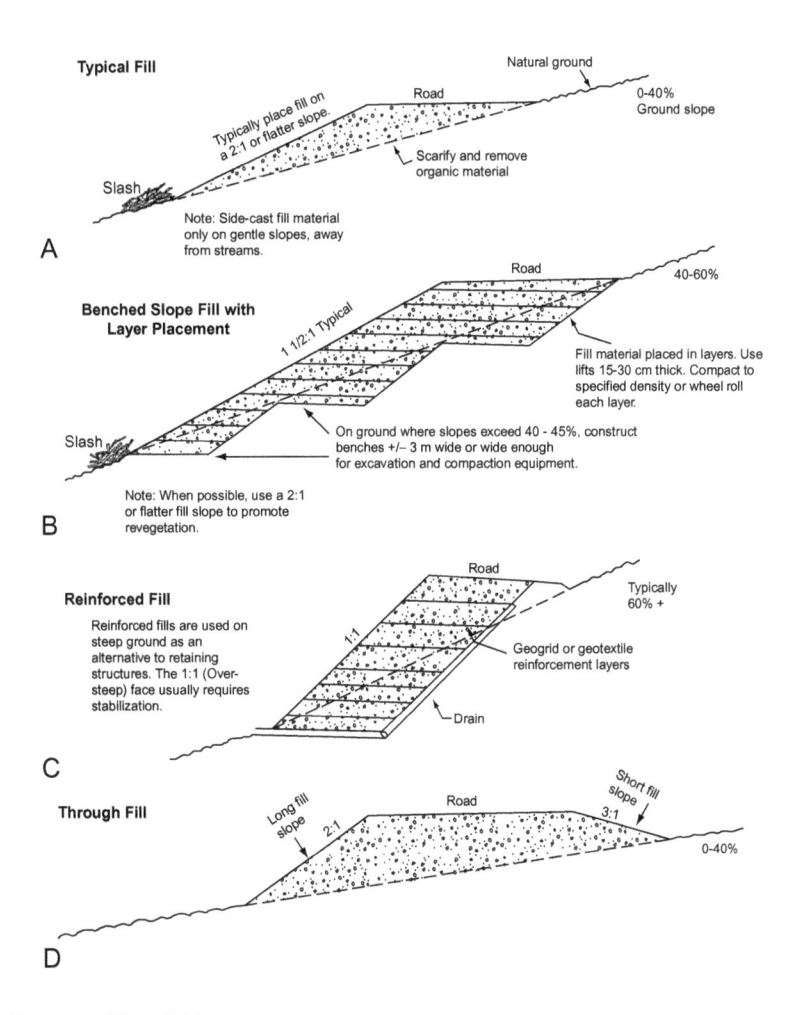

Typical Fill

Typically place fill on a 2:1 or flatter slope.
Natural ground
Road
0-40% Ground slope
Scarify and remove organic material
Slash
Note: Side-cast fill material only on gentle slopes, away from streams.

A

Benched Slope Fill with Layer Placement

1 1/2:1 Typical
Road
40-60%
Fill material placed in layers. Use lifts 15-30 cm thick. Compact to specified density or wheel roll each layer.
Slash
On ground where slopes exceed 40 - 45%, construct benches +/- 3 m wide or wide enough for excavation and compaction equipment.
Note: When possible, use a 2:1 or flatter fill slope to promote revegetation.

B

Reinforced Fill

Reinforced fills are used on steep ground as an alternative to retaining structures. The 1:1 (Oversteep) face usually requires stabilization.
Road
Typically 60% +
1:1
Geogrid or geotextile reinforcement layers
Drain

C

Through Fill

Long fill slope
2:1
Road
Short fill slope
3:1
0-40%

D

Source: U.S. Bureau of Land Management

4.5.4 Illustrationsn of Various Cut/Fill Configurations—Balanced Cut-Fill, Full Cut, Through Cut

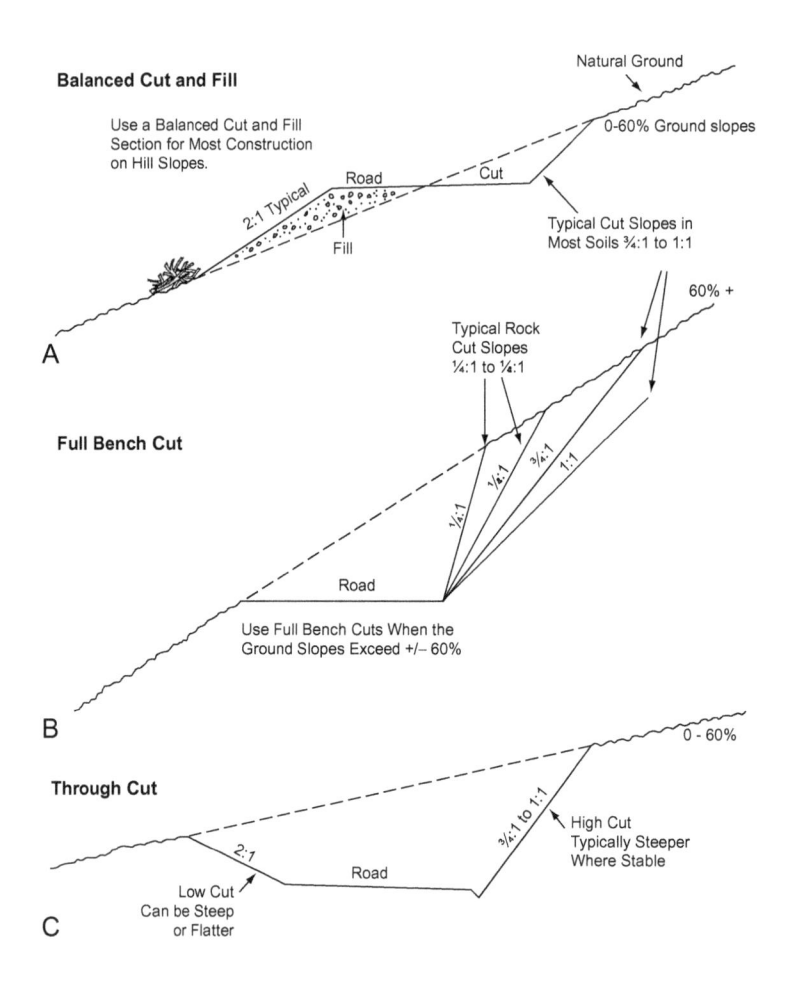

Balanced Cut and Fill

Use a Balanced Cut and Fill Section for Most Construction on Hill Slopes.

2:1 Typical — Road — Fill — Cut — Natural Ground — 0–60% Ground slopes

Typical Cut Slopes in Most Soils ¾:1 to 1:1

A

Full Bench Cut

60% +

Typical Rock Cut Slopes ¼:1 to ¼:1

¼:1, ¼:1, ¾:1, 1:1

Road

Use Full Bench Cuts When the Ground Slopes Exceed +/− 60%

B

Through Cut

0 – 60%

¾:1 to 1:1

High Cut Typically Steeper Where Stable

2:1 — Road

Low Cut Can be Steep or Flatter

C

Source: U.S. Bureau of Land Management

4.5.5 Calculating the Design of Gabion Retaining Walls to 20 Feet (6 Meters) in Height

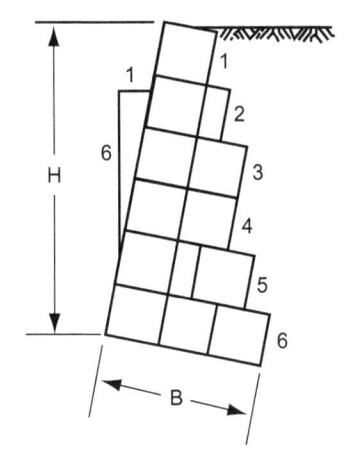

Flat Backfill (smooth face)

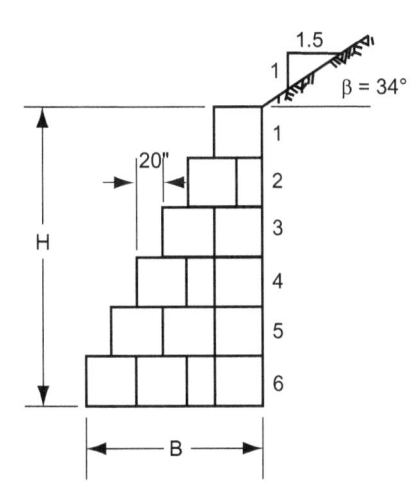

No. of levels	H	B	No. of gabions (per width)
1	3' 3"	3' 3"	1
2	6' 6"	4' 3"	1 1/2
3	9' 9"	5' 3"	2
4	13' 1"	6' 6"	2
5	16' 4"	8' 2"	2 1/2
6	19' 7"	9' 9"	3

Fill at 1 1/2:1 (face with steps)

$\beta = 34°$

No. of levels	H	B	No. of gabions (per width)
1	3' 3"	3' 3"	1
2	6' 6"	4' 11"	1 1/2
3	9' 9"	6' 6"	2
4	13' 1"	8' 2"	2 1/2
5	16' 4"	9' 9"	3
6	19' 7"	11' 5"	3 1/2

Note: Loading conditions are for silty sand to sand and gravel backfill. For finer or clay rich soils, earth pressure on the wall will increase, and the wall base width (B) will have to increase for each height. Backfill weight = 110 pcf (1.8 tons/m³) (1,762 kg/m³)

- Safe against overturning for soils with a minimum bearing capacity of 2 Tons/foot² (19,500 kg/m²)
- For flat or sloping backfills, either a flat or stepped face may be used.

Standard design for Gabion Retaining Structures up to 20 feet in height (6 meters) with flat or sloping backfill.

Source: U.S. Bureau of Land Management

4.5.6 Calculating the Design of Common Types of Retaining Structures

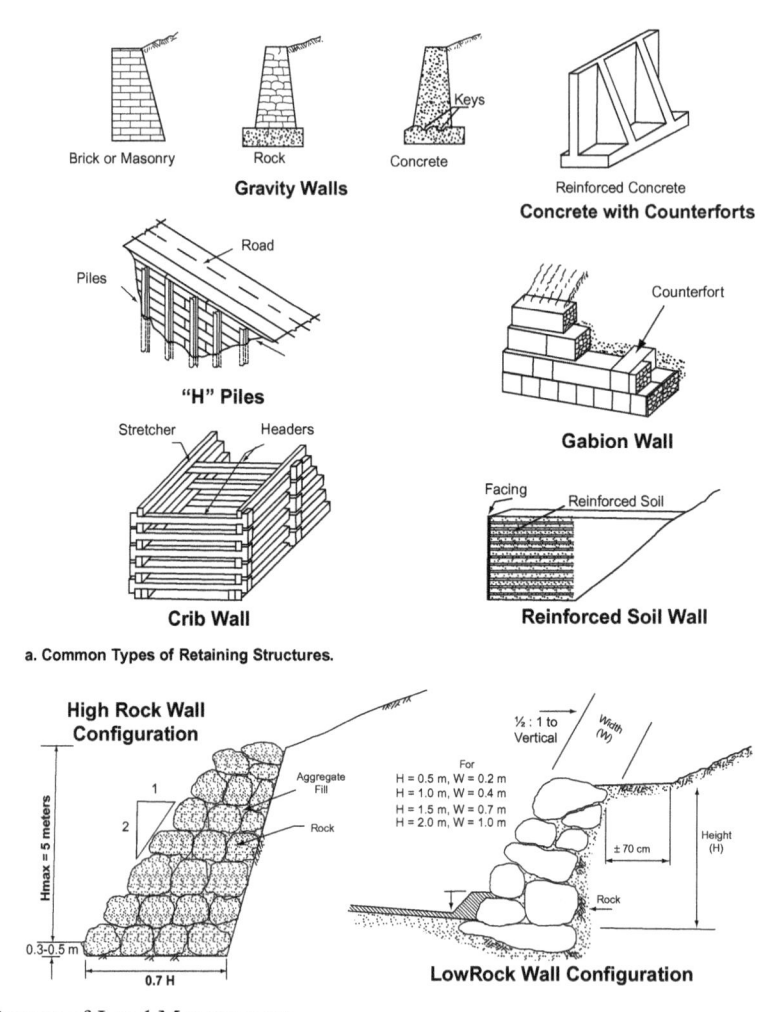

Gravity Walls: Brick or Masonry, Rock, Concrete (Keys)

Reinforced Concrete: Concrete with Counterforts

"H" Piles (Piles, Road)

Gabion Wall (Counterfort)

Crib Wall (Stretcher, Headers)

Reinforced Soil Wall (Facing, Reinforced Soil)

a. Common Types of Retaining Structures.

High Rock Wall Configuration
- Hmax = 5 meters
- 0.3–0.5 m
- 0.7 H
- Aggregate Fill
- Rock
- Slope 1:2

For
H = 0.5 m, W = 0.2 m
H = 1.0 m, W = 0.4 m
H = 1.5 m, W = 0.7 m
H = 2.0 m, W = 1.0 m

LowRock Wall Configuration
- ½ : 1 to Vertical
- Width (W)
- ± 70 cm
- Height (H)
- Rock

Source: U.S. Bureau of Land Management

4.7.0 Material Density Chart—Ashes to Wood

Material	Density (Loose)
Ashes	1100 lb/yd^3
Caliche	2100 lb/yd^3
Cement–Portland	2550 lb/yd^3
Cereals–Wheat, Bulk	1300 lb/yd^3
Clay (natural bed)	2800 lb/yd^3
Clay (Dry Lumps)	1820 lb/yd^3
Clay (Wet Lumps)	2700 lb/yd^3
Clay with Gravel (Dry)	2700 lb/yd^3

Material	Density (Loose)
Clay with Gravel (Wet)	3080 lb/yd³
Coal–Anthracite (Broken)	1850 lb/yd³
Coal–Bituminous (Broken)	1400 lb/yd³
Coke	875 lb/yd³
Earth–Dry, Packed	2550 lb/yd³
Earth–Wet, Excavated	2700 lb/yd³
Earth, Loam	2100 lb/yd³
Garbage–Wet	1350 lb/yd³
Granite–Broken	2800 lb/yd³
Gravel–Dry	2550 lb/yd³
Graveled Sand	3250 lb/yd³
Gravel–Dry (½" to 2")	2850 lb/yd³
Gravel–Wet (½" to 2")	3375 lb/yd³
Gypsum–Crushed	2700 lb/yd³
Kaolin	1730 lb/yd³
Limestone–Broken	2600 lb/yd³
Salt–Coarse	1350 lb/yd³
Sand–Dry	2700 lb/yd³
Sand–Wet	3500 lb/yd³
Sand with gravel–Dry	2900 lb/yd³
Sand with gravel–Wet	3400 lb/yd³
Sandstone–Broken	2550 lb/yd³
Sawdust	550 lb/yd³
Shale	2100 lb/yd³
Silage	865 lb/yd³
Slag–Broken	2950 lb/yd³
Snow–Dry	400 lb/yd³
Snow–Wet	600 lb/yd³
Stone–Crushed	2700 lb/yd³
Sugar–Raw, Refined	1750 lb/yd³
Topsoil	1600 lb/yd³
Wood Chips (Dry)	600 lb/yd³
Wood Chips (Wet)	900 lb/yd³

Actual material density may vary from these standard values.

Source: American Copier Systems

4.8.0 Calculating the Density of Rock, Sand, Till

Measuring the Density of Rock, Sand, Till, etc.

1 Summary

For measuring the density of a variety of geological materials, in particular oddly shaped samples of relatively consolidated material; density is important in the context of cosmogenic-nuclide measurements because the cosmic ray flux is attenuated according to mass depth below the surface; that is, it's necessary to think of depth of overburden or sample thickness in g cm^{-3}, a unit of mass per square area, rather than simply in length. This quantity is generally called mass depth and is equal to $z\rho$, where z is depth below the surface and ρ is the integrated density of overlying material between the surface and depth z. In order to compute this, you need to measure the density of your sample and/or its overburden.

1.1 References

Most of these methods are standard and can be found in any geological or soil science manual.

If you use the data, measurements, or conclusions in this document, however, please cite it as follows:

Balco, G., Stone, J.O., 2003. Measuring the density of rock, sand, till, etc. UW Cosmogenic Nuclide Laboratory, methods and procedures. *http://depts.washington.edu/cosmolab/chem.html.*

The glass bead method is not commonly described in manuals. We got the idea from Sheldrick (see below) and adapted it for our purposes. If you use it, please also cite:

Sheldrick, B.H., (Ed.), 1984. Analytical methods manual 1984. Land Resource Research Institute, Research Branch, Agriculture Canada. *http://sis.agr.gc.ca/cansis/publications/manuals/analytical.html.*

Source: University of Washington, Cosmogenic Nuclide Lab

2 Methods

2.1 A note on collecting samples

The idea of this whole procedure is to determine the density of the material in its natural condition. Thus, the most important thing is to try to get the sample to the measurement without disturbing it too much. If you can, measure the density in the field. For weakly consolidated material, try to collect the sample in some sort of rigid holder so that it won't be crushed during transport. For wet material, seal samples in something watertight so that water will not evaporate before you measure their density. Try to collect multiple samples, and try to make the samples representative. In general, larger samples are better.

2.2 Collecting a known volume in the field—unconsolidated sediment

In principle, it should be easy to measure the density of any material by cutting out a cube of the material, measuring the size of the cube to determine its volume, and then weighing the cube. In practice, it's nearly impossible to cut out a regular cube of any natural geological deposit.

For some unconsolidated sediments, it is possible to collect a known volume in the field. This works well for wet sand and silt. It sometimes works for glaciolacustrine sediment and wet, clay-rich glacial till. It generally doesn't work for gravel, dry sand and silt, or anything cemented. The preferred device is a section of aluminum pipe several inches long. It's helpful if the edges of the ends of the pipe have been beveled on the outside to make a sharp edge. The procedure is as follows:

1. Determine the volume of the pipe section by accurately measuring the inside diameter and length of the pipe. Measure as accurately as possible using calipers. If the pipe was cut by hand, measure the length at several locations around the circumference of the pipe and take the average. Determine the weight of the pipe.
2. In the field, push the pipe into the outcrop face until material starts to extrude out the near end. Hammer it as necessary. Be careful to ensure that there is no air space inside the pipe. Dig the pipe out and carefully slice the protruding sediment away from each end.
3. Weigh the pipe and sediment. It's best to bring the el-cheapo balance into the field with you and do this on site. If this is not possible, wrap the sample by placing something hard over either end (proper pipe caps are best) and then saran-wrapping and taping the whole thing to minimize water loss during transport. When disassembling it in the lab, make sure that all the sediment in the tube gets weighed and doesn't fall out during cap removal, or that you weigh the tube, sediment, and caps together and then the caps separately.
4. Subtract the weight of the pipe (and caps) from the total weight to determine the sample weight. Divide by the pipe volume to get the density.

It's good to do this a few times for each unit. The most important thing is to get the pipe entirely full of sediment.

2.3 Nonporous samples—weighing in water

The second easiest way to measure the density of material is to weigh it in air and then in water. If W_a is the weight of the sample in air and W_w is its weight when immersed in water, then its density is:

$$\rho = \frac{W_a}{W_a - W_w} \tag{1}$$

assuming, of course, that your water is pure H_2O at 25°C.

This is easy to do with most modern analytical balances, which generally have a hook on the bottom connected to the load cell so that one can weigh suspended objects. In our lab, use the Scout balance. Place it over one of the strategically located holes in the lab bench. Obtain a 1-meter length of thin steel wire. Weigh it. Affix the sample to one end of the wire. This may require some creativity. Make a loop in the other end of the wire. Fish it up through the hole and place it on the hook on the balance. Read off the weight. The sample weight in air is this weight less the weight of the wire. Fill a large beaker with DI water. Raise the beaker up from beneath the sample so that the sample is immersed about 1 inch below the surface of the water. Prop the beaker on something of the appropriate height. Read off the weight. The immersed sample weight is this weight less the weight of the wire (relatively little of the wire is immersed). Calculate the density with equation (1). Remember to dry the sample thoroughly before repeating the exercise, so as not to change the dry weight.

Obviously, this method is restricted to samples that are nonporous and will not absorb any of the water in which they are immersed. In practice, this means igneous and metamorphic rocks, and some limestones.

2.4 Oddly shaped and porous, but well-consolidated samples—glass bead method

This method is designed for oddly shaped samples of at-least-somewhat-consolidated material—for example, glacial till, compacted loess, saprolite, cemented sand, sandstone, shale. The procedure is as follows:

1. From the top left drawer in the sediment lab, select a stainless-steel tin slightly larger than your sample. Place it in one of the aluminum baking pans. Record the tare weight W_T and volume V_T of the tin (written on the side of most of the tins).
2. Using the Scout balance, weigh your sample and record the weight W_S.
3. Pour out enough beads to fill the tin about 5 mm deep. Bed a flat side of your sample in the beads. Make sure the sample does not stick out past the rim. Fill the rest of the tin with beads, making sure to tap the tin to settle the beads into all the nooks and crannies of the sample.
4. Overfill the tin with beads, then take the steel spatula and scrape the excess beads away, filling in gaps around the edges, until the surface of the beads is precisely flat with the rim. It's important to do this exactly the same way every time. Make sure to catch all the loose beads in the baking pan.
5. Weigh tin, sample, and packed beads and record the weight W_{TSP}. Dump the beads back into the baking pan. Remove the sample, trying not to break it up too much and get crud in the beads. Pour the beads from the baking pan back into the storage tin.
6. Calculate the weight of packing material, $W_P = W_{TSP} - W_T - W_S$. Calculate the volume of packing. $V_P = W_P/\rho_P$. ρ_P is the density of the packing material (see below). Calculate the sample density $\rho_S = W_S/(V_T - V_P)$.
7. Do the measurement a couple of times to make sure you really did fill all nooks and crannies with the beads the first time.

Notes:

- Metal vessels work best for this. Anything plastic will cause trouble with the beads owing to static electricity.
- We measure the volume of the tins by measuring the weight of water that will fit in them. It's important to make sure that the water surface is close to the actual top of the tin (to which you will grade the beads). It's possible to overfill due to surface tension.
- We use 1 mm glass beads (available from chemical supply companies). 0.5 mm beads also work OK but are a bit messier. We determine the density of packed beads by filling a tin of known volume and measuring the weight of the beads. The density of our 1 mm beads ρ_P is 1.53 g cm^{-3}. It's probably a good idea for each person to independently determine bead density with their own particular scraping/compacting technique.
- Occasionally it's necessary to clean the beads. We do this by sonicating them in water, rinsing thoroughly, and drying. If a lot of large chunks of foreign material build up in the beads, sieving might be needed.
- We also use clean beach sand (mostly quartz) in the 0.5–0.85 mm size range. We prepare it by sieving the sand to this size, sonicating it in water for approximately 1 hr, then rinsing it thoroughly and drying in the oven. The advantage of sand is that it is inexpensive and can be used in sacrificial applications, such as the wet density determination method described below, or when samples are very poorly compacted and are likely to break up during the process and make a mess. Sand does not compact as readily as glass beads (more angular grains), so it's very important to repeatedly tap the tin containing sand and sample on the bench as you are filling it, to make sure the sand is fully compacted. Also, each batch of sand will have a slightly different density that will need to be measured before starting. Our sand has a density ρ_P of 1.45 - 1.47 g · cm^{-3}.
- By repeatedly measuring the density of a variety of samples, including large quartz crystals, whose density, of course, we know exactly, we've determined the accuracy/precision of this technique to be $\pm$ 0.08 g · cm^{-3} for typical materials with densities of 1.2–2.7 g · cm^{-3}.

2.5 Unconsolidated samples—stuffing into a vial

If the sample is completely unconsolidated, for example, dry sand, there is one method remaining. Take a vial of known volume, for example, one of the small metal tins or plastic vials in the drawer. Using your fingers, press the sample into the vial, attempting to duplicate the natural compaction of the material. For most sands, this means squishing it in with some authority to ensure that the sand grains are well packed. Overfill the vial and blade off the excess with the steel spatula. Weigh the sample and vial, subtract the tare weight of the vial, and divide by the volume of the vial.

Despite the ad hoc nature of this technique, it probably does a fairly good job of measuring the density of sandy surficial sediments, because relatively well-sorted sand reaches its maximum compaction quickly and then does not compact any further until buried really deeply, like kilometers. This would also be the only way to measure the density of unconsolidated fluvial gravels, but the vessel would have to be much bigger, in keeping with the grain size of the gravel, to ensure a representative sample.

2.6 Notes on wet versus dry density

In reality, most geological materials are water-saturated below a few meters depth in most environments. In many cases, however, especially when working on drill core, the only samples available to measure overburden density have been dried during storage. Thus, we need some means of converting dry to wet density. The wet density of samples which are collected dry can be approximately measured by the following procedure:

1. For unconsolidated samples, pack a vial of known weight and volume with the sample as described in 2.5. For consolidated samples, pack a tin of known weight and volume with sample and packing material as described in 2.4. The packing material will be inseparable from the sample at the end of this procedure, so we suggest using sand as described in the notes to 2.4 above. Record the relevant weights.
2. Add distilled water to the vial or tin slowly and carefully until everything in the tin is completely saturated. Leave the sample to soak for at least 24 hours to ensure that well-compacted samples become fully saturated. For large samples of glacial till or the like, you may want to let them soak for a couple of days. Periodically add water to keep the sample fully saturated. When you are satisfied that the sample is fully saturated, record the total weight of vessel and contents.
3. For unconsolidated samples, subtract the weight of the vial from the total weight to obtain the wet weight of the sample, then divide by the volume of the vial to determine the wet density.
4. For consolidated samples, you have just measured W_{TSPW}. Calculate the total weight of water added $W_{WT} = W_{TSPW} - W_{TSP}$. Calculate the weight of water incorporated in the packing material $W_{WP} = V_P f_{WP} \rho_W$, where f_{WP} is the water content of saturated packing material (see below) and ρ_W is the density of water, i.e., $1 \text{ g} \cdot \text{cm}^{-3}$. Calculate the wet weight of the sample $W_{WS} = W_S + (W_{WT} - W_{WP})$. The wet density of the sample ρ_{wet} is then W_{WS}/V_S.

Our quartz sand has a saturated water content f_{WP} 0.45 by volume (0.24 by weight), which equates to a wet density of $1.90 \text{ g} \cdot \text{cm}^{-3}$.

This method is not always accurate, primarily because of the tendency of many materials that contain clays to expand when wet, but it is often the only option for determining the wet density of dry material obtained from old, cruddy drill core. It is always better to collect samples at natural moisture conditions in the field.

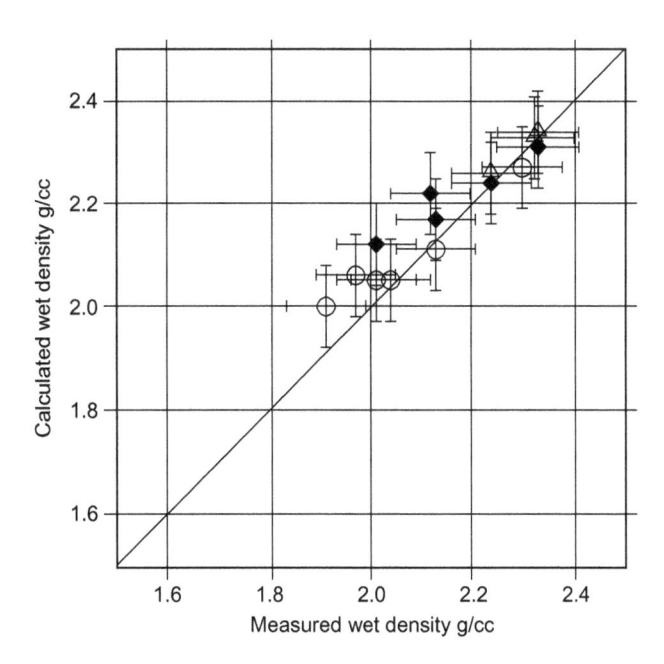

FIGURE 1 Measured wet densities compared with those calculated from dry densities using 2. Circles, wet density determined by saturation of dry sand, and triangles, of dry till, as described in 2.6. Diamonds, samples of glacial till collected wet and then oven-dried. Error-bars reflect what we believe to be measurement precision as described in 2.4.

In addition, this method is somewhat time consuming. A simple alternative is to assume that all of the grains in the sample are composed of quartz ($\rho = 2.65 \text{ g} \cdot \text{cm}^{-3}$) and that all the pore space is filled when wet. Under these assumptions:

$$\rho_{wet} = \left(1 - \frac{\rho_{dry}}{\rho_{quartz}}\right) + \rho_{dry} \qquad (2)$$

We tested this approximation with samples of unconsolidated sand and glacial till that we obtained from dried drillcore and whose wet density we measured as described above, as well as with samples of glacial till that were collected wet and whose density we measured before and after oven-drying (Figure 1). The results show that this approximation seems to be adequate within the resolution of our measurement technique.

Section 5

Calculations Relating to Concrete and Masonry

5.0.0	Standard American Concrete Institute (ACI) and Portland Cement Association (PCA) Divide the Production of Concrete into Seven (7) Basic Components and Ingredients	212	5.5.0	Structural Concrete Components—Calculations to Achieve High-Strength Concrete	233
5.0.1	Chemical Additives Provide Characteristics not Obtainable When Utilizing the Seven Basic Components	212	5.6.0	Lightweight Concrete Mix Design—Calculations Utilizing Perlite	235
5.0.1.1	Slump	213	5.7.0	Set-Retarding Admixtures Delay Hydration of Cement	237
5.0.1.2	Maximum Aggregate Size	214	5.8.0	Calculations for Mixing Small Batches of Concrete	238
5.0.1.3	Mixing Water and Air Content Estimation	214	5.9.0	In-Depth Concrete Inspection Guide as Published by U.S. DOT—FHWA	239
5.0.1.4	Water–Cement Ratio	215	5.10.0	Calculate the Size and Weight of Concrete Reinforcing Bars	243
5.0.1.5	Cement Content	215			
5.0.1.6	Adjustments for Aggregate Moisture	215	5.10.1	Reinforcing Bar Designations–Size and Diameter—U.S. and Metric	244
5.1.0	Portland Cement—ASTM Types	216			
5.1.1	Cement Composition	217	5.10.2	Welded Wire Mesh Designations—U.S and Metric	244
5.1.2	Physical Properties of Portland Cement	218	5.11.0	Embedded Anchor Bolts—Diameter, Length, Hook, and Thread Sizes—Plain Finish	245
5.1.3	Blended Portland Cement	219			
5.1.4	Modified Portland Cement (Expansive Cement)	220	5.11.1	Embedded Anchor Bolts—Diameter, Length, Hook, and Thread Sizes—Galvanized	247
5.2.0	Types of Cement and What They Do	221			
5.3.0	Concrete Compressive Strengths—U.S. and Metric	221	5.12.0	Brick Sizes—Nominal versus Actual Size	248
5.4.0	Sieve Analysis Defining Coarse and Fine Aggregates	222	5.12.1	Diagrams of Modular and Nonmodular Bricks	249
5.4.1	Air-Entraining Admixtures	223	5.12.2	Brick Positions in a Wall	250
5.4.2	Superplasticizers	225	5.12.3	Calculate the Number of Bricks in a Wall	250
5.4.3	Fly Ash	226			
5.4.4	Silica Fume	229	5.12.3.1	Calculate the Number of Bricks for Your Project	251
5.4.5	Ground Granulated Blast-Furnace Slag	231			

5.12.4	Percentages to Add for Various Bond Patterns	251	5.12.11	Horizontal Coursing—Soft and Hard Metric Dimensioning	257
5.12.5	Types of Brick—by Material	252	5.13.0	Profiles and Dimensions of Typical Concrete Masonry Units (CMUs)	258
5.12.6	Chart Reflecting Nominal Size, Joint Thickness, Actual Size—Modular/Nonmodular Bricks	252	5.14.0	Mortar Mixes—ASTM Minimums	260
			5.14.1	Mixture Calculations for Types N, M, S, O, K Mortar	260
5.12.7	Chart Reflecting Nominal size, Joint Thickness, Actual Size—Other Brick Sizes	254	5.14.2	Mixture Calculations for Straight Lime Mortar	262
5.12.8	Nominal Modular Size of Brick and Number of Courses in 16 Inches	254	5.14.3	Mixture Calculations for Glass Block Mortar	263
5.12.9	Calculate Vertical Coursing Height Based upon Number of Units	255	5.14.4	Mixture Calculations for Waterproof Portland Cement	263
5.12.10	Calculate Horizontal Coursing Based upon Number of Units	256	5.15.0	Typical Properties of Colorless Coatings for Brick Masonry	264

5.0.0 Standard American Concrete Institute (ACI) and Portland Cement Association (PCA) Divide the Production of Concrete into Seven (7) Basic Components and Ingredients

The standard ACI mix design procedure can be divided up into seven basic steps:

1. Choice of slump
2. Maximum aggregate size selection
3. Mixing water and air content selection
4. Water–cement ratio
5. Cement content
6. Coarse aggregate content
7. Fine aggregate content

   Source: Washington,edu

5.0.1 Chemical Additives Provide Characteristics not Obtainable When Utilizing the Seven Basic Components

Types of Concrete Admixtures

Posted by Civil Engineer

Chemical concrete admixtures are material in the form of powder or fluids that are added to concrete to give it certain characteristics not obtainable with plain concrete mixes. In normal use, admixture dosages are less than 5% by mass of cement and are added to the concrete at the time of batching/mixing. The most common types of concrete admixtures are:

1. **Accelerators** speed up the hydration (hardening) of the concrete.
2. **Retarders** slow the hydration (hardening) of the concrete and are used in large or difficult pours where partial setting before the pour is complete is undesirable.
3. **Air-entrainers** add and distribute tiny air bubbles in the concrete, which will reduce damage during freeze–thaw cycles, thereby increasing the concrete's durability.
4. **Plasticizers** (water-reducing admixtures) increase the workability of plastic of fresh concrete, allowing it to be placed more easily with less consolidating effort.

5. **Superplasticizers** (high-range water-reducing admixtures) are a class of plasticizers that have fewer deleterious effects when used to significantly increase workability. Alternatively; plasticizers can be used to reduce the water content of concrete (and have been called *water reducers* due to this application) while maintaining workability. This improves its strength and durability characteristics.
6. **Pigments** can be used to change the color of concrete, for aesthetics. Mainly they are ferrous oxides.
7. **Corrosion inhibitors** are used to minimize the corrosion of steel and steel bars in concrete.
8. **Bonding agents** are used to create a bond between old and new concrete.
9. **Pumping aids** improve pumpability, thicken the paste, and reduce dewatering of the paste.

Thus, chemical admixture is one ingredient creating concrete that provides the differentiation of concrete types.
Source: civilengineeringblog.com

5.0.1.1 Slump

The choice of slump is actually a choice of mix workability. Workability can be described as a combination of several different, but related, properties of portland cement concrete (PCC) related to its rheology:

- Ease of mixing
- Ease of placing
- Ease of compaction
- Ease of finishing

Generally, mixes of the stiffest consistency that can still be placed adequately should be used. Typically, slump is specified, but Table 5.14 shows general slump ranges for specific applications. Slump specifications are different for fixed-form paving and slip-form paving. Table 5.15 shows typical and extreme state Department of Transportation (DOT) slump ranges.
Source: washington.edu/PGI/html

TABLE 5.14 Slump Ranges for Specific Applications

	Slump	
Type of Construction	(mm)	(inches)
Reinforced foundation walls and footings	25 – 75	1 – 3
Plain footings, caissons, and substructure walls	25 – 75	1 – 3
Beams and reinforced walls	25 – 100	1 – 4
Building columns	25 – 100	1 – 4
Pavements and slabs	25 – 75	1 – 3
Mass concrete	25 – 50	1 – 2

TABLE 5.15 Typical State DOT Slump Specifications

	Fixed Form		Slip Form	
Specifications	(mm)	(inches)	(mm)	(inches)
Typical	25 – 75	1 – 3	0 – 75	0 – 3
Extremes	as low as 25 as high as 175	as low as 1 as high as 7	as low as 0 as high as 125	as low as 0 as high as 5

5.0.1.2 Maximum Aggregate Size

Maximum aggregate size will affect such PCC parameters as amount of cement paste, workability, and strength. In general, ACI (American Concrete Institute) recommends that maximum aggregate size be limited to one-third of the slab depth and three-fourths of the minimum clear space between reinforcing bars. Aggregate larger than these dimensions may be difficult to consolidate and compact, resulting in a honeycombed structure or large air pockets. Pavement PCC maximum aggregate sizes are on the order of 25 mm (1 in.) to 37.5 mm (1.5 in.).

5.0.1.3 Mxing Water and Air Content Estimation

Slump is dependent on nominal maximum aggregate size, particle shape, aggregate gradation, PCC temperature, amount of entrained air, and certain chemical admixtures. It is not generally affected by the amount of cementitious material. Therefore, ACI provides a table relating nominal maximum aggregate size, air entrainment, and desired slump to the desired mixing water quantity. Table 5.16 is a partial reproduction of ACI Table 6.3.3 (keep in mind that pavement PCC is almost always air-entrained, so air-entrained values are most appropriate). Typically, state agencies specify between about 4 and 8% air by total volume.

Note that the use of water-reducing and/or set-controlling admixtures can substantially reduce the amount of mixing water required to achieve a given slump.

Source: washington.edu/PGI/html

TABLE 5.16 Approximate Mixing Water and Air Content Requirements for Different Slumps and Maximum Aggregate Sizes

Slump	Mixing Water Quantity in kg/m^3 (lb/yd^3) for the listed Nominal Maximum Aggregate Size							
	9.5 mm (0.375 in.)	12.5 mm (0.5 in.)	19 mm (0.75 in.)	25 mm (1 in.)	37.5 mm (1.5 in.)	50 mm (2 in.)	75 mm (3 in.)	100 mm (4 in.)
Non–Air-Entrained PCC								
25 – 50 (1 – 2)	207 (350)	199 (335)	190 (315)	179 (300)	166 (275)	154 (260)	130 (220)	113 (190)
75 – 100 (3 – 4)	228 (385)	216 (365)	205 (340)	193 (325)	181 (300)	169 (285)	145 (245)	124 (210)
150 – 175 (6 – 7)	243 (410)	228 (385)	216 (360)	202 (340)	190 (315)	178 (300)	160 (270)	-
Typical entrapped air (percent)	3	2.5	2	1.5	1	0.5	0.3	0.2
Air-Entrained PCC								
25 – 50 (1 – 2)	181 (305)	175 (295)	168 (280)	160 (270)	148 (250)	142 (240)	122 (205)	107 (180)
75 – 100 (3 – 4)	202 (340)	193 (325)	184 (305)	175 (295)	165 (275)	157 (265)	133 (225)	119 (200)
150 – 175 (6 – 7)	216 (365)	205 (345)	197 (325)	184 (310)	174 (290)	166 (280)	154 (260)	-
Recommended Air Content (percent)								
Mild Exposure	4.5	4.0	3.5	3.0	2.5	2.0	1.5	1.0
Moderate Exposure	6.0	5.5	5.0	4.5	4.5	4.0	3.5	3.0
Severe Exposure	7.5	7.0	6.0	6.0	5.5	5.0	4.5	4.0

5.0.1.4 Water–Cement Ratio

The water–cement ratio is a convenient measurement whose value is well correlated with PCC strength and durability. In general, lower water–cement ratios produce stronger, more durable PCC. If natural pozzolans are used in the mix (such as fly ash), then the ratio becomes a water-cementitious material ratio (cementitious material = portland cement + pozzolonic material). The ACI method bases the water–cement ratio selection on desired compressive strength and then calculates the required cement content based on the selected water–cement ratio. Table 5.17 is a general estimate of 28-day compressive strength versus water–cement ratio (or water-cementitious ratio). Values in this table tend to be conservative. Most state DOTs tend to set a maximum water–cement ratio between 0.40 and 0.50.

5.0.1.5 Cement Content

Cement content is determined by comparing the following two items:

- The calculated amount based on the selected mixing water content and water–cement ratio.
- The specified minimum cement content, if applicable. Most state DOTs specify minimum cement contents in the range of 300–360 kg/m^3 (500–600 lbs/yd^3).

An older practice used to be to specify the cement content in terms of the number of 94 lb sacks of portland cement per cubic yard of PCC. This resulted in specifications such as a "6 sack mix" or a "5 sack mix." While these specifications are quite logical to a small contractor or an individual who buys portland cement in 94 lb sacks, they do not have much meaning to the typical pavement contractor or batching plant who buys portland cement in bulk. As such, specifying cement content by the number of sacks should be avoided.
Source: washington.edu/PGI/html

5.0.1.6 Adjustments for Aggregate Moisture

Unlike HMA (Hot Mix Asphalt), PCC batching does not require dried aggregate. Therefore, aggregate moisture content must be accounted for. Aggregate moisture affects the following parameters:

1. *Aggregate weights.* Aggregate volumes are calculated based on oven dry unit weights, but aggregate is typically batched based on actual weight. Therefore, any moisture in the aggregate will increase its weight, and stockpiled aggregates almost always contain some moisture. Without correcting for this, the batched aggregate volumes will be incorrect.
2. *Amount of mixing water.* If the batched aggregate is anything but saturated, surface drying it will absorb water (if oven dry or air dry) or give up water (if wet) to the cement paste. This causes a net change in the amount of water available in the mix and must be compensated for by adjusting the amount of mixing water added.
Source: washington.edu/PGI/html

TABLE 5.17 Water–Cement Ratio and Compressive Strength Relationship

	Water–Cement Ratio by Weight	
28-Day Compressive Strength in MPa (psi)	Non-Air-Entrained	Air-Entrained
41.4 (6000)	0.41	–
34.5 (5000)	0.48	0.40
27.6 (4000)	0.57	0.48
20.7 (3000)	0.68	0.59
13.8 (2000)	0.82	0.74

5.1.0 Portland Cement—ASTM Types

The properties of concrete depend on the quantities and qualities of its components. Because cement is the most active component of concrete and usually has the greatest unit cost, its selection and proper use are important in obtaining most economically the balance of properties desired for any particular concrete mixture.

Type I/II portland cements, which can provide adequate levels of strength and durability, are the most popular cements used by concrete producers. However, some applications require the use of other cements to provide higher levels of properties. The need for high-early strength cements in pavement repairs and the use of blended cements with aggregates susceptible to alkali-aggregate reactions are examples of such applications.

It is essential that highway engineers select the type of cement that will obtain the best performance from the concrete. This choice involves correct knowledge of the relationship between cement and performance and, in particular, between type of cement and durability of concrete.

Portland Cement (ASTM Types)

ASTM C 150 defines portland cement as "hydraulic cement (cement that not only hardens by reacting with water but also forms a water-resistant product) produced by pulverizing clinkers consisting essentially of hydraulic calcium silicates, usually containing one or more of the forms of calcium sulfate as an inter ground addition." Clinkers are nodules (diameters, 0.2–1.0 in. [5–25 mm]) of a sintered material that is produced when a raw mixture of predetermined composition is heated to high temperature. The low cost and widespread availability of the limestone, shales, and other naturally occurring materials make portland cement one of the lowest-cost materials widely used over the last century throughout the world. Concrete becomes one of the most versatile construction materials available in the world.

The manufacture and composition of portland cements, hydration processes, and chemical and physical properties have been repeatedly studied and researched, with innumerable reports and papers written on all aspects of these properties.

Types of Portland Cement

Different types of portland cement are manufactured to meet different physical and chemical requirements for specific purposes, such as durability and high-early strength. Eight types of cement are covered in ASTM C 150 and AASHTO M 85.

More than 92% of portland cement produced in the United States is Type I and II (or Type I/II); Type III accounts for about 3.5% of cement production. Type IV cement is only available on special request, and Type V may also be difficult to obtain (less than 0.5% of production).

Although IA, IIA, and IIIA (air-entraining cements) are available as options, concrete producers prefer to use an air-entraining admixture during concrete manufacture, where they can get better control in obtaining the desired air content. However, this kind of cement can be useful under conditions in which quality control is poor, particularly when no means of measuring the air content of fresh concrete is available.

If a given type of cement is not available, comparable results can frequently be obtained by using modifications of available types. High-early strength concrete, for example, can be made by using a higher content of Type I when Type III cement is not available, or by using admixtures such as chemical accelerators or high-range water reducers (HRWR). The availability of portland cements will be affected for years to come by energy and pollution requirements. In fact, the increased attention to pollution abatement and energy conservation has already greatly influenced the cement industry, especially in the production of low-alkali cements. Using high-alkali raw materials in the manufacture of low-alkali cement requires bypass systems to avoid

Portland Cement Types and Their Uses

Cement Type	Use
I[1]	General-purpose cement, when there are no extenuating conditions
II[2]	Aids in providing moderate resistance to sulfate attack
III	When high-early strength is required
IV[3]	When a low heat of hydration is desired (in massive structures)
V[4]	When high sulfate resistance is required
IA[4]	A type I cement containing an integral air-entraining agent
IIA[4]	A type II cement containing an integral air-entraining agent
IIIA[4]	A type III cement containing an integral air-entraining agent

[1] Cements that simultaneously meet requirements of Type I and Type II are also widely available.
[2] Type II low alkali (total alkali as Na2O < 0.6%) is often specified in regions where aggregates susceptible to alkali-silica reactivity are employed.
[3] Type IV cements are only available on special request.
[4] These cements are in limited production and not widely available.

concentrating alkali in the clinkers, which consumes more energy. It is estimated that 4% of energy used by the cement industry could be saved by relaxing alkali specifications. Limiting use of low-alkali cement to cases in which alkali-reactive aggregates are used could lead to significant improvement in energy efficiency.

Source: U.S. Department of Transportation—FHWA (Federal Highway Administration)

5.1.1 Cement Composition

Cement Composition. The composition of portland cements is what distinguishes one type of cement from another. ASTM C 150 and AASHTO M 85 present the standard chemical requirements for each type. The phase compositions in portland cement are denoted by ASTM as tricalcium silicate (C_3S), dicalcium silicate (C_2S), tricalcium aluminate (C_3A), and tetracalcium aluminoferrite (C_4AF). However, it should be noted that these compositions would occur at a phase equilibrium of all components in the mix and do not reflect the effects of burn temperatures, quenching, oxygen availability, and other real-world kiln conditions. The actual components are often complex chemical crystalline and amorphous structures, denoted by cement chemists as "elite" (C_3S), "belite" (C_2S), and various forms of aluminates. The behavior of each type of cement depends on the content of these components. Characterization of these compounds, their hydration, and their influence on the behavior of cements are presented in full detail in many texts. Different analytical techniques such as X-ray diffraction and analytical electron microscopy are used by researchers in order to understand fully the reaction of cement with water (hydration process) and to improve its properties.

In simplest terms, results of these studies have shown that early hydration of cement is principally controlled by the amount and activity of C_3A, balanced by the amount and type of sulfate interground with the cement. C_3A hydrates very rapidly and will influence early bonding characteristics. Abnormal hydration of (C_3A) and poor control of this hydration by sulfate can lead to such problems as flash set, false set, slump loss, and cement-admixture incompatibility.

Development of the internal structure of hydrated cement (referred to by many researchers as the microstructure) occurs after the concrete has set and continues for months (and even years) after placement. The microstructure of the cement hydrates will determine the mechanical behavior and durability of the concrete. In terms of cement composition, the C_3S and C_2S will have the primary influence on long-term development of structure, although aluminates may contribute to formation of compounds such as ettringite (sulfoaluminate hydrate), which can cause expansive disruption of concrete. Cements high in C_3S (especially those that are finely ground) will hydrate more rapidly and lead to higher early strength. However, the hydration products formed will, in effect, make it more difficult for hydration to proceed at later ages, leading to an ultimate strength lower than desired in some cases. Cements high in C_2S will hydrate much more slowly, leading to a denser ultimate structure and a higher long-term strength. The relative ratio of C_3S to C_2S, and the overall fineness of cements, have been steadily increasing over the past few decades. This ability to achieve desired strengths at a higher workability (and hence a higher w/c) may account for many durability problems, as it is now established that higher w/c invariably leads to higher permeability in the concrete.

One of the major aspects of cement chemistry that concern cement users is the influence of chemical admixtures on portland cement. Since the early 1960s, most states have permitted or required the use of water-reducing and other admixtures in highway pavements and structures. A wide variety of chemical admixtures have been introduced to the concrete industry over the last three decades, and engineers are increasingly concerned about the positive and negative effects of these admixtures on cement and concrete performance.

Considerable research dealing with admixtures has been conducted in the United States. Air-entraining agents are widely used in the highway industry in North America, where concrete will be subjected to repeated freeze–thaw cycles. Air-entraining agents have no appreciable effect on the rate of hydration of cement or on the chemical composition of hydration products. However, an increase in cement fineness or a decrease in cement alkali content generally increases the amount of an admixture required for a given air content. Water reducers or retarders influence cement compounds and their hydration. Lignosulfonate-based admixtures affect the hydration of C_3A, which controls the setting and early hydration of cement. C_3S and C_4AF hydration is also influenced by water reducers.

Test results showed that alkali and C_3A contents influence the required admixtures to achieve the desired mix. It appears that set retarders, for example, are more effective with cement of low alkali and low C_3A content, and that water reducers seem to improve the compressive strength of concrete-containing cements of low alkali content more than that of the concrete-containing cements of high alkali content.

Source: U.S. Department of Transportation–FHWA

5.1.2 Physical Properties of Portland Cement

<u>Physical Properties of Portland Cements.</u> ASTM C 150 and AASHTO M 85 have specified certain physical requirements for each type of cement. These properties include (1) fineness, (2) soundness, (3) consistency, (4) setting time, (5) compressive strength, (6) heat of hydration, (7) specific gravity, and (8) loss of ignition. Each of these properties has an influence on the performance of cement in concrete. The fineness of the cement, for example, affects the rate of hydration. Greater fineness increases the surface available for hydration, causing greater early strength and more rapid generation of heat (the fineness of Type III is higher than that of Type I cement).

ASTM C 150 and AASHTO M 85 specifications are similar except with regard to fineness of cement. AASHTO M 85 requires coarser cement, which will result in higher ultimate strengths and lower early-strength gain. The Wagner Turbidimeter and the Blaine air permeability test for measuring cement fineness are both required by the American Society for Testing Materials (ASTM) and the American Association for State Highway Transportation Officials (AASHTO). Average Blaine fineness of modern cement ranges from 3,000 to 5,000 cm^2/g (300 to 500 m^2/kg).

Soundness, which is the ability of hardened cement paste to retain its volume after setting, can be characterized by measuring the expansion of mortar bars in an autoclave (ASTM C 191, AASHTO T 130). The compressive strength of 2-in. (50-mm) mortar cubes after 7 days (as measured by ASTM C 109) should not be less than 2800 psi (19.3 MPa) for Type I cement. Other physical properties included in both ASTM C 150 and AASHTO M 95 are specific gravity and false set. False set is a significant loss of plasticity shortly after mixing

TABLE 1.2 Effects of cements on concrete properties.

Cement Property	Cement Effects
Placeability	Cement amount, fineness, setting characteristics
Strength	Cement composition (C_3S, C_2S and C_3A), loss on ignition, fineness
Drying Shrinkage	SO_3 content, cement composition
Permeability	Cement composition, fineness
Resistance to sulfate	C_3A content
Alkali Silica Reactivity	Alkali content
Corrosion of embedded steel	Cement Composition (esp. C_3A content)

due to the formation of gypsum or the formation of ettringite after mixing. In many cases, workability can be restored by remixing concrete before it is cast.

The effects of cement on the most important concrete properties are presented in Table 1.2.

Cement composition and fineness play a major role in controlling concrete properties. Fineness of cement affects the placeability, workability, and water content of a concrete mixture much like the amount of cement used in concrete does.

Cement composition affects the permeability of concrete by controlling the rate of hydration. However, the ultimate porosity and permeability are unaffected. The coarse cement tends to produce pastes with higher porosity than that produced by finer cement. Cement composition has only a minor effect on freeze–thaw resistance. Corrosion of embedded steel has been related to C_3A content. The higher the C_3A, the more chloride can be tied into chloroaluminate complexes—and thereby be unavailable for catalysis of the corrosion process.

Source: U.S. Department of Transportation—FHWA

5.1.3 Blended Portland Cement

Blended cement, as defined in ASTM C 595, is a mixture of portland cement and blast-furnace slag (BFS) or a "mixture of portland cement and a pozzolan (most commonly fly ash)."

The use of blended cements in concrete reduces mixing water and bleeding, improves finishability and workability, enhances sulfate resistance, inhibits the alkali-aggregate reaction, and lessens heat evolution during hydration, thus moderating the chances for thermal cracking on cooling.

Blended cement types and blended ratios are presented in Table 1.3.

TABLE 1.3 Blended cement types and blended ratios.

Type	Blended Ingredients
IP	15-40% by weight of pozzolan (fly ash)
I(PM)	0-15% by weight of Pozzolan (fly ash) (modified)
P	15-40% by weitth of pozzolan (fly ash)
IS	25-70% by weight of blast-furnace slag
I(SM)	0-25% by weight of blast-furnace slag (modified)
S	70-100% by weight of blast-furnace slag

The advantages to using mineral admixtures added at the batch plant are as follows.

- Mineral admixture replacement levels can be modified on a day-to-day and job-to-job basis to suit project specifications and needs.
- Cost can be decreased substantially while performance is increased (taking into consideration the fact that the price of blended cement is at least 10% higher than that of Type I/II cement [U.S. Dept. Int. 1989]).
- GGBFS can be ground to its optimum fineness.
- Concrete producers can provide specialty concretes in the concrete product markets.

At the same time, several precautions must be considered when mineral admixtures are added at the batch plant.

- Separate silos are required to store the different hydraulic materials (cements, pozzolans, slags). This might slightly increase the initial capital cost of the plant.
- There is a need to monitor variability in the properties of the cementitious materials, often enough to enable operators to adjust mixtures or obtain alternate materials if problems arise.
- Possibilities of cross-contamination or batching errors are increased as the number of materials that must be stocked and controlled is increased.

Source: U.S.Department of Transportation—FHWA

5.1.4 Modified Portland Cement (Expansive Cement)

Expansive cement, as well as expansive components, is a cement-containing hydraulic calcium silicate (such as those characteristic of portland cement) that, upon being mixed with water, forms a paste. During the early hydrating period occurring after setting, it increases in volume significantly more than does portland cement paste. Expansive cement is used to compensate for volume decrease due to shrinkage and to induce tensile stress in reinforcement.

Expansive cement concrete used to minimize cracking caused by drying shrinkage in concrete slabs, pavements, and structures is termed shrinkage-compensating concrete.

Self-stressing concrete is another expansive cement concrete in which the expansion, if restrained, will induce a compressive stress high enough to result in a significant residual compression in the concrete after drying shrinkage has occurred.

Types of Expansive Cements. Three kinds of expansive cement are defined in ASTM C 845.

- Type K: Contains anhydrous calcium aluminate
- Type M: Contains calcium aluminate and calcium sulfate
- Type S: Contains tricalcium aluminate and calcium sulfate

Only Type K is used in any significant amount in the United States.

Concrete placed in an environment where it begins to dry and lose moisture will begin to shrink. The amount of drying shrinkage that occurs in concrete depends on the characteristics of the materials, mixture proportions, and placing methods. When pavements or other structural members are restrained by subgrade friction, reinforcement, or other portions of the structure, drying shrinkage will induce tensile stresses. These drying shrinkage stresses usually exceed the concrete tensile strengths, causing cracking. The advantage of using expansive cements is to induce stresses large enough to compensate for drying shrinkage stresses and minimize cracking.

Physical and mechanical properties of shrinkage compensating concrete are similar to those of portland cement concrete (PCC). Tensile, flexural, and compressive strengths are comparable to those in PCC. Air-entraining admixtures are as effective with shrinkage-compensating concrete as with portland cement in improving freeze–thaw durability.

Some water-reducing admixtures may be incompatible with expansive cement. Type A water-reducing admixture, for example, may increase the slump loss of shrinkage-compensating concrete. Fly ash and other pozzolans may affect expansion and may also influence strength development and other physical properties.

Structural design considerations and mix proportioning and construction procedures are available in ACI 223-83. This report contains several examples of using expansive cements in pavements.

In Japan, admixtures containing expansive compounds are used instead of expansive cements. Described using expansive admixtures in building chemically prestressed precast concrete box culverts. Bending characteristics of chemically prestressed concrete box culverts were identical to those of reinforced concrete units of greater thickness. Expansive compounds are also available in the United States. They can be added to the mix in a way similar to how fly ash is added to concrete mixes.

Source: U.S. Department of Transportation—FHWA

5.2.0 Types of Cement and What They Do

Portland cement is a type of cement, not a brand name. Many cement manufacturers make portland cement. To find what concrete is made of and to learn about concrete mix designs, admixtures, and water-to-cement ratios, read our section "What Is Concrete".

Type 1—Normal portland cement. Type 1 is a general use cement.

Type 2—Used for structures in water or soil containing moderate amounts of sulfate, or when heat buildup is a concern.

Type 3—High-early strength. Used when high strengths are desired at very early periods.

Type 4—Low-heat portland cement. Used where the amount and rate of heat generation must be kept to a minimum.

Type 5—Sulfate-resistant portland cement. Used where the water or soil is high in alkali.

Types IA, IIA, and IIIA are cements used to make air-entrained concrete. They have the same properties as types I, II, have small quantities of air-entrained materials combined with them.

These are very short descriptions of the basic types of cement. There are other types for various purposes such as masonry cements, just to name two examples.

Your ready-mix company will know what the requirements are for your area and for your particular use. Simply ask the of cement is and if that will work for your conditions.

Source: concretenet work.com

5.3.0 Concrete Compressive Strengths—U.S. and Metric

Nominal MPa Values of Equivalent psi Concrete Strengths

In metric, concrete strength is denominated in megapascals (**MPa**).

In imperial, concrete strength is denominated in pounds per square inch (**psi**).

2500 psi = 18 MPa (17.23 MPa exact)
3000 psi = 20 MPa (20.67 MPa exact)
3500 psi = 25 MPa (24.12 MPa exact)
4000 psi = 30 mpa (27.57 mpa exact)
5000 psi = 35 mpa (34.46 mpa exact)
6000 psi = 40 mpa (41.35 mpa exact)

Use 0.0068915 to convert psi to MPa.

Newtons, psi, Concrete Strength, and Prestressed Slabs

Concrete strengths are customarily denominated in **psi** (pounds per square inch) in the imperial system and in **MPa** (megapascals) in metric. These are units of **pressure**.

The **newton** (N) is a measure of **force**. 1 newton is that force which pushes 1 gram of matter with an acceleration of 1 centimeter per second per second (or per second ^{2}) or, equivalently, the force that accelerates 1 kilogram of matter to 1 meter per second ^{2}.

$$\text{Force} = \text{mass} \times \text{acceleration}$$

Velocity is a measure of constant speed (i.e., meters per second, miles per hour, furlongs per fortnight)
  Velocity is speed in a certain direction
  Acceleration is the rate of change in velocity over time
  Acceleration can be either positive or negative (deceleration)

$$1\,N = 1\,kg \times (1\,\text{meter/second}^2) \longrightarrow 1\,N = 1\,kg.\text{meter/second}^2$$
$$1\,N = 1\,g \times (1\,\text{cm/second}^2)$$

When you apply the force of 1 newton to a 1 meter ^{2} area, you have **pressure**.

$$\text{Pressure} = \text{force per area}$$

By permission: mc2-ice.com

5.4.0 Sieve Analysis Defining Coarse and Fine Aggregates

Sieve Analysis Table									
Sieve Size	Coarse Aggregate			Fine Aggregate		Cumulative Combined		Combined
	1% Passing	2% Passing	3% Passing	1% Passing	2% Passing	% Passing	% Retained	% Retained
2 in.	100	100	-	100	-	100	0	0
1–1/2 in.	100	100	-	100	-	100	0	0
1 in.	95	100	-	100	-	98	2	2
3/4 in.	62	100	-	100	-	81	19	17
1/2 in.	35	100	-	100	-	67	33	14
3/8 in.	20	95	-	100	-	59	41	8
No. 4	1	65	-	100	-	46	54	13
No. 8	-	1	-	96	-	36	64	10
No. 16	-	-	-	79	-	29	71	7
No. 30	-	-	-	45	-	17	83	12
No. 50	-	-	-	17	-	6	94	11
No. 100	-	-	-	7	-	3	97	3
No. 200	-	-	-	2	-	1	99	3
Pan	-	-	-	0	-	0	100	1
% of Aggregate	50%	13%	0%	37%	0%	-	-	-

Charts

Coarseness Factor Chart—Use the cumulative combined sieve analysis to determine the coarseness and workability factors. Plot the coarseness and workability factors on the Coarseness Factor Chart.

Determine the coarseness factor using the following equation:

$$\text{Coarseness Factor} = \left(\frac{S}{T}\right) \times 100$$

where:

S = % Cumulative Retained on 3/8 in. Sieve and
T = % Cumulative Retained on No. 8 Sieve.

The workability factor is the cumulative combined percentage passing the No. 8 sieve.
Increase the workability factor by 2.5 percentage points for every 94 lb per cubic yard of cementitious material used in excess of 564 lb per cubic yard in the mix design.
Decrease the workability factor by 2.5 percentage points for every 94 lb per cubic yard

Source: Texas DOT

5.4.1 Air-Entraining Admixtures

Air Entrainment

Air entrainment is the process whereby many small air bubbles are incorporated into concrete and become part of the matrix that binds the aggregate together in the hardened concrete. These air bubbles are dispersed throughout the hardened cement paste but are not, by definition, part of the paste (Dolch 1984). Air entrainment has now been an accepted fact in concrete technology for more than 45 years. Although historical references indicate that certain archaic and early twentieth-century concretes were indeed inadvertently air entrained, the New York State Department of Public Works and the Universal Atlas Cement Company were among the first to recognize that controlled additions of certain naturally occurring organic substances derived from animal and wood by-products could materially increase the resistance of concrete in roadways to attack brought on by repeated freeze–thaw cycles and the application of deicing agents.

Extensive laboratory testing and field investigation concluded that the formation of minute air bubbles dispersed uniformly through the cement paste increased the freeze–thaw durability of concrete. This formation can be achieved through the use of organic additives, which enable the bubbles to be stabilized or entrained within the fresh concrete. These additives are called air-entraining agents.

Besides the increase in freeze–thaw and scaling resistances, air-entrained concrete is more workable than nonentrained concrete. The use of air-entraining agents also reduces bleeding and segregation of fresh concrete.

Materials and Specifications. The most commonly used chemical surfactants can be categorized into four groups: (1) salts of wood resins, (2) synthetic detergents, (3) salts of petroleum acids, and (4) fatty and resinous acids and their salts (Dolch 1984).

Until the early 1980s, the majority of concrete air entrainers were based solely on salts of wood resins or neutralized Vinsol resin (Edmeades and Hewlett 1986), and most concrete highway structures and pavements were air entrained by Vinsol resin. Today, a wider variety of air-entraining agents is available and competes with Vinsol resins.

Requirements and specifications of air-entraining agents to be used in concrete are covered in ASTM C 260 and AASHTO M 154. According to these specifications, each admixture to be used as an air-entraining agent should cause a substantial improvement in durability, and none of the essential properties of the concrete should be seriously impaired. This provides a means to evaluate air-entraining admixtures on a performance basis.

Factors Affecting Air Entrainment. The air-void system created by using air-entraining agents in concrete is also influenced by concrete materials and construction practice. Concrete materials such as cement, sand, aggregates, and other admixtures play an important role in maintaining the air-void system in concrete. It has been found that air content will increase as cement alkali levels increase and decrease as cement fineness increases significantly.

Fine aggregate serves as a three-dimensional screen and traps the air; the more median sand there is in the total aggregate, the greater the air content of the concrete will be (Dolch 1984). Gradation has more influence in leaner mixes. Median sand ranging from the No. 30 sieve to the No. 100 is the most effective at entraining air. Excessive fines, minus No. 100 material, causes a reduction in air entrainment.

Because the use of chemical and mineral admixtures in addition to air-entraining agents has become common practice, concrete users are always concerned about the effects of these admixtures on the air-void system and durability of concrete. The effects of water reducers, retarders, and accelerators were widely investigated by many researchers. In regards to gross air content obtained when water-reducing and retarding admixtures are used in concrete, numerous studies have shown that for most of the materials, less air-entraining agent is needed to achieve a given specified air content. Chemical admixtures should be added separately from air-entraining additives.

Source: U.S. Department of Transportation—FHWA

When lignosulfonate water reducers are used, less air-entraining agent is required because the lignosulfonates have a moderate air-entraining capacity, although alone they do not react as air-entraining agents (Dolch 1984). For a fixed amount of air-entraining agent, the effect of added calcium chloride is to slightly increase the air content (Edmeades and Hewlett 1986). The effect is more pronounced as amounts greater than 1% of the weight of cement are used. Some HRWR (superplasticizers) interact with cements and air-entraining agents, resulting in reductions in specific surfaces and increases in air-void spacing factors.

Mineral admixtures such as fly ash and silica fume also affect the formation of void systems in concrete. In their study on the effect of fly ash on air-void stability of concrete, showed that concretes containing fly ash produced relatively stable air-void systems. However, the volume of air retained is affected by fly ash types. In mixtures containing fly ashes, the amount of air-entraining agent required to produce a given percentage of entrained air is higher, and sometimes much higher, than it is in comparable mixtures without fly ash. In a series of papers, researchers presented the results of a study on factors that affect the air-void stability in concretes (Pigeon, Aitcin, and LaPlante 1987; Pigeon and Plante 1989). They found that silica fume has no significant influence on the production and stability of the air-void system during mixing and agitation. Also indicated that silica fume has no detrimental effects on the air-void system.

Temperature can also have a significant effect on air entrainment. Air entrainment varies inversely with temperature. The same mix will entrain more air at 50°F (10°C) than at 100°F (38°C).

Air Content Control. Measurement of air content is an important checking "sensor" for the concrete user to know whether concrete will resist freeze–thaw damage. Because average void spacing decreases as air content increases, an "optimum" air content at which void spacing will prevent the development of excessive pressure due to freezing and thawing will exist.

It is important to check the air content of fresh concrete regularly for control purposes. Air content should be tested not only at the mixer but also at the point of discharge into the forms, because of losses of air content due to handling and transportation.

Recommendations

1. Air-entraining admixtures should be specified when concrete will be exposed to freeze–thaw conditions, deicing salt applications, or sulfate attack.
2. Although air-entraining admixtures are compatible with most other admixtures, care should be taken to prevent them from coming in contact during the mixing process.

References

Sections of this document were obtained from Synthesis of Current and Projected Concrete Highway Technology, David Whiting et al., SHRP-C-345, Strategic Highway Research Program, National Research Council.

Dolch, W.L., 1984. Air-entraining admixtures. In: Ramachandran, V.S. (Ed.), Concrete admixtures handbook: Properties, science, and technology. Noyes Publications, Park Ridge, NJ, pp. 269–300.

Edmeades, R.M., Hewlett, P.C., 1986. Admixtures—Present and future trends. Concrete 20 (8), 4–7 (August).

Pigeon, M., Aitcin, P.C., LaPlante, P., 1987. Comparative study of the air-void stability in a normal and a condensed silica fume field concrete. ACI Journal 84 (3), 194–199 (May-June).

Pigeon, M., Plante, M., 1989. Air-void stability part I: Influence of silica fume and other parameters. ACI Journal 86 (5), 482–490.

www.fhwa.dot.gov/.../airentr.htm.

5.4.2 Superplasticizers

The use of superplasticizers (high-range water reducer) has become a quite common practice. This class of water reducers was originally developed in Japan and Germany in the early 1960s; it was introduced in the United States in the mid-1970s.

Superplasticizers are linear polymers containing sulfonic acid groups attached to the polymer backbone at regular intervals. Most of the commercial formulations belong to one of four families:

- Sulfonated melamine-formaldehyde condensates (SMF)
- Sulfonated naphthalene-formaldehyde condensates (SNF)
- Modified lignosulfonates (MLS)
- Polycarboxylate derivatives

The sulfonic acid groups are responsible for neutralizing the surface charges on the cement particles and causing dispersion, thus releasing the water tied up in the cement particle agglomerations and thereafter reducing the viscosity of the paste and concrete.

ASTM C 494 was modified to include high-range water-reducing admixtures in the edition published in July 1980. The admixtures were designated Type F water-reducing, high-range admixtures and Type G water-reducing, high-range, and retarding admixtures.

Effect of Superplasticizers on Concrete Properties. The main purpose of using superplasticizers is to produce flowing concrete with very high slump in the range of 7–9 in. (175–225 mm) to be used in heavily reinforced structures and in placements where adequate consolidation by vibration cannot be readily achieved. The other major application is the production of high-strength concrete at w/c's ranging from 0.3 to 0.4.

The ability of superplasticizers to increase the slump of concrete depends on such factors as the type, dosage, and time of addition of superplasticizer; w/c; and the nature or amount of cement. It has been found that for most types of cement, superplasticizer improves the workability of concrete. For example, incorporation of 1.5% SMF to a concrete containing Type I, II, and V cements increases the initial slump of 3 in. (76 mm) to 8.7, 8.5, and 9 in. (222, 216, and 229 mm), respectively.

The capability of superplasticizers to reduce water requirements 12–25% without affecting the workability leads to production of high-strength concrete and lower permeability. Compressive strengths greater than 14,000 psi (96.5 mpa) at 28 days have been attained. Use of superplasticizers in air-entrained concrete can produce coarser-than-normal air-void systems. The maximum recommended spacing factor for air-entrained concrete to resist freezing and thawing is 0.008 in. (0.2 mm). In superplasticized concrete, spacing factors in many cases exceed this limit. Even though the spacing factor is relatively high, the durability factors are above 90 after 300 freeze–thaw cycles for the same cases. A study indicated that high-workability concrete containing superplasticizer can be made with a high freeze–thaw resistance, but air content must be increased relative to concrete without superplasticizer. This study also showed that the type of superplasticizer has nearly no influence on the air-void system.

One problem associated with using a high-range water reducer in concrete is slump loss. In a study of the behavior of fresh concrete containing conventional water reducers and high-range water reducer, found that with time slump loss is very rapid in spite of the fact that second-generation high-range water reducers are claimed not to suffer as much from the slump loss phenomenon as the first-generation conventional water reducers do. However, slump loss of flowing concrete was found to be less severe, especially for newly developed admixtures based on copolymeric formulations.

The slump loss problem can be overcome by adding the admixture to the concrete just before the concrete is placed.

Source: U.S. Department of Transportation—FHWA

However, there are disadvantages to such a procedure. The dosage control, for example, might not be adequate, and it requires ancillary equipment such as truck-mounted admixture tanks and dispensers. Adding admixtures at the batch plant, besides dosage control improvement, reduces wear of truck mixers and lessens the tendency to add water onsite. New admixtures now being marketed can be added at the batch plant and can hold the slump above 8 in. (204 mm) for more than 2 hours.

Recommendations

1. Verification tests should be performed on liquid admixtures to confirm that the material is the same as that which was approved. The identifying tests include chloride and solids content, ph, and infrared spectrometry.
2. If transit mix trucks are used to mix high-slump concrete, it is recommended that a 75 mm slump concrete be used at a full mixing capacity to ensure uniform concrete properties.
3. If transit mix trucks are used to mix low w/c ratio concrete, it is recommended that the load size be reduced to ½ to ⅔ of the mixing capacity to ensure uniform concrete properties.
4. If freeze–thaw testing as described by ASTM C 666 indicates this to be a problem, it is recommended that the air content be increased by 1.5%.

5.4.3 Fly Ash

Infrastructure

Fly ashes are finely divided residue resulting from the combustion of ground or powdered coal. They are generally finer than cement and consist mainly of glassy-spherical particles as well as residues of hematite and magnetite, char, and some crystalline phases formed during cooling. Use of fly ash in concrete started in the United States in the early 1930s. The first comprehensive study was that described in 1937, by R. E. Davis at the University of California. The major breakthrough in using fly ash in concrete was the construction of Hungry Horse Dam in 1948, utilizing 120,000 metric tons of fly ash. This decision by the U.S. Bureau of Reclamation paved the way for using fly ash in concrete constructions.

In addition to economic and ecological benefits, the use of fly ash in concrete improves its workability; reduces segregation, bleeding, heat evolution, and permeability; inhibits alkali-aggregate reaction; and enhances sulfate resistance. Even though the use of fly ash in concrete has increased in the last 20 years, less than 20% of the fly ash collected was used in the cement and concrete industries.

One of the most important fields of application for fly ash is PCC pavement, where a large quantity of concrete is used and economy is an important factor in concrete pavement construction. The FHWA has been encouraging the use of fly ash in concrete. When the price of fly ash concrete is equal to, or less than, the price of mixes with only portland cement, fly ash concretes are given preference if technically appropriate under FHWA guidelines.

Classifications and Specifications

Two major classes of fly ash are specified in ASTM C 618 on the basis of their chemical composition resulting from the type of coal burned; these are designated Class F and Class C. Class F is fly ash normally produced from burning anthracite or bituminous coal, and Class C is normally produced from the burning of subbituminous coal and lignite (as are found in some of the western states of the United States) (Halstead 1986). Class C fly ash usually has cementitious properties in addition to pozzolanic properties due to free lime, whereas Class F is rarely cementitious when mixed with water alone. All fly ashes used in the United States before 1975 were Class F.

Fly ash that is produced at base-loaded electric generating plants is usually very uniform. Base-loaded plants are those plants that operate continuously. The only exception to uniformity is in the start-up and the shutdown of these plants. Contamination may occur from using other fuels to start the plant, and inconsistencies in carbon content occur until the plant reaches full operating efficiency. The ash produced from the start-up and shutdown must be separated from what is produced when the plant is running efficiently. In addition, when sources of coal are changed, it is necessary to separate the two types of fly ashes. Peak-load plants are subjected to many start-up and shutdown cycles. Because of this, these plants may not produce much uniform fly ash.

The most often used specifications for fly ash are ASTM C 618 and AASHTO M 295. While some differences exist, these two specifications are essentially equivalent. Some state transportation agencies have specifications that differ from the standards. The general classification of fly ash by the type of coal burned does not adequately define the type of behavior to be expected when the materials are used in concrete.

There are also wide differences in characteristics within each class. Despite the reference in ASTM C 618 to the classes of coal from which Class F and Class C fly ashes are derived, there was no requirement that a given class of fly ash must come from a specific type of coal. For example, Class F ash can be produced from coals that are not bituminous, and bituminous coals can produce ash that is not Class F (Halstead 1986). It should be noted that current standards contain numerous physical and chemical requirements that do not serve a useful purpose. Whereas some requirements are needed for ensuring batch-to-batch uniformity, many are unnecessary.

Source: U.S. Department of Transportation–FHWA

Mix Design

The substitution rate of fly ash for portland cement will vary depending on the chemical composition of both the fly ash and the portland cement. The rate of substitution typically specified is a minimum of 1 to 1½ pounds of fly ash to 1 pound of cement. It should be noted that the amount of fine aggregate will have to be reduced to accommodate the additional volume of fly ash. This is due to fly ash being lighter than the cement.

The amount of substitution is also dependent on the chemical composition of the fly ash and the portland cement. Currently, states allow a maximum substitution in the range of 15 to 25%.

The effects of fly ash, especially Class F, on fresh and hardened concrete properties have been extensively studied by many researchers in different laboratories, including the U.S. Army Corps of Engineers, PCA (Portland Cement Association), and the Tennessee Valley Authority. The two properties of fly ash that are of most concern are its carbon content and fineness. Both of these properties will affect the air content and water demand of the concrete.

The finer the material, the higher the water demand due to the increase in surface area. The finer material requires more air-entraining agent to give the mix the desired air content. The important thing to remember is uniformity. If fly ash is uniform in size, the mix design can be adjusted to give a good uniform mix.

The carbon content, which is indicated by the loss of ignition, also affects the air-entraining agents and reduces the entrained air for a given amount of air-entraining agent. More air-entraining agent will need to be added to get the desired air content. The carbon content will also affect water demand since the carbon will absorb water. Again, uniformity is important since the differences from non-fly ash concrete can be adjusted in the mix design.

Fresh Concrete Workability. Use of fly ash increases the absolute volume of cementitious materials (cement plus fly ash) compared to non-fly-ash concrete. Therefore, the paste volume is increased, leading to a reduction in aggregate particle interference and enhancement in concrete workability. The spherical particle shape of fly ash also participates in improving the workability of fly ash concrete because of the so-called ball-bearing effect. It has been found that both classes of fly ash improve concrete workability.

Bleeding. Using fly ash in air-entrained and non–air-entrained concrete mixtures usually reduces bleeding by providing greater fines volume and lower water content for a given workability. Although increased fineness usually increases the water demand, the spherical particle shape of the fly ash lowers particle friction and offsets such effects. Concrete with relatively high fly ash content will require less water than non–fly-ash concrete of equal slump.

Time of Setting. All Class F and most Class C fly ashes increase the time of setting of concrete. Time of setting of fly ash concrete is influenced by the characteristics and amounts of fly ash used in concrete. For highway construction, changes in time of setting of fly ash concrete from non–fly-ash concrete using similar materials will not usually introduce a need for changes in construction techniques; the delays that occur may be considered advantageous.

Strength and Rate of Strength of Hardened Concrete. Strength of fly ash concrete is influenced by type of cement, quality of fly ash, and curing temperature compared to that of non–fly-ash concrete proportioned for equivalent 28-day compressive strength. Concrete containing typical Class F fly ash may develop lower strength at 3 or 7 days of age when tested at room temperature. However, fly ash concretes usually have higher ultimate strengths when properly cured. The slow gain of strength is the result of the relatively slow pozzolanic reaction of fly ash. In cold weather, the strength gain in fly ash concretes can be more adversely affected than the strength gain in non–fly-ash concrete. Therefore, precautions must be taken when fly ash is used in cold weather.

Freeze–Thaw Durability of Hardened Concrete. On the basis of a comparative experimental study of freeze–thaw durability of conventional and fly ash concrete, it has been observed that the addition of fly ash has no major effect on the freeze–thaw resistance of concrete if the strength and air content are kept constant. The addition of fly ash may have a negative effect on the freeze–thaw resistance of concrete when a major part of the cement is replaced by it. The use of fly ash in air-entrained concrete will generally require an increase in the dosage rate of the air-entraining admixture to maintain constant air. Air-entraining admixture dosage depends on carbon content, loss of ignition, fineness, and amount of organic material in the fly ash. ACI Carbon content of fly ash, which is related to the coal burned by the producing utility of the type and condition of furnaces in the production process of fly ash, influences the behavior of admixtures in concrete. It has been found that high-carbon-content fly ash reduces the effectiveness of admixtures such as air-entraining agents.

Alkali-Silica Reaction of Hardened Concrete. One important reason for using fly ash in highway construction is to inhibit the expansion resulting from alkali-silica reaction (ASR). It has been found that (1) the alkalies released by the cement preferentially combine with the reactive silica in the fly ash rather than in the aggregate, and (2) the alkalies are tied up in nonexpansive calcium-alkali-silica gel. Thus hydroxyl ions remaining in the solution are insufficient to react with the material in the interior of the larger reactive aggregate particles, and disruptive osmotic forces are not generated.

In a paper presented at the 8th International Conference on alkali-aggregate reactivity held in Japan in 1989, Swamy and Al-Asali indicated that ASR expansion is generally not proportional to the percentage of cement replacement by fly ash. The rate of reactivity, the replacement level, the method of replacement, and the environment all have a profound influence on the protection against ASR afforded by fly ash. Several investigators have stated that ASR expansions correlated better with water-soluble alkali-silica content than with total alkali content. The addition of some high-calcium fly ash containing large amounts of soluble alkali sulfate might increase rather than decrease the alkali-aggregate reactivity. The effectiveness of different fly ashes in reducing long-term expansion varied widely; for each fly ash, this may be dependent on its alkali content or fineness.

Blended Cements

The following will discuss the Type "IP," "P," and "I(PM)" cements. The specifications for these cements are in AASHTO M-240 and ASTM C-595. Blended cements can be manufactured by either intimate blending of portland cement and pozzolan or intergrinding of the pozzolan with the cement clinker in the kiln. Type "I(PM)" (pozzolan-modified cement) allows up to 15% replacement of cement with fly ash. The Type "IP" and Type "P" are pozzolan-modified portland cements that allow 15-40% replacement with pozzolans. The difference in the two types of cements is in the ultimate strength and the rate of strength gain of the concretes. Most states specify limits on the pozzolanic content on Type "IP" cement. These limits are between 15 and 25%.

Restraints on the Use of Fly Ash Concrete in Highway Construction

It is well known now that both classes of fly ash improve the properties of concrete, but several factors and cautions should be considered when using fly ashes, especially in highway construction, where fly ash is heavily used. A report prepared by the Virginia Highway and Transportation Research Council (VHTRC) and discussed several restraints relating to the use of fly ash concrete for construction of highways and other highway structures. These restraints include the following: (1) special precautions may be necessary to ensure that the proper amount of entrained air is present; (2) not all fly ashes have sufficient pozzolanic activity to provide good results in concrete; (3) suitable fly ashes are not always available near the construction site, and transportation costs may nullify any cost advantage; and (4) mix proportions might have to be modified for any change in the fly ash composition.

Since the cement–fly ash reaction is influenced by the properties of the cement, it is important for a transportation agency not only to test and approve each fly ash source but also to investigate the properties of the specific fly ash–cement combination to be used for each project.

5.4.4 Silica Fume

Silica fume, also known as microsilica, is a by-product of the reduction of high-purity quartz with coal in electric furnaces in the production of silicon and ferrosilicon alloys. Silica fume is also collected as a by-product of other silicon alloys such as ferrochromium, ferromanganese, ferromagnesium, and calcium silicon. Before the mid-1970s, nearly all silica fume was discharged into the atmosphere. After environmental concerns necessitated the collection and landfilling of silica fume, it became economically justified to use it in various applications.

Silica fume consists of very fine vitreous particles with a surface area ranging from 60,000 to 150,000 ft^2/lb or 13,000 to 30,0000 m^2/kg when measured by nitrogen absorption techniques, with particles approximately 100 times smaller than the average cement particle. Because of its extreme fineness and high silica content, silica fume is a highly effective pozzolanic material. Silica fume is used in concrete to improve its properties. It has been found that it improves compressive strength, bond strength, and abrasion resistance; reduces permeability; and therefore helps in protecting reinforcing steel from corrosion.

Specifications

The first national standard for use of silica fume. The AASHTO and ASTM C 1240 covers microsilica for use as a mineral admixture in PCC and mortar to fill small voids and in cases in which pozzolanic action is desired. It provides the chemical and physical requirements, specific acceptance tests, and packaging and package marking.

Mix Design

Silica fume has been used as an addition to concrete up to 15% by weight of cement, although the normal proportion is 7 to 10%. With an addition of 15%, the potential exists for very strong, brittle concrete. It increases the water demand in a concrete mix; however, dosage rates of less than 5% will not typically require a water reducer. High replacement rates will require the use of a high-range water reducer.

Effects on Air Entrainment and Air-Void System of Fresh Concrete. The dosage of air-entraining agents needed to maintain the required air content when using silica fume is slightly higher than that for conventional concrete because of the high surface area and the presence of carbon. This dosage is increased with increasing amounts of silica fume content in concrete.

Effects on Water Requirements of Fresh Concrete. Silica fume added to concrete by itself increases water demands, often requiring one additional pound of water for every pound of added silica fume. This problem can be easily compensated for by using HRWR.

Effects on Consistency and Bleeding of Fresh Concrete. Concrete incorporating more than 10% silica fume becomes sticky; in order to enhance workability, the initial slump should be increased. It has been found that silica fume reduces bleeding because of its effect on rheologic properties.

Effects on Strength of Hardened Concrete. Silica fume has been successfully used to produce very high-strength, low-permeability, and chemically resistant concrete. Addition of silica fume by itself, with other factors being constant, increases the concrete strength.

Incorporation of silica fume into a mixture with HRWR also enables the use of a lower water-to-cementitious-materials ratio than may have been possible otherwise. The modulus of rupture of silica fume concrete is usually either about the same as or somewhat higher than that of conventional concrete at the same level of compressive strength.

Effects on Freeze–thaw Durability of Hardened Concrete. Air-void stability of concrete incorporating silica fume was studied by Pigeon, Aitcin, and LaPlante (1987) and Pigeon and Plante (1989). Their test results indicated that the use of silica fume has no significant influence on the production and stability of the air-void system. Freeze–thaw testing (ASTM C 666) on silica fume concrete showed acceptable results; the average durability factor was greater than 99%.

Source: U.S. Department of Transportation–FHWA

Effects on Permeability of Hardened Concrete. It has been shown by several researchers that addition of silica fume to concrete reduces its permeability. Rapid chloride permeability testing (AASHTO 277) conducted on silica fume concrete showed that addition of silica fume (8% silica fume) significantly reduces the chloride permeability. This reduction is primarily the result of the increased density of the matrix due to the presence of silica fume.

Effects on ASR of Hardened Concrete. Silica fume, like other pozzolans, can reduce ASR and prevent deleterious expansion due to ASR.

Availability and Handling

Silica fume is available in two conditions: dry and wet. Dry silica can be provided as produced or densified with or without dry admixtures and can be stored in silos and hoppers. Silica fume slurry with low or high dosages of chemical admixtures is available. Slurried products are stored in tanks with capacities ranging from a few thousand to 400,000 gallons (15100 m^3).

5.4.5 Ground Granulated Blast-Furnace Slag

Although portland blast-furnace slag cement, which is made by intergrinding the granulated slag with portland cement clinker (blended cement), has been used for more than 60 years, the use of separately ground slag combined with portland cement at the mixer as a mineral admixture did not start until the late 1970s. Ground granulated blast-furnace slag is the granular material formed when molten iron blast-furnace slag is rapidly chilled (quenched) by immersion in water. It is a granular product with very limited crystal formation, is highly cementitious in nature and, ground to cement fineness, and hydrates like portland cement.

Specifications

ASTM C 989-82 and AASHTO M 302 were developed to cover ground granulated blast-furnace slag for use in concrete and mortar. The three grades are 80, 100, and 120.

Mix Design

The use of grade 80 ground granulated blast-furnace slag should be avoided unless warranted in special circumstances. The grade of a ground granulated blast-furnace slag is based on its activity index, which is the ratio of the compressive strength of a mortar cube made with a 50% ground granulated blast-furnace slag-cement blend to that of a mortar cube made with a reference cement. For a given mix, the substitution of grade 120 ground granulated blast-furnace slag for up to 50% of the cement will generally yield a compressive strength at 7 days and beyond equivalent to or greater than that of the same concrete made without ground granulated blast-furnace slag. Substitution of grade 100 ground granulated blast-furnace slag will generally yield an equivalent or greater strength at 28 days. However, concrete made with grade 80 ground granulated blast-furnace slag will have a lower compressive strength at all ages. To provide a product with equivalent or greater compressive strengths, only grades 100 and 120 ground granulated blast-furnace slag should be used. However, in mass concrete, the heat of hydration may be an overriding factor, and the use of grade 80 slag may be appropriate.

Ground granulated blast-furnace slag is a cementitious material and can be substituted for cement on a 1:1 basis. In the absence of special circumstances or mix specific data, the substitution of ground granulated blast-furnace slag should be limited to 50% for areas not exposed to deicing salts and to 25% for concretes that will be exposed to deicing salts. While substitution of ground granulated blast-furnace slag for up to 70% of the portland cement in a mix has been used, there appears to be an optimum substitution percentage that produces the greatest 28-day strength. This is typically 50% of the total cementitious material but depends on the grade of ground granulated blast-furnace slag used. Also, research has shown that the scaling resistance of concretes decreases with ground granulated blast-furnace slag substitution rates greater than 25%.

These guidelines on ground granulated blast-furnace slag substitution rates are intended to provide a starting point for designers with little or no experience in the use of cement and concrete containing ground granulated blast-furnace slag. If local data shows good performance at greater percentages, this information can be used in lieu of the recommended guidelines. Section 4.2.3.2 of ACI 318-89, "Building Code Requirements for

Reinforced Concrete," indicates that substitution rates of up to 50% may be acceptable for concretes exposed to deicing chemicals. In addition, in mass concreting operations, the heat of hydration may be an overriding factor, and substitution rates greater than 50% may be appropriate.

Source: U.S. Department of Transportation—FHWA

Effects of Slags on Properties of Fresh Concrete. Use of slag or slag cements usually improves workability and decreases the water demand due to the increase in paste volume caused by the lower relative density of slag. The higher strength potential of Grade 120 slag may allow for a reduction of total cementitious material. In such cases, further reductions in water demand may be possible.

The setting times of concretes containing slag increase as the slag content increases. An increase of slag content from 35 to 65% by mass can extend the setting time by as much as 60 minutes. This delay can be beneficial, particularly in large pours and in hot weather conditions in which this property prevents the formation of "cold joints" in successive pours.

The rate and quantity of bleeding in concrete containing slag or slag cements are usually less than that in concrete containing no slag because of the relatively higher fineness of slag. The higher fineness of slag also increases the air-entraining agent required, compared to conventional concrete. However, unlike fly ash, slag does not contain carbon, which may cause instability and air loss in concrete.

Effect on Strength of Hardened Concrete. The compressive strength development of slag concrete depends primarily on the type, fineness, activity index, and proportions of slag used in concrete mixtures. In general, the strength development of concrete incorporating slags is slow at 1–5 days compared with that of the control concrete. Between 7 and 28 days, the strength approaches that of the control concrete; beyond this period, the strength of the slag concrete exceeds the strength of control concrete (Admixtures and ground slag 1990). Flexural strength is usually improved by the use of slag cement, which makes it beneficial to concrete paving applications where flexural strengths are important. It is believed that the increased flexural strength is the result of the stronger bonds in the cement-slag-aggregate system because of the shape and surface texture of the slag particles.

Problems occur when slag concrete is used in cold weather applications. At low temperatures, the strengths are substantially reduced up to 14 days, and the percentage of slag is usually reduced to 25–30% of replacement levels; when saw cutting of joints is required, the use of slag is discontinued.

Effects on Permeability of Hardened Concrete. Incorporation of granulated slags in cement paste helps in the transformation of large pores in the paste into smaller pores, resulting in decreased permeability of the matrix and of the concrete. Indicated that significant reduction in permeability is achieved as the replacement level of the slag increases from 40 to 65% of total cementitious material by mass. Because of the reduction in permeability, concrete containing granulated slag may require less depth of cover than conventional concrete requires to protect the reinforcing steel.

Effects on Freeze–Thaw Durability of Hardened Concrete. Freeze–thaw durability of slag concrete has been studied by many researchers. It has been reported that resistance of air-entrained concrete incorporating ground granulated blast-furnace slag is comparable to that of conventional concrete. Reported results of freeze–thaw tests on concrete incorporating 25–65% slag. Test results indicate that regardless of the water-to-(cement + slag) ratio, air-entrained slag concrete specimens performed excellently in freeze–thaw tests, with relative durability factors greater than 91%.

Effect on ASR of Hardened Concrete. Effectiveness of slag in preventing damage due to ASR is attributed to the reduction of total alkalies in the cement-slag blend, the lower permeability of the system, and the tying up of the alkalies in the hydration process. There have been many studies of ground granulated blast-furnace slag that has been used as partial replacement for portland cement in concrete to reduce expansion caused by alkali-aggregate reaction.

Section | 5 Calculations Relating to Concrete and Masonry

Handling, Storage, and Batching

Ground granulated blast-furnace slag should be stored in separate watertight silos (such as those used for cement) and should be clearly marked to avoid confusion with cement. In batching, it is recommended that portland cement be weighed first and then followed by the slag. Slag is like cement in that normal valves are adequate to stop the flow of material.

5.5.0 Structural Concrete Components—Calculations to Achieve High-Strength Concrete

Concrete Class	Required Field Compressive Strength	Cement Content: Minimum or Range	Air Content: Percent Range (Total)	Water Cement Ratio: Maximum or Range
B	3000 psi at 28 days	565 lbs/cu yd	5 – 8	N/A
BZ	4000 psi at 28 days	610 lbs/cu yd	N/A	N/A
D	4500 psi at 28 days	615 to 660 lbs./cu yd	5 – 8	0.44
DT	4500 psi at 28 days	700 lbs/cu yd	5 – 8	0.44
E	4200 psi at 28 days	660 lbs/cu yd	4 – 8	0.44
H	4500 psi at 56 days	580 to 640 lbs/cu yd	5 – 8	0.38–0.42
HT	4500 psi at 56 days	580 to 640 lbs/cu yd	5 – 8	0.38–0.42
P	4200 psi at 28 days	660 lbs/cu yd	4 – 8	0.44
S35	5000 psi at 28 days	615 to 720 lbs/cu yd	5 – 8	0.42
S40	5800 psi at 28 days	615 to 760 lbs/cu yd	5 – 8	0.40
S50	7250 psi at 28 days	615 to 800 lbs/cu yd	5 – 8	0.38

Class B concrete is an air-entrained concrete for general use. Class D or H concrete may be substituted for Class B concrete. Additional requirements for Class B concrete are: Class B concrete shall have a nominal coarse aggregate size of 37.5 mm (1½ in.) or smaller. Approved fly ash may be substituted for portland cement up to a maximum of 20% Class C or 30% Class F by weight.

Class BZ concrete is concrete for drilled piers. Additional requirements for class BZ concrete are: Entrained air is not required unless specified in the contract. High-range water reducers may be added at the job site to obtain desired slump and retardation. Slump shall be a minimum of 5 in. and a maximum of 8 in. Class BZ caisson concrete shall be made with 19.0 mm (¾ in.) nominal-sized coarse aggregate. Approved fly ash may be substituted for portland cement up to a maximum of 20% Class C or 30% Class F by weight.

Class D concrete is a dense medium strength structural concrete. Class H may be substituted for Class D concrete. Additional requirements for Class D concrete are: An approved water-reducing admixture shall be incorporated in the mix. Class D concrete shall be made with 19.0 mm (¾ in.) nominal-sized coarse aggregate. When placed in a bridge deck, Class D concrete shall contain a minimum of 55% AASHTO M 43 size No. 67 coarse aggregate. Approved fly ash may be substituted for portland cement up to a maximum of 20% Class C or 30% Class F by weight.

Class DT concrete may be used for deck resurfacing and repairs. Class HT may be substituted for Class DT concrete. Additional requirements for Class DT concrete are: An approved water-reducing admixture shall be incorporated in the mix. Class DT concrete shall contain a minimum of 50% AASHTO M 43 size No. 7 or No. 8 coarse aggregate. Approved fly ash may be substituted for portland cement up to a maximum of 20% Class C or 30% Class F by weight.

Class E concrete may be used for fast-track pavements needing early strength in order to open a pavement to service soon after placement. Additional requirements for Class E concrete are: Type III cement may be used. Class E concrete shall contain a minimum of 55% AASHTO M 43 size No. 357 or No. 467 coarse aggregate. If all transverse joints are doweled, then Class E concrete shall contain a minimum of 55% AASHTO M 43 sizes No. 57, No. 67, No. 357, or No. 467 coarse aggregate. Unless stated otherwise on the plans, Class E concrete shall achieve a field compressive strength of 2500 psi within 12 hours. Laboratory trial mix for Class E concrete must produce an average 28-day flexural strength of at least 650 psi. Approved fly ash may be substituted for portland cement up to a maximum of 30% Class F by weight.

Class H concrete is used for bare concrete bridge decks that will not receive a waterproofing membrane. Additional requirements for Class H concrete are: An approved water-reducing admixture shall be incorporated in the mix. Class H concrete shall contain a minimum of 55% AASHTO M 43 size No. 67 coarse aggregate. Class H concrete shall contain cementitious materials in the following ranges: 450 to 500 pounds per cubic yard Type II portland cement, 90 to 125 pounds per cubic yard fly ash, and 20 to 30 pounds per cubic yard silica fume. The total content of Type II portland cement, fly ash, and silica fume shall be 580 to 640 pounds per cubic yard. Laboratory trial mix for Class H concrete must not exceed permeability of 2000 coulombs at 56 days (ASTM C 1202). Laboratory trial mix for Class H concrete must not exhibit a crack at or before 14 days in the cracking tendency test (AASHTO PP 34).

Class HT concrete is used as the top layer for bare concrete bridge decks that will not receive a waterproofing membrane. Additional requirements for Class HT concrete are: An approved water-reducing admixture shall be incorporated in the mix. Class HT concrete shall contain a minimum of 50% AASHTO M 43 size No. 7 or No. 8 coarse aggregate. Class HT concrete shall contain cementitious materials in the following ranges: 450 to 500 pounds per cubic yard Type II portland cement, 90 to 125 pounds per cubic yard fly ash, and 20 to 30 pounds per cubic yard silica fume. The total content of Type II portland cement, fly ash, and silica fume shall be 580 to 640 pounds per cubic yard. Laboratory trial mix for Class HT concrete must not exceed permeability of 2000 coulombs at 56 days (ASTM C 1202). Laboratory trial mix for Class HT concrete must not exhibit a crack at or before 14 days in the cracking tendency test (AASHTO PP 34).

Class P concrete is used in pavements. Additional requirements for Class P concrete are: Class P concrete shall contain a minimum of 55% AASHTO M 43 size No. 357 or No. 467 coarse aggregate. If all transverse joints are doweled, then Class P concrete shall contain a minimum of 55% AASHTO M 43 sizes No. 57, No. 67, No. 357, or No. 467 coarse aggregate. Laboratory trial mix for Class P concrete must produce an average 28-day flexural strength of at least 650 psi. Class P concrete shall contain 70 to 80% portland cement and 20 to 30% Class F fly ash in the total weight of cement plus fly ash. Unless acceptance is based on flexural strength, the total weight of cement plus Class F fly ash shall not be less than 660 pounds per cubic yard. If acceptance is based on flexural strength, the total weight of cement plus Class F fly ash shall not be less than 520 pounds per cubic yard.

Class S35 concrete is a dense high-strength structural concrete. Additional requirements for Class S35 concrete are: An approved water-reducing admixture shall be incorporated in the mix. Class S35 concrete shall be made with 19 mm (¾ inch) nominal-sized coarse aggregate, that is, 100% passing the 25.0 mm (1 in.) sieve and 90 to 100% passing the 19 mm (¾ in.) sieve. When placed in a bridge deck, Class S35 concrete shall contain a minimum of 55% AASHTO M 43 size No. 67 coarse aggregate. Approved fly ash may be substituted for portland cement up to a maximum of 20% Class C or 30% Class F by weight.

Class S40 concrete is a dense high-strength structural concrete. Additional requirements for Class S40 concrete are: An approved water-reducing admixture shall be incorporated in the mix. Class S40 concrete shall be made with 19 mm (¾ in.) nominal-sized coarse aggregate. When placed in a bridge deck, Class S40 concrete shall contain a minimum of 55% AASHTO M 43 size No. 67 coarse aggregate. Approved fly ash may be substituted for portland cement up to a maximum of 20% Class C or 30% Class F by weight.

Class S50 concrete is a dense high-strength structural concrete. Additional requirements for Class S50 concrete are: An approved water-reducing admixture shall be incorporated in the mix. Class S50 concrete shall be made with 19 mm (¾ in.) nominal-sized coarse aggregate. When placed in a bridge deck, Class S50 concrete shall contain a minimum of 55% AASHTO M 43 size No. 67 coarse aggregate. Approved fly ash may be substituted for portland cement up to a maximum of 20% Class C or 30% Class F by weight. Laboratory trial mix for Class S50 concrete must not exhibit a crack at or before 14 days in the cracking tendency test (AASHTO PP 34).

5.6.0 Lightweight Concrete Mix Design—Calculations Utilizing Perlite

Perlite product guide

Perlite Concrete

Lightweight/Insulating/Fireproof

This product guide contains various mix designs for lightweight concrete, utilizing perlite as the primary aggregate. The basic mix designs presented may be used as stated or as a starting point for your own custom mixes.

Perlite lightweight concrete is used in many different applications, including lightweight tile mortar, garden sculpture, decorative brick, gas-fireplace logs and floor fills.

Some Perlite Concrete Applications

Chimney Lining	Statuary
Floor Systems	Tank Bases
Fuel Tanks	Tank Insulation
Pool Base	Tile Mortars
Sound/Firewalls	Underground Pipe

Perlite concrete, while not usually suited for structural or load bearing uses, offers many advantages beyond its lightweight. Perlite concrete provides sound deadening properties and is thermal insulating as well, depending on mix design. Generally speaking, the lighter the weight, the greater the insulative properties.

Mix Designs

Cement (sack)	Perlite (ft³)	Expanded Shale	Washed Concrete Sand (ft³)	Water (Gal)	Admix (Fl. Oz.)	Dry Density	Wet Density	Compressive Strength (lb/in²)	Thermal Conductivity ("k")	Yield (Cu. Ft.)
1	8	-	-	16	28^A	22	37	90-125	0.54	8
1	6	-	-	13	24^A	27	42	125-200	0.64	6
1	5	-	-	11	20^A	30	46	230-300	0.71	5
1	4	-	-	10	16^A	36	50	350-500	0.83	4
1	3	2^1	-	9	1^A & 3^B	54	72	1400-1700	n/a	3.8
1	3	2^2	-	10	2^B & 3^B	62	78	2000-2100	n/a	3.5
1	3	2^2	-	10	3^B	65	90	2500-2800	n/a	3.2
1	3	-	-	7.5	7	45	58	800-1100	n/a	3
1	1.6	-	2.5	9.2	3^A	82	98	1100-1300	n/a	5.1
1	2	-	-	5.5	3^A	60	74	1600-1900	n/a	2
1	1.1	-	2.1	7.8	3^A	88	105	2300-2500	n/a	3.5

EXPANDED SHALE ADMIXTURE
[1] 5/16" [A] Air Entrainment
[2] 1/2" [B] Pozzolith 300-N

Mix Instructions

Proper mixing will assure the maximum yield and uniformity. Low shear, low RPM mixers (similar to plaster mixers) are recommended for best results.

1. Add all materials except perlite to mixer; then mix until this slurry is fairly uniform. Two minutes will usually suffice.
2. Add all perlite; then mix again only long enough for a uniform mix, probably another 2 to 3 minutes. Excess water and undermixing may reduce yield and workability. Overmixing may degrade the perlite and increase concrete density, reducing yield. Optimum mixing cycle can usually be determined with one or two trial batches.

Source: Perlite Institute, Harrisburg, PA

General Considerations

- Addition of limited mason sand to a perlite/cement mix increases the compressive strength (to a point) and also the weight by approximately 100 lbs per cubic foot of sand.
- Addition of expanded shale also increases the compressive strength (to a point) and weight, but at about one-third the weight of sand, at higher cost.
- Addition of fibers increases the tensile and flexural strength of perlite concrete, thereby reducing shrink cracking.
- Addition of air-entraining agents reduces the weight and compressive strength of the mix and improves freeze–thaw performance.
- A range of aggregate size is desirable for increasing compressive strength. Superplasticizers and water reducers can also be used to increase strength.
- For detailed product finishes, finer aggregate particles can be used.

Typical Mix Data for Perlite Sand Concrete

Cement (sacks)	1	1
Perlite (ft^3)	3	2.4
Sand (ft^3)	2	1.5
AEA* (oz)		
Water (gal/sack)	8.2	8.1
Cement (factor/yd-100% yield)	5.87	7.44
Density (Wet)	83	84
Density (Dry)	69	74
Compressive Strength (28-day, lb/in^2)	1000	1200

*Air Entraining Agent as recommended by perlite manufacturer.

AMERICAN SOCIETY FOR TESTING
AND MATERIALS
1916 Race St., Philadelphia, PA 19103
PERLITE INSTITUTE
1924 N. Second Street, Harrisburg, PA 17102
717/238-9723. FAX: 717/238-9985
website: http://www.perlite.org. email: info@perlite.org

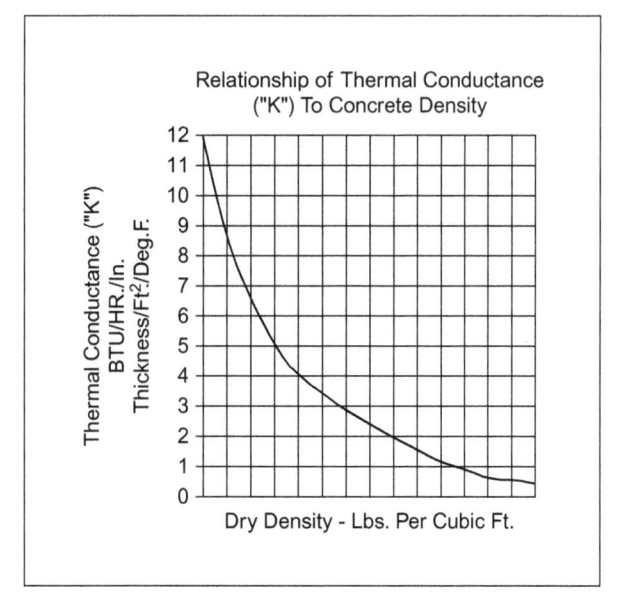

Grading Requirements for Lightweight Aggregates for Insulating Concrete

Size Designation	Weight % Passing Sieves								
	19.0-mm (¾-in.)	12.5-mm (½-in.)	9.5-mm (3/8-in.)	4.75-mm (No. 4)	2.36-mm (No. 8)	1.18-mm (No. 16)	600-μm (No. 30)	300-μm (No. 50)	150-μm (No. 100)
Group 1									
Perlite	'"	'"	'"	100	85 to 100	45 to 85	20 to 60	5 to 25	0 to 10

5.7.0 Set-Retarding Admixtures Delay Hydration of Cement

Set-Retarding

Retarding admixtures (retarders) are known to delay hydration of cement without affecting the long-term mechanical properties. They are used in concrete to offset the effect of high temperatures, which decrease setting times, or to avoid complications when unavoidable delays between mixing and placing occur. Use of set retarders in concrete pavement construction (1) enables farther hauling, thus eliminating the cost of relocating central mixing plants; (2) allows more time for texturing or plastic grooving of concrete pavements; (3) allows more time for hand finishing around the headers at the start and end of the production day; and (4) helps eliminate cold joints in two-course paving and in the event of equipment breakdown. Retarders can also be used to resist cracking due to form deflection that can occur when horizontal slabs are placed in sections (Mindess and Young 1981). Because of these advantages, set retarders are considered the second most commonly used admixtures in the highway industry, especially in the construction of bridge decks (U.S. Dept. Trans. 1990).

Composition and Mechanism of Retardation. Many water reducers have a retarding tendency. Therefore, some of the ingredients in water reducers, such as lignosulfate acids and hydroxycarboxylic acids, are also a basis for set-retarding admixtures. Other important materials used in producing set retarders are sugars and their derivatives.

Mechanisms of set retardation were studied by many researchers. Several theories have been offered to explain this mechanism. The role of retarding admixtures can be explained in a simple way: the admixtures form a film around the cement compounds (e.g., by absorption), thereby preventing or slowing the reaction with water. The thickness of this film will dictate how much the rate of hydration is retarded. After a while, this film breaks down, and normal hydration proceeds. However, in some cases when the dosage of admixtures exceeds a certain critical point, hydration of cement compounds will never proceed beyond a certain stage, and the cement paste will never set. Thus, it is important to avoid overdosing a concrete with a retarding admixture.

Other factors influencing the degree of retardation include the w/c, cement content, C3A and alkali contents in cement, the type and dosage of the admixture, and the stage at which the retarder is added to the mix. The effectiveness of retarder is increased if its addition to the fresh concrete is delayed for a few minutes.

Effect on Concrete Properties and Application. In addition to their role in controlling setting time, retarders—like any other admixtures—influence the properties of fresh and hardened concrete. Air entrainment of concrete is affected and fewer air-entraining agents need to be used because some retarders entrain air (see water reducers). Slump loss might increase even when abnormal setting behavior does not occur.

Because of retarding action, the 1-day strength of the concrete is reduced. However, ultimate strength is reported to be improved by using set-controlling admixtures. Rates of drying shrinkage and creep could increase by using retarders, but the ultimate values cannot increase.

One of the most important applications of retarding admixtures is hot-weather concreting, when delays between mixing and placing operations may result in early stiffening. Another important application is in pre-stressed concrete, where retarders prevent the concrete that is in contact with the strand from setting before vibrating operations are completed. Set retarders also allow use of high-temperature curing in prestressed concrete production without affecting the ultimate strength of the concrete.

Recommendations

1. Verification tests should be performed on liquid admixtures to confirm that the material is the same as that which was approved. The identifying tests include chloride and solids content, ph, and infrared spectrometry.
2. Water reducers and retarders may be used in bridge deck concrete to extend the time of set. This is especially important when the length of placement may result in flexural cracks created by dead load deflections during placement.
3. Increased attention needs to be placed on curing and protection due to the potential for shrinkage cracks and bleeding when water reducers are used.

5.8.0 Calculations for Mixing Small Batches of Concrete

Mixing Small Batches of Concrete on Site

Ready-mix concrete requires coarse aggregate (stone), fine aggregate (sand), cement, and water. These components can be measured by weight or by volume; the following charts provide the proportions.

To mix one cubic yard of concrete by volume

Maximum Size Aggregate	Cement	Sand	Stone	Water
3/8" (9.52 mm)	1	2½	1½	½
½" (12.6 mm)	1	2½	2	½
¾" (19.05 mm)	1	2½	2½	½
1" (25.39 mm)	1	2½	2¼	½
1½" (37.99 mm)	1	2½	3	½

To mix one cubic yard of concrete by weight

Maximum size Aggregate	Cement	Sand	Stone	Water
3/8" (9.52 mm)	29 lbs (13.15 kg)	59 lbs (26.76 kg)	46 lbs (20.87 kg)	11 lbs 4.99 kg)
½" (12.6 mm)	27 lbs (12.25 kg)	53 lbs (24.04 kg)	55 lbs (24.95 kg)	11 lbs (4.99 kg)
¾" (19.05 mm)	25 lbs (11.34kg)	47 lbs (21.32 kg)	65 lbs (29.66 kg)	10 lbs 4.54 kg)
1" (25.39 mm)	24 lbs (10.89 kg)	45 lbs (20.41 kg)	70 lbs (31.756 kg)	10 lbs (4.54 kg)
1½"	23 lbs (10.43 kg)	43 lbs (19.50 kg)	75 lbs (34.02 kg)	9 lbs (4.08 kg).

Other Simple Calculations for Small Batches

Cement—one bucket
Sand—two and one-quarter buckets
Stone—one and one-half buckets
Water—one-half bucket

Equivalents: One cubic foot of water is equal to 7.48 gallons
One gallon of water weighs 8.33 pounds
One gallon equals 3.787 liters
One liter equals 0.2642 gallons
One liter equals 1.0567 quarts

5.9.0 In-Depth Concrete Inspection Guide as Published by U.S. DOT—FHWA

. . . . Division Office*
1999 Inspection Guide (Metric Version)
Inspection-In-Depth: Major Structures—Concrete
Plant and Placement
PROJECT DATA

```
Project Number:
County:
Inspection Made By:
In Company With:
Date of Inspection:
Percent Complete:
Percent Time:
```

Scope of inspection:

The overall purpose of this inspection is to evaluate project enforcement of established concrete inspection procedures as it relates to Structural Concrete. The items of interest are concrete plant operations, form work, reinforcing steel, and concrete placement. As part of the inspection review, a determination should be conducted on the existing procedures and/or specifications as to whether they are practical and easily understood and implemented by construction personnel and the contractor and whether field personnel have sufficient training, ability, and interest to thoroughly understand and enforce existing procedures and specifications. The Area Engineer is provided the flexibility of using the guideline in its entirety or portions depending on job conditions and time limitations. This guide may be supplemented as deemed necessary by the Area Engineer for items distinct to the individual project. It is suggested that prior to the inspection, a review of all applicable provisions be made in order that a broad knowledge of the provisions can be achieved and utilized during the inspection.

REFERENCES

- Standard Specifications and Supplemental Specifications,
- Special Provisions,
- Construction Directives,

---

*This Division Office is to be filled in by the DOT-FHWA division using this form. The form can be accsessed on www.fhwa.dot.gov/construction/reviews/revconc1.cfm

- Construction Manual
  I. Concrete Plant Operation
     1. a. Are mix designs available for each class of concrete?
        b. Are they approved?
        c. Dates approved?
     2. a. Is the concrete plant approval available?
        b. Is the certification still valid?
        c. Expiration date?
        d. Is the plant inspection checklist used by .... available?
        e. Plant Type? (Central Mix, Transit Mix, Automatic, Manual, Manufacturer)
        f. Manufactures Rating?
        g. Has the mixer been checked for condition and wear (Section 601.5.3)?
     3. a. Have the scales been inspected and sealed?
        b. Expiration date?
        c. Are the Ten - 20 kg weights available?
        d. Have the scales been zero balanced and sensitivity checks been conducted?
     4. a. Is the Quality Control Plan available?
        b. Is it approved?
        c. Date approved?
        d. Approved by whom?
     5. a. Are the aggregates properly stored?
        b. Are intermixing, segregation, and/or contamination problems present?
        c. Where are the aggregate samples being taken (Section 601.5.2.7)?
        d. Are moisture tests being performed?
        e. Are the tests being documented?
     6. a. What is the water source used?
        b. Is the source approved?
        c. Comment on the water quality. Is it clean and free of oil, salt, acid, alkali, sugar, vegetation, etc.?
        d. Is the water added to the mix adjusted for moisture in aggregate?
        e. Is water added at the project site, or at the plant?
        f. Comment on the accuracy of the water measuring equipment (within 1%)?
        g. Is additional water being added at the project site?
        h. What is allowable Water/Cement (W/C) ratio from mix design?
        i. What is the W/C ratio that is being used (Section 601.7)?
     7. a. Is the cement bin and weight hopper properly sealed?
        b. Is the cement bin clean and dry?
        c. Is the fly ash in a separate bin?
        d. Is a Material Certification for the fly ash available?
        e. Are adequate records being kept of cement being delivered to the batch plant to properly track quantity received vs. quantity used for test coverage?
     8. a. Note the Field Laboratory location.
        b. Is the lab properly equipped?
        c. Is the lab in a reasonable proximity of plant?
     9. a. Are laboratory tests documented and available for water, cement, air entraining agent, retarder, curing compound, aggregate gradations, etc.?
        b. Where are the control charts being maintained?
           Are they complete and up to date?

Are there any undesirable trends noted?
Comments:
10. a. Is the Plant Inspector's Diary up to date?
    b. What entries are being made in the Plant Diary? (Materials received, tests, checks, etc.)
    c. Does the State or Contractor's Quality Control Personnel maintain a Plant Diary?
    d. Comment on the documentation of instructions to the contractor:
11. a. Comment on the knowledge of plant personnel:
    b. Comment on Personnel Certification:
12. General/Additional Comments on the Concrete Plant Operation:

II. Concrete Placement:

1. Form Work (Section 601.8)
   a. Type of form materials used?
   b. Is the mortar tight, and clean?
   c. What form treatment is being utilized (Section 601.8.5)?
   d. Are "Telltales," used for settlement monitoring, in-place?
   e. Is the form work adequately supported and true to line and grade?
   f. Has the overhang support been sufficiently checked (SD-7)?
   g. Is form removal (Section 601.8.7) based on early cylinder breaks or specification guidance?
   h. Slip forming (Section 601.8.8)?
   i. Any field welding noted?

2. **Reinforcing Steel**
   a. Are rebars being properly stored at the job site?
   b. Are the rebars clean?
   c. In the case of epoxy coated rebars, is the coating sound?
   d. Note bar sizes, spacing, clearances and general layout of steel mat.

	Plan	As Built
Rebar Size		
Bar Spacing		
Clearances		

   Are they in accordance with the plans?
   e. Is the reinforcing steel adequately supported?
   f. Are bar splices in compliance with the 30 bar diameters minimum overlap?
   g. Comment on welding (only if shown on the plans)?
   h. Field bending of reinforcing bars noted (Section 602.5)?

3. **Concrete Operation**
   a. Form work and reinforcing steel checked and approved to concrete placement?
   b. Was a preplacement conference held?

      Note items discussed (type of equipment, pour sequence, schedule, etc.)

   c. Was a dry run prior to concrete placement conducted?
   d. What is the minimum required placement rate for the operation observed?
      _____ cubic meters/hr
   e. Note the time placement commenced and was completed.

f. What placement rate was actually achieved?
   _____ cubic meters/hr
g. How was the concrete delivered (Chute, bucket, trough)?
h. Was proper technique executed?
i. Was concrete dropped through the air less than 1.5 meters?
j. Is concrete rehandling being avoided as much as possible?
k. Is concrete placed against previously placed concrete at all times?
l. Are concrete trucks clean of build-up and agitator blades not worn?
m. Note the condition of the contractor's equipment.
n. Were vibrators sufficient in quantity to do the work that was needed?
o. Were vibrators being used for a sufficient duration but not to the point of segregation?
p. Were vibrators being used to move concrete?
q. Note the type of placement/finishing machine used?
r. Was screed support adequate to maintain line and grade?
s. Was water applied to the surface to aid finishing?
t. Was a straight edge being used?
u. Was a rolling straight edge used?
   Any surface deficiencies found?
   Corrective action taken?
v. Were texture grooves 3 mm to 5 mm deep? (Section 601.11.4)
w. Comment on curing (Section 601.12) as to type, adequacy, timeliness, cold weather conditions, and maintenance of curing?
x. Note general weather conditions during concrete placement operation.
y. Comment on the establishment of adverse weather condition plans (Section 601.9) and their applicability.

4. **Field Testing**
   a. Observe State or Contractor's personnel conducting sample and field testing procedures. Comment on the following:
      - sample source
      - performance of tests in accordance with accepted test procedures
      - equipment adequacy
   b. Any material rejected?
   c. Record a sampling of test results for specification compliance:

Air (%)	Slump (mm)	Temperature (Celsius)		Depth Check		In Compliance (Yes/No)
		Air	Mix	Full Depth of Steel	To Top Mat	

5. General/additional comments on concrete placement:

III. **Project Records**
   1. Review inspectors daily reports (IDRs), supervisor's daily reports, HL-440s, and progress estimates. Comment on the documentation as to quantities used, work performed, test results recorded, problems encountered, and proper cross referencing.

2. a. How does the project monitor acceptance sampling and testing, independent assurance sampling and testing, concrete plant operations, manufactures certification of appropriate items (ex. curing compound)?
   b. Are these materials records found to be orderly?
   c. Review a sample of completed concrete cylinder strength test results:

Class of Concrete (A,B,K,ETC,)	Strength Tests (MPa)	Minimum Specification Requirements (MPa)	Meet Specs. (Yes/No)

   d. Any problems noted?
3. Is the contractor meeting the requirements of the Quality Control Plan?

IV. Closeout Conference:
1. Discuss all findings and come to an agreement on corrective actions when required.
2. Any recommendations from the review or from the project personnel?

PDF files can be viewed with the Acrobat® Reader®
Word files can be viewed with the Word Viewer 2003
This page last modified on 06/05/07
FHWA Home | Engineering | Construction

5.10.0 Calculate the Size and Weight of Concrete Reinforcing Bars

Concrete reinforcing bars are produced in straight lengths and in coils. The bars are referred to as "deformed" because they are made with ridges that allow them to "grip" the concrete into which they are embedded. Rebars are designated by size (diameter) in both U.S. and metric, and they are also designated by grade or minimum yield strength as set forth by ASTM standards.

Three ASTM grades are the most commonly manufactured

Inch-pound grade	Metric Grade	Yield in Pounds per Square Inch	Yield in Metric (Megapascals)
Grade 40	Grade 280	40,000	280
Grade 60	Grade 420	60,000	420
Grade A706	Grade 520	75,000	520

Coiled Rebar is produced in ASTM Grades 60 and A706 in the following diameters and coil size:

Size—in Diameter—Inches	Size—in Diameter—Metric	Weight per Coil
#3	10	4500 lbs
#4	13	4500 lbs
#5	16	3800 lbs
#6	19	3800 lbs

5.10.1 Reinforcing Bar Designations–Size and Diameter—U.S. and Metric

Bar Size.U.S.	Bar Size-Metric	Diameter-inches	Diameter Metric	Weight/Ft.
#3	#10	0.375	9.52 mm	0.376
#4	#13	0.500	12.7 mm	0.668
#5	#16	0.620	15.8 mm	1.043
#6	#19	0.750	19.05 mm	1.502
#7	#22	0.875	22.23 mm	2.044
#8	#25	1.000	25.4 mm	2.670
#9	#29	1.128	28.65 mm	3.400
#10	#32	1.270	32.25 mm	4.303
#11	#36	1.410	35.81 mm	5.313
#14	#43	1.693	43.0 mm	7.650
#18	#57	2.257	57.33 mm	13.60
#2*	#6	0.250	6.35 mm	0.167

*Number 2 bars are often referred to as "temperature bars" and are rarely used as concrete reinforcement.

5.10.2 Welded Wire Mesh Designations—U.S. and Metric

Welded Wire Mesh

Here are the metric equivalents of some common mesh sizes:

When you see a mesh size, what do the numbers mean?

6 × 6 W1.4/1.4 is a designated mesh size. The **6 × 6** is the horizontal and vertical spacing of the strands in inches. This is the size of the squares of space bounded by the wire strands in the mesh (the equivalent in metric is in millimeters, and so **6 × 6** inches is **152 × 152** mm). The **1.4/1.4** is the "W" number. This is the wire size (longitudinal/transverse) in cross-sectional area measured in 1/100's of a square inch (1.4 hundredths of a square inch). (The equivalent in metric is in square millimeters, and so 1.4 hundredths of a square inch is 9.1 mm^2.)

In earlier times, the **wire size** was given in **gauge** rather than in **cross-sectional area** by hundredths of a square inch. What is now called **6 × 6 W1.4/1.4** used to be called **6 × 6 W10/10**.

"**MW**" is for "Metric W number."

Current Mesh Name (wire size)	Former Mesh Name (wire gauge)	Metric Name
2×2 W4.0/4.0	2×2 - 4/4	50×50 MW25.8/25.8
2×2 W2.9/2.9	2×2 - 6/6	50×50 MW18.7/18.7
2×2 W2.1/2.1	2×2 - 8/8	50×50 MW13.3/13.3
2×2 W1.4/1.4	2×2 - 10/10	50×50 MW9.1/9.1
2×2 W0.9/0.9	2×2 - 12/12	50×50 MW5.6/5.6
2×2 W0.5/0.5	2×2 - 14/14	50×50 MW3.2/3.2
2×2 W0.3/0.3	2×2 - 16/16	50×50 MW2.0/2.0
3×3 W2.1/2.1	3×3 - 8/8	76×76 MW13.3/13.3
3×3 W1.4/1.4	3×3 - 10/10	76×76 MW9.1/9.1
3×3 W0.9/0.9	3×3 - 12/12	76×76 MW5.6/5.6

Section | 5 Calculations Relating to Concrete and Masonry   245

Current Mesh Name (wire size)	Former Mesh Name (wire gauge)	Metric Name
3×3 W0.5/0.5	3×3 - 14/14	76×76 MW3.2/3.2
4×4 W4.0/4.0 MW25.8/25.8	4×4 -4/4	102×102
4×4 W2.9/2.9 MW18.7/18.7	4×4 - 6/6	102×102
4×4 W2.1/2.1 MW13.3/13.3	4×4 - 8/8	102×102
4×4 W1.7/1.7 MW11.1/11.1	4×4 - 9/9	102×102
4×4 W1.4/1.4 MW9.1/9.1	4×4 - 10/10	102×102
4×4 W0.9/0.9 MW5.6/5.6	4×4 - 12/12	102×102
4×4 W0.7/0.7 MW4.2/4.2		102×102
4×4 W0.5/0.5 MW3.2/3.2	4×4 - 14/14	102×102
6×6 W7.4/7.4 MW47.6/47.6	6×6 - 0/0	152×152
6×6 W6.3/6.3 MW40.6/40.6	6×6 - 1/1	152×152
6×6 W5.4/5.4 MW34.9/34.9	6×6 - 2/2	152×152
6×6 W4.7/4.7 MW30.1/30.1	6×6 - 3/3	152×152
6×6 W4.0/4.0 MW25.8/25.8	6×6 - 4/4	152×152
6×6 W4.0/2.9 MW25.8/18.7	6×6 - 4/6	152×152
6×6 W3.4/3.4 MW21.7/21.7	6×6 - 5/5	152×152
6×6 W2.9/2.9 MW18.7/18.7	6×6 - 6/6	152×152
6×6 W2.5/2.5 MW15.9/15.9	6×6 - 7/7	152×152
6×6 W2.1/2.1 MW13.3/13.3	6×6 - 8/8	152×152
6×6 W1.7/1.7 MW11.1/11.1	6×6 - 9/9	152×152
6×6 W1.4/1.4 MW9.1/9.1	6×6 - 10/10	152×152
12×12 W5.4/5.4 MW34.9/34.9	12×12 - 2/2	305×305

5.11.0 Embedded Anchor Bolt—Diameter, Length, Hook, and Thread Sizes—Plain Finish

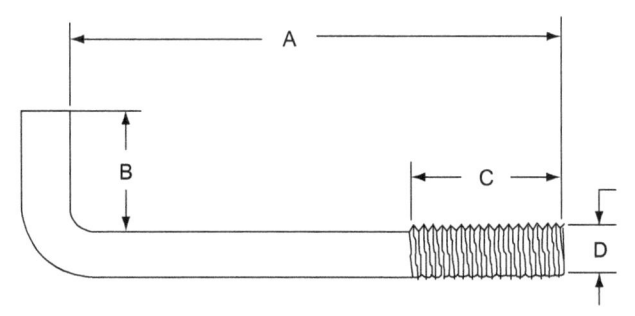

Anchor Bolts in Stock Plain Finish

Specifications: Steel to Astm F1554 (A36, GR. 55 & 105)		(New Anchor Bolt Specification)	
DIA. (D)	Length (A)	Hook (B)	Thread (C)
1/2"	6"	1-1/2"	1-1/2"
1/2"	8"	1-1/2"	1-1/2"

DIA. (D)	Length (A)	Hook (B)	Thread (C)
1/2"	10"	1-1/2"	1-1/2"
1/2"	12"	1-1/2"	1-1/2"
1/2"	14"	1-1/2"	1-1/2"
1/2"	16"	1-1/2"	1-1/2"
1/2"	18"	1-1/2"	1-1/2"
1/2"	20"	1-1/2"	1-1/2"

DIA. (D)	Length (A)	Hook (B)	Thread (C)
5/8"	8"	3"	3"
5/8"	10"	3"	3"
5/8"	12"	3"	3"
5/8"	14"	3"	3"
5/8"	15"	3"	3"
5/8"	16"	3"	3"
5/8"	18"	3"	3"
5/8"	20"	3"	3"
5/8"	21"	3"	3"
5/8"	24"	3"	3"
5/8"	30"	3"	3"

DIA. (D)	Length (A)	Hook (B)	Thread (C)
3/4"	8"	3"	4"
3/4"	10"	3"	4"
3/4"	12"	3"	4"
3/4"	14"	3"	4"
3/4"	15"	3"	4"
3/4"	16"	3"	4"
3/4"	18"	3"	4"
3/4"	20"	3"	4"
3/4"	21"	3"	4"
3/4"	24"	3"	4"
3/4"	30"	3"	4"

More Anchor Bolts in Stock Plain Finish
Specifications: Steel to Astm F1554 (A36, GR. 55 & 105)
(New Anchor Bolt Specification)

DIA. (D)	Length (A)	Hook (B)	Thread (C)
7/8"	12"	4"	4"
7/8"	14"	4"	4"
7/8"	16"	4"	4"

DIA. (D)	Length (A)	Hook (B)	Thread (C)
7/8"	18"	4"	4"
7/8"	20"	4"	4"
7/8"	22"	4"	4"
7/8"	24"	4"	4"
7/8"	30"	4"	4"

DIA. (D)	Length (A)	Hook (B)	Thread (C)
1"	12"	4"	4"
1"	14"	4"	4"
1"	15"	4"	4"
1"	16"	4"	4"
1"	18"	4"	4"
1"	20"	4"	4"
1"	22"	4"	4"
1"	24"	4"	4"
1"	30"	4"	6"
1"	36"	6"	7"

DIA. (D)	Length (A)	Hook (B)	Thread (C)
1-1/4"	24"	6"	6"
1-1/4"	30"	6"	6"
1-1/4"	36"	6"	6"
1-1/4'	44"	6"	8"

Source: St. Louis Screw and Bolt

5.11.1 Embedded Anchor Bolts—Diameter, Length, Hook, and Thread Sizes—Galvanized

Anchor Bolts in Stock   Hot Dip Galvanized   Specifications: Steel to Astm F1554 (A36, GR. 55 & 105)
(New Anchor Bolt Specification)

DIA. (D)	Length (A)	Hook (B)	Thread (C)
1/2"	6"	1-1/2"	1-1/2"
1/2"	8"	1-1/2"	1-1/2"
1/2"	10"	1-1/2"	1-1/2"
1/2"	12"	1-1/2"	1-1/2"
1/2"	14"	1-1/2"	1-1/2"
1/2"	16"	1-1/2"	1-1/2"
1/2"	18"	1-1/2"	1-1/2"
1/2"	20"	1-1/2"	1-1/2"

DIA. (D)	Length (A)	Hook (B)	Thread (C)
5/8"	8"	3"	3"
5/8"	10"	3"	3"
5/8"	12"	3"	3"
5/8"	14"	3"	3"
5/8"	15"	3"	3"
5/8"	16"	3"	3"
5/8"	18"	3"	3"
5/8"	20"	3"	3"
5/8"	21"	3"	3"
5/8"	24"	3"	3"
5/8"	30"	3"	3"

DIA. (D)	Length (A)	Hook (B)	Thread (C)
3/4"	8"	3"	4"
3/4"	10"	3"	4"
3/4"	12"	3"	4"
3/4"	14"	3"	4"
3/4"	15"	3"	4"
3/4"	16"	3"	4"
3/4"	18"	3"	4"
3/4"	20"	3"	4"
3/4"	21"	3"	4"
3/4"	24"	3"	4"
3/4"	30"	3"	4"

5.12.0 Brick Sizes—Nominal versus Actual Size

Standard and modular brick dimensions
   Face Brick

	Nominal Size	Actual Size	No. per sq ft*
Standard	4" × 2 2/3" × 8"	3 5/8" × 2 1/4" × 8"	6.27
Modular	4" × 2 2/3" × 8"	3 5/8" × 2 1/4" × 7 5/8"	6.86
King	3 3/8" × 3" × 10"	3" × 2 5/8" × 9 5/8"	4.80
Queen	2 3/4" × 3" × 10"	3 1/8" × 2 3/4" × 9 5/8"	4.61
Engineer	4" × 3 1/5" × 8"	3 5/8" × 2 13/16" × 7 5/8"	5.65
Economy	4" × 4" × 8"	3 5/8" × 3 5/8" × 7 5/8"	4.50
Utility	4" × 4" × 12"	3 5/8" × 3 5/8" × 11 1/2"	3.03
Jumbo	4" × 3" × 8"	3 5/8" × 2 3/4" × 8"	5.50
Norman	4" × 2 2/3" × 12"	3 5/8" × 2 1/4" × 11 5/8"	4.57
Norwegian	3 1/2" × 3" × 12"	3 1/2" × 2 3/4" × 11 5/8"	3.84

*With 3/8" mortar joints (bed and vertical).

There is no true standard of face brick sizes versus names. The same-name brick may vary in size among different manufacturers. It is best to specify face bricks by size first and then by name.

By permission:mc2-ice.com

5.12.1 Diagrams of Modular and Nonmodular Bricks

Modular Brick Sizes (Nominal Dimensions)

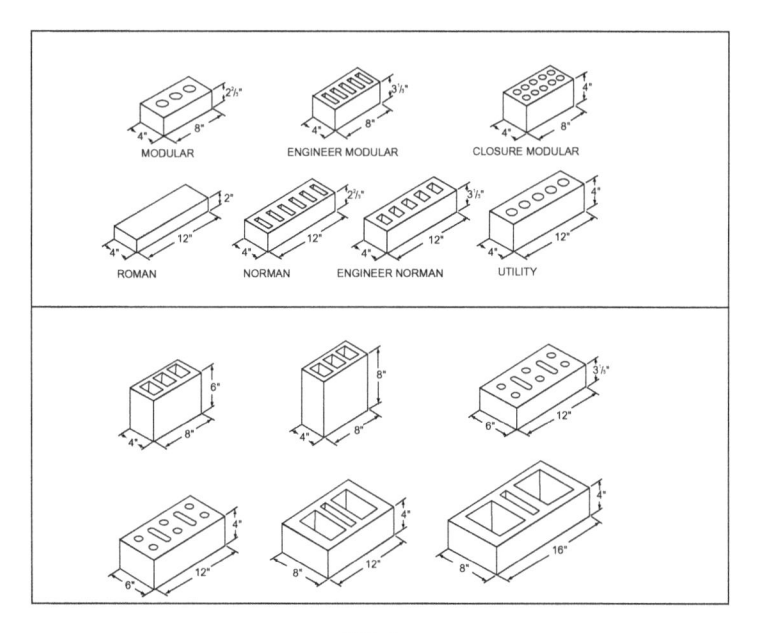

FIG. 1

Nonmodular Brick Sizes (Specified Dimensions)

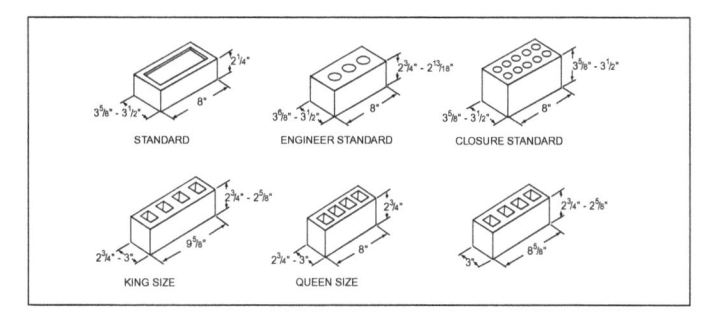

FIG. 2

5.12.2 Brick Positions in a Wall

Brick Positions

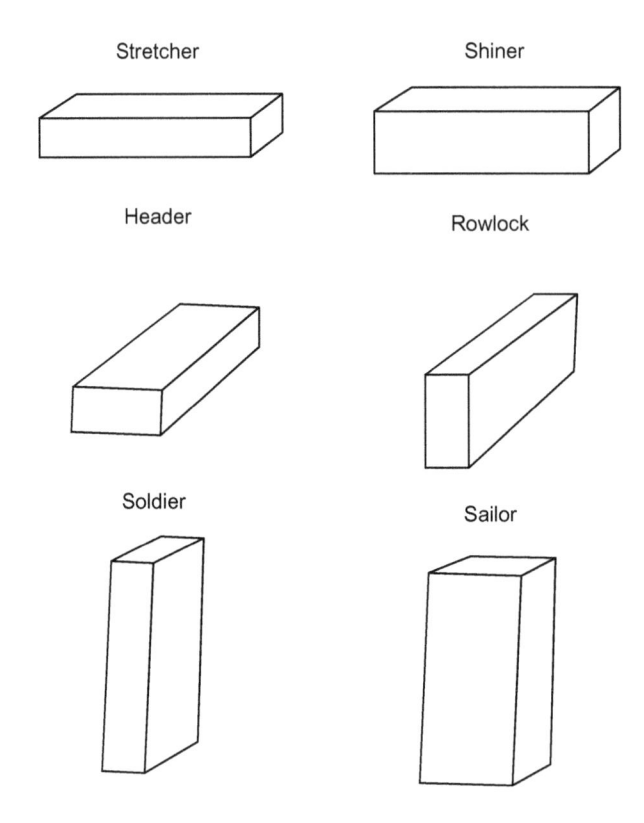

Stretcher

Shiner

Header

Rowlock

Soldier

Sailor

5.12.3 Calculate the Number of Bricks in a Wall

No. of Standard Bricks in One Sq Ft of Brick Wall								
Wall Thickness		No. of Bricks Thick	Thickness of Horizontal Mortar Joint					
5/8"	3/4"		1/8"	1/4"	3/8"	1/2"	5/8"	3/4"
4"	4 ½"	1	7.33	7.00	6.67	6.33	6.08	5.80
8"	9"	2	14.67	14.00	13.33	12.67	12.17	11.60
12"	13"	3	22.00	21.00	20.00	19.00	18.25	17.40
16"	17"	4	29.33	28.00	26.67	25.33	24.33	23.20
20"	21"	5	36.67	35.00	33.33	31.67	30.42	29.00
24"	25"	6	44.00	42.00	40.00	38.00	36.50	34.80

8"×2 ¼"×3 ¾" standard brick with ¼" vertical mortar joints in running bond

No. of square inches per standard brick with mortar joints
8"×2 ¼" face with ¼" vertical mortar joints
Thickness of horizontal mortar joints

1/8"	¼"	3/8"	½"	5/8"	¾"	7/8"	1"
19.30	20.625	21.656	22.69	23.72	24.75	25.78	26.81

5.12.3.1 Calculate the Number of Bricks for Your Project

1. Add up the total square footage of brick wall area.
    (Multiply height × length of brick wall area.)
2. Add up the total square footage of masonry openings.
    (Window and door areas not requiring brick.)
3. Subtract masonry opening square footage from brick wall area square footage.
4. Multiply this number of 7 if using modular size brick.
5. Multiply this number by 5.8 if using engineer size brick.

Example:
   1530 square feet of wall area. 154 square feet of windows and doors (Masonry Openings)
   1530 minus 154 – 1376 square feet of brick wall
   1376 × 7 = 9976 modular size brick
   1376 × 5.8 – 8,394 engineer size brick

How to Figure Pavers:
   1. Add up the total square footage of the horizontal area.
   2. Multiply this number by 5.

Example:
   10 × 10 area – 100 square feet of paving area.
   100 square feet × 5 = 500 pavers needed for a 100 square foot area.

5.12.4 Percentages to Add for Various Bond Patterns

Bonds

Face brick quantities

Percentages to add to numbers calculated for running bond walls

   English bond (with a full header course every other course)   50.000 %
   English bond (with a full header course every 6th course)   16.667
   English cross bond (with a full header course every other course)   50.000
   English cross bond (with a full header course every 6th course)   16.667
   Common bond* (with a full header course every 5th course)   20.000
   Common bond (with a full header course every 6th course)   16.667
   Common bond (with a full header course every 7th course)   14.333
   Dutch bond (with a full header course every other course)   50.000

***Common** bond is also called **American** bond.
Do not confuse **common** bond with everyday **running** bond.

Dutch bond (with a full header course every 6th course)   16.667
Dutch cross bond (with a full header course every other course)   50.000
Dutch cross bond (with a full header course every 6th course)   16.667
Double header bond (two headers and a stretcher every 5th course)   10.000
Double header bond (two headers and a stretcher every 6th course)   8.333
Flemish bond (with a full header course every other course)   33.333
Flemish bond (with a full header course every 6th course)   5.600
Double Flemish bond (with a full header course every other course)   10.000
Double Flemish bond (with a full header course every 3rd course)   6.667
3 stretcher Flemish bond (with a full header course every other course)   7.143
3 stretcher Flemish bond (with a full header course every 3rd course)   4.800
4 stretcher Flemish bond (with a full header course every other course)   5.600
4 stretcher Flemish bond (with a full header course every 3rd course)   3.704

5.12.5 Types of Brick—by Material

Adobe
Kiln burned (your common red brick)
Sand lime

By permission: mc2-ice.com

5.12.6 Chart Reflecting Nominal Size, Joint Thickness, Actual Size—Modular/Nonmodular Bricks

	MODULAR BRICK SIZES							
Unit Designation	Nominal Dimensions, in.			Joint Thickness[1], in.	Specified Dimensions[2], in.			Vertical Coursing
	w	h	l		w	h	l	
Modular	4	2⅔	8	⅜	3⅝	2¼	7⅝	3C = 8 in.
				½	3½	2¼	7½	
Engineer Modular	4	3⅕	8	⅜	3⅝	2¾	7⅝	5C = 16 in.
				½	3½	2¹³⁄₁₈	7½	
Closure Modular	4	4	8	⅜	3⅝	3⅝	7⅝	1C = 4 in.
				½	3½	3½	7½	
Roman	4	2	12	⅜	3⅝	1⅝	11⅝	2C = 4 in.
				½	3½	1½	11½	
Norman	4	2⅔	12	⅜	3⅝	2¼	11⅝	3C = 8 in.
				½	3½	2¼	11½	
Engineer Norman	4	3⅕	12	⅜	3⅝	2¾	11⅝	5C = 16 in.
				½	3½	2¹³⁄₁₈	11½	
Utility	4	4	12	⅜	3⅝	3⅝	11⅝	1C = 4 in.
				½	3½	3½	11½	

MODULAR BRICK SIZES—Cont'd

Unit Designation	Nominal Dimensions, in.			Joint Thickness, in.	Specified Dimensions, in.			Vertical Coursing
	w	h	l		w	h	l	
NONMODULAR BRICK SIZES								
Standard				3/8	3 5/8	2 1/4	8	3C = 8 in.
				1/2	3 1/2	2 1/4	8	
Engineer Standard				3/8	3 5/8	2 3/4	8	5C = 16 in.
				1/2	3 1/2	2 13/16	8	
Closure Standard				3/8	3 5/8	3 5/8	8	1C = 4 in.
				1/2	3 1/2	3 1/2	8	
King				3/8	3	2 3/4	9 5/8	5C = 16 in.
					2 3/4	2 5/8	9 5/8	
Queen				3/8	3	2 3/4	8	5C = 16 in.
					2 3/4	2 3/4	8	

[1] Common joint sizes used with length and width dimensions. Joint thicknesses of bed joints vary based on vertical coursing and specified unit height.
[2] Specified dimensions may vary within this range from manufacturer to manufacturer.

Brick dimensions

- Brick are identified by three dimensions: width, height, and length. Height and length are sometimes called face dimensions, for these are the dimensions showing when the brick is laid as a stretcher. The shaded areas indicate the surfaces of the brick that are exposed. Specifications and purchase orders should list brick dimensions in the standard order of **width** first, followed by **height**, then **length**.
- When specifying or designing with brick, it is important to understand the difference between nominal, specified, and actual dimensions. Nominal dimensions are most often used by the architect in modular construction. In modular construction, all dimensions of the brick and other building elements are multiples of a given module. Such dimensions are known as nominal dimensions. For brick masonry the *nominal* dimension is equal to the specified unit dimension plus the intended mortar joint thickness. The intended mortar joint thickness is the thickness required so that the unit plus joint thickness match the coursing module. In the inch-pound system of measurement, nominal brick dimensions are based on multiples (or fractions) of 4 in. In the SI (metric) system, nominal brick dimensions are based on multiples of 100 mm. For more information on modular construction see *Technical Notes* 10A Revised.
- As the name implies, the *specified* dimension is the anticipated manufactured dimension. It should be stated in project specifications and purchase orders. Specified dimensions are used by the structural engineer in the rational design of brick masonry.

By permission: Brick Industry Association, Reston, VA

5.12.7 Chart Reflecting Nominal Size, Joint Thickness, Actual Size—Other Brick Sizes

MODULAR BRICK SIZES

Nominal Dimensions, in.			Joint Thickness[1], in.	Specified Dimensions[2], in.			Vertical Coursing
w	h	l		w	h	l	
4	6	8	3/8	3 5/8	5 5/8	7 5/8	2C = 12 in.
			1/2	3 1/2	5 1/2	7 1/2	
4	8	8	3/8	3 5/8	7 5/8	7 5/8	1C = 8 in.
			1/2	3 1/2	7 1/2	7 1/2	
6	3 1/3	12	3/8	5 5/8	2 3/4	11 5/8	5C = 16 in.
			1/2	5 1/2	2 13/18	11 1/2	
6	4	12	3/8	5 5/8	3 5/8	11 5/8	1C = 4 in.
			1/2	5 1/2	3 1/2	11 1/2	
8	4	12	3/8	7 5/8	3 5/8	11 5/8	1C = 4 in.
			1/2	7 1/2	3 1/2	11 1/2	
8	4	16	3/8	7 5/8	3 5/8	15 5/8	1C = 4 in.
			1/2	7 1/2	3 1/2	15 1/2	

NONMODULAR BRICK SIZES

			Joint Thickness	w	h	l	Vertical Coursing
			3/8	3	2 3/4	8 5/8	5C = 16 in.
			3/8	3	2 5/8	8 5/8	

[1] Common joint sizes used with length and width dimensions. Joint thicknesses of bed joints vary based on vertical coursing and specified unit height.
[2] Specified dimensions may vary within this range from manufacturer to manufacturer.

5.12.8 Nominal Modular Size of Brick and Number of Courses in 16 Inches

Unit Designation	Thickness in.	Face Dimensions		Number of Courses in 16 in.
		Height in.	Length in.	
Standard	4	2 2/3	8	6
Engineer	4	3 1/5	8	5
Economy 8 or Jumbo Closure	4	4	8	4
Double	4	5 1/3	8	3
Roman	4	2	12	8
Norman	4	2 2/3	12	6
Norwegian	4	3 1/5	12	5
Economy 12 or Jumbo Utility	4	4	12	4
Triple	4	5 1/3	12	3
SCR brick[c]	6	2 2/3	12	6
6-in. Norwegian	6	3 1/5	12	5
6-in. Jumbo	6	4	12	4
8-in. Jumbo	8	4	12	4

Available as solid units conforming to ASTM C 216 or ASTM C 62, or, in some cases. as hollow brick conforming to ASTM C 652

By permission: Brick Industry Association, Reston VA

5.12.9 Calculate Vertical Coursing Height Based upon Number of Units

No. of Courses	Vertical Coursing of Unit			
	2C = 4 in.	3C = 8 in.	5C = 16 in.	1C = 4 in.
1	0' - 2"	0' - 2⅔"	0' - 3⅕"	0' - 4"
2	0' - 4"	0' - 5⅓"	0' - 6⅖"	0' - 8"
3	0' - 6"	0' - 8"	0' - 9⅗"	1' - 0"
4	0' - 8"	0' - 10⅔"	1' - 0⅘"	1' - 4"
5	0' - 10"	1' - 1⅓"	1' - 4"	1' - 8"
6	1' - 0"	1' - 4"	1' - 7⅕"	2' - 0"
7	1' - 2"	1' - 6⅔"	1' - 10⅖"	2' - 4"
8	1' - 4"	1' - 9⅓"	2' - 1⅗"	2' - 8"
9	1' - 6"	2' - 0"	2' - 4⅘"	3' - 0"
10	1' - 8"	2' - 2⅔"	2' - 8"	3' - 4"
11	1' - 10"	2' - 5⅓"	2' - 11⅕"	3' - 8"
12	2' - 0"	2' - 8"	3' - 2⅖"	4' - 0"
13	2' - 2"	2' - 10⅔"	3' - 5⅗"	4' - 4"
14	2' - 4"	3' - 1⅓"	3' - 8⅘"	4' - 8"
15	2' - 6"	3' - 4"	4' - 0"	5' - 0"
16	2' - 8"	3' - 6⅔"	4' - 3⅕"	5' - 4"
17	2' - 10"	3' - 9⅓"	4' - 6⅖"	5' - 8"
18	3' - 0"	4' - 0"	4' - 9⅗"	6' - 0"
19	3' - 2"	4' - 2⅔"	5' - 0⅘"	6' - 4"
20	3' - 4"	4' - 5⅓"	5' - 4"	6' - 8"
21	3' - 6"	4' - 8"	5' - 7⅕"	7' - 0"
22	3' - 8"	4' - 10⅔"	5' - 10⅖"	7' - 4"
23	3' - 10"	5' - 1⅓"	6' - 1⅗"	7' - 8"
24	4' - 0"	5' - 4"	6' - 4⅘"	8' - 0"
25	4' - 2"	5' - 6⅔"	6' - 8"	8' - 4"
26	4' - 4"	5' - 9⅓"	6' - 11⅕"	8' - 8"
27	4' - 6"	6' - 0"	7' - 2⅖"	9' - 0"
28	4' - 8"	6' - 2⅔"	7' - 5⅗"	9' - 4"
29	4' - 10"	6' - 5⅓"	7' - 8⅘"	9' - 8"
30	5' - 0"	6' - 8"	8' - 0"	10' - 0"
31	5' - 2"	6' - 10⅔"	8' - 3⅕"	10' - 4"
32	5' - 4"	7' - 1⅓"	8' - 6⅖"	10' - 8"
33	5' - 6"	7' - 4"	8' - 9⅗"	11' - 0"
34	5' - 8"	7' - 6⅔"	9' - 0⅘"	11' - 4"
35	5' - 10"	7' - 9⅓"	9' - 4"	11' - 8"
36	6' - 0"	8' - 0"	9' - 7⅕"	12' - 0"
37	6' - 2"	8' - 2⅔"	9' - 10⅖"	12' - 4"
38	6' - 4"	8' - 5⅓"	10' - 1⅗"	12' - 8"
39	6' - 6"	8' - 8"	10' - 4⅘"	13' - 0"
40	6' - 8"	8' - 10⅔"	10' - 8"	13' - 4"
41	6' - 10"	9' - 1⅓"	10' - 11⅕"	13' - 8"
42	7' - 0"	9' - 4"	11' - 2⅖"	14' - 0"
43	7' - 2"	9' - 6⅔"	11' - 5⅗"	14' - 4"
44	7' - 4"	9' - 9⅓"	11' - 8⅘"	14' - 8"
45	7' - 6"	10' - 0"	12' - 0"	15' - 0"

No. of Courses	Vertical Coursing of Unit 2C = 4 in.	3C = 8 in.	5C = 16 in.	1C = 4 in.
46	7′ - 8″	10′ - 2⅔″	12′ - 3⅕″	15′ - 4″
47	7′ - 10″	10′ - 5⅓″	12′ - 6⅖″	15′ - 8″
48	8′ - 0″	10′ - 8″	12′ - 9⅗″	16′ - 0″
49	8′ - 2″	10′ - 10⅔″	13′ - 0⅘″	16′ - 4″
50	8′ - 4″	11′ - 1⅓″	13′ - 4″	16′ - 8″
100	16′ - 8″	22′ - 2⅔″	26′ - 8″	33′ - 4″

1 in. = 25.4 mm; 1 ft = 0.3 m
Brick positioned in wall as stretchers or headers.

By permission: Brick Industry Association, Reston, VA

5.12.10 Calculate Horizontal Coursing Based upon Number of Units

	Unit Length					
	Nominal Dimensions, in.		*Specified Dimensions, in.*			
			8		*8⅝*	*9⅝*
Number of Units	*8*	*12*	*½ in. jt.*	*⅜ in. jt.*	*⅜ in. jt.*	*⅜ in. jt.*
1	0′ - 8″	1′ - 0″	0′ - 8½″	0′ - 8⅜″	0′ - 9″	0′ - 10″
2	1′ - 4″	2′ - 0″	1′ - 5″	1′ - 4¾″	1′ - 6″	1′ - 8″
3	2′ - 0″	3′ - 0″	2′ - 1½″	2′ - 1⅛″	2′ - 3″	2′ - 6″
4	2′ - 8″	4′ - 0″	2′ - 10″	2′ - 9½″	3′ - 0″	3′ - 4″
5	3′ - 4″	5′ - 0″	3′ - 6½″	3′ - 5⅛″	3′ - 9″	4′ - 2″
6	4′ - 0″	6′ - 0″	4′ - 3″	4′ - 2¼″	4′ - 6″	5′ - 0″
7	4′ - 8″	7′ - 0″	4′ - 11½″	4′ - 10⅝″	5′ - 3″	5′ - 10″
8	5′ - 4″	8′ - 0″	5′ - 8″	5′ - 7″	6′ - 0″	6′ - 8″
9	6′ - 0″	9′ - 0″	6′ - 4½″	6′ - 3⅜″	6′ - 9″	7′ - 6″
10	6′ - 8″	10′ - 0″	7′ - 1″	6′ - 11¾″	7′ - 6″	8′ - 4″
11	7′ - 4″	11′ - 0″	7′ - 9½″	7′ - 8⅛″	8′ - 3″	9′ - 2″
12	8′ - 0″	12′ - 0″	8′ - 6″	8′ - 4½″	9′ - 0″	10′ - 0″
13	8′ - 8″	13′ - 0″	9′ - 2½″	9′ - 0⅞″	9′ - 9″	10′ - 10″
14	9′ - 4″	14′ - 0″	9′ - 11″	9′ - 9¼″	10′ - 6″	11′ - 8″
15	10′ - 0″	15′ - 0″	10′ - 7½″	10′ - 5⅝″	11′ - 3″	12′ - 6″
16	10′ - 8″	16′ - 0″	11′ - 4″	11′ - 2″	12′ - 0″	13′ - 4″
17	11′ - 4″	17′ - 0″	12′ - 0½″	11′ - 10⅜″	12′ - 9″	14′ - 2″
18	12′ - 0″	18′ - 0″	12′ - 9″	12′ - 6¾″	13′ - 6″	15′ - 0″
19	12′ - 8″	19′ - 0″	13′ - 5½″	13′ - 3⅛″	14′ - 3″	15′ - 10″
20	13′ - 4″	20′ - 0″	14′ - 2″	13′ - 11½″	15′ - 0″	16′ - 8″
21	14′ - 0″	21′ - 0″	14′ - 10½″	14′ - 7⅞″	15′ - 9″	17′ - 6″
22	14′ - 8″	22′ - 0″	15′ - 7″	15′ - 4¼″	16′ - 6″	18′ - 4″
23	15′ - 4″	23′ - 0″	16′ - 3½″	16′ - 0⅝″	17′ - 3″	19′ - 2″
24	16′ - 0″	24′ - 0″	17′ - 0″	16′ - 9″	18′ - 0″	20′ - 0″
25	16′ - 8″	25′ - 0″	17′ - 8½″	17′ - 5⅜″	18′ - 9″	20′ - 10″

	Unit Length—Cont'd					
	Nominal Dimensions, in.		Specified Dimensions, in.			
			8		8⅝	9⅝
Number of Units	8	12	½ in. jt.	⅜ in. jt.	⅜ in. jt.	⅜ in. jt.
26	17' - 4"	26' - 0"	18' - 5"	18' - 1¾"	19' - 6"	21' - 8"
27	18' - 0"	27' - 0"	19' - 1½"	18' - 10⅛"	20' - 3"	22' - 6"
28	18' - 8"	28' - 0"	19' - 10"	19' - 6½"	21' - 0"	23' - 4"
29	19' - 4"	29' - 0"	20' - 6½"	20' - 2⅞"	21' - 9"	24' - 2"
30	20' - 0"	30' - 0"	21' - 3"	20' - 11¼"	22' - 6"	25' - 0"
31	20' - 8"	31' - 0"	21' - 11½"	21' - 7⅝"	23' - 3"	25' - 10"
32	21' - 4"	32' - 0"	22' - 8"	22' - 4"	24' - 0"	26' - 8"
33	22' - 0"	33' - 0"	23' - 4½"	23' - 0⅜"	24' - 9"	27' - 6"
34	22' - 8"	34' - 0"	24' - 1"	23' - 8¾"	25' - 6"	28' - 4"
35	23' - 4"	35' - 0"	24' - 9½"	24' - 5⅛"	26' - 3"	29' - 2"
36	24' - 0"	36' - 0"	25' - 6"	25' - 1½"	27' - 0"	30' - 0"
37	24' - 8"	37' - 0"	26' - 2½"	25' - 9⅞"	27' - 9"	30' - 10"
38	25' - 4"	38' - 0"	26' - 11"	26' - 6¼"	28' - 6"	31' - 8"
39	26' - 0"	39' - 0"	27' - 7½"	27' - 2⅝"	29' - 3"	32' - 6"
40	26' - 8"	40' - 0"	28' - 4"	27' - 11"	30' - 0"	33' - 4"
41	27' - 4"	41' - 0"	29' - 0½"	28' - 7⅜"	30' - 9"	34' - 2"
42	28' - 0"	42' - 0"	29' - 9"	29' - 3¾"	31' - 6"	35' - 0"
43	28' - 8"	43' - 0"	30' - 5½"	30' - 0⅛"	32' - 3"	35' - 10"
44	29' - 4"	44' - 0"	31' - 2"	30' - 8½"	33' - 0"	36' - 8"
45	30' - 0"	45' - 0"	31' - 10½"	31' - 4⅞"	33' - 9"	37' - 6"
46	30' - 8"	46' - 0"	32' - 7"	32' - 1¼"	34' - 6"	38' - 4"
47	31' - 4"	47' - 0"	33' - 3½"	32' - 9⅝"	35' - 3"	39' - 2"
48	32' - 0"	48' - 0"	34' - 0"	33' - 6"	36' - 0"	40' - 0"
49	32' - 8"	49' - 0"	34' - 8½"	34' - 2⅜"	36' - 9"	40' - 10"
50	33' - 4"	50' - 0"	35' - 5"	34' - 10¾"	37' - 6"	41' - 8"
100	66' - 8"	100' - 0"	70' - 10"	69' - 9½"	75' - 0"	83' - 4"

1 in. = 25.4 mm; 1 ft = 0.3 m

By permission: Brick Industry Association, Reston, VA

5.12.11 Horizontal Coursing—Soft and Hard Metric Dimensioning

Hard metric equivalent of $8'' \times 8'' \times 16'' = 400$ mm
$= 390$ mm $+ 10$ mm mortar joint

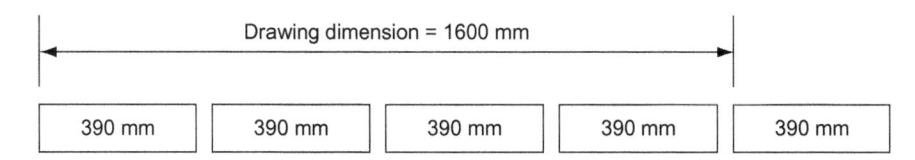

Drawing dimension = 1600 mm

| 390 mm | 390 mm | 390 mm | 390 mm | 390 mm |

Hard metric is based on 100 mm modules.

Detail 1 — Hard Metric Dimensioning

Soft metric equivalent of $8'' \times 8'' \times 16'' = 406.4$ mm
$= 396.9$ mm $+ 9.5$ mm mortar joint

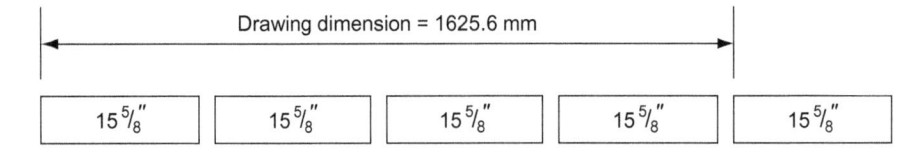

Drawing dimension = 1625.6 mm

| $15\,5/8''$ | $15\,5/8''$ | $15\,5/8''$ | $15\,5/8''$ | $15\,5/8''$ |

Soft metric is based on $4''$ modules.

Detail 2 — Soft Metric Dimensioning

5.13.0 Profiles and Dimensions of Typical Concrete Masonry Units (CMUs)

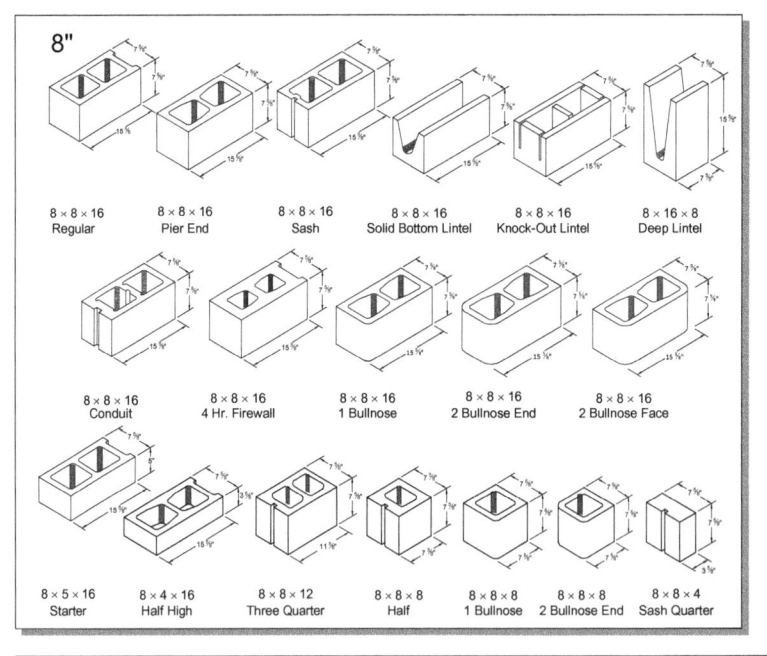

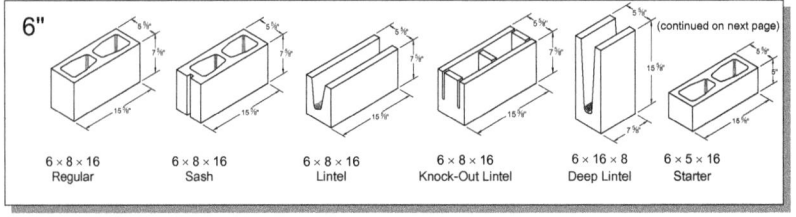

8"

- 8 × 8 × 16 Regular
- 8 × 8 × 16 Pier End
- 8 × 8 × 16 Sash
- 8 × 8 × 16 Solid Bottom Lintel
- 8 × 8 × 16 Knock-Out Lintel
- 8 × 16 × 8 Deep Lintel
- 8 × 8 × 16 Conduit
- 8 × 8 × 16 4 Hr. Firewall
- 8 × 8 × 16 1 Bullnose
- 8 × 8 × 16 2 Bullnose End
- 8 × 8 × 16 2 Bullnose Face
- 8 × 5 × 16 Starter
- 8 × 4 × 16 Half High
- 8 × 8 × 12 Three Quarter
- 8 × 8 × 8 Half
- 8 × 8 × 8 1 Bullnose
- 8 × 8 × 8 2 Bullnose End
- 8 × 8 × 4 Sash Quarter

6"

- 6 × 8 × 16 Regular
- 6 × 8 × 16 Sash
- 6 × 8 × 16 Lintel
- 6 × 8 × 16 Knock-Out Lintel
- 6 × 16 × 8 Deep Lintel
- 6 × 5 × 16 Starter

(continued on next page)

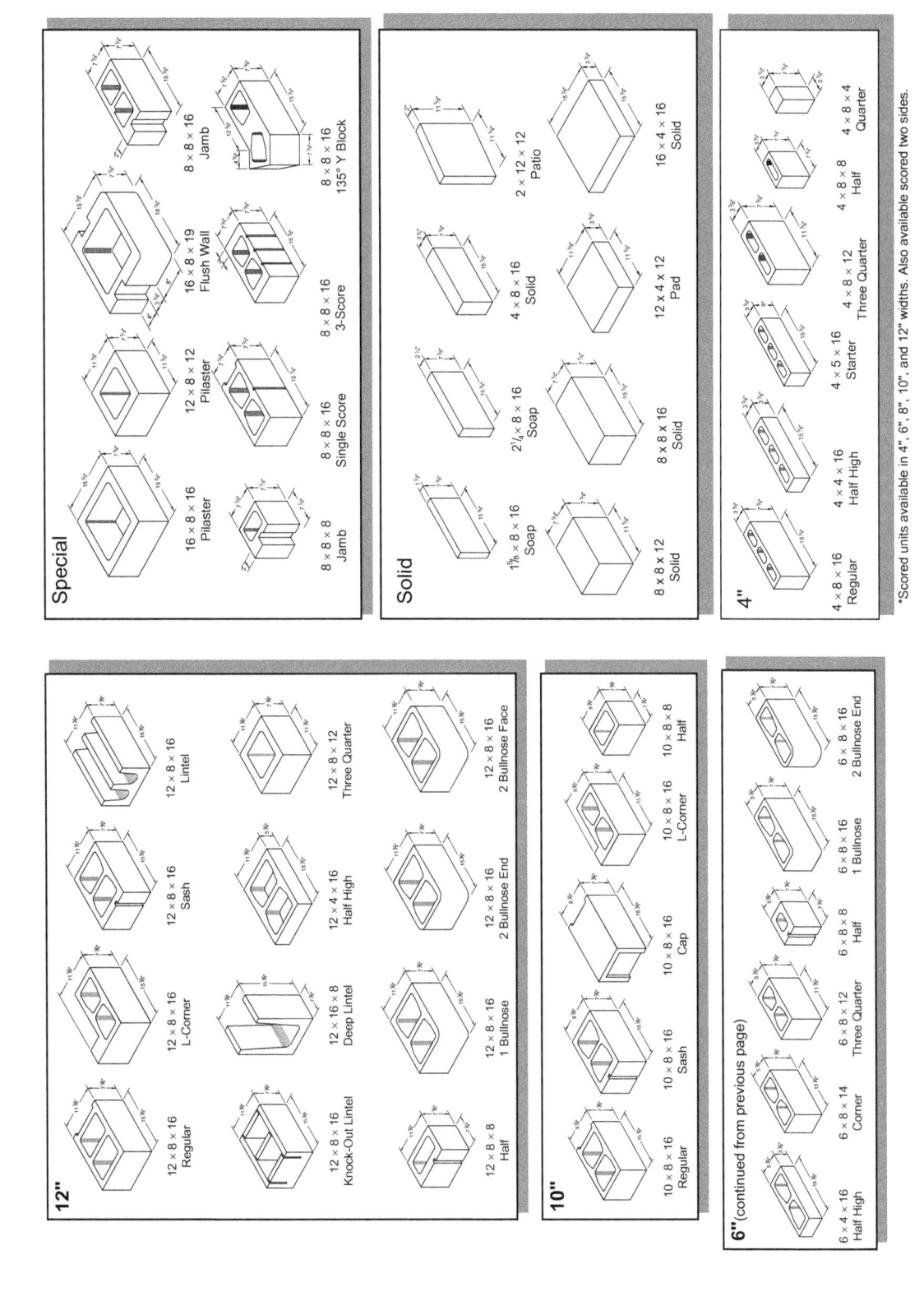

Section | 5 Calculations Relating to Concrete and Masonry

Source: Palestine Concrete Tile Co., Palestine, TX

5.14.0 Mortar Mixes—ASTM Minimums

Type M	2500 psi
Type S	1800 psi
Type N	750 psi
Type O	350 psi
Type K	75 psi

But be aware that the mix listed for type N mortar typically achieves a 28-day strength in the range of 1500 to 2400 psi. This meets and beats the ASTM requirement of 750 psi by a great deal.

Another example is the mix listed for type O mortar that provides a usual psi in the range of 750 to 1200 and higher, sometimes up to 2000. Again, this meets the minimum psi of 350 by a large percentage.

Typical type M mixes have strengths of 3000 to 3800 psi and so exceed the ASTM minimum compressive strength requirement of 2500 psi.

Type S mortars are required to have a minimum of 1800 psi, and their mixes usually give you strengths of from 2300 to 3000 psi.

5.14.1 Mixture Calculations for Types N, M, S, O, K Mortar

Here are seven common and uncommon mortar mixes. They are types N, M, S, and O. There is also mortar for glass block, straight lime mortar, and type K. Type K is used solely in historic preservation. Each one has a certain proportion of portland cement, hydrated lime, and sand. Mortar proportions are always expressed in that order. In addition, these proportions always refer to volumes, not to weight or a combination of volumes and weights. But then the components of these mixes are usually purchased by weight, but that's not how the mixes are measured.

A mix designated as 3/1/12 has three parts of portland cement, 1 part hydrated lime, and 12 parts sand. Now let's say that you want to compute mortar by the cubic yard. So how much of each mortar component is in a cubic yard? Let's go through all seven of the mixes and see.

Be aware that the proportions of lime, cement, and sand in each mix type can vary a bit by geographic regions or by contractors within a region. However, we are showing you the commonly used proportions and if you are used to something a little different, then you are simply using a regional or personal variation on the standard.

Also, these amounts are designed to add up to exactly one cu yd of material. Field amounts can show other quantities of components due to the realities of outdoor mortar mixing. Much of the literature on mortar proportions and mixes show greater or different quantities due to the great amount of waste in the actual preparation, transportation within the job site, and handling during the use of a batch of mortar. The numbers shown here reflect computed amounts. These are exact mathematical measurements down to the spoonful (though we give you final amounts of sand in tons and the other parts in bags). The tons and bags are finely measured. The terms **hydrated lime** and **lime putty** mean the same thing since lime putty is simply wet hydrated lime (you added some water to it and stirred it up), whereas in hydrated lime all of the water molecules are stoichiometrically bonded to the calcium and magnesium in the lime and the lime remains a dry powder. Lime putty is just wet hydrated lime.

The mix calculations use densities set out by the **ASTM**. These are:

Portland cement	94 lbs/cu ft
Hydrated lime	40 lbs/cu ft
Sand	80 lbs/cu ft

The purchased items are by these:

Portland cement	94 lb bags

Hydrated lime   50 lb bags
Sand   by the ton
Component amounts
Type N mortar

This uses a 1/1/6 mix and results in a mortar with a 750 psi compressive strength. Type N is the normal, general-purpose mortar mix and can be used in above-grade work in both exterior and interior load-bearing installations.

To get 1 cu yd of N mortar, you need 27 cubic feet of the components in a 1 to 1 to 6 proportion.

Portland cement   3.375 cu ft
Hydrated lime   3.375 cu ft
Sand   20.25 cu ft
Total   27 cu ft

Based on the ASTM densities, this gives you 317.25 lbs of portland cement, 135 lbs of hydrated lime, and 1620 lbs of sand.

To put together a single cubic yard of type N mortar, you need to buy and mix together:

3.375 bags of portland cement (94 lb bags)
2.7 bags of hydrated lime (50 lb bags)
0.81 tons of sand

Type M mortar

This uses a 3/1/12 mix and results in a mortar with a 2500 psi compressive strength. Type M is used for below-grade load-bearing masonry work and for chimneys and brick manholes.

To get 1 cu yd of M mortar, you need 27 cubic feet of the components in a 3 to 1 to 12 proportion.

Portland cement   5.0625 cu ft
Hydrated lime   1.6875 cu ft
Sand   20.25 cu ft
Total   27 cu ft

Based on the ASTM densities, this gives you 475.875 lbs of portland cement, 67.5 lbs of hydrated lime, and 1620 lbs of sand.

To put together a single cubic yard of type M mortar, you need to buy and mix:

5.0625 bags of portland cement (94 lb bags)
1.35 bags of hydrated lime (50 lb bags)
0.81 tons of sand

Type S mortar

This uses a 2/1/9 mix and results in a mortar with a 1800 psi compressive strength. Type S is used for below-grade work and in such areas as masonry foundation walls, brick manholes, retaining walls, sewers, brick walkways, brick pavement, and brick patios.

To get 1 cu yd of S mortar, you need 27 cubic feet of the components in a 2 to 1 to 9 proportion.

Portland cement   4.5 cu ft
Hydrated lime   2.25 cu ft
Sand   20.25 cu ft
Total   27 cu ft

Based on the ASTM densities, this gives you 423 lbs of portland cement, 90 lbs of hydrated lime, and 1620 lbs of sand.

To put together a single cubic yard of type S mortar, you need to buy and mix:

4.5 bags of portland cement (94 lb bags)
1.8 bags of hydrated lime (50 lb bags)
0.81 tons of sand

Type O mortar

This uses a 1/ 2/9 mix and results in a mortar with a 350 psi compressive strength. Type O is a lime-rich mortar and is also referred to as "pointing" mortar. It is used in **above-grade, non–load-bearing** situations in both interior and exterior environments.

To get 1 cu yd of O mortar, you need 27 cubic feet of the components in a 1 to 2 to 9 proportion.

Portland cement	2.25 cu ft
Hydrated lime	4.5 cu ft
Sand	20.25 cu ft
Total	27 cu ft

Based on the ASTM densities, this gives you 211.5 lbs of portland cement, 180 lbs of hydrated lime, and 1620 lbs of sand.

To put together a single cubic yard of type O mortar, you need to buy and mix together:

2.25 bags of portland cement (94 lb bags)
3.6 bags of hydrated lime (50 lb bags)
0.81 tons of sand

Type K mortar

This uses a 1 / 3 / 10 mix and results in a mortar with but a 75 psi compressive strength. Type K is useful only in historic preservation situations where load-bearing strength is not of importance and the porous qualities of this mortar allows very little movement due to temperature and moisture fluctuations. This aids in prolonging the integrity of the old or even ancient bricks in historic structures.

To get 1 cu yd of K mortar, you need 27 cubic feet of the components in a 1 to 3 to 10 proportion.

Portland cement	1.93 cu ft
Hydrated lime	5.79 cu ft
Sand	19.29 cu ft
Total	27 cu ft

Based on the ASTM densities, this gives you 181.42 lbs of portland cement, 231.6 lbs of hydrated lime, and 1543.2 lbs of sand.

To put together a single cubic yard of type K mortar, you need to buy:

1.93 bags of portland cement
4.632 bags of hydrated lime
0.7716 tons of sand

5.14.2 Mixture Calculations for Straight Lime Mortar

This uses a 0/ 1/ 3 mix and is used now only to re-create the construction and review the methods of times past or maybe for purely visual purposes. This mortar was made before portland cement was available in many areas, and so this is what was used. Sometimes you'll see straight lime mortar called "L" mortar (for

Section | 5 Calculations Relating to Concrete and Masonry

lime), but this is not designating it as "type L" mortar as in the MSNOK types. There is no "type L" mortar.

To get 1 cu yd of lime mortar, you need 27 cubic feet of the components in a 0 to 1 to 3 proportion.

Portland cement	none
Hydrated lime	6.75 cu ft
Sand	20.25 cu ft
Total	27 cu ft

Based on the ASTM densities, this gives you no portland cement, 270 lbs of hydrated lime, and 1620 lbs of sand.

To put together a single cubic yard of lime mortar, you need to buy:

No bags of portland cement
5.4 bags of hydrated lime (50 lb bags)
0.81 tons of sand

5.14.3 Mixture Calculations for Glass Block Mortar

This uses a 1/1/4 mix and is used with as little water as possible. This is a mix designed specifically for glass block. Also, note that it uses **waterproof** Portland cement in place of "regular" Portland cement.

To get 1 cu yd of glass block mortar, you need 27 cubic feet of the components in a 1 to 1 to 4 proportion.

5.14.4 Mixture Calculations for Waterproof Portland Cement

Waterproof Portland cement	4.5 cu ft
Hydrated lime	4.5 cu ft
Sand	18 cu ft
Total	27 cu ft

Based on the ASTM densities, this gives you 423 lbs of waterproof portland cement, 180 lbs of hydrated lime, and 1,440 lbs of sand.

To put together a single cubic yard of glass block mortar, you need to buy and mix:

4.5 bags of portland cement (94 lb bags)
3.6 bags of hydrated lime (50 lb bags)
0.72 tons of sand

Note

Lime Types versus Mortar Mix Designations

Limestone formed by nature contains varying proportions of calcium to magnesium. No large scientist with a giant beaker and a set of stoppered test tubes measured out the things that make up rocks beforehand. Some of it has more magnesium, while other limestone rock has more calcium. For making mortar, it is desirable to have from a third to a half of the rock from which the mortar lime is derived composed of magnesium carbonate. The remainder then would be from one-half to two-thirds calcium carbonate. A limestone whose composition falls within these percentages is **dolomitic limestone,** and from it is made **Type S** lime hydrate. Masonry lime made from limestone that is composed of less than 5% magnesium carbonate (called **high calcium limestone** since it is 95 to 99% calcium carbonate) is labeled Type N lime hydrate. Type S lime is used to make masonry mortar. **Type N** lime can be used only if it is tested and proven on a batch-by-batch basis. The type

S lime designation stands for **S**pecial and the type **N** stands for **N**ormal. The special lime hydrate is the one normally used, and the normal lime hydrate is used only with special testing. These lime "types" have absolutely nothing to with mortar mixes type N and type S. You must never, ever confuse these lime hydrate types with mortar mixes. They have nothing to do with one another. Why "they" should label them with the same designations, we have no idea.

Minimum compressive mortar strengths, ASTM and its psi requirements

The ASTM assigns minimum required compressive strengths to the various mortar types.

5.15.0 Typical Properties of Colorless Coatings for Brick Masonry

	Water Vapor Transmission	Water Repellency	Life-Span, Years	Available with Glossy Finish	Graffiti Resistance
Film Formers					
Acrylics	Poor	Very good	5 to 7	Yes	Yes
Stearates	Poor	Varies	1	No	No
Mineral gum waxes	Poor	Good	Varies	No	No
Urethanes	Poor	Very good	1 to 3	Yes	Yes
Penetrants					
Siloxanes	Very good	Very good	10+	No	No
Silanes	Very good	Very good	10+	No	No
Silicates	Poor	Poor	Varies	No	No
Methyl siliconates	Good	Fair	Varies	No	No
Silicone resins	Fair	Varies	1	Yes	No
RTV silicone rubber	Good	Good	5 to 10	No	Yes
Blends	Varies	Varies	Varies	No	No

By permission: Brick Industry Association, Reston, VA

Section 6

Calculating the Size/Weight of Structural Steel and Miscellaneous Metals

6.0.0	Ingredients of Steel	266
6.0.1	Structural Steel in the Construction Industry	267
6.0.1.1	AISC Shape Designations—Old and New	267
6.0.2	ASTM Designations for Most Common Types of Steel in Construction	268
6.0.3	Worldwide National Standards for Steel	274
6.0.3.1	Quick Review of U.S. Metric Conversions to Assist When Reviewing Steel Sizes	274
6.0.3.2	EN, DIN, JIS Standards	275
6.0.3.3	Tolerance on JIS Dimension and Shape of WF Beams	276
6.1.0	Approximate Minimum Mechanical Properties of Some Steels	277
6.2.0	Common Structural Shapes for U.S. Steel Sections	279
6.2.1	How Steel Wide-Flange Beams Are Identified	282
6.2.2	How Steel Channels Are Identified	282
6.2.3	How Steel Angles Are Identified	283
6.2.4	Cross Sections of Standard Structural Steel Members	283
6.3.0	Calculating the U.S. Weight and Size of Wide-Flange Beams-4″ × 4″ to 36″ × 36″	284
6.3.1	Calculating the Metric Weight and Size of Wide-Flange Beams W4s to W36s	289
6.3.2	Calculating the Weight and Size of I Beams and Junior Beams	293
6.4.0	Calculating the Weight and Size of U.S. Square High-Strength Steel Sections	295
6.4.1	Calculating the Weight and Size of Metric Square High-Strength Steel Sections	299
6.5.0	Calculating the Weight and Size of U.S. Rectangular High-Strength Steel Sections	302
6.5.1	Calculating the Weight and Size of Metric Rectangular High-Strength Steel Sections	306
6.6.0	Calculating the Weight and Size of U.S. Round High-Strength Steel Sections	311
6.6.1	Calculating the Weight and Size of Metric Round High-Strength Steel Sections	315
6.6.2	Calculating the Weight of Standard, Extra Strong, and Double Strong Steel Pipes	319
6.7.0	Calculating the Weight and Size of U.S. Steel C Channels	319
6.7.0.1	Calculating the Weight and Size of U.S. A-36 and a-36 Modified C Channels	321
6.7.0.2	Calculating the Weight and Size of U.S. Channels—Ship and Car	322
6.7.1	Calculating the Weight and Size of Metric Steel C Channels	323
6.7.1.1	Calculating the Weight and Size of Metric Channel, Box, Rectangular, and Square Tubing	324
6.8.0	Calculating the Weight and Size of Structural Steel Angles	326
6.9.0	Calculating the Weight and Size of Universal Mill Plates	327
6.10.0	Bar Size Tees—Calculating Their Weight and Size	328
6.11.0	Cold and Hot Rolled Rounds—Calculating Their Weight and Size	329
6.12.0	Aluminum Structural Angles—Calculate Their Weight and Size	330

6.12.1	Aluminum Channels—Calculate Their Weight and size	331	6.14.3	Converting Gauge Inches to Decimals for Sheet Steel, Aluminum, Stainless Steel 340
6.12.2	Aluminum Structural Beams—Calculate Their Weight and size	333	6.15.0	Carbon Steel Expanded Metal Grating—ASTM A1011 341
6.13.0	Plate Steel-3/16" to 6" Thickness—Calculate Their Weight and Size	334	6.15.1	Carbon Steel Catwalk Expanded Metal Grating—ASTMA569/569M 342
6.14.0	Sheet and Coil Steel—Types and Uses	336	6.15.2	Aluminum Expanded Metal Grating 342
6.14.1	Calculating the Weight of Various Types of Carbon, Stainless, and Galvanized Sheet Steel	337	6.16.0	Aluminum Rectangular Bar Grating 344
			6.16.1	Aluminum I Bar and Rec Bar Grating 346
6.14.2	Calculating the Weight of Low-Carbon, Hot-Dipped Galvanized Roof Deck	338	6.16.2	Aluminum Plank Sections and Pattern Availability 348

6.0.0 Ingredients of Steel

Iron Ore

Iron ore is a rock that contains iron combined with oxygen. It is sourced from mines around the world. Some of the world's highest quality iron ore comes from Australia.

Coke

Coke is made from coal. Once mined, the coal is crushed and washed. Coal is then baked in coke ovens for about 18 hours. During this process, by-products are removed and coke is produced.

Flux

Flux is a term that describes minerals used to collect impurities during iron and steelmaking. Fluxes used by BHP Steel include limestone and dolomite. The flux causes a chemical reaction, and the elements not needed for steelmaking combine to form slag.

Molten Iron

Iron is the main ingredient needed to make steel in the Basic Oxygen Steelmaking process. Molten iron is made from iron ore and other ingredients in a blast furnace.

Scrap Steel

Scrap steel comes from many different sources because it is very easily recycled. Some scrap comes from within the steelworks, where it might have been damaged or is at the end of a batch of one type of steel. It also comes from old car bodies, old ship containers, and buildings that have been demolished.

Another source of scrap can be found in our homes. Steel cans (food cans, pet food cans, aerosols, paint cans, etc.) are collected as part of council curbside collections and can be recycled an infinite number of times.

Alloying Materials

Alloying materials are used to give the steel special properties and make different types of steel. Alloying materials can be added as elements, like manganese, aluminium, and nickel, or as compounds of iron.

6.0.1 Structural Steel in the Construction Industry

There are four basic types of structural steel in general use in the construction industry today:

Carbon Steel

- A36-Structural Shapes and Plates
- A53-Structural Pipe and Tubing
- A500-Structural Pipe and Tubing
- A501-Structural Pipe and Tubing
- A529-Structural Shapes and Plate

High-Strength Low-Alloy Steel

- A441-Structural Shapes and Plates
- A572-Structural Shapes and Plates
- A618-Structural Plate and Tubing
- A992-W Shape Beams only
- A270-Structural Shapes and Plates

Corrosion-Resistant High-Strength Low-Alloy Steel

- A242-Structural Shapes and Plates
- A588-Structural Shapes and Plates

Quenched and Tempered Alloy Steel

- A514-Structural Shapes and Plates
- A517-Boiler and Pressure Vessel Steel

6.0.1.1 AISC Shape Designations—Old and New

Aisc Hot-Rolled Structural Designations Steel Shape		
New Designation	Type of Shape	Old Designation
W 24 × 76	W shape	24 WF 76
W14 × 26	W shape	14 B 26
S 24 × 100	S shape	24 I 100
M 8 × 18.5	M shape	8 M 18.5
M 10 × 9	M shape	10 JR 9.0
M 8 × 34.3	M shape	8 × 8 M 34.3
C 12 × 20.7	American Std. Channel	12 C 20.7
MC 12 × 45	Miscellaneous Channel	12 × 4 C 45.0
MC 12 × 10.6	Miscellaneous Channel	12 JR C 10.6
HP 14 × 73	HP shape	14 BP 73

(Continued)

Aisc Hot-Rolled Structural Designations Steel Shape—Cont'd		
New Designation	Type of Shape	Old Designation
L6 × 6 × 3/4	Equal Leg Angle	L 6 × 6 × 3/4
L6 × 4 × 5/8	Unequal Leg Angle	L 6 × 4 × 5/8
WT 12 × 38	Structural Tee	ST 12 WF 38
WT 7 × 13	Cut from W shape	ST 7 B 13
ST 12 × 50	Cut from S shape	ST 12/50
MT 4 × 9.25	Cut from M shape	ST 4 M 9.25
MT 5 × µ5	Cut from M shape	ST 5 Jµ4.5
MT 4 × 17.15	Cut from M shape	ST 4 M 17.15
PL ½ × 18	Plate	PL 18X7 ½
Bar 1	Square Bar	Bar 1
Bar 1-1/4 O	Round Bar	Bar 1-1/4 O
Bar 2-1/2 × 1/2	Flat Bar	Bar 2-1/2 × 1/2
Pipe 4 Std.	Pipe	Pipe 4 Std.
Pipe 4 × .Strong .	Pipe	Pipe 4 × -Strong
Pipe 4 × × Strong	Pipe	Pipe 4 × × -Strong.
TS 4 × 4 × .375	Structural Tubing: Square	Tube4 × 4 × .375
TS 5 × 3 × .375	Rectanqular	Tube 5 × 3 × .375
TS3 OD × .250	Circular	Tube 3 OD × .250

6.0.2 ASTM Designations for Most Common Types of Steel in Construction

American Society for Testing and Materials (ASTM) specifications that are common in steel design and construction for materials, preparation, and testing are given below. Just click the name to see its scope and most current date of revision. The links below get you the scope statements from ASTM. You can get nearly every one of them, though, in AISC's *2001 Selected ASTM Standards for Structural Steel Fabrication*.

- A6/A6M Standard Specification for General Requirements for Rolled Structural Steel Bars, Plates, Shapes, and Sheet Piling
- A27/A27M Standard Specification for Steel Castings, Carbon, for General Application
- A36/A36M Standard Specification for Carbon Structural Steel
- A53/A53M Standard Specification for Pipe, Steel, Black and Hot-Dipped, Zinc-Coated, Welded and Seamless
- A123/A123M Standard Specification for Zinc (Hot- Dip Galvanized) Coatings on Iron and Steel Products
- A148/A148M Standard Specification for Steel Castings, High Strength, for Structural Purposes
- A153/A153M Standard Specification for Zinc Coating (Hot- Dip) on Iron and Steel Hardware
- A193/A193M Standard Specification for Alloy-Steel and Stainless Steel Bolting Materials for High- Temperature Service
- A194/A194M Standard Specification for Carbon and Alloy Steel Nuts for Bolts for High Pressure or High- Temperature Service, or Both
- A242/A242M Standard Specification for High- Strength Low-Alloy Structural Steel
- A276 Specification for Stainless Steel Bars and Shapes
- A283/A283M Standard Specification for Low and Intermediate Tensile Strength Carbon Steel Plates

- A307 Standard Specification for Carbon Steel Bolts and Studs, 60 000 PSI Tensile Strength
- A325 Standard Specification for Structural Bolts, Steel, Heat Treated, 120/105 ksi Minimum Tensile Strength
- A325M Standard Specification for High- Strength Bolts for Structural Steel Joints (Metric)
- A354 Standard Specification for Quenched and Tempered Alloy Steel Bolts, Studs, and Other E×ternally Threaded Fasteners
- A370 Standard Test Methods and Definitions for Mechanical Testing of Steel Products
- A384 Standard Practice for Safeguarding Against Warpage and Distortion During Hot-Dip Galvanizing of Steel Assemblies
- A385 Standard Practice for Providing High-Quality Zinc Coatings (Hot-Dip)
- A435/A435M Standard Specification for Straight- Beam Ultrasonic Examination of Steel Plates
- A449 Standard Specification for Quenched and Tempered Steel Bolts and Studs
- A490 Standard Specification for Heat-Treated Steel Structural Bolts, 150 ksi Minimum Tensile Strength
- A490M Standard Specification for High- Strength Steel Bolts, Classes 10.9 and 10.9.3, for Structural Steel Joints (Metric)
- A500 Standard Specification for Cold-Formed Welded and Seamless Carbon Steel Structural Tubing in Rounds and Shapes
- A501 Standard Specification for Hot-Formed Welded and Seamless Carbon Steel Structural Tubing
- A514/A514M Standard Specification for High-Yield Strength, Quenched and Tempered Alloy Steel Plate, Suitable for Welding
- A529/A529M Standard Specification for High- Strength Carbon-Manganese Steel of Structural Quality
- A563 Standard Specification for Carbon and Alloy Steel Nuts
- A568/A568M Standard Specification for Steel, Sheet, Carbon, and High-Strength, Low-Alloy, Hot-Rolled and Cold-rolled
- A572/A572M Standard Specification for High-Strength Low-Alloy Columbium-Vanadium Structural Steel
- A578/A578M Standard Specification for Straight- Beam Ultrasonic Examination of Plain and Clad Steel Plates for Special Applications
- A588/A588M Standard Specification for High- Strength Low-Alloy Structural Steel with 50 ksi [345MPa] Minimum Yield Point to 4 in. [100mm] Thick
- A606 Standard Specification for Steel, Sheet and Strip, High-Strength, Low-Alloy, Hot-Rolled and Cold-Rolled, with Improved Atmospheric Corrosion Resistance
- A618 Standard Specification for Hot-Formed Welded and Seamless High-Strength Low-Alloy Structural Tubing
- A666 Standard Specification for Annealed or Cold- Worked Austenitic Stainless Steel Sheet, Strip, Plate, and Flat Bar
- A673/A673M Standard Specification for Sampling Procedure for Impact Testing of Structural Steel
- A706/A706M Standard Specification for Low-Alloy Steel Deformed and Plain Bars for Concrete Reinforcement
- A709/A709M Standard Specification for Carbon and High-Strength Low-Alloy Structural Steel Shapes, Plates, and Bars and Quenched-and- Tempered Alloy Structural Steel Plates for Bridges
- A759 Standard Specification for Carbon Steel Crane Rails
- A770/A770M Standard Specification for Through- Thickness Tension Testing of Steel Plates for Special Applications
- A780 Ptandard Sractice for Repair of Damaged and Uncoated Areas of Hot-Dip Galvanized Coatings
- A786/A786M Standard Specification for Rolled Steel Floor Plates
- A847 Standard Specification for Cold-Formed Welded and Seamless High-Strength, Low-Alloy Structural Tubing with Improved Atmospheric Corrosion Resistance
- A852/A852M Standard Specification for Quenched and Tempered Low-Alloy Structural Steel Plate with 70 ksi [485 MPa] Minimum Yield Strength to 4 in. [100 mm] Thick
- A913/A913M Standard Specification for High- Strength Low-Alloy Steel Shapes of Structural Quality, Produced by Quenching and Self-Tempering Process (QST)
- A931 Standard Test Method for Tension Testing of Wire Ropes and Strand
- A941 Standard Terminology Relating to Steel, Stainless Steel, Related Alloys, and Ferroalloys

- A949/A949M Standard Specification for Spray-Formed Seamless Ferritic/Austenitic Stainless Steel Pipe
- A992/A992M Standard Specification for Steel for Structural Shapes for Use in Building Framing
- A1011/A1011M Standard Specification for Steel, Sheet and Strip, Hot-Rolled, Carbon, Structural, High-Strength Low-Alloy and High-Strength Low-Alloy with Improved Formability
- B695 Standard Specification for Coatings of Zinc Mechanically Deposited on Iron and Steel
- D3359 Standard Test Methods for Measuring Adhesion by Tape Test
- D4541 Standard Test Method for Pull-Off Strength of Coatings Using Portable Adhesion Testers
- D4752 Standard Test Method for Measuring MEK Resistance of Ethyl Silicate (Inorganic) Zinc-Rich Primers by Solvent Rub
- D5402 Standard Practice for Assessing the Solvent Resistance of Organic Coatings Using Solvent Rubs
- E94 Standard Guide for Radiographic Examination
- E165 Standard Test Method for Liquid Penetrant Examination
- E709 Standard Guide for Magnetic Particle Examination
- E1032 Standard Test Method for Radiographic Examination of Weldments
- F436 Standard Specification for Hardened Steel Washers
- F606 Standard Test Methods for Determining the Mechanical Properties of Externally and Internally Threaded Fasteners, Washers, and Rivets
- F959 Standard Specification for Compressible-Washer- Type Direct Tension Indicators for Use with Structural Fasteners
- F1554 Standard Specification for Anchor Bolts, Steel, 36, 55, and 105 ksi Yield Strength
- F1852 Standard Specification for "Twist-Off" Type Tension Control Structural Bolt/Nut/Washer Assemblies, Steel, Heat Treated, 120/105 ksi Minimum Tensile Strength
- G101 Standard Guide for Estimating the Atmospheric Corrosion Resistance of Low-Alloy Steels

Here are a few brand-new and newer ASTM specifications (and some discontinued and replaced ones, too).

- ASTM A992/A992M—The new 50 ksi steel for wide-flange shapes (only) that replaces ASTMA36, ASTM A572 grade 50 and the similar dual-certified products for wide-flange shapes (only). Read about this in three places: *Are You Properly Specifying Materials? Part 1—Structural Shapes* from the January 1999 issue of AISC's *Modern Steel Construction* magazine, *Steel Industry Embraces ASTM A292* from the April 1999 issue of AISC's *Modern Steel Construction* magazine and AISC Technical Bulletin No.3, which was AISC's announcement of ASTM A992 before it had an ASTM number. Also, note that 50 ksi W-shapes are now less expensive than 36 ksi W-shapes, as explained here.
- ASTM F1554 —The new ASTM specification for anchor rods (what used to be called anchor bolts). Read about this in *Are You Properly Specifying Materials? Part 3—Fastening Products* from the March 1999 issue of AISC's *Modern Steel Construction* magazine.
- ASTM F1852—The new ASTM specification for twist-off-type tension-control bolt assemblies that meet mechanical and chemical requirements similar to ASTM A325 high-strength bolts. Read about this in *Are You Properly Specifying Materials? Part 3—Fastening Products* from the March 1999 issue of AISC's *Modern Steel Construction* magazine.

And here are ASTM Specifications that have been discontinued or replaced:

- ASTM A687 was discontinued (1999) without replacement.
- ASTM A570/A570M was discontinued (2000) and replaced by ASTM A1011/A1011M.
- ASTM E142 was discontinued (2000) and replaced by ASTM E94.

Home > Useful Information > Materials, Preparation, and Testing

Properly specify requirements for materials, preparation, and testing in your steel projects with the help in this feature. Here's a three-part article in AISC's *Modern Steel Construction* magazine that will help you to do so. Surely it's no coincidence that its title is "Are You Properly Specifying Materials?"

- Part 1—Structural Shapes (January 1999 issue)
- Part 2—Plate Products (February 1999 issue)
- Part 3—Fastening Products (March 1999 issue)

American Society for Testing and Materials (ASTM) specifications that are common in steel design and construction for materials, preparation, and testing are given below. Just click the name to see its scope and most current date of revision. The links below get you the scope statements from ASTM. You can get nearly every one of them, though, in AISC's *2001 Selected ASTM Standards Fabrication*.

- A6/A6M Standard Specification for General Requirements for Rolled Structural Steel Bars, Plates, Shapes, and Sheet Piling
- A27/A27M Standard Specification for Steel Castings, Carbon, for General Application
- A36/A36M Standard Specification for Carbon Structural Steel
- A53/A53M Standard Specification for Pipe, Steel, Black and Hot-Dipped, Zinc-Coated, Welded and Seamless
- A123/A123M Standard Specification for Zinc (Hot-Dip Galvanized) Coatings on Iron and Steel Products
- A148/A148M Standard Specification for Steel Castings, High Strength, for Structural Purposes
- Al53/A153M Standard Specification for Zinc Coating (Hot-Dip) on Iron and Steel Hardware
- A193/A193M Standard Specification for Alloy-Steel and Stainless Steel Bolting Materials for High-Temperature Service
- A194/A194M Standard Specification for Carbon and Alloy Steel Nuts for Bolts for High Pressure or High- Temperature Service, or Both
- A242/A242M Standard Specification for High-Strength Low-Alloy Structural Steel
- A276 Specification for Stainless Steel Bars and Shapes
- A283/A283M Standard Specification for Low and Intermediate Tensile Strength Carbon Steel Plates
- A307 Standard Specification for Carbon Steel Bolts and Studs, 60 000 PSI Tensile Strength
- A325 Standard Specification for Structural Bolts, Steel, Heat Treated, 120/105 ksi Minimum Tensile Strength
- A325M Standard Specification for High-Strength Bolts for Structural Steel Joints (Metric)
- A354 Standard Specification for Quenched and Tempered Alloy Steel Bolts, Studs, and Other Externally Threaded Fasteners
- A501 Standard Specification for Hot-Formed Welded and Seamless Carbon Steel Structural Tubing
- A514/A514M Standard Specification for High-Yield Strength, Quenched and Tempered Alloy Steel Plate, Suitable for Welding
- A529/A529M Standard Specification for High-Strength Carbon-Manganese Steel of Structural Quality
- A563 Standard Specification for Carbon and Alloy Steel Nuts
- A568/A568M Standard Specification for Steel, Sheet, Carbon, and High-Strength, Low-Alloy, Hot-Rolled and Cold-rolled, General Requirements for steel
- A572/A572M Standard Specification for High-Strength Low-Alloy Columbium- Vanadium Structural Steel
- A578/A578M Standard Specification for Straight-Beam Ultrasonic Examination of Plain and Clad Steel Plates for Special Applications

- A588/A588M Standard Specification for High-Strength Low-Alloy Structural Steel with 50 ksi [345MPa] Minimum Yield Point to 4 in. [100mm] Thick
- A606 Standard Specification for Steel, Sheet and Strip, High-Strength, Low-Alloy, Hot-Rolled and Cold-Rolled, with Improved Atmospheric Corrosion Resistance
- A618 Standard Specification for Hot-Formed Welded and Seamless High-Strength Low-Alloy Structural Tubing
- A666 Standard Specification for Annealed or Cold-Worked Austenitic Stainless Steel Sheet, Strip, Plate, and Flat Bar
- A673/A673M Standard Specification for Sampling Procedure for Impact Testing of Structural Steel
- A706/A1Q6M Standard Specification for Low-Alloy Steel Deformed and Plain Bars for Concrete Reinforcement
- A709/A709M Standard Specification for Carbon and High-Strength Low-Alloy Structural Steel Shapes, Plates, and Bars and Ferritic/Austenitic Stainless Steel Pipe
- A852/A852M Standard Specification for Quenched and Tempered Low-Alloy Structural Steel Plate with 70 ksi [485 MPa] Minimum Yield Strength to 4 in. [100 mm] Thick
- A913/A913M Standard Specification for High-Strength Low-Alloy Steel Shapes of Structural Quality, Produced by Quenching and Self-Tempering Process (QST)
- A931 Standard Test Method for Tension Testing of Wire Ropes and Strand
- A941 Standard Terminology Relating to Steel, Stainless Steel, Related Alloys, and Ferroalloys
- A949/A949M Standard Specification for Spray-Formed Seamless Ferritic/Austenitic Stainless Steel Pipe
- A992/A992M Standard Specification for Steel for Structural Shapes for Use in Building Framing.
- A1011/A1011M Standard Specification for Steel, Sheet and Strip, Hot-Rolled, Carbon, Structural, High-Strength Low-Alloy and High-Strength Low-Alloy with Improved Formability
- B695 Standard Specification for Coatings of Zinc Mechanically Deposited on Iron and Steel
- D3359 Standard Test Methods for Measuring Adhesion by Tape Test
- D4541 Standard Test Method for Pull-Off Strength of Coatings Using Portable Adhesion Testers
- D4752 Standard Test Method for Measuring MEK Resistance of Ethyl Silicate (Inorganic) Zinc-Rich Primers by Solvent Rub
- D5402 Standard Practice for Assessing the Solvent Resistance of Organic Coatings Using Solvent Rubs
- E94 Standard Guide for Radiographic Examination
- El65 Standard Test Method for Liquid Penetrant Examination
- E709 Standard Guide for Magnetic Particle Examination
- E1032 Standard Test Method for Radiographic Examination of Weldments

Here are a few brand-new and newer ASTM specifications (and some discontinued and replaced ones, too).

- ASTM A992/A992M – The new 50 ksi steel for wide-flange shapes (only) that replaces ASTM A36, ASTM A572 grade 50, and the similar dual-certified products for wide-flange shapes (only). Read about this in three places: *Are You Properly Specifying Materials? Part 1 – Structural Shapes* from the January 1999 issue of AISC's *Modern Steel Construction* magazine, *Steel Industry Embraces ASTM A992* from the April 1999 issue of AISC's *Modern Steel Construction* magazine, and AISC Technical Bulletin No. 3, which was AISC's announcement of ASTM A992 before it had

an ASTM number. Also, note that 50 ksi W-shapes are now less expensive than 36 ksi W-shapes, as explained here.
- ASTM F1554 – The new ASTM specification for anchor rods (what used to be called "anchor bolts"). Read about this in *Are You Properly Specifying Materials? Part 3 – Fastening Products* from the March 1999 issue of AISC's *Modern Steel Construction* magazine.
- ASTM F1852 – The new ASTM specification for twist-off-type tension-control bolt assemblies that met mechanicals and chemical requirements similar to ASTM A325 high-strength bolts. Read about this in *Are You Properly Specifying Materials? Part 3 – Fastening Products* from the March 1999 issue of AISC's *Modern Steel Construction* magazine

And here are ASTM specifications that have been discontinued or replaced:

- ASTM A687 was discontinued (1999) without replacement.
- ASTM A570/A570M was discontinued (2000) and replaced by ASTM A1011/A1011M.
- ASTM E142 was discontinued (2000) and replaced by ASTM E94.

- A370 Standard Test Methods and Definitions for Mechanical Testing of Steel Products
- A384 Standard Practice for Safeguarding Against Warpage and Distortion During Hot-Dip Galvanizing of Steel Assemblies
- A385 Standard Practice for Providing High-Quality Zinc. Coatings (Hot-Dip)
- A435/A435M Standard Specification for Straight-Beam Ultrasonic Examination of Steel Plates
- A449 Standard Specification for Quenched and Tempered Steel Bolts and Studs
- A490 Standard Specification for Heat-Treated Steel Structural Bolts, 150 ksi Minimum Tensile Strength
- A490M Standard Specification for High-Strength Steel Bolts, Classes 10.9 and 10.9.3, for Structural Steel Joints (Metric)
- A500 Standard Specification for Cold-Formed Welded and Seamless Carbon Steel Structural Tubing in Rounds and Shapes Quenched-and-Tempered Alloy Structural Steel Plates for Bridges
- A759 Standard Specification for Carbon Steel Crane Rails
- A770/A770M Standard Specification for Through-Thickness Tension Testing of Steel Plates for Special Applications
- A780 Standard Practice for Repair of Damaged and Uncoated Areas for Hot-Dip Galvanized Coatings
- A786/A786M Standard Specification for Rolled Steel Floor Plates
- A847 Standard Specification for Cold-Formed Welded and Seamless High-Strength, Low-Alloy Structural Tubing with Improved Atmospheric Corrosion Resistance
- F436 Standard Specification for Hardened Steel Washers
- F606 Standard Test Methods for Determining the Mechanical Properties of Externally and Internally Threaded Fasteners, Washers, and Rivets
- F959 Standard Specification for Compressible-Washer-Type Direct Tension Indicators for Use with Structural Fasteners
- F1554 Standard Specification for Anchor Bolts, Steel, 36, 55, and 105 ksi Yield Strength
- F1852 Standard Specification for "Twist-Off" type Tension Control Structural Bolt/Nut/Washer Asemblies, Steel, Heat Treated, 120/105 ksi Minimum Tensile Strength
- G101 Standard Guide for Estimating the Atmospheric Corrosion Resistance of Low-Alloy Steels

Can't find the specification you need in the above list? Give these materials web sites a try:

- ASTM – their standards search page.
- NSSN – "a national resource for global materials standards."
- Metal Basics – "digital solutions for marketing and sourcing metals."

6.0.3 Worldwide National Standards for Steel

Australia : AS standards
Austria : ONORM standards
Belgium : NBN standards
Bulgaria: BDS standards
Canada : CSA standards
China : GB standards
Czech/Slovak Republic : CSN standards
Finland : SFS standards
France : AFNOR standards
Great Britain : BS standards
Hungary : MSZ standards
Italy : UNI standards
Japan : JIS standards
Norway : NS standards
Poland PN standards
Romania : STAS standards
Russia : UNE standards
Spain : UNE standards
Sweden : SS standards
Switzerland : SNV/VSM standards
United States : ACI, AISI, AMS, ASME, ASTM, AWS, FED, MIL, SAE, UNS standards
European: Euronorm standards

Source: West Yorkshire Steel Ltd., Leeds, UK

6.0.3.1 Quick Review of U.S. Metric Conversions to Assist When Reviewing Steel Sizes

	psi	ksi	ksf	kPa	atm	bar
1 psi =	1	0.001	0.144	6.895	0.068	0.069
1 ksi =	1000	1	144.0	6894.757	68.045	68.948
1 ksf =	6.94	0.007	1	47.88	0.473	0.479
1 mPa =	145	0.145	20.885	1000	9.869	10
1 atm =	14.696	0.015	2.116	101.325	1	1.013
1 bar =	14.504	0.015	2.089	100	0.987	1

U.S. Customary Units	U.S. Units Equivalent	SI Units Equivalent
one square inch (sq in)	$= 1/144$ sq ft	$= 645.16$ mm^2
one square foot (sq ft)	$= 144$ sq in $= 1/9$ sq yd	$= 0.092903$ m^2
one square yard (sq yd)	$= 1296$ sq in $= 9$ sq ft	$= 0.83613$ m^2
one acre (ac)	$= 160$ sq rd $= 43560$ sq ft $= 1/640$ sq mi	$= 4046.85642$ m^2 $= 0.40469$ ha
one square rod (sq rd)	$= 272\ 1/4$ sq ft	$= 25.2929$ m^2
one rood	$= 40$ sq rd $= 10890$ sq ft	$= 1011.7141$ m^2
one square mile	$= 640$ acres	$= 2.58999$ km^2
Metric Units	**US units equivalent**	**SI units equivalent**
one acre (a)	$= \sim 1076^{3}/8$ sq ft	$= 100$ m^2 $= 0.01$ ha
one hectare (ha)	$= \sim 2\ 1/2$ acres	$= 10000$ m^2
one km^2	$= \sim 3/8$ sq mi	$= 1000000$ m^2 $= 100$ ha^2

By permission: Structural-Drafting-Net-Expert.com

6.0.3.2 EN, DIN, JIS Standards

The term *steel specification* is very often closely related to and used interchangeably with standards, although their meaning is not really identical. Hence, German steel specifications often start with the letters DIN, Japanese with JIS international with ISO, and so on.

The most widely used standard steel specifications in the United States are those published by ASTM; these steel specifications represent a consensus drawn from producers, fabricators, and users of steel mill products. In many cases, the dimensions, tolerances, limits, and restrictions in the ASTM specifications are the same as the corresponding items of the standard practices in the AISI steel product manuals.

Many of the ASTM specifications have been adopted by the American Society of Mechanical Engineers (ASME) with slight or no modifications. ASME uses the prefix S with the ASTM specifications; for example, ASME SA 213 and ASTMA 213 are the same. SAE/AISI designations for the chemical compositions of carbon and alloy steels are sometimes included in the ASTM specifications for bars, wires, and billets for forging. Some ASTM specifications for sheet products incorporate SAE-AISI designations for chemical composition.

EN (Euronorm) is a harmonized system of European countries. Although it is accepted and effectively used in all European countries, "obsolete" national systems, such as German DIN, British BS, French AFNOR, and Italian UNI, can still often be found in many documents.

DIN standards are developed by Deutsches Institute für Normung in Germany. All German steel standards and specifications are represented by the letters DIN and followed by an alphanumeric or a numeric code. For example, DIN 40NiCrMo66 or 1.6565 is a Ni-Cr-Mo steel that contains 0.35–0.45%C, 0.9-1.4%Cr, 0.5-0.7%Mn, 0.2-0.3%Mo, 1.4-1.7%Ni, 0.035%S; DIN 17200 1.1149 or DIN 17200 Cm22 is a nonresulfurized carbon steel containing 0.17-0.245C, 0.3-0.6%Mn, 0.02-0.035%S and 0.4% max Si.

JIS standards are developed by the Japanese Industrial Standards Committee (JISC) in Tokyo. The specifications begin with the prefix JIS, followed by a letter G for carbon and low-alloy steels. Examples: JIS G3445

STKM11A is a low-carbon tube steel containing 0.12%C, 0.35%Si, 0.60%Mn, 0.04%P, 0.04%S; JIS G4403 SKH2 (AISI T1Grade) is a tungsten high-speed tool steel containing 0.73-0.83%C, 3.8-4.5%Cr, 0.4%Mn, 0.4%Si, 0.8-1.2%V and 17-19%W.

The KEY to METALS database brings global metal specifications and properties together into one integrated and searchable database. Quick and easy access to the mechanical properties, chemical composition, cross-reference tables, and more provide users with an unprecedented wealth of information. Click the buttons below to learn more from the Guided Tour or to test drive the KEY to METALS database.

6.0.3.3 Tolerance on JIS Dimension and Shape of WF Beams

		Hot-rolled Wide Flange Shapes for Building Structure		
		Range	Tolerance	Remarks
Width (B)		$B \leq 400$mm	± 2.0mm	
		400 mm < B	± 3.0mm	
Depth (H)		H < 800 mm B ≤ 400 mm	± 2.0mm	
		400mm < B	± 3.0mm	
		800mm ≤ H	± 3.0mm	
Thickness	Flange (tz)*	t_2 < 16mm	± 1.0mm	
		16mm ≤ t_2 < 25mm	± 1.5mm	
		25mm ≤ t_2 < 40 mm	± 1.7mm	
		40mm ≤ t_2	± 2.0mm	
	Web (t1)	t_1 < 16mm	± 0.7mm	
		16mm ≤ t_1 < 25mm	± 1.0mm	
		25mm ≤ t_1 < 40mm	± 1.5mm	
		40mm ≤ t_1	± 2.0mm	
Length (L)		L ≤ 7m	+40mm – 0mm	
		7m < L	+ tolerance increases 5mm for the increment of every 1m or fraction thereof.	
Flange Out-of-squareness (T)		H ≤ 300mm	≤ B x 0.01 The minimum tolerance shall be 1.5 mm.	
		300mm < H	≤ B x 0.012 The minimum tolerance shall be 1.5mm.	
Bend		H ≤ 300mm	≤ L x 0.0015	Applies to both vertical and horizontal deviations
		300mm < H	≤ L x 0.001	

	Hot-rolled Wide Flange Shapes for Building Structure		
	Range	Tolerance	Remarks
Web off Center (S)	B ≦ 400mm	± 2.0mm	$S = \dfrac{b_1 - b_2}{2}$
	400mm < B	±3.5mm	
Comber of Web (δ)	H ≦ 350mm	≦ 2.0	
	350mm ≦ H < 550mm	≦ 2.5	
	550mm ≦ H	≦ 3.0	
Ends Out-of-square (e)	–	≦ B or H × 0.016 The minimum tolerance shall be 3.0 mm.	
Out-of-squareness (t)	B ≦ 400mm	≦ b × 0.015 The maximum tolerance shall be 1.5 mm	

*For the hot-rolled wide-flange shapes of JIS G3136, the following table should be used.

6mm ≦ t2 < 16mm	+1.7mm − 0.3mm
16mm ≦ t2 < 40mm	+2.3mm − 0.7mm
40mm ≦ t2 ≦ 100mm	+2.5mm − 1.5mm

6.1.0 Approximate Minimum Mechanical Properties of Some Steels

Hot Rolled—Cold Drawn—Annealed—Quenched & Tempered at 1000°F.

The following table represents an average of results obtained from a large number of tests and is offered only as a guide in accordance with standard procedure; the specimens used were 1" diameter. Under no condition do we guarantee these statistics to be accurate.

The section size, finishing temperature, and cooling rate during the rolling process influence the final mechanical properties of any steel in the As-Rolled condition. The amount of size reduction in cold drawing will affect the

Approximate Mechanical Properties
(Tensile and Yield Expressed in Thousands of Pounds Per Square Inch)

AISI GRADE	Condition	Machinability 1212 equals 100%	S.F.M.	Strength T.S.	Y.P.	Ductility El 2"%	RA	Hardness Brnl.	Rock-well
1018	H.R.	52	86	58	32	25	50	116	68B
1018	C.D.	65	107	64	54	15	40	126	72B
M1020	H.R.	50	83	55	30	25	50	110	66B
1035	H.R.	65	107	70	30	20	35	155	83B
1035	C.D.	67	111	90	75	10	40	170	87B
1035	Q&T			95	70	19	55	91	92B
1042	H.R.	61	101	80	50	15	35	175	88B
1042	C.D.	63	104	90	75	12	30	185	91B
1042	Q&T			105	80	15	40	215	96B
M1044	H.R.	53	87	82	49	15	30	170	87B
1045	H.R.	56	92	85	50	15	30	175	88B
1045	C.D.	60	99	90	80	10	30	195	93B
1095	HRA	45	74	90	55	15	40	190	90B
1117	H.R.	85	140	60	35	20	45	115	68B
1117	C.D.	90	149	75	60	15	40	143	79B
11L17	H.R.	92	152	60	35	20	45	115	89B
11L17	C.D.	100	165	75	60	15	40	143	93B
1137	H.R.	70	116	85	50	18	35	179	24C
1137	C.D.	75	121	100	85	10	30	197	89B
1137	Q&T			110	85	15	40	250	93B
1141	H.R.	65	107	90	60	15	25	180	28C
1141	C.D.	70	116	100	85	8	20	195	94B
1141	Q&T			120	100	10	35	270	96B
1144	H.R.	75	124	95	60	15	30	200	30C
1144	C.D.	85	110	100	90	7	20	210	87B
1144	Q&T			130	110	15	45	286	87B
1212	C.D.	100	165	80	70	10	40	170	87B
1213	C.D.	150	248	80	70	10	40	170	81B
B1113	C.D.	150	248	80	70	10	40	170	
12L14	C.D.	170	281	60	55	12	40	150	81B
(Type A Leaded)									
12L14	C.D.	215	355	60	55	12	40	150	81B
Selenium Treated									
1215	C.D.	150	248	80	70	10	40	170	87B
Jalcase 100	C.D.	80	132	120	100	10	25	248	24C
Jalcase 100L	C.D.	98	162	120	100	10	25	248	24C
(Leaded)									
4142	H.R.A.	56	92	85	55	20	45	170	87B
4142	C.D.A.	65	107	100	85	12	40	196	93B
4142	Q&T			150	130	15	45	300	32C
4147-50	H.R.A.	52	86	90	65	20	50	185	92B
4147-50	Q&T			170	145	15	50	350	37C

Approximate Mechanical Properties—Cont'd
(Tensile and Yield Expressed in Thousands of Pounds Per Square Inch)

		Machinability		Strength		Ductility		Hardness	
AISI GRADE	Condition	1212 equals 100%	S.F.M.	T.S.	Y.P.	El 2"%	RA	Brnl.	Rock-well
4340	H.R.A.	45	74	100	70	15	10	220	20C
4340	Q&T			175	155	12	18	370	38C
8620	H.R.	60	99	80	55	18	15	160	84B
8620	Q&T			130	92	25	55	218	24C
8620	C.D.	63	104	90	70	15	10	185	90B

mechanical properties of Cold Drawn Bars. Turned and Polished as well as Turned, Ground, and Polished Bars have approximately the same mechanical properties as the Hot-Rolled Bars from which they were produced.

6.2.0 Common Structural Shapes for U.S. Steel Sections

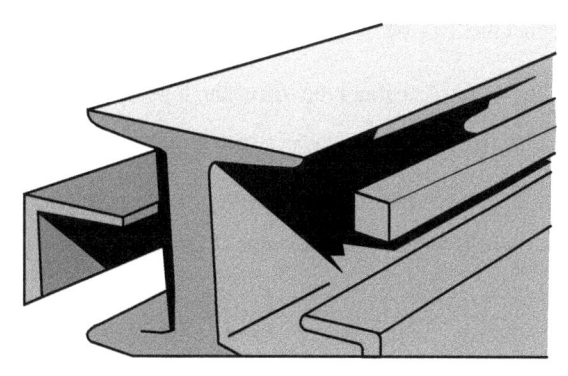

The beams are known by their profile meaning:

- The length of the beam
- The shape of the cross section
- The material used

The most commonly found steel beam is the I beam, or the wide-flanged beam, also known by the name of universal beam or stouter sections as the universal column. Such beams are commonly used in the construction of bridges and steel frame buildings.

Types of Beams

The most commonly found types of steel beams are varied, and they are as follows:

- I beams
- Wide-flange beams
- HP shape beams
- Special shape nonstandard beams
- H beams
- Junior beams

Typical Characteristics of Beams

Beams experience tensile, sheer, and compressive stresses internally due to the loads applied to them. Generally, in cinder gravity loads there is a slight reduction in the original length of the beam. This results in a smaller radius arc enclosure at the top of the beam, thus showing compression, while the same beam at the bottom is slightly stretched, enclosing a larger radius arc due to tension. The length of the beam midway and at the bends is the same as it is not under tension or compression and is defined as the neutral axis. The beam is completely exposed to shear stress above the support. There are some reinforced concrete beams that are completely under compression. These beams are called prestressed concrete beams and are built in such a manner as to produce a compression more than the expected tension under loading conditions.

Steel Channels, Stainless Steel Channels,...

- **J channels:** This kind of channel has two legs and a web. One leg is longer. This channel resembles the letter J.
- **Hat channels:** This channel has legs that are folded in the outward direction resembling an old-fashioned man's hat.
- **U channels:** This is the most common and basic channel variety. It has a base known as a web and two equal-length legs.
- **C channels:** In this channel the legs are folded back in the channel and resembles the letter C. C channels are known as rests.
- **Hemmed channels:** In this kind of channel the top of the leg is folded, hence forming double thickness.

There are other variations of channels that are available, which are customized according to the customer's needs.

Application

Steel channels are subjected to a wide array of applications. The application fields are:

- Construction
- Appliances
- Transportation
- Used in making signposts
- Used in wood flooring for athletic purposes
- Used in installing and making windows and doors

A major variant of the channel is the mild steel channel. Such channels are generally used in heavy industries. They are used in the heavy machinery industry and the automotive industry too. The mild steel channel is divided into major variants, namely:

- **Lipped channels:** The letters "LL" denote the Lipped channels. In the diagram, the number following the letters, the nominal web dimension of the channel is indicated by the first three digits. The measurements are in millimeters.

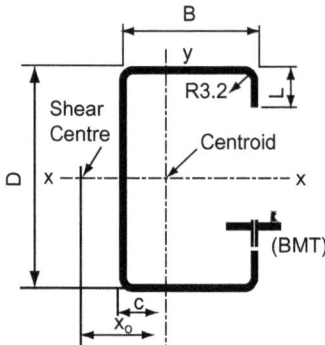

- **Plain channels:** Such channels are represented by the letters "LL". The numbers after the letters denote the web dimensions of the channel measured in millimeters. The thickness of the material is denoted by the last two digits and is measured in the tenth of millimeter.

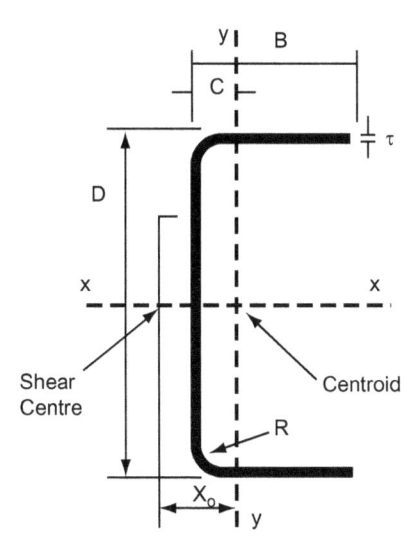

The steel angle finds an application in a number of things, such as the following:

- Used in framing
- Used in trims
- For reinforcement
- In brackets
- Used in transmission towers
- Bridges
- Lifting and transporting machinery
- Reactors
- Vessels
- Warehouses
- Industrial boilers
- Structural steel angles are used in rolling shutters for fabricating guides for strength and durability.

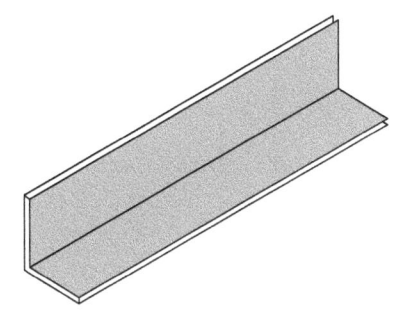

6.2.1 How Steel Wide-Flange Beams Are Identified

Steel Wide-Flange I Beams

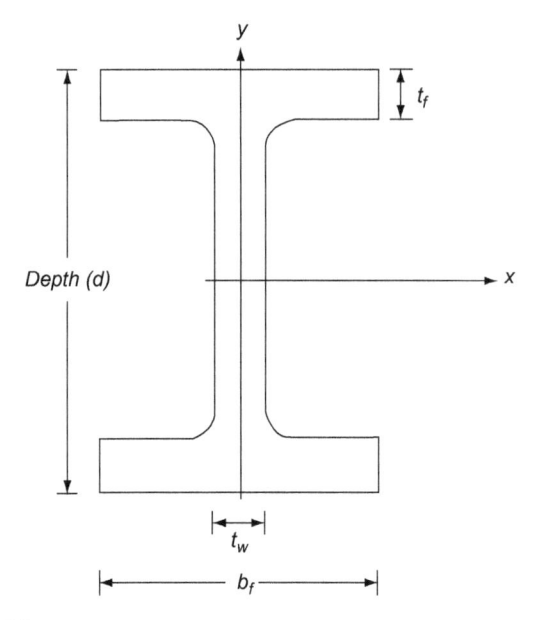

The I Beams are identified by:
W DEPTH (inches) × WEIGHT PER UNIT LENGTH (pound force per foot)
For example: **W27 × 161** is an I beam with a depth of 27 inches and a nominal weight per foot of 161 lbf/ft.

6.2.2 How Steel Channels Are Identified

American Standard Steel Channels

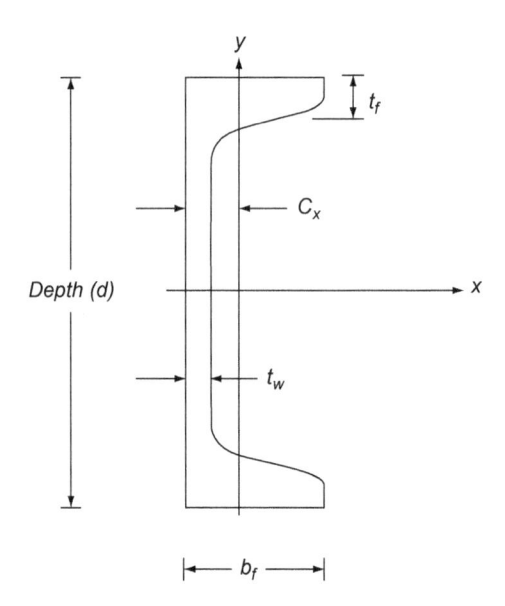

The channels are identified by:
C DEPTH (inches) × WEIGHT PER UNIT LENGTH (pound force per foot)
For example: **C12 × 30** is a channel with a depth of 12 inches and a nominal weight per foot of 30 lbf/ft.

6.2.3 How Steel Angles Are Identified

Steel Angles

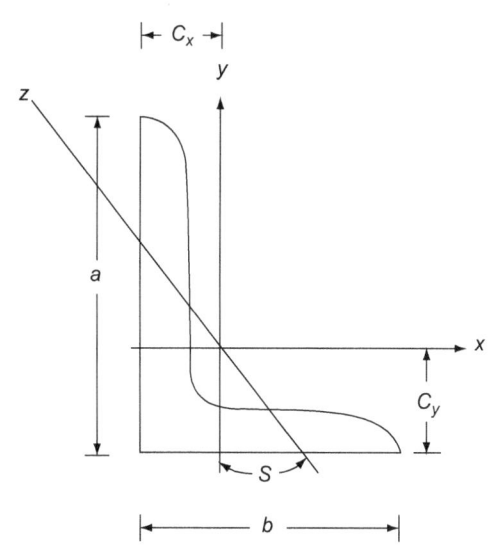

The angles are identified by:
LLEG$_a$ inches × LEG$_b$ inches × THICKNESS inches
For example, **L4 × 3 ×$^5/_8$** is an angle with one 4-inch leg and one 3-inch leg and having a nominal weight per foot of 161 lbf/ft.

6.2.4 Cross Sections of Standard Structural Steel Members

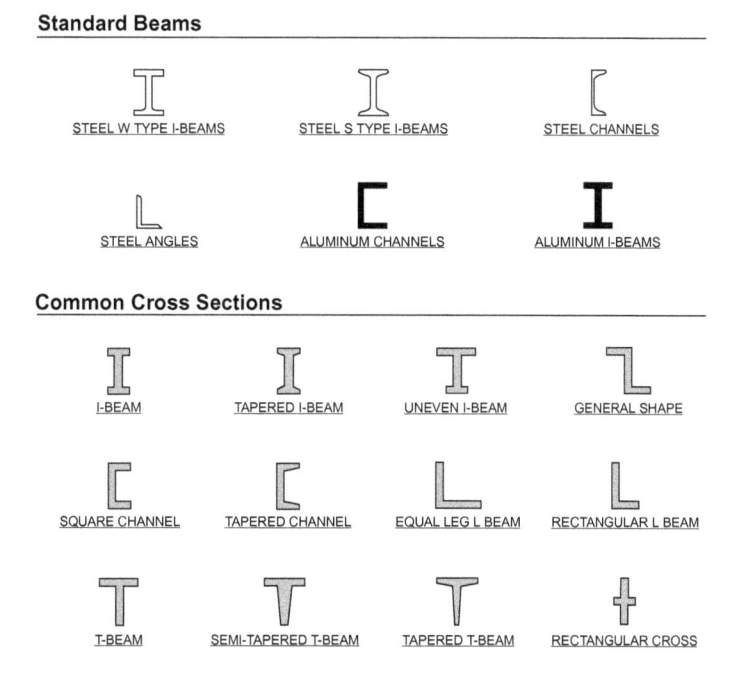

Standard Beams

STEEL W TYPE I-BEAMS   STEEL S TYPE I-BEAMS   STEEL CHANNELS

STEEL ANGLES   ALUMINUM CHANNELS   ALUMINUM I-BEAMS

Common Cross Sections

I-BEAM   TAPERED I-BEAM   UNEVEN I-BEAM   GENERAL SHAPE

SQUARE CHANNEL   TAPERED CHANNEL   EQUAL LEG L BEAM   RECTANGULAR L BEAM

T-BEAM   SEMI-TAPERED T-BEAM   TAPERED T-BEAM   RECTANGULAR CROSS

6.3.0 Calculating the U.S. Weight and Size of Wide-Flange Beams-4"×4" to 36"×16"

WIDE-FLANGE BEAMS
Conforms to A-36 and A-572-Gr50
Standard Lengths 40', 50' and 60'

Nominal Size in inches	Weight Per Foot	A Depth of Section	B Flange Width Inches	C Average Flange Thickness Inches	D Web Thickness Inches	Weight LBS. 40'	Weight LBS. 50'	Weight LBS. 60'
4 × 4	13	4.16	4.060	.375	.280	520	650	780
5 × 5	16	5.00	5.000	.360	.240	640	800	960
5 × 5	19	5.15	5.030	.430	.270	760	950	1140
6 × 4	9	5.90	3.940	.215	.170	360	450	540
6 × 4	12	6.00	4.000	.279	.230	480	600	720
6 × 4	16	6.25	4.030	.404	.260	640	800	960
6 × 6	15	5.99	5.990	.260	.230	600	750	900
6 × 6	20	6.20	6.018	.367	.258	800	1000	1200
6 × 6	25	6.37	6.080	.456	.320	1000	1250	1500
8 × 4	10	7.90	3.940	.204	.170	400	500	600
8 × 4	13	8.00	4.000	.254	.230	520	650	780
8 × 4	15	8.12	4.015	.314	.245	600	750	900
8 × 5 ½	18	8.14	5.250	.330	.230	720	900	1080
8 × 5 ½	21	8.28	5.270	.400	.250	840	1050	1260
8 × 6 ½	24	7.93	6.500	.398	.245	960	1200	1440
8 × 6 ½	28	8.06	6.540	.4.63	.285	1120	1400	1680
8 × 8	31	8.00	8.000	.433	.288	1240	1550	1860
8 × 8	35	8.12	8.027	.493	.315	1400	1750	2100
H	40	8.00	8.083	.521	.458	1600	2000	2400
8 × 8	40	8.25	8.077	.558	.365	1600	2000	2400
8 × 8	48	8.50	8.117	.683	.405	1920	2400	2880
8 × 8	58	8.75	8.222	.808	.510	2320	2900	3480
8 × 8	67	9.00	8.287	.933	.575	2680	3350	4020
10 × 4	12	9.87	3.960	.210	.190	480	600	720
10 × 4	15	10.00	4.000	.269	.230	600	750	900
10 × 4	17	10.12	4.010	.329	.240	680	850	1020
10 × 4	19	10.25	4.020	.394	.250	760	950	1140
10 × 5 3/4	22	10.17	5.75	.360	.240	880	1100	1320
10 × 5 3/4	26	10.33	5.750	.360	.240	1040	1300	1560
10 × 5 3/4	30	10.47	5.810	.510	.300	1200	1500	1800
10 × 8	33	9.75	7.964	.433	.292	1320	1650	1980
10 × 8	39	9.94	7.990	.528	.318	1560	1950	2340
10 × 8	45	10.12	8.022	.618	.350	1800	2250	2700
10 × 10	49	10.00	10.000	.558	.340	1960	2450	2940
10 × 10	54	10.12	10.028	.618	.368	2160	2700	3240
10 × 10	60	10.25	10.075	.683	.415	2400	3000	3600
10 × 10	68	10.40	10.130	.770	.470	2720	3400	4080
10 × 10	77	10.62	10.195	.868	.535	3080	3850	4620
10 × 10	88	10.84	10.265	.990	.605	3520	4400	5280

WIDE-FLANGE BEAMS—Cont'd

Conforms to A-36 and A-572-Gr50
Standard Lengths 40', 50' and 60'

Nominal Size in inches	Weight Per Foot	A Depth of Section	B Flange Width Inches	C Average Flange Thickness Inches	D Web Thickness Inches	Weight LBS. 40'	50'	60'
10 × 10	100	11.12	10.345	1.118	.685	4000	5000	6000
10 × 10	112	11.38	10.415	1.248	.755	4480	5600	6720
12 × 4	14	11.91	3.970	.224	.200	560	700	840
12 × 4	16	11.99	3.99	.265	.220	640	800	960
12 × 4	19	12.16	4.010	.349	.240	760	950	1140
12 × 4	22	12.31	4.030	.424	.260	880	1100	1320
12 × 6 1/2	26	12.22	6.490	.380	.230	1040	1300	1560
12 × 6 1/2	30	12.34	6.520	.440	.260	1200	1500	1800
12 × 6 1/2	35	12.50	6.560	.520	.300	1400	1750	2100
12 × 8	40	11.94	8.000	.516	.294	1600	2000	2400
12 × 8	45	12.06	8.042	.576	.336	1800	2250	2700
12 × 8	50	12.19	8.077	.640	.371	2000	2500	3000
12 × 10	53	12.06	10.000	5.76	.345	2120	2650	3180
12 × 10	58	12.19	10.014	.641	.359	2320	2900	3480
12 × 12	65	12.12	12.000	.606	.390	2600	3250	3900
12 × 12	72	12.25	12.040	.671	.430	2880	3600	4320
12 × 12	79	12.38	12.080	.736	.470	3160	3950	4740
12 × 12	87	12.53	12.125	.810	.515	3480	4350	5220
12 × 12	96	12.71	12.160	.900	.550	3840	4800	5760
12 × 12	106	12.88	12.230	.986	.620	4240	5300	6360
12 × 12	120	13.12	12.320	1.106	.710	4800	6000	7200
12 × 12	136	13.41	12.400	1.250	.790	5440	6800	8160
12 × 12	152	13.71	12.480	1.400	.870	6080	7600	9120
12 × 12	170	14.03	12.570	1.560	.960	6800	8500	10200
12 × 12	190	14.38	12.670	1.736	1.060	7600	9500	11400
14 × 5	22	13.72	5.000	.335	.230	880	1100	1320
14 × 5	26	13.89	5.025	.418	.255	1040	1300	1560
14 × 63/4	30	13.86	6.733	.383	.270	1200	1500	1800
14 × 63/4	34	14.00	6.750	.453	.287	1360	1700	2040
14 × 63/4	38	14.12	6.776	.513	.313	1520	1900	2280
14 × 8	43	13.68	8.000	.528	.308	1720	2150	2580
14 × 8	48	13.81	8.031	.593	.339	1920	2400	2880
14 × 8	53	13.94	8.062	.658	.370	2120	2650	3180
14 × 10	61	13.91	10.000	.643	.378	2440	3050	3660
14 × 10	68	14.06	10.040	.718	.418	2720	3400	4080
14 × 10	74	14.19	10.072	.783	.450	2960	3700	4440
14 × 10	82	14.31	10.130	.855	.510	3280	4100	4920
14 × 14 1/2	90	14.02	14.520	.710	.440	3600	4500	5400
14 × 14 1/2	99	14.16	14.565	.780	.485	3960	4950	5940
14 × 14 1/2	109	14.32	14.6050	.860	.525	4360	5450	6540
14 × 14 1/2	120	14.48	14.670	.940	.590	4800	6000	7200
14 × 14 1/2	132	14.66	14.725	1.030	.645	5280	6600	7920
14 × 16	145	14.78	15.500	1.090	.680	5800	7250	8700
14 × 16	159	14.98	15.565	1.190	.745	6360	7950	9540
14 × 16	176	15.25	15.640	1.313	.820	7040	8800	10560

(Continued)

WIDE-FLANGE BEAMS—Cont'd

Conforms to A-36 and A-572-Gr50
Standard Lengths 40′, 50′ and 60′

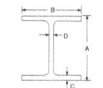

Nominal Size in inches	Weight Per Foot	A Depth of Section	B Flange Width Inches	C Average Flange Thickness Inches	D Web Thickness Inches	Weight LBS. 40′	50′	60′
14 × 16	193	15.50	15.710	1.438	.890	7720	9650	11580
14 × 16	211	15.75	15.800	1.563	.980	8440	10550	12660
14 × 16	233	16.04	15.890	1.720	1.070	9320	11650	13980
14 × 16	257	16.38	15.995	1.890	1.175	10280	12850	15420
14 × 16	283	16.74	16.110	2.070	1.290	11320	14150	16980
14 × 16	311	17.12	16.230	2.260	1.410	12440	15550	18660
14 × 16	342	17.56	16.365	2.468	1.545	13680	17100	20520
14 × 16	370	17.94	16.4575	2.658	1.655	14800	18500	22200
14 × 16	398	18.31	16.590	2.843	1.770	15920	19900	23880
14 × 16	426	18.69	16.695	3.033	1.875	17040	21300	25560
14 × 16	455	19.05	16.828	3.213	2.008	18200	22750	27300
14 × 16	500	19.63	17.008	3.501	2.188	20000	25000	30000
14 × 16	550	20.26	17.206	3.818	2.386	22000	27500	33000
14 × 16	605	20.94	17.418	4.157	2.598	24200	30250	36300
14 × 16	665	21.67	17.646	4.522	2.826	26600	33250	39900
14 × 16	730	22.44	17.889	4.910	3.069	29200	36500	43800
16 × 5 1/2	26	15.65	5.500	.345	.250	1040	1300	1560
16 × 5 1/2	31	15.84	5.525	.442	.275	1240	1550	1860
16 × 7	36	15.85	6.992	.428	.299	1440	1800	2160
16 × 7	40	16.00	7.000	.503	.307	1600	2000	2400
16 × 7	45	16.12	7.039	.563	.346	1800	2250	2700
16 × 7	50	16.25	7.073	.628	.380	2000	2500	3000
16 × 7	57	16.43	7.120	.715	.430	2280	2850	3420
16 × 10 1/4	67	16.33	10.235	.665	.395	2680	3350	4020
16 × 10 1/4	77	16.52	10.295	.760	.455	3080	3850	4620
16 × 10 1/4	89	16.75	10.365	.875	.525	3560	4450	5340
16 × 10 1/4	100	16.97	10.425	.985	.585	4000	5000	6000
18 × 6	35	17.71	6.000	.429	.298	1400	1750	2100
18 × 6	40	17.90	6.018	.524	.316	1600	2000	2400
18 × 6	46	18.06	6.060	.605	.360	1840	2300	2760
18 × 7 1/2	50	18.00	7.500	.570	.358	2000	2500	3000
18 × 7 1/2	55	18.12	7.532	.630	.390	2200	2750	3300
18 × 7 1/2	60	18.25	7.558	.695	.416	2400	3000	600
18 × 7 1/2	65	18.35	7.590	.750	.450	2600	3250	3900
18 × 7 1/2	71	18.47	7.635	.810	.495	2840	3550	4260
18 × 11	76	18.21	11.035	.680	.425	3040	3800	4560
18 × 11	86	18.39	11.090	.770	.480	3440	4300	5160
18 × 11	97	18.59	11.145	.870	.535	3880	4850	5820
18 × 11	106	18.73	11.200	.940	.590	4240	5300	6360
18 × 11	119	18.97	11.265	1.060	.655	4760	5950	7140
21 × 6 1/2	44	20.66	6.500	.451	.348	1760	2200	2640
21 × 6 1/2	50	20.83	6.530	.535	.380	2000	2500	3000
21 × 6 1/2	57	21.06	6.555	.650	.405	2280	2850	3420
21 × 8 1/4	62	20.99	8.240	.615	.400	2480	3100	3720
21 × 8 1/4	68	21.13	8.270	.685	.430	2720	3400	4080
21 × 8 1/4	73	21.24	8.295	.740	.455	2920	3650	4380
21 × 8 1/4	83	21.43	8.355	.835	.515	3320	4150	4980

WIDE-FLANGE BEAMS—Cont'd

Conforms to A-36 and A-572-Gr50
Standard Lengths 40', 50' and 60'

Nominal Size in inches	Weight Per Foot	A Depth of Section	B Flange Width Inches	C Average Flange Thickness Inches	D Web Thickness Inches	Weight LBS. 40'	50'	60'
21 × 8 1/4	93	21.62	8.420	.930	.580	3720	4650	5580
21 × 12 1/4	101	21.36	12.290	.800	.500	4040	5050	6060
14 × 16	605	20.94	17.418	4.157	2.598	24200	30250	36300
14 × 16	665	21.67	17.646	4.522	2.826	26600	33250	39900
14 × 16	730	22.44	17.889	4.910	3.069	29200	36500	43800
16 × 5 1/2	26	15.65	5.500	.345	.250	1040	1300	1560
16 × 5 1/2	31	15.84	5.525	.442	.275	1240	1550	1860
16 × 7	36	15.85	6.992	.428	.299	1440	1800	2160
16 × 7	40	16.00	7.000	.503	.307	1600	2000	2400
16 × 7	45	16.12	7.039	.563	.346	1800	2250	2700
16 × 7	50	16.25	7.073	.628	.380	2000	2500	3000
16 × 7	57	16.43	7.120	.715	.430	2280	2850	3420
16 × 10 1/4	67	16.33	10.235	.665	.395	2680	3350	4020
16 × 10 1/4	77	16.52	10.295	.760	.455	3080	3850	4620
16 × 10 1/4	89	16.75	10.365	.875	.525	3560	4450	5340
16 × 10 1/4	100	16.97	10.425	.985	.585	4000	5000	6000
18 × 6	35	17.71	6.000	.429	.298	1400	1750	2100
18 × 6	40	17.90	6.018	.524	.316	1600	2000	2400
18 × 6	46	18.06	6.060	.605	.360	1840	2300	2760
18 × 7 1/2	50	18.00	7.500	.570	.358	2000	2500	3000
18 × 7 1/2	55	18.12	7.532	.630	.390	2200	2750	3300
18 × 7 1/2	60	18.25	7.558	.695	.416	2400	3000	600
18 × 7 1/2	65	18.35	7.590	.750	.450	2600	3250	3900
18 × 7 1/2	71	18.47	7.635	.810	.495	2840	3550	4260
18 × 11	76	18.21	11.035	.680	.425	3040	3800	4560
18 × 11	86	18.39	11.090	.770	.480	3440	4300	5160
18 × 11	97	18.59	11.145	.870	.535	3880	4850	5820
18 × 11	106	18.73	11.200	.940	.590	4240	5300	6360
18 × 11	119	18.97	11.265	1.060	.655	4760	5950	7140
21 × 6 1/2	44	20.66	6.500	.451	.348	1760	2200	2640
21 × 6 1/2	50	20.83	6.530	.535	.380	2000	2500	3000
21 × 6 1/2	57	21.06	6.555	.650	.405	2280	2850	3420
21 × 8 1/4	62	20.99	8.240	.615	.400	2480	3100	3720
21 × 8 1/4	68	21.13	8.270	.685	.430	2720	3400	4080
21 × 8 1/4	73	21.24	8.295	.740	.455	2920	3650	4380
21 × 8 1/4	83	21.43	8.355	.835	.515	3320	4150	4980
21 × 8 1/4	93	21.62	8.420	.930	.580	3720	4650	5580
21 × 12 1/4	101	21.36	12.290	.800	.500	4040	5050	6060
21 × 12 1/4	111	21.51	12.340	.875	.550	4440	5550	6660
21 × 12 1/4	122	21.68	12.390	.960	.600	4880	6100	7320
21 × 12 1/4	132	21.83	12.440	1.035	.650	5280	6600	7920
21 × 12 1/4	147	22.06	12.510	1.150	.720	5880	7350	8820

(Continued)

WIDE-FLANGE BEAMS—Cont'd

Conforms to A-36 and A-572-Gr50
Standard Lengths 40′, 50′ and 60′

Nominal Size in inches	Weight Per Foot	A Depth of Section	B Flange Width Inches	C Average Flange Thickness Inches	D Web Thickness Inches	Weight LBS.		
						40′	50′	60′
24 × 7	55	23.55	7.000	.503	.396	2200	2750	3300
24 × 7	62	23.72	7.040	.590	.430	2480	3100	3720
24 × 9	68	23.71	8.961	.582	.416	2720	3400	4080
24 × 9	76	23.91	8.985	.682	.440	3040	3800	4560
24 × 9	84	24.09	9.015	.772	.470	3360	4200	5040
24 × 9	94	24.29	9.061	.872	.516	3760	4700	5640
24 × 12 3/4	104	24.06	12.750	.750	.500	4160	5200	6240
24 × 12 3/4	117	24.26	12.800	.850	.550	4680	5850	7020
24 × 12 3/4	131	24.48	12.855	.960	.605	5240	6550	7860
24 × 12 3/4	146	24.74	12.900	1.090	.650	5840	7300	8760
24 × 12 3/4	162	25.00	12.955	1.220	.705	6480	8100	9720
27 × 10	84	26.69	9.963	.636	.463	3360	4200	5040
27 × 10	94	26.91	9.990	.747	.490	3760	4700	5640
27 × 10	102	27.07	10.018	.827	.518	4080	5100	6120
27 × 10	114	27.28	10.070	.932	.570	4560	5700	6840
27 × 14	146	27.38	13.965	.975	.605	5840	7300	8760
27 × 14	161	27.59	14.020	1.080	.660	6440	8050	9660
27 × 14	178	27.81	14.085	1.190	.725	7120	8900	10680
30 × 10 1/2	99	29.64	10.458	.670	.522	3960	4950	5940
30 × 10 1/2	108	29.82	10.484	.760	.548	4320	5400	6480
30 × 10 1/2	116	30.00	10.500	.850	.564	4640	5800	6960
30 × 10 1/2	124	30.16	10.521	.930	.585	4960	6200	7440
30 × 10 1/2	132	30.30	10.551	1.000	.615	5280	6600	7920
30 × 15	173	30.44	14.985	1.065	.622	6920	8650	10380
30 × 15	191	30.68	15.040	1.185	.710	7640	9550	11460
30 × 15	211	30.94	15.105	1.315	.775	8440	10550	12660
33 × 11 1/2	118	32.86	11.484	.738	.554	4720	5900	7080
33 × 11 1/2	130	33.10	11.510	.855	.580	5200	6500	7800
33 × 11 1/2	141	33.31	11.535	.960	.605	5640	7050	8460
33 × 11 1/2	152	33.50	11.565	1.055	.635	6080	7600	9120
33 × 15 3/4	201	33.68	15.745	1.150	.715	8040	10050	12060
33 × 15 3/4	221	33.93	15.850	1.275	.775	8840	11050	13260
33 × 15 3/4	241	34.18	15.860	1.400	.830	9640	12050	14460
36 × 12	135	35.55	11.945	.794	.598	5400	6750	8100
36 × 12	150	35.84	11.972	.940	.625	6000	7500	9000
36 × 12	160	36.00	12.000	1.020	.653	6400	8000	9600
36 × 12	170	36.16	12.027	1.100	.680	6800	8500	10200
36 × 12	182	36.32	12.072	1.180	.725	7280	9100	10920
36 × 12	194	36.48	12.117	1.260	.770	7760	9700	11640
36 × 12	210	36.69	12.180	1.360	.830	8400	10500	12600
36 × 16 1/2	230	35.88	16.475	1.260	.765	9200	11500	13800
36 × 16 1/2	245	36.06	16.512	1.350	.802	9800	12250	14700

WIDE-FLANGE BEAMS—Cont'd

Conforms to A-36 and A-572-Gr50
Standard Lengths 40', 50' and 60'

Nominal Size in inches	Weight Per Foot	A Depth of Section	B Flange Width Inches	C Average Flange Thickness Inches	D Web Thickness Inches	Weight LBS. 40'	Weight LBS. 50'	Weight LBS. 60'
36 × 16 1/2	260	36.24	16.555	1.440	.845	10400	13000	15600
36 × 16 1/2	280	36.50	16.595	1.680	.885	11200	14000	16800
36 × 16 1/2	300	36.72	16.655	1.680	.945	12000	15000	18000

6.3.1 Calculating the Metric Weight and Size of Wide-Flange Beams W4s to W36s

Wide-Flange Beams
Imperial to Metric Conversions
W/D Imperial = M/D Metric * 0.017

Imperial Size (In.)	× Wt. (lb/ft)	Metric Size (mm)	× WT. (kg/m)	Imperial W/D (lb/ft/in.)	Metric M/D (kg/m/m)
W4	× 13	W100	× 19	0.65	39
W5	× 16	W130	× 24	0.65	39
W5	× 19	W130	× 28	0.76	45
W6	× 9	W150	× 14	0.39	23
W6	× 12	W150	× 18	0.51	30
W6	× 15	W150	× 22	0.51	30
W6	× 16	W150	× 24	0.66	39
W6	× 20	W150	× 30	0.67	40
W6	× 25	W150	× 37	0.82	49
W8	× 10	W200	× 15	0.37	22
W8	× 13	W200	× 19	0.47	28
W8	× 15	W200	× 22	0.54	32
W8	× 18	W200	× 27	0.57	34
W8	× 21	W200	× 31	0.66	39
W8	× 24	W200	× 36	0.69	41
W8	× 28	W200	× 42	0.8	48
W8	× 31	W200	× 46	0.79	47
W8	× 35	W200	× 52	0.88	52
W8	× 40	W200	× 59	1	59
W8	× 48	W200	× 71	1.18	70
W8	× 58	W200	× 86	1.41	83
W8	× 67	W200	× 100	1.61	95
W10	× 12	W250	× 18	0.38	23
W10	× 15	W250	× 2	0.48	29
W10	× 17	W250	× 25	0.54	32
W10	× 19	W250	× 28	0.59	35
W10	× 22	W250	× 33	0.59	35
W10	× 26	W250	× 39	0.69	41

(Continued)

Wide-Flange Beams—Cont'd
Imperial to Metric Conversions
W/D Imperial = M/D Metric * 0.017

Imperial		Metric		Imperial	Metric
Size (In.)	× Wt. (lb/ft)	Size (mm)	× WT. (kg/m)	W/D (lb/ft/in.)	M/D (kg/m/m)
W10	× 30	W250	× 45	0.79	47
W10	× 33	W250	× 49	0.77	46
W10	× 39	W250	× 58	0.9	53
W10	× 45	W250	× 87	1.03	61
W10	× 49	W250	× 73	0.99	59
W10	× 54	W250	× 80	1.09	65
W10	× 60	W250	× 89	1.2	71
W10	× 68	W250	× 101	1.35	80
W10	× 77	W250	× 115	1.52	90
W10	× 88	W250	× 131	1.72	102
W10	× 100	W250	× 149	1.93	114
W10	× 112	W250	× 167	2.14	126
W12	× 14	W310	× 21	0.4	24
W12	× 16	W310	× 24	0.45	27
W12	× 19	W310	× 28	0.53	32
W12	× 22	W310	× 33	0.61	36
W12	× 26	W310	× 39	0.6	36
W12	× 30	W310	× 45	0.69	41
W12	× 35	W310	× 52	0.79	47
W12	× 40	W310	× 60	0.85	50
W12	× 45	W310	× 67	0.95	56
W12	× 50	W310	× 74	1.04	62
W12	× 53	W310	× 79	0.99	59
W12	× 54	W310	× 80	1.04	62
W12	× 55	W310	× 82	1.04	62
W12	× 58	W310	× 86	1.08	64
W12	× 65	W310	× 97	1.09	65
W12	× 72	W310	× 107	1.2	71
W12	× 79	W310	× 118	1.32	78
W12	× 87	W310	× 129	1.44	85
W12	× 96	W310	× 143	1.57	93
W12	× 106	W310	× 158	1.73	102
W12	× 120	W310	× 179	1.94	115
W12	× 136	W310	× 202	2.17	128
W12	× 152	W310	× 226	2.4	142
W12	× 170	W310	× 253	2.66	157
W12	× 190	W310	× 283	2.93	173
W12	× 210	W310	× 313	3.21	189
W12	× 230	W310	× 342	3.47	205
W12	× 252	W310	× 375	3.76	222
W12	× 279	W310	× 415	4.1	242
W12	× 305	W310	× 454	4.41	260
W12	× 336	W310	× 500	4.78	282
W14	× 22	W360	× 33	0.52	31
W14	× 26	W360	× 39	0.61	36
W14	× 30	W360	× 45	0.63	38
W14	× 34	W360	× 51	0.71	42
W14	× 38	W360	× 57	0.79	47
W14	× 43	W360	× 64	0.85	50
W14	× 48	W360	× 72	0.94	56
W14	× 53	W360	× 79	1.03	61
W14	× 61	W360	× 91	1.07	63

Wide-Flange Beams—Cont'd
Imperial to Metric Conversions
W/D Imperial = M/D Metric * 0.017

Imperial		Metric		Imperial	Metric
Size (In.)	× Wt. (lb/ft)	Size (mm)	× WT. (kg/m)	W/D (lb/ft/in.)	M/D (kg/m/m)
W14	× 68	W360	× 101	1.19	70
W14	× 74	W360	× 110	1.28	76
W14	× 82	W360	× 122	1.41	83
W14	× 90	W360	× 134	1.27	75
W14	× 99	W360	× 147	1.39	82
W14	× 109	W360	× 162	1.53	90
W14	× 120	W360	× 179	1.67	99
W14	× 132	W360	× 196	1.83	108
W14	× 145	W360	× 216	1.94	115
W14	× 159	W360	× 237	2.11	125
W14	× 176	W360	× 262	2.32	137
W14	× 193	W360	× 287	2.53	149
W14	× 211	W360	× 314	2.74	162
W14	× 233	W360	× 347	3	177
W14	× 257	W360	× 382	3.27	193
W14	× 283	W360	× 421	3.57	210
W14	× 311	W360	× 463	3.88	229
W14	× 342	W360	× 509	4.21	248
W14	× 370	W360	× 551	4.51	266
W14	× 398	W360	× 592	4.8	283
W14	× 426	W360	× 634	5.09	300
W14	× 455	W360	× 677	5.38	317
W14	× 500	W360	× 744	5.82	343
W14	× 550	W360	× 816	6.3	371
W14	× 605	W360	× 900	6.8	400
W14	× 665	W360	× 990	7.34	432
W14	× 730	W360	× 1086	7.9	465
W16	× 26	W410	× 39	0.55	33
W16	× 31	W410	× 46	0.65	39
W16	× 36	W410	× 54	0.69	41
W16	× 40	W410	× 60	0.76	45
W16	× 45	W410	× 67	0.85	50
W16	× 50	W410	× 74	0.94	56
W16	× 57	W410	× 85	1.07	63
W16	× 67	W410	× 100	1.07	63
W16	× 77	W410	× 114	1.22	72
W16	× 89	W410	× 132	1.4	83
W16	× 100	W410	× 149	1.56	92
W18	× 35	W460	× 52	0.66	39
W18	× 40	W460	× 60	0.75	45
W18	× 46	W460	× 68	0.86	51
W18	× 50	W460	× 74	0.87	52
W18	× 55	W460	× 82	0.95	56
W18	× 60	W460	× 89	1.03	61
W18	× 65	W460	× 97	1.11	66
W18	× 71	W460	× 106	1.21	72
W18	× 76	W460	× 113	1.11	66
W18	× 86	W460	× 128	1.24	73
W18	× 97	W460	× 144	1.39	82
W18	× 106	W460	× 158	1.52	90
W18	× 119	W460	× 177	1.68	99

(Continued)

Wide-Flange Beams—Cont'd
Imperial to Metric Conversions
W/D Imperial = M/D Metric * 0.017

Imperial		Metric		Imperial	Metric
Size (In.)	× Wt. (lb/ft)	Size (mm)	× WT. (kg/m)	W/D (lb/ft/in.)	M/D (kg/m/m)
W21	× 44	W530	× 66	0.73	43
W21	× 50	W530	× 74	0.83	49
W21	× 57	W530	× 85	0.93	55
W21	× 62	W530	× 92	0.94	56
W21	× 68	W530	× 101	1.03	61
W22	× 73	W530	× 108	1.1	65
W21	× 83	W530	× 123	1.24	73
W21	× 93	W530	× 138	1.38	82
W21	× 101	W530	× 150	1.29	76
W21	× 111	W530	× 165	1.41	83
W21	× 122	W530	× 182	1.54	91
W21	× 132	W530	× 196	1.66	98
W21	× 147	W530	× 219	1.83	108
W24	× 55	W610	× 82	0.82	49
W24	× 62	W610	× 92	0.92	55
W24	× 68	W610	× 101	0.93	55
W24	× 76	W610	× 113	1.02	60
W24	× 84	W610	× 125	1.13	67
W24	× 94	W610	× 140	1.26	75
W24	× 104	W610	× 153	1.22	72
W24	× 117	W610	× 174	1.36	80
W24	× 131	W610	× 195	1.52	90
W24	× 146	W610	× 217	1.68	99
W24	× 162	W610	× 241	1.85	109
W27	× 84	W690	× 125	1.02	60
W27	× 94	W690	× 140	1.13	67
W27	× 102	W690	× 152	1.23	73
W27	× 114	W690	× 170	1.36	80
W27	× 146	W690	× 217	1.53	90
W27	× 161	W690	× 240	1.68	99
W27	× 178	W690	× 265	1.85	109
W30	× 99	W760	× 147	1.1	65
W30	× 108	W760	× 161	1.2	71
W30	× 116	W760	× 173	1.28	76
W30	× 124	W760	× 185	1.37	81
W30	× 132	W760	× 196	1.45	86
W30	× 173	W760	× 257	1.66	98
W30	× 191	W760	× 284	1.82	108
W30	× 211	W760	× 314	2	118
W33	× 118	W840	× 176	1.19	70
W33	× 130	W840	× 193	1.31	78
W33	× 141	W840	× 210	1.41	83
W33	× 152	W840	× 226	1.51	89
W33	× 201	W840	× 299	1.78	105
W33	× 221	W840	× 329	1.94	115
W33	× 241	W840	× 359	2.11	125

Wide-Flange Beams—Cont'd
Imperial to Metric Conversions
W/D Imperial = M/D Metric * 0.017

Imperial		Metric		Imperial	Metric
Size (In.)	× Wt. (lb/ft)	Size (mm)	× WT. (kg/m)	W/D (lb/ft/in.)	M/D (kg/m/m)
W36	× 135	W920	× 201	1.28	76
W36	× 150	W920	× 223	1.41	83
W36	× 160	W920	× 238	1.5	89
W36	× 170	W920	× 253	1.59	94
W36	× 182	W920	× 271	1.69	100
W36	× 194	W920	× 289	1.8	106
W36	× 210	W920	× 313	1.94	115
W36	× 230	W920	× 342	1.92	113
W36	× 245	W920	× 365	2.04	120
W36	× 260	W920	× 387	2.16	128
W36	× 280	W920	× 417	2.31	136
W36	× 300	W920	× 446	2.47	146

6.3.2 Calculating the Weight and Size of I Beams and Junior Beams

STANDARD JUNIOR

I BEAMS AND JUNIOR BEAMS
Structural Steel Shapes

Structural Steel Shapes are usually ordered as specification ASTM-36.

This Standard of the American Society for Testing and Materials is issued under the designation A-36.

The number immediately following the designation includes the year of original adoption, or, in the case of revision, the year of last revision.

A-36 Specification
Tensile strength, psi 56,000–80,000

Min. yield strength, psi 36,000
Carbon Content, .26 max.
Standard Lengths 40′, 50′, 60′

A-572 Specification
Tensile strength, psi 65,000
Min. yield strength, psi 36,000
Carbon Content, .26 max.
Standard Lengths 40′, 50′, 60′

Source: Cardinal Metals, St. Louis, MO

STANDARD I BEAMS

[S]

Conforms to A-36
Standard Lengths 20', 40', 60'

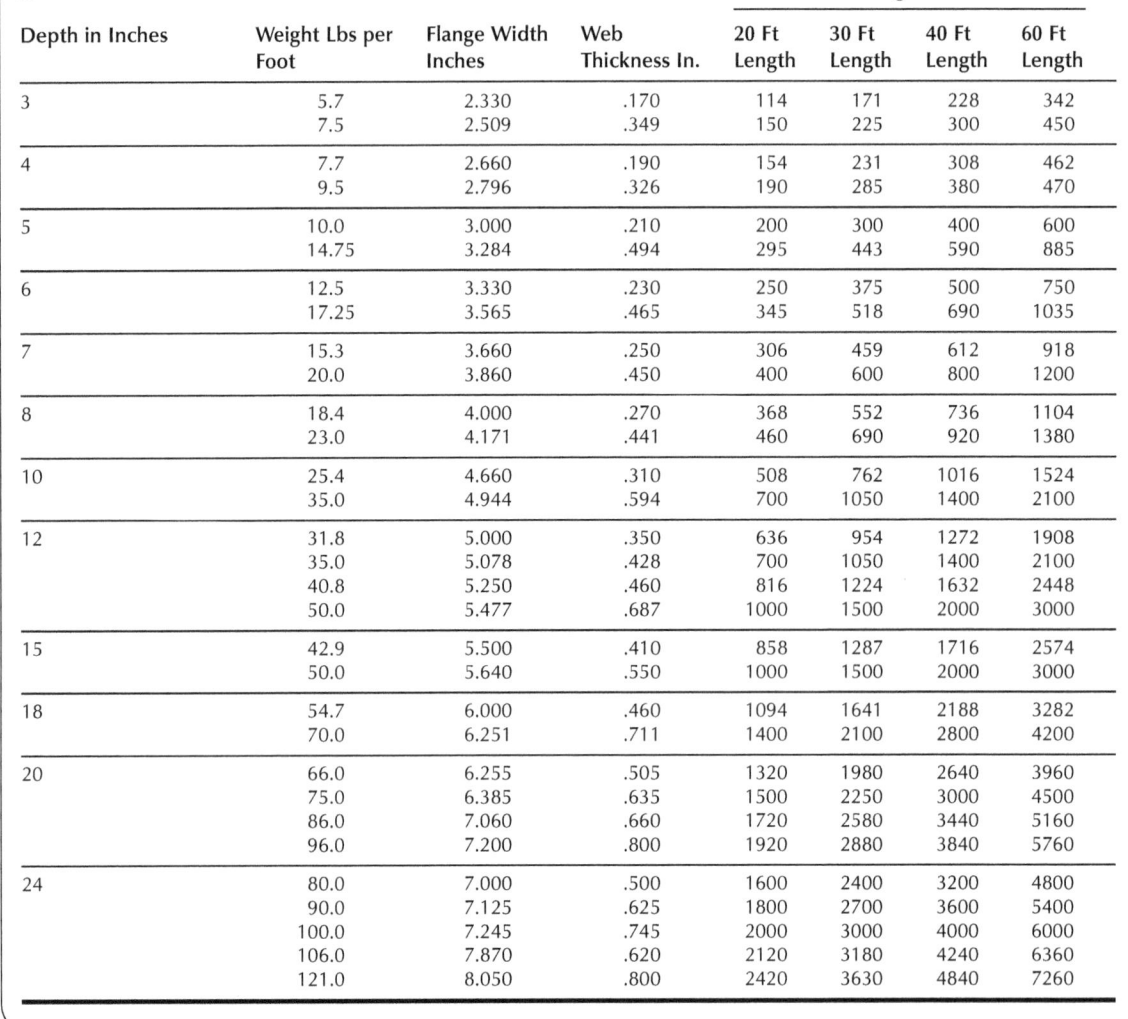

A	B	C	Weight Lbs			
Depth in Inches	Weight Lbs per Foot	Flange Width Inches	Web Thickness In.	20 Ft Length	30 Ft Length	40 Ft Length	60 Ft Length
3	5.7	2.330	.170	114	171	228	342
	7.5	2.509	.349	150	225	300	450
4	7.7	2.660	.190	154	231	308	462
	9.5	2.796	.326	190	285	380	470
5	10.0	3.000	.210	200	300	400	600
	14.75	3.284	.494	295	443	590	885
6	12.5	3.330	.230	250	375	500	750
	17.25	3.565	.465	345	518	690	1035
7	15.3	3.660	.250	306	459	612	918
	20.0	3.860	.450	400	600	800	1200
8	18.4	4.000	.270	368	552	736	1104
	23.0	4.171	.441	460	690	920	1380
10	25.4	4.660	.310	508	762	1016	1524
	35.0	4.944	.594	700	1050	1400	2100
12	31.8	5.000	.350	636	954	1272	1908
	35.0	5.078	.428	700	1050	1400	2100
	40.8	5.250	.460	816	1224	1632	2448
	50.0	5.477	.687	1000	1500	2000	3000
15	42.9	5.500	.410	858	1287	1716	2574
	50.0	5.640	.550	1000	1500	2000	3000
18	54.7	6.000	.460	1094	1641	2188	3282
	70.0	6.251	.711	1400	2100	2800	4200
20	66.0	6.255	.505	1320	1980	2640	3960
	75.0	6.385	.635	1500	2250	3000	4500
	86.0	7.060	.660	1720	2580	3440	5160
	96.0	7.200	.800	1920	2880	3840	5760
24	80.0	7.000	.500	1600	2400	3200	4800
	90.0	7.125	.625	1800	2700	3600	5400
	100.0	7.245	.745	2000	3000	4000	6000
	106.0	7.870	.620	2120	3180	4240	6360
	121.0	8.050	.800	2420	3630	4840	7260

JUNIOR BEAMS

[M]

Conforms to A-36
Standard Lengths 20' and 40'

In Inches	Per Foot	Width Inches	Thick Ness In.	20 FT. Length	30 FT. Length	40 FT. Length	60 FT. Length
6	4.4	1.844	.114	88	132	176	
8	6.5	2.281	.135	130	195	260	
10	9.0	2.688	.155	180	270	360	
12	11.8	3.063	.175	236	354	472	

6.4.0 Calculating the Weight and Size of U.S. Square High-Strength Steel Sections

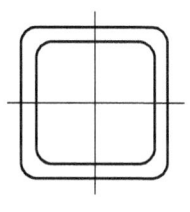

Nominal Size			Weight Per Foot	Wall Thickness t	b/t	h/t	Cross Section Area	I	S	r	Z	Torsional Stiffness Constant J	Torsional Shear Constant C	Surface Area Per Foot
in.		in.	lb.	in.			in.²	in.⁴	in.³	in.	in.³	in.⁴	in.³	ft.²
32	×	32												
		5/8*	259.83	0.625	48.2	48.2	76.4	12300	771	12.7	890	19700	1230	10.34
		1/2*	210.72	0.500	61.0	61.0	61.9	10100	634	12.8	727	15900	991	10.45
		3/8*	159.37	0.375	82.3	82.3	46.8	7750	485	12.9	553	12000	750	10.51
30	×	30												
		5/8*	242.82	0.625	45.0	45.0	71.4	10100	673	11.9	778	16200	1070	9.68
		1/2*	197.11	0.500	57.0	57.0	57.9	8320	555	12.0	637	13000	869	9.79
		3/8*	149.16	0.375	77.0	77.0	43.8	6370	424	12.1	485	9870	658	9.84
28	×	28												
		5/8*	225.80	0.625	41.8	41.8	66.4	8140	582	11.1	674	13100	933	9.01
		1/2*	183.50	0.500	53.0	53.0	53.9	6730	480	11.2	552	10600	755	9.12
		3/8*	138.95	0.375	71.7	71.7	40.8	5150	368	11.2	421	8010	572	9.17
26	×	26												
		5/8*	208.79	0.625	38.6	38.6	61.4	6460	497	10.3	577	10500	801	8.34
		1/2*	169.89	0.500	49.0	49.0	49.9	5350	411	10.4	474	8430	649	8.48
		3/8*	128.74	0.375	66.3	66.3	37.8	4110	316	10.4	362	6400	492	8.51
24	×	24												
		5/8*	191.78	0.625	35.4	35.4	56.4	5030	419	9.44	487	8180	679	7.68
		1/2*	156.28	0.500	45.0	45.0	45.9	4170	348	9.53	401	6610	551	7.79
		3/8*	118.53	0.375	61.0	61.0	34.8	3210	268	9.60	307	5020	418	7.84
22	×	22												
		5/8*	174.76	0.625	32.2	32.2	51.4	3820	347	8.62	406	6260	567	7.01
		1/2*	142.67	0.500	41.0	41.0	41.9	3190	290	8.72	335	5070	461	7.12
		3/8*	108.32	0.375	55.7	55.7	31.8	2460	223	8.78	256	3850	350	7.17

(Continued)

Nominal Size					Weight Per Foot	Wall Thickness t	b/t	h/t	Cross Section Area	I	S	r	Z	Torsional Stiffness Constant J	Torsional Shear Constant C	Surface Area Per Foot
in.		in.		in.	lb.	in.			in.²	in.⁴	in.³	in.	in.³	in.⁴	in.³	ft.²
20	×	20	×	5/8*	157.75	0.625	29.0	29.0	46.4	2830	283	7.81	331	4670	465	6.34
				1/2*	129.06	0.500	37.0	37.0	37.9	2370	237	7.90	275	3790	379	6.45
				3/8*	98.12	0.375	50.3	50.3	28.8	1830	183	7.97	211	2880	288	6.51
18	×	18	×	5/8*	140.73	0.625	25.8	25.8	41.4	2020	224	6.99	264	3370	373	5.68
				1/2*	115.45	0.500	33.0	33.0	33.9	1700	189	7.08	220	2740	305	5.79
				3/8*	87.91	0.375	45.0	45.0	25.8	1320	147	7.15	169	2090	232	5.84
16	×	16	×	5/8	127.37	0.581	24.5	24.5	35.0	1370	171	6.25	200	2170	276	5.17
				1/2	103.30	0.465	31.4	31.4	28.3	1130	141	6.31	164	1770	224	5.20
				3/8	78.52	0.349	42.8	42.8	21.5	873	109	6.37	126	1350	171	5.23
				5/16	65.87	0.291	52.0	52.0	18.1	739	92.3	6.39	106	1140	144	5.25
14	×	14	×	5/8	110.36	0.581	21.1	21.1	30.3	896	128	5.44	151	1430	208	4.50
				1/2	89.68	0.465	27.1	27.1	24.6	743	106	5.49	124	1170	170	4.53
				3/8	68.31	0.349	37.1	37.1	18.7	577	82.5	5.55	95.4	900	130	4.57
				5/16	57.36	0.291	45.1	45.1	15.7	490	69.9	5.58	80.5	759	109	4.58
12	×	12	×	5/8	93.34	0.581	17.7	17.7	25.7	548	91.3	4.62	109	885	151	3.83
				1/2	76.07	0.465	22.8	22.8	20.9	457	76.2	4.68	89.6	728	123	3.87
				3/8	58.10	0.349	31.4	31.4	16.0	357	59.5	4.73	69.2	561	94.6	3.90
				5/16	48.86	0.291	38.2	38.2	13.4	304	50.7	4.76	58.6	474	79.7	3.92
				1/4	39.43	0.233	48.5	48.5	10.8	248	41.4	4.79	47.6	384	64.5	3.93
10	×	10	×	5/8	76.33	0.581	14.2	14.2	21.0	304	60.8	3.80	73.2	498	102	3.17
				1/2	62.46	0.465	18.5	18.5	17.2	256	51.2	3.86	60.7	412	84.2	3.20
				3/8	47.90	0.349	25.7	25.7	13.2	202	40.4	3.92	47.2	320	64.8	3.23
				5/16	40.35	0.291	31.4	31.4	11.1	172	34.5	3.94	40.1	271	54.8	3.25
				1/4	32.63	0.233	39.9	39.9	8.96	141	28.3	3.97	32.7	220	44.4	3.27
				3/16	24.73	0.174	54.5	54.5	6.76	108	21.6	4.00	24.8	167	33.6	3.28
9	×	9	×	1/2	55.66	0.465	16.4	16.4	15.3	182	40.6	3.45	48.4	296	67.4	2.87
				3/8	42.79	0.349	22.8	22.8	11.8	145	32.2	3.51	37.8	231	52.1	2.90
				5/16	36.10	0.291	27.9	27.9	9.92	124	27.6	3.54	32.1	196	44.0	2.92
				1/4	29.23	0.233	35.6	35.6	8.03	102	22.7	3.56	26.2	159	35.8	2.93
				3/16	22.18	0.174	48.7	48.7	6.06	78.2	17.4	3.59	20.0	121	27.1	2.95
8	×	8	×	5/8	59.32	0.581	10.8	10.8	16.4	146	36.5	2.99	44.7	244	63.2	2.50
				1/2	48.85	0.465	14.2	14.2	13.5	125	31.2	3.04	37.5	204	52.4	2.53
				3/8	37.69	0.349	19.9	19.9	10.4	99.6	24.9	3.10	29.4	160	40.7	2.57
				5/16	31.84	0.291	24.5	24.5	8.76	85.6	21.4	3.13	25.1	136	34.5	2.58
				1/4	25.82	0.233	31.3	31.3	7.10	70.7	17.7	3.15	20.5	111	28.1	2.60

7	×	7	×	3/16	19.63	0.174	9.0	9.0	5.37	54.4	13.6	3.18	15.7	84.5	21.3	2.62
			×	5/8	5.81	0.581	9.0	9.0	14.0	93.3	26.7	2.58	33.1	158	47.1	2.17
				1/2	42.05	0.465	12.1	12.1	11.6	80.5	23.0	2.63	27.9	133	39.3	2.20
				3/8	32.58	0.349	17.1	17.1	8.97	64.9	18.6	2.69	22.1	105	30.7	2.23
				5/16	27.59	0.291	21.1	21.1	7.59	56.1	16.0	2.72	18.9	89.7	26.1	2.25
				1/4	22.42	0.233	27.0	27.0	6.17	46.5	13.3	2.75	15.5	73.5	21.3	2.27
				3/16	17.08	0.174	37.2	37.2	4.67	36.0	10.3	2.77	11.9	56.1	16.2	2.28
6	×	6	×	5/8	42.30	0.581	7.3	7.3	11.7	55.1	18.4	2.17	23.2	94.9	33.4	1.83
				1/2	35.24	0.465	9.9	9.9	9.74	48.2	16.1	2.23	19.8	81.1	28.1	1.87
				3/8	27.48	0.349	14.2	14.2	7.58	39.4	13.1	2.28	15.8	64.6	22.1	1.90
				5/16	23.34	0.291	17.6	17.6	6.43	34.3	11.4	2.31	13.6	55.4	18.9	1.92
				1/4	19.02	0.233	22.8	22.8	5.24	28.6	9.54	2.34	11.2	45.6	15.4	1.93
				3/16	14.53	0.174	31.5	31.5	3.98	22.3	7.42	2.37	8.63	35.0	11.8	1.95
				1/8	9.86	0.116	48.7	48.7	2.70	15.5	5.15	2.39	5.92	23.9	8.03	1.97
5 1/2	×	5 1/2	×	3/8	24.93	0.349	12.8	12.8	6.88	29.7	10.8	2.08	13.1	49.0	18.4	1.73
				5/16	21.21	0.291	15.9	15.9	5.85	25.9	9.43	2.11	11.3	42.2	15.7	1.75
				1/4	17.32	0.233	20.6	20.6	4.77	21.7	7.90	2.13	9.32	34.8	12.9	1.77
				3/16	13.25	0.174	28.6	28.6	3.63	17.0	6.17	2.16	7.19	26.7	9.85	1.78
				1/8	9.01	0.116	44.4	44.4	2.46	11.8	4.30	2.19	4.95	18.3	6.72	1.80
5	×	5	×	1/2	28.43	0.465	7.8	7.8	7.88	26.0	10.4	1.82	13.1	44.6	18.7	1.53
				3/8	22.37	0.349	11.3	11.3	6.18	21.7	8.67	1.87	10.6	36.1	14.9	1.57
				5/16	19.08	0.291	14.2	14.2	5.26	19.0	7.61	1.90	9.16	31.2	12.8	1.58
				1/4	15.62	0.233	18.5	18.5	4.30	16.0	6.41	1.93	7.61	25.8	10.5	1.60
				3/16	11.97	0.174	25.7	25.7	3.28	12.6	5.03	1.96	5.89	19.9	8.08	1.62
				1/8	8.16	0.116	40.1	40.1	2.23	8.80	3.52	1.99	4.07	13.7	5.53	1.63
4 1/2	×	4 1/2	×	1/2	25.03	0.465	6.7	6.7	6.95	18.0	8.02	1.61	10.2	31.3	14.8	1.37
				3/8	19.82	0.349	9.9	9.9	5.48	15.3	6.78	1.67	8.36	25.7	11.9	1.40
				5/16	16.96	0.291	12.5	12.5	4.68	13.5	5.99	1.70	7.27	22.3	10.2	1.42
				1/4	13.91	0.233	16.3	16.3	3.84	11.4	5.08	1.73	6.06	18.5	8.44	1.43
				3/16	10.70	0.174	22.9	22.9	2.93	9.02	4.01	1.75	4.71	14.4	6.49	1.45
				1/8	7.31	0.116	35.8	35.8	2.00	6.35	2.82	1.78	3.27	9.92	4.45	1.47
4	×	4	×	1/2	21.63	0.465	5.6	5.6	6.02	11.9	5.95	1.41	7.70	21.0	11.2	1.20
				3/8	17.27	0.349	8.5	8.5	4.78	10.3	5.13	1.46	6.39	17.5	9.14	1.23
				5/16	14.83	0.291	10.7	10.7	4.10	9.14	4.57	1.49	5.59	15.3	7.91	1.25
				1/4	12.21	0.233	14.2	14.2	3.37	7.80	3.90	1.52	4.69	12.8	6.56	1.27
				3/16	9.42	0.174	20.0	20.0	2.58	6.21	3.10	1.55	3.67	9.96	5.07	1.28

(Continued)

Nominal Size			Weight Per Foot	Wall Thickness t	b/t	h/t	Cross Section Area	I	S	r	Z	Torsional Stiffness Constant J	Torsional Shear Constant C	Surface Area Per Foot
in.		in.	lb.	in.			in.²	in.⁴	in.³	in.	in.³	in.⁴	in.³	ft.²
		1/8	6.46	0.116	31.5	31.5	1.77	4.40	2.20	1.58	2.56	6.91	3.49	1.30
3 1/2 ×	3 1/2	3/8	14.72	0.349	7.0	7.0	4.09	6.48	3.70	1.26	4.69	11.2	6.77	1.07
		5/16	12.70	0.291	9.0	9.0	3.52	5.84	3.34	1.29	4.14	9.89	5.90	1.08
		1/4	10.51	0.233	12.0	12.0	2.91	5.04	2.88	1.32	3.50	8.35	4.92	1.10
		3/16	8.15	0.174	17.1	17.1	2.24	4.05	2.31	1.35	2.76	6.56	3.83	1.12
		1/8	5.61	0.116	27.2	27.2	1.54	2.90	1.66	1.37	1.93	4.58	2.65	1.13
3 ×	3	3/8	12.17	0.349	5.6	5.6	3.39	3.77	2.51	1.05	3.25	6.64	4.74	0.90
		5/16	10.58	0.291	7.3	7.3	2.94	3.45	2.30	1.08	2.90	5.94	4.18	0.92
		1/4	8.81	0.233	9.9	9.9	2.44	3.02	2.01	1.11	2.48	5.08	3.52	0.93
		3/16	6.87	0.174	14.2	14.2	1.89	2.46	1.64	1.14	1.97	4.03	2.76	0.95
		1/8	4.75	0.116	22.9	22.9	1.30	1.78	1.19	1.17	1.40	2.84	1.92	0.97
2 1/2 ×	2 1/2	5/16	8.45	0.291	5.6	5.6	2.35	1.82	1.45	0.879	1.88	3.20	2.74	0.75
		1/4	7.11	0.233	7.7	7.7	1.97	1.63	1.30	0.908	1.63	2.79	2.35	0.77
		3/16	5.59	0.174	11.4	11.4	1.54	1.35	1.08	0.937	1.32	2.25	1.86	0.78
		1/8	3.90	0.116	18.6	18.6	1.07	0.998	0.798	0.965	0.947	1.61	1.31	0.80
2 1/4 ×	2 1/4	1/4	6.26	0.233	6.7	6.7	1.74	1.13	1.00	0.805	1.28	1.96	1.85	0.68
		3/16	4.96	0.174	9.9	9.9	1.37	0.952	0.847	0.835	1.04	1.60	1.48	0.70
		1/8	3.48	0.116	16.4	16.4	0.96	0.712	0.633	0.863	0.755	1.15	1.05	0.72
2 ×	2	1/4	5.41	0.233	5.6	5.6	1.51	0.745	0.745	0.703	0.964	1.31	1.41	0.60
		3/16	4.32	0.174	8.5	8.5	1.19	0.640	0.640	0.732	0.797	1.09	1.14	0.62
		1/8	3.05	0.116	14.2	14.2	0.84	0.486	0.486	0.761	0.584	0.796	0.817	0.63
1 3/4 ×	1 3/4	3/16	3.68	0.174	7.1	7.1	1.02	0.405	0.462	0.630	0.585	0.699	0.844	0.53
1 5/8 ×	1 5/8	3/16	3.36	0.174	6.3	6.3	0.93	0.312	0.384	0.579	0.491	0.544	0.712	0.49
		1/8	2.42	0.116	11.0	11.0	0.67	0.246	0.302	0.608	0.370	0.410	0.522	0.51
1 1/2 ×	1 1/2	3/16	3.04	0.174	5.6	5.6	0.84	0.235	0.314	0.528	0.406	0.414	0.592	0.45
		1/8	2.20	0.116	9.9	9.9	0.61	0.188	0.251	0.556	0.309	0.316	0.438	0.47
1 1/4 ×	1 1/4	3/16	2.40	0.174	4.2	4.2	0.67	0.121	0.194	0.425	0.259	0.218	0.383	0.37
		1/8	1.78	0.116	7.8	7.8	0.49	0.101	0.162	0.454	0.204	0.174	0.292	0.38

6.4.1 Calculating the Weight and Size of Metric Square High-Strength Steel Sections

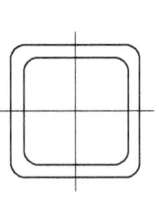

Nominal Size			Mass Per Meter	Weight Per Meter	Design Wall Thickness	b/t	h/t	Area	$1/10^6$		$S/10^3$	r	$Z/10^3$	Torsional Stiffness Constant $J/10^3$	Torsional Shear Constant $C/10^3$	Surface Area Per Meter
US Customary	SI/Metric															
Inches	Millimeters		kg	kN	mm			mm^2	mm^4	mm^4	mm^3	mm	mm^3	mm^4	mm^3	m^2
32×32×5/8*	812.8×812.8×15.9*		387.3	3.798	15.9	48.1	48.1	49300	5140		12700	323	14600	7960000	20100	3.15
1/2*	12.7*		313.6	3.076	12.7	61.0	61.0	40000	4220		10400	325	11900	6440000	16200	3.19
3/8*	9.5*		236.6	2.320	9.5	82.6	82.6	30100	3220		7920	327	9040	4890000	12300	3.20
30×30×5/8*	762.0×762.0×15.9*		361.9	3.549	15.9	44.9	44.9	46100	4210		11000	302	12800	6520000	17600	2.95
1/2*	12.7*		293.4	2.877	12.7	57.0	57.0	37400	3460		9090	304	10400	5280000	14200	2.98
3/8*	9.5*		221.4	2.171	9.5	77.2	77.2	28200	2640		6940	306	7920	4020000	10700	3.00
28×28×5.8*	711.2×711.2×15.9*		336.6	3.301	15.9	41.7	41.7	42900	3390		9550	281	11100	5270000	15300	2.75
1/2*	12.7*		273.1	2.678	12.7	53.0	53.0	34800	2800		7870	284	9050	4270000	12400	2.78
3/8*	9.5*		206.3	2.023	9.5	71.9	71.9	26300	2140		6020	285	6880	3250000	9350	2.80
26×26×5/8*	660.4×660.4×15.9*		311.2	3.052	15.9	38.5	38.5	39600	2690		8150	261	9460	4190000	13100	2.54
1/2*	12.7*		252.9	2.480	12.7	49.0	49.0	32200	2230		6740	263	7760	340000	10600	2.58
3/8*	9.5*		191.1	1.874	9.5	66.5	66.5	24300	1700		5160	265	5910	2590000	8040	2.59
24×24×5/8*	609.6×609.6×15.9*		285.8	2.803	15.9	35.3	35.3	36400	2090		6870	240	8000	3270000	11100	2.34
1/2*	12.7*		232.6	2.281	12.7	45.0	45.0	29600	1740		5700	242	6580	2660000	9030	2.37
3/8*	9.5*		176.0	1.726	9.5	61.2	61.2	22400	1330		4370	244	5010	2030000	6830	2.39
22×22×5/8*	558.8×558.8×15.9*		260.5	2.554	15.9	32.1	32.1	33200	1590		5700	219	6650	2490000	9310	2.14
1/2*	12.7*		212.3	2.082	12.7	41.0	41.0	27000	1330		4750	221	5490	2030000	7550	2.17
3/8*	9.5*		160.8	1.577	9.5	55.8	55.8	20500	1020		3650	223	4190	1560000	5720	2.19
20×20×5/8*	508.0×508.0×15.9*		235.1	2.306	15.9	28.9	28.9	30000	1180		4640	198	5430	1850000	7630	1.93
1/2*	12.7*		192.1	1.884	12.7	37.0	37.0	24500	985		3880	201	4500	1510000	6210	1.97
3/8*	9.5*		145.7	1.428	9.5	50.5	50.5	18600	760		2990	202	3440	1160000	4710	1.98
18×18×5/8*	457.2×457.2×15.9*		209.8	2.057	15.9	25.8	25.8	26700	842		3680	177	4340	1330000	6130	1.73
1/2*	12.7*		171.8	1.685	12.7	33.0	33.0	21900	708		3100	180	3610	1090000	5000	1.76
3/8*	9.5*		130.5	1.280	9.5	45.1	45.1	16600	548		2400	182	2770	840000	3800	1.78
16×16×5/8	406.4×406.4×15.9		189.9	1.862	14.8	24.5	24.5	22600	571		2810	159	3290	854000	4530	1.57
1/2*	12.7		153.7	1.508	11.8	31.4	31.4	18300	469		2310	160	2680	703000	3670	1.59
3/8	9.5		116.6	1.143	8.9	42.7	42.7	13900	365		1790	162	2070	547000	2810	1.60

(Continued)

Nominal Size		Mass Per Meter	Weight Per Meter	Design Wall Thickness	b/t	h/t	Area	$I/10^6$	$S/10^3$	r	$Z/10^3$	Torsional Stiffness Constant $J/10^3$	Torsional Shear Constant $C/10^3$	Surface Area Per Meter
US Customary	SI/Metric													
Inches	Millimeters	kg	kN	mm			mm^2	mm^4	mm^3	mm	mm^3	mm^4	mm^3	m^2
5/16	7.9	97.6	0.957	7.4	51.9	51.9	11700	308	1510	162	1740	461000	2350	1.60
14×14×5/8	355.6×355.6×15.9	164.5	1.613	14.8	21.0	21.0	19600	374	2100	138	2480	559000	3430	1.37
1/2	12.7	133.5	1.309	11.8	27.1	27.1	15900	309	1740	140	2030	463000	2780	1.38
3/8	9.5	101.4	0.995	8.9	37.0	37.0	12100	241	1360	141	1570	361000	2140	1.39
5/16	7.9	85.0	0.833	7.4	45.1	45.1	10200	204	1150	142	1320	306000	1790	1.40
12×12×5/8	304.8×304.8×15.9	139.1	1.364	14.8	17.6	17.6	16600	229	1500	117	1780	341000	2480	1.17
1/2	12.7	113.2	1.110	11.8	22.8	22.8	13500	190	1250	119	1470	284000	2020	1.18
3/8	9.5	86.3	0.846	8.9	31.2	31.2	10300	149	979	120	1140	224000	1560	1.19
5/16	7.9	72.4	0.710	7.4	38.2	38.2	8660	127	831	121	961	190000	1310	1.19
1/4	6.4	59.1	0.580	5.9	48.7	48.7	6960	103	676	122	777	154000	1050	1.20
10×10×5/8	254.0×254.0×15.9	113.8	1.116	14.8	14.2	14.2	13600	127	1000	96.6	1200	189000	1680	0.97
1/2	12.7	93.0	0.912	11.8	18.5	18.5	11100	106	838	98.0	994	159000	1380	0.98
3/8	9.5	71.1	0.697	8.9	25.5	25.5	8520	84.3	664	99.4	777	126000	1070	0.99
5/16	7.9	59.8	0.586	7.4	31.3	31.3	7160	71.8	566	100	657	108000	898	0.99
1/4	6.4	48.9	0.480	5.9	40.1	40.1	5770	58.7	462	101	534	88000	726	1.00
3/16	4.8	37.1	0.364	4.4	54.7	54.7	4340	44.8	353	102	405	67200	548	1.00
9×9×1/2	228.6×228.6×12.7	82.8	0.812	11.8	16.4	16.4	9870	75.9	664	87.7	793	113000	1100	0.87
3/8	9.5	63.5	0.623	8.9	22.7	22.7	7620	60.4	529	89.1	622	90400	856	0.88
5/16	7.9	53.5	0.525	7.4	27.9	27.9	6410	51.7	452	89.8	527	77300	723	0.89
1/4	6.4	43.8	0.430	5.9	35.7	35.7	5170	42.3	370	90.5	429	63400	584	0.89
3/16	4.8	33.3	0.326	4.4	49.0	49.0	3900	32.4	284	91.2	326	48600	442	0.90
8×8×5/8	203.2×203.2×15.9	88.4	0.867	14.8	10.7	10.7	10600	60.9	599	75.8	734	90100	1040	0.76
1/2	12.7	72.7	0.713	11.8	14.2	14.2	8680	51.8	510	77.3	614	77100	858	0.77
3/8	9.5	56.0	0.549	8.9	19.8	19.8	6710	41.6	409	78.7	484	62100	669	0.78
5/16	7.9	47.2	0.463	7.4	24.5	24.5	5650	35.7	351	79.4	412	53400	566	0.79
1/4	6.4	38.7	0.380	5.9	31.4	31.4	4570	29.3	289	80.1	336	43900	459	0.79
3/16	4.8	29.4	0.289	4.4	43.2	43.2	3450	22.6	222	80.9	256	33800	347	0.80
7×7×5/8	177.8×177.8×15.9	75.7	0.743	14.8	9.0	9.0	9090	38.9	438	65.4	543	57300	774	0.66
1/2	12.7	62.6	0.614	11.8	12.1	12.1	7480	33.5	377	66.9	457	49700	644	0.67
3/8	9.5	48.4	0.474	8.9	17.0	17.0	5810	27.1	305	68.3	363	40400	505	0.68
5/16	7.9	40.9	0.401	7.4	21.0	21.0	4900	23.4	263	69.0	310	34900	428	0.69
1/4	6.4	33.6	0.330	5.9	27.1	27.1	3970	19.3	217	69.8	254	28900	348	0.69
3/16	4.8	25.6	0.251	4.4	37.4	37.4	3000	14.9	168	70.5	194	22400	264	0.70
6×6×5/8	152.4×152.4×15.9	63.0	0.618	14.8	7.3	7.3	7580	23.0	301	55.0	381	33500	548	0.56
1/2	12.7	52.4	0.514	11.8	9.9	9.9	6280	20.1	263	56.5	324	29600	460	0.57
3/8	9.5	40.8	0.400	8.9	14.1	14.1	4900	16.5	216	57.9	260	24500	364	0.58
5/16	7.9	34.6	0.339	7.4	17.6	17.6	4150	14.3	188	58.7	223	21300	310	0.58
1/4	6.4	28.5	0.280	5.9	22.8	22.8	3370	11.9	156	59.4	183	17800	252	0.59

Size	in	mm													
	3/16	4.8	21.8	0.214	4.4	31.6	31.6	2550	9.23	121	60.1	141	13800	192	0.59
	1/8	3.2	14.8	0.145	3.0	47.8	47.8	1770	6.54	85.8	60.8	98.7	9800	134	0.60
5 1/2×5 1/2×3/8 139.7×139.7×9.5			37.0	0.363	8.9	12.7	12.7	4450	12.4	177	52.8	215	18400	302	0.53
	5/16	7.9	31.4	0.308	7.4	15.9	15.9	3780	10.8	155	53.5	185	16100	257	0.53
	1/4	6.4	26.0	0.255	5.9	20.7	20.7	3070	9.02	129	54.2	152	13500	210	0.54
	3/16	4.8	19.9	0.195	4.4	28.8	28.8	2330	7.04	101	54.9	117	10500	161	0.54
	1/8	3.2	13.5	0.132	3.0	43.6	43.6	1620	5.00	71.6	55.6	82.5	7500	112	0.55
5×5×1/2 127.0×127.0×12.7			42.3	0.415	11.8	7.8	7.8	5080	10.8	170	46.1	214	15800	307	0.47
	3/8	9.5	33.2	0.326	8.9	11.3	11.3	4000	9.05	143	47.6	174	13400	246	0.48
	5/16	7.9	28.3	0.277	7.4	14.2	14.2	3400	7.93	125	48.3	150	11800	210	0.48
	1/4	6.4	23.4	0.230	5.9	18.5	18.5	2770	6.65	105	49.0	124	9930	172	0.49
	3/16	4.8	18.0	0.176	4.4	25.9	25.9	2110	5.22	82.2	49.7	96.1	7810	132	0.49
	1/8	3.2	12.2	0.120	3.0	39.3	39.3	1460	3.72	58.6	50.4	67.7	5580	92.2	0.50
4 1/2×4 1/2×1/2 114.3×114.3×12.7			37.3	0.365	11.8	6.7	6.7	4480	7.51	131	40.9	167	10900	242	0.42
	3/8	9.5	29.4	0.289	8.9	9.8	9.8	3550	6.37	111	42.4	137	9400	195	0.43
	5/16	7.9	25.1	0.246	7.4	12.4	12.4	3020	5.62	98.3	43.1	119	8340	168	0.43
	1/4	6.4	20.9	0.205	5.9	16.4	16.4	2470	4.74	83.0	43.8	99.1	7070	138	0.44
	3/16	4.8	16.0	0.157	4.4	23.0	23.0	1880	3.74	65.5	44.6	76.9	5600	106	0.44
	1/8	3.2	11.0	0.107	3.0	35.1	35.1	1310	2.69	47.0	45.2	54.4	4020	74.2	0.45
4×4×1/2 101.6×101.6×12.7			32.2	0.316	11.8	5.6	5.6	3880	4.95	97.5	35.7	126	7120	184	0.37
	3/8	9.5	25.6	0.252	8.9	8.4	8.4	3100	4.28	84.2	37.2	105	6280	150	0.38
	5/16	7.9	22.0	0.216	7.4	10.7	10.7	2650	3.81	74.9	37.9	91.7	5630	130	0.38
	1/4	6.4	18.3	0.179	5.9	14.2	14.2	2170	3.24	63.8	38.6	76.7	4820	107	0.39
	3/16	4.8	14.1	0.139	4.4	20.1	20.1	1660	2.57	50.7	39.4	59.9	3850	82.8	0.39
	1/8	3.2	9.7	0.095	3.0	30.9	30.9	1160	1.86	36.6	40.0	42.6	2790	58.2	0.40
3 1/2×3 1/2×3/8 88.9×88.9×9.5			21.9	0.214	8.9	7.0	7.0	2640	2.70	60.8	32.0	77.1	3940	111	0.33
	5/16	7.9	18.8	0.185	7.4	9.0	9.0	2270	2.43	54.7	32.7	67.9	3580	96.7	0.33
	1/4	6.4	15.8	0.154	5.9	12.1	12.1	1870	2.09	47.1	33.5	57.2	3100	80.5	0.34
	3/16	4.8	12.2	0.120	4.4	17.2	17.2	1440	1.68	37.8	34.2	45.0	2510	62.5	0.34
	1/8	3.2	8.4	0.082	3.0	26.6	26.6	1010	1.22	27.6	34.9	32.2	1830	44.2	0.35
3×3×3/8 76.2×76.2×9.5			18.1	0.177	8.9	5.6	5.6	2190	1.57	41.2	26.8	53.4	2260	77.9	0.27
	5/16	7.9	15.7	0.154	7.4	7.3	7.3	1900	1.44	37.7	27.5	47.6	2090	68.5	0.28
	1/4	6.4	13.2	0.129	5.9	9.9	9.9	1570	1.25	32.9	28.3	40.5	1850	57.5	0.28
	3/16	4.8	10.3	0.101	4.4	14.3	14.3	1210	1.02	26.8	29.0	32.2	1520	45.0	0.29
	1/8	3.2	7.1	0.070	3.0	22.4	22.4	855	0.753	19.8	29.7	23.3	1130	32.0	0.29
2 1/2×2 1/2×5/16 63.5×63.5×7.9			12.5	0.123	7.4	5.6	5.6	1520	0.757	23.8	22.3	30.9	1090	45.0	0.23
	1/4	6.4	10.6	0.104	5.9	7.8	7.8	1270	0.676	21.3	23.1	26.7	988	38.4	0.23
	3/16	4.8	8.4	0.082	4.4	11.4	11.4	990	0.561	17.7	23.8	21.6	832	30.4	0.24
	1/8	3.2	5.9	0.057	3.0	18.2	18.2	703	0.421	13.3	24.5	15.8	629	21.9	0.24
2 1/4×2 1/4×1/4 57.2×57.2×6.4			9.4	0.092	5.9	6.7	6.7	1120	0.471	16.5	20.5	20.9	683	30.3	0.21
	3/16	4.8	7.4	0.073	4.4	10.0	10.0	879	0.396	13.9	21.2	17.1	585	24.2	0.21
	1/8	3.2	5.2	0.051	3.0	16.1	16.1	627	0.301	10.5	21.9	12.6	449	17.5	0.22

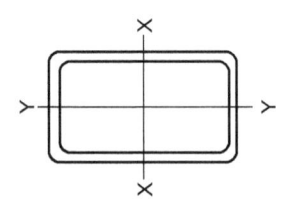

6.5.0 Calculating the Weight and Size of U.S. Rectangular High-Strength Steel Sections

Nominal Size			Weight per Foot	Wall Thickness t	b/t	h/t	Cross Sectional Area	x-x Axis				Y-Y Axis				Torsional Stiffness Constant J	Torsional Shear Constant C	Surface Area Per Foot
								I_x	S_x	r_x	Z_x	I_y	S_y	r_y	Z_y			
in.	in.	in.	lb.	in.			in.²	in.⁴	in.³	in.	in.³	in.⁴	in.³	in.	in.³	in.⁴	in.³	ft.²
32 × 24	×	5/8*	225.80	0.625	35.4	48.2	66.4	9880	617	12.2	733	6390	533	9.81	604	12600	913	9.01
		1/2*	183.50	0.500	45.0	61.0	53.9	8160	510	12.3	601	5280	440	9.98	495	10100	739	9.12
		3/8*	138.95	0.375	61.0	82.3	40.8	6250	391	12.4	458	4050	337	9.96	378	7670	560	9.17
30 × 24	×	5/8*	217.30	0.625	35.4	45.0	63.9	8480	565	11.5	668	6050	504	9.73	575	11400	854	8.68
		1/2*	176.70	0.500	45.0	57.0	51.9	7010	468	11.6	548	5000	417	9.82	472	9220	692	8.79
		3/8*	133.84	0.375	61.0	77.0	39.3	5380	359	11.7	418	3840	320	9.88	360	6990	524	8.84
28 × 24	×	5/8*	208.79	0.625	35.4	41.8	61.4	7210	515	10.8	605	5710	476	9.65	546	10300	796	8.34
		1/2*	169.89	0.500	45.0	53.0	49.9	5970	426	10.9	497	4730	394	9.73	448	8330	645	8.45
		3/8*	128.74	0.375	61.0	71.7	37.8	4580	327	11.0	379	3630	302	9.79	342	6320	489	8.51
26 × 24	×	5/8*	200.28	0.625	35.4	38.6	58.9	6060	466	10.1	545	5370	447	9.55	517	9240	737	8.01
		1/2*	163.08	0.500	45.0	49.0	47.9	5020	386	10.2	448	4450	371	9.64	425	7460	598	8.12
		3/8*	123.64	0.375	61.0	66.3	36.3	3860	297	10.3	342	3420	285	9.70	324	5660	453	8.17
24 × 22	×	5/8*	183.27	0.625	32.2	35.4	53.9	4680	390	9.33	458	4110	373	8.73	432	7150	621	7.34
		1/2*	149.47	0.500	41.0	45.0	43.9	3900	325	9.42	378	3420	311	8.82	356	5780	504	7.45
		3/8*	113.43	0.375	55.7	61.0	33.3	3000	250	9.49	289	2630	239	8.89	273	4390	383	7.51
22 × 20	×	5/8*	166.25	0.625	29.0	32.2	48.9	3530	321	8.51	379	3060	306	7.91	355	5400	514	6.68
		1/2*	135.86	0.500	37.0	41.0	39.9	2950	269	8.60	313	2560	256	8.00	294	4370	418	6.79
		3/8*	103.22	0.375	50.3	55.7	30.3	2280	207	8.67	240	1970	197	8.07	225	3330	318	6.84
20 × 18	×	5/8*	149.24	0.625	25.8	29.0	43.9	2590	259	7.69	307	2210	245	7.10	286	3960	417	6.01
		1/2*	122.25	0.500	33.0	37.0	35.9	2180	218	7.78	255	1850	206	7.19	238	3220	340	6.12
		3/8*	93.01	0.375	45.0	50.3	27.3	1690	169	7.85	196	1440	160	7.25	183	2450	259	6.17
20 × 16	×	5/8*	140.73	0.625	22.6	29.0	41.4	2360	236	7.55	283	1680	210	6.37	243	3280	368	5.68
		1/2*	115.45	0.500	29.0	37.0	33.9	1990	199	7.65	236	1410	177	6.46	203	2670	301	5.79
		3/8*	87.91	0.375	39.7	50.3	25.8	1540	154	7.72	181	1100	137	6.52	156	2040	229	5.84
20 × 12	×	5/8*	123.72	0.625	16.2	29.0	36.4	1890	189	7.20	234	864	144	4.87	166	2030	271	5.01
		1/2	103.30	0.465	22.8	40.0	28.3	1550	155	7.39	188	705	117	4.99	132	1540	209	5.20
		3/8	78.52	0.349	31.4	54.3	21.5	1200	120	7.45	144	547	91.1	5.04	102	1180	160	5.23
		5/16	65.87	0.291	38.2	65.7	18.1	1010	101	7.48	122	464	77.3	5.07	85.8	997	134	5.25

Size			t														
20	×	8	5/8	110.36	0.581	10.8	31.4	30.3	1440	144	6.89	185	338	84.6	3.34	96.4	916	167	4.50
			1/2	89.68	0.465	14.2	40.0	24.6	1190	119	6.96	152	283	70.8	3.39	79.5	757	137	4.53
			3/8	68.31	0.349	19.9	54.3	18.7	926	92.6	7.03	117	222	55.6	3.44	61.5	586	105	4.57
			5/16	57.36	0.291	24.5	65.7	15.7	786	78.6	7.07	98.6	189	47.4	3.47	52.0	496	88.3	4.58
20	×	4	1/2	76.07	0.465	5.6	40.0	20.9	838	83.8	6.33	115	58.7	29.3	1.68	34.0	195	63.8	3.87
			3/8	58.10	0.349	8.5	54.3	16.0	657	65.7	6.42	89.3	47.6	23.8	1.73	26.8	156	49.9	3.90
			5/16	48.86	0.291	10.7	65.7	13.4	560	56.0	6.46	75.6	41.2	20.6	1.75	22.9	134	42.4	3.92
18	×	12	5/8*	115.21	0.625	16.2	25.8	33.9	1450	161	6.55	199	783	131	4.81	152	1740	243	4.68
			1/2*	95.03	0.500	21.0	33.0	27.9	1240	138	6.67	168	668	111	4.89	127	1430	200	4.79
			3/8*	72.59	0.375	29.0	45.0	21.3	971	108	6.75	130	524	87.3	4.95	98.6	1100	153	4.84
18	×	6	5/8	93.34	0.581	7.3	28.0	25.7	923	103	6.00	135	158	52.6	2.48	61.0	462	109	3.83
			1/2	76.07	0.465	9.9	35.7	20.9	770	85.6	6.07	112	134	44.6	2.53	50.7	387	89.9	3.87
			3/8	58.10	0.349	14.2	48.6	16.0	602	66.9	6.15	86.4	106	35.5	2.58	39.5	302	69.5	3.90
			5/16	48.86	0.291	17.6	58.9	13.4	513	57.0	6.18	73.1	91.3	30.4	2.61	33.5	257	58.7	3.92
			1/4	39.43	0.233	22.8	74.3	10.8	419	46.5	6.22	59.4	75.1	25.0	2.63	27.3	210	47.7	3.93
16	×	12	5/8*	106.71	0.625	16.2	22.6	31.4	1090	136	5.89	167	702	117	4.73	137	1470	215	4.34
			1/2	89.68	0.465	22.8	31.4	24.6	904	113	6.06	135	581	96.8	4.86	111	1120	166	4.53
			3/8	68.31	0.349	31.4	42.8	18.7	702	87.7	6.12	104	452	75.3	4.91	85.5	862	127	4.57
			5/16	57.36	0.291	38.2	52.0	15.7	595	74.4	6.15	87.7	384	64.0	4.94	72.2	727	107	4.58
16	×	8	5/8	93.34	0.581	10.8	24.5	25.7	815	102	5.63	129	274	68.5	3.27	79.2	681	132	3.83
			1/2	76.07	0.465	14.2	31.4	20.9	679	84.9	5.70	106	230	57.6	3.32	65.5	563	108	3.87
			3/8	58.10	0.349	19.9	42.8	16.0	531	66.3	5.77	82.1	181	45.3	3.37	50.8	436	83.4	3.90
			5/16	48.86	0.291	24.5	52.0	13.4	451	56.4	5.80	69.4	155	38.7	3.40	43.0	369	70.4	3.92
16	×	4	1/2	62.46	0.465	5.6	31.4	17.2	455	56.9	5.15	77.3	47.0	23.5	1.65	27.4	150	50.7	3.20
			3/8	47.90	0.349	8.5	42.8	13.2	360	45.0	5.23	60.2	38.3	19.1	1.71	21.7	120	39.7	3.23
			5/16	40.35	0.291	10.7	52.0	11.1	308	38.5	5.27	51.1	33.2	16.6	1.73	18.5	103	33.8	3.25
14	×	12	1/2*	81.42	0.500	21.0	25.0	23.9	678	96.9	5.32	116	536	89.3	4.73	104	990	154	4.12
			3/8*	62.39	0.375	29.0	34.3	18.3	534	76.3	5.40	90.0	422	70.4	4.80	81.2	762	118	4.17
14	×	10	5/8	93.34	0.581	14.2	21.1	25.7	687	98.2	5.17	120	407	81.5	3.98	95.1	832	146	3.83
			1/2	76.07	0.465	18.5	27.1	20.9	573	81.8	5.23	98.8	341	68.1	4.04	78.5	685	120	3.87
			3/8	58.10	0.349	25.7	37.1	16.0	447	63.9	5.29	76.3	267	53.4	4.09	60.7	528	91.8	3.90
			5/16	48.86	0.291	31.4	45.1	13.4	380	54.3	5.32	64.6	227	45.5	4.12	51.4	446	77.4	3.92
			1/4	39.43	0.233	39.9	57.1	10.8	310	44.3	5.35	52.4	186	37.2	4.14	41.8	362	62.6	3.93
14	×	6	5/8	76.33	0.581	7.3	21.1	21.0	478	68.2	4.77	88.7	124	41.2	2.43	48.4	334	83.7	3.17
			1/2	62.46	0.465	9.9	27.1	17.2	402	57.4	4.84	73.6	105	35.1	2.48	40.4	279	69.3	3.20
			3/8	47.90	0.349	14.2	37.1	13.2	317	45.3	4.91	57.3	84.1	28.0	2.53	31.6	219	53.7	3.23
			5/16	40.35	0.291	17.6	45.1	11.1	271	38.7	4.94	48.6	72.3	24.1	2.55	26.9	186	45.5	3.25
			1/4	32.63	0.233	22.8	57.1	8.96	222	31.7	4.98	39.6	59.6	19.9	2.58	22.0	152	36.9	3.27
			3/16	24.73	0.174	31.5	77.5	6.76	170	24.3	5.01	30.1	45.9	15.3	2.61	16.7	116	28.0	3.28
14	×	4	5/8	67.82	0.581	3.9	21.1	18.7	373	53.3	4.47	73.1	47.1	23.6	1.59	28.5	148	52.6	2.83
			1/2	55.66	0.465	5.6	27.1	15.3	317	45.3	4.55	61.0	41.1	20.6	1.64	24.1	127	44.1	2.87

(Continued)

Nominal Size			Weight per Foot	Wall Thickness t	b/t	h/t	Cross Sectional Area	x–x Axis				Y-Y Axis				Torsional Stiffness Constant J	Torsional Shear Constant C	Surface Area Per Foot
								I_x	S_x	r_x	Z_x	I_y	S_y	r_y	Z_y			
in.	in.	in.	lb.	in.			in.²	in.⁴	in.³	in.	in.³	in.⁴	in.³	in.	in.³	in.⁴	in.³	ft.²
12 × 10		3/8	42.79	0.349	8.5	37.1	11.8	252	36.0	4.63	47.8	33.6	16.8	1.69	19.1	102	34.6	2.90
		5/16	36.10	0.291	10.7	45.1	9.92	216	30.9	4.67	40.6	29.2	14.6	1.72	16.4	87.7	29.5	2.92
		1/4	29.23	0.233	14.2	57.1	8.03	178	25.4	4.71	33.2	24.4	12.2	1.74	13.5	72.4	24.1	2.93
		3/16	22.18	0.174	20.0	77.5	6.06	137	19.5	4.74	25.3	19.0	9.48	1.77	10.3	55.8	18.4	2.95
12 × 10	×	1/2	69.27	0.465	18.5	22.8	19.0	395	65.9	4.56	78.8	298	59.7	3.96	69.6	545	102	3.53
		3/8	53.00	0.349	25.7	31.4	14.6	310	51.6	4.61	61.1	234	46.9	4.01	54.0	421	78.3	3.57
		5/16	44.60	0.291	31.4	38.2	12.2	264	44.0	4.64	51.7	200	40.0	4.04	45.7	356	66.1	3.58
		1/4	36.03	0.233	39.9	48.5	9.90	216	36.0	4.67	42.1	164	32.7	4.07	37.2	289	53.5	3.60
12 × 8	×	5/8	76.33	0.581	10.8	17.7	21.0	396	66.1	4.34	82.1	210	52.5	3.16	61.9	454	97.7	3.17
		1/2	62.46	0.465	14.2	22.8	17.2	333	55.5	4.40	68.1	177	44.4	3.21	51.5	377	80.4	3.20
		3/8	47.90	0.349	19.9	31.4	13.2	262	43.7	4.47	53.0	140	35.1	3.27	40.1	293	62.1	3.23
		5/16	40.35	0.291	24.5	38.2	11.1	224	37.4	4.50	44.9	120	30.1	3.29	34.1	248	52.4	3.25
		1/4	32.63	0.233	31.3	48.5	8.96	184	30.6	4.53	36.6	98.8	24.7	3.32	27.8	202	42.5	3.27
		3/16	24.73	0.174	43.0	66.0	6.76	140	23.4	4.56	27.8	75.7	18.9	3.35	21.1	153	32.2	3.28
12 × 6	×	5/8	67.82	0.581	7.3	17.7	18.7	321	53.4	4.14	68.8	106	35.5	2.39	42.1	271	71.1	2.83
		1/2	55.66	0.465	9.9	22.8	15.3	271	45.2	4.21	57.4	91.1	30.4	2.44	35.2	227	59.0	2.87
		3/8	42.79	0.349	14.2	31.4	11.8	215	35.8	4.28	44.8	72.9	24.3	2.49	27.7	178	45.8	2.90
		5/16	36.10	0.291	17.6	38.2	9.92	184	30.7	4.31	38.1	62.8	20.9	2.52	23.6	152	38.8	2.92
		1/4	29.23	0.233	22.8	48.5	8.03	151	25.2	4.34	31.1	51.9	17.3	2.54	19.3	124	31.6	2.93
		3/16	22.18	0.174	31.5	66.0	6.06	116	19.4	4.38	23.7	40.1	13.3	2.57	14.7	94.6	24.0	2.95
12 × 4	×	5/8	59.32	0.581	3.9	17.7	16.4	245	40.8	3.87	55.5	40.3	20.1	1.57	24.5	122	44.6	2.50
		1/2	48.85	0.465	5.6	22.8	13.5	209	34.9	3.95	46.7	35.3	17.6	1.62	20.9	105	37.5	2.53
		3/8	37.69	0.349	8.5	31.4	10.4	168	28.0	4.02	36.7	28.9	14.5	1.67	16.6	84.1	29.5	2.57
		5/16	31.84	0.291	10.7	38.2	8.76	144	24.0	4.06	31.3	25.2	12.6	1.70	14.2	72.4	25.2	2.58
		1/4	25.82	0.233	14.2	48.5	7.10	119	19.9	4.10	25.6	21.0	10.5	1.72	11.7	59.8	20.6	2.60
		3/16	19.63	0.174	20.0	66.0	5.37	91.8	15.3	4.13	19.6	16.4	8.20	1.75	9.00	46.1	15.7	2.62
12 × 3 1/2	×	3/8	36.41	0.349	7.0	31.4	10.0	156	26.0	3.94	34.7	21.3	12.2	1.46	14.0	67.7	25.5	2.48
		5/16	24.97	0.291	9.0	38.2	8.46	134	22.4	3.98	29.6	18.6	10.6	1.48	12.1	56.0	21.8	2.50
12 × 3	×	5/16	29.72	0.291	7.3	38.2	8.17	124	20.7	3.90	27.9	13.1	8.73	1.27	10.0	41.3	18.4	2.42
		1/4	24.12	0.233	9.9	48.5	6.63	103	17.2	3.94	22.9	11.1	7.38	1.29	8.28	34.5	15.1	2.43
		3/16	18.35	0.174	14.2	66.0	5.02	79.6	13.3	3.98	17.5	8.72	5.81	1.32	6.40	26.8	11.6	2.45
12 × 2	×	1/4	22.42	0.233	5.6	48.5	6.17	86.9	14.5	3.75	20.1	4.40	4.40	0.845	5.08	15.1	9.64	2.27
		3/16	17.08	0.174	8.5	66.0	4.67	67.4	12.2	3.80	15.5	3.55	3.55	0.872	3.97	12.0	7.49	2.28
10 × 8	×	1/2	55.66	0.465	14.2	18.5	15.3	214	42.7	3.73	51.9	151	37.8	3.14	44.5	288	66.4	2.87
		3/8	42.79	0.349	19.9	25.7	11.8	169	33.9	3.79	40.5	120	30.0	3.19	34.8	224	51.4	2.90
		5/16	36.10	0.291	24.5	31.4	9.92	145	29.0	3.82	34.4	103	25.7	3.22	296	190	43.5	2.92
		1/4	29.23	0.233	31.3	39.9	8.03	119	23.8	3.85	28.1	84.7	21.2	3.25	24.2	155	35.3	2.93
		3/16	22.18	0.174	43.0	54.5	6.06	91.4	18.3	3.88	21.4	65.1	16.3	3.28	18.4	118	26.7	2.95
10 × 6	×	5/8	59.32	0.581	7.3	14.2	16.4	201	40.2	3.50	51.3	89.4	29.8	2.34	35.8	209	58.6	2.50
		1/2	48.85	0.465	9.9	18.5	13.5	171	34.3	3.57	43.0	76.8	25.6	2.39	30.1	176	48.7	2.53
		3/8	37.69	0.349	14.2	25.7	10.4	137	27.3	3.63	33.8	61.8	20.6	2.44	23.7	139	37.9	2.57
		5/16	31.84	0.291	17.6	31.4	8.76	118	23.5	3.66	28.8	53.3	17.8	2.47	20.2	118	32.2	2.58
		1/4	25.82	0.233	22.8	39.9	7.10	96.9	19.4	3.69	23.6	44.1	14.7	2.49	16.6	96.7	26.2	2.60

Size	Thickness	Wt/ft															
	3/16	19.63	0.174		31.5	54.5	5.37	74.6	14.9	3.73	18.0	34.1	11.4	2.52	12.7	73.8	19.9	2.62
10 × 5	3/8	35.13	0.349	11.3	25.7	9.67	120	24.1	3.53	30.4	40.6	16.2	2.05	18.7	100	31.2	2.40
	5/16	29.72	0.291	14.2	31.4	8.17	104	20.8	3.56	26.0	35.2	14.1	2.07	16.0	86.0	26.5	2.42
	1/4	24.12	0.233	18.5	39.9	6.63	85.8	17.2	3.60	21.3	29.3	11.7	2.10	13.2	70.7	21.6	2.43
	3/16	18.35	0.174	25.7	54.5	5.02	66.2	13.2	3.63	16.3	22.7	9.09	2.13	10.1	54.1	16.5	2.45
10 × 4	5/8	50.81	0.581	3.9	14.2	14.0	149	29.9	3.26	40.3	33.4	16.7	1.54	20.6	95.7	36.7	2.17
	1/2	42.05	0.465	5.6	18.5	11.6	129	25.8	3.34	34.1	29.4	14.7	1.59	17.6	82.6	31.0	2.20
	3/8	32.58	0.349	8.5	25.7	8.97	104	20.8	3.41	27.0	24.3	12.1	1.64	14.0	66.5	24.4	2.23
	5/16	27.59	0.291	10.7	31.4	7.59	90.1	18.0	3.44	23.1	21.2	10.6	1.67	12.1	57.3	20.9	2.25
	1/4	22.42	0.233	14.2	39.9	6.17	74.7	14.9	3.48	19.0	17.7	8.87	1.70	9.96	47.4	17.1	2.27
	3/16	17.08	0.174	20.0	54.5	4.67	57.8	11.6	3.52	14.6	13.9	6.93	1.72	7.66	36.5	13.1	2.28
10 × 3 1/2	3/16	16.44	0.174	17.1	54.5	4.50	53.6	10.7	3.45	13.7	10.3	5.89	1.51	6.52	28.6	11.4	2.20
10 × 3	3/8	30.03	0.349	5.6	25.7	8.27	88.0	17.6	3.26	23.7	12.4	8.27	1.22	9.73	37.8	17.7	2.07
	5/16	25.46	0.291	7.3	31.4	7.01	76.3	15.3	3.30	20.3	11.0	7.30	1.25	8.42	33.0	15.2	2.08
	1/4	20.72	0.233	9.9	39.9	5.70	63.6	12.7	3.34	16.7	9.28	6.18	1.28	6.99	27.6	12.5	2.10
	3/16	15.80	0.174	14.2	54.5	4.32	49.4	9.87	3.38	12.8	7.33	4.89	1.30	5.41	21.5	9.64	2.12
	1/8	10.71	0.116	22.9	83.2	2.93	34.2	6.83	3.42	8.80	5.16	3.44	1.33	3.74	14.9	6.61	2.13
10 × 2	3/8	27.48	0.349	2.7	25.7	7.58	71.7	14.3	3.08	20.3	4.69	4.69	0.786	5.76	15.9	11.0	1.90
	5/16	23.34	0.291	3.9	31.4	6.43	62.6	12.5	3.12	17.5	4.24	4.24	0.812	5.06	14.2	9.56	1.92
	1/4	19.02	0.233	5.6	39.9	5.24	52.5	10.5	3.17	14.4	3.67	3.67	0.837	4.26	12.2	7.99	1.93
	3/16	14.53	0.174	8.5	54.5	3.98	41.0	8.19	3.21	11.1	2.97	2.97	0.864	3.34	9.74	6.22	1.95
9 × 7	5/8	59.32	0.581	9.0	12.5	16.4	174	38.7	3.26	48.3	117	33.5	2.68	40.5	235	62.0	2.50
	1/2	48.85	0.465	12.1	16.4	13.5	149	33.0	3.32	40.5	100	28.7	2.73	34.0	197	51.5	2.53
	3/8	37.69	0.349	17.1	22.8	10.4	119	26.4	3.38	31.8	80.4	23.0	2.78	26.7	154	40.0	2.57
	5/16	31.84	0.291	21.1	27.9	8.76	102	22.6	3.41	27.1	69.2	19.8	2.81	22.8	131	33.9	2.58
	1/4	25.32	0.233	27.0	35.6	7.10	84.1	18.7	3.44	22.2	57.2	16.3	2.84	18.7	107	27.6	2.60
	3/16	19.63	0.174	37.2	48.7	5.37	64.7	14.4	3.47	16.9	44.1	12.6	2.87	14.3	81.7	20.9	2.62
9 × 5	5/8	50.81	0.581	5.6	12.5	14.0	133	29.6	3.08	38.5	51.9	20.8	1.92	25.3	128	42.5	2.17
	1/2	42.05	0.465	7.8	16.4	11.6	115	25.5	3.14	32.5	45.2	18.1	1.97	21.5	109	35.6	2.20
	3/8	32.58	0.349	11.3	22.8	8.97	92.5	20.5	3.21	25.7	36.8	14.7	2.03	17.1	86.9	27.9	2.23
	5/16	27.59	0.291	14.2	27.9	7.59	79.8	17.7	3.24	22.0	32.0	12.8	2.05	14.6	74.4	23.8	2.25
	1/4	22.42	0.233	18.5	35.6	6.17	66.1	14.7	3.27	18.1	26.6	10.6	2.08	12.0	61.2	19.4	2.27
	3/16	17.08	0.174	25.7	48.7	4.67	51.1	11.4	3.31	13.8	20.7	8.28	2.10	9.25	46.9	14.8	2.28
9 × 3	1/12	35.24	0.465	3.5	16.4	9.74	80.8	17.9	2.88	24.6	13.2	8.79	1.16	10.8	40.0	19.7	1.87
	3/8	27.48	0.349	5.6	22.8	7.58	66.3	14.7	2.96	19.7	11.2	7.45	1.21	8.80	33.1	15.8	1.90
	5/16	23.34	0.291	7.3	27.9	6.43	57.7	12.8	3.00	16.9	9.88	6.59	1.24	7.63	28.9	13.6	1.92
	1/4	19.02	0.233	9.9	35.6	5.24	48.2	10.7	3.04	14.0	8.38	5.59	1.27	6.35	24.2	11.3	1.93
	3/16	14.53	0.174	14.2	48.7	3.98	37.6	8.35	3.07	10.8	6.63	4.42	1.29	4.92	18.9	8.66	1.95

6.5.1 Calculating the Weight and Size of Metric Rectangular High-Strength Steel Sections

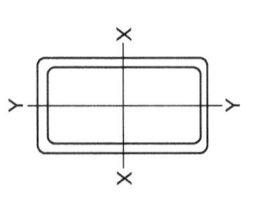

Nominal Size			Mass per Meter	Weight per Meter	Design Wall Thickness t	b/t	h/t	Area	X-X Axis					Y-Y Axis					Torsional Stiffness Constant	Torsional Shear Constant	Surface Area Per Meter
US Customary Inches	SI/Metric	Millimeters	kg	kN	mm			mm²	$I_x/10^6$ mm⁴	$S_x/10^3$ mm³	r_x mm	$Z_x/10^3$ mm³	$I_y/10^6$ mm⁴	$S_y/10^3$ mm³	r_y mm	$Z_y/10^3$ mm³	$J/10^3$	$C/10^3$ mm³	m²
32 × 24 × 5/8*	812.8 × 609.6 × 15.9*		336.6	3.301	15.9	35.3	48.1	42900	4120	10100	310	12000	2660	8740	249	9920	5050000	1500	2.75
1/2*	12.7*		273.1	2.678	12.7	45.0	61.0	34800	3400	8360	313	9850	2200	7210	251	8120	4090000	12100	2.78
3/8*	9.5*		206.3	2.023	9.5	61.2	82.6	26300	2600	6390	314	7490	1680	5510	253	6170	3120000	9150	2.80
30 × 24 × 5/8*	762.0 × 609.6 × 15.9*		323.9	3.176	15.9	35.3	44.9	41300	3530	9280	293	11000	2520	8270	247	9440	4590000	1400	2.64
1/2*	12.7*		263.0	2.579	12.7	45.0	57.0	33500	2920	7660	295	8980	2080	6830	249	7730	3720000	11300	2.68
3/8*	9.5*		198.7	1.949	9.5	61.2	77.2	25300	2230	5860	297	6830	1590	5230	251	5880	2840000	8570	2.69
28 × 24 × 5/8*	711.2 × 609.6 × 15.9*		311.2	3.052	15.9	35.3	41.7	39600	3000	8450	275	9930	2380	7810	245	8960	4140000	13100	2.54
1/2*	12.7*		252.9	2.480	12.7	45.0	53.0	32200	2480	6990	278	8150	1970	6450	247	7350	3360000	10600	2.58
3/8*	9.5*		191.1	1.874	9.5	61.2	71.9	24300	1900	5350	280	6200	1510	4940	249	5590	2560000	7990	2.59
26 × 24 × 5/8*	660.4 × 609.6 × 15.9*		298.5	2.928	15.9	35.3	38.5	38000	2520	7650	258	8940	2240	7340	243	8480	370000	12100	2.44
1/2*	12.7*		242.7	2.380	12.7	45.0	49.0	30900	2090	6330	260	7350	1850	6080	245	6960	3010000	9800	2.47
3/8*	9.5*		183.5	1.800	9.5	61.2	66.5	23400	1600	4850	262	5600	1420	4660	246	5300	2290000	7410	2.49
24 × 22 × 5/8*	609.6 × 558.8 × 15.9*		273.2	2.679	15.9	32.1	35.3	34800	1950	6400	237	7520	1710	6130	222	7090	2850000	10200	2.24
1/2*	12.7*		222.5	2.182	12.7	41.0	45.0	28300	1620	5320	239	6190	1420	5090	224	5840	2320000	8260	2.27
38/*	9.5*		168.4	1.651	9.5	55.8	61.2	21500	1250	4090	241	4720	1090	3910	226	4460	1780000	6250	2.29
22 × 20 × 5/8*	558.8 × 508.0 × 15.9*		247.8	2.430	15.9	28.9	32.1	31600	1470	5270	216	6220	1280	5020	201	5830	2140000	8430	2.04
1/2*	12.7*		202.2	1.983	12.7	37.0	41.0	25800	1230	4400	218	5140	1060	4190	203	4820	1750000	6850	2.07
3/8*	9.5*		153.2	1.503	9.5	50.5	55.8	19500	947	3390	220	3930	820	3230	205	3680	1340000	5190	2.08
20 × 18 × 5/8*	508.0 × 457.2 × 15.9*		224.4	2.181	15.9	25.8	28.9	28300	1080	4250	195	5040	920	4030	180	4690	1560000	6840	1.83
1/2*	12.7*		182.0	1.784	12.7	33.0	37.0	23200	906	3570	198	4180	772	3380	183	3890	1280000	5570	1.86

Size																
20 × 16 × 5/8*	508.0 × 406.4 × 15.9*	209.8	2.057	15.9	22.6	28.9	26700	982	3870	192	4640	700	3440	162	4000	1290000	6040	1.73
1/2*	12.7*	171.8	1.685	12.7	29.0	37.0	21900	827	3250	194	3860	589	2900	164	3320	1060000	4930	1.76
3/8*	9.5*	130.5	1.280	9.5	39.8	50.5	16600	640	2520	196	2960	456	2250	166	2550	81800	3750	1.78
20 × 12 × 5/8*	508.0 × 304.8 × 15.9*	184.4	1.808	15.9	16.2	28.9	23500	786	3090	183	3850	360	2360	124	2720	791000	4450	1.53
1/2	12.7	153.7	1.508	11.8	22.8	40.1	18300	644	2530	188	3080	293	1920	127	2170	613000	3420	1.59
3/8	9.5	116.6	1.143	8.9	31.2	54.1	13900	500	1970	189	2370	228	1500	128	1670	477000	2630	1.60
5/16	7.9	97.6	0.957	7.4	38.2	65.6	11700	422	1660	190	1990	193	1270	129	1410	404000	2200	1.60
20 × 8 × 5/8	508.0 × 203.2 × 15.9	164.5	1.613	14.8	10.7	31.3	19600	600	2360	175	3040	141	1390	84.8	1580	357000	2740	1.37
1/2	12.7	133.5	1.309	11.8	14.2	40.1	15900	496	1950	177	2480	118	1160	86.2	1300	298000	2240	1.38
3/8	9.5	101.4	0.995	8.9	19.8	54.1	12100	387	1520	179	1920	92.8	914	87.4	1010	235000	1720	1.39
5/16	7.9	85.0	0.833	7.4	24.5	65.6	10200	327	1290	179	1620	78.9	777	88.1	853	200000	1450	1.40
20 × 4 × 1/2	508 × 101.6 × 12.7	113.2	1.110	11.8	5.6	40.1	13500	348	1370	161	1890	24.4	480	42.6	557	76200	1050	1.18
3/8	9.5	86.3	0.846	8.9	8.4	54.1	10300	274	1080	163	1470	19.9	391	43.9	440	62200	821	1.19
5/16	7.9	72.4	0.710	7.4	10.7	65.6	8660	233	918	164	1240	17.2	338	44.5	375	53800	696	1.19
18 × 12 × 5/8*	457.2 × 304.8 × 15.9*	171.7	1.684	15.9	16.2	25.8	21900	606	2650	166	3270	326	2140	122	2490	678000	3990	1.43
1/2*	12.7*	141.4	1.387	12.7	21.0	33.0	18000	517	2260	169	2750	278	1830	124	2090	564000	3280	1.46
3/8*	9.5*	107.8	1.057	9.5	29.1	45.1	13700	403	1760	171	2120	217	1430	126	1610	438000	2500	1.48
18 × 6 5/8	457.2 × 152.4 × 15.9	139.1	1.364	14.8	7.3	27.9	16600	385	1680	152	2220	65.8	864	63.0	1000	178000	1790	1.17
1/2	12.7	113.2	1.110	11.8	9.9	35.7	13500	320	1400	154	1830	55.7	731	64.3	830	151000	1470	1.18
3/8	9.5	86.3	0.846	8.9	14.1	48.4	10300	252	1100	156	1420	44.4	583	65.6	649	15000	1470	1.18
5/16	7.9	72.4	0.710	7.4	17.6	58.8	8660	214	934	157	1200	38.0	499	66.3	550	103000	964	1.19
1/4	6.4	59.1	0.580	5.9	22.8	74.5	6960	174	760	158	971	31.2	409	66.9	447	84600	779	1.20
16 × 12 × 5/8*	406.4 × 304.8 × 15.9*	159.0	1.560	15.9	16.2	22.6	20300	453	2230	150	2730	293	1920	120	2250	568000	3520	1.32
1/2	12.7	133.5	1.309	11.8	22.8	31.4	15900	376	1850	154	2210	242	1590	123	1820	443000	2720	1.38
3/8	9.5	101.4	0.995	8.9	31.2	42.7	12100	293	1440	155	1710	189	1240	125	1410	346000	2090	1.39
5/16	7.9	85.0	0.833	7.4	38.2	51.9	10200	248	1220	156	1440	160	1050	125	1180	293000	1750	1.40
16 × 8 5/8	406.4 × 203.2 × 15.9	139.1	1.364	14.8	10.7	24.5	16600	340	1670	143	2120	114	1130	83.0	1300	262000	2170	1.17
1/2	12.7	113.2	1.110	11.8	14.2	31.4	13500	283	1390	145	1740	95.8	943	84.3	1070	220000	1780	1.18
3/8	9.5	86.3	0.846	8.9	19.8	42.7	10300	222	1090	146	1350	75.7	745	85.6	835	174000	1370	1.19
5/16	7.9	72.4	0.710	7.4	24.5	51.9	8660	188	925	147	1140	64.5	635	86.3	706	148000	1150	1.19
16 × 4 × 1/2	406.4 × 101.6 × 12.7	93.0	0.912	11.8	5.6	31.4	11100	189	931	131	1270	19.5	385	42.0	449	57600	830	0.98
3/8	9.5	71.1	0.697	8.9	8.4	42.7	8520	150	739	133	990	16.0	314	43.3	357	47300	653	0.99
5/16	7.9	59.8	0.586	7.4	10.7	51.9	7160	128	631	134	839	13.8	272	44.0	304	41000	555	0.99
14 × 12 × 1/2*	355.6 × 304.8 × 12.7*	121.2	1.188	12.7	21.0	25.0	15400	282	1590	135	1900	223	1460	120	1710	387000	2520	1.26
3/8*	9.5*	92.6	0.908	9.5	29.1	34.4	11800	222	1250	137	1470	175	1150	122	1330	302000	1930	1.27
14 × 10 × 5/8	355.6 × 254.0 × 15.9	139.1	1.364	14.8	14.2	21.0	16600	287	1610	.131	1970	170	*1340	101	1560	320000	2400	1.17
1/2	12.7	113.2	1.110	11.8	18.5	27.1	13500	238	1340	133	1620	142	1120	103	1280	267000	1960	1.18
3/8	9.5	86.3	0.846	8.9	25.5	37.0	10300	187	1050	134	1260	111	878	104	998	211000	1510	1.19
5/16	7.9	72.4	0.710	7.4	31.3	45.1	8660	158	891	135	1060	94.7	746	105	843	179000	1270	1.19
1/4	6.4	59.1	0.580	5.9	40.1	57.3	6960	129	724	136	857	77.1	607	105	682	146000	1020	1.20

(Continued)

Nominal Size			Mass per Meter	Weight per Meter	Design Wall Thickness t	b/t	h/t	Area	X-X Axis					Y-Y Axis					Torsional Stiffness Constant J/10³	Torsional Shear Constant C/10³	Surface Area Per Meter
US Customary Inches	SI/Metric Millimeters		kg	kN	mm			mm²	$I_x/10^6$ mm⁴	$S_x/10^3$ mm³	r_x mm	$Z_x/10^3$ mm³		$I_y/10^6$ mm⁴	$S_y/10^3$ mm³	r_y mm	$Z_y/10^3$ mm³		mm⁴	mm³	m²
14 × 6 5/8	355.6 × 152.4 × 15.9		13.8	1.116	14.8	7.3	21.0	13600	199	1120	121	1460		51.6	677	61.6	794		126000	1380	0.97
1/2		12.7	93.0	0.912	11.8	9.9	27.1	11100	167	941	123	1210		43.8	575	62.9	661		108000	1130	0.98
3/8		9.5	71.1	0.697	8.9	14.1	37.0	8520	132	745	125	942		35.1	461	64.2	520		86400	883	0.99
5/16		7.9	59.8	0.586	7.4	17.6	45.1	7160	113	634	126	798		30.1	395	64.9	441		74100	746	0.99
1/4		6.4	48.9	0.480	5.9	22.8	57.3	5770	92.1	518	126	647		24.8	325	65.5	359		60900	604	1.00
3/16		4.8	37.1	0.364	4.4	31.6	77.8	4340	70.4	396	127	491		19.0	250	66.2	273		46800	457	1.00
14 × 4 5/8	355.6 × 101.6 × 15.9		101.1	0.991	14.8	3.9	21.0	12100	156	875	113	1200		19.6	387	40.3	468		55300	863	0.86
1/2		12.7	82.8	0.812	11.8	5.6	27.1	9870	132	741	116	1000		17.1	337	41.6	395		48500	722	0.87
3/8		9.5	63.5	0.623	8.9	8.4	37.0	7620	105	592	118	785		14.0	276	42.9	315		39900	569	0.88
5/16		7.9	53.5	0.525	7.4	10.7	45.1	6410	90.0	506	119	667		12.2	239	43.6	269		34700	484	0.89
1/4		6.4	43.8	0.430	5.9	14.2	57.3	5170	73.8	415	120	542		10.1	199	44.2	220		28800	394	0.89
3/16		4.8	33.3	0.326	4.4	20.1	77.8	3900	56.6	318	121	413		7.86	155	44.9	169		22400	300	0.90
12 × 10 1/2	304.8 × 254.0 × 12.7		103.1	1.011	11.8	18.5	22.8	12300	164	1080	116	1290		124	977	101	1140		212000	1670	1.08
3/8		9.5	78.7	0.772	8.9	25.5	31.2	9430	129	849	117	1000		97.9	771	102	888		167000	1290	1.09
5/16		7.9	66.1	0.648	7.4	31.3	38.2	7910	110	722	118	849		83.3	656	103	750		142000	1080	1.09
1/4		6.4	54.0	0.530	5.9	40.1	48.7	6360	89.6	588	119	688		67.9	535	103	608		116000	874	1.10
12 × 8 5/8	304.8 × 203.2 × 15.9		113.8	1.116	14.8	10.7	17.6	13600	165	1090	110	1350		87.6	863	80.3	1020		172000	1600	0.97
1/2		12.7	93.0	0.912	11.8	14.2	22.8	11100	139	910	112	1120		73.8	727	81.6	843		145000	1320	0.98
3/8		9.5	71.1	0.697	8.9	19.8	31.2	8520	110	719	113	871		58.7	577	83.0	660		116000	1020	0.99
5/16		7.9	59.8	0.586	7.4	24.5	38.2	7160	93.4	613	114	737		50.1	493	83.6	559		98600	860	0.99
1/4		6.4	48.9	0.480	5.9	31.4	48.7	5770	76.2	500	115	598		41.0	404	84.3	454		80700	695	1.00
3/16		4.8	37.1	0.364	4.4	43.2	66.3	4340	58.2	382	116	454		31.4	309	85.0	345		61800	525	1.00
12 × 6 5/8	304.8 × 152.4 × 15.9		101.1	0.991	14.8	7.3	17.6	12100	134	877	105	1130		44.4	583	60.6	691		101000	1170	0.86
1/2		12.7	82.8	0.812	11.8	9.9	22.8	9870	113	741	107	940		37.9	497	61.9	577		86800	966	0.87
3/8		9.5	63.5	0.623	8.9	14.1	31.2	7620	89.8	589	109	737		30.5	400	63.2	455		69900	753	0.88
5/16		7.9	53.5	0.525	7.4	17.6	38.2	6410	76.7	504	109	625		26.2	343	63.9	387		60000	637	0.89
1/4		6.4	43.8	0.430	5.9	22.8	48.7	5170	62.9	412	110	508		21.5	283	64.6	315		49400	516	0.89
3/16		4.8	33.3	0.326	4.4	31.6	66.3	3900	48.1	316	111	387		16.6	218	65.2	240		38000	391	0.90
12 × 4 5/8	304.8 × 101.6 × 15.9		88.4	0.867	14.8	3.9	17.6	10600	102	670	98.2	912		16.8	331	39.8	403		44900	733	0.76
1/2		12.7	72.7	0.713	11.8	5.6	22.8	8680	87.1	572	100	764		14.7	289	41.1	341		39600	615	0.77
3/8		9.5	56.0	0.549	8.9	8.4	31.2	6710	70.0	460	102	603		12.1	238	42.4	273		32700	486	0.78
5/16		7.9	47.2	0.463	7.4	10.7	38.2	5650	60.1	394	103	513		10.5	207	43.1	233		28400	413	0.79
1/4		6.4	38.7	0.380	5.9	14.2	48.7	4570	49.5	325	104	419		8.74	172	43.7	191		23700	337	0.79
3/16		4.8	29.4	0.289	4.4	20.1	66.3	3450	38.0	250	105	319		6.80	134	44.4	147		18400	257	0.80
12 × 3 1/2	304.8 × 88.9 × 935		54.1	0.530	8.9	7.0	31.2	6490	65.1	427	100	570		8.88	200	37.0	231		25100	419	0.76
3/8																					
5/16		7.9	45.6	0.447	7.4	9.0	38.2	5470	56.0	367	101	486		7.75	174	37.7	198		22000	357	0.76
12 × 3 5/16	304.8 × 76.2 × 7.9		44.0	0.432	7.4	7.3	38.2	5280	51.8	340	99.1	458		5.45	143	32.1	164		16100	301	0.74
1/4		6.4	36.2	0.355	5.9	9.9	48.7	4270	42.8	281	100	374		4.59	121	32.8	135		13600	247	0.74
3/16		4.8	27.5	0.270	4.4	14.3	66.3	3230	33.0	217	101	286		3.62	94.9	33.5	104		10700	189	0.75

12× 2× 1/4	304.8× 50.8× 6.4	33.6		5.9		5.6	48.7	3970	36.1	237	95.4	329	1.83	72.0	21.5	83.1	5930	158	0.69
	3/16	4.8	25.6		0.330	4.4	8.5	66.3	3000	28.0	183	96.5	252	1.47	58.0	22.1	64.9	4790	122	0.70
10× 8× 1/2	254.0× 203.2× 12.7	82.8			0.812	11.8	14.2	18.5	9870	88.9	700	94.9	849	62.8	618	79.8	728	110000	1090	0.87
	3/8	9.5	63.5	8.9	0.623	8.9	19.8	25.5	7620	70.7	557	96.3	666	50.1	493	81.1	572	88000	845	0.88
	5/16	7.9	53.5	7.4	0.525	7.4	24.5	31.3	6410	60.4	476	97.1	565	42.9	422	81.8	485	75300	713	0.89
	1/4	6.4	43.8	5.9	0.430	5.9	31.4	40.1	5170	49.5	389	97.8	459	35.2	346	82.5	395	61800	577	0.89
	3/16	4.8	33.3	4.4	0.326	4.4	43.2	54.7	3900	37.9	298	98.6	349	27.0	265	83.2	300	47400	436	0.90
10× 6× 5/8	254.0× 152.4× 15.9	88.4		14.8	0.867	14.8	7.3	14.2	10600	83.8	660	89.0	842	37.3	489	59.3	587	77300	962	0.76
	1/2	12.7	72.7	11.8	0.713	11.8	9.9	18.5	8680	71.3	561	90.6	704	31.9	419	60.7	493	66600	797	0.77
	3/8	9.5	56.0	8.9	0.549	8.9	14.1	25.5	6710	57.1	450	92.2	555	25.8	338	62.0	390	53960	623	0.78
	5/16	7.9	47.2	7.4	0.463	7.4	17.6	31.3	5650	49.0	385	93.0	472	22.2	291	62.7	332	46400	528	0.79
	1/4	6.4	38.7	5.9	0.380	5.9	22.8	40.1	4570	40.2	317	93.9	385	18.3	240	63.3	271	38300	428	0.79
	3/16	4.8	29.4	4.4	0.289	4.4	31.6	54.7	3450	30.9	243	94.7	293	14.1	185	64.0	207	29500	325	0.80
10× 5× 3/8	254.0× 127.0× 9.5	52.2		8.9	0.512	8.9	11.3	25.5	6260	50.3	396	89.6	500	16.9	267	52.0	307	38900	513	0.73
	5/16	7.9	44.0	7.4	0.432	7.4	14.2	31.3	5280	43.2	340	90.5	426	14.7	231	52.7	263	33600	435	0.74
	1/4	6.4	36.2	5.9	0.355	5.9	18.5	40.1	4270	35.6	280	91.4	348	12.2	191	53.4	215	27900	354	0.74
	3/16	4.8	27.5	4.4	0.270	4.4	25.9	54.7	3230	27.4	216	92.2	265	9.42	148	54.0	165	21600	269	0.75
10× 4× 5/8	254.0× 101.6× 15.9	75.7		14.8	0.743	14.8	3.9	14.2	9090	62.3	490	82.8	662	13.9	274	39.2	338	34600	602	0.66
	1/2	12.7	62.6	11.8	0.614	11.8	5.6	18.5	7480	53.7	423	84.7	559	12.2	241	40.5	288	30800	507	0.67
	3/8	9.5	48.4	8.9	0.474	8.9	8.4	25.5	5810	43.5	343	86.6	444	10.1	199	41.7	231	25600	402	0.68
	5/16	7.9	40.9	7.4	0.401	7.4	10.7	31.3	4900	37.5	295	87.5	379	8.82	174	42.4	198	22300	342	0.69
	1/4	6.4	33.6	5.9	0.330	5.9	14.2	40.1	3970	31.0	244	88.4	310	7.36	145	43.1	163	18600	279	0.69
	3/16	4.8	25.6	4.4	0.251	4.4	20.1	54.7	3000	23.9	189	89.3	238	5.74	113	43.7	125	14500	213	0.70
10× 31/2×	254.0× 88.9× 4.8	24.7		4.4	0.242	4.4	17.2	54.7	2890	22.2	175	87.6	224	4.28	96.2	38.5	106	11400	185	0.67
	3/16																			
10× 3× 3/8	254.0× 76.2× 9.5	44.6		8.9	0.437	8.9	5.6	25.5	5360	36.7	289	82.8	389	5.18	136	31.1	160	14500	291	0.63
	5/16	7.9	37.7	7.4	0.370	7.4	7.3	31.3	4530	31.8	250	83.8	333	4.56	120	31.7	138	12800	250	0.63
	1/4	6.4	31.1	5.9	0.305	5.9	9.9	40.1	3670	26.4	208	84.8	273	3.85	101	32.4	114	10800	205	0.64
	3/16	4.8	23.7	4.4	0.232	4.4	14.3	54.7	2780	20.5	161	85.8	210	3.04	79.8	33.1	88.4	8540	157	0.65
	1/8	3.2	16.1	3.0	0.158	3.0	22.4	81.7	1920	14.5	114	86.7	147	2.18	57.3	33.7	62.3	6130	110	0.65
10× 2× 3/8	254.0× 50.8× 9.5	40.8		8.9	0.400	8.9	2.7	25.5	4900	29.9	236	78.1	334	1.95	76.9	20.0	94.7	6000	180	0.58
	5/16	7.9	34.6	7.4	0.339	7.4	3.9	31.3	4150	26.1	205	79.3	287	1.76	69.5	20.6	83.0	5470	157	0.58
	1/4	6.4	28.5	5.9	0.280	5.9	5.6	40.1	3370	21.8	171	80.4	236	1.53	60.0	21.3	69.6	4760	131	0.59
	3/16	4.8	21.8	4.4	0.214	4.4	8.5	54.7	2550	17.0	134	81.5	182	1.23	48.5	22.0	54.5	3860	102	0.59
9× 7× 5/8	228.6× 177.8× 15.9	88.4		14.8	0.867	14.8	9.0	12.4	10600	72.7	636	82.8	793	489	550	68.0	665	86800	1020	0.76
	1/2	12.7	72.7	11.8	0.713	11.8	12.1	16.4	8680	61.8	541	84.4	663	11.7	470	69.4	557	74400	843	0.77
	3/8	9.5	56.0	8.9	0.549	8.9	17.0	22.7	6710	49.5	433	85.9	522	33.6	378	70.7	440	60000	658	0.78
	5/16	7.9	47.2	7.4	0.463	7.4	21.0	27.9	5650	42.5	371	86.7	444	28.8	324	71.4	374	51600	556	0.79
	1/4	6.4	38.7	5.9	0.380	5.9	27.1	35.7	4570	34.9	305	87.4	362	23.7	267	72.1	305	42500	451	0.79
	3/16	4.8	29.4	4.4	0.289	4.4	37.4	49.0	3450	26.8	235	88.2	276	18.3	206	72.8	233	32700	342	0.80
9× 5× 5/8	228.6× 127.0× 15.9	75.7		14.8	0.743	14.8	5.6	12.4	9090	55.5	485	78.1	632	21.6	341	48.8	416	46500	698	0.66
	1/2	12.7	62.6	11.8	0.614	11.8	7.8	16.4	7480	47.7	417	79.9	533	18.8	296	50.1	352	40700	583	0.67
	3/8	9.5	48.4	8.9	0.474	8.9	11.3	22.7	5810	38.6	338	81.5	423	15.4	242	51.4	281	33400	459	0.68
	5/16	7.9	40.9	7.4	0.401	7.4	142	27.9	4900	33.3	291	82.4	361	13.3	210	52.1	240	29000	390	0.69
	1/4	6.4	33.6	5.9	0.330	5.9	18.5	35.7	3970	27.5	240	83.2	295	11.1	174	52.8	197	24100	317	0.69
	3/16	4.8	25.6	4.4	0.251	4.4	25.9	49.0	3000	21.2	185	84.0	226	8.58	135	53.4	151	18700	242	0.70

(Continued)

Nominal Size			Mass per Meter	Weight per Meter	Design Wall Thickness t	b/t	h/t	Area	X-X Axis				Y-Y Axis				Torsional Stiffness Constant	Torsional Shear Constant	Surface Area Per Meter
US Customary	SI/Metric								$I_x/10^6$	$S_x/10^3$	r_x	$Z_x/10^3$	$I_y/10^6$	$S_y/10^3$	r_y	$Z_y/10^3$	$J/10^3$	$C/10^3$	
Inches	Millimeters		kg	kN	mm			mm^2	mm^4	mm^3	mm	mm^3	mm^4	mm^3	mm	mm^3	mm^4	mm^3	m^2
9 × 3 × 1/2	228.6 × 76.2 × 12.7		52.4	0.514	11.8	3.5	16.4	6280	33.6	294	73.2	403	5.49	144	29.6	177	14600	323	0.57
3/8	9.5		40.8	0.400	8.9	5.6	22.7	4900	27.7	242	75.1	324	4.66	122	30.8	145	12600	260	0.58
5/16	7.9		34.6	0.339	7.4	7.3	27.9	4150	24.1	210	76.1	278	4.12	108	31.5	125	11100	224	0.58
1/4	6.4		28.5	0.280	5.9	9.9	35.7	3370	20.0	175	77.1	229	3.48	91.4	32.1	104	9430	184	0.59
3/16	4.8		21.8	0.214	4.4	14.3	49.0	2550	15.6	136	78.1	176	2.75	72.2	32.8	80.4	7460	141	0.59
8 × 6 × 5/8	203.2 × 152.4 × 15.9		75.7	0.743	14.8	7.3	10.7	9090	47.5	468	72.3	592	30.1	395	57.6	484	54500	755	0.66
1/2	12.7		62.6	0.614	11.8	9.9	14.2	7480	40.8	402	73.9	499	26.0	341	59.0	408	47300	629	0.67
3/8	9.5		48.4	0.474	8.9	14.1	19.8	5810	33.0	325	75.4	396	21.1	277	60.3	325	38600	494	0.68
5/16	7.9		40.9	0.401	7.4	17.6	24.5	4900	28.5	280	76.2	338	18.2	239	61.0	278	33400	419	0.69
1/4	6.4		33.6	0.330	5.9	22.8	31.4	3970	23.5	231	77.0	276	15.1	198	61.7	227	27700	340	0.69
3/16	4.8		25.6	0.251	4.4	31.6	43.2	3000	18.1	178	77.7	211	11.7	153	62.4	174	21400	259	0.70
8 × 4 × 5/8	203.2 × 101.6 × 15.9		63.0	0.618	14.8	3.9	10.7	7580	34.1	336	67.1	451	11.1	218	38.2	272	24800	472	0.56
1/2	12.7		52.4	0.514	11.8	5.6	14.2	6280	29.8	294	68.9	384	9.81	193	39.5	234	22300	399	0.57
3/8	9.5		40.8	0.400	8.9	8.4	19.8	4900	24.5	241	70.7	308	8.18	161	40.8	189	18700	318	0.58
5/16	7.9		34.6	0.339	7.4	10.7	24.5	4150	21.3	209	71.6	264	7.15	141	41.5	163	16400	271	0.58
1/4	6.4		28.5	0.280	5.9	14.2	31.4	3370	17.7	174	72.4	217	5.99	118	42.2	134	13700	222	0.59
3/16	4.8		21.8	0.214	4.4	20.1	43.2	2550	13.7	135	73.3	167	4.39	92.3	42.8	103	10800	170	0.59
1/8	3.2		14.8	0.145	3.0	30.9	64.7	1770	9.71	95.5	74.1	117	3.34	65.8	43.5	72.7	7670	118	0.60
8 × 3 × 1/2	203.2 × 76.2 × 12.7		47.4	0.465	11.8	3.5	14.2	5680	24.3	240	65.5	327	4.86	127	29.2	158	12400	285	0.52
3/8	9.5		37.0	0.363	8.9	5.6	19.8	4450	20.2	199	67.4	264	4.15	109	30.5	129	10700	230	0.53
5/16	7.9		31.4	0.308	7.4	7.3	24.5	3780	17.6	174	68.4	228	3.67	96.3	31.2	112	9500	198	0.53
1/4	6.4		26.0	0.255	5.9	9.9	31.4	3070	14.7	145	69.3	188	3.11	81.6	31.8	93.2	8070	163	0.54
3/16	4.8		19.9	0.195	4.4	14.3	43.2	2330	11.5	113	70.3	145	2.46	84.6	32.5	72.3	6400	125	0.54
1/8	3.2		13.5	0.132	3.0	22.4	64.7	1620	8.18	80.5	71.1	102	1.77	46.6	33.1	51.1	4610	87.8	0.55
8 × 2 × 3/8	203.2 × 50.8 × 9.5		33.2	0.326	8.9	2.7	19.8	4000	16.0	157	63.2	220	1.55	61.1	19.7	75.8	4490	142	0.48
5/16	7.9		28.3	0.277	7.4	3.9	24.5	3400	14.0	138	64.3	191	1.41	55.4	20.3	66.7	4120	124	0.48
1/4	6.4		23.4	0.230	5.9	5.6	31.4	2770	11.8	116	65.4	158	1.22	48.1	21.0	56.1	3600	104	0.49
3/16	4.8		18.0	0.176	4.4	8.5	43.2	2110	9.30	91.5	66.4	123	0.990	39.0	21.7	44.1	2930	80.8	0.49
1/8	3.2		12.2	0.120	3.0	13.9	64.7	1460	6.55	65.5	67.4	86.5	0.728	28.7	22.3	31.6	2160	57.3	0.50

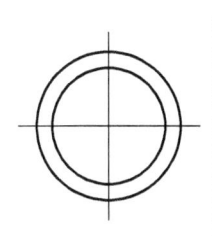

6.6.0 Calculating the Weight and Size of U.S. Round High-Strength Steel Sections

Nominal Size		Weight per Foot	Wall Thickness t	D/t	Cross-Sectional Area	I	S	r	Z	Torsional Stiffness Constant	Torsional Shear Constant	Surface Area per Foot
Outside Diameter	Wall											
in.	in.	lb.	in.		in.²	in.⁴	in.³	in.	in.³	in.⁴	in.³	ft.²
20.000 ×	0.500	104.23	0.465	43.0	28.5	1360	136	6.91	177	2720	272	5.24
	0.375	78.67	0.349	57.3	21.5	1040	104	6.95	135	2080	208	5.24
18.000 ×	0.500	93.54	0.465	38.7	25.6	985	109	6.20	143	1970	219	4.71
	0.375	70.65	0.349	51.6	19.4	754	83.8	6.24	109	1510	168	4.71
16.000 ×	0.500	82.85	0.465	34.4	22.7	685	85.7	5.49	112	1370	171	4.19
	0.438	72.86	0.407	39.3	19.9	606	75.8	5.51	99.0	1210	152	4.19
	0.375	62.64	0.349	45.8	17.2	526	65.7	5.53	85.5	1050	131	4.19
	0.312	52.32	0.291	55.0	14.4	443	55.4	5.55	71.8	886	111	4.19
14.000 ×	0.500	72.16	0.465	30.1	19.8	453	64.8	4.79	85.2	907	130	3.67
	0.375	54.62	0.349	40.1	15.0	349	49.8	4.83	65.1	698	99.7	3.67
	0.312	45.65	0.291	48.1	12.5	295	42.1	4.85	54.7	589	84.2	3.67
12.750 ×	0.500	65.48	0.465	27.4	17.9	339	53.2	4.35	70.2	678	106	3.34
	0.375	49.61	0.349	36.5	13.6	262	41.0	4.39	53.7	523	82.1	3.34
	0.250	33.41	0.233	54.7	9.16	180	28.2	4.43	36.5	359	56.6	3.34
12.500 ×	0.625	79.34	0.581	21.5	21.8	387	62.0	4.22	82.6	774	124	3.27
	0.500	64.14	0.465	26.9	17.6	319	51.0	4.26	67.4	638	102	3.27
	0.375	48.61	0.349	35.8	13.3	246	39.4	4.30	51.5	492	78.7	3.27
	0.312	40.65	0.291	43.0	11.2	208	33.3	4.32	43.4	416	66.6	3.27
	0.250	32.74	0.233	53.6	8.98	169	27.0	4.34	35.1	338	54.1	3.27
	0.188	24.74	0.174	71.8	6.74	128	20.5	4.36	26.4	256	41.0	3.27
12.313 ×	0.625	78.09	0.581	21.2	21.4	369	60.0	4.15	80.0	739	120	3.22
	0.500	63.14	0.465	26.5	17.3	304	49.4	4.19	65.3	608	98.8	3.22
	0.375	47.86	0.349	35.3	13.1	235	38.2	4.23	50.0	470	76.3	3.22
	0.312	40.03	0.291	42.3	11.0	199	32.3	4.25	42.1	397	64.5	3.22
	0.250	32.24	0.233	52.8	8.84	161	26.2	4.27	34.0	323	52.4	3.22
	0.188	24.37	0.174	70.8	6.64	122	19.9	4.29	25.6	244	39.7	3.22
12.250 ×	0.625	77.67	0.581	21.1	21.3	363	59.3	4.13	79.2	727	119	3.21

(Continued)

Nominal Size		Weight per Foot	Wall Thickness t	D/t	Cross-Sectional Area	I	S	r	Z	Torsional Stiffness Constant	Torsional Shear Constant	Surface Area per Foot
Outside Diameter	Wall											
in.	in.	lb.	in.		in.²	in.⁴	in.³	in.	in.³	in.⁴	in.³	ft.²
	0.500	62.80	0.465	26.3	17.2	299	48.9	4.17	64.6	599	97.7	3.21
	0.375	47.60	0.349	35.1	13.0	231	37.7	4.21	49.4	462	75.5	3.21
	0.312	39.82	0.291	42.1	10.9	196	31.9	4.23	41.6	391	63.9	3.21
	0.250	32.07	0.233	52.6	8.80	159	25.9	4.25	33.7	318	51.9	3.21
	0.188	24.24	0.174	70.4	6.60	120	19.6	4.27	25.4	241	39.3	3.21
11.250 ×	0.625	70.99	0.581	19.4	19.5	278	49.4	3.78	66.2	556	98.8	2.95
	0.500	57.46	0.465	24.2	15.8	229	40.8	3.82	54.1	459	81.6	2.95
	0.375	43.60	0.349	32.2	12.0	178	31.6	3.86	41.5	355	63.2	2.95
	0.312	36.48	0.291	38.7	10.0	151	26.8	3.88	35.0	301	53.5	2.95
	0.250	29.40	0.233	48.3	8.06	122	21.8	3.90	28.3	245	43.5	2.95
	0.188	22.23	0.174	64.7	6.05	92.9	16.5	3.92	21.3	186	33.0	2.95
10.750 ×	0.500	54.79	0.465	23.1	15.0	199	37.0	3.64	49.2	398	74.1	2.81
	0.365	40.52	0.340	31.6	11.1	151	28.1	3.68	36.9	36.9	56.1	2.81
	0.250	28.06	0.233	46.1	7.70	106	19.8	3.72	25.8	25.8	39.6	2.81
10.000 ×	0.625	62.64	0.581	17.2	17.2	191	38.3	3.34	51.6	383	76.6	2.62
	0.500	50.78	0.465	21.5	13.9	159	37.1	3.38	42.3	317	63.5	2.62
	0.375	38.58	0.349	28.7	10.6	123	24.7	3.41	32.5	247	49.3	2.62
	0.312	32.31	0.291	34.4	8.88	105	20.9	3.43	27.4	209	41.9	2.62
	0.250	26.06	0.233	42.9	7.15	85.3	17.1	3.45	22.2	171	34.1	2.62
	0.188	19.72	0.174	57.5	5.37	64.8	13.0	3.47	16.8	130	25.9	2.62
9.625 ×	0.500	48.77	0.465	20.7	13.4	141	29.2	3.24	39.0	281	58.5	2.52
	0.375	37.08	0.349	27.6	10.2	110	22.8	3.28	30.0	219	45.5	2.52
	0.312	31.06	0.291	33.1	8.53	93.0	19.3	3.30	25.4	186	38.7	2.52
	0.250	25.05	0.233	41.3	6.87	75.9	15.8	3.32	20.6	152	31.5	2.52
	0.188	18.97	0.174	55.3	5.17	57.7	12.0	3.34	15.5	115	24.0	2.52
8.750 ×	0.500	44.10	0.465	18.8	12.1	104	23.8	2.93	32.0	208	47.6	2.29
	0.375	33.57	0.349	25.1	9.21	81.4	18.6	2.97	24.6	163	37.2	2.29
	0.312	28.14	0.291	30.1	7.73	69.3	15.8	2.99	20.8	139	31.7	2.29
	0.250	22.72	0.233	37.6	6.23	56.6	12.9	3.01	16.9	113	25.9	2.29
	0.188	17.21	0.174	50.3	4.69	43.1	9.86	3.03	12.8	86.2	19.7	2.29
8.625 ×	0.500	43.43	0.465	18.5	11.9	99.5	23.1	2.89	31.0	199	46.2	2.26
	0.375	33.07	0.349	24.7	9.07	77.8	18.0	2.93	23.9	156	36.1	2.26
	0.322	28.58	0.300	28.7	7.85	68.1	15.8	2.95	20.8	136	31.6	2.26
	0.250	22.38	0.233	37.0	6.14	54.1	12.5	2.97	16.4	108	25.1	2.26
	0.188	16.96	0.174	49.6	4.62	41.3	9.57	2.99	12.4	82.5	19.1	2.26
7.625 ×	0.375	29.06	0.349	21.8	7.98	52.9	13.9	2.58	18.5	106	27.8	2.00
	0.328	25.59	0.305	25.0	7.01	47.1	12.3	2.59	16.4	94.1	24.7	2.00
	0.125	10.02	0.116	65.7	2.74	19.3	5.06	2.66	6.54	38.6	10.1	2.00

7.500	×	0.500	37.42	0.465	16.1	10.3	63.9	17.0	2.49	23.0	128	34.1	1.96
		0.375	28.56	0.349	21.5	7.84	50.2	13.4	2.53	17.9	100	26.8	1.96
		0.312	23.97	0.291	25.8	6.59	42.9	11.4	2.55	15.1	85.8	22.9	1.96
		0.250	19.38	0.233	32.2	5.32	35.2	9.37	2.57	12.3	70.3	18.7	1.96
		0.188	14.70	0.174	43.1	4.00	26.9	7.17	2.59	9.34	53.8	14.3	1.96
7.000	×	0.500	34.74	0.465	15.1	9.55	51.2	14.6	2.32	19.9	102	29.3	1.83
		0.375	26.56	0.349	20.1	7.29	40.4	11.6	2.35	15.5	80.9	23.1	1.83
		0.312	22.31	0.291	24.1	6.13	34.6	9.88	2.37	13.1	69.1	19.8	1.83
		0.250	18.04	0.233	30.0	4.95	28.4	8.11	2.39	10.7	56.8	16.2	1.83
		0.188	13.69	0.174	40.2	3.73	21.7	6.21	2.41	8.11	43.5	12.4	1.83
		0.125	9.19	0.116	60.3	2.51	14.9	4.25	2.43	5.50	29.7	8.49	1.83
6.875	×	0.500	34.07	0.465	14.8	9.36	48.3	14.1	2.27	19.1	96.7	28.1	1.80
		0.375	26.06	0.349	19.7	7.16	38.2	11.1	2.31	14.9	76.4	22.2	1.80
		0.312	21.89	0.291	23.6	6.02	32.7	9.51	2.33	12.6	65.4	19.0	1.80
		0.250	17.71	0.233	29.5	4.86	26.8	7.81	2.35	10.3	53.7	15.6	1.80
		0.188	13.44	0.174	39.5	3.66	20.6	5.99	2.37	7.81	41.1	12.0	1.80
6.625	×	0.500	32.74	0.465	14.2	9.00	42.9	13.0	2.18	17.7	85.9	25.9	1.73
		0.432	28.60	0.403	16.4	7.88	38.3	11.6	2.20	15.6	76.6	23.1	1.73
		0.375	25.05	0.349	19.0	6.88	34.0	10.3	2.22	13.8	68.0	20.5	1.73
		0.312	21.06	0.291	22.8	5.79	29.1	8.79	2.24	11.7	58.2	17.6	1.73
		0.280	18.99	0.261	25.4	5.22	26.5	7.99	2.25	10.6	52.9	16.0	1.73
		0.250	17.04	0.233	28.4	4.68	23.9	7.22	2.26	9.52	47.9	14.4	1.73
		0.188	12.94	0.174	38.1	3.53	18.4	5.54	2.28	7.24	36.7	11.1	1.73
		0.125	8.69	0.116	57.1	2.37	12.6	3.79	2.30	4.92	25.1	7.59	1.73
6.125	×	0.500	30.07	0.465	13.2	8.27	33.3	10.9	2.01	14.9	66.7	21.8	1.60
		0.375	23.05	0.349	17.6	6.33	26.5	8.66	02.05	11.7	53.0	17.3	1.60
		0.312	19.39	0.291	21.0	5.33	22.7	7.43	2.07	9.91	45.5	14.9	1.60
		0.250	15.70	0.233	26.3	4.31	18.7	6.12	2.08	8.09	37.5	12.2	1.60
		0.188	11.93	0.174	35.2	3.25	14.4	4.71	2.10	6.16	28.8	9.41	1.60
6.000	×	0.500	29.40	0.465	12.9	8.09	31.2	10.4	1.96	14.3	62.4	20.8	1.57
		0.375	22.55	0.349	17.2	6.20	24.8	8.28	2.00	11.2	49.7	16.6	1.57
		0.312	18.97	0.291	20.6	5.22	21.3	7.11	2.02	9.49	42.6	14.2	1.57
		0.280	17.12	0.261	23.0	4.71	19.4	6.47	2.03	8.60	38.8	12.9	1.57
		0.250	15.37	0.233	25.8	4.22	17.6	5.86	2.04	7.75	35.2	11.7	1.57
		0.188	11.68	0.174	34.5	3.18	13.5	4.51	2.06	5.91	27.0	9.02	1.57
		0.125	7.85	0.116	51.7	2.14	9.28	3.09	2.08	4.02	18.6	6.19	1.57

Nominal Size		Weight per Foot	Wall Thickness t	D/t	Cross-Sectional Area	I	S	r	Z	Torsional Stiffness Constant	Torsional Shear Constant	Surface Area per Foot
Outside Diameter	Wall											
in.	in.	lb	in.		in.²	in.⁴	in.³	in.	in.³	in.⁴	in.³	ft.²
5.563 ×	0.375	20.80	0.349	15.9	5.72	19.5	7.02	1.85	9.50	39.0	14.0	1.46
	0.258	14.63	0.241	23.1	4.03	14.3	5.14	1.88	6.83	28.6	10.3	1.46
	0.188	10.80	0.174	32.0	2.95	10.7	3.85	1.91	5.05	21.4	7.70	1.46
	0.134	7.78	0.125	44.5	2.14	7.90	2.84	1.92	3.70	15.8	5.68	1.46
5.500 ×	0.500	26.73	0.465	11.8	7.36	23.5	8.55	1.79	11.8	47.0	17.1	1.44
	0.375	20.54	0.349	15.8	5.65	18.8	6.84	1.83	9.27	37.6	13.7	1.44
	0.258	14.46	0.241	22.8	3.98	13.8	5.02	1.86	6.67	27.6	10.0	1.44
5.000 ×	0.500	24.05	0.465	10.8	6.62	17.2	6.88	1.61	9.60	34.4	13.8	1.31
	0.375	18.54	0.349	14.3	5.10	13.9	5.55	1.65	7.56	27.7	11.1	1.31
	0.312	15.64	0.291	17.2	4.30	12.0	4.79	1.67	6.46	24.0	9.58	1.31
	0.258	13.08	0.241	20.7	3.60	10.2	4.09	1.68	5.46	20.5	8.18	1.31
	0.250	12.69	0.233	21.5	3.49	9.94	3.97	1.69	5.30	19.9	7.95	1.31
	0.188	9.67	0.174	28.7	2.64	7.69	3.08	1.71	4.05	15.4	6.15	1.31
	0.125	6.51	0.116	43.1	1.78	5.31	2.12	1.73	2.77	10.6	4.25	1.31
4.500 ×	0.337	15.00	0.315	14.3	4.14	9.12	4.05	1.48	5.53	18.2	8.11	1.18
	0.237	10.80	0.221	20.4	2.97	6.82	3.03	1.51	4.05	13.6	6.06	1.18
	0.188	8.67	0.174	25.9	2.36	5.54	2.46	1.53	3.26	11.1	4.93	1.18
	0.125	5.85	0.116	38.8	1.60	3.84	1.71	1.55	2.23	7.68	3.41	1.18
4.000 ×	0.337	13.20	0.315	12.7	3.65	6.24	3.12	1.31	4.29	12.5	6.24	1.05
	0.313	12.34	0.291	13.7	3.39	5.87	2.93	1.32	4.01	11.7	5.87	1.05
	0.250	10.02	0.233	17.2	2.76	4.91	2.45	1.33	3.31	9.82	4.91	1.05
	0.237	9.53	0.221	18.1	2.62	4.70	2.35	1.34	3.16	9.40	4.70	1.05
	0.226	9.12	0.211	19.0	2.51	4.52	2.26	1.34	3.03	9.04	4.52	1.05
	0.220	8.89	0.205	19.5	2.44	4.41	2.21	1.34	2.96	8.83	4.41	1.05
	0.188	7.66	0.174	23.0	2.09	3.83	1.92	1.35	2.55	7.67	3.83	1.05
	0.125	5.18	0.116	34.5	1.42	2.67	1.34	1.37	1.75	5.34	2.67	1.05

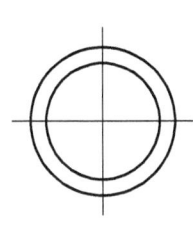

6.6.1 Calculating the Weight and Size of Metric Round High-Strength Steel Sections

Normal Size			Mass per Meter	Design Weight per Meter	Wall Thickness t		D/t	Area	I/106		S/103	r	Z/103	Torsional Stiffness Constant J/103	Torsional Shear Constant C/103	Surface Area per Meter
U.S. Customary Outside Diameter Wall	SI/Metric Outside Diameter Wall															
Inches	Millimeters		kg	kN	mm			mm²	mm⁴	mm*	mm³	mm	mm³	mm⁴	mm³	m²
20.000 × 0.500	508.0 × 12.7		155.1	1.521	11.8		43.1	18400	566		2230	175	2910	1130000	4460000	1.60
0.375	9.5		116.8	1.145	8.9		57.1	14000	435		1710	176	2220	869000	3420000	1.60
18.000 × 0.500	457.2 × 12.7		139.2	1.365	11.8		38.7	16500	410		1790	158	2340	819000	3580000	1.44
0.375	9.5		104.9	1.029	8.9		51.4	12500	315		1380	159	1790	630000	2760000	1.44
16.000 × 0.500	406.4 × 12.7		123.3	1.209	11.8		34.4	14600	285		1400	140	1840	570000	2800000	1.28
0.438	11.1		108.2	1.061	10.3		39.5	12800	252		1240	140	1620	503000	2480000	1.28
0.375	9.5		93.0	0.912	8.9		45.7	11100	220		1080	141	1410	439000	2160000	1.28
0.312	7.9		77.6	0.761	7.4		54.9	9280	185		909	141	1180	369000	1820000	1.28
14.000 × 0.500	355.6 × 12.7		107.4	1.053	11.8		30.1	12700	189		1060	122	1400	377000	2120000	1.12
0.375	9.5		81.1	0.795	8.9		40.0	9690	146		820	123	1070	291000	1640000	1.12
0.312	7.9		67.7	0.664	7.4		48.1	8090	123		690	123	897	245000	1380000	1.12
12.750 × 0.500	323.9 × 12.7		97.5	0.956	11.8		27.4	11600	141		871	110	1150	282000	1740000	1.02
0.375	9.5		73.7	0.722	8.9		36.4	8810	109		675	111	883	219000	1350000	1.02
0.250	6.4		50.1	0.491	5.9		54.9	5890	74.5		460	112	597	149000	920000	1.02
12.500 × 0.625	317.5 × 15.9		118.3	1.160	14.8		21.5	14100	162		1020	107	1360	323000	2040000	1.00
0.500	12.7		95.5	0.936	11.8		26.9	11300	133		835	108	1100	265000	1670000	1.00
0.375	9.5		72.2	0.708	8.9		35.7	8630	103		648	109	848	206000	1300000	1.00
0.312	7.9		60.3	0.592	7.4		42.9	7210	86.7		546	110	712	173000	1090000	1.00
0.250	6.4		49.1	0.482	5.9		53.8	5780	70.1		442	110	573	140000	883000	1.00
0.188	4.8		37.0	0.363	4.4		72.2	4330	53.0		334	111	431	106000	668000	1.00
12.313 × 0.625	312.8 × 15.9		116.4	1.141	14.8		21.1	13900	154		985	105	1310	308000	1970000	0.98
0.500	12.7		94.0	0.922	11.8		26.5	11200	127		809	106	1070	253000	1620000	0.98
0.375	9.5		71.0	0.697	8.9		35.1	8500	98.1		628	107	822	196000	1260000	0.98
0.312	7.9		59.4	0.582	7.4		42.3	7100	82.8		529	108	690	166000	1060000	0.98
0.250	6.4		48.4	0.474	5.9		53.0	5690	67.0		428	109	556	134000	856000	0.98
0.188	4.8		36.5	0.357	4.4		71.1	4260	50.7		324	109	418	101000	648000	0.98

(Continued)

Normal Size			Mass per Meter	Design Weight per Meter	Wall Thickness t		D/t	Area	I/10⁶	S/10³	r	Z/10³	Torsional Stiffness Constant J/10³	Torsional Shear Constant C/10³	Surface Area per Meter
U.S. Customary Outside Diameter Wall	SI/Metric Outside Diameter Wall														
Inches	Millimeters		kg	kN	mm			mm²	mm⁴	mm³	mm	mm³	mm⁴	mm³	m²
12.250 × 0.625	311.2 × 15.9		115.8	1.135	14.8		21.0	13800	152	975	105	1300	303000	1950000	0.98
0.500	12.7		93.5	0.917	11.8		26.4	11100	124	800	106	1060	249000	1600000	0.98
0.375	9.5		70.7	0.693	8.9		35.0	8450	96.6	621	107	813	193000	1240000	0.98
0.312	7.9		59.1	0.579	7.4		42.0	7060	81.5	524	107	683	163000	1050000	0.98
0.250	6.4		48.1	0.472	5.9		52.7	5660	65.9	424	108	550	132000	847000	0.98
0.188	4.8		36.3	0.356	4.4		70.7	4240	49.9	321	108	414	100000	641000	0.98
11.250 × 0.625	285.8 × 15.9		105.8	1.038	14.8		19.3	12600	116	812	96.0	1090	232000	1620000	0.90
0.500	12.7		85.5	0.839	11.8		24.2	10200	95.5	668	97.0	886	191000	1340000	0.90
0.375	9.5		64.7	0.635	8.9		32.1	7740	74.3	520	97.9	683	149000	1040000	0.90
0.312	7.9		54.1	0.531	7.4		38.6	6470	62.7	439	98.5	574	125000	878000	0.90
0.250	6.4		44.1	0.432	5.9		48.4	5190	50.8	356	99.0	462	102000	711000	0.90
0.188	4.8		33.3	0.326	4.4		65.0	3890	38.5	270	99.5	348	77000	539000	0.90
10.750 × 0.500	273.1 × 12.7		81.6	0.800	11.8		23.1	9690	82.8	607	92.5	806	166000	1210000	0.86
0.365	9.3		60.5	0.593	8.6		31.8	7150	62.6	458	93.6	602	125000	916000	0.86
0.250	6.4		42.1	0.413	5.9		46.3	4950	44.2	324	94.5	421	88400	648000	0.86
10.000 × 0.625	254.0 × 15.9		93.4	0.916	14.8		17.2	11100	79.8	629	84.7	848	160000	1260000	0.80
0.500	12.7		75.6	0.741	11.8		21.5	8980	66.0	520	85.7	693	132000	1040000	0.80
0.375	9.5		57.3	0.562	8.9		28.5	6850	51.5	406	86.7	535	103000	811000	0.80
0.312	7.9		47.9	0.470	7.4		34.3	5730	43.6	343	87.2	450	87200	687000	0.80
0.250	6.4		39.1	0.383	5.9		43.1	4600	35.4	279	87.7	363	70800	558000	0.80
0.188	4.8		29.5	0.289	4.4		57.7	3450	26.9	212	88.3	274	53800	423000	0.80
9.625 × 0.500	244.5 × 12.7		72.6	0.712	11.8		20.7	8630	58.5	479	82.4	640	117000	958000	0.77
0.375	9.5		55.1	0.540	8.9		27.5	6590	45.8	374	83.4	494	91500	749000	0.77
0.312	7.9		46.1	0.452	7.4		33.0	5510	38.8	317	83.9	416	77500	634000	0.77
0.250	6.4		37.6	0.369	5.9		41.4	4420	31.5	258	84.4	336	63000	515000	0.77
0.188	4.8		28.4	0.278	4.4		55.6	3320	23.9	196	84.9	254	47800	391000	0.77
8.750 × 0.500	222.3 × 12.7		65.6	0.644	11.8		18.8	7800	43.4	390	74.5	523	86700	780000	0.70
0.375	9.5		49.9	0.489	8.9		25.0	5970	34.0	306	75.5	406	68000	612000	0.70
0.312	7.9		41.8	0.410	7.4		30.0	5000	28.9	260	76.0	342	57700	520000	0.70
0.250	6.4		34.1	0.334	5.9		37.7	4010	23.5	211	76.5	276	47000	423000	0.70
0.188	4.8		25.7	0.252	4.4		50.5	3010	17.9	161	77.1	209	35800	322000	0.70
8.625 × 0.500	219.1 × 12.7		64.6	0.634	11.8		18.6	7680	41.4	378	73.4	508	82800	756000	0.69
0.375	9.5		49.1	0.482	8.9		24.6	5880	32.5	297	74.4	393	65000	594000	0.69
0.322	8.2		42.6	0.418	7.6		28.8	5050	28.3	258	74.8	340	56500	516000	0.69
0.250	6.4		33.6	0.329	5.9		37.1	3950	22.5	205	75.4	268	44900	410000	0.69
0.188	4.8		25.4	0.249	4.4		49.8	2970	17.1	156	75.9	203	34200	312000	0.69
7.625 × 0.375	193.7 × 9.5		43.2	0.423	8.9		21.8	5170	22.1	228	65.4	304	44200	457000	0.61
0.328	8.3		37.9	0.372	7.7		25.2	4500	19.5	201	65.8	267	39000	402000	0.61
0.125	3.2		15.0	0.147	2.9		66.8	1740	7.91	81.7	67.5	106	15800	163000	0.61

Metric Dimensions and Section Properties of Round HSS

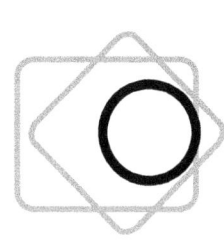

Normal Size		Mass per Meter	Design Weight per Meter	Wall Thickness t	D/t	Area	I/10⁶	S/10³	r	Z/10³	Torsional Stiffness Constant J/10³	Torsional Shear Constant C/10³	Surface Area per Meter
U.S. Customary Outside Diameter Wall	SI/Metric Outside Diameter Wall												
Inches	Millimeters	kg	kN	mm		mm²	mm⁴	mm³	mm	mm³	mm⁴	mm³	m²
7.500 × 0.500	190.5 × 12.7	55.7	0.546	11.8	16.1	6620	26.6	279	63.3	377	53100	558000	0.60
0.375	9.5	42.4	0.416	8.9	21.4	5080	21.0	220	64.3	294	42000	441000	0.60
0.312	7.9	35.6	0.349	7.4	25.7	4260	17.9	188	64.8	248	35700	375000	0.60
0.250	6.4	29.1	0.285	5.9	32.3	3420	14.6	153	65.3	201	29200	306000	0.60
0.188	4.8	22.0	0.216	4.4	43.3	2570	11.1	117	65.8	152	22300	234000	0.60
7.000 × 0.500	177.8 × 12.7	51.7	0.507	11.8	15.1	6150	21.3	240	58.8	326	42600	479000	0.56
0.375	9.5	39.4	0.387	8.9	20.0	4720	16.9	190	59.8	254	33800	380000	0.56
0.312	7.9	33.1	0.325	7.4	24.0	3960	14.4	162	60.3	215	28800	324000	0.56
0.250	6.4	27.1	0.265	5.9	30.1	3190	11.8	133	60.8	174	23600	265000	0.56
0.188	4.8	20.5	0.201	4.4	40.4	2400	9.01	101	61.3	132	18000	203000	0.56
0.125	3.2	13.8	0.135	2.9	61.3	1590	6.09	68.6	61.8	88.7	12200	137000	0.56
6.875 × 0.500	174.6 × 12.7	50.7	0.497	11.8	14.8	6040	20.1	230	57.7	313	40200	460000	0.55
0.375	9.5	38.7	0.379	8.9	19.6	4630	15.9	183	58.7	245	31900	365000	0.55
0.312	7.9	32.5	0.318	7.4	23.6	3890	13.6	156	59.2	207	27200	312000	0.55
0.250	6.4	26.5	0.260	5.9	29.6	3130	11.1	128	59.7	168	22300	255000	0.55
0.188	4.8	20.1	0.197	4.4	39.7	2350	8.52	97.6	60.2	127	17000	195000	0.55
6.625 × 0.500	168.3 × 12.7	48.7	0.478	11.8	14.3	5800	17.9	212	55.5	290	35700	425000	0.53
0.432	11.0	42.7	0.418	10.2	16.5	5070	15.9	189	56.0	255	31800	378000	0.53

(Continued)

Normal Size		Mass per Meter	Design Weight per Meter	Wall Thickness t	D/t	Area	I/106	S/103	r	Z/103	Torsional Stiffness Constant J/103	Torsional Shear Constant C/103	Surface Area per Meter
U.S. Customary Outside Diameter Wall	SI/Metric Outside Diameter Wall												
Inches	Millimeters	kg	kN	mm		mm²	mm⁴	mm³	mm	mm³	mm⁴	mm³	m²
0.375	9.5	37.2	0.365	8.9	18.9	4460	14.2	169	56.4	226	28400	337000	0.53
0.312	7.9	31.3	0.306	7.4	22.7	3740	12.1	144	56.9	192	24300	288000	0.53
0.280	7.1	28.2	0.277	6.6	25.5	3350	11.0	130	57.2	173	22000	261000	0.53
0.250	6.4	25.6	0.251	5.9	28.5	3010	9.94	118	57.5	156	19900	236000	0.53
0.188	4.8	19.4	0.190	4.4	38.3	2270	7.61	90.5	58.0	118	15200	181000	0.53
0.125	3.2	13.0	0.128	2.9	58.0	1510	5.15	61.3	58.5	79.3	10300	123000	0.53
6.125 × 0.500	155.6 × 12.7	44.8	0.439	11.8	13.2	5330	13.9	178	51.0	245	27700	357000	0.49
0.375	9.5	34.2	0.336	8.9	17.5	4100	11.1	142	52.0	192	22100	285000	0.49
0.312	7.9	28.8	0.282	7.4	21.0	3450	9.48	122	52.5	163	19000	244000	0.49
0.250	6.4	23.5	0.231	5.9	26.4	2770	7.78	100	53.0	132	15600	200000	0.49
0.188	4.8	17.9	0.175	4.4	35.4	2090	5.98	76.8	53.5	101	12000	154000	0.49
6.000 × 0.500	152.4 × 12.7	43.8	0.429	11.8	12.9	5210	13.0	170	49.9	234	25900	340000	0.48
0.375	9.5	33.5	0.328	8.9	17.1	4010	10.4	136	50.8	184	20700	272000	0.48
0.312	7.9	28.2	0.276	7.4	20.6	3370	8.88	117	51.3	156	17800	233000	0.48
0.280	7.1	25.4	0.249	6.6	23.1	3020	8.05	106	51.6	140	16100	211000	0.48
0.250	6.4	23.0	0.226	5.9	25.8	2720	7.30	95.8	51.8	127	14600	192000	0.48
0.188	4.8	17.5	0.171	4.4	34.6	2050	5.61	73.6	52.3	96.4	11200	147000	0.48
0.125	3.2	11.8	0.115	2.9	52.6	1360	3.81	50.0	52.9	64.8	7610	99900	0.48

6.6.2 Calculating the Weight of Standard, Extra Strong, and Double Strong Steel Pipes

Steel Pipe

Standard Weight Pipe		Extra Strong Pipe		Double Extra Strong Pipe	
½"	0.85 lbs/lnft	½"	1.09 lbs/lnft	2"	9.03
¾"	1.13	¾"	1.47	2 ½"	13.69
1"	1.68	1"	2.17	3"	18.58
1 ¼"	2.27	1 ¼"	3.00	4"	27.54
1 ½"	2.72	1 ½"	3.63	5"	38.55
2"	3.65	2"	5.02	6"	53.16
2 ½"	5.79	2 ½"	7.66	8"	74.42
3"	7.58	3"	10.25		
3 ½"	9.11	3 ½"	12.50		
4"	10.79	4"	14.98		
5"	14.62	5"	20.78		
6"	18.97	6"	28.57		
8"	28.55	8"	43.39		
10"	40.48	10"	54.74		
12"	49.56	12"	65.42		

6.7.0 Calculating the Weight and Size of U.S. Steel C Channels

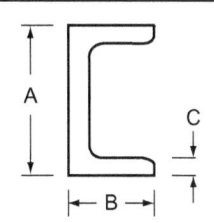

Channels are available in Carbon, Stainless Steel, and Aluminum. Sizes shown are for Carbon only.

MC-Shapes	A Depth in Inches	B Width in Inches	C Web in Inches	Per Ft.	Weight Lbs 20-Ft.	30-Ft.	40-Ft.
MC 3 × 7.1	3	1.938	.312	7.1	142	213	284
MC 4 × 13.8	4	2.500	.500	13.8	276	414	552
MC 6 × 12	6	2.497	.310	12.0	240	360	480
MC 6 × 15.3	6	3.500	.340	15.3	306	459	612
MC 6 × 16.3	6	3.000	.375	16.3	326	489	652
MC 6 × 18	6	3.504	.379	18.0	360	540	720
MC 7 × 17.6	7	3.000	.375	17.6	352	528	704
MC 7 × 19.1	7	3.452	.352	19.1	382	573	764
MC 7 × 22.7	7	3.603	.503	22.7	454	681	908
MC 8 × 8.5	8	1.875	.188	8.5	170	225	340

(Continued)

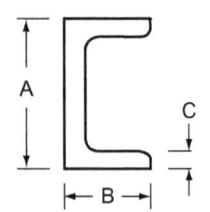

Channels are available in Carbon, Stainless Steel, and Aluminum. Sizes shown are for Carbon only.

MC-Shapes	A Depth in Inches	B Width in Inches	C Web in Inches	Weight Lbs Per Ft.	20-Ft.	30-Ft.	40-Ft.
MC 8 × 18.7	8	2.978	.353	18.7	374	561	748
MC 8 × 20	8	3.025	.400	20	400	600	800
MC 8 × 21.4	8	3.450	.375	21.4	428	642	856
MC 8 × 22.8	8	3.502	.353	22.8	456	684	912
MC 9 × 23.9	9	3.450	.400	23.9	478	717	956
MC 9 × 25.4	9	3.500	.450	25.4	508	762	1016
MC 10 × 6.5	10	1.125	.150	6.5	130	195	260
MC 10 × 8.4	10	1.500	.170	8.4	168	252	336
MC 10 × 22	10	3.315	.290	22	440	660	880
MC 10 × 25	10	3.405	.380	25	500	750	1000
MC 10 × 28.5	10	3.950	.425	28.5	570	855	1140
MC 10 × 33.6	10	4.100	.575	33.6	672	1008	1344
MC 12 × 10.6	12	1.500	1.90	10.6	212	318	424
MC 12 × 31	12	3.670	.370	31.0	620	930	1240
MC 12 × 35	12	3.767	.467	35.0	700	1050	1400
MC 12 × 37	12	3.600	.600	37.0	740	1110	1480
MC 12 × 40	12	3.890	.590	40.0	800	1200	1600
MC 12 × 45	12	4.012	.712	45.0	900	1350	1800
MC 12 × 50	12	4.135	.835	50.0	1000	1500	2000
MC 13 × 31.8	13	4.000	.375	31.8	636	954	1272
MC 13 × 40	13	4.185	.560	40.0	800	1200	1600
MC 13 × 50	13	4.412	.787	50.0	1000	1500	2000
MC 18 × 42.7	18	3.950	.450	42.7	854	1281	1708
MC 18 × 45.8	18	4.000	.500	45.8	916	1374	1832
MC 18 × 51.9	18	4.100	.600	51.9	1038	1557	2076
MC 18 × 58	18	4.200	.700	58.0	1160	1740	2320

6.7.0.1 Calculating the Weight and Size of U.S. A-36 and A-36 Modified C Channels

Conforms to A-36 and A-36 Modified Standard Lengths 2,' 40,' and 60'

A	B		C	Weight Lbs			
Depth in Inches	Weight Lbs per Foot	Flange Width Inches	Web Thickness In.	20 Ft Length	30 Ft Length	40 Ft Length	60 Ft Length
3	4.1	1.410	.170	82	123	164	246
3	5.0	1.498	.258	100	150	200	300
3	6.0	1.596	.356	120	180	240	360
4	5.4	1.580	.180	108	162	216	324
4	6.25	1.647	.247	125	188	250	375
4	7.25	1.720	.320	145	218	290	435
5	6.7	1.750	.190	134	201	268	402
5	9	1.885	.325	180	270	360	540
6	8.2	1.920	.200	164	246	328	492
6	10.5	2.034	.314	210	315	420	630
6	13.0	2.157	.437	260	390	520	780
7	9.8	2.090	.210	196	294	392	588
7	12.25	2.194	.314	245	368	490	735
7	14.75	2.299	.419	295	443	590	885
8	8.55	1.875	.180	170	255	340	510
8	11.5	2.260	.220	230	345	460	690
8	13.75	2.343	.303	275	413	550	825
9	18.75	2.527	.487	375	563	750	1125
9	1.3.4	2.430	.230	268	402	536	801
9	15.0	2.485	.285	300	450	600	900
10	20.0	2.648	.448	400	600	800	1200
10	15.3	2.600	.240	306	459	612	918
10	20.0	2.739	.379	400	600	800	1200
10	25.0	2.886	.526	500	750	1000	1500
12	30.0	3.033	.673	600	900	1200	1800
12	20.7	2.940	.280	414	621	828	1242
12	25.0	3.047	.387	500	750	1000	1500
12	30.0	3.170	.510	600	900	1200	1800
15	33.9	3.400	.400	678	1017	1356	2034
15	40.0	3.520	.520	800	1200	1600	2400
15	50.0	3.716	.716	1000	1500	2000	3000

6.7.0.2 Calculating the Weight and Size of U.S. Channels—Ship and Car

MC CHANNELS–SHIP AND CAR
Conforms to A-36 and A-36 Standard Lengths 20′, 40′ and 60′

A	B	C		Weight Lbs.		
Depth in Inches	Weight Lbs per Foot	Flange Width Inches	Web Thickness In.	20 Ft. Length	30 Ft. Length	40 fT. Length	60 Ft. Length
3	7.1	1.938	.313	142	213	248	426
4	13.8	2.500	.500	276	414	552	828
6	12.0	2.500	.313	240	360	480	720
6	15.1	2.941	.316	302	453	604	906
6	15.3	3.500	.340	306	459	612	918
6	16.3	3.000	.375	326	489	652	978
6	18.0	3.500	.375	360	540	720	1080
7	17.6	3.000	.375	352	528	704	1056
7	19.1	3.450	.350	382	573	764	1146
7	22.7	3.600	.500	454	681	908	1362
8	18.7	2.975	.350	374	561	748	1122
8	20.0	3.025	.400	400	600	800	1200
8	21.4	3.450	.375	428	642	856	1284
8	22.8	3.500	.425	456	684	912	1368
9	23.9	3.450	.400	478	717	956	1434
9	25.4	3.500	.450	508	762	1016	1524
10	21.9	3.450	.325	438	657	876	1314
10	22.0	3.315	.290	440	660	880	1320
10	24.9	3.400	.375	498	747	996	1494
10	25.0	3.405	.380	500	750	1000	1500
10	28.3	3.502	.477	566	849	1132	1698
10	28.5	3.950	.425	570	855	1140	1710
10	33.6	4.100	.575	672	1008	1344	2016
10	41.1	4.321	.796	822	1233	1644	2466
12	31.0	3.670	.370	620	930	1240	1860
12	32.9	3.500	.500	658	987	1316	1974
12	35.0	3.767	.467	700	1050	1400	2100
12	40.0	3.890	.590	800	1200	1600	2400
12	45.0	4.012	.712	900	1350	1800	2700
12	50.0	4.135	.835	1000	1500	2000	3000
13	31.8	4.000	.375	636	954	1272	1908
13	35.0	4.072	.447	700	1050	1400	2100
13	40.0	4.185	.560	800	1200	1600	2400
13	50.0	4.412	.787	1000	1500	2000	3000
18	42.7	3.950	.450	854	1281	1708	2562
18	45.8	4.000	.500	916	1374	1832	2748
18	51.9	4.100	.600	1038	1557	2076	3114
18	51.9	4.100	.600	1038	1557	2076	3114

6.7.1 Calculating the Weight and Size of Metric Steel C Channels

Disclaimer: The information on this page has not been checked by an independent person. Use this information at your own risk.

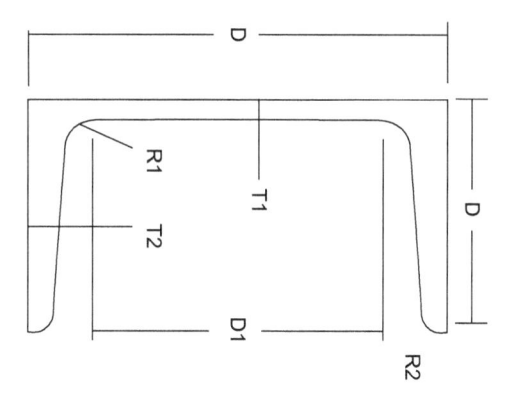

Dimensions
M = Mass per m, D = Depth of Section, B = Width of Section, T1 = Web Thickness, T2 = Flange Thickness, R1 = Root Radius, R2 = Toe Rad, Area = Area of Section

Serial Size mm	M kg	D mm	B mm	T1 mm	T2 mm	R1 mm	R2 mm	DT mm	D/T	Area cm²
432 × 102	65.54	431.8	101.6	12.2	16.8	15.2	4.8	362.5	25.7	83.49
381 × 102	55.1	381.0	101.6	10.4	16.3	15.2	4.8	312.4	23.4	70.19
305 × 102	46.18	304.8	101.6	10.2	14.8	15.2	4.8	239.3	20.6	58.83
305 × 89	41.69	304.8	88.9	10.2	13.7	13.7	3.2	245.4	22.2	53.11
254 × 89	35.74	254.0	88.9	9.1	13.6	13.7	3.2	194.8	18.7	45.52
254 × 76	28.29	254.0	76.2	8.1	10.9	12.2	3.2	203.7	23.3	36.03
229 × 89	32.76	228.6	88.9	8.6	13.3	13.7	3.2	169.9	17.2	41.73
229 × 76	26.06	228.6	76.2	7.6	11.2	12.2	3.2	178.1	20.4	33.20
203 × 89	29.78	203.2	88.9	8.1	12.9	13.7	3.2	145.3	15.8	37.94
203 × 76	23.82	203.2	76.2	7.1	11.2	12.2	3.2	152.4	18.1	30.34
178 × 89	26.81	177.8	88.9	7.6	12.3	13.7	3.2	120.9	14.5	34.15
178 × 76	20.84	177.8	76.2	6.6	10.3	12.2	3.2	128.8	17.3	26.54
152 × 89	23.84	152.4	88.9	7.1	11.6	13.7	3.2	97.0	13.1	30.36
152 × 76	17.88	152.4	76.2	6.4	9.0	12.2	2.4	105.9	16.9	22.77
127 × 64	14.90	127.0	63.5	6.4	9.2	10.7	2.4	84.1	13.8	18.98
102 × 51	10.42	101.6	50.8	6.1	7.6	9.1	2.4	65.8	13.4	13.28

6.7.1.1 Calculating the Weight and Size of Metric Channel, Box, Rectangular, and Square Tubing

Web O.D	Web I.D	Leg O.D	Leg I.D	Returns O.D.	Gap/Opening I.D.	Metal Thickness
.425"	.377"	.679"	.631"	.191"	.011"	0.024
.500"	.342"	.500"	.342"	.125"	.250"	0.079
.500"	.380"	.500"	.380"	.125"	.250"	0.060
.500"	.428"	.500"	.428"	.125"	.250"	0.036
.600"	.528"	.148"	.076"	.135"	.330"	0.036
.609"	.535"	.487"	.413"	.145"	.319"	0.037
.745"	.625"	.620"	.500"	.185"	.375"	0.060
.750"	.702"	.281"	.233"	.250"	.250"	0.024
.750"	.600"	.750"	.600"	.250"	.250"	0.075
.750"	.630"	.750"	.630"	.250"	.250"	0.060
.780"	.742"	.575"	.537"	.269"	.242"	0.019
.813"	.733"	.406"	.272"	.188"	.437"	0.040
.860"	.800"	.160"	.100"	.115"	.630"	0.030
.904"	.784"	1.004"	.793"	.226"	.452"	0.060
.922"	.850"	.178"	.106"	.156"	.610"	0.036
.961"	.913"	.704"	.656"	.303"	.355"	0.024
.970"	.898"	.187"	.115"	.145"	.680"	0.036
.975"	.903"	.178"	.106"	.219"	.537"	0.036
1.000"	.964"	.200"	.164"	.095"	.810"	0.018
1.000"	.952"	.500"	.452"	.343"	.315"	0.024
1.000"	.904"	1.000"	.904"	.375"	.250"	0.048
1.000"	.880"	1.000"	.880"	.375"	.250"	0.060
1.000"	.896"	1.500"	1.396"	.171"	.658"	0.052
1.038"	.982"	.923"	.867"	.288"	.462"	0.028
1.042"	.982"	.471"	.411"	.117" & .283"	.642"	0.030
1.050"	.984"	.780"	.714"	.240"	.570"	0.033
1.062"	.990"	.156"	.084"	.156"	.750"	0.036
1.215"	1.143"	.188"	.116"	.145"	.925"	0.036
1.236"	1.176"	.866"	.806"	.278"	.680"	0.030
1.249"	1.199"	.219"	.169"	.156"	.937"	0.025
1.250"	1.156"	.437"	.312"	.156"	.938"	0.047
1.250"	1.124"	1.250"	1.124"	.250"	.750"	0.063
1.250"	1.130"	2.000"	1.880"	.595"	.060"	0.060

Web O.D	Web I.D	Leg O.D	Leg I.D	Returns O.D.	Gap/Opening I.D.	Metal Thickness
1.253"	1.133"	.209"	.089"	.470"	.313"	0.060
1.259"	1.199"	.197"	.137"	.100"	1.060"	0.030
1.375"	1.225"	.480"	.330"	.250"	.875"	0.075
1.403"	1.291"	.468"	.256"	.468"	.467"	0.056
1.491"	1.431"	.221"	.161"	.221"	1.049"	0.030
1:491"	1.423"	.246"	.178"	.280"	.931"	0.034
1.494"	1.444"	.219"	.169"	.161"	1.172"	0.025
1.500"	1.404"	.500"	.404"	.500"	.500"	0.048
1.500"	1.380"	.500"	.380"	.500"	.500"	0.060
1.500"	1.380"	.625"	.505"	.312"	.876"	0.060
1.500"	1.350"	1.000"	.850"	.312"	.876"	0.075
1.500"	1.380"	1.500"	1.380"	.500"	.500"	0.060
1.725"	1.653"	.188"	.116"	.145"	1.435"	0.036
1.998"	1.902"	1.498"	1.402"	.655"	.688"	0.048
2.000"	.760"	2.000"	1.760"	.735"	.531"	0.120
2.000"	.880"	2.000"	.880"	.735"	.531"	0.060
2.004"	.954"	.219"	.169"	.161"	1.682"	0.025
2.125"	2.029"	.313"	.217"	.313"	1.499"	0.048
2.250"	2.122"	2.000"	1.872"	.638"	.975"	0.064
2.500"	2.380"	1.750"	1.630"	.563"	1.375"	0.060
2.500"	2.452"	.563"	.469"	.312"	1.875"	0.024
3.438"	3.318"	1.750"	.630"	.532"	2.375"	0.060
4.000"	3.950"	.100"	.050"	.100"	3.800"	0.025
5.000"	4.952"	.563"	.469"	.281"	4.438"	0.024

- Gap/Openings can be larger than shown; narrower requires tooling modification.
- Materials used: Steel, Stainless, Aluminum, Brass, Copper, Bronze, Tempered, Textured, Embossed, Perforated, Alloys, Clad.
- Prefinishes Used: Plain, Dull, Prepolished, Prepainted, Preanodized, Strippable-PVC Coated, Prelaminated Vinyl, Preplated, Prelaquered, Galvanized (Hot Dip: G30, G60, G90, G210, Bonderized, Electro, Chromate) Galvannealled, Aluminized, Galvanized.
- Many other sizes not shown here are possible.
- Thickness can vary slightly using same tooling.
- Inline Press Fabricating: Holes, Notching, End Fabricating, Perforating, Embossing, Cut-to-Length, and More.
- I.D's will vary with different thicknesses used.

Source: Johnsonrollforming.com

6.8.0 Calculating the Weight and Size of Structural Steel Angles

Angles

Weight per Lnft

Size	Weight	Size	Weight	Size	Weight
L9 × 4×5/8	26.3 lbs/lnft	L6 × 6×9/16	21.9	L4 × 3½×½	11.9
L9 × 4×9/16	23.8	L6 × 6×½	19.6	L4 × 3½×7/16	10.6
L9 × 4×½	21.3	L6 × 6×7/16	17.2	L4 × 3½×3/8	9.1
		L6 × 6×3/8	14.9	L4 × 3½×5/16	7.7
L8 × 8×1-1/8	56.9	L6 × 6×5/16	12.4	L4 × 3½×¼	6.2
L8 × 8×1	51.0	L6 × 4×7/8	27.2	L4 × 3×½	11.1
L8 × 8×7/8	45.0	L6 × 4×¾	23.6	L4 × 3×7/16	9.8
L8 × 8×¾	38.9	L6 × 4×5/8	20.0	L4 × 33/8	8.5
L8 × 8×5/8	32.7	L6 × 4×9/16	18.1	L4 × 3×5/16	7.2
L8 × 8×9/16	29.6	L6 × 4×½	16.2	L4 × 3×¼	5.8
L8 × 8×1/2	26.4	L6 × 4×7/16	14.3		
		L6 × 4×3/8	12.3	L3½×3½×½	11.1
L8 × 6×1	44.2	L6 × 4×5/16	10.3	L3½×3½×7/16	9.8
L8 × 6×7/8	39.1			L3½×3½×3/8	8.5
L8 × 6×¾	33.8			L3½×3½×5/16	7.2
L8 × 6×5/8	28.5	L6 × 3½×½	15.3	L3½×3½×¼	5.8
L8 × 6×9/16	25.7	L6 × 3½×3/8	11.7		
L8 × 6×½	23.0	L6 × 3½×5/16	9.8	3½×3×½	10.2
L8 × 6×7/16	20.2			L3½×3 × 7/16	9.1
		L5 × 5×7/8	27.2	L3½×3 × 3/8	7.9
L8 × 4×1	37.4	L5 × 5×¾	23.6	L3½×3 × 5/16	6.6
L8 × 4×¾	28.7	L5 × 3×5/8	20.0	L3½×3×¼	5.4
L8 × 4×9/16	21.9	L5 × 3×½	16.2		
L8 × 4×½	19.6	L5 × 3×7/16	14.3	L3½×2½×½	9.4
		L5 × 5×3/8	12.3	L3½×2½×7/16	8.3
L7 × 4×¾	26.2	L5 × 5×5/16	10.3	L3½×2½×3/8	7.2
L7 × 4×5/8	22.1			L3½×2½×5/16	6.1
L7 × 4×½	17.9	L5 × 3½×¾	19.8	L3½×2½×¼	4.9
L7 × 4×3/8	13.6	L5 × 3½×5/8	16.8		
		L5 × 3½×½	13.6	L3 × 3×½	9.4
L6 × 6×1	37.4	L5 × 3½×7/16	12.0	L3 × 3×7/16	8.3
L6 × 6×7/8	33.1	L5 × 3½×3/8	10.4	L3 × 3×3/8	7.2
L6 × 6×¾	28.7	L5 × 3½×5/16	8.7	L3 × 3×5/16	6.1
L6 × 6×5/8	24.2	L5 × 3½×¼	7.0	L3 × 3×¼	4.9
				L3 × 3×3/16	3.71
		L5 × 3×5/8	15.7		
		L5 × 3×½	12.8	L3 × 2½×½	8.5
		L5 × 3×7/16	11.3	L3 × 2½×7/16	7.6
		L5 × 3×3/8	9.8	L3 × 2½×3/8	6.6
		L5 × 3×5/16	8.2	L3 × 2½×5/16	5.6
		L5 × 3×¼	6.6	L3 × 2½×¼	4.5
				L3 × 2½×3/16	3.39
		L4 × 4×¾	18.5		
		L4 × 4×5/8	15.7	L3 × 2×½	7.7
		L4 × 4×½	12.8	L3 × 2×7/16	6.8
		L4 × 4×7/16	11.3	L3 × 2×3/8	5.9
		L4 × 4×3/8	9.8	L3 × 2×5/16	5.0
		L4 × 4×5/16	8.2	L3 × 2×¼	4.1
		L4 × 4×¼	6.6	L3 × 2×3/16	3.0
				L2½×2½×½	7.7
				L2½×2½×3/8	5.9
				L2½×2½×5/16	5.0
				L2½×2½×¼	4.1

L2½×2½×3/16	3.07
L2½×2 × 3/8	5.3
L2½×2 × 5/16	4.5
L2½×2×¼	3.62
L2½×2 × 3/16	2.75
L2 × 2×3/8	4.7
L2 × 2×5/16	3.92
L2 × 2×¼	3.19
L2 × 2×3/16	2.44
L2 × 2×1/8	1.65
L1¾×1¾×¼	2.77
L1¾×1¾×3/16	2.12
L1½×1½×¼	2.34
L1½×1½×3/16	1.80
L1¼×1¼×¼	1.92
L1¼×1¼×3/16	1.48
L1-1/8 × 1-1/8 × 1/8	0.90
L1 × 1×1/8	0.80

Source: Mc2-ice.com-Wt of Steel Angles, p. 2

6.9.0 Calculating the Weight and Size of Universal Mill Plates

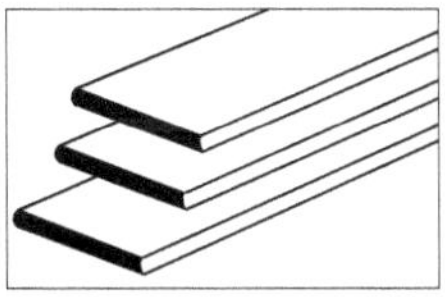

Hot Rolled ASTM-A36

Stock lengths: 20′, 25′, 30′, & 40′

Size in Inches	Wt. per Ft, Lb	Size in Inches	Wt. Per Ft, Lb
1/4 × (10.20 Lb. per Sq Ft)		3/4 × (30.60 Lb. per Sq Ft)	
9	7.65	9	22.95
10	8.50	10	25.50
11 1/2	9.78	12	30.60
12	10.20	14	35.70
14	11.89	16	40.80
24	20.40	24	61.20
5/16 × (12.75 Lb per Sq Ft)		7/8 × (35.70 Lb per Sq Ft)	
9	9.56	9	26.78
10	10.62	10	29.75
11 1/2	12.22	12	35.70
12	12.75	14	41.65
16	17.00	16	45.60
24	25.50	24	71.40
3/8 × (15.30 Lb per Sq Ft)		1 × (40.80 Lb per Sq Ft)	
9	11.48	9	30.60
10	12.75	10	34.00
11 1/2	14.66	12	40.80
12	15.30	14	47.60
14	17.85	16	54.40
16	20.40	18	61.20
24	30.60	24	81.60
7/16 × (17.85 lb. per sq ft)		1 1/4 × (51.00 Lb per Sq Ft)	
9	13.39	10	42.50
10	14.87	12	51.00
12	17.85	18	76.50
24	35.70	24	102.00

(Continued)

Size in Inches	Wt. per Ft, Lb	Size in Inches	Wt. Per Ft, Lb
1/2 × (20.40 Lb per Sq Ft)		1 1/2 × (61.20 per Sq Ft)	
9	15.30	10	51.00
10	17.00	12	61.20
11	18.70	24	122.40
12	20.60	2 × (81.60 Lb per Sq Ft)	
14	23.80	10	68.00
16	27.20	12	81.60
18	30.60	24	163.20
24	40.80		
5/8 × (25.50 Lb per Sq Ft)			
9	19.13		
10	21.25		
12	25.50		
16	34.00		
18	38.25		
24	51.00		

Source: Illinois Steel Service

6.10.0 Bar Size Tees—Calculating Their Weight and Size

BAR SIZE TEES
Hot Rolled
M-1020

Specify tees by flange first, then stem and thickness

F Flange Width, Inches	D Depth of Tee, Inches	S Stem Thickness, Inches	Weight, Pounds per Ft	Weight, Pounds of 20' Bar	Stock Lengths
3/4	3/4	1/8	0.62	12.40	20
1	1	1/8	0.85	17.00	20
1-1/4	1-1/4	1/8	1.09	21.80	20
1-1/4	1-1/4	3/16	1.55	31.00	20
1-1/4	1-1/4	1/4	1.93	38.60	20
1-1/2	1-1/2	3/16	1.90	38.00	20
1-1/2	1-1/2	1/4	2.43	48.60	20
1-3/4	1-3/4	3/16	2.26	45.20	20
1-3/4	1-3/4	1/4	2.90	58.00	20
2	1/1/2	1/4	3.10	62.00	20
2	2	1/4	3.62	72.40	20
2	2	5/16	4.30	86.00	20
2	2	3/8	4.78	95.60	20
2-1/4	2-1/4	1/4	4.10	82.00	20
2-1/2	2-1/2	1/4	4.60	92.00	20

Source: Illinois Steel Service

6.11.0 Cold and Hot Rolled Rounds—Calculating Their Weight and Size

Cold Finished

Hot Rolled

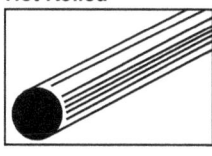

Merchant Quality
AISI M-1020-C-1045
ASTM-A-36

Merchant Quality is a term applied to a grade of steel in common use for production and repair work when forging or heat-treating or intricate machining is not required.

C1018
C1040
C1141
1/8" THRU 6" DIAMETER
12 FT. AND 20 FT. LENGTHS

B1112
B1113
B1117

C1212
C1213

Size in Inches	Weight, Pounds Per Ft.	Weight, Pounds Per 20' Bar	Stock Lengths, Ft.
5/32	0.07	12' - .78	12
3/16	0.09	12' - 1.13	12
1/4	0.17	3.34	20
5/16	0.26	5.22	20
3/8	0.38	7.52	20
7/16	0.51	10.22	20
1/2	0.67	13.36	20
9/16	0.85	16.90	20
5/8	1.04	20.80	20
11/16	1.26	25.20	20
3/4	1.50	30.00	20
13/16	1.76	35.20	20
7/8	2.04	40.08	20
15/16	2.35	47.00	20
1	2.67	53.40	20
1 1/16	3.01	60.20	20
1 1/8	3.38	67.60	20
1 3/16	3.77	75.40	20
1 1/4	4.17	83.40	20
1 5/16	4.60	92.00	20
1 3/8	5.05	101.00	20
1 7/16	5.52	110.40	20
1 1/2	6.01	120.20	20
1 5/8	7.05	141.00	20
1 3/4	8.18	163.60	20
1 7/8	9.39	187.80	20
2	10.68	213.60	20
2 1/8	12.06	241.20	20
2 1/4	13.52	270.40	20
2 3/8	15.06	301.20	20
2 1/2	16.69	333.80	20
2 5/8	18.40	368.00	20
2 3/4	20.19	403.80	20
2 7/8	22.07	441.40	20

Source: Illinois Steel Service

6.12.0 Aluminum Structural Angles—Calculate Their Weight and Size

Structural Angles

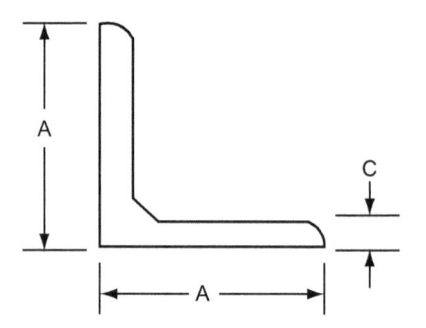

6061-T6 Aluminum Structural Angles
25 Ft Lengths—Equal Leg—ASTM-B221-or- QQ-A-200/16

A		Size A		C	Approx. Wt. per Lin. Ft	Approx. Wt. per Length	Group/Size/Grade
3/4	×	3/4	×	1/8	.201	5.03	6061A\.75-.125\T6
1	×	1	×	1/8	.275	6.8	6061A\1-.125\T6
				3/16	.400	10.00	6061A\1-.1875\T6
				1/4	.514	12.85	6061A\1-.25\T6
1-1/4	×	1-1/4	×	1/8	.343	8.58	6061A\1.25-.125\T6
				3/16	.510	12.75	6061A\1.25-.1875\T6
				1/4	.656	16.40	6061A\1.25-.25\T6
1-1/2	×	1-1/2	×	1/8	.423	10.58	6061A\1.5-.125\T6
				3/16	.619	15.48	6061A\1.5-.1875\T6
				1/4	.809	20.23	6061A\1.5-.25\T6
1-3/4	×	1-3/4	×	1/4	.980	24.50	6061A \ 1.75-.25 \ T6
2	×	2	×	1/8	.577	14.43	6061A\2-.125\T6
				3/16	.850	21.25	6061A\2-.1875\T6
				1/4	1.110	27.75	6061A\2-.25\T6
				3/8	1.606	40.15	6061A\2-.375\T6
2-1/2	×	2-1/2	×	3/16	1.070	26.75	6061A\2.5-.1875\T6
				1/4	1.404	35.10	6061A\2.5-.25\T6
				3/8	2.047	51.18	6061A\2.5-.375\T6
3	×	3	×	3/16	1.275	31.88	6061A\3-.1875\T6
				1/4	1.684	42.10	6061A\3-.25\T6
				3/8	2.474	61.85	6061A\3-.375\T6
				1/2	3.227	80.68	6061A\3-.5\T6
3-1/2	×	3-1/2	×	1/4	1.989	49.73	6061A\3.5-.25\T6
4	×	4	×	1/4	2.283	57.08	6061A\4-.25\T6
				3/8	3.366	84.15	6061A\4-.375\T6
				1/2	4.414	110.35	6061A\4-.5\T6
5	×	5	×	3/8	4.237	105.93	6061A\5-.375\T6
6	×	6	×	1/2	6.754	168.85	6061A\6-.5\T6
8	×	8	×	1/2	9.141	228.53	6061A\8-.5\T6

Source: Pacificmetral.com

Section | 6 Calculating the Size/Weight of Structural Steel and Miscellaneous Metals   331

6061-T6 Aluminum Structural Angles
25 Ft. Lengths—Unequal Leg—ASTM-B221-or- QQ-A-200/16

A	Size B		C	Approx. Wt. Per Lin. Ft.	Approx. Wt. Per Length	Group/Size/Grade
1-1/4	×	1	×	1/8	.31	7.75	6061A\1.25-1.125\T6
1-1/2	×	1-1/4	×	3/16	.57	14.25	6061A\1.5-1.25-.18\T6
2	×	1-1/2	×	1/8	.496	12.40	6061A\2-1.5-.125\T6
				3/16	.731	18.28	6061A\2-1.5-.1875\T6
				1/4	.956	23.90	6061A\2-1.5-.25\T6
2-1/2	×	2	×	3/16	.961	24.03	6061A\2.5-2-.1875\T6
				1/4	1.257	31.43	6061A\2.5-2-.25\T6
3	×	1-1/2	×	3/16	.971	24.28	6061A\3-1.5-1875\T6
		2	×	3/16	1.071	26.78	6061A\3-2-.1875\T6
				1/4	1.403	35.08	6061A\3-2-.25\T6
				3/8	2.046	51.15	6061A\3-2-.375\T6
		2-1/2	×	1/4	1.537	38.43	6061A\3-2.5-25\T6
3-1/2	×	2-1/2	×	1/4	1.684	42.10	6061A\3.5-2.5-.25\T6
4	×	3	×	1/4	1.988	49.70	6061A\4-3-.25\T6
				3/8	2.926	73.15	6061A\4-3-.375\T6
5	×	3	×	3/8	3.450	86.30	6061A\5-3-.375\T6
6	×		×	3/8	4.237	105.93	6061A\6-4-.375\T6

6.12.1 Aluminum Channels—Calculate Their Weight and Size

6061-T6 Aluminum Structural Channels
American Standard—25 Ft Lengths—ASTM-B221-or-QQ-A-200/16

Size (Inches) A	Web Thick C	Flange Width B	Approx. Wt. per Lin. Ft	Approx. Wt. per Length	Group/Size/Grade
3	.170	1.410	1.417	35.43	6061N\3-.17\T6-2
	.258	1.498	1.729	43.23	6061N\3-.258\T6-2
	.356	1.596	2.074	51.85	6061N\3-.356\T6-2
4	.180	1.580	1.846	46.15	6061N\4-.18 \ T6-2
	.247	1.647	2.161	54.03	6061N\4-.247\T6-2
	.320	1.720	2.504	62.60	6061N\4-.32\T6-2
5	.190	1.750	2.316	57.90	6061N\5-/19\T6-2
	.325	1.885	3.108	77.70	6061N\5-325\T6-2
	.472	2.032	3.975	99.38	6061N\5-.472\T6-2
6	.200	1.920	2.826	70.65	6061N\6-.2\T6-2
	.225	1.945	3.002	7.05	6061N\6-.225\T6-2
	.314	2.034	3.631	90.78	6061N\6-314\T6-2
7	.230	2.110	3.541	88.53	6061N\7-.23\T6-2
8	.250	2.290	4.252	106.30	6061N\8-.25\T6-2
	.488	2.527	6.484	162.10	6061N\8-.488\T6-2

(Continued)

6061-T6 Aluminum Structural Channels—Cont'd

Size (Inches) A	Web Thick C	Flange Width B	Approx. Wt. per Lin. Ft	Approx. Wt. per Length	Group/Size/Grade
9	.230	2.430	4.604	115.10	6061N\9-.23\T6-2
10	.240	2.600	5.278	131.95	6061N\10-.24\T6-2
	.526	2.886	8.641	216.30	6061N\10-.526\T6-2
12	.300	2.960	7.415	185.38	6061N\12-.3\T6-2
12	.510	3.170	10.374	259.35	6061N\12-.51\T6-2

6061-T6 Aluminum Structural Channels
Aluminum Association (AA)—25 Ft Lengths—ASTM-B221-or-QQ-A-200/16

Size (Inches) A	Web Thick C	Flange Width B	Flange Thick C-1	Radius R	Approx. Wt. per Lin. Ft	Approx. Wt. per Length	Group/Size/Grade
2	.13	1.00	.13	.10	.577	14.43	6061N\2-.13\T6-1
	.17	1.25	.26	.15	1.071	26.78	6061N\2-.17\T6-1
3	.13	1.50	.20	.25	1.135	28.93	6061N\3-.13\T6-1
	.17	1.75	.26	.25	1.597	39.93	6061N\3-.17\T6-1
4	.15	2.00	.23	.25	1.738	43.45	6061N\4-.15\T6-1
	.19	2.25	.29	.25	2.331	58.28	6061N\4-.19\T6-1
5	.15	2.25	.26	.30	2.212	55.30	6061N\5-.15\T6-1
	.19	2.75	.32	.30	3.089	77.23	6061N\5-.19\T6-1
6	.17	2.50	.29	.30	2.834	70.85	6061N\6-.17\T6-1
	.21	3.25	.35	.30	4.060	100.75	6061N\6-.21\T6-1
7	.21	3.50	.38	.30	4.715	103.68	6061N\7-.21\T6-1
8	.19	3.00	.35	.30	4.147	103.68	6061N\8-.19\T6-1
	.25	3.75	.41	.35	5.789	144.73	6061N\8-.25\T6-1
10	.25	3.50	.41	.35	6.139	153.40	6061N\10-.25\T6-1
12	.29	4.00	.41	.35	8.274	206.90	6061N\12-.29\T69-1

6061-T6 Special Aluminum Channel
Sharp Corners—QQ-A-200/8—Die # 40517—22 Ft Length—ASTM-B221

Size (Inches) A	Web Thick C	Flange Width B	Approx. Wt. per Lin. Ft	Approx. Wt. per Length	Group/Size/Grade
2.50	.125	1.50	.787	17.31	6061N\2.5-1.5-.125\T6-3

Section | 6 Calculating the Size/Weight of Structural Steel and Miscellaneous Metals

6.12.2 Aluminum Structural Beams—Calculate Their Weight and Size

Beams

6061-T6 Aluminum Structural I Beam
American Standard—25 Ft Length—ASTM-B221-or-QQ-A-200/16

Size (Inches) A	Web Thick C	Flange Width B	Approx. Wt. per Lin. Ft	Approx. Wt. per Length	Group/Size/Grade
3	.170	2.330	1.963	49.80	6061\I3-.17\T6-2
	.349	2.509	2.591	64.78	6061I\3-.349\T6-2
4	.190	2.660	2.664	66.60	6061I\4-.19\T6-2
	.326	2.796	3.283	82.08	6061I\4-.326\T6-2
5	.210	3.000	3.430	85.75	6061I\5-.21\T6-2
	.494	3.284	5.000	127.50	6061I\5-.494\T6-2
6	.230	3.330	4.302	107.55	6061I\6-.23\T6-2

6061-T6 Aluminum Structural I Beam
Aluminum Association (AA)—25 Ft—Length—ASTM-B221-or- QQ-A-200/16

Size (Inches) A	Web Thick C	Flange Width B	Flange Thick C-1	Radius R	Approx. Wt. per Lin. Ft	Approx. Wt. per Length	Group/Size/Grade
4	.15	3.00	.23	.25	2.310	57.75	60611\4-.15\T6-1
	.17	3.00	.29	.25	2.793	69.83	6061I\4-.17\T6-1
5	.19	3.50	.32	.30	3.700	92.50	60611\5-.19\T6-1
6	.19	4.00	.29	.30	4.030	100.75	60611\6-.19\T6-1
	.21	4.00	.35	.30	4.693	117.33	60611\6-.21\T6-1
8	.23	5.00	.35	.30	6.181	154.53	60611\8-.23\T6-1

6061-T6 Aluminum Structural H Beams
25 PL Length—ASTM-B221—or—QQ-A-200/16

Size (in.) A	Web Thick C	Flange Width B	Approx. Wt. per Lin. Ft	Approx. Wt. per Length	Group/Size/Grade
4	.312	4.000	4.760	119.0	6061H\4-.313\T6
5	.312	5.000	6.490	162.3	6061H\5-.312\T6

6061-T6 Aluminum Structural Wide Flange Beams
Aluminum Association (AA)—25 Ft Length—ASTM-B221-or- QQ-A-200/16

Size (in.) A	Web Thick C	Flange Width B	Approx. Wt. per Lin. Ft	Approx. Wt. per Length	Group/Size/Grade
6	.230	4.000	4.160	104.0	6061L\6-4-.23\T6-1
	.240	6.000	5.400	135.0	6061L\6-.24\T6-1
8	.288	8.000	10.725	268.1	6061L\8-.288\T6-1

6.13.0 Plate Steel-3/16″ to 6″ Thickness—Calculate Their Weight and Size

Plate Steel
Hot Rolled C-33 MAX, ASTM-A36

Size in Inches	Wt. Per Ft, Lb	Size in Inches	Wt. per Ft, Lb
*3/16 × (7.65 Lb per Sq Ft)		1 1/8 × (45.9 Lb per Sq Ft)	
84	53.55	72	275.40
96	61.20	96	367.20
1/4 × (10.20 Lb Per Sq Ft)		1 1/4 × (51.00 Lb. per Sq Ft)	
30	25.50	36	153.00
36	30.60	48	204.00
42	35.70	60	255.00
48	40.80	72	306.00
54	45.90	84	357.00
60	51.00	96	408.00
72	61.20	1 3/8 × (56.10 Lb per Sq Ft)	
84	71.40	60	280.50
96	81.60	72	336.60
120	102.00	84	392.70
5/16 × (12.75 Lb per Sq Ft)		96	448.80
30	31.88	1 1/2 × (61.20 per Sq Ft)	
36	38.25	48	244.80
48	51.00	60	306.00
60	63.75	72	367.20
72	76.50	84	428.40
84	89.25	1 5/8 × (66.3 Lb per Sq Ft)	
96	102.00	72	397.80
120	127.50	1 3/4 × (71.40 Lb per Sq Ft)	
3/8 × (15.30 Lb per Sq Ft)		48	285.60
30	38.25	60	357.00
36	45.90	72	428.40
42	53.55	96	571.20
48	61.20	2 × (81.60 Lb per Sq Ft)	
60	76.50	48	326.40
72	91.80	60	408.00
84	107.10	72	489.60
96	122.40	84	571.20
120	153.00	96	652.80

Plate Steel—Cont'd

Size in Inches	Wt. Per Ft, Lb	Size in Inches	Wt. per Ft, Lb
7/16 × (17.85 Lb Per Sq Ft)		2 1/4 × (91.80 Lb per Sq Ft)	
30	44.63	72	550.80
36	53.55	2 1/2 × (102.0 Lb per Sq Ft.)	
42	62.48	48	408.00
48	71.40	60	510.00
60	89.25	72	612.00
72	107.10	2 3/4 × (112.20 Lb per Sq Ft)	
84	124.95	48	448.80
96	142.80	60	561.00
120	178.50	72	673.20
1/2 × (20.40 Lb per Sq Ft)		3 × (122.40 Lb per Sq Ft)	
30	51.00	48	489.60
36	61.20	60	612.00
42	71.40	72	734.40
48	81.60	3 1/4 × (132.60 Lb per Sq Ft)	
60	102.00	48	530.40
72	122.40	60	663.00
84	142.80	72	795.60
96	163.20	3 1/2 × (142.80 Lb. per Sq Ft)	
120	204.00	48	571.00
9/16 × (22.95 Lb per Sq Ft)		60	714.00
72	137.70	72	856.80
5/8 × (25.75 Lb per Sq Ft)		3 3/4 × (153.00 Lb per Sq Ft)	
30	64.38	48	612.00
36	77.25	60	765.00
42	90.13	72	918.00
48	103.00	4 × (163.20 Lb per Sq Ft)	
60	128.75	48	652.80
72	154.50	60	816.00
84	180.25	72	979.20
96	206.00	4 1/2 × (183.60 Lb per Sq Ft)	
120	257.50	48	734.40
11/16 × (28.05 Lb per Sq Ft)		72	1101.60
72	168.30	5 × (204.0 Lb per Sq Ft)	
96	224.40	48	816.00
3/4 × (30.60 Lb per Sq Ft)		60	1020.00
30	76.50	72	1224.00
36	91.80	84	1428.00
42	107.10	5 1/2 × (224.40 Lb per Sq Ft)	
48	122.40	48	897.00
60	153.00	60	1122.00
72	183.60	72	1346.40
84	214.20	6 × (244.80 Lb per Sq Ft)	
96	244.80	48	979.20

6.14.0 Sheet and Coil Steel—Types and Uses

HOT COLD ROLLED STEEL SHEETS
STANDARD SHEET GAUGE & WEIGHTS

AISI THICKNESS TOLERANCE RANGE
GALVANIZED

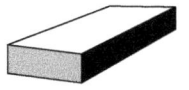

Sheet and Coil Selection Guide

Low carbon

A low-carbon sheet is a low-cost steel sheet, soft enough to bend flat on itself in any direction, without cracking, ductile enough for shallow drawing. Carbon is .10 max. (instead of .15 max.) for improved welding and forming. Surface has normal mill oxide. It conforms to ASTM A569 and is used for tanks, barrels, farm implements, and other applications where finish is secondary.

Pickled and oiled

Acid pickling provides a smoother, more uniform surface. Paint and enamel adhere well. Properties and characteristics are the same as low-carbon sheets. It conforms to ASTM A569. Typical applications include auto parts, appliances, and toys.

Abrasions resisting

Medium carbon content plus higher manganese greatly improve resistance to abrasion. 210 min. Brinell. For scrapers, liners, chutes, conveyors–outlasting low-carbon steel by two to ten times. Moderate formability.

Cold rolled

Low carbon

This continuous mill product is made with a high degree of gauge accuracy and uniformity of physical characteristics. The smooth deoxidized matte finish gives an excellent base for paint, lacquer, and enamel. Box annealing and absence of scale permit stamping and moderate drawing operations. Sheets bend flat on themselves without cracking. Oiling protects surface against rust. It conforms to ASTM A366—with carbon content held to .10 max. (instead of .15 max. as the spec permits) for improved welding and forming.

Galvanized

The term g*alvanized* has long been used to describe steel sheets coated with zinc. It is usually associated with the hot-dipped process, but the zinc coating can also be applied by electroplating. Our hot-dipped galvanized sheets are produced to conform with ASTM A525. Electrogalvanized conforms to ASTM A591.

In recent years, the producing mills have accomplished many technological advancements in the manufacture of both hot-dipped and electrogalvanized sheets. The end use should determine the type of coated product and surface condition required regardless of the coating weight, spangle size, method of manufacture, or surface preparation required. Following are descriptions of the grades and conditions of galvanized sheets used most often.

Coating	
ASTM Designation	Minimum Check Limit by Triple Spot Test (oz/sq ft)
G-60 Light Commercial	.060
G-90 Commercial up to	0.90
G-210	2.10

6.14.1 Calculating the Weight of Various Types of Carbon, Stainless, and Galvanized Sheet Steel

Standard Sheet Gauge and Weights
For Accuracy
Specify Thickness by Decimal Part of an Inch

	Carbon Sheets to U.S.S. or Mfrs. Gauge		Galvanized Sheets to Galvanized Sheet Gauge		Stainless Sheets to Stainless Sheet Gauge		
						Pounds per Sq Ft	
Gauge No.	Thickness in Inches	Pounds per Sq Ft	Thickness in Inches	Pounds per Sq. Ft	Thickness in Inches	200 & 300 Series	400 Series
3	.2391	10.00					
4	.2242	9.375					
5	.2092	8.750					
6	.1943	8.125					
7	.1793	7.500					
8	.1644	6.875	.1681	7.031	.171875	7.2187	7.0813
9	.1495	6.250	.1532	6.406	.156250	6.5625	6.4375
10	.1345	5.625	.1382	5.781	.140625	5.9062	5.7937
11	.1196	5.000	.1233	5.516	.125000	5.2500	5.1500
12	.1046	4.375	.1084	4.531	.109375	4.5937	4.5063
13	.0897	3.750	.0934	3.906	.09375	3.9375	3.8625
14	.0747	3.125	.0785	3.281	.078125	3.2812	3.2187
15	.0673	2.813	.0710	2.969	.070313	2.9531	2.8968
16	.0598	2.500	.0635	2.656	.062500	2.6250	2.5750
17	.0538	2.250	.0575	2.406	.056250	2.3625	2.3175
18	.0478	2.000	.0516	2.156	.050000	2.1000	2.0600
19	.0418	1.750	.0456	1.906	.043750	1.8375	1.8025

(Continued)

Standard Sheet Gauge and Weights—Cont'd

	Carbon Sheets to U.S.S. or Mfrs. Gauge		Galvanized Sheets to Galvanized Sheet Gauge		Stainless Sheets to Stainless Sheet Gauge		
						Pounds per Sq Ft	
Gauge No.	Thickness in Inches	Pounds per Sq Ft	Thickness in Inches	Pounds per Sq. Ft	Thickness in Inches	200 & 300 Series	400 Series
20	.0359	1.500	.0396	1.656	.037500	1.5750	1.5450
21	.0329	1.375	.0366	1.531	.034375	1.4437	1.4160
22	.0299	1.250	.0336	1.406	.031250	1.3125	1.2875
23	.0269	1.125	.0306	1.281	.028125	1.1813	1.1587
24	.0239	1.000	.0276	1.156	.025000	1.0500	1.0300
25	.0209	.8750	.0247	1.031	.021875	.9187	.9013
26	.0179	.7500	.0217	.9063	.018750	.7875	.7725
27	.0164	.6875	.0202	.8438	.017188	.7218	.7081
28	.0149	.6250	.0187	.7813	.015625	.6562	.6438
29	.0135	.5625	.0172	.7188	.014063	.5906	.5794
30	.0120	.5000	.0157	.6563	.012500	.5250	.5150

¼" Thick & Heavier classed as plates.

6.14.2 Calculating the Weight of Low-Carbon, Hot-Dipped, Galvanized Roof Deck

Galvanized Steel Sheet   Galvanized Corrugated Roofing   Abrasion Resisting Steel Sheets

Galvanized Steel Sheet Commercial Quality – Low Carbon

Hot-Dipped G60-G90 ASTM A525

Gauge	Width and Length		Estimated Wt. Lbs per Sheet	Gauge	Width and Length		Estimated Wt. Lbs per Sheet
26 Ga. – 906 lbs / sq ft .02174				16 Ga. – 2.66 lbs / sq ft .0635			
	36 ×	96	21.7		36 ×	96	63.7
	36 ×	120	27.2		36 ×	120	79.7
	48 ×	96	29.0		48 ×	96	85.0
	48 ×	120	36.2		48 ×	120	106.2
24 Ga. – 1.16 lbs / sq ft .0276					60 ×	120	132.8
	36 ×	96	27.7	14 Ga. – 3.28 lbs / sq ft .0785			
	36 ×	120	34.7		36 ×	96	78.7
	48 ×	96	37.0		36 ×	120	98.4
	48 ×	120	46.2		48 ×	96	105.0

Galvanized Steel Sheet

Galvanized Steel Sheet
Commercial Quality – Low Carbon

Hot-Dipped G60-G90 ASTM A525

Gauge	Width and Length			Estimated Wt. Lbs per Sheet
22 Ga. – 1.41 lbs / sq ft .0336				
	36	×	96	33.7
	36	×	120	42.2
	48	×	96	45.0
	48	×	120	56.2
	60	×	120	70.3
20 Ga. – 1.66 lbs / sq ft 0396				
	36	×	96	39.7
	36	×	120	49.7
	48	×	96	53.0
	48	×	120	66.2
	60	×	120	82.8
18 Ga. – 2.16 lbs / sq ft .0516				
	36	×	96	51.7
	36	×	120	64.7
	48	×	96	69.0
	48	×	120	85.2
	60	×	120	107.8

Abrasion Resisting Steel Sheets—Cont'd

Gauge	Width and Length			Estimated Wt. Lbs per Sheet
	48	×	120	131.2
	60	×	120	164.1
12 Ga. – 4.53 lbs / sq ft 1084				
	48	×	96	145.0
	48	×	120	181.2
	60	×	120	226.6
10 Ga. – 53.78 lbs / sq ft .1382				
	48	×	96	185.0
	48	×	120	231.2
	60	×	120	289.1

GALVANIZED CORRUGATED ROOFING

Hot-Dipped G90 Coating	Wt. per Sheet	Sheets per Square	Wt per Square
24 Ga. × 27 1/2 × 96	23.1	5.4	126
24 Ga. × 27 1/2 × 120	28.9	4.3	126
24 Ga. × 27 1/2 × 144	34.7	3.6	126
26 Ga. × 27 1/2 × 96	22.6	4.3	98
26 Ga. × 27 1/2 × 120	28.2	3.4	98
26 Ga. × 27 1/2 × 144	33.8	2.8	98
28 Ga. × 27 1/2 × 96	15.6	5.4	85
28 Ga. × 27 1/2 × 120	19.5	4.3	85
28 Ga. × 27 1/2 × 144	23.4	3.6	85

Electrogalvanized sheets are offered subject to inquiry

ABRASION RESISTING STEEL SHEETS

Thickness Inches	Size Inches			Est Weight per Sheet Lbs
No. 14 (3.125 lbs/sq ft)				
.0747	36	×	120	93.80
No. 12 (4.375 lbs/sq ft.)				
.1046	36	×	120	131.30
1/8" (5.10 lbs/sq ft)				
.125	48	×	96	163.20

For heavier gauges see plate section page

6.14.3 Converting Gauge Inches to Decimals for Sheet Steel, Aluminum, Stainless Steel

Galvanized Steel

Specified Width, Inches		Thickness tolerance, inch over only. No tolerance under specified minimum thickness					
		over .101 thru .177	over .075 thru .101	over .061 thru .075	over .043 thru .061	over .023 thru .043	.023 and thinner
	thru 32	.016	.014	.012	.010	.008	.006
over 32	thru 40	.016	.016	.012	.010	.008	.006
over 40	thru 60	.018	.016	.012	.010	.008	.006
over 60	thru 72	.018	.018	.012	.010	.008	—

Hot-Rolled Steel

Specified Width, Inches		Thickness tolerance, inch over only. No tolerance under specified minimum thickness			
		over .179 thru .230	over .097 thru .179	over .082 thru .097	over .071 thru .082
—	thru 20	.016	.016	.014	.014
over 20	thru 32	.018	.016	.014	.014
over 32	thru 40	.018	.018	.016	.014
over 40	thru 48	.020	.020	.016	.014
over 48	thru 60	—	.020	.016	.014
over 60	thru 72	—	.022	.018	.016
over 72		—	.024	.018	.016

Section | 6 Calculating the Size/Weight of Structural Steel and Miscellaneous Metals   341

Cold-Rolled Steel

Specified Width, Inches	Thickness tolerance, inch over only. No tolerance under specified minimum thickness					
	over .098 thru .142	over .071 thru .098	over .057 thru .071	over .039 thru .057	over .019 thru .039	.over .014 thru .019
24 thru 72	.012	.010	.010	.008	.006	.004
over 72	.014	.012	.010	.008	.006	—

By permission: Corrugated Metals, Inc., Belvidere, ILL

6.15.0 Carbon Steel Expanded Metal Grating—ASTM A1011

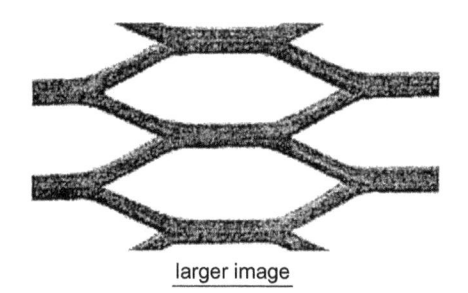

Expanded Metal Grating - ASTM A1011

larger image

Results 1 -8 of 8

Item #	Item Name	-	Opening Width	-	Opening Length	-	Open Area	-	Overall Thickness	-
2 lb -	Expanded Metal Grating–Carbon Steel	-	1.000 in.	-	3.600 in.	-	77%	-	0.460 in.	-
3 lb -	Expanded Metal Grating–Carbon Steel	-	0.938 in.	-	3.438 in.	-	73%	-	0.500 in.	-
3. 14 lb -	Expanded Metal Grating–Carbon Steel	-	1.625 in.	-	4.875 in.	-	74%	-	0.562 in.	-
4 lb -	Expanded Metal Grating–Carbon Steel	-	0.938 in.	-	3.438 in.	-	65%	-	0.625 in.	-
4.27 lb -	Expanded Metal Grating–Carbon Steel	-	1.000 in.	-	2.875 in.	-	58%	-	0.625 in.	-
5 lb -	Expanded Metal Grating–Carbon Steel	-	0.813 in.	-	3.375 in.	-	52%	-	0.625 in.	-
6.25 lb -	Expanded Metal Grating–Carbon Steel	-	0.813 in.	-	3.375 in.	-	55%	-	0.750 in.	-
7 lb -	Expanded Metal Grating–Carbon Steel	-	0.813 in.	-	3.375 in.	-	60%	-	0.750 in.	-

Source: Direct Metals, Inc., Kennesaw, GA

6.15.1 Carbon Steel Catwalk Expanded Metal Grating—ASTMA569/569M

Carbon Steel Catwalk Expanded Metal Grating

Check up to five results to perform an action.

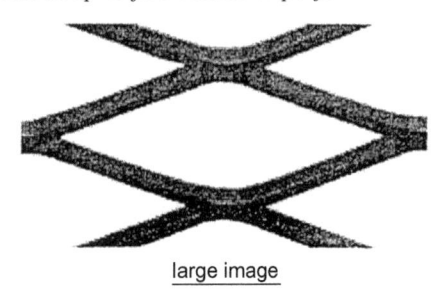

Catwalk Grating-ASTMA 569/569M

large image

Results 1-7 of 7

Item #	Item Name	Opening Width	Opening Length	Open Area	Overall Thickness
2 lb	Catwalk Expanded Metal Grating	1.000 in.	3.600 in.	77%	0.460 in.
3 lb	Catwalk Expanded Metal Grating	0.938 in.	3.438 in.	73%	0.500 in.
3. 14 lb	Catwalk Expanded Metal Grating	1.625 in.	4.875 in.	74%	0.562 in.
4 lb	Catwalk Expanded Metal Grating	0.938 in.	3.438 in.	65%	0.625 in.
4.27 lb	Catwalk Expanded Metal Grating	1.000 in	2.875 in.	58%	0.625 in.
5 lb	Catwalk Expanded Metal Grating	0.813 in.	3.375 in.	52%	0.625 in.
6.25 lb	Catwalk Expanded Metal Grating	0.813 in.	3.375 in.	55%	0.750 in.

6.15.2 Aluminum Expanded Metal Grating

Item # 2 lb, Aluminum Expanded Metal Grating

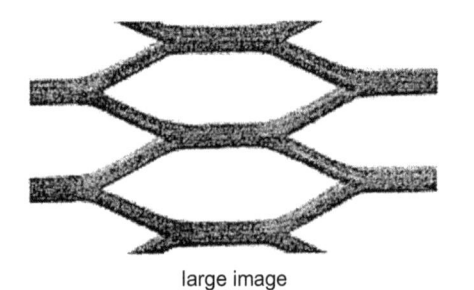

Aluminum Expanded Metal Grating
Aluminum Expanded Metal Grating—Alloy 5052 H32

large image

Specifications

Style	2 lb
Weight Plain per CSF	200 lb
Standard Width	4 ft
Standard Length	8 ft
Opening Width	1.000 in.
Opening Length	3.600 in.
Center to Center of Bond Width	1.333 in.
Center to Center to Bond Length	5.330 in.
Stands Width	0.235 in.
Open Area	77 %
No. Diamonds per ft SWD	9
Overall Thickness	0.460 in.

Source: Direct Metals, Inc., Kennesaw, GA

6.16.0 Aluminum Rectangular Bar Grating

Bar Grating
Aluminum Rectangular Bar (SG Series)

U- Safe uniform load in pounds/sq ft
C-Safe concentrated load in pounds/ft grating width
D-Deflection in inches

Loads and deflections given in this table are theoretical and are based on a unit stress of 12,000 psi.

Grating Types	Bar Size	Wt. *Lbs Sq Ft	Sect. Prop. Ft of Width		2'-0"	2'-6"	3'-0"	3'-6"	4'-0"	4'-6"	5'-0"	5'-6"	6'-0"	6'-6"	7'-0"
19 SG 4 1-3/16"	× 1/8	1.85	Sx = .228	U	458	293	203	149	114						
				D	.144	.225	.324	.441	.576						
			1x = .114	C	458	366	305	261	229						
				D	.115	.180	.259	.352	.461						
15 SG 4 15/16"	× 1/4 × 1/8	2.25	Sx = .358	U	686	439	305	224	172	136					
				D	.144	.225	.324	.441	.576	.728					
			1x = .223	C	686	549	458	392	343	305					
				D	.115	.190	.259	.352	.461	.583					
11 SG 4 11/16"	× 1/4 × 1/8	2.25	Sx = .358	U	715	458	318	233	179	141	114				
				D	.115	.180	.259	.351	.480	.581	.717				
			1x = .223	C	715	572	477	408	358	318	286				
				D	.092	.144	.207	.282	.369	.466	.575				
7 SG 4 7/16"	1/4 × 3/16	3.25	Sx = .536	U	1074	687	477	350	268	212	172				
				D	.115	.180	.259	.351	.460	.581	.717				
			1x = .335	C	1074	859	716	614	537	477	429				
				D	.092	.144	.207	.282	.369	.486	.575				
19 SG 2 1-3/16"	1/2 × 1/8	2.65	Sx = .515	U	1030	659	458	336	257	203	165	136			
				D	.096	.150	.216	.294	.383	.485	.599	.724			
			1x = .387	C	1030	824	686	588	515	458	412	374			
				D	.077	.120	.173	.235	.307	.389	.480	.579			
15 SG 2 15/16"	1/2 × 3/16	3.86	Sx = .773	U	1547	990	687	505	387	306	247	204	172		
				D	.096	.150	.216	.294	.383	.485	.599	.724	.861		
			1x = .579	C	1547	1237	1031	884	773	687	619	562	516		
				D	.077	.120	.173	.235	.307	.389	.480	.579	.690		
11 SG 2 11/16"	3/4 × 3/16	4.48	Sx = 1.052	U	2105	1347	936	687	526	416	337	278	234	199	
				D	.082	.128	.185	.252	.629	.417	.515	.622	.741	.868	
			1x = .902	C	2105	1684	1404	1203	1053	936	842	766	702	648	
				D	.066	.103	.148	.202	.264	.334	.412	.498	.593	.696	
7 SG 2 7/16"	× 3/16	5.08	Sx = 1.375	U	2750	1760	1222	898	688	543	440	364	306	260	224
				D	.072	.112	.162	.220	.288	.364	.450	.545	.649	.759	.880
			1x = 1.375	C	2750	2200	1833	1571	1375	1222	1100	1000	917	846	786
				D	.058	.090	.130	.176	.230	.292	.360	.436	.518	.608	.706

Bar Grating—Cont'd

Bar Size	Wt. Lbs Sq Ft	Sect. Prop. Ft of Width		Clear Span											
				2'-0"	2'-6"	3'-0"	3'-6'	4'-0"	4'-6"	5'-0"	5'-6"	6'-0"	6'-6"	7'-0"	8'-0"
2-1/4 × 3/16	5.69	Sx =1.740	U	3480	2227	1547	1136	870	687	557	460	387	330	284	218
			D	.064	.100	.144	.196	.256	.324	.400	.484	.577	.677	.784	1.027
		1x=1.958	C	3480	2784	2320	1989	1740	1547	1392	1266	1160	1071	994	870
			D	.051	.080	.115	.157	.205	.259	.320	.387	.461	.541	.627	.819
2-1/2 × 3/16	6.29	Sx= 1.148	U	4297	2750	1910	1403	1074	849	687	568	477	407	351	268
			D	.058	.090	.130	.176	.230	.292	.360	.436	.518	.609	.706	.920
		1x = 2.685	C	4297	3437	2864	2455	2148	1910	1719	1562	1432	1322	1228	1074
			D	.046	.072	.104	.141	.184	.233	.288	.348	.415	.487	.565	.737

Based on 11 bars/ft. of grating width. Bearing bars 1-3/16 c.c. Add. 4 lbs/sq ft for 19.SG.2.
NOTE: Grating for spans to the left of the heavy line have a deflection less than 1/4" for uniform loads of 100 lbs/sq/ft. This is the maximum deflection to afford pedestrian comfort and can be exceeded for other types of load at the discretion of the engineer. When serrated grating is specified, the depth of grating required for a specified load will be 1/4" greater than that shown in these tables. 1" serrated grating not available.

19-SG-4/19-SG-2 Panel Width Chart (in.)

Dimensions Are Out-to-Out of Bearing Bars**

No. of Bars	2	3	4	5	6	7	8	9	10	11	12	13	14	15	16
3/16" Bar	1-3/8	2-9/16	3-3/4	4-15/16	6-1/8	7-5/16	8-1/2	9-11/16	10-7/8	12-1/16	13-1/4	14-7/16	15-5/8	16-13/16	18
No. of Bars	17	18	19	20	21	22	23	24	25	26	27	28	29	30	31
3/16" Bars	19-3/16	20-3/8	21-9/16	22-3/4	23-15/16	25-1/8	26-5/16	27-1/2	28-11/16	29-7/8	31-1/16	32-1/4	33-7/16	34-5/8	35-15/16

**Add 1/4* for extended cross bars. Deduct 1/16" for 1/8* bearing bars. Standard panel widths indicated with white numbers.

6.16.1 Aluminium I Bar and Rec Bar Grating

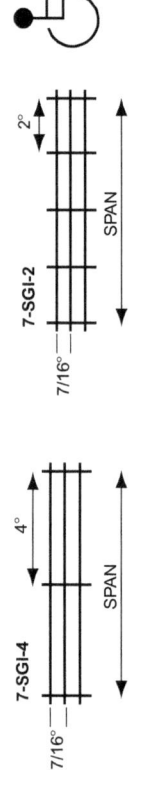

Bar Size, In.	Ped Span, In.	Wt. *Lbs SqFt	Sec. Prop Sx*, in³		2'-0"	2'-6"	3'-0"	3'-6"	4'-0"	4'-6"	5'-0"	5'-6"	6'-0"	6'-6"	7'-0"	8'-0"
1 × 3/16	56	6.30	0.857	U	1714	1097	762	560	429	339	274	U-Safe uniform load in pounds/sq ft				
				D	0.144	0.225	0.324	0.441	0.577	0.730	0.899	C-Safe concentrated load in pounds/ft ftating width				
1" 1-Bar		4.79	0.429	C	1714	1371	1143	980	857	762	686	D- Deflection in inches				
				D	0.115	0.180	0.259	0.353	0.461	0.583	0.720					
1-1/4 × 3/16	66	7.78	1.339	U	2679	1714	1190	875	670	529	429	354	298		Loads and deflections given in this table	
				D	0.115	0.180	0.259	0.353	0.461	0.583	0.721	0.871	1.038		are theoretical and are based on a unit	
1-1/4" 1-Bar		5.75	0.837	C	2679	2143	1786	1531	1339	1190	1071	974	893		stress of 12,000 psi.	
				D	0.092	0.144	0.207	0.282	0.369	0.466	0.576	0.697	0.830			
1-1/2 × 3/16	76	9.28	1.929	U	3857	2469	1714	1259	964	762	617	510	429	365		
				D	0.096	0.150	0.216	0.294	0.384	0.486	0.600	0.726	0.865	1.014		
1-1/2" 1-Bar		6.74	1.446	C	3857	3086	2571	2204	1929	1714	1543	1403	1286	1187		
				D	0.077	0.120	0.173	0.235	0.307	0.389	0.480	0.581	0.691	0.811		
1-3/4 × 3/16	85	10.80	2.625	U	5250	3360	2333	1714	1313	1037	840	694	583	497	429	328
				D	0.082	0.129	0.185	0.252	0.329	0.417	0.514	0.622	0.740	0.869	1.009	1.316
1-3/4" 1-Bar		7.70	2.297	C	5250	4200	3500	3000	2625	2333	2100	1909	1750	1615	1500	1313
				D	0.066	0.103	0.148	0.202	0.263	0.333	0.411	0.498	0.592	0.695	0.806	1.054
2 × 3/16	94	12.32	3.429	U	6857	4389	3048	2239	1714	1355	1097	907	762	649	560	429
				D	0.072	0.113	0.162	0.220	0.288	0.365	0.450	0.545	0.648	0.760	0.882	1.153
2" 1-Bar		8.71	3.429	C	6857	5486	4572	3918	3429	3048	2743	2494	2286	2110	1959	1714
				D	0.058	0.090	0.130	0.176	0.230	0.292	0.360	0.436	0.518	0.608	0.706	0.921
2-1/4 × 3/16	103	13.83	4.339	U	8679	5554	3857	2834	2170	1714	1389	1148	864	822	708	542
				D	0.064	0.100	0.144	0.196	0.256	0.324	0.400	0.484	0.576	0.676	0.783	1.023
2-1/4" 1-Bar		9.59	4.882	C	8679	6943	5786	4959	4339	3857	3471	3156	2893	2670	2480	2170
				D	0.051	0.080	0.115	0.157	0.205	0.259	0.320	0.387	0.461	0.541	0.627	0.819
2-1/2 × 3/16	111	15.33	5.357	U	10714	6857	4762	3499	2679	2116	1714	1417	1190	1014	875	670
				D	0.058	0.090	0.130	0.176	0.230	0.292	0.360	0.436	0.518	0.608	0.706	0.922
2-1/2" 1-Bar		10.66	6.697	C	10714	8572	7143	6123	5357	4762	4286	3896	3571	3297	3061	2679
				D	0.046	0.072	0.104	0.141	0.184	0.233	0.288	0.348	0.415	0.487	0.564	0.737

*Based on 27.429 bars/ft of grating width. Bearing bars 7/16" c.c. Add 3 lbs/sq ft for 7-SG-2. 1/8" bearing bars available by inquiry. Note: Grating for spans to the left of the heavy line has a deflection less than 1/4" for uniform loads of 100 lbs/sqft. This is the maximum deflection to afford pedestrian comfort and can be exceeded for other types of load at the discretion of the engineer.

7-SGI-4 7-SGI-2 Panel Width Chart (in.) — Dimensions Are Out-to-Out of Bearing Bars**

No. of Bars	2	3	4	5	6	7	8	9	10	11	12	13	14	15	16
1/4" Flange	11/16	1-1/8	1-9/16	2	2-7/16	2-7/8	3-5/16	3-3/4	4-3/16	4-5/8	5-1/16	5-15/2	6-3/16	6-3/8	6-13/16
No. of Bars	17	18	19	20	21	22	23	24	25	26	27	28	29	30	31
1/4" Flange	7-1/4	7-11/16	8-1/8	8-9/16	9	9-7/16	9-7/8	10-5/16	10-3/4	11-3/16	11-5/8	12-1/16	12-1/2	12-15/16	13-3/8
No. of Bars	32	33	34	35	36	37	38	39	40	41	42	43	44	45	46
1/4" Bars	13-13/16	14-1/4	14-11/16	15-1/8	15-9/16	16	16-7/16	16-7/8	17-5/16	17-3/4	18-3/16	18-8/8	19-1/16	19-1/2	19-15/16
No. of Bars	47	48	49	50	51	52	53	54	55	56	57	58	59	60	61
1/4" Bars	20-3/8	20-13/16	21-1/4	21-11/16	22-1/8	22-9/16	23	23-7/16	23-7/8	24-5/16	24-3/4	25-3/16	25-5/8	26-1/16	26-1/2
No. of Bars	62	63	64	65	66	67	68	69	70	71	72	73	74	75	76
1/4" Bars	26-15/16	27-3/8	27-13/16	28-1/4	28-11/16	29-1/8	29-9/16	30	30-7/16	30-7/8	31-5/16	31-3/4	32-3/16	32-5/8	33-1/16
No. of Bars	77	78	79	80	81	82	83								
1/4" Flange	33-1/2	33-15/16	34-3/8	34-13/16	35-1/4	35-11/16	36-1/8								

**Bars thickness is 1/4" at top and bottom. Add 1/4" for extended cross bars.

6.16.2 Aluminum Plank Sections and Pattern Availability

Bar grating

Aluminum Plank Section Availability

Aluminum plank is structurally sound and cosmetically attractive. Plank grating is nonsparking, nonmagnetic, nonskid, and relatively maintenance free. It is durable and corrosion resistant and possesses a high strength-to-weight ratio. The surface can be provided unpunched or with a variety of punch/patterns for the passage of air, light, heat, or moisture. The interconnecting webs offer a flush top walking surface. Aluminum plank grating has found application in sewage and wastewater treatment plants, as well as in the marine refrigerator (reefer), freezer, and cargo-hold flooring market. Aluminum plank grating is available in five cross-sectional designs: Heavy Duty (plain sides), Heavy Duty (interlocking sides), Light Series (plain sides), Reefer (plain sides), and Reefer (interlocking sides).

The Heavy Duty sections are used primarily in water and waste treatment and the marine markets, while the Light Series and Reefer sections are exclusively in the marine refrigerated stores application. Interlocking Heavy Duty, Reefer sections, and edge sections are available in 1″ deep grating only.

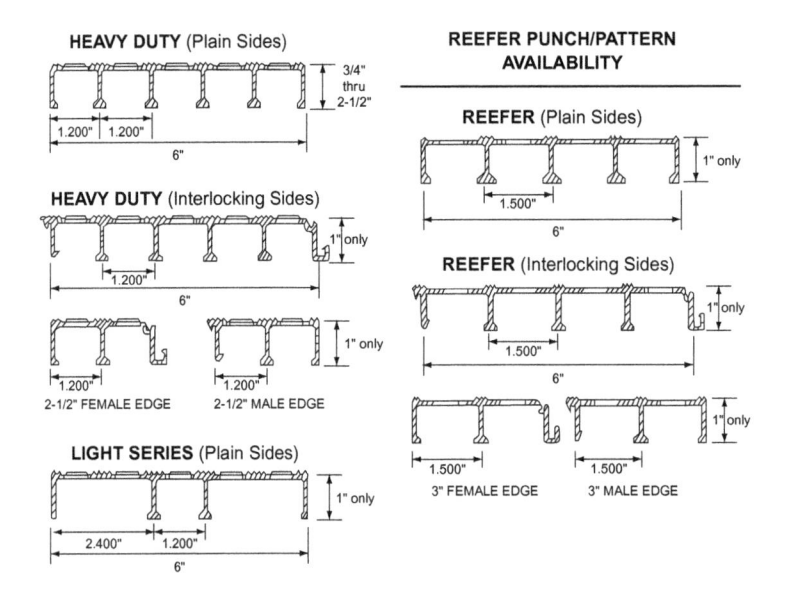

Plank Punch/Pattern Availability

Aluminum plank grating is available unpunched or with a variety of punch/patterns as shown below. Rectangular or square punched holes are most commonly used for water and waste treatment plants and in marine applications, while the round holes find application primarily in the marine market. The surface of plank grating can be specified as plain or with one of two styles of upsets (OGI or WACO) designed to promote a slip-resistant walkway, especially in the presence of moisture, oil, or other spilled substances.

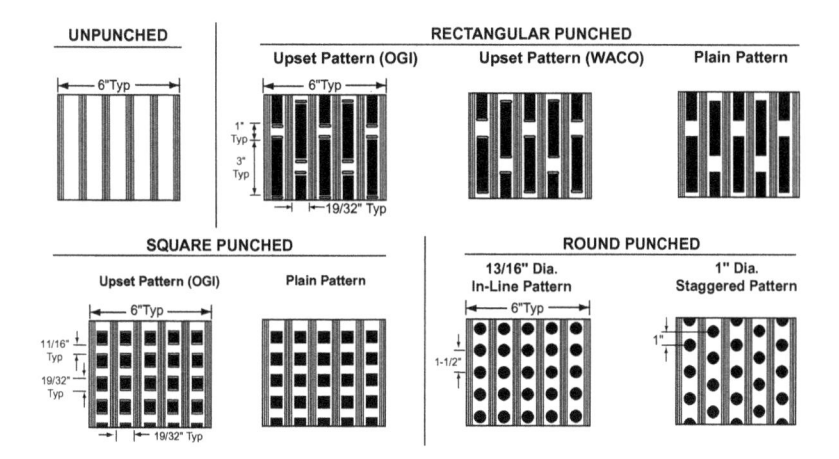

UNPUNCHED

← 6"Typ →

RECTANGULAR PUNCHED

Upset Pattern (OGI) | **Upset Pattern (WACO)** | **Plain Pattern**

← 6"Typ →
1" Typ
3" Typ
← 19/32" Typ

SQUARE PUNCHED

Upset Pattern (OGI) | **Plain Pattern**

← 6"Typ →
11/16" Typ
19/32" Typ
← 19/32" Typ

ROUND PUNCHED

13/16" Dia. In-Line Pattern | **1" Dia. Staggered Pattern**

← 6"Typ →
1-1/2"
1"

Section 7

Lumber—Calculations to Select Framing and Trim Materials

7.0.1	How Lumber Is Cut from a Log Affects Its Grain, Drying Process, and Waste Factor	352	7.2.4	Weights of Green and Kiln-Dried Hardwoods	386
7.0.2	The Physical Properties of Wood–Illustrated	353	7.2.5	National Hardwood Lumber Association Grading Rules	388
7.0.3	Lumber Industry Abbreviations	356	7.3.0	Various Strengths—Static Bending and Compression of Softwoods and Hardwoods	398
7.1.0	Softwood Lumber—Commercial Names of Principal Softwood Species	358	7.4.0	Equilibrium Moisture Content of Wood in Each of 50 States	404
7.1.1	Standard Sizes for Framing Lumber	360			
7.1.2	Nominal and Minimum Dressed Sizes of Finish Lumber	361	7.4.1	Coefficients for Dimensional Change as result of Shrinkage	405
7.1.3	Appearance Lumber—Grades, Nominal, and Dressed Sizes	363	7.4.2	Drying of Wood—Recommended Moisture Content	407
7.1.4	Section Properties of Joists, Beams, and Timbers	365	7.5.0	Mechanical Properties of Some Common Woods Grown in the United States—Inch-Pound	410
7.1.4.1	Stress-Graded Lumber—Nominal and Dry Dressed Dimensions	369	7.5.1	Mechanical Properties of Some Common Woods Grown in the United States—Metric	416
7.1.5	Average Moisture Content of Green Word, by Species	370			
7.1.6	Notching and Boring Guides for Softwood Floor Joist and Stud Walls	373	7.6.0	Mechanical Properties of Wood Imported from Other Countries—Inch-Pound	420
7.1.7	Profiles of Typical "Worked" Lumber	375			
7.1.7.1	Nominal and Dressed Sizes of Worked Lumber	377	7.7.0	Mechanical Properties of Wood Imported from Other Countries—Metric	426
7.1.7.2	Nominal and Minimum Sizes of 19% Moisture Content Dressed Lumber	378	7.8.0	Thermal Qualities of Wood	430
7.2.0	No. I and No. 2 Common Board Designations	380	7.9.0	Machine Stress-Rated Lumber—Calculating Design Values	431
7.2.1	Hardwood—Measurement System	381	7.10.0	Grade Names for Interior and Exterior Plywood	433
7.2.2	Hardwoods—Selects and No. I and No. 2 Standards	384	7.10.1	Softwood Plywood Species Grouped by Stiffness and Strength	434
7.2.3	American Standard Lumber Sizes for Yard and Structural Lumber for Construction	385	7.10.2	Typical Grade Stamps for Plywood and Oriented Strand Board (OSB)	435

7.11.0	Classification of Wood Composite Boards by Particle Size and Process Type 436	7.12.0	Medium-Density Fiberboard and Hardboard Property Requirements 438
7.11.1	Particleboard Grade Requirements 436	7.13.0	Physical and Mechanical Properties of Hardboard Siding 439
7.11.2	Particleboard Flooring Grade Requirements 437		

7.0.1 How Lumber Is Cut from a Log Affects Its Grain, Drying Process, and Waste Factor

Quarter Sawn

Rift sawing at a 30-degree or greater angle to the growth rings produces narrow boards with accentuated vertical grain pattern. Rift-sawn boards are often favored for fine furniture and other applications where matching grain is important. This type of lumber is available in limited quantities and species and, as can be seen, produces more wasted material per log.

Rift Sawn

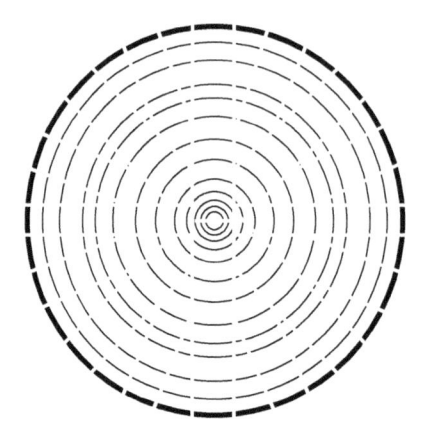

To summarize: if the growth rings are at an angle of 0° to 45° relative to the face of the board, the board is flat sawn, and if the growth rings are 45° to 90° the board is quarter sawn. Quarter-sawn boards are less likely to distort or crack during the drying process and are more stable in service.

Wood shrinks and expands when it rains a lot with high moisture content. Some wooden doors may stick a bit because the wood has expanded. This can become an issue when using wood.

Plain Sawn

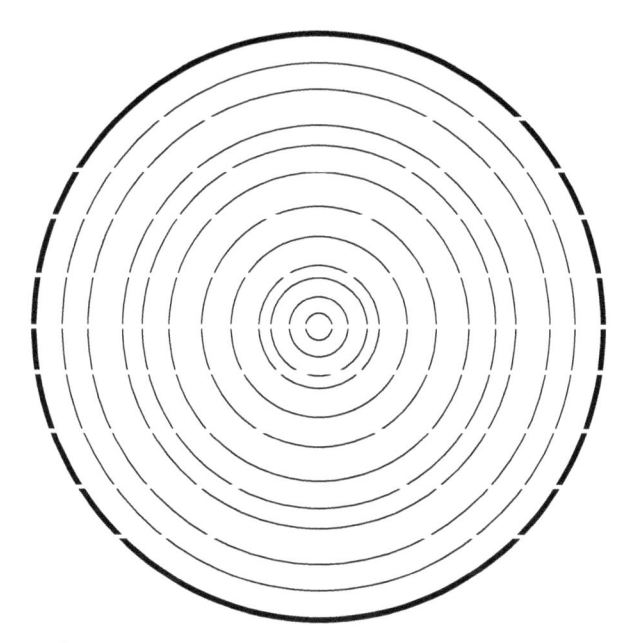

Quarter sawing means cutting a log radially (at a 90-degree angle) to the growth rings to produce a vertical grain pattern. This pattern produces fewer and narrower boards per log than plain sawing. However, the quality of each board is much better. Quarter-sawn boards will expand and contract less than boards sawn by other methods. Quarter-sawn logs require a lot of processing and produce lots of quarter-sawn boards of various grades. There is also little waste in the process.

7.0.2 The Physical Properties of Wood–Illustrated

Physical Properties Illustrated

Extreme Fiber Stress in Bending – F_b (Fig. 1)

When loads are applied, structural members bend, producing tension in the fibers along the faces farthest from the applied load and compression in the fibers along the face nearest to the applied load. These induced stresses in the fibers are designated as "extreme fiber stress in bending" (F_b).

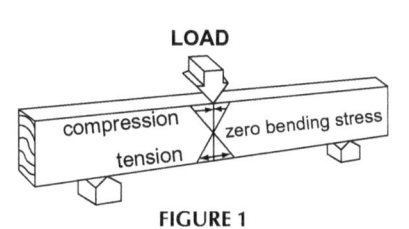

FIGURE 1

Single Member F_b

Design values are used in design where the strength of an individual piece, such as a beam, may be solely responsible for carrying a specific design load.

Repetitive Member F_b

Design values are used in design when three or more load-sharing members, such as joists, rafters, or studs, are spaced no more than 24" apart and are joined by flooring, sheathing, or other load-distributing elements. Repetitive members are also used where pieces are adjacent, such as decking.

Fiber Stress in Tension - F_t (Fig. 2)

Tensile stresses are similar to compression parallel to grain in that they act across the full cross section and tend to stretch the piece. Length does not affect tensile stresses.

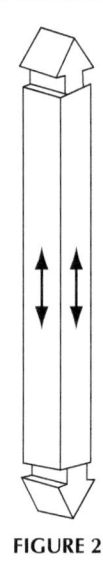

FIGURE 2

Horizontal Shear - F_v (Fig. 3)

Horizontal shear stresses tend to slide fibers over each other horizontally. Most predominate in short, heavily loaded deep beams. Increasing beam cross section decreases shear stresses.

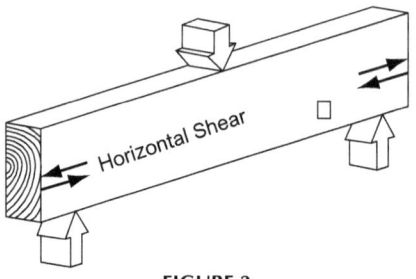

FIGURE 3

Compression Perpendicular to Grain - $F_{c\perp}$ (Fig. 4)

Where a joist, beam, or similar piece of lumber bears on supports, the load tends to compress the fibers. It is necessary that the bearing area be sufficient to prevent side-grain crushing.

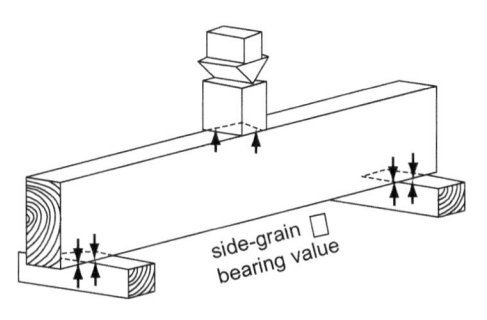

FIGURE 4 — side-grain bearing value

Compression Parallel to Grain - F_c *(Fig. 5)*

In many parts of a structure, stress grades are used where the loads are supported on the ends of the pieces. Such uses are as studs, posts, columns and struts. The internal stress induced by this kind of loading is the same across the whole cross section, and the fibers are uniformly stressed parallel to and along the full length of the piece.

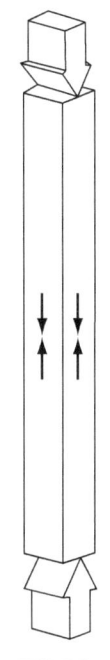

FIGURE 5

Modulus of Elasticity - E *(Fig. 6)*

The modulus of elasticity is a ratio of the amount a material will deflect in proportion to an applied load.

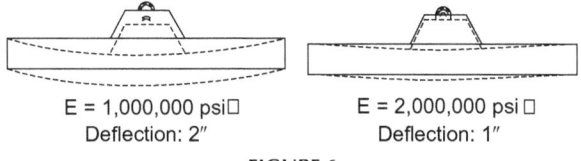

E = 1,000,000 psi
Deflection: 2"

E = 2,000,000 psi
Deflection: 1"

FIGURE 6

COURTEST: WESTERN WOOD PRODUCTS ASSOCIATION

7.0.3 Lumber Industry Abbreviations

The following abbreviations are commonly used for softwood lumber, although not all of them are necessarily applicable to all species. Additional abbreviations that are applicable to a particular region or species shall not be used unless included in certified grading rules.

Abbreviations are commonly used in the forms indicated, but variations such as the use of upper- and lower-case type, and the use or omission of periods and other forms of punctuation, are not required.

AD	Air-dried
ADF	After deducting freight sides
ALS	American Softwood Lumber Standard
AV or AVG	Average
Bd	Board
Bd ft	Board foot or feet
Bdl	Bundle
Bev	Beveled
B/L	Bill of lading
BM	Board Measure
Btr	Better
B&B or B&Btr	B and better
B&S	Beams and stringers
CB1S	Center bead one side
CB2S	Center bead two sides
CF	Cost and freight
CG2E	Center groove two edges
CIF	Cost, insurance, and freight
CIFE	Cost, insurance, freight, and exchange
Clg	Ceiling
Clr	Clear
CM	Center matched
Com	Common
CS	Caulking seam
Csg	Casing
Cu Ft	Cubic foot or feet
CV1S	Center Vee one side
CV2S	Center Vee two sides
D&H	Dressed and headed
D&M	Dressed and matched
DB Clg	Double-beaded ceiling (E&CB1S)
DB Part	Double-beaded partition (E&CB2S)
DET	Double end trimmed
Dim	Dimension
Dkg	Decking
D/S or D/Sdg	Drop siding
EB1S	Edge bead one side
EB2S	Edge bead two sides
E&CB1S	Edge and center bead one side
E&CB2S	Edge and center bead two sides
E&CV1S	Edge and center Vee one side
E&CV2S	Edge and center Vee two sides
EE	Eased edges
EG	Edge (vertical) grain
EM	End matched
EV1S	Edge Vee one side
EV2S	Edge Vee two sides
Fac	Factory
FAS	Free alongside (named vessel)

FBM	Foot or board measure
FG	Flat (slash) grain
Fig	Flooring
FOB	Free on board (named point)
FOHC	Free of heart center or centers
FOK	Free of knots
Frt	Freight
Ft	Foot or feet
GM	Grade marked
G/R or G/Rfg	Grooved roofing
HB	Hollow back
H&M	Hit-and-miss
H or M	Hit-or-miss
Hrt	Heart
Hrt CC	Heart cubical content
Hrt FA	Heart facial area
Hrt G	Heart girth
IN	Inch or inches
J&P	Joists and planks
KD	Kiln-dried
Lbr	Lumber
LCL	Less than carload
LFT or Lin Ft	Linear foot or feet
Lgr	Longer
Lgth	Length
Lin	Linear
Lng	Lining
M	Thousand
MBM	Thousand (feet) board measure
MC	Moisture content
Merch	Merchantable
Mldg	Molding
Mm	Millimeter
No	Number
N1E	Nosed one edge
N2E	Nosed two edges
Og	Ogee
Ord	Order
Par	Paragraph
Part	Partition
Pat	Pattern
Pc	Piece
Pcs	Pieces
PE	Plain end
PO	Purchase order
P&T	Post and timbers
Reg	Regular
Res	Resawed or resawn
Rfg	Roofing
Rgh	Rough
TUL	Random lengths
R/W	Random widths
R/W&L	Random widths and lengths
Sdg	Siding
Sel	Select
S&E	Side and Edge (surfaced on)
SE Sdg	Square edge siding
SE & S	Square edge and sound
S/L or S/LAP	Shiplap
SL&C	Shipper's load and count

SM or Std M	Standard matched
Specs	Specifications
Std	Standard
Stpg	Stepping
Str or Struc	Structural
S1E	Surfaced one edge
S1S	Surfaced one side
S1S1E	Surfaced one side and one edge
S1S2E	Surfaced one side and two edges
S2E	Surfaced two edges
S2E	Surfaced two sides
S2S	Surfaced two sides and one edge
S2S1E	Surfaced two sides and center matched
S2S&CM	Surfaced two sides and standard matched
S2S&SM	Surfaced four sides
S4S	Surfaced four sides and caulking seam
S4S&CS	Tongued and grooved
T&G	Vertical grain
VG	Wider
Wdr	Weight
Wt	Pattern

7.1.0 Softwood Lumber—Commercial Names of Principal Softwood Species

The commercial names listed below are intended to provide a correlation between commercial names for lumber and the botanical names of the species from which the lumber is to be manufactured. In some instances, more than one species is associated with a single commercial name. For stress-graded lumber, the species to be associated with a commercial name will be determined in accordance with 6.3.2.1. These commercial names are to be used in grading rule descriptions and in specifications [see 2.15]. The provisions of this Standard apply to lumber manufactured from hardwood species or lumber manufactured from foreign species when the species is included in rules certified by the Board of Review. The information contained herein is a partial list of commercial names of the principal softwood species and species groups. Additional species and species groups are provided in ASTM Standard D 1165-03 and the rules certified by the Board of Review.

Commercial Species or Species Group Names[1]	Official Common Tree Names[2]	Botanical Names
Cedar:		
Alaska Cedar	Alaska-cedar	*Chamaecyparis nootkatensis*
Incense Cedar	incense-cedar	*Libocedrus decurrens*
Port Orford Cedar	Port-Orford-cedar	*Chamaecyparis lawsoniana*
Eastern Red Cedar	eastern redcedar	*Juniperus virginiana*
	southern redcedar	*J. silicicola*
Western Red Cedar	western redcedar	*Thuja plicata*
Northern White Cedar	northern white-cedar	*T. occidentalis*
Southern White Cedar	Atlantic white-cedar	*Chamaecyparis thyoides*
Cypress[3]		
Baldcypress	Baldcypress	*Taxodium distichum*
Pond cypress	Pond cypress	*T. distichum var. nutans*
Fir:		
Balsam Fir[4]	balsam fir	*Abies balsamea*
Fraser Fir	Fraser fir	*A. fraseri*
Douglas Fir[5]	Douglas-fir	*Pseudotsuga menziesii*

Commercial Species or Species Group Names[1]	Official Common Tree Names[2]	Botanical Names
Bigcone Douglas fir	Bigcone Douglas fir	P. macrocarpa
Noble Fir	noble fir	Abies procera
Alpine Fir	subalpine fir (alpine fir)	A. lasiocarpa
California Red Fir	California red fir	A. magnifica
Grand Fir	grand fir	A. grandis
Pacific Grand Fir	Pacific silver fir	A. amabilis
White Fir	white fir	A. concolor
Hemlock:		
Carolina Hemlock	Carolina hemlock	Tsuga caroliniana
Eastern Hemlock	eastern hemlock	T. Canadensis
Mountain Hemlock	mountain hemlock	T. mertensiana
Western Hemlock	western hemlock	T. heterophylla
Juniper:		
Western Juniper	alligator juniper	Juniperus deppeana
	Rocky Mountain juniper	J. scopulorum
	Utah juniper	J. osteosperma
	western juniper	J. occidentalis
Larch:		
Western Larch	western larch	Larix occidentalis
Pine:		
Bishop Pine	Bishop pine	Pinus muricata
Digger Pine	Digger pine	P. sabiniana
Knobcone Pine	knobcone pine	P. attenuata
Coulter Pine	Coulter pine	P. coulteri
Jeffrey Pine	Jeffrey pine	P. jeffreyi
Jack Pine	jack pine	P. banksiana
Limber Pine	limber pine	P. flexilis
Lodgepole Pine	lodgepole pine	P. contorta
Norway Pine	red pine	P. resinosa
Pitch Pine	pitch pine	P. rigida
Ponderosa Pine	ponderosa pine	P. ponderosa
Radiata/Monterey Pine	Monterey pine	P. radiate
Sugar Pine	sugar pine	P. lambertiana
Whitebark Pine	whitebark pine	P. albicaulis
Idaho White Pine	western white pine	P. monticola
Northern White Pine	eastern white pine	P. strobes
Longleaf Pine[6]	longleaf pine	P. palustris
	slash pine	P. elliottii
Southern Pine (Major)	loblolly pine	P. taeda
	longleaf pine	P. palustris
	shortleaf pine	P. echinata
	slash pine	P. elliottii
Southern Pine (Minor)	pond pine	P. serotina
	Virginia pine	P. virginiana
	sand pine	P. clausa
	spruce pine	P. glabra
Redwood:		
Redwood	Redwood	Sequoia sempervirens
Spruce:		
Black spruce	black spruce	Picea mariana

(Continued)

Commercial Species or Species Group Names[1]	Official Common Tree Names[2]	Botanical Names
Red spruce	red spruce	P. rubens
White spruce	white spruce	P. glauca
Blue Spruce	blue spruce	P. pungens
Engelmann Spruce	Engelmann spruce	P. engelmannii
Sitka Spruce	Sitka spruce	P. sitchensis
Norway Spruce	Norway spruce	P. abies
Tamarack:		
Tamarack	Tamarack	Larix laricina
Yew:		
Pacific Yew	Pacific yew	Taxus brevifolia

[1] The commercial names for species represent those commonly accepted. Some grading rules certified by the Board provide for the inclusion of additional species under the established names.
[2] The official common tree names conform to the Checklist of United States Trees (Native and Naturalized), Agriculture Handbook No. 541 (1979), and are sometimes used as names for lumber. In addition to the official common names for a species, the Handbook lists other names by which the species and the lumber produced from it are sometimes designated.
[3] Cypress includes types designated as Red Cypress, White Cypress, and Yellow Cypress. Red Cypress is frequently classified and sold separately from the other types.
[4] Balsam fir lumber is sometimes designated either as Eastern fir or as Balsam.
[5] Douglas fir from Arizona, Colorado, Nevada, New Mexico, and Utah is recognized as Douglas fir-South.
[6] The commercial requirements for Longleaf Pine lumber are that not only must it be produced from trees of the botanical species of Pinus elliottii and Pinus palustris, but each piece in addition must average either on one end or the other not less than six annual rings per inch and not less than one-third summerwood. Longleaf Pine lumber is sometimes designated as Pitch Pine in the export trade.

7.1.1 Standard Sizes for Framing Lumber

Nominal and Dressed (based on *Western Lumber Grading Rules*)

Product	Description	Nominal Size Thickness (inch)	Nominal Size Width (inch)	Dressed Dimensions – Surfaced Dry (inch)	Dressed Dimensions – Surfaced Dry (mm)	Dressed Dimensions – Surfaced Unseasoned (inch)	Dressed Dimensions – Surfaced Unseasoned (mm)	Length (feet)
DIMENSION	S4S	2	2	1 ½	38	1 9/16	40	6' (183 cm) and longer, generally shipped in multiples of 2' (61 cm)
		3	3	2 ½	61	2 9/16	65	
		4	4	3 ½	89	3 9/16	90	
			5	4 ½	114	4 5/8	117	
			6	5 ½	140	5 5/8	143	
			8	7 ¼	184	7 ½	191	
			10	9 ¼	235	9 ½	241	
			12	11 ¼	289	11 ½	292	
			over 12	¾ off nominal	19 off nominal	½ off nominal	13 off nominal	
				Thickness unseasoned		**Width unseasoned**		6' (183 cm) and longer, generally shipped in multiples of 2' (61 cm)
TIMBERS	Rough or S4S (shipped unseasoned)	5 and larger	5 and larger	½" (13mm) off nominal (S4S). See 3.20 of WWPA Grading Rules for Rough.				

Nominal and Dressed (based on *Western Lumber Grading Rules*)—Cont'd

Product	Description	Thickness	Width	Thickness Dry		Width Dry		Length feet
				inch	mm	inch	mm	
DECKING	2" (Single T&G)	2	5	1 ½	38	4	102	6' (183 cm) and longer, generally shipped in multiples of 2' (61 cm)
			6			5	127	
			8			6 ¾	172	
			10			8 ¾	222	
			12			10 ¾	273	
	3" and 4" (Double T&G)	3	6	2 ½	64	5 ¼	133	
		4		3 ½	89			

Abbreviations: T&F—Tongued and grooved   Rough—Unsurfaced   S4S—Surfaced four sides
Note on Metrics: Metric equivalents are provided for surfaces (actual) sizes.

7.1.2 Nominal and Minimum Dressed Sizes of Finish Lumber

The thicknesses apply to all widths, and all widths apply to all thicknesses except as modified. Sizes are given in millimeters and inches. Metric units are based on dressed size.

Nominal and minimum-dressed dry sizes of finish, flooring, ceiling, partition, and stepping at 19% maximum-moisture content

Item	Thicknesses			Widths		
	Nominal Inch	Minimum Dressed		Nominal (inch)	Minimum Dressed	
		mm	inch		mm	inch
Finish	3/8	8	5/16	2	38	1-1/2
	1/2	11	7/16	3	64	2-1/2
	5/8	14	9/16	4	89	3-1/2
	3/4	16	5/8	5	114	4-1/2
	1	19	3/4	6	140	5-1/2
	1-1/4	25	1	7	165	6-1/2
	1-1/2	32	1-1/4	8	184	7-1/4
	1-3/4	35	1-3/8	9	210	8-1/4
	2	38	1-1/2	10	235	9-1/4
	2-1/2	51	2	11	260	10-1/4
	3	64	2-1/2	12	286	11-1/4
	3-1/2	76	3	14	337	13-1/4
	4	89	3-1/2	16	387	15-1/4
Flooring[a]	3/8	8	5/16	2	29	1-1/8
	1/2	11	7/16	3	54	2-1/8
	5/8	14	9/16	4	79	3-1/8
	1	19	3/4	5	105	4-1/8
	1-1/4	25	1	6	130	5-1/8
	1-1/2	32	1-1/4			
Ceiling[a]	3/8	8	5/16	3	54	2-1/8
	1/2	11	7/16	4	79	3-1/8
	5/8	14	9/16	5	105	4-1/8

(Continued)

Nominal and minimum-dressed dry sizes of finish, flooring, ceiling, partition, and stepping at 19% maximum-moisture content—Cont'd

Item	Thicknesses			Nominal (inch)	Widths	
	Nominal Inch	Minimum Dressed			Minimum Dressed	
		mm	inch		mm	inch
	3/4	17	11/16	6	130	5-1/8
Partition[a]				3	54	2-1/8
				4	79	3-1/8
	1	18	23/32	5	105	4-1/8
				6	130	5-1/8
Stepping	1	19	3/4	8	184	7-1/4
	1-1/4	25	1	10	235	9-1/4
	1-1/2	32	1/4	12	286	11-1/4
	2	38	1-1/2			

[a] In tongued-and-grooved flooring and in tongued-and-grooved and shiplapped ceiling of 8 mm (5/16 inch), 11 mm (7/16 inch), and 14 mm (9/16 inch) dressed thicknesses, the tongue or lap shall be 5 mm (3/16 inch) wide, with the overall widths 5 mm (3/16 inch) wider than the face widths shown in the above table. In all other worked lumber shown in this table of dressed thicknesses of 16 mm (5/8 inch) to 32 mm (1-1/4 inches), the tongue shall be 6 mm (1/4 inch) wide or wider in tongued-and-grooved lumber, and the lap shall be 10 mm (3/8 inch) wide or wider in shiplapped lumber, and the overall widths shall be not less than the dressed face widths shown in the above table plus the width of the tongue or lap.

The thicknesses apply to all widths, and all widths apply to all thicknesses. Sizes are given in millimeters and inches. Metric units are based on dressed size. See B1, Appendix B, for rounding rule for metric units.

Nominal and minimum-dressed sizes of worked lumber

Item	Nominal Inch	Thicknesses				Nominal Inch	Widths			
		Minimum Dressed					Minimum Dressed			
		Dry[a]		Green[a]			Dry[a]		Green[a]	
		mm	inch	mm	inch		mm	inch	mm	inch
Shiplap, 10 mm (3/8 inch) lap	1	19[c]	3/4[c]	20	25/32	4	79	3-1/8	81	3-3/16
						6	130	5-1/8	133	5-1/4
						8	175	6-7/8	181	7-1/8
						10	225	8-7/8	232	9-1/8
						12	276	10-7/8	283	11-1/8
						14	327	12-7/8	333	13-1/8
						16	378	14-7/8	384	15-1/8
Shiplap, 13 mm (1/2 inch) lap[b]	1	19[c]	3/4[c]	20	25/32	4	76	3	78	3-1/16
	2	38	1-1/2	40	1-9/16	6	127	5	130	5-1/8
	2-1/2	51	2	52	2-1/16	8	171	6-3/4	178	7
	3	64	2-1/2	65	2-9/16	10	222	8-3/4	229	9
	3-1/2	76	3	78	3-1/16	12	273	10-3/4	279	11
	4	89	3-1/2	90	3-9/16	14	324	12-3/4	330	13
	4-1/2	102	4	103	4-1/16	16	375	14-3/4	381	15

Nominal and minimum-dressed sizes of worked lumber—Cont'd

Item	Nominal Inch	Thicknesses Minimum Dressed Dry[a] mm	Dry[a] inch	Green[a] mm	Green[a] inch	Nominal Inch	Widths Minimum Dressed Dry[a] mm	Dry[a] inch	Green[a] mm	Green[a] inch
Centermatch (Tongue & Groove), 6 mm (1/4 inch) tongue	1	19[c]	3/4[c]	20	25/32	4	79	3-1/8	81	3-3/16
	1-1/4	25	1	26	1-1/32	5	105	4-1/8	108	4-1/4
	1-1/2	32	1-1/4	33	1-9/32	6	130	5-1/8	133	5-1/4
						8	175	6-7/8	181	7-1/8
						10	225	8-7/8	232	9-1/8
						12	276	10-7/8	283	11-1/8
Centermatch (Tongue & Groove), 10 mm (3/8 inch) tongue[b]	2	38	1-1/2	40	1-9/16	4	76	3	78	3-1/16
	2-1/2	51	2	52	2-1/16	6	127	5	130	5-1/8
	3	64	2-1/2	65	2-9/16	8	171	6-3/4	178	7
	3-1/2	76	3	78	3-1/16	10	222	8-3/4	229	9
	4	89	3-1/2	90	3-9/16	12	273	10-3/4	279	11
	4-1/2	102	4	103	4-1/16					
Grooved-for-Splines	2-1/2	51	2	52	2-1/16	4	89	3-1/2	90	3-9/16
	3	64	2-1/2	65	2-9/16	6	140	5-1/2	143	5-5/8
	3-1/2	76	3	78	3-1/16	8	184	7-1/4	190	7-1/2
	4	89	3-1/2	90	3-9/16	10	235	9-1/4	241	9-1/2
	4-1/2	102	4	103	4-1/16	12	286	11-1/4	292	11-1/2

7.1.3 Appearance Lumber—Grades, Nominal, and Dressed Sizes

	Product	Grades[1]	Equivalent Grades in Idaho White Pine	WWPA Grading Rules Section Number
Highest Quality Appearance Grades	Selects (all species)	B & BTR SELECT	SUPREME	10.11
		C SELECT	CHOICE	10.12
		D SELECT	QUALITY	10.13
	Finish (usually available only in Doug Fir and Hem-Fir)	SUPERIOR		10.51
		PRIME		10.52
		E		10.53
	Special Western Red Cedar Pattern[2] Grades	CLEAR HEART		20.11
		A GRADE		20.12
		B GRADE		20.13

(Continued)

	Product	Grades[1]	Equivalent Grades in Idaho White Pine	WWPA Grading Rules Section Number
General Purpose Grades	Common Boards (WWPA Rules) (primarily in pines, spruces, and cedars)	1 COMMON 2 COMMON 3 COMMON 4 COMMON 5 COMMON	COLONIAL STERLING STANDARD UTILITY INDUSTRIAL	30.11 30.12 30.13 30.14 30.15
	Alternate Boards (WCLIB Rules) (primarily in Dough Fir and Hem-Fir)	SELECT MERCHANTABLE CONSTRUCTION STANDARD UTILITY ECONOMY		**WCLIB[3,4]** 118-a 118-b 118-c 118-d 118-e
	Special Western Red Cedar Pattern[2] Grades	SELECT KNOTTY QUALITY KNOTTY		**WCLIB[3]** 111-e 111-f

[1] Refer to WWPA's Vol. 2, Western Wood Species book for full-color photography and to WWPA's Natural Wood Siding for complete information on siding grades, specification, and installation.
[2] "PATTERN" includes finish, paneling, ceiling and siding grades.
[3] West Coast Lumber Inspection Bureau's West Coast Lumber Standard Grading Rules.
[4] Also found in WWPA's Western Lumber Grading Rules.

Standard Signs—Appearance Lumber
Nominal & Dressed (Based on Western Lumber Grading Rules)

Product	Description	Nominal Size		Dry Dressed Dimensions				Lengths (feet)
		Thickness	Width	Thickness		Width		
		inches	inches	inches	mm	inches	mm	
SELECTS AND COMMONS	S1S, S2S, S4S, S1S1E, S1S2E	4/4 5/4 6/4 7/4 8/4 9/4 10/4 11/4 12/4 16/4	2 3 4 5 6 7 8 & wider	¾ 1 5/32 1 13/32 1 19/32 1 13/16 2 3/32 2 ⅜ 2 9/16 2 ¾ 3 ¾	19 29 36 40 46 53 60 65 70 95	1 ½ 2 ½ 3 ½ 4 ½ 5 ½ 6 ½ ¾ off nominal	38 64 89 114 140 165 19 off nominal	6' (183 cm) and longer in multiples of 1' (31 cm), except Douglas Fir and Larch Selects shall be 4' (122 cm) and longer with 3% of 4' (122 cm) and 5' (152 cm) permitted.

Section | 7   Lumber—Calculations to Select Framing and Trim Materials   365

Standard Signs—Appearance Lumber—Cont'd

Product	Description	Nominal Size		Dry Dressed Dimensions				Lengths (feet)
		Thickness	Width	Thickness		Width		
		inches	inches	inches	mm	inches	mm	
FINISH AND ALTERNATE BOARD GRADES	S1S, S2S, S4S, S1S1E, S1S2E	⅜ ½ ⅝ ¾ [1] 1 [1] 1¼ [1] 1½ [1] 1¾ 2 2½ 3 3½ 4	2 3 4 5 6 7 8 & wider	5⁄16 7⁄16 9⁄16 ⅝ ¾ 1 1¼ 1⅜ 1½ 2 2½ 3 3½	2 11 14 16 19 25 32 35 38 51 64 76 89	1½ 2½ 3½ 4½ 5½ 6½ ¾ off nominal	38 64 89 114 140 165 16 off nominal	3′ (91 cm) and longer. In SUPERIOR grade, 3% of 3′ (91 cm) and 4′ (122 cm) and 7% of 5′ (152 cm) and 6′ (183 cm) are permitted. In PRIME grade 20% of 3′. (91 cm) to 6′ (183 cm) is permitted.

[1] These sizes apply only to WCLIB Alternate Board grades.

Abbreviations: S1S—Surfaced one side   S1S1E—Surfaced one side, one edge
 S2S—Surfaced two side   S1S2E—Surfaced one side, two edges
 S4S—Surfaced four sides

Note on Metrics: Metric equivalents are provided for surfaced (actual) sizes.

7.1.4 Section Properties of Joists, Beams, and Timbers

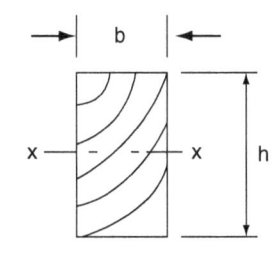

Nominal Size in Inches $b \times h$	Surfaced Size for Design in Inches $b \times h$	Area (A) $A = bh$ (in²)	Section Modulus (S) $S = \dfrac{bh^2}{6}$ (in³)	Moment of Inertia (I) $I = \dfrac{bh^3}{12}$ (in⁴)	Board Feet per Lineal Foot of Piece
2 × 2	1.5 × 1.5	2.25	0.562	0.422	0.33
2 × 3	1.5 × 2.5	3.75	1.56	1.95	0.50
2 × 4	1.5 × 3.5	5.25	3.06	5.36	0.67
2 × 6	1.5 × 5.5	8.25	7.56	20.80	1.00

(Continued)

Nominal Size in Inches b × h	Surfaced Size for Design in Inches b × h	Area (A) A = bh (in²)	Section Modulus (S) $S = \dfrac{bh^2}{6}$ (in³)	Moment of Inertia (I) $I = \dfrac{bh^3}{12}$ (in⁴)	Board Feet per Lineal Foot of Piece
2 × 8	1.5 × 7.25	10.88	13.14	47.63	1.33
2 × 10	1.5 × 9.25	13.88	21.39	98.93	1.67
2 × 12	1.5 × 11.25	16.88	31.64	177.98	2.00
2 × 14	1.5 × 13.25	19.88	43.89	290.78	2.33
3 × 3	2.5 × 2.5	6.25	2.60	3.26	0.75
3 × 4	2.5 × 3.5	8.75	5.10	8.93	1.00
3 × 6	2.5 × 5.5	13.75	12.60	34.66	1.50
3 × 8	2.5 × 7.25	18.12	21.90	79.39	2.00
3 × 10	2.5 × 9.25	23.12	35.65	164.89	2.50
3 × 12	2.5 × 11.25	28.12	52.73	296.63	3.00
3 × 14	2.5 × 13.25	33.12	73.15	484.63	3.50
3 × 16	2.5 × 15.25	38.12	96.90	738.87	4.00
4 × 4	3.5 × 3.5	12.25	7.15	12.51	1.33
4 × 6	3.5 × 5.5	19.25	17.65	48.53	2.00
4 × 8	3.5 × 7.25	25.38	30.66	111.15	2.67
4 × 10	3.5 × 9.25	32.38	49.91	230.84	3.33
4 × 12	3.5 × 11.25	39.38	73.83	415.28	4.00
4 × 14	3.5 × 13.25	46.38	102.41	678.48	4.67
4 × 16	3.5 × 15.25	53.38	135.66	1034.42	5.33
6 × 6	5.5 × 5.5	30.25	27.73	76.26	3.00
6 × 8	5.5 × 7.5	41.25	51.56	193.36	4.00
6 × 10	5.5 × 9.5	52.25	82.73	392.96	5.00
6 × 12	5.5 × 11.5	63.25	121.23	697.07	6.00
6 × 14	5.5 × 13.5	74.25	167.06	1127.67	7.00
6 × 16	5.5 × 15.5	85.25	220.23	1706.78	8.00
6 × 18	5.5 × 17.5	96.25	280.73	2456.38	9.00
6 × 20	5.5 × 19.5	107.25	348.56	3398.48	10.00
8 × 8	7.5 × 7.5	56.25	70.31	263.67	5.33
8 × 10	7.5 × 9.5	71.25	112.81	535.86	6.67
8 × 12	7.5 × 11.5	86.25	165.31	950.55	8.00
8 × 14	7.5 × 13.5	101.25	227.81	1537.73	9.33
8 × 16	7.5 × 15.5	116.25	300.31	2327.42	10.67
8 × 18	7.5 × 17.5	131.25	382.81	3349.61	12.00
8 × 20	7.5 × 19.5	146.25	475.31	4634.30	13.33
8 × 22	7.5 × 21.5	161.25	577.81	6211.48	14.67
8 × 24	7.5 × 23.5	176.25	690.31	8111.17	16.00
10 × 10	9.5 × 9.5	90.25	142.90	678.76	8.33
10 × 12	9.5 × 11.5	109.25	209.40	1204.03	10.00
10 × 14	9.5 × 13.5	128.25	288.56	1947.80	11.67
10 × 16	9.5 × 15.5	147.25	380.40	2948.07	13.33
10 × 18	9.5 × 17.5	166.25	484.90	4242.84	15.00
10 × 20	9.5 × 19.5	185.25	602.06	5870.11	16.67
10 × 22	9.5 × 21.5	204.25	731.90	7867.88	18.33
12 × 12	11.5 × 11.5	132.25	253.48	1457.51	12.00
12 × 14	11.5 × 13.5	155.25	349.31	2357.86	14.00
12 × 16	11.5 × 15.5	178.25	460.48	3568.71	16.00
12 × 18	11.5 × 17.5	201.25	586.98	5136.07	18.00

Section | 7 Lumber—Calculations to Select Framing and Trim Materials   367

Nominal Size in Inches b × h	Surfaced Size for Design in Inches b × h	Area (A) $A = bh$ (in²)	Section Modulus (S) $S = \dfrac{bh^2}{6}$ (in³)	Moment of Inertia (I) $I = \dfrac{bh^3}{12}$ (in⁴)	Board Feet per Lineal Foot of Piece
12 × 20	11.5 × 19.5	224.25	728.81	7105.92	20.00
12 × 22	11.5 × 21.5	247.25	885.98	9524.28	22.00
12 × 24	11.5 × 23.5	270.25	1058.48	12437.13	24.00

Used with Courtesy of Western Wood Products Association.

FRAMING LUMBER

Beams and Stringers Design Values
5″ and thicker, width more than 2″ greater than thickness   Grades described in Section 53.00 and 70.00 of *Western Lumber Grading Rules*

Species of Group	Grade	Extreme Fiber Stress in Bending Single Member F_b	Tension Parallel to Grain F_t	Horizontal Shear³ F_v	Compression Perpendicular $F_{c\perp}$	Compression Parallel to Grain F_c	Modulus of Elasticity E
Douglas Fir-Larch	Dense Selected Structural	1900	1100	170	730	1300	1,700,000
	Dense No. 1	1550	775	170	730	1100	1,700,000
	Dense No. 2	1000	500	170	730	700	1,400,000
	Select Structural	1600	950	170	625	1100	1,600,000
	No. 1	1350	675	170	625	925	1,600,000
	No. 2	875	425	170	625	600	1,300,000
Douglas Fir-South	Select Structural	1550	900	165	520	1000	1,200,000
	No. 1	1300	625	165	520	850	1,200,000
	No. 2	825	425	165	520	550	1,000,000
Hem-Fir	Select Structural	1300	750	140	405	925	1,300,000
	No. 1	1050	525	140	405	750	1,300,000
	No. 2	675	350	140	405	500	1,100,000
Mountain Hemlock	Select Structural	1350	775	170	570	875	1,100,000
	No. 1	1100	550	170	570	725	1,100,000
	No. 2	725	375	170	570	475	900,000
Sitka Spruce	Select Structural	1200	675	140	435	825	1,300,000
	No. 1	1000	500	140	435	675	1,300,000
	No. 2	650	325	140	435	450	1,100,000
Spruce-Pine-Fir (South)	Select Structural	1050	625	125	335	675	1,200,000
	No. 1	900	450	125	335	550	1,200,000
	No. 2	575	300	125	335	375	1,000,000
Western Cedars	Select Structural	1400	825	170	410	1000	1,400,000
	No. 1	1150	575	170	410	850	1,400,000
	No. 2	750	375	170	410	550	1,100,000

(Continued)

Beams and Stringers Design Values—Cont'd

Species of Group	Grade	Extreme Fiber Stress in Bending Single Member F_b	Tension Parallel to Grain F_t	Horizontal Shear[3] F_V	Compression Perpendicular $F_{c\perp}$	Compression Parallel to Grain F_c	Modulus of Elasticity E
Western Woods (and White Woods)	Select Structural	1050	625	125	345	750	1,100,000
	No. 1	900	450	125	345	625	1,100,000
	No. 2	575	300	125	345	425	900,000

Posts and Timbers Design Values[1]

5" × 5" and larger, width not more than 2" greater than thickness[2]

Grades described in Section 53.00 and 80.00 of *Western Lumber Grading Rules*

Species of Group	Grade	Extreme Fiber Stress in Bending Single Member F_b	Tension Parallel to Grain F_t	Horizontal Shear[3] F_V	Compression Perpendicular $F_{c\perp}$	Compression Parallel to Grain F_c	Modulus of Elasticity E
Douglas Fir-Larch	Dense Selected Structural	1750	1150	170	730	1350	1,700,000
	Dense No. 1	1400	950	170	730	1200	1,700,000
	Dense No. 2	850	550	170	730	825	1,400,000
	Select Structural	1500	1000	170	625	1150	1,600,000
	No. 1	1200	825	170	625	1000	1,600,000
	No. 2	750	475	170	625	700	1,300,000
Douglas Fir-South	Select Structural	1450	950	165	520	1050	1,200,000
	No. 1	1150	775	165	520	925	1,200,000
	No. 2	675	450	165	520	650	1,000,000
Hem-Fir	Select Structural	1200	800	140	405	975	1,300,000
	No. 1	975	650	140	405	850	1,300,000
	No. 2	575	375	140	405	575	1,100,000
Mountain Hemlock	Select Structural	1250	825	170	570	925	1,100,000
	No. 1	1000	675	170	570	800	1,100,000
	No. 2	625	400	170	570	550	900,000
Sitka Spruce	Select Structural	1150	750	140	435	875	1,300,000
	No. 1	925	600	140	435	750	1,300,000
	No. 2	550	350	140	435	525	1,100,000
Spruce-Pine-Fir (South)	Select Structural	1000	675	125	335	700	1,200,000
	No. 1	875	550	125	335	625	1,200,000
	No. 2	475	325	125	335	425	1,000,000

Posts and Timbers Design Values[1] —Cont'd

Species of Group	Grade	Extreme Fiber Stress in Bending Single Member F_b	Tension Parallel to Grain F_t	Horizontal Shear[3] F_v	Compression Perpendicular $F_{c\perp}$	Compression Parallel to Grain F_c	Modulus of Elasticity E
Western Cedars	Select Structural	1100	725	140	425	925	1,000,000
	No. 1	875	600	140	425	800	1,000,000
	No. 2	550	350	140	425	550	800,000
Western Hemlock	Select Structural	1300	875	170	410	1100	1,400,000
	No. 1	1050	700	170	410	950	1,400,000
	No. 2	650	425	170	410	650	1,100,000
Western Woods (and White Woods)	Select Structural	1000	675	125	345	800	1,100,000
	No. 1	800	525	125	345	700	1,100,000
	No. 2	475	325	125	345	475	900,000

[1] Design Values in pounds per square inch. See Sections 100.00 through 180.00 in the Western Lumber Grading Rules for additional information on these values.
[2] When the depth of a sawn lumber member exceeds 12 inches, the design value for extreme fiber stress in bending (F_b) shall be multiplied by the size factor in Table J.
[3] All horizontal shear values are assigned in accordance with ASTM standards, which include a reduction to compensate for any degree of shake, check, or split that might develop in a piece.

7.1.4.1 Stress-Graded Lumber —Nominal and Dry Dressed Dimensions

Standard Sizes[1]
Stress-Rated Boards

	Nominal	Surfaced Unseasoned		Surfaced Dry	
		inches	mm	inches	mm
Thickness	1"	25/32	20	3/4	19
	1 1/4"	1 1/32	26	1	25
	1 1/2"	1 9/32	33	1 1/4	32
Widths	2"	1 9/16	40	1 1/2	38
	3"	2 9/16	65	2 1/2	64
	4"	3 9/16	90	3 1/2	89
	5"	4 5/8	117	4 1/2	114
	6"	5 5/8	143	5 1/2	140
	8" and wider	1/2 off nominal	13 off nominal	3/4 off nominal	19 off nominal

[1] Standard lengths are 6' (183 cm) and longer in multiples of 1' (31 cm).
Note on Metrics: Metric equivalents are provided for surfaced (actual) sizes.

7.1.5 Average Moisture Content of Green Wood, by Species

Species	Moisture Content[a] (%) Heartwood	Sapwood	Species	Moisture Content[a] (%) Heartwood	Sapwood
Hardwoods			**Softwoods**		
Alder, red	—	97	Baldcypress	121	171
Apple	81	74	Cedar, eastern red	33	—
Ash, black	95	—	Cedar, incense	40	213
Ash, green	—	58	Cedar, Port-Orford	50	96
Ash, white	45	44	Cedar, western red	58	249
Aspen	95	113	Cedar, yellow	32	166
Basswood, American	81	133	Douglas-fir, coast type	37	115
Beech, American	55	72	Fir, balsam	88	173
Birch, paper	89	72	Fir, grand	91	136
Birch, sweet	75	70	Fir, noble	34	115
Birch, yellow	74	72	Fir, Pacific silver	55	164
Cherry, black	58	—	Fir, white	98	160
Chestnut, American	120	—	Hemlock, eastern	97	119
Cottonwood	162	146	Hemlock, western	85	170
Elm, American	95	92	Larch, western	54	119
Elm, cedar	66	61	Pine, loblolly	33	110
Elm, rock	44	57	Pine, lodgepole	41	120
Hackberry	61	65	Pine, longleaf	31	106
Hickory, bitternut	80	54	Pine, ponderosa	40	148
Hickory, mockernut	70	52	Pine, red	32	134
Hickory, pignut	71	49	Pine, shortleaf	32	122
Hickory, red	69	52	Pine, sugar	93	219
Hickory, sand	68	50	Pine, western white	62	148
Hickory, water	97	62	Redwood, old growth	86	210
Magnolia	80	104	Spruce, black	52	113
Maple, silver	58	97	Spruce, Engelmann	51	173
Maple, sugar	65	72	Spruce, Sitka	41	142
Oak, California black	76	75	Tamarack	49	—
Oak, northern red	80	69			
Oak, southern red	83	75			
Oak, water	81	81			
Oak, white	64	78			
Oak, willow	82	74			
Sweetgum	79	137			
Sycamore, American	114	130			
Tupelo, black	87	115			
Tupelo, swamp	101	108			
Tupelo, water	150	116			
Walnut, black	90	73			
Yellow-poplar	83	106			

[a] Based on weight when oven dry.

Section | 7 Lumber—Calculations to Select Framing and Trim Materials   371

Density of wood as a function of specific gravity and moisture content (metric)

Density (kg/m³) when the specific gravity G_m is

Moisture Content of Wood (%)	0.30	0.32	0.34	0.36	0.38	0.40	0.42	0.44	0.46	0.48	0.50	0.52	0.54	0.56	0.58	0.60	0.62	0.64	0.66	0.68	0.70
0	300	320	340	360	380	400	420	440	460	480	500	520	540	560	580	600	620	640	660	680	700
4	312	333	354	374	395	416	437	458	478	499	520	541	562	582	603	624	645	666	686	707	728
8	324	36	367	389	410	432	454	475	497	518	540	562	583	605	626	648	670	691	713	734	756
12	336	358	381	403	426	448	470	493	515	538	560	582	605	627	650	672	694	717	739	762	784
16	348	371	394	418	441	464	487	510	534	557	580	603	626	650	673	696	719	742	766	789	812
20	360	384	408	432	456	480	504	528	552	576	600	624	648	672	696	720	744	768	792	816	840
24	372	397	422	446	471	496	521	546	570	595	620	645	670	694	719	744	769	794	818	843	868
28	384	410	435	461	486	512	538	563	589	614	640	666	691	717	742	768	794	819	845	870	896
32	396	422	449	475	502	528	554	581	607	634	660	686	713	739	766	792	818	845	871	898	924
36	408	435	462	490	517	544	571	598	626	653	680	707	734	762	789	816	843	870	898	925	952
40	420	448	476	504	532	560	88	616	644	672	700	728	756	784	812	840	868	896	924	952	890
44	432	461	490	518	547	576	605	634	662	691	720	749	778	806	835	864	893	922	950	979	1,008
48	444	474	503	533	562	592	622	651	681	710	740	770	899	829	858	888	918	947	977	1,006	1,036
52	456	486	517	547	578	608	638	669	699	730	760	790	821	851	882	912	942	973	1,003	1,034	1,064
56	468	499	530	562	593	624	655	686	718	749	790	811	842	874	905	936	967	998	1,030	1,061	1,092
60	480	512	544	576	608	640	672	704	736	768	800	832	864	896	928	960	992	1,024	1,056	1,088	1,120
64	492	525	558	590	623	656	689	722	754	787	820	853	886	918	951	984	1,017	1,050	1,082	1,115	1,148
68	504	538	571	605	638	672	706	739	773	806	840	874	907	941	974	1,008	1,042	1,075	1,109	1,142	1,176
72	516	550	585	619	654	688	722	757	791	826	860	894	929	963	998	1,032	1,066	1,101	1,135	1,170	1,204
76	528	563	598	634	669	704	739	774	810	845	880	915	950	986	1,021	1,056	1,091	1,126	1,162	1,197	
80	540	576	612	648	684	720	756	792	828	864	900	936	972	1,008	1,044	1,080	1,116	1,152	1,188		
84	552	589	626	662	699	736	773	810	846	883	920	957	994	1,030	1,067	1,104	1,141	1,178			
88	564	602	639	677	714	752	790	827	865	902	940	978	1,015	1,053	1,090	1,128	1,166				
92	576	614	653	691	730	768	806	845	883	922	960	998	1,037	1,075	1,114	1,152	1,190				
96	588	627	666	706	745	784	823	862	902	941	980	1,019	1,058	1,098	1,137	1,176					
100	600	640	680	720	760	800	840	880	920	960	1,000	1,040	1,080	1,120	1,160	1,200					
110	630	672	714	756	798	840	832	924	922	1,008	1,050	1,092	1,134	1,176	1,218						
120	660	704	748	792	836	880	924	968	1,012	1,056	1,100	1,144	1,188	1,232							
130	690	736	782	828	874	920	966	1,012	1,058	1,104	1,150	1,196	1,242	1,288							
140	720	868	816	864	912	960	1,008	1,056	1,104	1,152	1,200	1,248	1,296								
150	750	900	850	900	950	1,000	1,050	1,100	1,150	1,200	1,250	1,300	1,350								

Density of wood as a function of specific gravity and moisture content (inch-pound)

Moisture Content of Wood (%)	Density (lb/ft³) when the specific gravity G_m is																				
	0.30	0.32	0.34	0.36	0.38	0.40	0.42	0.44	0.46	0.48	0.50	0.52	0.54	0.56	0.58	0.60	0.62	0.64	0.66	0.68	0.70
0	18.7	20.0	21.2	22.5	23.7	25.0	26.2	27.5	28.7	30.0	31.2	32.4	33.7	34.9	36.2	37.4	38.7	39.9	41.2	42.4	43.7
4	19.5	20.8	22.1	2.4	24.7	26.0	27.2	29.6	29.8	31.2	32.4	33.7	35.0	36.6	37.6	38.9	40.2	41.5	42.8	44.1	45.4
8	20.2	21.6	22.9	24.3	25.6	27.0	28.3	29.6	31.0	32.3	33.7	35.0	36.4	37.7	39.1	40.4	41.8	43.1	44.5	45.8	47.2
12	21.0	22.4	23.8	25.2	36.6	38.0	39.4	30.8	32.2	33.5	34.9	36.3	37.7	39.1	40.5	41.9	43.3	44.7	46.1	47.5	48.9
16	21.7	23.2	24.6	26.0	27.5	29.0	30.4	31.8	33.3	34.7	36.2	37.6	39.1	40.5	42.0	43.4	44.9	46.3	47.8	49.2	50.7
20	22.5	24.0	25.5	27.0	28.4	30.0	31.4	32.9	34.4	35.9	37.4	38.9	40.4	41.9	43.4	44.9	46.4	47.9	49.4	50.9	52.4
24	23.2	24.8	26.3	27.8	29.4	31.0	32.5	34.0	35.6	37.1	38.7	40.2	41.8	43.3	44.9	46.4	48.0	49.5	54.1	52.6	54.2
28	24.0	25.6	27.2	28.8	30.4	31.9	33.5	35.1	36.7	38.3	39.9	41.5	43.1	44.7	46.3	47.9	49.5	51.1	52.7	54.3	55.9
32	24.7	26.4	38.0	39.7	31.3	32.9	34.6	36.2	37.9	39.5	41.2	42.8	44.5	46.1	47.8	49.4	51.1	52.7	54.4	56.0	57.7
36	25.5	27.2	28.9	30.6	32.2	33.9	35.6	37.3	39.0	40.7	42.4	44.1	45.8	47.5	49.2	50.9	52.6	54.3	56.0	57.7	59.4
40	26.2	28.0	29.7	31.4	33.2	34.9	36.7	38.4	40.2	41.9	43.7	45.4	47.2	48.9	50.7	52.4	54.2	55.9	57.7	59.5	61.2
44	27.0	28.8	30.6	32.3	34.1	35.9	37.7	39.5	41.3	43.1	44.9	46.7	48.5	50.3	52.1	53.9	55.7	57.5	59.3	61.1	62.9
48	27.7	29.6	31.4	33.2	35.1	36.9	38.8	40.6	42.5	44.3	46.2	48.0	49.9	51.7	53.6	55.4	57.3	59.1	61.0	62.8	64.6
52	28.5	30.4	32.2	34.1	36.0	37.9	39.8	41.7	43.6	45.5	47.4	49.3	51.2	53.1	55.0	56.9	58.8	60.7	62.6	54.5	66.4
56	29.2	31.2	33.1	35.0	37.0	38.9	40.9	42.8	44.8	46.7	48.7	50.6	52.6	54.5	56.5	58.4	60.4	62.3	64.2	66.2	68.1
60	30.0	31.9	33.9	35.9	37.9	39.9	41.9	43.9	45.9	47.9	49.9	51.9	53.9	55.9	57.9	59.9	61.9	63.9	65.9	67.9	69.9
64	30.7	32.7	34.8	36.8	38.9	40.9	43.0	45.0	47.1	49.1	51.2	53.2	55.3	57.3	59.4	61.4	63.4	65.5	67.5	69.6	71.6
68	31.4	33.5	35.6	37.7	39.8	41.9	44.0	46.1	48.2	50.3	52.4	54.5	56.6	58.7	60.8	62.9	65.0	67.1	69.2	71.3	73.4
72	32.2	34.3	36.5	38.6	40.8	42.9	45.1	47.2	49.4	51.5	53.7	55.8	58.0	60.1	62.3	64.4	66.5	68.7	70.8	73.0	75.1
76	32.9	35.1	37.3	39.5	41.7	43.9	46.1	48.3	50.5	52.7	54.9	57.1	59.3	61.5	63.7	65.9	68.1	70.3	72.5		
80	33.7	35.9	38.2	40.4	42.7	44.9	47.2	49.4	51.7	53.9	56.2	58.4	60.7	62.9	65.1	67.4	69.6	71.9	74.1		
84	34.4	36.7	39.0	41.3	43.6	45.9	48.2	50.5	52.8	55.1	57.4	59.7	62.0	64.3	66.6	68.9	71.2	73.5			
88	35.2	37.5	39.9	42.2	44.6	46.9	49.3	51.6	54.0	56.3	58.7	61.0	63.3	65.7	68.0	70.4	72.7				
92	35.9	38.3	40.7	43.1	45.5	47.9	50.3	52.7	55.1	57.5	59.9	62.3	64.7	67.1	69.4	71.9	74.3				
96	36.7	39.1	41.6	44.0	46.5	48.9	51.4	53.8	56.3	58.7	61.2	63.6	66.0	68.5	70.9	73.4					
100	37.4	39.9	42.4	44.9	47.4	49.9	52.4	54.9	57.4	59.9	62.4	64.9	67.4	69.9	72.4	74.9					
110	39.3	41.9	44.6	47.2	49.8	52.4	55.0	57.7	60.3	62.9	65.5	68.1	70.8	73.4	76.0						
120	41.2	43.9	46.7	49.4	52.2	54.9	57.7	60.4	63.1	65.9	68.6	71.4	74.1	76.9							
130	43.1	45.9	48.8	51.7	54.5	57.4	60.3	63.1	66.0	68.9	71.8	74.6	77.5	80.4							
140	44.9	47.9	50.9	53.9	56.9	59.9	62.9	65.9	68.9	71.9	74.9	77.9	80.9								
150	46.8	49.9	53.0	56.2	59.3	62.4	65.5	68.6	71.8	74.9	78.0	81.1	84.2								

7.1.6 Notching and Boring Guides for Softwood Floor Joist and Stud Walls

Notching and boring guide for floor joists and stud walls in conventional light-frame construction

Notching and Boring Guidelines

Intended for use by residential builders, this WWPA TIP Sheet serves as a guide to code-allowed size and placement of cuts (notching and boring) in floor-joist and stud-wall framing members.

A number of problems can occur if cuts are made through framing members to make room for plumbing or electrical runs, ductwork, or other mechanical elements such as sound or security systems.

Whenever a hole or notch is cut into a member, the structural capacity of the piece is weakened and a portion of the load supported by the cut member must be transferred properly to other joists.

It is best to design and frame a project to accommodate mechanical systems from the outset, as notching and boring should be avoided whenever possible; however, unforeseen circumstances sometimes arise during construction.

If it is necessary to cut into a framing member, the following diagrams provide a guide for doing so in the least destructive manner.

Diagrams comply with the requirements of the three major model building codes: Uniform (UBC), Standard (SBC), and National (BOCA), and the CABO One- & Two-Family Dwelling Code.

Floor Joists

The following references are to actual, not nominal, dimensions. (See Figure 1: Placement of Cuts in Floor Joists and Table 1: Maximum Sizes for Cuts in Floor Joists.)

> **Holes:** Do not bore holes closer than 2" from joist edges, nor make them larger than 1/3 the depth of the joist.
> **Notches:** Do not make notches in the middle third of the span where the bending forces are greatest.

Notches should be no deeper than 1/6 the depth of the joist. Notches at the end of the joist should be no deeper than 1/4 the depth. Limit the length of notches to 1/3 of the joist's depth.

When a Notch Becomes a Rip

Codes do not address the maximum allowable length of a notch; however, the 1991 National Design Specification (NDS) does limit the maximum length of a notch to 1/3 the depth of a member.

It is important to recognize the point at which a notch becomes a rip, such as when floor joists at the entry of a home are ripped down to allow underlayment for a tile floor.

Ripping wide-dimension lumber lowers the grade of the material and is unacceptable under all building codes.

When a sloped surface is necessary, a nonstructural member can be ripped to the desired slope and fastened to the structural member in a position above the top edge. Do not rip the structural member.

Stud Walls

When structural wood members are used vertically to carry loads in compression, the same engineering procedure is used for both studs and columns. However, differences between studs and columns are recognized in the model building codes for conventional light-frame residential construction.

The term "column" describes an individual major structural member subjected to axial compression loads, such as columns in timber-frame or post-and-beam structures.

The term "stud" describes one of the members in a wall assembly or wall system carrying axial compression loads, such as 2 x 4 studs in a stud wall that includes sheathing or wall board. The difference between columns and studs can be further described in terms of the potential consequences of failure.

Columns function as individual major structural members; consequently failure of a column is likely to result in partial collapse of a structure (or complete collapse in extreme cases due to the domino effect). However, studs function as members in a system. Due to the system effects (load sharing, partial composite action, redundancy, load distribution, etc.), studs are much less likely to fail and result in a total collapse than are columns.

Notching or boring into columns is not recommended and rarely acceptable; however, model codes establish guidelines for allowable notching and boring into studs used in a stud-wall system.

Figures 2 and 3 illustrate the maximum allowable notching and boring of 2 × 4 studs under all model codes except BOCA. BOCA allows a hole one-third the width of the stud in all cases.

Bored holes shall not be located in the same cross section of a stud as a cut or notch.

For additional information on framing (and common framing errors), contact WWPA for reprints of the following articles written by Association field staff.

Field Guide to Common Framing Errors (JLC-2) reprinted from *Journal of Light Construction*: article focuses on most commonly encountered job-site errors.

Common Roof-Framing Errors (JLC-3) reprinted from *Journal of Light Construction*: focuses on problems and solutions with trusses, rafters, collar ties, and structural ridges.

Picture Perfect Framing (B-1) reprinted from *Builder Magazine*: discusses cantilevers, joist hangers, blocking, notching and boring, cathedral ceilings, and overcutting tapers. Article reprints are 75 cents each to cover postage and handling.

Maximum Sizes for Cuts in Floor Joists

Joist Size	Max. Hole	Max Notch Depth	Max. End Notch
2 × 4	None	none	none
2 × 6	1-1/2"	7/8"	1-3/8"
2 × 8	2-3/8"	1-1/4"	1-7/8"
2 × 10	3"	1-1/2"	2-3/8"
2 × 12	3-3/4"	1-7/8"	2-7/8"

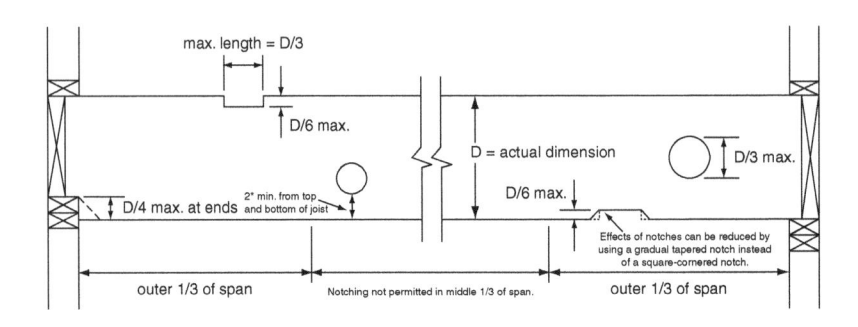

FIGURE 1 Placement of Cuts in Floor Joists

FIGURE 2 Notches in 2×4 Studs

FIGURE 3 Bored Holes in 2×4 Studs

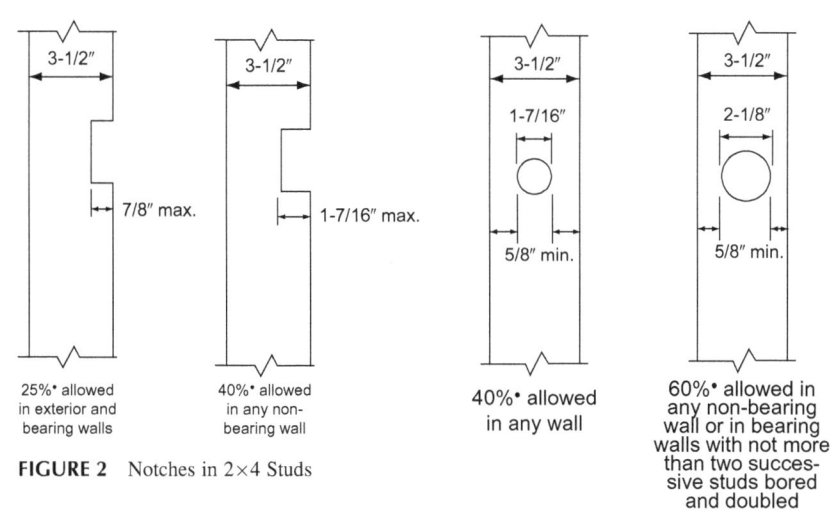

Western Wood Products Association 522 SW Fifth Avenue Suite 400 Portland, OR 97204-2122 503/224-3930 Fax: 503/224-3934 e-mail: mailto: info@wwpa.org web site: http://www.wwpa.org

7.1.7 Profiles of Typical "Worked" Lumber

Lumber Standard and further detailed in the grading rules. Classifications of manufacturing imperfections (combinations of imperfections allowed) are established in the rules as Standard A, Standard B, and so on. For example, Standard A admits very light torn grain, occasional slight chip marks, and very slight knife marks. These classifications are used as part of the grade rule description of some lumber products to specify the allowable surface quality.

Patterns

Lumber that has been matched, shiplapped, or otherwise patterned, in addition to being surfaced, is often classified as "worked lumber." Figure 5–3 shows typical patterns.

Softwood Lumber Species

The names of lumber species adopted by the trade as standard may vary from the names of trees adopted as official by the USDA Forest Service. Refer to USDA Forest service website for American Softwood Lumber Standard Commercial names, tree names, botanical names and Southern Pine and Hem-Fire combinations.

Softwood Lumber Grading

Most lumber is graded under the supervision of inspection bureaus and grading agencies. These organizations supervise lumber mill grading and provide re-inspection services to resolve disputes concerning lumber shipments. Some of these agencies also write grading rules that reflect the species and products in the geographic regions they represent. These grading rules follow the American Softwood Lumber Standard (PS–20). Names and addresses of rules-writing organizations in the United States and Canadian softwood lumber imported into the United States may be obtained from the American Lumber Standard Committee, P.O. Box 210, Germantown, MD 20879.

Flooring (standard match)

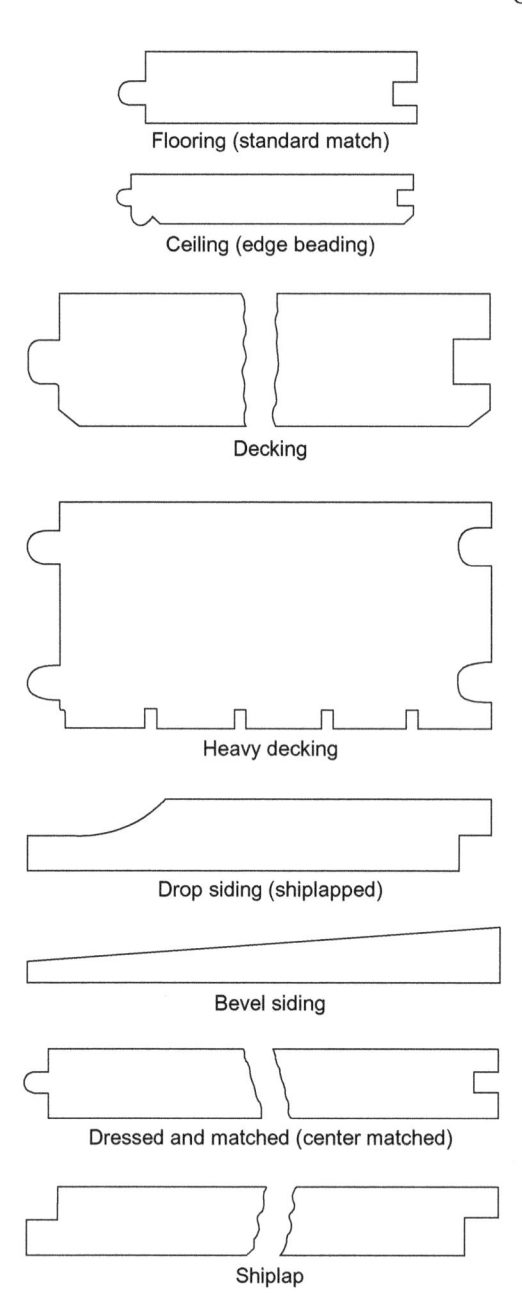

Ceiling (edge beading)

Decking

Heavy decking

Drop siding (shiplapped)

Bevel siding

Dressed and matched (center matched)

Shiplap

Typical patterns of worked lumber.

Purchase of Lumber

After primary manufacture, most lumber products are marketed through wholesalers to remanufacturing plants or retail outlets. Because of the extremely wide variety of lumber products, wholesaling is very specialized—some organizations deal with only a limited number of species or products. Where the primary manufacturer can readily identify the customers, direct sales may be made. Primary manufacturers often sell directly to large retail-chain contractors, manufacturers of mobile and modular housing, and truss fabricators.

7.1.7.1 Nominal and Dressed Sizes of Worked Lumber

The thicknesses apply to all widths, and all widths apply to all thicknesses. Sizes are given in millimeters and inches. Metric units are based on dressed size. See Bl, Appendix B, for rounding rule for metric units.

Item	Thicknesses					Widths				
	Nominal Inch	Minimum Dressed				Nominal Inch	Minimum Dressed			
		Dry		Green			Dry		Green	
		mm	inch	Mm	inch		mm	inch	mm	inch
Shiplap, 10 mm (3/8 inch) lap	1	19	3/4	20	25/32	4	79	3-1/8	81	3-3/16
						6	130	5-1/8	133	5-1/4
						8	175	6-7/8	181	7-1/8
						10	225	8-7/8	232	9-1/8
						12	276	10-7/8	283	11-1/8
						14	327	12-7/8	333	13-1/8
						16	378	14-7/8	384	15-1/8
Shiplap, 13 mm (1/2 inch) lap	1	19	3/4	20	25/32	4	76	3	78	3-1/16
	2	38	1-1/2	40	1-9/16	6	127	5	130	5-1/8
	2-1/2	51	2	52	2-1/16	8	171	6-3/4	178	7
	3	64	2-1/2	65	2-9/16	10	222	8-3/4	229	9
	3-1/2	76	3	78	3-1/16	12	273	10-3/4	279	11
	4	89	3-1/2	90	3-9/16	14	324	12-3/4	330	13
	4-1/2	102	4	103	4-1/16	16	375	14-3/4	381	15
Centermatch (Tongue & Groove), 6 mm (1/4 inch) tongue	1	19	3/4	20	25/32	4	79	3-1/8	81	3-3/16
	1-1/4	25	1	26	1-1/32	5	105	4-1/8	108	4-1/4
	1-1/2	32	1-1/4	33	1-9/32	6	130	5-1/8	133	5-1/4
						8	175	6-7/8	181	7-1/8
						10	225	8-7/8	232	9-1/8
						12	276	10-7/8	283	11-1/8
Centermatch (Tongue & Groove), 10 mm (3/8 inch) tongue	2	38	1-1/2	40	1-9/16	4	76	3	78	3-1/16
	2-1/2	51	2	52	2-1/16	6	127	5	130	5-1/8
	3	64	2-1/2	65	2-9/16	8	171	6-3/4	178	7
	3-1/2	76	3	78	3-1/16	10	222	8-3/4	229	9
	4	89	3-1/2	90	3-9/16	12	273	10-3/4	279	11
	4-1/2	102	4	103	4-1/16					
Grooved-for-Splines	2-1/2	51	2	52	2-1/16	4	89	3-1/2	90	3-9/16
	3	64	2-1/2	65	2-9/16	6	140	5-1/2	143	5-5/8
	3-1/2	76	3	78	3-1/16	8	184	7-1/4	190	7-1/2
	4	89	3-1/2	90	3-9/16	10	235	9-1/4	241	9-1/2
	4-1/2	102	4	103	4-1/16	12	286	11-1/4	292	11-1/2

7.1.7.2 Nominal and Minimum Sizes of 19% Moisture Content Dressed Lumber

The thicknesses apply to all widths and all widths apply to all thicknesses except as modified. Sizes are given in millimeters and inches. Metric units are based on dressed size. See Bl, Appendix B, for rounding rule for metric units.

Item	Thicknesses Nominal (inch)	Minimum Dressed mm	Minimum Dressed inch	Widths Nominal (inch)	Minimum Dressed mm	Minimum Dressed inch
Finish	3/8	8	5/16	2	38	1-1/2
	1/2	11	7/16	3	64	2-1/2
	5/8	14	9/16	4	89	3-1/2
	3/4	16	5/8	5	114	4-1/2
	1	19	3/4	6	140	5-1/2
	1-1/4	25	1	7	165	6-1/2
	1-1/2	32	1-1/4	8	184	7-1/4
	1-3/4	35	1-3/8	9	210	8-1/4
	2	38	1-1/2	10	235	9-1/4
	2-1/2	51	2	11	260	10-1/4
	3	64	2-1/2	12	286	11-1/4
	3-1/2	76	3	14	337	13-1/4
	4	89	3-1/2	16	387	15-1/4
Flooring[a]	3/8	8	5/16	2	29	1-1/8
	1/2	11	7/16	3	54	2-1/8
	5/8	14	9/16	4	79	3-1/8
	1	19	3/4	5	105	4-1/8
	1-1/4	25	1	6	130	5-1/8
	1-1/2	32	1-1/4			
Ceiling[a]	3/8	8	5/16	3	54	2-1/8
	1/2	11	7/16	4	79	3-1/8
	5/8	14	9/16	5	105	4-1/8
	3/4	17	11/16	6	130	5-1/8
Partition[a]	1	18	23/32	3	54	2-1/8
				4	79	3-1/8
				5	105	4-1/8
				6	130	5-1/8
Stepping	1	19	3/4	8	184	7-1/4
	1-1/4	25	1	10	235	9-1/4
	1-1/2	32	1-1/4	12	286	11-1/4
	2	38	1-1/2			

[a] In tongued-and-grooved flooring and in tongued-and-grooved and shiplapped ceiling of 8 mm (5/16 inch), 11 mm (7/16 inch), and 14 mm (9/16 inch) dressed thicknesses, the tongue or lap shall be 5 mm (3/16 inch) wide, with the overall widths 5 mm (3/16 inch) wider than the face widths shown in the above table. In all other worked lumber shown in this table of dressed thicknesses of 16 mm (5/8 inch) to 32 mm (1-1/4 inches), the tongue shall be 6 mm (1/4 inch) wide or wider in tongued-and-grooved lumber, and the lap shall be 10 mm (3/8 inch) wide or wider in shiplapped lumber, and the overall widths shall be not less than the dressed face widths shown in the above table plus the width of the tongue or lap.

Nominal and Minimum-Dressed Dry Sizes of Siding at 19% Maximum-Moisture Content

		Thicknesses		Widths		
		Minimum Dressed			Minimum Dressed	
Item	Nominal (inch)	mm	inch	Nominal (inch)	mm	inch
Plain Bevel	1/2	11 butt, 5 tip	7/16 butt, 3/16 tip	4	89	3-1/2
	9/16	12 butt, 5 tip	15/32 butt, 3/16 tip	5	114	4-1/2
	5/8	14 butt, 5 tip	9/16 butt, 3/16 tip	6	140	5-1/2
	3/4	17 butt, 5 tip	11/16 butt, 3/16 tip	8	184	7-1/4
	1	19 butt, 5 tip	3/4 butt, 3/16 tip	10	235	9-1/4
				12	286	11-1/4
Rabbeted Bevel	1/2	11 butt, 5 tip	7/16 butt, 3/16 tip	4	89	3-1/2
	3/4	17 butt, 7 tip	11/16 butt, 9/32 tip	6	140	5-1/2
				8	184	7-1/4
				10	235	9-1/4
				12	286	11-1/4
Bungalow	3/4	17 butt, 5 tip	11/16 butt, 3/16 tip	6	140	5-1/2
				8	184	7-1/4
				10	235	9-1/4
				12	286	11-1/4
Shiplap (10 mm [3/8 in.] lap)	5/8	14	9/16	4	76	3
	1	18[a]	23/32[a]	5	102	4
				6	127	5
				8	171	6-3/4
				10	222	8-3/4
				12	273	10-3/4
Shiplap (13 mm [1/2 in.] lap)	5/8	14	9/16	4	73	2-7/8
	1	18[a]	23/32[a]	5	98	3-7/8
	2	38	1-1/2	6	124	4-7/8
				8	168	6-5/8
				10	219	8-5/8
				12	270	10-5/8
Dressed and Matched (Tongue and Grooved, 6 mm [1/4 in.] tongue)	5/8	14	9/16	4	79	3-1/8
	1	18[a]	23/32[a]	5	105	4-1/8
				6	130	5-1/8
				8	175	6-7/8
				10	225	8-7/8
				12	276	10-7/8
Dressed and Matched (Tongue and Grooved, 10 mm [3/8 in.) tongue)	1	18[a]	23/32[a]	4	76	3
				6	127	5
				8	171	6-3/4
				10	222	8-3/4
				12	273	10-3/4

[a]Minimum dressed thickness for 1-inch nominal redwood and western red cedar shiplap and tongue and groove siding patterns is 17 mm (11/16 inch).

7.2.0 No. I and No. 2 Common Board Designations

No. 1 Common and No. 2A Common

Number 1 Common (No. 1C)

The Number 1 Common grade is often referred to as the Cabinet grade in the United States because of its adaptability to the standard sizes of kitchen cabinet doors used throughout the United States. Number 1 Common is widely used in the manufacture of furniture parts as well for this same reason. The Number 1 Common grades include boards that are a minimum of 3″ wide and 4′ long and will yield clear face cuttings from 66 2/3% (8/12ths) up to, but not including, the minimum requirement for FAS (831/3%). The smallest clear cuttings allowed are 3″ by 3′ and 4″ by 2′. The number of these clear cuttings is determined by the size of the board. Both faces of the board must meet the minimum requirement for Number 1 Common.

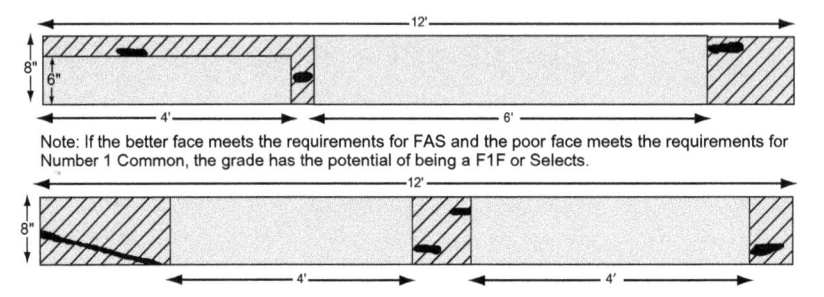

Note: If the better face meets the requirements for FAS and the poor face meets the requirements for Number 1 Common, the grade has the potential of being a F1F or Selects.

Number 2A Common (No. 2AC)

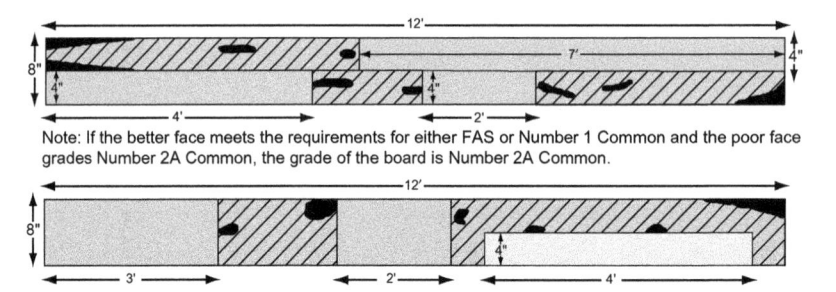

Note: If the better face meets the requirements for either FAS or Number 1 Common and the poor face grades Number 2A Common, the grade of the board is Number 2A Common.

There are lower NHLA (National Hardwood Lumber Association) grades than Number 2A Common, but they are usually converted into dimension parts, flooring parts, or used domestically in the United States.

The Number 2A Common grade is often referred to as the Economy grade because of its price and suitability for a wide range of furniture parts. It is also the grade of choice for the U.S. hardwood flooring industry. The Number 2A Common grade includes boards that are a minimum of 3″ wide and 4′ long that yield from 50% (6/12ths) up to, but not including, the minimum requirement for Number 1 Common (662/3%). The smallest clear cutting allowed is 3″ by 2′, and the number of these cuttings depends on the size of the board. If the poorest face meets the minimum requirements for Number 2A Common, it does not matter what the grade of the better face is.

These Standard Grades form the framework by which all American hardwoods are traded. It is important to note that between buyer and seller any exception to these rules is permissible and even encouraged. For a complete description of the NHLA grades, consult the NHLA's "Rules for the Measurement and Inspection of Hardwoods and Cypress."

7.2.1 Hardwood—Measurement System

The NHLA lumber grading rules adopted by the U.S. hardwood industry are based on an imperial measurement system using inches and feet. In contrast, most export markets are more familiar with a metric standard. In addition, the grade rules were developed with random width and length lumber in mind. Any selection for particular specifications should be discussed prior to ordering.

Board Foot

A board foot (BF) is the unit of measurement for hardwood lumber. A board foot is 1 foot long × 1 foot wide × 1 inch thick. (1 foot = 0.305 meters, 1 inch = 25.4 mm)
   The formula for determining board feet in a board is:
   (Width in inches × length in feet × thickness in inches) divided by 12
   The percentages of clear wood required for each grade are based on this 12′ unit of measure.

Surface Measure

Surface measure (SM) is the surface area of a board in square feet. To determine surface measure, multiply the width of the board in inches by the length of the board in feet and divide the sum by 12 rounding up or down to the nearest whole number. The percentage of clear wood required for each grade is based on the surface measure, not the board feet, and because of this all boards, no matter what the thickness, are graded in the same way.
   Some examples for surface measure calculations are as follows:

6½ × 8′ ÷ 12 = 4⅓ = 4′ SM

8″ × 12′ ÷ 12 = 8′ SM

10″ × 13′ ÷ 12 = 10% = 11′ SM

Example of SM and BF:
The board is 2″ thick, 6¼″ wide, and 8′ long.
6¼″ × 8′ ÷ 12 = 4¼, thus the SM is 4′. Multiply the SM by the thickness 2″ and the BF is 8′.
   When preparing a bundle tally for export, the boards are recorded by their width and length. Random widths above or below the half inch are rounded to the nearest whole inch. Board widths falling exactly on the half inch are alternatively rounded up or down. Lengths that fall between whole foot increments are always rounded down to the nearest whole foot. For example, a board 5¼″ width and 8½′ long is tallied 5″ and 8′.

Standard Thickness for Rough Sawn Lumber

Standard thickness for rough sawn lumber is expressed in quarters of an inch. For example, 1″ = ¼. The majority of U.S. hardwood lumber production is sawn between 1″ and 2″, although other thicknesses are available in more limited volumes. The standard thicknesses and their exact metric equivalent are shown below.

3/4	(3/4″ = 19.0mm)	8/4	(2″ = 50.8mm)
4/4	(1″ = 25.4mm)	10/4	(2½″ = 63.5mm)
5/4	(1¼″ = 31.8mm)	12/4	(3″ = 76.2mm)
6/4	(1½″ = 38.1mm)	16/4	(4″ = 10.1mm)

Standard Thickness for Surfaced (Planed) Lumber

When rough sawn lumber is surfaced (planed) to a finished thickness, defects such as checks, stain, and warp are not considered when establishing the grade of a board, **if they can be removed in the surfacing (planning) process**. The finished thickness for lumber of 1½" and less can be determined by subtracting 3/16" from the nominal thickness. For lumber 13/4" and thicker, subtract ¼".

Measurement of Kiln-dried Lumber

Net tally: The actual board feet of kiln-dried lumber measured after kiln drying.
Gross or green tally: The actual board feet measured before kiln drying. When kiln-dried lumber is sold on this basis, the buyer can expect to receive approximately 7% less board feet because of shrinkage in the kiln drying process.

Estimating Board Feet in a Bundle of Lumber

To determine the board feet of one board, the procedure is to multiply the surface measure by the thickness. A bundle of lumber can be estimated in much the same manner. First, calculate the surface measure of one layer of boards. Do this by multiplying the width of the bundle, minus gaps, by the length of the bundle and divide the sum by 12. If there are several lengths in the bundle, use an average length. Once one layer is estimated, multiply this sum by the total number of layers.

Example:

Average width of unit 40"
(lumber only, after allowing for gaps between boards)
Length of unit 10'

$40'' \times 10' = 400 \div 12 \qquad = 33.33$
Thickness of lumber 8/4 $\qquad \times 2$
$\qquad\qquad\qquad\qquad\qquad = 66.66$
Number of layers : $\qquad\qquad \times 10$
$\qquad\qquad\qquad\qquad\qquad = 666.67$

Estimated board feet of the bundle 667 BF

Conversion Factors
1":	25.4 millimeters (mm)	1m³: 424 board feet (BF)
1m:	3.281 feet	1m³: 35.315 cubic feet (cu ft)
1.000BF:(1MBF)	2.36 cubic meters (m³)	

Additional Guidance

Regional exceptions to the standard NHLA grades

The NHLA grades cover the majority of commercial hardwood species growing in the United States. The following is a brief summary of various species and color sorting that can be ordered from the American supplier.

Red Alder

Grows exclusively in the Pacific Northwest between the vast stands of softwood timber such as Douglas fir and pine and is the most important commercial hardwood in this region. The grading rules for red alder are geared more toward specific end uses and appearance. The rules were developed on the West Coast of the United States with those manufacturers and exports in mind. An exceptional cabinet wood typically sold surfaced (planed) and often cut to specific lengths and widths. Consult with your local supplier for a more detailed explanation of the alder grades and products available.

Walnut

Considered the elite of the American hardwoods, walnut is the favorite of the darker woods for fine furniture, interiors, and gunstocks. Walnut grows in widely scattered stands throughout the eastern half of the United States, primarily in the Midwest. Historically, the grading rules for FAS walnut have been refined to encourage better use of this valuable species. Because of this, FAS Walnut grades allow for smaller boards, in both width and length. Natural characteristics are also admitted to a greater extent than the standard NHLA grade rules for other species. A detailed explanation can be found in the NHLA rules book. Consult with your local supplier for the walnut grades and products available.

Color Sorting

In addition to sorting for grades or selecting specific widths, various species are commercially sold at an added value when color is also considered. It is important to note that color in this explanation refers to sapwood and heartwood.

Number 1 and 2 White

A color selection is typically made on hard maple, but can be applied to any species where sapwood clear cuttings are desired, such as ash, birch, and soft maple.

Number 1 white means both faces and edges of the clear cuttings must be all sapwood.

Number 2 white means that one face and both edges of the clear cuttings must be sapwood and not less than 50% sapwood on the reverse face.

Sap and Better

Commercially sold when only one face of the board needs to be sapwood. Usually applied to the same species as Number 1 and 2 White, although just a little less stringent. In Sap and Better, every board should have a minimum of one sapwood face in the clear cuttings.

Red One Face and Better

Commercially sold when a minimum of one face of the board needs to be heartwood. Usually applied to species such as cherry, oak, walnut, gum, and even birch and maple in certain applications. What the producer is looking for in this specification is that all clear cuttings must have a minimum of one heartwood face.

A wide range of additional options is open to American hardwood producers in sorting and selecting specific lengths, widths, and even grain patterns. If these options can be agreed upon individually between producers and buyers, there can be benefits by making modifications to the standard grades shown in this guide. This may also assist with improving the yield from each log and thus contribute to the sustainability of the forest. It may also reduce costs to both sides or add value to the delivery.

The Steps in Determining Grade

1. Determine species.
2. Calculate the surface measure (SM).
3. Determine the poor side of the board.
4. From this poor face, calculate the percentage of clear wood available.
   Note: If Number 1 Common is the grade of the poor face, check the better face to see if it will grade FAS for the F1F or Selects grades to be achieved.

5. Once the grade is determined, check for any special features such as sapwood or heartwood cuttings for special color sorts.
6. Sort to bundles according to buyer and seller specifications.

7.2.2 Hardwoods—Selects and No. 1 and No. 2 Standards

FAS

The FAS grade, which derives from an original grade "First And Seconds," will provide the user with long, clear cuttings best suited for high-quality furniture, interior joinery, and solid wood moldings. Minimum board size is $6''$ and wider and $8'$ and longer. The FAS grade includes a range of boards that yield from $83 1/3\%$ (10/12ths) to 100% clear-wood cuttings over the entire surface of the board. The clear cuttings must be a minimum size of $3''$ wide by $7'$ long or $4''$ wide by $5'$ long. The number of these cuttings permitted depends on the size of the board, with most boards permitting one to two. The minimum width and length will vary, depending on species and whether the board is green or kiln dried. Both faces of the board must meet the minimum requirement for FAS.

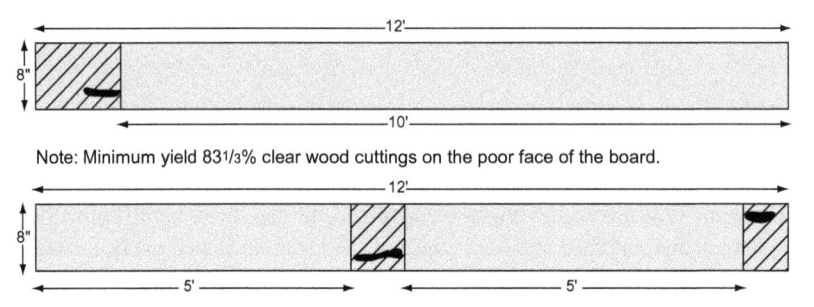

Note: Minimum yield 83 1/3% clear wood cuttings on the poor face of the board.

FAS One Face (F1F)

This grade is nearly always shipped with FAS. The better face must meet all FAS requirements, while the poor face must meet all the requirements of the Number 1 Common grade, thus ensuring the buyer with at least one FAS face. Often export shipments are assembled with an 80–20 mix, 80% being the percentage of FAS boards and 20% being the percentage of F1F boards. These percentages are strictly left to individual buyer and seller agreement.

Selects

This grade is virtually the same as F1F except for the minimum board size required. Selects allow boards $4''$ and wider and $6'$ and longer in length. The Selects grade is generally associated with the northern regions of the United States and is also shipped in combination with the FAS grade.

Often export shipments of upper grades are simply referred to as FAS. The conventional business practice for American hardwoods is to ship these upper grades in some combination. Working closely with the supplier will enable the buyer to be sure that the expected quality will be received. Whether FAS is combined with F1F (Face And Better) or Selects (Sel And Better), every board in the shipment must have a minimum of one FAS face.

Prime grade: This grade has evolved from the NHLA grade of FAS for the export market. It is square edged and virtually wane free. The minimum clear yield will be select and better, with appearance being a major factor. Minimum size of the boards varies, depending on the species, region, and supplier.

Comsel grade: This grade has evolved from the NHLA grades of Number 1 Common and Selects. For the export market the minimum clear yield should be Number 1 Common or slightly better with appearance a main factor. Minimum size of the boards varies, depending on the species, region, and supplier.

Note: The terms Prime and Comsels are not standard NHLA definitions and therefore fall outside the official range of the NHLA grading rules.

7.2.3 American Standard Lumber Sizes for Yard and Structural Lumber for Construction

American Standard Lumber Sizes for Yard and Structural Lumber for Construction

Item	Nominal (in.)	Thickness Minimum dressed Dry (mm)	(in.)	Green (mm)	(in.)	Nominal (in.)	Face Width Minimum dressed Dry (mm)	(in.)	Green (mm)	(in.)
Boards	1	19	(3/4)	20	(25/32)	2	38	(1-1/2)	40	(1-9/16)
	1-1/4	25	(1)	26	(1-1/32)	3	64	(2-1/2)	65	(2-9/16)
	1-1/2	32	(1-1/4)	33	(1-9/32)	4	89	(3-1/2)	90	(3-9/16)
						5	114	(4-1/2)	117	(4-5/8)
						6	140	(5-1/2)	143	(5-5/8)
						7	165	(6-1/2)	168	(6-5/8)
						8	184	(7-1/4)	190	(7-1/2)
						9	210	(8-1/4)	216	(8-1/2)
						10	235	(9-1/4)	241	(9-1/2)
						11	260	(10-1/4)	267	(10-1/2)
						12	286	(11-1/4)	292	(11-1/2)
						14	337	(13-1/4)	343	(13-1/2)
						16	387	(15-1/4)	394	(15-1/2)
Dimension	2	38	(1-1/2)	40	(1-9/16)	2	38	(1-1/2)	40	(1-9/16)
	2-1/2	51	(2)	52	(2-1/16)	3	64	(2-1/2)	65	(2-9/16)
	3	64	(2-1/2)	65	(2-9/16)	4	89	(3-1/2)	90	(3-9/16)
	3-1/2	76	(3)	78	(3-1/16)	5	114	(4-1/2)	117	(4-5/8)
	4	89	(3-1/2)	90	(3-9/16)	6	140	(5-1/2)	143	(5-5/8)
	4-1/2	102	(4)	103	(4-1/16)	8	184	(7-1/4)	190	(7-1/2)
						10	235	(9-1/4)	241	(9-1/2)
						12	286	(11-1/4)	292	(11-1/2)
						14	337	(13-1/4)	343	(13-1/2)
						16	387	(15-1/4)	394	(15-1/2)
Timbers	≥5	13 mm off	(1/2 in. off)	13 mm off	(1/2 in. off)	≥5	13 mm off	(1/2 in. off)	13 mm off	(1/2 in. off)

Factory and Shop lumber for remanufacture is offered in specified sizes to fit end-product requirements. Factory (Shop) grades for general cuttings are offered in thickness from standard 19 to 89 mm (nominal 1 to 4 in.). Thicknesses of door cuttings start at 35 mm (nominal 1-3/8 in.). Cuttings are of various lengths and widths. Laminating stock is sometimes offered oversize, compared with standard dimension sizes, to permit resurfacing prior to laminating. Industrial Clears can be offered rough or surfaced in a variety of sizes, starting from standard 38 mm (nominal 2 in.) and thinner and as narrow as standard 64 mm (nominal 3 in.). Sizes for special product grades such as molding stock and ladder stock are specified in appropriate grading rules or handled by purchase agreements.

Surfacing

Lumber can be produced either rough or surfaced (dressed). Rough lumber has surface imperfections caused by the primary sawing operations. It may be greater than target size by variable amounts in both thickness and width, depending on the type of sawmill equipment. Rough lumber serves as a raw material for further manufacture and also for some decorative purposes. A rough-sawn surface is common in post and timber products. Because of surface roughness, grading of rough lumber is generally more difficult.

Surfaced lumber has been surfaced by a machine on one side (S1S), two sides (S2S), one edge (S1E), two edges (S2E), or combinations of sides and edges (S1S1E, S2S1E, S1S2, S4S). Lumber is surfaced to attain smoothness and uniformity of size.

Imperfections or blemishes defined in the grading rules and caused by machining are classified as "manufacturing imperfections." For example, chipped and torn grain are surface irregularities in which surface fibers have been torn out by the surfacing operation. Chipped grain is a "barely perceptible" characteristic, while torn grain is classified by depth. Raised grain, skip, machine burn and gouge, chip marks, and wavy surfacing are other manufacturing imperfections. Manufacturing imperfections are defined in the American Softwood.

7.2.4 Weights of Green and Kiln-Dried Hardwoods

		Weights of Green and Kiln-Dried Lumber[a]							
		Green				Kiln-Dried[b]			
Common Name	Latin Name	lb/ft3	kg/m3	lb/MBF	kg/MBF	lb/ft3	kg/m3	lb/MBF	kg/MBF
Alder, Red	Alnus rubra	46	737	3833	1739	27.5	440	2288	1038
Ash									
Black	Fraxinus nigra	52	833	4333	1966	34.1	546	2843	1290
Green	Fraxinus pennsylvanica	49	785	4083	1852	39.3	629	3274	1485
White	Fraxinus americana	48	769	4000	1814	41.0	657	3420	1551
Aspen	Populus tremuloides	43	689	3583	1625	26.5	425	2211	1003
Basswood, Amer.	Tilia americana	42	673	3500	1588	24.4	390	2031	921
Beech, Amer.	Fagus grandifolia	54	865	4500	2041	43.2	691	3597	1632
Birch									
Sweet	Betula lenta	57	913	4750	2155	45.6	731	3803	1725
White	Betula papyrifera	50	801	4167	1890	36.8	590	3067	1391
Yellow	Betula alleghaniensis	57	913	4750	2155	42.2	677	3520	1597
Black Gum/Tupelo	Nyssa spp.	45	721	3750	1701	34.6	555	2887	1310
Black Locus	Robinia pseudoacacia	58	929	4833	2192	48.0	769	4003	1816
Black Walnut	Juglans nigra	58	929	4833	2192	37.9	607	3159	1433
Boxelder[c]	Acer negundo	32	513	2667	1210	27.5	508	2642	1198
Buckeye, Yellow	Aesculus octandra	49	785	4083	1852	24.5	392	2039	925
Butternut	Juglans cinerea	46	737	3833	1739	26.3	421	2191	994
Cherry, Black	Prunus serotina	45	721	3750	1701	34.6	554	2881	1307
Chestnut, Amer.	Castanea dentate	55	881	4583	2079	29.4	472	2453	1113
Cottonwood[c]	Populus deltoids	49	785	4083	1852	27.5	444	2308	1047
Cypress	Taxodium distichum	60	961	5000	2268	30.6	491	2553	1158

Weights of Green and Kiln-Dried Lumber[a]—Cont'd

Common Name	Latin Name	Green				Kiln-Dried[b]			
		lb/ft3	kg/m3	lb/MBF	kg/MBF	lb/ft3	kg/m3	lb/MBF	kg/MBF
Elm									
Hard	Ulmus thomasil	53	849	4417	2003	43.1	690	3591	1629
Soft	Ulmus rubra	56	897	4667	2117	36.0	576	2997	1359
Hackberry	Celtis spp.	50	801	4167	1890	36.7	588	3060	1388
Hickory									
True (average)	Carya spp.	64	1025	5333	2449	49.6	795	4135	1876
Pecan (bitternut)	Carya spp.	61	977	5083	2306	27.5	724	3767	1709
Honeylocust[c]	Gleditsia triacanthos	61	977	5083	2306	27.5	702	3650	1656
Madrone	Arbutus spp.	60	961	5000	2268	45.0	722	3754	1703
Magnolia, So.[c]	Magnolia grandiflora	59	945	4917	2230	27.5	485	2525	1145
Maple									
Hard (sugar)	Acer spp.	56	897	4667	2117	42.3	677	3523	1598
Soft (red)	Acer spp.	50	801	4167	1890	36.4	582	3030	1374
Red Oak Group									
Black	Quercus velutina	62	993	5167	2344	42.4	679	3534	1603
Cherrybark[c]	Q. falcate v. pagodifolia	68	1089	5667	2570	46.5	745	3875	1758
Laurel	Q. laurifolia	65	1041	5417	2457	43.8	702	3652	1657
Northern red	Q. rubra	63	1009	5250	2381	41.9	672	3494	1585
Pin	Q. palustris	63	1009	5250	2381	43.7	700	3642	1652
Scarlet	Q. coccinea	62	993	5167	2344	45.3	725	3774	1712
Southern red	Q. faleata	62	993	5167	2344	39.7	636	3309	1501
Water	Q. nigra	63	1009	5250	2381	42.8	685	3564	1617
Willow	Q. phellos	67	1073	5583	2533	43.8	701	3649	1655
White Oak Group									
Bur	Quercus macrocarpa	62	993	5167	2344	43.1	690	3589	1628
Chestnut	Q. prinus	61	977	5083	2306	43.6	699	3636	1649
Post	Q. stellata	63	1009	5250	2381	45.9	734	3821	1733
Swamp chestnut	Q. michauxii	65	1041	5417	2457	45.9	736	3828	1736
Swamp white[c]	Q. bicolor	69	1105	5750	2608	27.5	799	4158	1886
White	Q. alba	62	993	5167	2344	45.9	735	3825	1735
Live Oak	Quercus virginiana	76	1218	6333	2873	60.4	967	5032	2283
Redcedar, Eastern	Juniperus virginia	36	583	3033	1376	31.4	504	2620	1188
Sassafras	Sassafras albidum	44	705	3667	1663	27.5	497	2583	1172
Sweetgum	Liquidambar styracifulua	50	801	4167	1890	35.0	561	2920	1325
Sycamore	Platanus occidentalis	52	833	4333	1966	34.0	545	2836	1286
Tanoak	Lithocarpus densiflorus	65	1041	5417	2457	44.7	717	3728	1691

(Continued)

Common Name	Latin Name	Weights of Green and Kiln-Dried Lumber[a]—Cont'd							
		Green				Kiln-Dried[b]			
		lb/ft3	kg/m3	lb/MBF	kg/MBF	lb/ft3	kg/m3	lb/MBF	kg/MBF
Yellow-poplar	Liriodendron tulipifera	38	609	3167	1436	29.7	476	2475	1123
Willow, Black	Salix nigra	50	801	4167	1890	27.0	432	2249	1020

[a] Green weights directly from Hardwoods of North America, USDA Forest Service, FPL-GTR-83; and Wood Handbook, USDA Forest Service, Ag Handbook 72. Kiln-dried weights from Dry Kiln Operator's Manual, USDA Forest Service, except where noted. Weights are calculated based on average densities for each species. Thus, the weights given here should always be considered approximations because of the natural variation in anatomy, moisture content, and the ratio of heartwood to sapwood that occurs in a load. Board foot weights are based on exact board footage (1BF = 1' × 1' × 1"). No allowance is given for over thickness.
[b] Kiln-dried weights are calculated at 8% moisture content.
[c] Kiln-dried weights extrapolated from Hardwoods of North America, USDA Forest Service, FPL-GTR-83.

7.2.5 National Hardwood Lumber Association Grading Rules

Grading Rules

SELECTS and BETTER:
Widths: 4" and wider, of which 5% of 3" width is admitted.
Lengths: Random 4' and longer. Minimum cuttings: 4" wide by 3' long, or 3" wide by 6' long.

There is no limit to the number of cuttings. This grade admits all boards of 1' and over, surface measure, that will yield not less than 83-1/3% of clear-face cuttings, the reverse side of the cuttings sound as defined in "SOUND CUTTING," except that boards of 1' and over surface measure yielding not less than 83-1/3% clear-face cuttings on one face, the reverse side of the board grading not below No. 1 Shop.

Pith: No piece shall be admitted which contains pith exceeding in the aggregate in inches in length twice the surface measure in feet.

Splits: No piece shall be admitted which contains splits exceeding in the aggregate in inches in length twice the surface measure of the piece in feet, nor when diverging more than one inch to the foot in length, except when one foot or shorter and covered by Paragraph 59 of Standard Grades.

Wane: On the face side of Selects and Better, wane or its equivalent shall be limited to one-twelfth the surface measure of the piece.

On the No. 1 Shop side of Selects and Better, wane or its equivalent shall not exceed one-fourth the width by three-fourths the length in the aggregate, or pieces may alternately have wane one-third the width by one-half the length in the aggregate. Width of the wane may be divided and show on both edges. The reverse side of the cuttings in Selects and Better and No. 1 Shop are not required to be sound.

NO. 1 SHOP: Widths: 4" and wider, admitting 5% of 3" width.
Lengths: Random 4' and longer.

Minimum cuttings: 3" wide by 3' long, or 4" wide by 2' 1' and over surface measure shall yield not less than 66-2/3% clear-face cuttings, the reverse side of the cuttings sound as defined in "Sound Cutting."

NO. 2 SHOP: Widths: 4" and wider, admitting 5% of 3" width.
Lengths: Random 4' and longer.

Minimum cuttings: 3" wide by 2' long. There is no limit to the number of cuttings. Admits boards of 1' and over surface measure that will yield not less than 50% clear-face cuttings, the reverse side of the cuttings sound as defined in "Sound Cutting."

NO. 3 SHOP: Widths: 3″ and wider.
Lengths: Random 4′ and longer.

Minimum cuttings: 3″ wide by 2′ long. There is no limit to the number of cuttings. Admits boards of 1′ and over surface measure that will yield not less than 33-1/3% of sound cuttings or better.

NOTE: *Lumber poorer in cutting percentage, or less in width or length than admitted in No. 3 Shop described above, shall be tallied and reported below grade.*

FRAME GRADE: Widths: 4″ and wider
Lengths: 7 and longer
Minimum cuttings: Each piece must contain at least one cutting 4″ × 7″; other cuttings, minimum size 4″ × 2′. Each piece shall yield not less than 83-1/3%. There is no limit to the number of cuttings.

Wane: Wane shall not exceed one-fourth the width and one-half the length of the piece. Width of the wane may be divided and show on both edges.

Each cutting shall be reasonably flat and straight; will admit bark pockets, season checks, slight surface shake that does not impair the strength of the cutting, firm tight pith, stain, worm holes, and other holes or unsound knots that do not exceed in their greatest dimension 1-1/2″ in 4″ to 5″ wide cuttings and 2″ in 6″ and wider cuttings. Sound knots that do not exceed in their greatest dimension one-half the width of the cutting and other defects that do not impair the strength of the cutting more than the above mentioned defects are admitted.

Rules apply to both faces of the piece.

Quartered Sap Gum, Quartered Black Gum, and Quartered Tupelo

FAS: Standard. Except:
Widths 5″ and wider; pieces 5″ wide containing 3′ and 4′ surface measure shall be clear, pieces 5″ wide containing 5′ to 7′ surface measure shall yield 11/12 (91-2/3%) clear-face in one cutting.

F1F:
SELECTS:
NO. 1 COMMON:
NO. 2A COMMON:
No figure is required.
Stain is admitted in all grades.
Pieces below the grade of No. 2A Common shall be graded as Sap Gum or Black Gum.

Ribbon Stripe

When ribbon stripe figure is specified each piece shall be-selected for the stripe effect caused by the wavy grain brought out in quarter sawing. One face of each required cutting shall show 90% in the aggregate of such ribbon stripe figure.

Plain Red Gum

Red Gum is lumber produced from the Sweet Gum tree, containing sufficient heartwood to be admitted into the grades defined under the caption of Red Gum.

Stain is admitted in the sapwood in all grades. Any part of the sapwood allowed may be included in the cuttings.

FAS: Standard, except:
FAS will admit 1″ of sapwood in the aggregate on one face and one-fifth of the surface in the aggregate on the reverse side.
F1F: Standard.
SELECTS: Standard, except:

Pieces 4″ and 5″ wide and pieces 6′ and 7′ long shall be free of sapwood on one face; pieces 6″ and wider 8′ and longer will admit 1″ of sapwood in the aggregate on one face; such faces shall meet the grading requirements of Standard Selects. Unlimited sapwood is admitted on the reverse side.

NO. 1 COMMON: Standard, except:
Each cutting shall have one clear heartwood face.
NO. 2A COMMON: Standard, except:
Each cutting shall have one clear heartwood face.
Pieces below the grade of No. 2A Common shall be graded as Sap Gum.

Plain Sawn Red Gum, Figured Wood

Each piece shall be especially selected for markings and color tones of spots and streaks producing a variegated effect on the surface.

One piece of each required cutting shall show 90% in the aggregate of such markings and color tones, with the exception that unfigured spaces not exceeding 1″ by 24″ or its equivalent in area between spots and streaks, shall be disregarded.

Otherwise the rules for Plain Red Gum shall apply.

Quartered Red Gum

(No figure is required. Stain is admitted in the sapwood in all grades.)

FAS: Standard, except:

Widths 5″ and wider; pieces 5″ wide containing 3′ and 4′ surface measure shall be clear, pieces 5″ wide containing 5′ to 7′ surface measure shall yield 11/12 (91-2/3%) clear-face in one cutting.

In FAS, pieces 5″ wide shall be free of sapwood on one face; pieces 6″ and 7″ wide may have 3/4″ of sapwood in the aggregate on one face; pieces 8″ and wider may have 1″ of sapwood in the aggregate on one face. The reverse side of any piece will admit sapwood aggregating one-fifth of its surface. Any part of the sapwood allowed may be included in the cuttings.

SELECTS: Standard, except:
Pieces 4″ and 5″ wide shall be free of sapwood on one face; pieces 6″ and 7″ wide will admit 3/4″ of sapwood and pieces 8″ and wider 1″ of sapwood in the aggregate on one face, which faces shall meet the grading requirements of Standard Selects. Unlimited sapwood is admitted on the reverse side.
NO. 1 COMMON: Standard, except:
Each cutting shall have one clear heartwood face.

Bridge Plank and Crossing Plank

Widths: 6″ and wider.
Lengths: 8′ to 16′, 25% of the pieces in a shipment may be 1/4″ scant in thickness.
Will admit pin, shot, and spot worm holes; an occasional grub or knot hole; sound knots; split in each end not exceeding in length the width of the piece.
PITH: Firm pith may be admitted on one face in pieces 2-1/4″ or less in thickness. Firm pith may be admitted, either boxed or on one face, in pieces 2-1/2″ and thicker.
SHAKE: Shake may be admitted on the pith face not to exceed one-third the length of the piece in the aggregate.

In this grade no shake shall be admitted that extends from edge to edge; from edge to either face; or from one face to the opposite face.

In planking 18′ and longer, pith may be admitted on one face. On the opposite face it may be admitted up to one-sixth of the length of the piece in the aggregate.

In planking 18′ and longer, shake may be admitted to the extent permitted in standard lengths.

WANE: One face and two edges shall be sound except that wane not exceeding one-third the length, one-third the width, and one-third the thickness of the piece will be admitted.

Mine Lumber and Timber Products

Mixed Hardwood

Cribbing Blocks, Mine Caps, Wedges, Mine Rails, Mine Ties, Headers, Bars

Will admit pith, boxed or showing on one face and one edge; knots; season checks; splits and other defects that do not impair the strength or prevent the use of the piece in its full size for purposes of strength. Wane not exceeding one-third the width or the thickness is admitted on one corner, or its aggregate equivalent on two or more corners, or it may extend across only one face for one-third the length to a depth not exceeding one-twelfth the distance to the opposite face.

Sheet Piling, Sewer Sheathing, Hardwood Hearts

Will admit pith, boxed or showing on the surface; knots; checks; splits and defects commonly found in heart stock that do not seriously impair the strength or prevent the use of the piece in its full size. Wane not exceeding one-third the width or the thickness is admitted on one corner, or its aggregate equivalent on two or more corners, or it may extend across only one face for one-third the length to a depth not exceeding one-twelfth the distance to the opposite face.

Military or Commercial Timbers and Planking

This designation of quality shall consist of the grades of "Select Car Stock"; " Dimension"; and "Sound Square Edge." Timbers or planking sold in accordance with this designation must contain not less than 50% of the quality of Common Dimension and Better, of which 50%, one-half, must be of the grade of Select Car Stock. No material lower in quality than that defined under the caption of SOUND SQUARE EDGE shall be admitted.

Hardwood Construction and Utility Boards

Finish and Dimension

GRADES: "A" finish—"B" finish, NO. 1, NO. 2, and NO. 3 Construction Boards and Utility Boards, No. 1 and No. 2 Dimension. Rough or dressed as specified.

Nominal Rough and Dressed Thicknesses						
Nominal Rough	1″	1-1/4″	1-1/2″	2″	1-1/2″	3″
S1S or S2S	25/32″	1-1/16″	1-5/16″	1-5/8″	2-1/8″	2-5/8″

Nominal and Dressed Widths			
Nominal	Finish SIE or S2E	Construction and Utility Boards SIE or S2E	Dimension SIE or S2E
3 in.	2-5/8 in.	2-5/8 in.	
4 in.	3-5/8 in.	3-5/8 in.	3-5/8 in.
5 in.	4-5/8 in.	4-5/8 in.	
6 in.	5-5/8 in.	5-5/8 in.	5-5/8 in.
7 in.	6-5/8 in.	6-5/8 in.	
8 in.	7-1/2 in.	7-1/2 in.	7-1/2 in.
9 in.	8-1/2 in.	8-1/2 in.	
10 in.	9-1/2 in.	9-1/2 in.	9-1/2 in.
11 in.	10-1/2 in.	10-1/2 in.	
12 in.	11-1/2 in.	11-1/2 in.	11-1/2 in.
Over 12 in.	off 5/8 in.	off 5/8 in.	off 5/8 in.

No. 1 Common admits pieces that will yield clear-face cuttings as follows:

Surface Measure of Piece	Percentage of Yield	Number of Cuttings
2'	75	1
3' and 4'	66-2/3	1
	75	2
5' to 7'	66-2/3	2
	75	3
8' to 11"	66-2/3	3
12' and over	66-2/3	4

NO. 2A COMMON: Standard, except
Lengths: 6 ft. and longer.

There is no limit to the number of cuttings.

NO. 2B COMMON: All the requirements for No. 2A Common Mahogany shall apply except cuttings to be sound as defined in Sound Cutting.
NO. 3 COMMON: Standard, to include No. 3A Common and No. 3B Common as one grade, except lengths are 6 ft. and longer.

Tropical American, African Mahogany, and Spanish Cedar FAS 61 to 7-11" long (When Specified)

Widths: 6" and wider.

Pieces 3' surface measure shall be clear; 4' and over surface measure shall grade the same as Standard lengths in these woods.

Pin Wormy Mahogany

SELECTED FAS PIN WORMY (N.O. GRADE):
Widths: 6" and wider.
Lengths: 6' and longer.

One face of each board shall yield 75% clear of pin worm holes and other defects, in cuttings of not less than 144 square inches each and without limit to the number of cuttings. The reverse side shall grade No. 1 Common Pin Wormy (N Wormy) or better.

FAS PIN WORMY (A WORMY):
Widths: 6″ and wider.
Lengths: 6′ and longer.

Shall grade FAS Mahogany except as to lengths, and except that pin worm holes or grooves, sapwood, and stain will be admitted in the cuttings. Black track worm grooves shall not be admitted in the cuttings.

NO. 1 COMMON PIN WORMY (N WORMY);
Widths: 4″ and wider.
Lengths: 6′ and longer.

Shall grade No. 1 Common except that pin worm holes or grooves, burls, stain, 3/4″ sound knots, and equivalent defects are admitted in the cuttings, and that 50% of the required cuttings in the aggregate shall be free of black track worm grooves.

NO. 2 COMMON PIN WORMY (B WORMY):
Widths: 3″ and wider.
Lengths: 6′ and longer.

Shall grade No. 2 Common except that pin worm holes or grooves, burls, stain, small checks, 3/4″ sound knots, and equivalent defects are admitted in the cuttings. Black track worm grooves are admitted without limit in the cuttings.

Mahogany Shorts

GRADES: FAS Shorts, Common Shorts and Pin Wormy Shorts.
Standard Lengths: 2″, 2-1/4′, 2-1/2′, 2-3/4′, 3′, 3-1/4″, 3-1/2′, 3-3/4′, 4′, 4-1/4′, 4-1/2′, 4-3/4′, 5″, 5-1/4′, 5-1/2′.

Lengths other than standard shall be measured as of the next lower standard length.
Shorts shall be measured and tallied as if four times the actual standard length and the resulting total divided by four.

FAS:
Widths: 4″ and wider.
Pieces 4″ and 5″ wide shall be clear.

Pieces 6″ and wider will admit standard defects or their equivalent according to the above basis of surface measure (four times the actual surface measure) as follows:

8′, 1; 16′, 2; 22′, 3; 26′, 4.
COMMON SHORTS:
Widths: 3″ and wider.

Shall yield 50% clear face in not over two cuttings.
No cutting containing less than 36 square inches shall be considered.

PIN WORMY SHORTS: Widths: 3″ and wider.

Shall grade First and Seconds Shorts except as to minimum width, and pin worm holes or grooves, burls and stain are admitted.

Mahogany Strips

Inspection shall be made from the better face of the piece.

Odd lengths are admitted without limit. Fractions over one-half foot in length shall be counted up, and fractions of one-half foot or less in length shall be dropped. This does not change the minimum length requirement of Strips.

The widths in Clear and No. 1 Common Strips are 2″, 2-1/2″, 3″, 3-1/2″, 4″, 4-1/2″, 5″ and 5-1/2″.

Strips may be 1/8″ scant in width when shipping dry. In Clear Strips, tapering pieces shall be measured at the narrow end. In the grades of No. 1 Common and Wormy Strips, tapering pieces shall be measured one-third the length of the piece from the narrow end.

Sapwood is admitted without limit in all grades.
CLEAR:
Lengths: 6′ and longer.

Shall have one clear face, the reverse side will admit wane or its equivalent in other defects, not exceeding one-third the length, one-third the width and one-third the thickness of the piece and shall otherwise be sound as defined in "Sound Cutting."
NO. 1 COMMON:
Lengths: 6′ and longer.

Both edges of pieces 6′ and 7′ long and both edges of each cutting in 8′ and longer shall be clear. In addition to the above requirements, pieces 6′ and 7′ long will admit one standard defect; 8′ and longer shall yield 66-2/3% clear face in not over two cuttings in 8′ to 11′ and not over three cuttings in 12′ and longer. No cutting shall be less than 2′ long nor less than 2″ wide in pieces 2″ and 2-1/2″ wide, nor less than 3″ wide in pieces 3″ and wider.

The reverse side of the cuttings to be sound as defined in "Sound Cutting."

WORMY:
Lengths: 6′ and longer.
Widths: 1-1/2″, 2″, 3″, 3-1/2″, and 3-3/4″.

Shall yield 50% sound, no cutting to be considered that is less than 1-1/2″ wide by 2′ long. The number of cuttings not limited. Pin worm holes, clear or stained pin worm grooves, burls, stain, small checks, sound knots not over 3/4″ in average diameter, or other sound defects not exceeding in extent or damage the defects described will be admitted without limit.

One edge of each piece shall be square; the other edge will admit wane not exceeding in thickness or width the thickness of the piece and not exceeding one-third the length of the piece or the equivalent of such aggregate wane at one or both ends.

Philippine Mahogany

"Philippine Red Mahogany" includes Tanguile, Red Lauan, and Tiaong.

"Light Red Philippine Mahogany" includes Almond, Bagtican, Mayapis, and White Lauan.

NOTE: *National Hardwood Lumber Association inspectors will undertake to make distinction between Philippine Red Mahogany and Light Red Philippine Mahogany when required, but the Association does not assume financial liability with respect to color.*

Odd lengths are admitted without limit.

In FAS, bright sapwood not exceeding in the aggregate one-third the width of the piece will be admitted on one face. Any part of the sapwood may be included in the cuttings.

COUNTERS (when specified):
Widths: 18″ to 24″.
Lengths: 12′ to 40′.

Counters shall be free of all defects on one face; the reverse side shall grade not below FAS. Splits shall be measured out.

FAS: Standard, except:
Pieces of 4′ and 5′ surface measure shall yield 11/12 (91-2/3%) clear face in one cutting.
F1F: Standard.
SELECTS: Standard, except:
Widths: 6″ and wider.
Lengths: 8′ and longer.
NO. 1 COMMON: Standard, except:
Widths: 4″ and wider
Lengths: 6′ and longer.

Splits exceeding in the aggregate in inches in length twice the surface measure of the piece in feet shall not be admitted.

NO. 2A COMMON: Standard, except:
Lengths: 6′ and longer.
There is no limit to the number of cuttings.
NO. 2B COMMON: All the requirements for No. 2A Common Philippine Mahogany shall apply except cuttings to be sound as defined in Sound Cutting.
NO. 3 COMMON: Standard, to include No. 3A Common and No. 3B Common as one grade, except lengths are 2′ and longer.

Pin Wormy Philippine

FAS PIN WORMY: Same as FAS, except:
Pin worm holes are admitted without limit.
Widths: 6″ and wider.
Lengths: 8′ and longer.

Scattered stained pin worm grooves not exceeding 25% of the required cutting area are admitted.

NO. 1 COMMON WORMY: Same as No. 1 Common except:

Pin worm holes, stained or otherwise, pin worm grooves, burls, sound pin knots not exceeding 1/2″ in diameter or other sound defects that do not exceed in extent of damage the defects described, are admitted.

Philippine Mahogany Shorts

FAS SHORTS:
Widths: 3″ and wider.
Standard Lengths: 2′ 2-1/2′, 3′, 3-1/2′, 4′, 4-1/2′, 5′ and 5-1/2′.

Lengths other than standard shall be measured as of the next lower standard length.
Shorts shall be measured and tallied as if four times the actual standard length, and the resulting total divided by four.

Pieces 3″ to 5″ wide shall be clear.

Pieces 6″ and wider will admit standard defects or their equivalent according to the above basis of surface Measure (four times the actual surface measure) as follows:

8′, 1; 16′, 2; 22′, 3; 26′, 4.
COMMON SHORTS:

Widths: 3" and wider.
Standard Lengths: 2', 2-1/2', 3', 3-1/2', 4', 4-1/2', 5' and 5-1/2'.

Lengths other than standard shall be measured as of the next lower standard length.
Shorts shall be measured and tallied as if four times the actual length and the resulting total divided by four.
Shall yield 50% clear face in not over two cuttings. No cutting containing less than 36 square inches shall be considered.

Strips

STRIPS:
CLEAR: Standard Strip Grade, except:
Lengths: 6' and longer, admitting 10% of 6' and 7'.
NO. 1 COMMON: Standard Strip Grade, except:
6' and 7' lengths may have one standard defect. Each cutting shall have clear edges.

Flitches

CLEAR VENEER FLITCHES:
Thicknesses: 6" and thicker.
Widths: 8" and wider.
Lengths: 8' and longer.

Shall be clear on one face and two edges; the reverse side shall grade not below FAS. Knots admitted on the reverse side shall be sound and not exceeding 3/4" in diameter. Sapwood and worm defects are not admitted.

PIN WORMY FLITCHES:
Thicknesses: 6" and thicker.
Widths: 8" and wider.
Lengths: 8' and longer.

Will admit stained pin worm holes and grooves, pin knots not exceeding 5/8" in diameter without limit. Sapwood not exceeding one-sixth the thickness and showing on one face will be admitted.

NO. 1 COMMON FLITCHES:
Thicknesses: 6" and thicker.
Widths: 8" and wider.
Lengths: 8' and longer.
Shall yield 66-2/3% in clear-face cuttings.
Minimum cutting 4" wide by 3' long.

Ribbon Stripe

When ribbon stripe figure is specified in lumber or flitches, each piece shall be selected for the stripe effect brought out in quarter sawing. One face of each required cutting shall show 90% or more in the aggregate of such ribbon stripe figure.

Apitong and Other Philippine Hardwoods

All other Philippine hardwoods shall be graded under the rules for the inspection of Philippine Mahogany.

NOTE: National Hardwood Lumber Association inspectors will not accept responsibility for the distinction between Apitong and other Philippine hardwoods included in this classification and when issuing Association certificates for the grading of these woods will use the term "Said to be."

North American, Tropical American, and African Hardwoods (Other than Mahogany and Spanish Cedar)

Unless otherwise specified, North American, Tropical American and African hardwoods for which there are no established grading rules, shall be graded under the STANDARD GRADES.

NOTE: National inspectors will not accept responsibility for distinguishing the species of woods included in this classification and when issuing certificates for the grading will use the term "Said to be."

Aromatic Red Cedar

GRADES: NO. 1 COMMON AND BETTER and NO. 2A COMMON:
Will admit sound knots, white streaks and firm, tight pith in the cuttings, which otherwise shall be sound.
No cutting may contain sapwood in the aggregate exceeding one-sixth the heartwood side. Unlimited sapwood is admitted on the reverse side.

Variation in thickness may be 1/2″ on 4/4″ to 8/4″.
Thicknesses: Standard.
NO. 1 COMMON AND BETTER:
Widths: 3″ and wider, admitting 25% of 3″ width.
Lengths: 3′ and longer.
Minimum cutting: 3″ wide by 2′ long or 2″ wide by 3′ long.
Each piece shall yield not less than 66-2/3% of cuttings.
There is no limit to the number of cuttings.
NO. 2A COMMON:
Widths: 2″ and wider, admitting-35% of 2″ width.
Lengths: 2′ and longer.

Minimum Cutting: 2″ or wider containing not less than 48 square inches.
Each piece shall yield not less than 50% of cuttings.
There is no limit to the number of cuttings.

Pacific Coast Red Alder, Pacific Coast Maple

When Pacific Coast Alder or Maple is sold and specified "Pin knots no defect," knots or their equivalent, not exceeding 1/4″ in their greatest dimension, sound or containing unsound centers not over 1/8″ in diameter, shall be admitted in the cuttings.
The General Instructions and Standard Grades (pages 6–21) shall govern the measurement and inspection of all commercial hardwoods indigenous to the Northwest hardwood belt, with the exceptions as set forth under the respective species.
These rules shall apply to green, dry, rough, or surfaced lumber.
The better face of boards in all thicknesses shall yield not less than the minimum percentage of cuttings required for the grade; the reverse side of the cuttings in all thicknesses shall be sound as defined in "SOUND CUTTING," or better, unless otherwise specified.
No exception shall be made to these rules unless agreed to by the seller and the buyer and specifically stated in the purchase order.

7.3.0 Various Strengths—Static Bending and Compression of Softwoods and Hardwoods

Strength Properties of Some Commercially Important Woods Grown in the United States (inch-pound)[a]

Common species names	Moisture content	Specific gravity[b]	Modulus of rupture (lbf/in^2)	Static bending			Compression parallel to grain (lbf/in^2)	Compression perpendicular to grain (lbf/in^2)	Shear parallel to grain (lbf/in^2)	Tension perpendicular to grain (lbf/in^2)	Side hardness (lbf)
				Modulus of elasticity[c] ($\times 10^6$ lbf/in^2)	Work to maximum load (in-lbf/in^3)	Impact bending (in.)					
Hickory, pecan											
Bitternut	Green	0.60	10,300	1.40	20.0	66	4,570	800	1,240	—	—
	12%	0.66	17,100	1.79	18.2	66	9,040	1,680	—	—	—
Nutmeg	Green	0.56	9,100	1.29	22.8	54	3,980	760	1,030	—	—
	12%	0.60	16,600	1.70	25.1	—	6,910	1,570	—	—	—
Pecan	Green	0.60	9,800	1.37	14.6	53	3,990	780	1,480	680	1,310
	12%	0.66	13,700	1.73	13.8	44	7,850	1,720	2,080	—	1,820
Water	Green	0.61	10,700	1.56	18.8	56	4,660	880	1,440	—	—
	12%	0.62	17,800	2.02	19.3	53	8,600	1,550	—	—	—
Hickory, true											
Mockernut	Green	0.64	11,100	1.57	26.1	88	4,480	810	1,280	—	—
	12%	0.72	19,200	2.22	22.6	77	8,940	1,730	1,740	—	—
Pignut	Green	0.66	11,700	1.65	31.7	89	4,810	920	1,370	—	—
	12%	0.75	20,100	2.26	30.4	74	9,190	1,980	2,150	—	—
Shagbark	Green	0.64	11,000	1.57	23.7	74	4,580	840	1,520	—	—
	12%	0.72	20,200	2.16	25.8	67	9,210	1,760	2,430	—	—
Shellbark	Green	0.62	10,500	1.34	29.9	104	3,920	810	1,190	—	—
	12%	0.69	18,100	1.89	23.6	88	8,000	1,800	2,110	—	—
Honeylocust	Green	0.60	10,200	1.29	12.6	47	4,420	1,150	1,660	930	1,390
	12%	—	14,700	1.63	13.3	47	7,500	1,840	2,250	900	1,580
Locust, black	Green	0.66	13,800	1.85	15.4	44	6,800	1,160	1,760	770	1,570
	12%	6.69	19,400	2.05	18.4	57	10,180	1,830	2,480	640	1,700
Magnolia											
Cucumbertree	Green	0.44	7,400	1.56	10.0	30	3,140	330	990	440	520
	12%	0.48	12,300	1.82	12.2	35	6,310	570	1,340	660	700
Southern	Green	0.46	6,800	1.11	15.4	54	2,700	460	1,040	610	740
	12%	0.50	11,200	1.40	12.8	29	5,460	860	1,530	740	1,020
Maple											
Bigleaf	Green	0.44	7,400	1.10	8.7	23	3,240	450	1,110	600	620
	12%	0.48	10,700	1.45	7.8	28	5,950	750	1,730	540	850
Black	Green	0.52	7,900	1.33	12.8	48	3,270	600	1,130	720	840
	12%	0.57	13,300	1.62	12.5	40	6,680	1,020	1,820	670	1,180

Red	Green	0.49	7,700	1.39	11.4	32	3,280	400	1,150	—	700
	12%	0.54	13,400	1.64	12.5	32	6,540	1,000	1,850	—	950
Silver	Green	0.44	5,800	0.94	11.0	29	2,490	370	1,050	560	590
	12%	0.47	8,900	1.14	8.3	25	5,220	740	1,480	500	700
Sugar	Green	0.56	9,400	1.55	13.3	40	4,020	640	1,460	—	970
	12%	0.63	15,800	1.83	16.5	39	7,830	1,470	2,330	—	1,450
Oak, red											
Black	Green	0.56	8,200	1.18	12.2	40	3,470	710	1,220	—	1,060
	12%	0.61	13,900	1.64	13.7	41	6,520	930	1,910	—	1,210
Cherrybark	Green	0.61	10,800	1.79	14.7	54	4,620	760	1,320	800	1,240
	12%	0.68	18,100	2.28	18.3	49	8,740	1,250	2,000	840	1,480
Laurel	Green	0.56	7,900	1.39	11.2	39	3,170	570	1,180	770	1,000
	12%	0.63	12,600	1.69	11.8	39	6,980	1,060	1,830	790	1,210
Northern red	Green	0.56	8,300	1.35	13.2	44	3,440	610	1,210	750	1,000
	12%	0.63	14,300	1.82	14.5	43	6,760	1,010	1,780	800	1,290
Pin	Green	0.58	8,300	1.32	14.0	48	3,680	720	1,290	800	1,070
	12%	0.63	14,000	1.73	14.8	45	6,820	1,020	2,080	1,050	1,510
Scarlet	Green	0.60	10,400	1.48	15.0	54	4,090	830	1,410	700	1,200
	12%	0.67	17,400	1.91	20.5	53	8,330	1,120	1,890	870	1,400
Southern red	Green	0.52	6,900	1.14	8.0	29	3,030	550	930	480	860
	12%	0.59	10,900	1.49	9.4	26	6,090	870	1,390	510	1,060
Port-Orford	Green	0.39	6,600	1.30	7.4	21	3,140	300	840	180	380
	12%	0.43	12,700	1.70	9.1	28	6,250	720	1,370	400	630
Western redcedar	Green	0.31	5,200	0.94	5.0	17	2,770	240	770	230	260
	12%	0.32	7,500	1.11	5.8	17	4,560	460	990	220	350
Yellow	Green	0.42	6,400	1.14	9.2	27	3,050	350	840	330	440
	12%	0.44	11,100	1.42	10.4	29	6,310	620	1,130	360	580
Douglas-fir[d]											
Coast	Green	0.45	7,700	1.56	7.6	26	3,780	380	900	300	500
	12%	0.48	12,400	1.95	9.9	31	7,230	800	1,130	340	710
Interior West	Green	0.46	7,700	1.51	7.2	26	3,870	420	940	290	510
	12%	0.50	12,600	1.83	10.6	32	7,430	760	1,290	350	660
Interior North	Green	0.45	7,400	1.41	8.1	22	3,470	360	950	340	420
	12%	0.48	13,100	1.79	10.5	26	6,900	770	1,400	390	600
Interior South	Green	0.43	6,800	1.16	8.0	15	3,110	340	950	250	360
	12%	0.46	11,900	1.49	9.0	20	6,230	740	1,510	330	510
Fir											
Balsam	Green	0.33	5,500	1.25	4.7	16	2,630	190	662	180	290
	12%	0.35	9,200	1.45	5.1	20	5,280	404	944	180	400
California red	Green	0.36	5,800	1.17	6.4	21	2,760	330	770	380	360
	12%	0.38	10,500	1.50	8.9	24	5,460	610	1,040	390	500
Grand	Green	0.35	5,800	1.25	5.6	22	2,940	270	740	240	360
	12%	0.37	8,900	1.57	7.5	28	5,290	500	900	240	490
Noble	Green	0.37	6,200	1.38	6.0	19	3,010	270	800	230	290

(Continued)

Strength Properties of Some Commercially Important Woods Grown in the United States (inch-pound)[a]—Cont'd

Common species names	Moisture content	Specific gravity[b]	Modulus of rupture (lbf/in²)	Static bending Modulus of elasticity[c] (×10⁶ lbf/in²)	Work to maximum load (in-lbf/in³)	Impact bending (in.)	Compression parallel to grain (lbf/in²)	Compression perpendicular to grain (lbf/in²)	Shear parallel to grain (lbf/in²)	Tension perpendicular to grain (lbf/in²)	Side hardness (lbf)
Pacific silver	12%	0.39	10,700	1.72	8.8	23	6,100	520	1,050	220	410
	Green	0.40	6,400	1.42	6.0	21	3,140	220	750	240	310
	12%	0.43	11,000	1.76	9.3	24	6,410	450	1,220	—	430
Subalpine	Green	0.31	4,900	1.05	—	—	2,300	190	700	—	260
	12%	0.32	8,600	1.29	—	—	4,860	390	1,070	—	350
White	Green	0.37	5,900	1.16	5.6	22	2,900	280	760	300	340
	12%	0.39	9,800	1.50	7.2	20	5,800	530	1,100	300	480
Hemlock											
Eastern	Green	0.38	6,400	1.07	6.7	21	3,080	360	850	230	400
	12%	0.40	8,900	1.20	6.8	21	5,410	650	1,060	—	500
Mountain	Green	0.42	6,300	1.04	11.0	32	2,880	370	930	330	470
	12%	0.45	11,500	1.33	10.4	32	6,440	860	1,540	—	680
Western	Green	0.42	6,600	1.31	6.9	22	3,360	280	860	290	410
	12%	0.45	11,300	1.63	8.3	23	7,200	550	1,290	340	540
Larch, western	Green	0.48	7,700	1.46	10.3	29	3,760	400	870	330	510
	12%	0.52	13,000	1.87	12.6	35	7,620	930	1,360	430	830
Pine											
Eastern white	Green	0.34	4,900	0.99	5.2	17	2,440	220	680	250	290
	12%	0.35	8,600	1.24	6.8	18	4,800	440	900	310	380
Jack	Green	0.40	6,000	1.07	7.2	26	2,950	300	750	360	400
	12%	0.43	9,900	1.35	8.3	27	5,660	580	1,170	420	570
Loblolly	Green	0.47	7,300	1.40	8.2	30	3,510	390	860	260	450
	12%	0.51	12,800	1.79	10.4	30	7,130	790	1,390	470	690
Lodgepole	Green	0.38	5,500	1.08	5.6	20	2,610	250	680	220	330
	12%	0.41	9,400	1.34	6.8	20	5,370	610	880	290	480
Longleaf	Green	0.554	8,500	1.59	8.9	35	4,320	480	1,040	330	590
	12%	0.59	14,500	1.98	11.8	34	8,470	960	1,510	470	870
Pitch	Green	0.47	6,800	1.20	9.2	—	2,950	360	860	—	—
	12%	0.52	10,800	1.43	9.2	—	5,940	820	1,360	—	—
Hardwoods											
Alder, red	Green	0.37	6,500	1.17	8.0	22	2,960	250	770	390	440
	12%	0.41	9,800	1.38	8.4	20	5,820	440	1,080	420	590
Ash											
Black	Green	0.45	6,000	1.04	12.1	33	2,300	350	860	490	520

Species	Condition									
Blue	Green	0.53	9,600	1.24	14.7	—	4,180	810	1,540	—	—
	12%	0.58	13,800	1.40	14.4	—	6,980	1,420	2,030	—	—
Green	Green	0.53	9,500	1.40	11.8	35	4,200	730	1,260	590	870
	12%	0.56	14,100	1.66	13.4	32	7,080	1,310	1,910	700	1,200
Oregon	Green	0.50	7,600	1.13	12.2	39	3,510	530	1,190	590	790
	12%	0.55	12,700	1.36	14.4	33	6,040	1,250	1,790	720	1,160
White	Green	0.55	9,500	1.44	15.7	38	3,990	670	1,350	590	960
	12%	0.60	15,000	1.74	16.6	43	7,410	1,160	1,910	940	1,320
Aspen											
Bigtooth	Green	0.36	5,400	1.12	5.7	—	2,500	210	730	—	—
	12%	0.39	9,100	1.43	7.7	—	5,300	450	1,080	—	—
Quaking	Green	0.35	5,100	0.86	6.4	22	2,140	180	660	230	300
	12%	0.38	8,400	1.18	7.6	21	4,250	370	850	260	350
Basswood, American	Green	0.32	5,000	1.04	5.3	16	2,220	170	600	280	250
	12%	0.37	8,700	1.46	7.2	16	4,730	370	990	350	410
Beech, American	Green	0.56	8,600	1.38	11.9	43	3,550	540	1,290	720	850
	12%	0.64	14,900	1.72	15.1	41	7,300	1,010	2,010	1,010	1,300
Birch											
Paper	Green	0.48	6,400	1.17	16.2	49	2,360	270	840	380	560
	12%	0.55	12,300	1.59	16.0	34	5,690	600	1,210	—	910
Sweet	Green	0.60	9,400	1.65	15.7	48	3,740	470	1,240	430	970
	12%	0.65	16,900	2.17	18.0	47	8,540	1,080	2,240	950	1,470
Yellow	Green	0.55	8,300	1.50	16.1	48	3,380	430	1,110	430	780
	12%	0.62	16,600	2.01	20.8	55	8,170	970	1,880	920	1,260
Butternut	Green	0.36	5,400	0.97	8.2	24	2,420	220	760	430	390
	12%	0.38	8,100	1.18	8.2	24	5,110	460	1,170	440	490
Cherry, black	Green	0.47	8,000	1.31	12.8	33	3,540	360	1,130	570	660
	12%	0.50	12,300	1.49	11.4	29	7,110	690	1,700	560	950
Chestnut, American	Green	0.40	5,600	0.93	7.0	24	2,470	310	800	440	420
	12%	0.43	8,600	1.23	6.5	19	5,320	620	1,080	460	540
Cottonwood											
Balsam, poplar	Green	0.31	3,900	0.75	4.2	—	1,690	140	500	—	—
	12%	0.34	6,800	1.10	5.0	—	4,020	300	790	—	—
Black	Green	0.31	4,900	1.08	5.0	20	2,200	160	610	270	250
	12%	0.35	8,500	1.27	6.7	22	4,500	300	1,040	330	350
Eastern	Green	0.37	5,300	1.01	7.3	21	2,280	200	680	410	340
	12%	0.40	8,500	1.37	7.4	20	4,910	380	930	580	430

(Continued)

Strength Properties of Some Commercially Important Woods Grown in the United States (inch-pound)[a]—Cont'd

Common species names	Moisture content	Specific gravity[b]	Static bending				Compression parallel to grain (lbf/in²)	Compression perpendicular to grain (lbf/in²)	Shear parallel to grain (lbf/in²)	Tension perpendicular to grain (lbf/in²)	Side hardness (lbf)
			Modulus of rupture (lbf/in²)	Modulus of elasticity[c] (×10⁶ lbf/in²)	Work to maximum load (in-lbf/in³)	Impact bending (in.)					
Elm											
American	Green	0.46	7,200	1.11	11.8	38	2,910	360	1,000	590	620
	12%	0.50	11,800	1.34	13.0	39	5,520	690	1,510	660	830
Rock	Green	0.57	9,500	1.19	19.8	54	3,780	610	1,270	—	940
	12%	0.63	14,800	1.54	19.2	56	7,050	1,230	1,920	—	1,320
Slippery	Green	0.48	8,000	1.23	15.4	47	3,320	420	1,110	640	660
	12%	0.53	13,000	1.49	16.9	45	6,360	820	1,630	530	860
Hackberry	Green	0.49	6,500	0.95	14.5	48	2,650	400	1,070	630	700
	12%	0.53	11,000	1.19	12.8	43	5,440	890	1,590	580	880
Pond	Green	0.51	7,400	1.28	7.5	—	3,660	440	940	—	—
	12%	0.56	11,600	1.75	8.6	—	7,540	910	1,380	—	—
Ponderosa	Green	0.38	5,100	1.00	5.2	21	2,450	280	700	310	320
	12%	0.40	9,400	1.29	7.1	19	5,320	580	1,130	420	460
Red	Green	0.41	5,800	1.28	6.1	26	2,730	260	690	300	340
	12%	0.46	11,000	1.63	9.9	26	6,070	600	1,210	460	560
Sand	Green	0.46	7,500	1.02	9.6	—	3,440	450	1,140	—	—
	12%	0.48	11,600	1.41	9.6	—	6,920	836	—	—	—
Shortleaf	Green	0.47	7,400	1.39	8.2	30	3,530	350	910	320	440
	12%	0.51	13,100	1.75	11.0	33	7,270	820	1,390	470	690
Slash	Green	0.54	8,700	1.53	9.6	—	3,820	530	960	—	—
	12%	0.59	16,300	1.98	13.2	—	8,140	1020	1,680	—	—
Spruce	Green	0.41	6,000	1.00	—	—	2,840	280	900	—	450
	12%	0.44	10,400	1.23	—	—	5,650	730	1,490	—	660
Sugar	Green	0.34	4,900	1.03	5.4	17	2,460	210	720	270	270
	12%	0.36	8,200	1.19	5.5	18	4,460	500	1,130	350	380
Virginia	Green	0.45	7,300	1.22	10.9	34	3,420	390	890	400	540
	12%	0.48	13,000	1.52	13.7	32	6,710	910	1,350	380	740
Western white	Green	0.35	4,700	1.19	5.0	19	2,430	190	680	260	260
	12%	0.38	9,700	1.46	8.8	23	5,040	470	1,040	—	420
Redwood											
Old-growth	Green	0.38	7,500	1.18	7.4	21	4,200	420	800	260	410
	12%	0.40	10,000	1.34	6.9	19	6,150	700	940	240	480
Young-growth	Green	0.34	5,900	0.96	5.7	16	3,110	270	890	300	350
	12%	0.35	7,900	1.10	5.2	15	5,220	520	1,110	250	420

Spruce										
Black	Green	0.38	6,100	1.38	7.4	24	2,840	240	739	100	370
	12%	0.42	10,800	1.61	10.5	23	5,960	550	1,230	—	520
Engelmann	Green	0.33	4,700	1.03	5.1	16	2,180	200	640	240	260
	12%	0.35	9,300	1.30	6.4	18	4,480	410	1,200	350	390
Red	Green	0.37	6,000	1.33	6.9	18	2,720	260	750	220	350
	12%	0.40	10,800	1.61	8.4	25	5,540	550	1,290	350	490
Sitka	Green	0.37	5,700	1.23	6.3	24	2,670	280	760	250	350
	12%	0.40	10,200	1.57	9.4	25	5,610	580	1,150	370	510
White	Green	0.33	5,000	1.14	6.0	22	2,350	210	640	220	320
	12%	0.36	9,400	1.43	7.7	20	5,180	430	970	360	480
Tamarack	Green	0.49	7,200	1.24	7.2	28	3,480	390	860	260	380
	12%	0.53	11,600	1.64	7.1	23	7,160	800	1,280	400	590

[a] Results of tests on small clear specimens in the green and air-dried conditions. Definition of properties: impact bending is height of drop that causes complete failure, using 0.71-kg (50-lb) hammer; compression parallel to grain is also called maximum crushing strength; compression perpendicular to grain is fiber stress at proportional limit; shear is maximum shearing strength; tension is maximum tensile strength; and side hardness is hardness measured when load is perpendicular to grain.
[b] Specific gravity is based on weight when oven dry and volume when green or at 12% moisture content.
[c] Modulus of elasticity measured from a simply supported, center-loaded beam, on a span depth ratio of 14/1. To correct for shear deflection, the modulus can be increased by 10%.
[d] Coast Douglas-fir is defined as Douglas-fir growing in Oregon and Washington State west of the Cascade Mountains summit. Interior West includes California and all counties in Oregon and Washington east of, but adjacent to, the Cascade summit; Interior North, the remainder of Oregon and Washington plus Idaho, Montana, and Wyoming; and Interior South, Utah, Colorado, Arizona, and New Mexico.

7.4.0 Equilibrium Moisture Content of Wood in Each of 50 States

Equilibrium Moisture Content of Wood, Exposed to Outdoor Atmosphere, in Several U.S. Locations in 1997

		Equilibrium moisture content[a] (%)											
State	City	Jan.	Feb.	Mar.	Apr.	May	June	July	Aug.	Sept.	Oct.	Nov.	Dec.
AK	Juneau	16.5	16.0	15.1	13.9	13.6	13.9	15.1	16.5	18.1	18.0	17.7	18.1
AL	Mobile	13.8	13.1	13.3	13.3	13.4	13.3	14.2	14.4	13.9	13.0	13.7	14.0
AZ	Flagstaff	11.8	11.4	10.8	9.3	8.8	7.5	9.7	11.1	10.3	10.1	10.8	11.8
AZ	Phoenix	9.4	8.4	7.9	6.1	5.1	4.6	6.2	6.9	6.9	7.0	8.2	9.5
AR	Little Rock	13.8	13.2	12.8	13.1	13.7	13.1	13.3	13.5	13.9	13.1	13.5	13.9
CA	Fresno	16.4	14.1	12.6	10.6	9.1	8.2	7.8	8.4	9.2	10.3	13.4	16.6
CA	Los Angeles	12.2	13.0	13.8	13.8	14.4	14.8	15.0	15.1	14.5	13.8	12.4	12.1
CO	Denver	10.7	10.5	10.2	9.6	10.2	9.6	9.4	9.6	9.5	9.5	11.0	11.0
DC	Washington	11.8	11.5	11.3	11.1	11.6	11.7	11.7	12.3	12.6	12.5	12.2	12.2
FL	Miami	13.5	13.1	12.8	12.3	12.7	14.0	13.7	14.1	14.5	13.5	13.9	13.4
GA	Atlanta	13.3	12.3	12.0	11.8	12.5	13.0	13.8	14.2	13.9	13.0	12.9	13.2
HI	Honolulu	13.3	12.8	11.9	11.3	10.8	10.6	10.6	10.7	10.8	11.3	12.1	12.9
ID	Boise	15.2	13.5	11.1	10.0	9.7	9.0	7.3	7.3	8.4	10.0	13.3	15.2
IL	Chicago	14.2	13.7	13.4	12.5	12.2	12.4	12.8	13.3	13.3	12.9	14.0	14.9
IN	Indianapolis	15.1	14.6	13.8	12.8	13.0	12.8	13.9	14.5	14.2	13.7	14.8	15.7
IA	Des Moines	14.0	13.9	13.3	12.6	12.4	12.6	13.1	13.4	13.7	12.7	13.9	14.9
KS	Wichita	13.8	13.4	12.4	12.4	13.2	12.5	11.5	11.8	12.6	12.4	13.2	13.9
KY	Louisville	13.7	13.3	12.6	12.0	12.8	13.0	13.3	13.7	14.1	13.3	13.5	13.9
LA	New Orleans	14.9	14.3	14.0	14.2	14.1	14.6	15.2	15.3	14.8	14.0	14.2	15.0
ME	Portland	13.1	12.7	12.7	12.1	12.6	13.0	13.0	13.4	13.9	13.8	14.0	13.5
MA	Boston	11.8	11.6	11.9	11.7	12.2	12.1	11.9	12.5	13.1	12.8	12.6	12.2
MI	Detroit	14.7	14.1	13.5	12.6	12.3	12.3	12.6	13.3	13.7	13.5	14.4	15.1
MN	Minneapolis-St. Paul	13.7	13.6	13.3	12.0	11.9	12.3	12.5	13.2	13.8	13.3	14.3	14.6
MS	Jackson	15.1	14.4	13.7	13.8	14.1	13.9	14.6	14.6	14.6	14.1	14.3	14.9
MO	St. Louis	14.5	14.1	13.2	12.4	12.8	12.6	12.9	13.3	13.7	13.1	14.0	14.9
MT	Missoula	16.7	15.1	12.8	11.4	11.6	11.7	10.1	9.8	11.3	12.9	16.2	17.6
NE	Omaha	14.0	13.8	13.0	12.1	12.6	12.9	13.3	13.8	14.0	13.0	13.9	14.8
NV	Las Vegas	8.5	7.7	7.0	5.5	5.0	4.0	4.5	5.2	5.3	5.9	7.2	8.4
NV	Reno	12.3	10.7	9.7	8.8	8.8	8.2	7.7	7.9	8.4	9.4	10.9	12.3
NM	Albuquerque	10.4	9.3	8.0	6.9	6.8	6.4	8.0	8.9	8.7	8.6	9.6	10.7
NY	New York	12.2	11.9	11.5	11.0	11.5	11.8	11.8	12.4	12.6	12.3	12.5	12.3
NC	Raleigh	12.8	12.1	12.2	11.7	13.1	13.4	13.8	14.5	14.5	13.7	12.9	12.8
ND	Fargo	14.2	14.6	15.2	12.9	11.9	12.9	13.2	13.2	13.7	13.5	15.2	15.2
OH	Cleveland	14.6	14.2	13.7	12.6	12.7	12.7	12.8	13.7	13.8	13.3	13.8	14.6
OK	Oklahoma City	13.2	12.9	12.2	12.1	13.4	13.1	11.7	11.8	12.9	12.3	12.8	13.2
OR	Pendleton	15.8	14.0	11.6	10.6	9.9	9.1	7.4	7.7	8.8	11.0	14.6	16.5
OR	Portland	16.5	15.3	14.2	13.5	13.1	12.4	11.7	11.9	12.6	15.0	16.8	17.4
PA	Philadelphia	12.6	11.9	11.7	11.2	11.8	11.9	12.1	12.4	13.0	13.0	12.7	12.7
SC	Charleston	13.3	12.6	12.5	12.4	12.8	13.5	14.1	14.6	14.5	13.7	13.2	13.2
SD	Sioux Falls	14.2	14.6	14.2	12.9	12.6	12.8	12.6	13.3	13.6	13.0	14.6	15.3
TN	Memphis	13.8	13.1	12.4	12.2	12.7	12.8	13.0	13.1	13.2	12.5	12.9	13.6
TX	Dallas-Ft.Worth	13.6	13.1	12.9	13.2	13.9	13.0	11.6	11.7	12.9	12.8	13.1	13.5
TX	El Paso	9.6	8.2	7.0	5.8	6.1	6.3	8.3	9.1	9.3	8.8	9.0	9.8
UT	Salt Lake City	14.6	13.2	11.1	10.0	9.4	8.2	7.1	7.4	8.5	10.3	12.8	14.9
VA	Richmond	13.2	12.5	12.0	11.3	12.1	12.4	13.0	13.7	13.8	13.5	12.8	13.0
WA	Seattle-Tacoma	15.6	14.6	15.4	13.7	13.0	12.7	12.2	12.5	13.5	15.3	16.3	16.5

Equilibrium Moisture Content of Wood, Exposed to Outdoor Atmosphere, in Several U.S. Locations in 1997—Cont'd

State	City	\multicolumn{12}{c}{Equilibrium moisture content[a] (%)}
		Jan.	Feb.	Mar.	Apr.	May	June	July	Aug.	Sept.	Oct.,	Nov.	Dec.
WI	Madison	14.5	14.3	14.1	12.8	12.5	12.8	13.4	14.4	14.9	14.1	15.2	15.7
WV	Charleston	13.7	13.0	12.1	11.4	12.5	13.3	14.1	14.3	14.0	13.6	13.0	13.5
WY	Cheyenne	10.2	10.4	10.7	10.4	10.8	10.5	9.9	9.9	9.7	9.7	10.6	10.6

[a] EMC values were determined from the average of 30 or more years of relative humidity and temperature data available from the National Climatic Data Center of the National Oceanic and Atmospheric Administration.

7.4.1 Coefficients for Dimensional Change as result of Shrinkage

Coefficients for Dimensional Change as a Result of Shrinking or Swelling within Moisture Content Limits of 6% to 14% (C_T = dimensional change coefficient for tangential direction; C_R = radial direction)

Species	Dimensional change coefficient[a]		Species	Dimensional change coefficient[a]	
	C_R	C_T		C_R	C_T
Hardwoods					
Alder, red	0.00151	0.00256	Honeylocust	0.00144	0.00230
Apple	0.00205	0.00376	Locust, black	0.00158	0.00252
Ash, black	0.00172	0.00274	Madrone, Pacific	0.00194	0.00451
Ash, Oregon	0.00141	0.00285	Magnolia, cucumbertree	0.00180	0.00312
Ash, pumpkin	0.00126	0.00219	Magnolia, southern	0.00187	0.00230
Ash, white	0.00169	0.00274	Magnolia, sweetbay	0.00162	0.00293
Ash, green	0.00169	0.00274	Maple, bigleaf	0.00126	0.00248
Aspen, quaking	0.00119	0.00234	Maple, red	0.00137	0.00289
Basswood, American	0.00230	0.00330	Maple, silver	0.00102	0.00252
Beech, American	0.00190	0.00431	Maple, black	0.00165	0.00353
Birch, paper	0.00219	0.00304	Maple, sugar	0.00165	0.00353
Birch, river	0.00162	0.00327	Oak, black	0.00123	0.00230
Birch, yellow	0.00256	0.00338	Red Oak, commercial	0.00158	0.00369
Birch, sweet	0.00256	0.00338	Red oak, California	0.00123	0.00230
Buckeye, yellow	0.00123	0.00285	Red oak: water, laurel, willow	0.00151	0.00350
Butternut	0.00116	0.00223	White Oak, commercial	0.00180	0.00365
Catalpa, northern	0.00085	0.00169	White oak, live	0.00230	0.00338
Cherry, black	0.00126	0.00248	White oak, Oregon white	0.00144	0.00327
Chestnut, American	0.00116	0.00234	White oak, overcup	0.00183	0.00462
Cottonwood, black	0.00123	0.00304	Persimmon, common	0.00278	0.00403
Cottonwood, eastern	0.00133	0.00327	Sassafras	0.00137	0.00216
Elm, American	0.00144	0.00338	Sweet gum	0.00183	0.00365
Elm, rock	0.00165	0.00285	Sycamore, American	0.00172	0.00296
Elm, slippery	0.00169	0.00315	Tanoak	0.00169	0.00423
Elm, winged	0.00183	0.00419	Tupelo, black	0.00176	0.00308
Elm, cedar	0.00183	0.00419	Tupelo, water	0.00144	0.00267
Hackberry	0.00165	0.00315	Walnut, black	0.00190	0.00274
Hickory, pecan	0.00169	0.00315	Willow, black	0.00112	0.00308
Hickory, true	0.00259	0.00411	Willow, Pacific	0.00099	0.00319

(Continued)

Coefficients for Dimensional Change as a Result of Shrinking or Swelling within Moisture Content Limits of 6% to 14% (C_T = dimensional change coefficient for tangential direction; C_R = radial direction)—Cont'd

Species	Dimensional change coefficient[a]		Species	Dimensional change coefficient[a]	
	C_R	C_T		C_R	C_T
Holly, American	0.00165	0.00353	Yellow-poplar	0.00158	0.00289
Softwoods					
Baldcypress	0.00130	0.00216	Pine, eastern white	0.00071	0.00212
Cedar, yellow	0.00095	0.00208	Pine, jack	0.00126	0.00230
Cedar, Atlantic white	0.00099	0.00187	Pine, loblolly	0.00165	0.00259
Cedar, eastern red	0.00106	0.00162	Pine, pond	0.00165	0.00259
Cedar, Incense	0.00112	0.00180	Pine, lodgepole	0.00148	0.00234
Cedar, Northern white"	0.00101	0.00229	Pine, Jeffrey	0.00148	0.00234
Cedar, Port-Orford	0.00158	0.00241	Pine, longleaf	0.00176	0.00263
Cedar, western red"	0.00111	0.00234	Pine, ponderosa	0.00133	0.00216
Douglas-fir, Coast-type	0.00165	0.00267	Pine, red	0.00130	0.00252
Douglas-fir, Interior north	0.00130	0.00241	Pine, shortleaf	0.00158	0.00271
Douglas-fir, Interior west	0.00165	0.00263	Pine, slash	0.00187	0.00267
Fir, balsam	0.00099	0.00241	Pine, sugar	0.00099	0.00194
Fir, California red	0.00155	0.00278	Pine, Virginia	0.00144	0.00252
Fir, noble	0.00148	0.00293	Pine, western white	0.00141	0.00259
Fir, Pacific silver	0.00151	0.00327	Redwood, old-growth[b]	0.00120	0.00205
Fir, subalpine	0.00088	0.00259	Redwood, second-growth[b]	0.00101	0.00229
Fir, grand	0.00112	0.00245	Spruce, black	0.00141	0.00237
Fir, white	0.00112	0.00245	Spruce, Engelmann	0.00130	0.00248
Hemlock, eastern	0.00102	0.00237	Spruce, red	0.00130	0.00274
Hemlock, western	0.00144	0.00274	Spruce, white	0.00130	0.00274
Larch, western	0.00155'	0.00323	Spruce, Sitka	0.00148	0.00263
			Tamarack	0.00126	0.00259
Imported Woods					
Andiroba, crabwood	0.00137	0.00274	Light .red "Philippine mahogany"	0.00126	0.00241
Angelique	0.00180	0.00312	Limba	0.00151	0.00187
Apitong, keruing[b] (all *Dipterocarpus* spp.)	0.00243	0.00527	Mahogany[b]	0.00172	0.00238
			Meranti	0.00126	0.00289
Avodire	0.00126	0.00226	Obeche	0.00106	0.00183
Balsa	0.00102	0.00267	Okoume	0.00194	0.00212
Banak	0.00158	0.00312	Parana, pine	0.00137	0.00278
Cativo	0.00078	0.00183	Paumarfim	0.00158	0.00312
Cuangare	0.00183	0.00342	Primavera	0.00106	0.00180
Greenheart[b]	0.00390	0.00430	Ramin	0.00133	0.00308
Iroko"	0.00153	0.00205	Santa Maria	0.00187	0.00278
Khaya	0.00141	0.00201	Spanish-cedar	0.00141	0.00219
Kokrodua"	0.00148	0.00297	Teak[b]	0.00101	0.00186
Lauans: dark red "Philippine mahogany"	0.00133	0.00267			

[a] Per 1% change in moisture content, based on dimension at 10% moisture content and a straight-line relationship between moisture content at which shrinkage starts and total shrinkage. (Shrinkage assumed to start at 30% for all species except those indicated by footnote b.)
[b] Shrinkage assumed to start at 22% moisture content.

Calculation Based on Green Dimensions

Approximate dimensional changes associated with moisture content changes greater than 6% to 14%, or when one moisture value is outside of those limits, can be calculated by

$$\Delta D = \frac{D_I(M_F - M_I)}{30(100)/S_T - 30 + M_I} \tag{12-3}$$

where S_T is tangential shrinkage (%) from green to oven dry (Ch. 3, Tables 3–5 and 3–6) (use radial shrinkage S_R when appropriate).

Neither M_I nor M_F should exceed 30%, the assumed moisture content value when shrinkage starts for most species.

7.4.2 Drying of Wood—Recommended Moisture Content

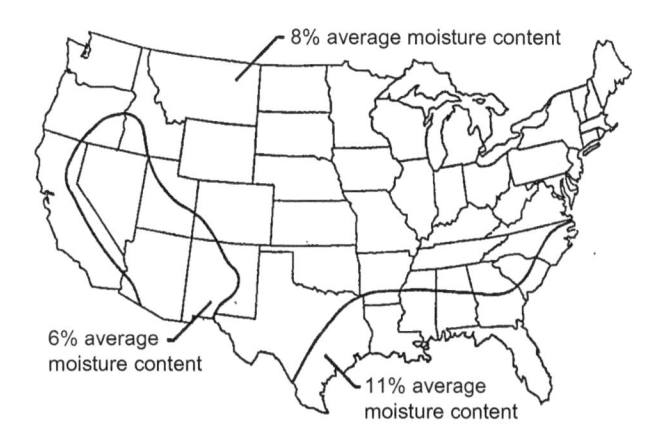

FIGURE 12–1 Recommended Average Moisture Content for Interior Use of Wood Products in Various Areas of the United States.

Hot-pressed plywood and other board products, such as particleboard and hardboard, usually do not have the same moisture content as lumber. The high temperatures used in hot presses cause these products to assume a lower moisture content for a given relative humidity. Because this lower equilibrium moisture content varies widely, depending on the specific type of hot-pressed product, it is recommended that such products be conditioned at 30 to 40% relative humidity for interior use and 65% for exterior use.

Lumber used in the manufacture of large laminated members should be dried to a moisture content slightly less than the moisture content expected in service so that moisture absorbed from the adhesive will not cause the moisture content of the product to exceed the service value. The range of moisture content between laminations assembled into a single member should not exceed 5 percentage points. Although laminated members are often massive and respond rather slowly to changes in environmental conditions, it is desirable to follow the recommendations in Table 12-2 for moisture content at time of installation.

Recommended Moisture Content Values for Various Wood Items at Time of Installation

	Recommended moisture content (%) in various climatological regions					
	Most areas of the United States		Dry southwestern area[a]		Damp, warm coastal area[a]	
Use of wood	Average[b]	Individual pieces	Average[b]	Individual pieces	Average[b]	Individual pieces
Interior: woodwork, flooring, furniture, wood trim	8	6-10	6	4-9	11	8-13
Exterior: siding, wood trim, sheathing, laminated timbers	12	9-14	9	7-12	12	9-14

[a]Major areas are indicated in Figure 12-1.
[b]To obtain a realistic average, test at least 10% of each item. If the quantity of a given item is small, make several tests. For example, in an ordinary dwelling having about 60 floor joists, at least 10 tests should be made on joists selected at random.
Source: U.S. Department of Agriculture.

Drying of Wood

Drying is required for wood to be used in most products. Dried lumber has many advantages over green lumber for producers and consumers. Removal of excess water reduces weight, thus lowering shipping and handling costs. Proper drying confines shrinking and swelling of wood in use to manageable amounts under all but extreme conditions of relative humidity or flooding. As wood dries, most of its strength properties increase, as well as its electrical and thermal insulating properties. Properly dried lumber can be cut to precise dimensions and machined more easily and efficiently; wood parts can be more securely fitted and fastened together with nails, screws, bolts, and adhesives; warping, splitting, checking, and other harmful effects of uncontrolled drying are largely eliminated; and paint, varnish, and other finishes are more effectively applied and maintained. Wood must be relatively dry before it can be glued or treated with decay-preventing and fire-retardant chemicals.

The key to successful and efficient drying is control of the drying process. Timely application of optimum or at least adequate temperature, relative humidity, and air circulation conditions is critical. Uncontrolled drying leads to drying defects that can adversely affect the serviceability and economics of the product. The usual strategy is to dry as fast as the particular species, thickness, and end-product requirements will allow without damaging the wood. Slower drying can be uneconomical as well as introduce the risk of stain.

Softwood lumber intended for framing in construction is usually targeted for drying to an average moisture content of 15%, not to exceed 19%. Softwood lumber for many other uses is dried to a low moisture content, 10 to 12% for many appearance grades to as low as 7 to 9% for furniture, cabinets, and millwork. Hardwood lumber for framing in construction, though not in common use, should also be dried to an average moisture content of 15%, not to exceed 19%. Hardwood lumber for furniture, cabinets, and millwork is usually dried to 6 to 8% moisture content.

A typical hardwood schedule might begin at 49°C (120°F) and 80% relative humidity when the lumber is green. By the time the lumber has reached 15% moisture content, the temperature is as high as 82°C (180°F). A typical hardwood drying schedule is shown in Table 12-3. Some method of monitoring moisture content during drying is required for schedules based on moisture content. One common method is the use of short kiln samples that are periodically weighed, usually manually but potentially remotely with load cells. Alternatively, electrodes are embedded in sample boards to sense the change in electrical conductivity with moisture content. This system is limited to moisture content values less than 30%.

Softwood kiln schedules generally differ from hardwood schedules in that changes in kiln temperature and relative humidity are made at predetermined times rather than moisture content levels. Examples of time-based schedules, both conventional temperature ($<100°C$ ($<212°F$)) and high temperature ($>110°C$ ($>230°F$)), are given in Table 12-3.

Drying Defects

Most drying defects or problems that develop in wood products during drying can be classified as fracture or distortion, warp, or discoloration. Defects in any one of these categories are caused by an interaction of wood properties with processing factors. Wood shrinkage is mainly responsible for wood ruptures and distortion of shape. Cell structure and chemical extractives in wood contribute to defects associated with uneven moisture content, undesirable color, and undesirable surface texture. Drying temperature is the most important processing factor because it can be responsible for defects in each category.

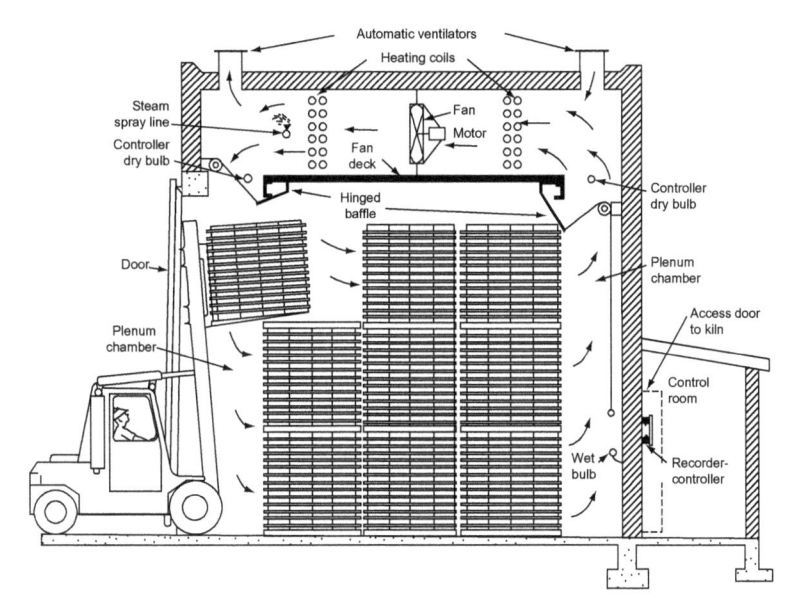

FIGURE 12-6 Package-loaded kiln with fans connected directly to motors.

7.5.0 Mechanical Properties of Some Common Woods Grown in the United States—Inch-Pound

Strength properties of some commercially important woods grown in the United States (Inch-Ponud)

Common species names	Moisture content	Specific gravity[b]	Static bending — Modulus of rupture (lbf/in²)	Static bending — Modulus of elasticity[c] (×10⁶ lbf/in²)	Static bending — Work to maximum load (in-lbf/in³)	Impact bending (in.)	Compression parallel to grain (lbf/in²)	Compression perpendicular to grain (lbf/in²)	Shear parallel to grain (lbf/in²)	Tension perpendicular to grain (lbf/in²)	Side hardness (lbf)
Hardwoods											
Alder, red	Green	0.37	6,500	1.17	8.0	22	2,960	250	770	390	440
	12%	0.41	9,800	1.38	8.4	20	5,820	440	1,080	420	590
Ash											
Black	Green	0.45	6,000	1.04	12.1	33	2,300	350	860	490	520
	12%	0.49	12,600	1.60	14.9	35	5,970	760	1,570	700	850
Blue	Green	0.53	9,600	1.24	14.7	—	4,180	810	1,540	—	—
	12%	0.58	13,800	1.40	14.4	—	6,980	1,420	2,030	—	—
Green	Green	0.53	9,500	1.40	11.8	35	4,200	730	1,260	590	870
	12%	0.56	14,100	1.66	13.4	32	7,080	1,310	1,910	700	1,200
Oregon	Green	0.50	7,600	1.13	12.2	39	3,510	530	1,190	590	790
	12%	0.55	12,700	1.36	14.4	33	6,040	1,250	1,790	720	1,160
White	Green	0.55	9,500	1.44	15.7	38	3,990	670	1,350	590	960
	12%	0.60	15,000	1.74	16.6	43	7,410	1,160	1,910	940	1,320
Aspen											
Bigtooth	Green	0.36	5,400	1.12	5.7	—	2,500	210	730	—	—
	12%	0.39	9,100	1.43	7.7	—	5,300	450	1,080	—	—
Quaking	Green	0.35	5,100	0.86	6.4	22	2,140	180	660	230	300
	12%	0.38	8,400	1.18	7.6	21	4,250	370	850	260	350
Basswood, American	Green	0.32	5,000	1.04	5.3	16	2,220	170	600	280	250
	12%	0.37	8,700	1.46	7.2	16	4,730	370	990	350	410
Beech, American	Green	0.56	8,600	1.38	11.9	43	3,550	540	1,290	720	850
	12%	0.64	14,900	1.72	15.1	41	7,300	1,010	2,010	1,010	1,300
Birch											
Paper	Green	0.48	6,400	1.17	16.2	49	2,360	270	840	380	560
	12%	0.55	12,300	1.59	16.0	34	5,690	600	1,210	—	910
Sweet	Green	0.60	9,400	1.65	15.7	48	3,740	470	1,240	430	970
	12%	0.65	16,900	2.17	18.0	47	8,540	1,080	2,240	950	1,470
Yellow	Green	0.55	8,300	1.50	16.1	48	3,380	430	1,110	430	780
	12%	0.62	16,600	2.01	20.8	55	8,170	970	1,880	920	1,260

Species	Condition								
Butternut	Green	0.36	5,400	0.97	8.2	24	2,420	220	760	430	390
	12%	0.38	8,100	1.18	8.2	24	5,110	460	1,170	440	490
Cherry, black	Green	0.47	8,000	1.31	12.8	33	3,540	360	1,130	570	660
	12%	0.50	12,300	1.49	11.4	29	7,110	690	1,700	560	950
Chestnut, American	Green	0.40	5,600	0.93	7.0	24	2,470	310	800	440	420
	12%	0.43	8,600	1.23	6.5	19	5,320	620	1,080	460	540
Cottonwood											
Balsam, poplar	Green	0.31	3,900	0.75	4.2	—	1,690	140	500	—	—
	12%	0.34	6,800	1.10	5.0	—	4,020	300	790	—	—
Black	Green	0.31	4,900	1.08	5.0	20	2,200	160	610	270	250
	12%	0.35	8,500	1.27	6.7	22	4,500	300	1,040	330	350
Eastern	Green	0.37	5,300	1.01	7.3	21	2,280	200	680	410	340
	12%	0.40	8,500	1.37	7.4	20	4,910	380	930	580	430
Elm											
American	Green	0.46	7,200	1.11	11.8	38	2,910	360	1,000	590	620
	12%	0.50	11,800	1.34	13.0	39	5,520	690	1,510	660	830
Rock	Green	0.57	9,500	1.19	19.8	54	3,780	610	1,270	—	940
	1.2%	0.63	14,800	1.54	19.2	56	7,050	1,230	1,920	—	1,320
Slippery	Green	0.48	8,000	1.23	15.4	47	3,320	420	1,110	640	660
	12%	0.53	13,000	1.49	16.9	45	6,360	820	1,630	530	860
Hackberry	Green	0.49	6,500	0.95	14.5	48	2,650	400	1,070	630	700
	12%	0.53	11,000	1.19	12.8	43	5,440	890	1,590	580	880
Hickory, pecan											
Bittternut	Green	0.60	10,300	1.40	20.0	66	4,570	800	1,240	—	—
	12%	0.66	17,100	1.79	18.2	66	9,040	1,680	—	—	—
Nutmeg	Green	0.56	9,100	1.29	22.8	54	3,980	760	1,030	—	—
	12%	0.60	16,600	1.70	25.1	—	6,910	1,570	—	—	—
Pecan	Green	0.60	9,800	1.37	14.6	53	3,990	780	1,480	680	1,310
	12%	0.66	13,700	1.73	13.8	44	7,850	1,720	2,080	—	1,820
Water	Green	0.61	10,700	1.56	18.8	56	4,660	880	1,440	—	—
	12%	0.62	17,800	2.02	19.3	53	8,600	1,550	—	—	—
Hickory, true											
Mockernut	Green	0.64	11,100	1.57	26.1	88	4,480	810	1,280	—	—
	12%	0.72	19,200	2.22	22.6	77	8,940	1,730	1,740	—	—
Pignut	Green	0.66	11,700	1.65	31.7	89	4,810	920	1,370	—	—
	12%	0.75	20,100	2.26	30.4	74	9,190	1,980	2,150	—	—
Shagbark	Green	0.64	11,000	1.57	23.7	74	4,580	840	1,520	—	—
	12%	0.72	20,200	2.16	25.8	67	9,210	1,760	2,430	—	—
Shellbark	Green	0.62	10,500	1.34	29.9	104	3,920	810	1,190	—	—
	12%	0.69	18,100	1.89	23.6	88	8,000	1,800	2,110	—	—
Honeylocust	Green	0.60	10,200	1.29	12.6	47	4,420	1,150	1,660	930	1,390
	12%		14,700	1.63	13.3	47	7,500	1,840	2,250	900	1,580
Locust, black	Green	0.66	13,800	1.85	15.4	44	6,800	1,160	1,760	770	1,570
	12%	0.69	19,400	2.05	18.4	57	10,180	1,830	2,480	640	1,700

(Continued)

Strength properties of some commercially important woods grown in the United States (Inch-Ponud)—Cont'd

Common species names	Moisture content	Specific gravity[b]	Modulus of rupture (lbf/in²)	Modulus of elasticity[c] (×10⁶ lbf/in²)	Work to maximum load (in-lbf/in³)	Impact bending (in.)	Compression parallel to grain (lbf/in²)	Compression perpendicular to grain (lbf/in²)	Shear parallel to grain (lbf/in²)	Tension perpendicular to grain (lbf/in²)	Side hardness (lbf)
Magnolia											
Cucumbertree	Green	0.44	7,400	1.56	10.0	30	3,140	330	990	440	520
	12%	0.48	12,300	1.82	12.2	35	6,310	570	1,340	660	700
Southern	Green	0.46	6,800	1.11	15.4	54	2,700	460	1,040	610	740
	12%	0.50	11,200	1.40	12.8	29	5,460	860	1,530	740	1,020
Maple											
Bigleaf	Green	0.44	7,400	1.10	8.7	23	3,240	450	1,110	600	620
	12%	0.48	10,700	1.45	71.8	28	5,950	750	1,730	540	850
Black	Green	0.52	7,900	1.33	12.8	48	3,270	600	1,130	720	840
	12%	0.57	13,300	1.62	12.5	40	6,680	1,020	1,820	670	1,180
Red	Green	0.49	7,700	1.39	11.4	32	3,280	400	1,150	—	700
	12%	0.54	13,400	1.64	12.5	32	6,540	1,000	1,850	—	950
Silver	Green	0.44	5,800	0.94	11.0	29	2,490	370	1,050	560	590
	12%	0.47	8,900	1.14	8.3	25	5,220	740	1,480	500	700
Sugar	Green	0.56	9,400	1.55	13.3	40	4,020	640	1,460	—	970
	12%	0.63	15,800	1.83	16.5	39	7,830	1,470	2,330	—	1,450
Oak, red											
Black	Green	0.56	8,200	1.18	12.2	40	3,470	710	1,220	—	1,060
	12%	0.61	13,900	1.64	13.7	41	6,520	930	1,910	—	1,210
Cherrybark	Green	0.61	10,800	1.79	14.7	54	4,620	760	1,320	800	1,240
	12%	0.68	18,100	2.28	18.3	49	8,740	1,250	2,000	840	1,480
Laurel	Green	0.56	7900	1.39	11.2	39	3,370	570	1,180	770	1,000
	12%	0.63	12,600	1.69	11.8	39	6,980	1,060	1,830	790	1,210
Northern red	Green	0.56	8,300	1.35	13.2	44	3,440	610	1,210	750	1,000
	12%	0.63	14,300	1.82	14.5	43	6,760	1,010	1,780	800	1,290
Pin	Green	0.58	8,300	1.32	14.0	48	3,680	720	1,290	800	1,070
	12%	0.63	14,000	1.73	14.8	45	6,820	1,020	2,080	1,050	1,510
Scarlet	Green	0.60	10,400	1.48	15.0	54	4,090	830	1,410	700	1,200
	12%	0.67	17,400	1.91	20.5	53	8,330	1,120	1,890	870	1,400
Southern red	Green	0.52	6,900	1.14	8.0	29	3,030	550	930	480	860
	12%	0.59	10,900	1.49	9.4	26	6,090	870	1,390	510	1,060
Water	Green	0.56	8,900	1.55	11.1	39	3,740	620	1,240	820	1,010
	12%	0.63	15,400	2.02	21.5	44	6,770	1,020	2,020	920	1,190

Section 7 Lumber—Calculations to Select Framing and Trim Materials

Species	Moisture										
Willow	Green	0.56	7,400	1.29	8.8	35	3,000	610	1,180	760	980
	12%	0.69	14,500	1.90	14.6	42	7,040	1,130	1,650	—	1,460
Oak, white											
Bur	Green	0.58	7,200	0.88	10.7	44	3,290	680	1,350	800	1,110
	12%	0.64	10,300	1.03	9.8	29	6,060	1,200	1,820	680	1,370
Chestnut	Green	0.57	8,000	1.37	9.4	35	3,520	530	1,210	690	890
	12%	0.66	13,300	1.59	11.0	40	6,830	840	1,490	—	1,130
Live	Green	0.80	11,900	1.58	12.3	—	5,430	2,040	2,210	—	—
	12%	0.88	18,400	1.98	18.9	—	8,900	2,840	2,660	—	—
Overcup	Green	0.57	8,000	1.15	12.6	44	3,370	540	1,320	730	960
	12%	0.63	12,600	1.42	15.7	38	6,200	810	2,000	940	1,190
Post	Green	0.60	8,100	1.09	11.0	44	3,480	860	1,280	790	1,130
	12%	0.67	13,200	1.51	13.2	46	6,600	1,430	1,840	780	1,360
Swamp chestnut	Green	0.60	8,500	1.35	12.8	45	3,540	570	1,260	670	1,110
	12%	0.67	13,900	1.77	12.0	41	7,270	1,110	1,990	690	1,240
Swamp white	Green	0.64	9,900	1.59	14.5	50	4,360	760	1,300	860	1,160
	12%	0.72	17,700	2.05	19.2	49	8,600	1,190	2,000	830	1,620
White	Green	0.60	8,300	1.25	11.6	42	3,560	670	1,250	770	1,060
	12%	0.68	15,200	1.78	14.8	37	7,440	1,070	2,000	800	1,360
Sassafras	Green	0.42	6,000	0.91	7.1	—	2,730	370	950	—	—
	12%	0.46	9,000	1.12	8.7	—	4,760	850	1,240	—	—
Sweetgum	Green	0.46	7,100	1.20	10.1	36	3,040	370	990	540	600
	12%	0.52	12,500	1.64	11.9	32	6,320	620	1,600	760	850
Sycamore, American	Green	0.46	6,500	1.06	7.5	26	2,920	360	1,000	630	610
	12%	0.49	10,000	1.42	8.5	26	5,380	700	1,470	720	770
Tanoak	Green	0.58	10,500	1.55	13.4	—	4,650	—	—	—	—
	12%	—	—	—	—	—	—	—	—	—	—
Tupelo											
Black	Green	0.46	7,000	1.03	8.0	30	3,040	480	1,100	570	640
	12%	0.50	9,600	1.20	6.2	22	5,520	930	1,340	500	810
Water	Green	0.46	7,300	1.05	8.3	30	3,370	480	1,190	600	710
	12%	0.50	9,600	1.26	6.9	23	5,920	870	1,590	700	880
Walnut, Black	Green	0.51	9,500	1.42	14.6	37	4,300	490	1,220	570	900
	12%	0.55	14,600	1.68	10.7	34	7,580	1,010	1,370	690	1,010
Willow, Black	Green	0.36	4,800	0.79	11.0	—	2,040	180	680	—	—
	12%	0.39	7,800	1.01	8.8	—	4,100	430	1,250	—	—
Yellow-poplar	Green	0.40	6,000	1.22	7.5	26	2,660	270	790	510	440
	12%	0.42	10,100	1.58	8.8	24	5,540	500	1,190	540	540
Softwoods											
Baldcypress	Green	0.42	6,600	1.18	6.6	25	3,589	400	810	300	390
	12%	0.46	10,600	1.44	8.2	24	6,360	730	1,000	270	510

(Continued)

Strength properties of some commercially important woods grown in the United States (Inch-Ponud)—Cont'd

Common species names	Moisture content	Specific gravity[b]	Static bending				Impact bending (in.)	Compression parallel to grain (lbf/in^2)	Compression perpendicular to grain (lbf/in^2)	Shear parallel to grain (lbf/in^2)	Tension perpendicular to grain (lbf/in^2)	Side hardness (lbf)
			Modulus of rupture (lbf/in^2)	Modulus of elasticity[c] (×10^6 lbf/in^2)	Work to maximum load (in-lbf/in^3)							
Cedar												
Atlantic white	Green	0.31	4,700	0.75	5.9		18	2,390	240	690	180	290
	12%	0.32	6,800	0.93	4.1		13	4,700	410	800	220	350
Eastern redcedar	Green	0.44	7,000	0.65	15.0		35	3,570	700	1,010	330	650
	12%	0.47	8,800	0.88	8.3		22	6,020	920	—	—	—
Incense	Green	0.35	6,200	0.84	6.4		17	3,150	370	830	280	390
	12%	0.37	8,000	1.04	5.4		17	5,200	590	880	270	470
Northern White	Green	0.29	4,200	0.64	5.7		15	1,990	230	620	240	230
	12%	0.31	6,500	0.80	4.8		12	3,960	310	850	240	320
Port-Orford	Green	0.39	6,600	1.30	7.4		21	3,140	300	840	180	380
	12%	0.43	12,700	1.70	9.1		28	6,250	720	1,370	400	630
Western redcedar	Green	0.31	5,200	0.94	5.0		17	2,770	240	770	230	260
	12%	0.32	7,500	1.11	5.8		17	4,560	460	990	220	350
Yellow	Green	0.42	6,400	1.14	9.2		27	3,050	350	840	330	440
	12%	0.44	11,100	1.42	10.4		29	6,310	620	1,130	360	580
Douglas-fir[d]												
Coast	Green	0.45	7,700	1.56	7.6		26	3,780	380	900	300	500
	12%	0.48	12,400	1.95	9.9		31	7,230	800	1,130	340	710
Interior West	Green	0.46	7,700	1.51	7.2		26	3,870	420	940	290	510
	12%	0.50	12,600	1.83	10.6		32	7,430	760	1,290	350	660
Interior North	Green	0.45	7,400	1.41	8.1		22	3,470	360	950	340	420
	12%	0.48	13,100	1.79	10.5		26	6,900	770	1,400	390	600
Interior South	Green	0.43	6,800	1.16	8.0		15	3,110	340	950	250	360
	12%	0.46	11,900	1.49	9.0		20	6,230	740	1,510	330	510
Fir												
Balsam	Green	0.33	5,500	1.25	4.7		16	2,630	190	662	180	290
	12%	0.35	9,200	1.45	5.1		20	5,280	404	944	180	400
California red	Green	0.36	5,800	1.17	6.4		21	2,760	330	770	380	360
	12%	0.38	10,500	1.50	8.9		24	5,460	610	1,040	390	500
Grand	Green	0.35	5,800	1.25	5.6		22	2,940	270	740	240	360
	12%	0.37	8,900	1.57	7.5		28	5,290	500	900	240	490
Noble	Green	0.37	6,200	1.38	6.0		19	3,010	270	800	230	290
	12%	0.39	10,700	1.72	8.8		23	6,100	520	1,050	220	410

Section 7 Lumber—Calculations to Select Framing and Trim Materials

Species	Moisture									
Pacific silver	Green	0.40	6,400	1.42	6.0	21	3,140	220	750	240	310
	12%	0.43	11,000	1.76	9.3	24	6,410	450	1,220	—	430
Subalpine	Green	0.31	4,900	1.05	—	—	2,300	190	700	—	260
	12%	0.32	8,600	1.29	—	—	4,860	390	1,070	—	350
White	Green	0.37	5,900	1.16	5.6	22	2,900	280	760	300	340
	12%	0.39	9,800	1.50	1.1	20	5,800	530	1,100	300	480
Hemlock											
Eastern	Green	0.38	6,400	1.07	6.7	21	3,080	360	850	230	400
	12%	0.40	8,900	1.20	6.8	21	5,410	650	1,060	—	500
Mountain	Green	0.42	6,300	1.04	11.0	32	2,880	370	930	330	470
	12%	0.45	11,500	1.33	10.4	32	6,440	860	1,540	—	680
Western	Green	0.42	6,600	1.31	6.9	22	3,360	280	860	290	410
	12%	0.45	11,300	1.63	8.3	23	7,200	550	1,290	340	540
Larch, western	Green	0.48	7,700	1.46	10.3	29	3,760	400	870	330	510
	12%	0.52	13,000	1.87	12.6	35	7,620	930	1,360	430	830
Pine											
Eastern white	Green	0.34	4,900	0.99	5.2	17	2,440	220	680	250	290
	12%	0.35	8,600	1.24	6.8	18	4,800	440	900	310	380
Jack	Green	0.40	6,000	1.07	7.2	26	2,950	300	750	360	400
	12%	0.43	9,900	1.35	8.3	27	5,660	580	1,170	420	570
Loblolly	Green	0.47	7,300	1.40	8.2	30	3,510	390	860	260	450
	12%	0.51	12,800	1.79	10.4	30	7,130	790	1,390	470	690
Lodgepole	Green	0.38	5,500	1.08	5.6	20	2,610	250	680	220	330
	12%	0.41	9,400	1.34	6.8	20	5,370	610	880	290	480
Longleaf	Green	0.554	8,500	1.59	8.9	35	4,320	480	1,040	330	590
	12%	0.59	14,500	1.98	11.8	34	8,470	960	1,510	470	870
Pitch	Green	0.47	6,800	1.20	9.2	—	2,950	360	860	—	—
	12%	0.52	10,800	1.43	9.2	—	5,940	820	1,360	—	—

7.5.1 Mechanical Properties of Some Common Woods Grown in the United States—Metric

Strength properties of some commercially important woods grown in the United States (Metric)

Common species names	Moisture content	Specific gravity[b]	Static bending			Impact bending (mm)	Compression parallel to grain (kPa)	Compression perpendicular to grain (kPa)	Shear parallel to grain (kPa)	Tension perpendicular to grain (kPa)	Side hardness (N)
			Modulus of rupture (kPa)	Modulus of elasticity[c] (MPa)	Work to maximum load (kJ/m³)						
Hardwoods											
Alder, red	Green	0.37	45,000	8,100	55	560	20,400	1,700	5,300	2,700	2,000
	12%	0.41	68,000	9,500	58	510	40,100	3,000	7,400	2,900	2,600
Ash											
Black	Green	0.45	41,000	7,200	83	840	15,900	2,400	5,900	3,400	2,300
	12%	0.49	87,000	11,000	103	890	41,200	5,200	10,800	4,800	3,800
Blue	Green	0.53	66,000	8,500	101	—	24,800	5,600	10,600	—	—
	12%	0.58	95,000	9,700	99	—	48,100	9,800	14,000	—	—
Green	Green	0.53	66,000	9,700	81	890	29,000	5,000	8,700	4,100	3,900
	12%	0.56	97,000	11,400	92	810	48,800	9,000	13,200	4,800	5,300
Oregon	Green	0.50	52,000	7,800	84	990	24,200	3,700	8,200	4,100	3,500
	12%	0.55	88,000	9,400	99	840	41,600	8,600	12,300	5,000	5,200
White	Green	0.55	66,000	9,900	108	970	27,500	4,600	9,300	4,100	4,300
	12%	0.60	103,000	12,000	115	1,090	51,100	8,000	13,200	6,500	5,900
Aspen											
Bigtooth	Green	0.36	37,000	7,700	39	—	17,200	1,400	5,000	—	—
	12%	0.39	63,000	9,900	53	—	36,500	3,100	7,400	—	—
Quaking	Green	0.35	35,000	5,900	44	560	14,800	1,200	4,600	1,600	1,300
	12%	0.38	58,000	8,100	52	530	29,300	2,600	5,900	1,800	1,600
Basswood, American	Green	0.32	34,000	7,200	37	410	15,300	1,200	4,100	1,900	1,100
	12%	0.37	60,000	10,100	50	410	32,600	2,600	6,800	2,400	1,800
Beech, American	Green	0.56	59,000	9,500	82	1,090	24,500	3,700	8,900	5,000	3,800
	12%	0.64	103,000	11,900	104	1,040	50,300	7,000	13,900	7,000	5,800
Birch											
Paper	Green	0.48	44,000	8,100	112	1,240	16,300	1,900	5,800	2,600	2,500
	12%	0.55	85,000	11,000	110	860	39,200	4,100	8,300	—	4,000
Sweet	Green	0.60	65,000	11,400	108	1,220	25,800	3,200	8,500	3,000	4,300
	12%	0.65	117,000	15,000	124	1,190	58,900	7,400	15,400	6,600	6,500
Yellow	Green	0.55	57,000	10,300	111	1,220	23,300	3,000	7,700	3,000	3,600
	12%	0.62	114,000	13,900	143	1,400	56,300	6,700	13,000	6,300	5,600
Butternut	Green	0.36	37,000	6,700	57	610	16,700	1,500	5,200	3,000	1,700
	12%	0.38	56,000	8,100	57	610	36,200	3,200	8,100	3,000	2,200
Cherry, black	Green	0.47	55,000	9,000	88	840	24,400	2,500	7,800	3,900	2,900
	12%	0.50	85,000	10,300	79	740	49,000	4,800	11,700	3,900	4,200
Chestnut, American	Green	0.40	39,000	6,400	48	610	17,000	2,100	5,500	3,000	1,900
	12%	0.43	59,000	8,500	45	480	36,700	4,300	7,400	3,200	2,400
Cottonwood											
Balsam poplar	Green	0.31	27,000	5,200	29	—	11,700	1,000	3,400	—	—
	12%	0.34	47,000	7,600	34	—	27,700	2,100	5,400	—	—
Black	Green	0.31	34,000	7,400	34	510	15,200	1,100	4,200	1,900	1,100
	12%	0.35	59,000	8,800	46	560	31,000	2,100	7,200	2,300	1,600
Eastern	Green	0.37	37,000	7,000	50	530	15,700	1,400	4,700	2,800	1,500
	12%	0.40	59,000	9,400	51	510	33,900	2,600	6,400	4,000	1,900

Strength properties of some commercially important woods grown in the United States (Metric)—Cont'd

Common species names	Moisture content	Specific gravity[b]	Modulus of rupture (kPa)	Modulus of elasticity[c] (MPa)	Work to maximum load (kJ/m³)	Impact bending (mm)	Compression parallel to grain (kPa)	Compression perpendicular to grain (kPa)	Shear parallel to grain (kPa)	Tension perpendicular to grain (kPa)	Side hardness (N)
Elm											
American	Green	0.46	50,000	7,700	81	970	20,100	2,500	6,900	4,100	2,800
	12%	0.50	81,000	9,200	90	990	38,100	4,800	10,400	4,600	3,700
Rock	Green	0.57	66,000	8,200	137	1,370	26,100	4,200	8,800	—	—
	12%	0.63	102,000	10,600	132	1,420	48,600	8,500	13,200	—	—
Slippery	Green	0.48	55,000	8,500	106	1,190	22,900	2,900	7,700	4,400	2,900
	12%	0.53	90,000	10,300	117	1,140	43,900	5,700	11,200	3,700	3,800
Hackberry	Green	0.49	45,000	6,600	100	1,220	18,300	2,800	7,400	4,300	3,100
	12%	0.53	76,000	8,200	88	1,090	37,500	6,100	11,000	4,000	3,900
Hickory, pecan											
Bitternut	Green	0.60	71,000	9,700	138	1,680	31,500	5,500	8,500	—	—
	12%	0.66	118,000	12,300	125	1,680	62,300	11,600	—	—	—
Nutmeg	Green	0.56	63,000	8,900	157	1,370	27,400	5,200	7,100	—	—
	12%	0.60	114,000	11,700	173	—	47,600	10,800	—	—	—
Pecan	Green	0.60	68,000	9,400	101	1,350	27,500	5,400	10,200	4,700	5,800
	12%	0.66	94,000	11,900	95	1,120	54,100	11,900	14,300	—	8,100
Water	Green	0.61	74,000	10,800	130	1,420	32,100	6,100	9,900	—	—
	12%	0.62	123,000	13,900	133	1,350	59,300	10,700	—	—	—
Hickory, true											
Mockrnut	Green	0.64	77,000	10,800	180	2,240	30,900	5,600	8,800	—	—
	12%	0.72	132,00	15,300	156	1,960	61,600	11,900	12,000	—	—
Pignut	Green	0.66	81,000	11,400	219	2,260	33,200	6,300	9,400	—	—
	12%	0.75	139,000	15,600	210	1,880	63,400	13,700	14,800	—	—
Shagbark	Green	0.64	76,000	10,800	163	1,880	31,600	5,800	10,500	—	—
	12%	0.72	139,000	14,900	178	1,700	63,500	12,100	16,800	—	—
Shellbark	Green	0.62	72,000	9,200	206	2,640	27,000	5,600	8,200	—	—
	12%	0.69	125,000	13,000	163	2,240	55,200	12,400	14,500	—	—
Honeylocust	Green	0.60	70,000	8,900	87	1,190	30,500	7,900	11,400	6,400	6,200
	12%	—	101,000	11,200	92	1,190	51,700	12,700	15,500	6,200	7,000
Locust, black	Green	0.66	95,000	12,800	106	1,120	46,900	8,000	12,100	5,300	7,000
	12%	0.69	134,000	14,100	127	1,450	70,200	12,600	17,100	4,400	7,600
Magnolia											
Cucumber tree	Green	0.44	51,000	10,800	69	760	21,600	2,300	6,800	3,000	2,300
	12%	0.48	85,000	12,500	84	890	43,500	3,900	9,200	4,600	3,100
Southern	Green	0.46	47,000	7,700	106	1,370	18,600	3,200	7,200	4,200	3,300
	12%	0.50	77,000	9,700	88	740	37,600	5,900	10,500	5,100	4,500
Maple											
Bigleaf	Green	0.44	51,000	7,600	60	580	22,300	3,100	7,700	4,100	2,800
	12%	0.48	74,000	10,000	54	710	41,000	5,200	11,900	3,700	3,800
Black	Green	0.52	54,000	9,200	88	1,220	22,500	4,100	7,800	5,000	3,700
	12%	0.57	92,000	11,200	86	1,020	46,100	7,000	12,500	4,600	5,200
Red	Green	0.49	53,000	9,600,	79	810	22,600	2,800	7,900	—	3,100
	12%	0.54	92,000	11,300	86	810	45,100	6,900	12,800	—	4,200
Silver	Green	0.44	40,000	6,500	76	740	17,200	2,600	7,200	3,900	2,600
	12%	0.47	61,000	7,900	57	640	36,000	5,100	10,200	3,400	3,100
Sugar	Green	0.56	65,000	10,700	92	1,020	27,700	4,400	10,100	—	4,300
	12%	0.63	109,000	12,600	114	990	54,000	10,100	16,100	—	6,400

(Continued)

Strength properties of some commercially important woods grown in the United States (Metric)—Cont'd

Common species names		Moisture content	Specific gravity[b]	Static bending			Impact bending (mm)	Compression parallel to grain (kPa)	Compression perpendicular to grain (kPa)	Shear parallel to grain (kPa)	Tension perpendicular to grain (kPa)	Side hardness (N)
				Modulus of rupture (kPa)	Modulus of elasticity[c] (MPa)	Work to maximum load (kJ/m³)						
Oak, red												
	Black	Green	0.56	57,000	8,100	84	1,020	23,900	4,900	8,400	—	4,700
		12%	0.61	96,000	11,300	94	1,040	45,000	6,400	13,200	—	5,400
	Cherrybark	Green	0.61	74,000	12,300	101	1,370	31,900	5,200	9,100	5,500	5,500
		12%	0.68	125,000	15,700	126	1,240	60,300	8,600	13,800	5,800	6,600
	Laurel	Green	0.56	54,000	9,600	77	990	21,900	3,900	8,100	5,300	4,400
		12%	0.63	87,0000	11,700	81	990	48,100	7,300	12,600	5,400	5,400
	Northern red	Green	0.56	57,000	9,300	91	1,120	23,700	4,200	8,300	5,200	4,400
		12%	0.63	99,000	12,500	100	1,090	46,600	7,000	12,300	5,500	5,700
	Pin	Green	0.58	57,000	9,100	97	1,220	25,400	5,000	8,900	5,500	4,800
		12%	0.63	97,000	11,900	102	1,140	47,000	7,000	14,300	7,200	6,700
	Scarlet	Green	0.60	72,000	10,200	103	1,370	28,200	5,700	9,700	4,800	5,300
		12%	0.67	120,000	13,200	141	1,350	57,400	7,700	13,000	6,000	6,200
	Southern red	Green	0.52	48,000	7,900	55	740	20,900	3,800	6,400	3,300	3,800
		12%	0.59	75,000	10,300	65	660	42,000	6,000	9,600	3,500	4,700
	Water	Green	0.56	61,000	10,700	77	990	25,800	4,300	8,500	5,700	4,500
		12%	0.63	106,000	13,900	148	1,120	46,700	7,000	13,900	6,300	5,300
	Willow	Green	0.56	51,000	8,900	61	890	20,700	4,200	8,100	5,200	4,400
		12%	0.69	100,000	13,100	101	1,070	48,500	7,800	11,400	—	6,500
Oak, white												
	Bur	Green	0.58	50,000	6,100	74	1,120	22,700	4,700	9,300	5,500	4,900
		12%	0.64	71,000	7,100	68	740	41,800	8,300	12,500	4,700	6,100
	Chestnut	Green	0.57	55,000	9,400	65	890	24,300	3,700	8,300	4,800	4,000
		12%	0.66	92,000	11,000	76	1,020	47,100	5,800	10,300	—	5,000
	Live	Green	0.80	82,000	10,900	85	—	37,400	14,100	15,200	—	—
		12%	0.88	127,000	13,700	130	—	61,400	19,600	18,300	—	—
	Overcup	Green	0.57	55,000	7,900	87	1,120	23,200	3,700	9,100	5,000	4,300
		12%	0.63	87,000	9,800	108	970	42,700	5,600	13,800	6,500	5,300
	Post	Green	0.60	56,000	7,500	76	1,120	24,000	5,900	8,800	5,400	5,000
		12%	0.67	91,000	10,400	91	1,170	45,300	9,900	12,700	5,400	6,000
	Swamp chestnut	Green	0.60	59,000	9,300	88	1,140	24,400	3,900	8,700	4,600	4,900
		12%	0.67	96,000	12,200	83	1,040	50,100	7,700	13,700	4,800	5,500
	Swamp white	Green	0.64	68,000	11,000	100	1,270	30,100	5,200	9,000	5,900	5,200
		12%	0.72	122,000	14,100	132	1,240	59,300	8,200	13,800	5,700	7,200
	White	Green	0.60	57,000	8,600	80	1,070	24,500	4,600	8,600	5,300	4,700
		12%	0.68	105,000	12,300	102	940	51,300	7,400	13,800	5,500	6,000
Sassafras		Green	0.42	41,000	6,300	49	—	18,800	2,600	6,600	—	—
		12%	0.46	62,000	7,700	60	—	32,800	5,900	8,500	—	—
Sweetgum		Green	0.46	49,000	8,300	70	910	21,000	2,600	6,800	3,700	2,700
		12%	0.52	86,000	11,300	82	810	43,600	4,300	11,000	5,200	3,800
Sycamore, American		Green	0.46	45,000	7,300	52	660	20,100	2,500	6,900	4,300	2,700
		12%	0.49	69,000	9,800	59	660	37,100	4,800	10,100	5,000	3,400
Tanoak		Green	0.58	72,000	10,700	92	—	32,100	—	—	—	—
		12%	—	—	—	—	—	—	—	—	—	—
Tupelo Black		Green	0.46	48,000	7,100	55	760	21,000	3,300	7,600	3,900	2,800
		12%	0.50	66,000	8,300	43	560	38,100	6,400	9,200	3,400	3,600

Section | 7 Lumber—Calculations to Select Framing and Trim Materials

Strength properties of some commercially important woods grown in the United States (Metric)—Cont'd

Common species names	Moisture content	Specific gravity[b]	Static bending			Impact bending (mm)	Compression parallel to grain (kPa)	Compression perpendicular to grain (kPa)	Shear parallel to grain (kPa)	Tension perpendicular to grain (kPa)	Side hardness (N)
			Modulus of rupture (kPa)	Modulus of elasticity[c] (MPa)	Work to maximum load (kJ/m³)						
Water	Green	0.46	50,000	7,200	57	760	23,200	3,300	8,200	4,100	3,200
	12%	0.50	66,000	8,700	48	580	40,800	6,000	11,000	4,800	3,900
Walnut,	Green	0.51	66,000	9,800	101	940	29,600	3,400	8,400	3,900	4,000
black	12%	0.55	101,000	11,600	74	860	52,300	7,000	9,400	4,800	4,500
Willow,	Green	0.36	33,000	5,400	76	—	14,100	1,200	4,700	—	—
black	12%	0.39	54,000	7,000	61	—	28,300	3,000	8,600	—	—
Yellow-	Green	0.40	41,000	8,400	52	660	18,300	1,900	5,400	3,500	2,000
poplar	12%	0.42	70,000	10,900	61	610	38,200	3,400	8,200	3,700	2,400
Softwoods											
Baldcypress	Green	0.42	46,000	8,100	46	640	24,700	2,800	5,600	2,100	1,700
	12%	0.46	73,000	9,900	57	610	43,900	5,000	6,900	1,900	2,300
Cedar											
Atlantic white	Green	0.31	32,000	5,200	41	460	16,500	1,700	4,800	1,200	1,300
	12%	0.32	47,000	6,400	28	330	32,400	2,800	5,500	1,500	1,600
Eastern redcedar	Green	0.44	48,000	4,500	103	890	24,600	4,800	7,000	2,300	2,900
	12%	0.47	61,000	6,100	57	560	41,500	6,300	—	—	4,000
Incense	Green	0.35	43,000	5,800	44	430	21,700	2,600	5,700	1,900	1,700
	12%	0.37	55,000	7,200	37	430	35,900	4,100	6,100	1,900	2,100
Northern white	Green	0.29	29,000	4,400	39	380	13,700	1,600	4,300	1,700	1,000
	12%	0.31	45,000	5,500	33	300	27,300	2,100	5,900	1,700	1,400
Port-Orford	Green	0.39	45,000	9,000	51	530	21,600	2,100	5,800	1,200	1,700
	12%	0.43	88,000	11,700	63	710	43,100	5,000	9,400	2,800	2,800
Western redcedar	Green	0.31	35,900	6,500	34	430	19,100	1,700	5,300	1,600	1,200
	12%	0.32	51,700	7,700	40	430	31,400	3,200	6,800	1,500	1,600
Yellow	Green	0.42	44,000	7,900	63	690	21,000	2,400	5,800	2,300	2,000
	12%	0.44	77,000	9,800	72	740	43,500	4,300	7,800	2,500	2,600
Douglas-fir[d] Coast	Green	0.45	53,000	10,800	52	660	26,100	2,600	6,200	2,100	2,200
	12%	0.48	85,000	13,400	68	790	49,900	5,500	7,800	2,300	3,200
Interior West	Green	0.46	53,000	10,400	50	660	26,700	2,900	6,500	2,000	2,300
	12%	0.50	87,000	12,600	73	810	51,200	5,200	8,900	2,400	2,900
Interior North	Green	0.45	51,000	9,700	56	560	23,900	2,500	6,600	2,300	1,900
	12%	0.48	90,000	12,300	72	660	47,600	5,300	9,700	2,700	2,700
Interior South	Green	0.43	47,000	8,000	55	380	21,400	2,300	6,600	1,700	1,600
	12%	0.46	82,000	10,300	62	510	43,000	5,100	10,400	2,300	2,300
Fir											
Balsam	Green	0.33	38,000	8,600	32	410	18,100	1,300	4,600	1,200	1,300
	12%	0.35	63,000	10,000	35	510	36,400	2,800	6,500	1,200	1,800
California red	Green	0.36	40,000	8,100	44	530	19,000	2,300	5,300	2,600	1,600
	12%	0.38	72,400	10,300	61	610	37,600	4,200	7,200	2,700	2,200
Grand	Green	0.35	40,000	8,600	39	560	20,300	1,900	5,100	1,700	1,600
	12%	0.37	61,400	10,800	52	710	36,500	3,400	6,200	1,700	2,200
Noble	Green	0.37	43,000	9,500	41	480	20,800	1,900	5,500	1,600	1,300
	12%	0.39	74,000	11,900	61	580	42,100	3,600	7,200	1,500	1,800
Pacific silver	Green	0.40	44,000	9,800	41	530	21,600	1,500	5,200	1,700	1,400
	12%	0.43	75,800	12,100	64	610	44,200	3,100	8,400	—	1,900

(Continued)

Strength properties of some commercially important woods grown in the United States (Metric)—Cont'd

Common species names	Moisture content	Specific gravity[b]	Static bending			Impact bending (mm)	Compression parallel to grain (kPa)	Compression perpendicular to grain (kPa)	Shear parallel to grain (kPa)	Tension perpendicular to grain (kPa)	Side hardness (N)
			Modulus of rupture (kPa)	Modulus of elasticity[c] (MPa)	Work to maximum load (kJ/m³)						
Subalpine	Green	0.31	34,000	7,200	—	—	15,900	1,300	4,800	—	1,200
	12%	0.32	59,000	8,900	—	—	33,500	2,700	7,400	—	1,600
White	Green	0.37	41,000	8,000	39	560	20,000	1,900	5,200	2,100	1,500
	12%	0.39	68,000	10,300	50	510	40,000	3,700	7,600	2,100	2,100
Hemlock											
Eastern	Green	0.38	44,000	7,400	46	530	21,200	2,500	5,900	1,600	1,800
	12%	0.40	61,000	8,300	47	530	37,300	4,500	7,300	—	2,200
Mountain	Green	0.42	43,000	7,200	76	810	19,900	2,600	6,400	2,300	2,100
	12%	0.45	79,000	9,200	72	810	44,400	5,900	10,600	—	3,000
Western	Green	0.42	46,000	9,000	48	560	23,200	1,900	5,900	2,000	1,800
	12%	0.45	78,000	11,300	57	580	49,000	3,800	8,600	2,300	2,400
Larch, western	Green	0.48	53,000	10,100	71	740	25,900	2,800	6,000	2,300	2,300
	12%	0.52	90,000	12,900	87	890	52,500	6,400	9,400	3,000	3,700
Pine											
Eastern white	Green	0.34	34,000	6,800	36	430	16,800	1,500	4,700	1,700	1,300
	12%	0.35	59,000	8,500	47	460	33,100	3,000	6,200	2,100	1,700
Jack	Green	0.40	41,000	7,400	50	660	20,300	2,100	5,200	2,500	1,800
	12%	0.43	68,000	9,300	57	690	39,000	4,000	8,100	2,900	2,500
Loblolly	Green	0.47	50,000	9,700	57	760	24,200	2,700	5,900	1,800	2,000
	12%	0.51	88,000	12,300	72	760	49,200	5,400	9,600	3,200	3,100
Lodgepole	Green	0.38	38,000	7,400	39	510	18,000	1,700	4,700	1,500	1,500
	12%	0.41	65,000	9,200	47	510	37,000	4,200	6,100	2,000	2,100
Longleaf	Green	0.54	59,000	11,000	61	890	29,800	3,300	7,200	2,300	2,600
	12%	0.59	1,00,000	13,700	81	860	58,400	6,600	10,400	3,200	3,900
Pitch	Green	0.47	47,000	8,300	63	—	20,300	2,500	5,900	—	—
	12%	0.52	74,000	9,900	63	—	41,000	5,600	9,400	—	—

7.6.0 Mechanical Properties of Wood Imported from Other Countries—Inch-Pound

Mechanical Properties of Some Woods Imported into the United States other than Canadian Imports (Inch-Pound)[a]

Common and botanical names of species	Moisture content	Specific gravity	Static bending			Compression parallel to grain (lbf/in²)	Shear parallel to grain (lbf/in²)	Side hardness (lbf)	Sample origin[b]
			Modulus of rupture (lbf/in²)	Modulus of elasticity (x16⁶ lbf/in³)	Work to maximum load (in-lbf/in³)				
Afrormosia (Pericopsis elata)	Green	0.61	14,800	1.77	19.5	7,490	1,670	1,600	AF
	12%		18,400	1.94	18.4	9,940	2,090	1,560	
Albarco (Cariniana spp.)	Green	0.48	—	—	—	—	—	—	AM
	12%		4,500	1.5	13.8	6,820	2,310	1,020	

Mechanical Properties of Some Woods Imported into the United States other than Canadian Imports (Inch-Pound)[a] — Cont'd

Common and botanical names of species	Moisture content	Specific gravity	Static bending — Modulus of rupture (lbf/in²)	Static bending — Modulus of elasticity (x10⁶ lbf/in³)	Static bending — Work to maximum load (in-lbf/in³)	Compression parallel to grain (lbf/in²)	Shear parallel to grain (lbf/in²)	Side hardness (lbf)	Sample origin[b]
Andiroba (*Carapa guianensis*)	Green	0.54	10,000	1.69	9.8	4,780	1,220	880	AM
	12%	—	15,500	2	14	8,120	1,510	1,130	
Angelin (*Andira inermis*)	Green	0.6.5	—	—	—	—	—	—	AF
	12%		18,000	2.49	—	9,200	1,840	1,750	
Angelique (*Dicorynia guianensis*)	Green	0.6	11,400	1.84	12	5,590	1,340	1,100	AM
	12%	—	17,400	2.19	15.2	8,770	1,660	1,290	
Avodire (*Turraeanthus africanus*)	Green	0.48	—	—	—	—	—	—	AF
	12%		12,700	1.49	9.4	7,150	2,030	1,080	
Azobe (*Lophira alata*)	Green	0.87	16,900	2.16	12	9,520	2,040	2,890	AF
	12%		24,500	2.47	—	12,600	2,960	3,350	
Balsa (*Ochroma pyramidale*)	Green	0.16	—	—	—	—	—	—	AM
	12%		3,140	0.49	2.1	2,160	300	—	
Banak (*Virola spp.*)	Green	0.42	5,600	1.64	4.1	2,390	720	320	AM
	12%	—	10,900	2.04	10	5,140	980	510	
Benge (*Guibourtia arnoldiana*)	Green	0.65	—	—	—	—	—	—	AF
	12%		21,400	2.04	—	11,400	2,090	1,750	
Bubinga (*Guibourtia spp.*)	Green	0.71	—	—	—	—	—	—	AF
	12%		22,600	2.48	—	10,500	3,110	2,690	
Bulletwood (*Manilkara bidentata*)	Green	0.85	17,300	2.7	13.6	8,690	1,900	2,230	AM
	12%		27,300	3.45	28.5	11,640	2,500	3,190	
Cativo (*Prioria copaifera*)	Green	0.4	5,900	0.94	5.4	2,460	860	440	AM
	12%	—	8,600	1.11	7.2	4,290	1,060	630	
Ceiba (*Ceiba pentandra*)	Green	0.25	2,200	0.41	1,2	1,060	350	220	AM
	12%		4,300	0.54	2.8	2,380	550	240	
Courbaril (*Hymenaea courbaril*)	Green	0.71	12,900	1.84	14.6	5,800	1,770	1,970	AM
	12%	—	19,400	2.16	17.6	9,510	2,470	2,350	
Cuangare (*Dialyanthera spp.*)	Green	0.31	4,000	1.01	—	2,080	590	230	AM
	12%		7,300	1.52	—	4,760	830	380	
Cypress, Mexican (*Cupressus lustianica*)	Green	0.93	6,200	0.92	—	2,880	950	340	AF
	12%		10,300	1.02	—	5,380	1,580	460	
Degame (*Calycophyllum candidissimum*)	Green	0.67	14,300	1.93	18.6	6,200	1,660	1,630	AM
	12%		22,300	2.27	27	9,670	2,120	1,940	
Determa (*Ocotea rubra*)	Green	0.52	7,800	1.46	4.8	3,760	860	520	AM
	12%		10,500	1.82	6.4	5,800	980	660	
Ekop (*Tetraberlinia tubmaniana*)	Green	0.6	—	—	—	—	—	—	AF
	12%		16,700	2.21	—	9,010	—	—	
Goncalo alves (*Astronium graveolens*)	Green	0.84	12,100	1.94	6.7	6,580	1,760	1,910	AM
	12%	—	16,600	2.23	10.4	10,320	1,960	2,160	

(Continued)

Mechanical Properties of Some Woods Imported into the United States other than Canadian Imports (Inch-Pound)[a]—Cont'd

Common and botanical names of species	Moisture content	Specific gravity	Static bending			Compression parallel to grain (lbf/in²)	Shear parallel to grain (lbf/in²)	Side hardness (lbf)	Sample origin[b]
			Modulus of rupture (lbf/in²)	Modulus of elasticity (×16⁶ lbf/in³)	Work to maximum load (in-lbf/in³)				
Greenheart (Chlorocardium rodiei)	Green	0.8	19,300	2.47	10.5	9,380	1,930	1,880	AM
	12%		24,900	3.25	25.3	12,510	2,620	2,350	
Hura (Hura crepitans)	Green	0.38	6,300	1.04	5.9	2,790	830	440	AM
	12%		8,700	1.17	6.7	4,800	1,080	550	
Ilomba (Pycnanthus angolensis)	Geen	0.4	5,500	1.14	—	2,900	840	470	AF
	12%		9,900	1.59	—	5,550	1,290	610	
Ipe (Tabebuia spp., lapacho group)	Green	0.92	22,600	2.92	27.6	10,350	2,120	3,060	AM
	12%		25,400	3.14	22	13,010	2,060	3,680	
Iroko (Chlorophora spp.)	Green	0.54	10,200	1.29	10.5	4,910	1,310	1,080	AF
	12%		12,400	1.46	9	7,590	1,800	1,260	
Jarrah (Eucalyptus marginata)	Green	0.67	9,900	1.48	—	5,190	1,320	1,290	AS
	12%	—	16,200	1.88	—	8,870	2,130	1,910	
Jelutong (Dyera costulata)	Green	0.36	5,600	1.16	5.6	3,050	760	330	AS
	15%		7,300	1.18	6.4	3,920	840	390	
Kaneelhart (Licaria spp.)	Green	0.96	22,300	3.82	13.6	13,390	1,680	2,210	AM
	12%		29,900	4.06	17.5	17,400	1,970	2,900	
Kapur (Dryobalanops spp.)	Green	0.64	12,800	1.6	15.7	6,220	1,170	980	AS
	12%		18,300	1.88	18.8	10,090	1,990	1,230	
Karri (Eucalyptus diversicolor)	Green	0.82	11,200	1.94	11.6	5,450	1,510	1,360	AS
	12%		20,160	2.6	25.4	10,800	2,420	2,040	
Kempas (Koompassia malaccensis)	Green	0.71	14,500	2.41	12.2	7,930	1,460	1,480	AS
	12%		17,700	2.69	15.3	9,520	1,790	1,710	
Keruing (Dipterocarpus spp.)	Green	0.69	11,900	1.71	13.9	5,680	1,170	1,060	AS
	12%		19,900	2.07	23.5	10,500	2,070	1,270	
Lignumvitae (Guaiacum spp.)	Green	1.05	—	—	—	—	—	—	AM
	12%	—	—	—	—	11,400	—	4,500	
Limba (Terminate superba)	Green	0.38	6,000	0.77	7.7	2,780	88	400	AF
	12%		8,800	1.01	8.9	4,730	1,410	490	
Macawood (Platymiscium spp.)	Green	0.94	22,300	3.02	—	10,540	1,840	3,320	AM
	12%		27,600	3.2	—	16,100	2,540	3,150	
Mahogany, African (Khaya spp.)	Green	0.42	7,400	1.15	7.1	3,730	931	640	AF
	12%		10,700	1.4	8.3	6,460	1,500	830	
Mahogany, true (Swietenia macrophylla)	Green	0.45	9,000	1.34	9.1	4,340	1,240	740	AM
	12%	—	11,500	1.5	7.5	6,780	1,230	800	
Manbarklak (Eschweilera spp.)	Green	0.87	17,100	2.7	17.4	7,340	1,630	2,280	AM
	12%		26,500	3.14	33.3	11,210	2,070	3,480	
Manni (Symphonia globulifera)	Green	0.58	11,200	1.96	11.2	5,160	1,140	940	AM
	12%		16,900	2.46	16.5	8,820	1,420	1,120	

Mechanical Properties of Some Woods Imported into the United States other than Canadian Imports (Inch-Pound)[a]—Cont'd

| Common and botanical names of species | Moisture content | Specific gravity | Static bending | | | Compression parallel to grain (lbf/in²) | Shear parallel to grain (lbf/in²) | Side hardness (lbf) | Sample origin[b] |
			Modulus of rupture (lbf/in²)	Modulus of elasticity (x10⁶ lbf/in³)	Work to maximum load (in-lbf/in³)				
Marishballi (*Lincania* spp.)	Green	0.88	17,100	2.93	13.4	7,580	1,620	2,250	AM
	12%		27,700	3.34	14.2	13,390	1,750	3,570	
Merbau (*Intsia* spp.)	Green	0.64	12,900	2.02	12.8	6,770	1,560	1,380	AS
	15%	—	16,800	2.23	14.8	8,440	1,810	1,500	
Mersawa (*Anisoptera* spp.)	Green	0.52	8,000	1.77	—	3,960	740	880	AS
	12%		13,800	2.28	—	7,370	890	1,290	
Mora (*Mora* spp.)	Green	0.78	12,600	2.33	13.5	6,400	1,400	1,450	AM
	12%		22,100	2.96	18.5	11,840	1,900	2,300	
Oak (*Quercus* spp.)	Green	0.76	—	—	—	—	—	—	AM
	12%		23,000	3.02	16.5	—	—	2,500	
Obeche (*Triplochiton scleroxylon*)	Green	0.3	5,100	0.72	6.2	2,570	660	420	AF
	12%		7,400	0.86	6.9	3,930	990	430	
Okoume (*Aucoumea klaineana*)	Green	0.33	—	—	—	—	—	—	AF
	12%		7,400	1.14	—	3,970	970	380	
Opepe (*Nauclea diderrichii*)	Green	0.63	13,600	1.73	12.2	7,480	1,900	1,520	AF
	12%		17,400	1.94	14.4	10,400	2,480	1,630	
Ovangkol (*Guibourtia ehie*)	Green	0.67	—	—	—	—	—	—	AF
	12%		16,900	2.56	—	8,300	—	—	
Para-angelim (*Hymenolobium excelsum*)	Green	0.63	14,600	1.95	12.8	7,460	1,600	1,720	AM
	12%		17,600	2.05	15.9	8,990	2,010	1,720	
Parana-pine (*Araucaria augustifolia*)	Green	0.46	7,200	1.35	9.7	4,010	970	560	AM
	1.2%	—	13,500	1.61	12.2	7,660	1,730	780	
Pau marfim (*Balfourodendron riedelianum*)	Green	0.73	14,400	1.66	—	6,070	—	—	AM
	15%		18,900	—	—	8,190	—	—	
Peroba de campos (*Paratecoma peroba*)	Green	0.62	—	—	—	—	—	—	AM
	12%		15,400	1.77	10.1	8,880	2,130	1,600	
Peroba rosa (*Aspidosperma* spp., peroba group)	Green	0.66	10,900	1.29	10.5	5,540	1,880	1,580	AM
	12%		12,100	1.53	9.2	7,920	2,490	1,730	
Pilon (*Hyeronima* spp.)	Green	0.65	10,700	1.88	8.3	4,960	1,200	1,220	AM
	12%		18,200	2.27	12.1	9,620	1,720	1,700	
Pine, Caribbean (*Pinus caribaea*)	Green	0.68	11,200	1.88	10.7	4,900	1,170	980	AM
	12%	—	16,700	2.24	17.3	8,540	2,090	1,240	
Pine, ocote (*Pinus oocarpa*)	Green	0.55	8,000	1.74	6.9	3,690	1,040	580	AM
	12%	—	14,900	2.25	10.9	7,680	1,720	910	
Pine, radiata (*Pinus radiata*)	Green	0.42	6,100	1.18	—	2,790	750	480	AS
	12%	—	11,700	1.48	—	6,080	1,600	750	

(Continued)

Mechanical Properties of Some Woods Imported into the United States other than Canadian Imports (Inch-Pound)[a] — Cont'd

| Common and botanical names of species | Moisture content | Specific gravity | Static bending | | | Compression parallel to grain (lbf/in²) | Shear parallel to grain (lbf/in²) | Side hardness (lbf) | Sample origin[b] |
			Modulus of rupture (lbf/in²)	Modulus of elasticity (x10⁶ lbf/in³)	Work to maximum load (in-lbf/in³)				
Piquia (*Caryocar* spp.)	Green	0.72	12,400	1.82	8.4	6,290	1,640	1,720	AM
	12%		17,000	2.16	15.8	8,410	1,990	1,720	
Primavera (*Tabebuia donnell-smithii*)	Green	0.4	7,200	0.99	7.2	3,510	1,030	700	AM
	12%		9,500	1.04	6.4	5,600	1,390	660	
Purpleheart (*Peltogyne* spp.)	Green	0.67	1,370	2	14.8	7,020	1,640	1,810	AM
	12%		19,200	2.27	17.6	10,320	2,220	1,860	
Ramin (*Gonystylus bancanus*)	Green	0.52	9,800	1.57	9	5,390	990	640	AS
	12%	—	18,500	2.17	17	10,080	1,520	1,300	
Robe (*Tabebuia* spp., roble group)	Green	0.52	10,800	1.45	11.7	4,910	1,250	910	AM
	12%		13,800	1.6	12.5	7,340	1,450	960	
Rosewood, Brazilian	Green	0.8	14,100	1.84	13.2	5,510	2,360	2,440	AM
	12%	—	19,000	1.88	—	9,600	2,110	2,720	
Rosewood, Indian (*Dalbergia latifolia*)	Green	0.75	9,200	1.19	11.6	4,530	1,400	1,560	AS
	12%		16,900	1.78	13.1	9,220	2,090	3,170	
Sande (*Brosimum* spp., *utile* group)	Green	0.49	8,500	1.94	—	4,490	1,040	600	AM
	12%		14,300	2.39	—	8,220	1,290	900	
Santa Maria (*Calophyllum brasiliense*)	Green	0.52	10,500	1.59	12.7	4,560	1,260	890	AM
	12%	—	14,600	1.83	16.1	6,910	2,080	1,150	
Sapele (*Entandrophragma cylindricum*)	Green	0.55	10,200	1.49	10.5	5,010	1,250	1,020	AF
	12%	—	15,300	1.82	15.7	8,160	2,260	1,510	
Sepetir (*Pseudosindora palustris*)	Green	0.56	11,200	1.57	13.3	5,460	1,310	950	AS
	12%		17,200	1.97	13.3	8,880	2,030	1,410	
Shorea (*Shorea* spp., *bullau* group)	Green	0.68	11,700	2.1		5,380	1,440	1,350	AS
	12%		18,800	2.61	—	10,180	2,190	1,780	
Shorea, lauan-meranti group									
Dark red meranti	Green	0.46	9,400	1.5	8.6	4,720	1,110	700	AS
	12%		12,700	1.77	13.8	7,360	1,450	780	
Light red meranti	Green	0.34	6,600	1.04	6.2	3,330	710	440	AS
	12%		9,500	1.23	8.6	5,920	970	460	
White meranti	Green	0.55	9,800	1.3	8.3	5,490	1,320	1,000	AS
	15%		12,400	1.49	11.4	6,350	1,540	1,140	
Yellow meranti	Green	0.46	8,000	1.3	8.1	3,880	1,030	750	AS
	12%		11,400	1.55	10.1	5,900	1,520	770	
Spanish-cedar (*Cedrela* spp.)	Green	0.41	7,500	1.31	7.1	3,370	990	550	AM
	12%	—	11,500	1.44	9.4	6,210	1,100	600	
Sucupira (*Bowdichia* spp.)	Green	0.74	17,200	2.27	—	9,730	—	—	AM
	15%		19,400	—	—	11,100	—	—	

Mechanical Properties of Some Woods Imported into the United States other than Canadian Imports (Inch-Pound)[a]—Cont'd

Common and botanical names of species	Moisture content	Specific gravity	Static bending			Compression parallel to grain (lbf/in^2)	Shear parallel to grain (lbf/in^2)	Side hardness (lbf)	Sample origin[b]
			Modulus of rupture (lbf/in^2)	Modulus of elasticity (x10^6 lbf/in^3)	Work to maximum load (in-lbf/in^3)				
Sucupira (Diplotropis purpurea)	Green	0.78	17,400	2.68	13	8,020	1,800	1,980	AM
	12%		20,600	2.87	14.8	12,140	1,960	2,140	
Teak (Tectona grandis)	Green	0.55	11,600	1.37	13.4	5,960	1,290	930	AS
	12%		14,600	1.55	12	8,410	1,890	1,000	
Tornillo (Cedrelinga cateniformis)	Green	0.45	8,400	—	—	4,100	1,170	870	AM
	12%	—	—	—	—	—	—	—	
Wallaba (Eperua spp.)	Green	0.78	14,300	2.33	—	8,040	—	1,540	AM
	12%	—	19,100	2.28	—	10,760	—	2,040	

[a] Results of tests on small, clear, straight-grained specimens. Property values were taken from world literature (not obtained from experiments conducted at the Forest Products Laboratory). Other species may be reported in the world literature, as well as additional data on many of these species. Some property values have been adjusted to 12% moisture content.
[b] AF is Africa; AM, America; AS, Asia.

Average Coefficients of Variation for Some Mechanical Properties of Clear Wood

Property	Coefficient of Variation[a] (%)
Static bending	
Modulus of rupture	16
Modulus of elasticity	22
Work to maximum load	34
Impact bending	25
Compression parallel to grain	18
Compression perpendicular to grain	28
Shear parallel to grain, maximum shearing strength	14
Tension parallel to grain	25
Side hardness	20
Toughness	34
Specific gravity	10

[a] Values based on results of tests of green wood from approximately 50 species. Values for wood adjusted to. 12% moisture content may be assumed to be approximately of the same magnitude.

7.7.0 Mechanical Properties of Wood Imported from Other Countries—Metric

Mechanical Properties of Some Woods Imported into the United States other than Canadian Imports (Metric)[a]

| Common and botanical names of species | Moisture content | Specific gravity | Static bending | | | Compression parallel to grain (kPa) | Shear parallel to grain (kPa) | Side hardness (N) | Sample origin[b] |
			Modulus of rupture (kPa)	Modulus of elasticity (MPa)	Work to maximum load (kJ/m³)				
Afrormosia (*Pericopsis elata*)	Green	0.61	102,000	12,200	135	51,600	11,500	7,100	AF
	12%		126,900	13,400	127	68,500	14,400	6,900	
Albarco (*Cariniana* spp.)	Green	0.48	—	—	—	—	—	—	AM
	12%		100,000	10,300	95	47,000	15,900	4,500	
Andiroba (*Carapa guianensis*)	Green	0.54	71,000	11,700	68	33,000	8,400	3,900	AM
	12%	—	106,900	13,800	97	56,000	10,400	5,000	
Angelin (*Andira inermis*)	Green	0.65	—	—	—	—	—	—	AF
	12%		124,100	17,200	—	63,400	12,700	7,800	
Angelique (*Dicorynia guianensis*)	Green	0.6	78,600	12,700	83	38,500	9,200	4,900	AM
	12%	—	120,000	15,100	105	60,500	11,400	5,700	
Avodire (*Turraeanthus africanus*)	Green	0.48	—	—	—	—	—	—	AF
	12%		87,600	10,300	65	49,300	14,000	4,800	
Azobe (*Lophira alata*)	Green	0.87	116,500	14,900	83	65,600	14,100	12,900	AF
	12%		168,900	17,000	—	86,900	20,400	14,900	
Balsa (*Ochrpma pyramidale*)	Green	0.16	—	—	—	—	—	—	AM
	12%		21,600	3,400	14	14,900	2,100	—	
Banak (*Virola* spp.)	Green	0.42	38,600	11,300	28	16,500	5,000	1,400	AM
	12%	—	75,200	14,100	69	35,400	6,800	2,300	
Benge (*Guibourtia arnoldiana*)	Green	0.65	—	—	—	—	—	—	AF
	12%		147,500	14,100	—	78,600	14,400	7,800	
Bubinga (*Guibourtia* spp.)	Green	0.71	—	—	—	—	—	—	AF
	12%		155,800	17,100	—	72,400	21,400	12,000	
Bulletwood (*Manilkara bidentata*)	Green	0.85	119,300	18,600	94	59,900	13,100	9,900	AM
	12%		188,200	23,800	197	80,300	17,200	14,200	
Cativo (*Prioria copaifera*)	Green	0.4	40,700	6,500	37	17,000	5,900	2,000	AM
	12%	—	59,300	7,700	50	29,600	7,300	2,800	
Ceiba (*Ceiba pentandra*)	Green	0.25	15,200	2,800	8	7,300	2,400	1,000	AM
	12%		29,600	3,700	19	16,400	3,800	1,100	
Courbaril (*Hymenaea courbaril*)	Green	0.71	88,900	12,700	101	40,000	12,200	8,800	AM
	12%	—	133,800	14,900	121	65,600	17,000	10,500	
Cuangare (*Dialyanthera* spp.)	Green	0.31	27,600	7,000	—	14,300	4,100	1,000	AM
	12%		50,300	10,500	—	32,800	5,700	1,700	
Cypress, Mexican (*Cupressus lustianica*)	Green	0.93	42,700	6,300	—	19,900	6,600	1,500	AF
	12%		71,000	7,000	—	37,100	10,900	2,000	
Degame (*Calycophyllum candidissimum*)	Green	0.67	98,600	13,300	128	42,700	11,400	7,300	AM
	12%		153,800	15,700	186	66,700	14,600	8,600	
Determa (*Ocotea rubra*)	Green	0.52	53,800	10,100	33	25,900	5,900	2,300	AM
	12%		72,400	12,500	44	40,000	6,800	2,900	
Ekop (*Tetraberlinia tubmaniana*)	Green	0.6	—	—	—	—	—	—	AF
	12%		115,100	15,200	—	62,100	—	—	
Goncalo alves (*Astronium graveolens*)	Green	0.84	83,400	13,400	46	45,400	12,100	8,500	AM
	12%	—	114,500	15,400	72	71,200	13,500	9,600	

Mechanical Properties of Some Woods Imported into the United States other than Canadian Imports (Metric)[a]—Cont'd

| Common and botanical names of species | Moisture content | Specific gravity | Static bending | | | Compression parallel to grain (kPa) | Shear parallel to grain (kPa) | Side hardness (N) | Sample origin[b] |
			Modulus of rupture (kPa)	Modulus of elasticity (MPa)	Work to maximum load (kJ/m³)				
Greenheart (*Chlorocardium rodiei*)	Green	0.8	133,100	17,000	72	64,700	13,300	8,400	AM
	12%		171,700	22,400	175	86,300	18,100	10,500	
Hura (*Hura crepitans*)	Green	0.38	43,400	7,200	41	19,200	5,700	2,000	AM
	12%		60,000	8,100	46	33,100	7,400	2,400	
Ilomba (*Pycnanthus angolensis*)	Geen	0.4	37,900	7,900		20,000	5,800	2,100	AF
	12%		68,300	11,000	—	38,300	8,900	2,700	
Ipe (*Tabebuia* spp., lapacho group)	Green	0.92	155,800	20,100	190	71,400	14,600	13,600	AM
	12%		175,100	21,600	152	89,700	14,200	16,400	
Iroko (*Chlorophora* spp.)	Green	0.54	70,300	8,900	72	33,900	9,000	4,800	AF
	12%		85,500	10,100	62	52,300	12,400	5,600	
Jarrah (*Eucalyptus marginata*)	Green	0.67	68,300	10,200	—	35,800	9,100	5,700	AS
	12%	—	111,700	13,000	—	61,200	14,700	8,500	
Jelutong (*Dyera costulata*)	Green	0.36	38,600	8,000	39	21,000	5,200	1,500	AS
	15%		50,300	8,100	44	27,000	5,800	1,700	
Kaneelhart (*Licaria* spp.)	Green	0.96	153,800	26,300	94	92,300	11,600	9,800	AM
	12%		206,200	28,000	121	120,000	13,600	12,900	
Kapur (*Dryobalanops* spp.)	Green	0.64	88,300	11,000	108	42,900	8,100	4,400	AS
	12%		126,200	13,000	130	69,600	13,700	5,500	
Karri (*Eucalyptus diversicolor*)	Green	0.82	77,200	13,400	80	37,600	10,400	6,000	AS
	12%		139,000	17,900	175	74,500	16,700	9,100	
Kempas (*Koompassia malaccensis*)	Green	0.71	100,000	16,600	84	54,700	10,100	6,600	AS
	12%		122,000	18,500	106	65,600	12,300	7,600	
Keruing (*Dipterocarpus* spp.)	Green	0.69	82,000	11,800	96	39,200	8,100	4,700	AS
	12%		137,200	14,300	162	72,400	14,300	5,600	
Lignumvitae (*Guaiacum* spp.)	Green	1.05	—	—	—	—			AM
	12%	—	—	—	—	78,600	—	20,000	
Limba (*Terminalia superba*)	Green	0.38	41,400	5,300	53	19,200	600	1,800	AF
	12%		60,700	7,000	61	32,600	9,700	2,200	
Macawood (*Platymiscium* spp.)	Green	0.94	153,800	20,800	—	72,700	12,700	14,800	AM
	12%		190,300	22,100	—	111,000	17,500	14000	
Mahogany, African (*Khaya* spp.)	Green	0.42	51,000	7,900	49	25,700	6,400	2,800	AF
	12%.		73,800	9,700	57	44,500	10,300	3,700	
Mahogany, true (*Swietenia macrophylla*)	Green	0.45	62,100	9,200	63	29,900	8,500	3,300	AM
	12%	—	79,300	10,300	52	46,700	8,500	3,600	
Manbarklak (*Eschweilera* spp.)	Green	0.87	117,900	18,600	120	50,600	11,200	10,100	AM
	12%		182,700	21,600	230	77,300	14,300	15,500	
Manni (*Symphonia globulifera*)	Green	0.58	77,200	13,500	77	35,600	7,900	4,200	AM
	12%		116,500	17,000	114	60,800	9,800	5,000	
Marishballi (*Lincania* spp.)	Green	0.88	117,900	20,200	92	52,300	11,200	10,000	AM
	12%		191,000	23,000	98	92,300	12,100	15,900	
Merbau (*Intsia* spp.)	Green	0.64	88,900	13,900	88	46,700	10,800	6,100	AS
	15%	—	115,800	15,400	102	58,200	12,500	6,700	
Mersawa (*Anisoptera* spp.)	Green	0.52	55,200	12,200	—	27,300	5,100	3,900	AS
	12%		95,100	15,700	—	50,800	6,100	5,700	

(Continued)

Mechanical Properties of Some Woods Imported into the United States other than Canadian Imports (Metric)[a]—Cont'd

Common and botanical names of species	Moisture content	Specific gravity	Static bending Modulus of rupture (kPa)	Static bending Modulus of elasticity (MPa)	Static bending Work to maximum load (kJ/m^3)	Compression parallel to grain (kPa)	Shear parallel to grain (kPa)	Side hardness (N)	Sample origin[b]
Mora (*Mora* spp.)	Green	0.78	86,900	16,100	93	44,100	9,700	6,400	AM
	12%		152,400	20,400	128	81,600	13,100	10,200	
Oak (*Quercus* spp.)	Green	0.76	—	—	—	—	—	—	AM
	12%		158,600	20,800	114			11,100	
Obeche (*Triplochiton scleroxylon*)	Green	0.3	35,200	5,000	43	17,700	4,600	1,900	AF
	12%		51,000	5,900	48	27,100	6,800	1,900	
Okoume (*Aucoumea klaineana*)	Green	0.33	—	—	—	—	—	—	AF
	12%		51,000	7,900	—	27,400	6,700	1,700	
Opepe (*Nauclea diderrichii*)	Green	0.63	93,800	11,900	84	51,600	13,100	6,800	AF
	12%		120,000	13,400	99	71,700	17,100	7,300	
Ovangkol (*Guibourtia ehie*)	Green	0.67	—	—	—	—	—	—	AF
	12%		116,500	17,700	—	57,200	—	—	
Para-angelim (*Hymenolobium excelsum*)	Green	0.63	100,700	13,400	88	51,400	11,000	7,700	AM
	12%		121,300	14,100	110	62,000	13,900	7,700	
Parana-pine (*Araucaria augustifolia*)	Green	0.46	49,600	9,300	67	27,600	6,700	2,500	AM
	12%	—	93,100	11,100	84	52,800	11,900	3,500	
Pau marfim (*Balfourodendron riedelianum*)	Green	0.73	99,300	11,400	—	41,900	—	—	AM
	15%		130,300	—	—	56,500			
Peroba de campos (*Paratecoma peroba*)	Green	0.62	—	—	—	—	—	—	AM
	12%		106,200	12,200	70	61,200	14,700	7,100	
Peroba rosa (*Aspidosperma* spp., peroba group)	Green	0.66	75,200	8,900	72	38,200	13,000	7,000	AM
	12%		83,400	10,500	63	54,600	17,200	7,700	
Pilon (*Hyeronima* spp.)	Green	0.65	73,800	13,000	57	34,200	8,300	5,400	AM
	12%	—	125,500	15,700	83	66,300	11,900	7,600	
Pine, Caribbean (*Pinus caribaea*)	Green	0.68	77,200	13,000	74	33,800	8,100	4,400	AM
	12%	—	115,100	15,400	119	58,900	14,400	5,500	
Pine, ocote (*Pinus oocarpa*)	Green	0.55	55,200	12,000	48	25,400	7,200	2,600	AM
	12%	—	102,700	15,500	75	53,000	11,900	4,000	
Pine, radiata (*Pinus radiata*)	Green	0.42	42,100	8,100	—	19,200	5,200	2,100	AS
	12%	—	80,700	10,200	—	41,900	11,000	3,300	
Piquia (*Caryocar* spp.)	Green	0.72	85,500	12,500	58	43,400	11,300	7,700	AM
	12%		117,200	14,900	109	58,000	13,700	7,700	
Primavera (*Tabebuia donnell-smithii*)	Green	0.4	49,600	6,800	50	24,200	7,100	3,100	AM
	12%		65,500	7,200	44	38,600	9,600	2,900	
Purpleheart (*Peltogyne* spp.)	Green	0.67	9,400	13,800	102	48,400	11,300	8,100	AM
	12%		132,400	15,700	121	71,200	15,300	8,300	
Ramin (*Gonystylus bancanus*)	Green	0.52	67,600	10,800	62	37,200	6,800	2,800	AS
	12%	—	127,600	15,000	117	69,500	10,500	5,800	
Robe (*Tabebuia* spp., roble group)	Green	0.52	74,500	10,000	81	33,900	8,600	4,000	AM
	12%	—	95,100	11,000	86	50,600	10,000	4,300	
Rosewood, Brazilian (*Dalbergia nigra*)	Green	0.8	97,200	12,700	91	38,000	16,300	10,900	AM
	12%	—	131,000	13,000	—	66,200	14,500	12,100	

Mechanical Properties of Some Woods Imported into the United States other than Canadian Imports (Metric)[a]—Cont'd

| Common and botanical names of species | Moisture content | Specific gravity | Static bending | | | Compression parallel to grain (kPa) | Shear parallel to grain (kPa) | Side hardness (N) | Sample origin[b] |
			Modulus of rupture (kPa)	Modulus of elasticity (MPa)	Work to maximum load (kJ/m³)				
Rosewood, Indian (*Dalbergia latifolia*)	Green	0.75	63,400	8,200	80	31,200	9,700	6,900	AS
	12%		116,500	12,300	90	63,600	14,400	14,100	
Sande (*Brosimum* spp., utile group)	Green	0.49	58,600	13,400	—	31,000	7,200	2,700	AM
	12%		98,600	16,500	—	56,700	8,900	4,000	
Santa Maria (*Calophyllum brasiliense*)	Green	0.52	72,400	11,000	88	31,400	8,700	4,000	AM
	12%	—	100,700	12,600	111	47,600	14,300	5,100	
Sapele (*Entandrophragma cylindricum*)	Green	0.55	70,300	10,300	72	34,500	8,600	4,500	AF
	12%	—	105,500	12,500	108	56,300	15,600	6,700	
Sepetir (*Pseudosindora palustris*)	Green	0.56	77,200	10,800	92	37,600	9,000	4,200	AS
	12%	—	118,600	13,600	92	61,200	14,000	6,300	
Shorea (*Shorea* spp., baulau group)	Green	0.68	80,700	14,500	—	37,100	9,900	6,000	AS
	12%	—	129,600	18,000	—	70,200	15,100	7,900	
Shorea, lauan-meranti group									
Dark red meranti	Green	0.46	64,800	10,300	59	32,500	7,700	3,100	AS
	12%		87,600	12,200	95	50,700	10,000	3,500	
Light red meranti	Green	0.34	45,500	7,200	43	23,000	4,900	2,000	AS
	12%		65,500	8,500	59	40,800	6,700	2,000	
White meranti	Green	0.55	67,600	9,000	57	37,900	9,100	4,400	AS
	15%		85,500	10,300	79	43,800	10,600	5,100	
Yellow meranti	Green	0.46	55,200	9,000	56	26,800	7,100	3,300	AS
	12%		78,600	10,700	70	40,700	10,500	3,400	
Spanish-cedar (*Cedrela* spp.)	Green	0.41	51,700	9,000	49	23,200	6,800	2,400	AM
	12%	—	79,300	9,900	65	42,800	7,600	2,700	
Sucupira (*Bowdichia* spp.)	Green	0.74	118,600	15,700	—	67,100	—	—	AM
	15%		133,800	—	—	76,500	—	—	
Sucupira (*Diplotropis purpurea*)	Green	0.78	120,000	18,500	90	55,300	12,400	8,800	AM
	12%		142,000	19,800	102	83,700	13,500	9,500	
Teak (*Tectona grandis*)	Green	0.55	80,000	9,400	92	41,100	8,900	4,100	AS
	12%		100,700	10,700	83	58,000	13,000	4,400	
Tornillo (*Cedrelinga cateniformis*)	Green	0.45	57,900	—	—	28,300	8,100	3,900	AM
	12%	—	—	—	—	—			
Wallaba (*Eperua* spp.)	Green	0.78	98,600	16,100	—	55,400		6,900	AM
	12%		131,700	15,700	—	74,200	—	9,100	

[a] Results of tests on small, clear, straight-grained specimens. Property values were taken from world literature (not obtained from experiments conducted at the Forest Products Laboratory). Other species may be reported in the world literature, as well as additional data on many of these species. Some property values have been adjusted to 12% moisture content.
[b] AF is Africa; AM, America; AS, Asia.

7.8.0 Thermal Qualities of Wood

Species	Specific gravity	Conductivity (W/m-K (Btuin/h-ft^2·°F))		Resistivity (W/m.K (h.ft^2·°F/Btu.in))	
		Ovendry	12% MC	Ovendry	12% MC
Softwoods					
Baldcypress	0.47	0.11 (0.76)	0.13 (0.92)	9.1 (1.3)	7.5 (1.1)
Cedar					
Atlantic white	0.34	0.085 (0.59)	0.10 (0.70)	12 (1.7)	9.9 (1.4)
Eastern red	0.48	0.11 (0.77)	0.14 (0.94)	8.9 (1.3)	7.4 (1.1)
Northern white	0.31	0.079 (0.55)	0.094 (0.65)	13 (1.8)	11 (1.5)
Port-Orford	0.43	0.10 (0.71)	0.12 (0.85)	9.8 (1.4)	8.1 (1.2)
Western red	0.33	0.083 (0.57)	0.10 (0.68)	12 (1.7)	10 (1.5)
Yellow	0.46	0.11 (0.75)	0.13 (0.90)	9.3 (1.3)	7.7 (1.1)
Douglas-fir					
Coast	0.51	0.12 (0.82)	0.14 (0.99)	8.5 (1.2)	7.0 (1.0)
Interior north	0.50	0.12 (0.80)	0.14 (0.97)	8.6 (1.2)	7.1 (1.0)
Interior west	0.52	0.12 (0.83)	0.14 (1.0)	8.4 (1.2)	6.9 (1.0)
Fir					
Balsam	0.37	0.090 (0.63)	0.11 (0.75)	11 (1.6)	9.2 (1.3)
White	0.41	0.10 (0.68)	0.12 (0.82)	10 (1.5)	8.5 (1.2)
Hemlock					
Eastern	0.42	0.10 (0.69)	0.12 (0.84)	10 (1.4)	8.3 (1.2)
Western	0.48	0.11 (0.77)	0.14 (0.94)	8.9 (1.3)	7.4 (1.1)
Larch, western	0.56	0.13 (0.88)	0.15 (1.1)	7.9 (1.1)	6.5 (0.93)
Pine					
Eastern white	0.37	0.090 (0.63)	0.11 (0.75)	11 (1.6)	9.2 (1.3)
Jack	0.45	0.11 (0.73)	0.13 (0.89)	9.4 (1.4)	7.8 (1.1)
Loblolly	0.54	0.12 (0.86)	0.15 (1.0)	8.1 (1.2)	6.7 (0.96)
Lodgepole	0.43	0.10 (0.71)	0.12 (0.85)	9.8 (1.4)	8.1 (1.2)
Longleaf	0.62	0.14 (0.96)	0.17 (1.2)	7.2 (1.0)	5.9 (0.85)
Pitch	0.53	0.12 (0.84)	0.15 (1.0)	8.2 (1.2)	6.8 (0.98)
Ponderosa	0.42	0.10 (0.69)	0.12 (0.84)	10 (1.4)	8.3 (1.2)
Red	0.46	0.11 (0.75)	0.13 (0.90)	9.3 (1.3)	7.7 (1.1)
Shortleaf	0.54	0.12 (0.86)	0.15 (1.0)	8.1 (1.2)	6.7 (0.96)
Slash	0.61	0.14 (0.95)	0.17 (1.2)	7.3 (1.1)	6.0 (0.86)
Western white	0.37	0.090 (0.63)	0.11 (0.75)	11 (1.6)	9.2 (1.3)
Redwood	0.40	0.10 (0.67)	0.12 (0.80)	10 (1.5)	8.6 (1.2)
Old growth	0.41	0.10 (0.68)	0.12 (0.82)	10 (1.5)	8.5 (1.2)
Young growth	0.37	0.090 (0.63)	0.11 (0.75)	11 (1.6)	9.2 (1.3)
Spruce					
Black	0.43	0.10 (0.71)	0.12 (0.85)	9.8 (1.4)	8.1 (1.2)
Engelmann	0.37	0.090 (0.63)	0.11 (0.75)	11 (1.6)	9.2 (1.3)
Red	0.42	0.10 (0.69)	0.12 (0.84)	10 (1.4)	8.3 (1.2)
Sitka	0.42	0.10 (0.69)	0.12 (0.84)	10 (1.4)	8.3 (1.2)
White	0.37	0.090 (0.63)	0.11 (0.75)	11 (1.6)	9.2 (1.3)

Values in this table are approximate and should be used with caution; actual conductivities may vary by as much as 20%. The specific gravities also do not represent species averages.
Source: U.S. Department of Agriculture.

7.9.0 Machine Stress-Rated Lumber—Calculating Design Values

Design Values of MSR Lumber

MSR lumber assures the performance and reliability of your engineered components and structures. The machine grading process sorts dimension lumber by strength and stiffness to improve consistency. Daily quality control testing for strength and stiffness ensures that products meet structural requirements. Machine graded lumber certification and quality control procedures are based on approved American or Canadian lumber standards (ALS or CLS).

Design values in pounds per square inch (psi)

Grade Designation	Bending F_b	Tension Parallel to Grain F_t	Compression Parallel to Grain $F_{c//}$	Modulus of Elasticity E
1650f-1.5E	1650	1020	1700	1,500,000
1800f-1.6E	1800	1175	1750	1,600,000
2100f-1.8E	2100	1575	1875	1,800,000
2400f-2.0E	2400	1925	1975	2,000,000

Notes:

(1) Other grades: The grades listed above are meant as examples of commonly produced MSR grades. This is not intended to be a complete list. The *Supplement to the National Design Specification (NDS) for Wood Construction* (AF&PA, 1997) provides a good summary of the established grades in Machine Stress-Rated (MSR) and Machine Evaluated Lumber (MEL) in Table 4C on page 35.

(2) Quality Control: QC testing takes place on a daily basis for all machine-graded lumber products. Depending on the grade requirements, testing takes place for one or more of the following properties: E, F_b, F_t. This testing process is established to verify that production meets design requirements for all products shipped to customers as MSR, MEL, or E-rated laminating grades.

(3) Common Species: The *MSR Lumber Production Survey* (MSR Lumber Producers Council, 1998) identifies Spruce-Pine-Fir (SPF) as the species in which the majority of MSR is produced (69% of total MSR in 1998). Douglas Fir-Larch (DFL), Hem-Fir (HF), and Southern Pine (SYP) share significant volumes as well (5 to 14% of total MSR in 1998). Consult the MSR Lumber Production Survey for more information.

(4) Species-specific design values: MSR lumber simplifies many design considerations since a grade like 1650f-1.5E maintains the same F_b, F_t, $F_{c//}$, and E values no matter what species or size is considered. Other properties, however, change by species as they relate to the specific gravity (density) of wood. A couple of examples are listed below. Please refer to NDS Supplement footnote #2 for Table 4C for more detailed information. Grade rules writing agencies are another good source of up-to-date information on this topic (see NDS Supplement Section 1.1 on page 2).

Grade Designation	Species	SG	Shear Parallel to Grain F_v	Compression Perpendicular to Grain $F_{c\perp}$
1650f-1.5E	SPF	0.42	135	425
1650f-1.5E	HF	0.46	145	405
2400f-2.0E	SPF	0.50	170	615
2400f-2.0E	SYP	0.57	190	805

(5) Adjustment Factors: Adjustment factors must be applied to the allowable design values presented here. Adjustments to values are taken for duration of load, repetitive member situations, beam and column stability, and other factors as summarized in NDS Table 2.3.1 on page 9. Machine graded lumber enjoys specific advantages in beam stability, column stability, and buckling stiffness factors as a result of the consistency of E compared to visually graded lumber.

Lumber Producers Council
Educational Information
Design Values of Machine Stress-Rated Lumber

Machine Stress-Rated lumber (MSR Lumber) assures the performance and reliability of your engineered components and structures. The machine grading process sorts dimension lumber by strength and stiffness to improve consistency. Daily quality control testing for strength and stiffness ensures that products meet structural requirements. Machine-graded lumber certification and quality control procedures are based on approved American or Canadian lumber standards (ALS or CLS).

Design Values in Pounds per Square Inch (psi)				
Grade Designation	Bending F_b	Tension Parallel to Grain F_t	Compression Parallel to Grain $F_{c//}$	Modulus of Elasticity E
1650f-1.5E	1650	1020	1700	1,500,000
1800f-1.6E	1800	1175	1750	1,600,000
2100f-1.8E	2100	1575	1875	1,800,000
2400f-2.0E	2400	1925	1975	2,000,000

Notes:

(1) **Other grades of Machine Stress-Rated Lumber:** The grades listed above are meant as examples of commonly produced Machine Stress-Rated (MSR) grades. This is not intended to be a complete list. The Supplement to the National Design Specification (NDS) for Wood Construction (AF& PA, 1997) provides a good summary of the established grades in Machine Stress-Rated (MSR) and Machine Evaluated Lumber (MEL) in Table 4C on page 35.

(2) **Quality Control of Machine Stress-Rated Lumber:** QC testing takes place on a daily basis for all machine-graded lumber products. Depending on the grade requirements, testing takes place for one or more of the following properties: E, F_b, F_t. This testing process is established to verify that production meets design requirements for all products shipped to customers as MSR, MEL, or E-rated laminating grades.

(3) **Common Species used in Machine Stress-Rated Lumber:** The Machine Stress-Rated Lumber Production Survey (MSR Lumber Producers Council, 1998) identifies Spruce-Pine-Fir (SPF) as the species in which the majority of Machine Stress-Rated Lumber (MSR Lumber) is produced (69% of total Machine Stress Rated in 1998). Douglas Fir-Larch (DFL), Hem-Fir (HF), and Southern Pine (SYP) and share significant volumes as well (5% to 14% of total Machine Stress-Rated Lumber (MSR Lumber) in 1998). Consult the MSR Lumber Production Survey for more information.

(4) **Species-specific design values for Machine Stress-Rated Lumber:** Machine Stress-Rated lumber (MSR lumber) simplifies many design considerations since a grade like 1650M.5E maintains the same F_b, F_t, $F_{c//}$,

and E values no matter what species or size is considered. Other properties, however, change by species as they relate to the specific gravity (density) of wood. A couple of examples are listed below. Please refer to NDS Supplement footnotes #2 for Table 4C for more detailed information. Grade rules writing agencies are another good source of up-to-date information on this topic (see NDS Supplement Section 1.1 on page 2).

Grade Designation	Species	SG	Shear Parallel to Grain Fv	Compression Perpendicular to Grain Fc.L
1650f-1.5E	SPF	0.42	135	425
1650f-1.5E	HF	0.43	145	405
2400f-2.0E	SPF	0.50	170	615
2400f-2.0E	SYP	0.57	190	805

(5) **Adjustment Factors:** Adjustment factors must be applied to the allowable design values presented here. Adjustments to values are taken for duration of load, repetitive member situations, beam and column stability, and other factors as summarized in NDS Table 2.3.1 on page 9. Machine-graded lumber enjoys specific advantages in beam stability, column stability, and buckling stiffness factors as a result of the consistency of E compared to visually graded lumber.

7.10.0 Grade Names for Interior and Exterior Plywood

Panel Grade Designation	Minimum Face	Veneer Back	Quality Inner Plies	Surface
N–N	N	N	C	S2S[a]
N–A	N	A	C	S2S
N–B	N	B	C	S2S
N–D	N	D	D	S2S
A–A	A	A	D	S2S
A–B	A	B	D	S2S
A–D	A	D	D	S2S
B–B	B	B	D	S2S
B–D	B	D	D	S2S
Underlayment	C plugged	D	C & D	Touch sanded
C–D plugged	C plugged	D	D	Touch sanded
Structural 1 C–D				Unsanded
Structural 1 C–D plugged, underlayment				Touch sanded
C–D	C	D	D	Unsanded
C–D with exterior adhesive	C	D	D	Unsanded

[a] Sanded on two sides.

Grade Names for Exterior Plywood Grades[a]

Panel Grade Designation	Minimum Face	Veneer Back	Quality inner Plies	Surface
Marine, A–A, A–B. B–B, HDO, MDO				See regular grades
Special exterior, A–A, A–B, B–B, HDO, MDO				See regular grades
A–A	A	A	C	S2S[b]
A–B	A	B	C	S2S
A–C	A	C	C	S2S
B–B (concrete form)				
B–B	B	B	C	S2S
B–C	B	C	C	S2S
C–C plugged	C plugged	C	C	Touch sanded
C–C	C	C	C	Unsanded
A–A high-density overlay	A	A	C plugged	—
B–B high-density overlay	B	B	C plugged	—
B–B high-density concrete form overlay	B	B	C plugged	—
B–B medium-density overlay	B	B	C	—
Special overlays	C	C	C	—

[a]NIST 1995.
[b]Sanded on two sides.

7.10.1 Softwood Plywood Species Grouped by Stiffness and Strength

Group 1	Group 2	Group 3	Group 4	Group 5
Apitong	Cedar, Port Orford	Alder, red	Aspen	Basswood
Beech, American	Cypress	Birch, paper	Bigtooth	Poplar
Birch	Douglas-fir	Cedar, yellow	Quaking	Balsam
Sweet	Fir	Fir, subalpine	Cativo	
Yellow	Balsam	Hemlock, eastern	Cedar	
Douglas-fir	California red	Maple, bigleaf	Incense	
Kapur	Grand	Pine	Western	
Keruing	Noble	Jack	Red	
Larch, western	Pacific silver	Lodgepole	Cottonwood	
Maple, sugar	White	Ponderosa	Eastern	
Pine	Hemlock, western	Spruce	Black	
Caribbean	Lauan	Redwood	(Western Poplar)	
Ocofe	Almon	Spruce	Pine, eastern	
Pine, Southern	Bagtikan	Engelman	White, sugar	
Loblolly	Mayapis	White		
Longleaf	Red lauan			
Shortleaf	Tangile			
Slash	White lauan			
Tanoak	Maple, black			
	Mengkulang			
	Meranti, red			
	Mersawa			
	Pine			
	Pond			
	Red			
	Virginia			
	Western white			

Group 1	Group 2	Group 3	Group 4	Group 5

	Spruce
	  Black
	  Red
	  Sitka
	Sweetgum
	Tamarack
	Yellow poplar			

7.10.2 Typical Grade Stamps for Plywood and Oriented Stand Board (OSB)

1. Product Standard that governs specifics of production for construction and industrial plywood.
2. Nominal panel thickness subject to acceptable tolerances.
3. Panel grade designation indicating minimum veneer grade used for panel face and back, or grade name based on panel use.
4. Performance-rated panel standard indicating structural-use panel test procedure recognized by National Evaluation Service (NES).
5. NES report number from Council of American Building Officials (CABO).
6. Exposure durability classification: Exposure 1 indicates interior panel bonded with exterior glue suitable for uses not permanently exposed to weather.
7. Span rating indicating maximum spacing of roof and floor supports for ordinary residential construction applications; 32/16 rating identifies a panel rated for use on roof supports spaced up to 813 mm (32 in.) o.c, or floor supports spaced up to 406 mm (16 in.) o.c.
8. Sized for spacing denotes panels that have been sized to allow for spacing of panel edges during installation to reduce the possibility of buckling.

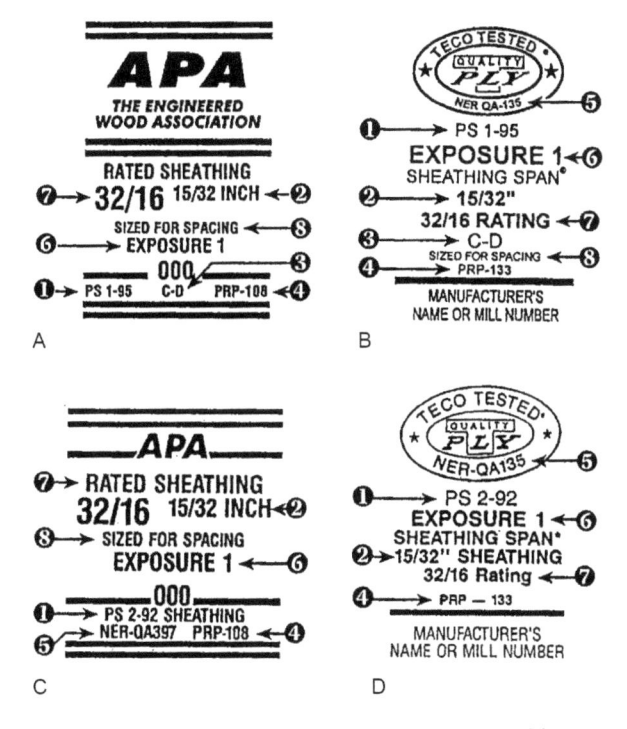

FIGURE 10–4 Typical grade stamps for plywood and OSB.

7.11.0 Classification of Wood Composite Boards by Particle Size and Process Type

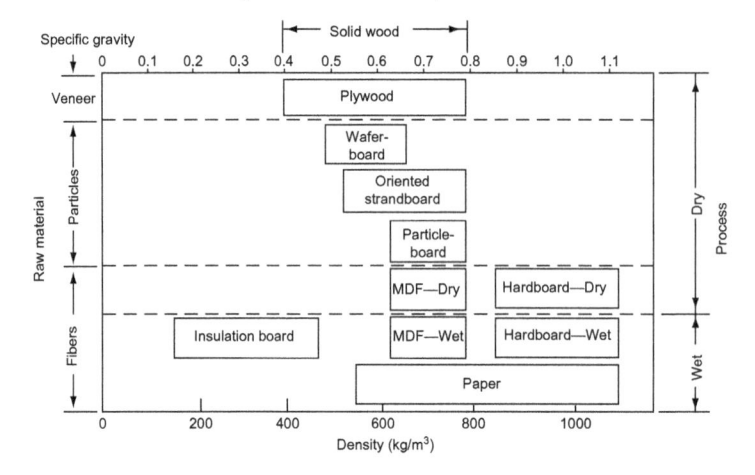

FIGURE 10–2 Classification of wood composite boards by particle size, density, and process type (Suchsland and Woodson 1986).

7.11.1 Particleboard Grade Requirements

Particleboard grade requirement[a,b,c]

Grade[d]	MOR (MPa)	MOE (MPa)	Internal Bond (MPa)	Hardness (N)	Linear Expansion max avg (%)	Screw-holding (N) Face	Screw-holding (N) Edge	Formaldehyde Maximum Emission (ppm)
H–1	16.5	2,400	0.90	2,225	NS	1,800	1,325	0.30
H–2	20.5	2,400	0.90	4,450	NS	1,900	1,550	0.30
H–3	23.5	2,750	1.00	6,675	NS	2,000	1,550	0.30
H–1	11.0	1,725	0.40	2,225	0.35	NS	NS	0.30
H–S	12.5	1,900	0.40	2,225	0.35	900	800	0.30
M–2	14.5	2,225	0.45	2,225	0.35	1,000	900	0.30
M–3	16.5	2,750	0.55	2,225	0.35	1,100	1,000	0.30
LD–1	3.0	550	0.10	NS	0.35	400	NS	0.30
LD–2	5.0	1,025	0.15	NS	0.35	550	NS	0.30

[a]From NPA (1993). Particleboard made with phenol-formaldehyde-based resins does not emit significant quantities of formaldehyde. Therefore, such products and other particleboard products made with resin without formaldehyde are not subject to formaldehyde emission conformance testing.
[b]Panels designated as "exterior adhesive" must maintain 50% MOR after ASTM D1037 accelerated aging.
[c]MOR = modulus of rupture; MOE = modulus of elasticity. NS = not specified. 1 MPa = 145 lb/in^2; 1 N = 0.22 lb.
[d]H = density > 800 kg/m^3 (> 50 lb/ft^3), M = density 640 to 800 kg/m^3 (40 to 50 lb/ft^3). LD = density < 640 kg/m^3 (< 40 lb/ft^3). Grade M–S refers to medium density; "special" grade added to standard after grades M–1, M–2, and M–3. Grade M–S falls between M–1 M–2 in physical properties.

7.11.2 Particleboard Flooring Grade Requirements

Grade[a]	MOR (MPa)	MOE (MPa)	Internal Bond (MPa)	Hardness (N)	Linear Expansion max avg (%)	Formaldehyde Maximum Emission (ppm)
PBU	11.0	1,725	0.40	2,225	0.35	0.20
D–2	16.5	2,750	0.55	2,225	0.30	0.20
D–3	19.5	3,100	0.55	2,225	0.30	0.20

[a]PBU = underlayment; D = manufactured home decking.

Attrition milling is an age-old concept whereby material is fed between two disks, one rotating and the other stationary. As the material is forced through the preset gap between the disks, it is sheared, cut, and abraded into fibers and fiber bundles. Grain has been ground in this way for centuries.

Attrition milling, or refining as it is commonly called, can be augmented by water soaking, steam cooking, or chemical treatments. Steaming the lignocellulosic weakens the lignin bonds between the cellulosic fibers. As a result, the fibers are more readily separated and usually are less damaged than fibers processed by dry processing methods. Chemical treatments, usually alkali, are also used to weaken the lignin bonds. All of these treatments help increase fiber quality and reduce energy requirements, but they may reduce yield as well. Refiners are available with single- or double-rotating disks, as well as steam-pressurized and unpressurized configurations. For medium-density fiberboard (MDF), steam-pressurized refining is typical.

Fiberboard is normally classified by density and can be made by either dry or wet processes (Fig. 10–2). Dry processes are applicable to boards with high density (hardboard) and medium density (MDF). Wet processes are applicable to both high-density hardboard and low-density insulation board. The following subsections briefly describe the manufacturing of high-and medium-density dry-process fiberboard, wet-process hardboard, and wet-process low-density insulation board. Suchsland and Woodson (1986) and Maloney (1993) provide more detailed information.

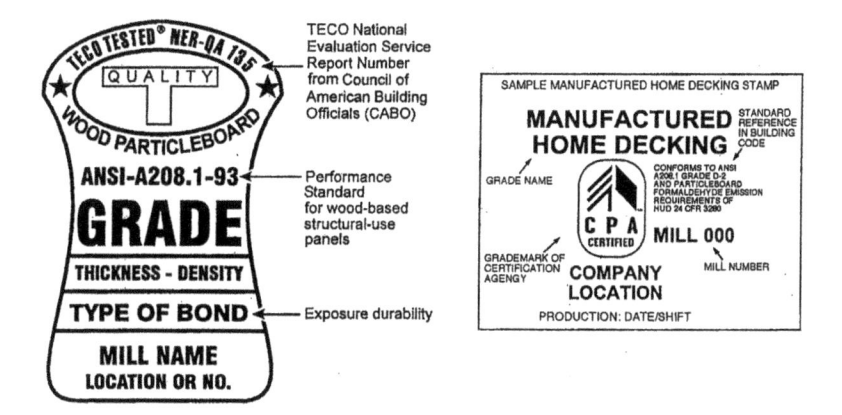

Examples of grade stamps for particleboard.

7.12.0 Medium-Density Fiberboard and Hardboard Property Requirements

Product Class[a]	Nominal Thickness (mm)	MOR (MPa)	MOE (MPa)	Internal Bond (MPa)	Screw-holding (N) Face	Screw-holding (N) Edge	Formaldehyde Emission[b] (ppm)
Interior MDF							
HD		34.5	3,450	0.75	1,555	1,335	0.30
MD	≤21	24.0	2,400	0.60	1,445	1,110	0.30
	>21	24.0	2,400	0.55	1,335	1,000	0.30
LD		14.0	1,400	0.30	780	670	0.30
Exterior MDF							
MD–Exterior	≤21	34.5	3,450	0.90	1,445	1,110	0.30
adhesive	>21	31.0	3,100	0.70	1,335	1,000	0.30

[a] MD–Exterior adhesive panels shall maintain at least 50% of listed MOR after ASTM D1037–1991, accelerated aging (3.3.4). HD = density > 800 kg/m^3 (>50 lb/ft^3), MD = density 640 to 800 kg/m^3 (40 to 50 lb/ft^3), LD = density < 640 kg/m^3 (< 40 lb/ft^3).
[b] Maximum emission when tested in accordance with ASTM E1333–1990. Standard test method for determining formaldehyde levels from wood products under defined test conditions using a larger chamber (ASTM).

Hardboard Physical Property Requirements[a]

Product Class	Normal Thickness (mm)	Water Resistance (max avg/panel) Water Absorption Based on Weight (%)	Water Resistance (max avg/panel) Thickness Swelling (%)	MOR (min avg/ Panel) (MPa)	Tensile Strength (min avg/panel) (MPa) Parallel to Surface	Tensile Strength (min avg/panel) (MPa) Perpendicular to Surface
Tempered	2.1	30	25	41.4	20.7	0.90
	2.5	25	20	41.4	20.7	0.90
	3.2	25	20	41.4	20.7	0.90
	4.8	25	20	41.4	20.7	0.90
	6.4	20	15	41.4	20.7	0.90
	7.9	15	10	41.4	20.7	0.90
	9.5	10	9	41.4	20.7	0.90
Standard	2.1	40	30	31.0	15.2	0.62
	2.5	35	25	31.0	15.2	0.62
	3.2	35	25	31.0	15.2	0.62
	4.8	35	25	31.0	15.2	0.62
	6.4	25	20	31.0	15.2	0.62
	7.9	20	15	31.0	15.2	0.62
	9.5	15	10	31.0	15.2	0.62
Service-tempered	3.2	35	30	31.0	3.8	0.52
	4.8	30	30	31.0	3.8	0.52
	6.4	30	25	31.0	3.8	0.52
	9.5	20	15	31.0	3.8	0.52

[a] AHA 1995a.

7.13.0 Physical and Mechanical Properties of Hardboard Siding

Property[a]	Requirement	
Water absorption (based on weight)	12% (max avg/panel)	
Thickness swelling	8% (max avg/panel)	
Weatherability of substrate (max residual swell)	20%	
Weatherability of primed substrate	No checking, erosion, flaking, or objectionable fiber raising; adhesion, less than 3.2 mm (0.125 in.) of coating picked up	
Linear expansion 30% to 90% RH (max)	Thickness range (cm)	Maximum linear expansion (%)
	0.220–0.324	0.36
	0.325–0.375	0.38
	0.376–0.450	0.40
	>0.451	0.40
Nail-head pull-through	667 N (150 lb) (min avg/panel)	
Lateral nail resistance	667 N (150 lb) (min avg/panel)	
Modulus of rupture	12.4 MPa (1,800 lb/in^2) for 9.5, 11, and 12.7 mm (3/8, 7/16, and 1/2 in.) thick (min avg/panel) 20.7 MPa (3,000 lb/in^2) for 6.4 mm (1/4 in.) thick (min avg/panel)	
Hardness	2002 N (450 lb) (min avg/panel)	
Impact	229 mm (9 in.) (min avg/panel)	
Moisture contend[b]	4% to 9% included, and not more than 3% variance between any two boards in any one shipment or order	

[a]Refer to ANSI/AHA A135.6 1-1990 for test method for determining information on properties.
[b]Since hardboard is a wood-based material, its moisture content varies with environmental humidity conditions. When the environmental humidity conditions in the area of intended use are a critical factor, the purchaser should specify a moisture content range more restrictive than 4 to 9% so that fluctuation in the moisture content of the siding will be kept to a minimum.

Properties of hardboard siding; hardboard siding products come in a great variety of finishes and textures (smooth or embossed) and in different sizes. For application purposes, the AHA classifies siding into three basic types:

Lap siding—boards applied horizontally, with each board overlapping the board below it

Square edge panels—siding intended for vertical application in full sheets

Shiplap edge panel siding—siding intended for vertical application, with the long edges incorporating shiplap joints

The type of panel dictates the application method. The AHA administers a quality conformance program for hardboard for both panel and lap siding. Participation in this program is voluntary and is open to all (not restricted to AHA members). Under this program, hardboard siding products are tested by an independent laboratory in accordance with product standard ANSI/AHA A135.6. Figure 10-13a provides an example of a grade stamp for a siding product meeting this standard.

Insulation Board—Physical and mechanical properties of insulation board are published in the ASTM C208 standard specification for cellulosic fiber insulation board.

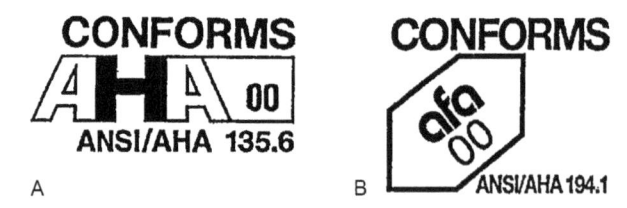

FIGURE 10–13 Examples of grade stamps: (a) grade stamp for siding conforming to ANSI/AHA A135.6 standard, and (b) grade mark stamp for cellulosic fiberboard products conforming to ANSI/AHA A194.1 standard.

Section 8

Fasteners for Wood and Steel—Calculations for Selection

8.1.0 Nail Sizes—Common Wire Nails	442	
8.1.1 Bright Common Nails, Box Nails, Annular Nails—Length and Diameter in United States and Metric	442	
8.1.2 Nail Sizes—Common Wire Spikes	444	
8.1.3 Nail Sizes—Casing Nails	444	
8.1.4 Nail Sizes—Finishing Nails	445	
8.1.5 Nail Sizes—Deformed Shank Nails	445	
8.1.6 Nail Sizes—Roofing Nails	446	
8.1.7 Nail Sizes—Joist Hanger Nails	447	
8.1.8 Cut Nails—Various Configurations	448	
8.2.0 Instructions on Nail Selection and Usage for Exposed Wood Structures	449	
8.3.0 How Did the "Penny Weight" Nail Designation Originate?	454	
8.4.0 About Nails—Historic and Otherwise	455	
8.4.1 Cut Floor Brads	456	
8.4.2 Palm Holdfast	457	
8.4.3 Moulder Brad	457	
8.4.4 Flat Countersunk Head Spike	458	
8.4.5 Rosehead Fine Shank	458	
8.4.6 Rosehead Square Shank Spike	459	
8.4.7 Décor Nail	460	
8.4.8 Boat Nail	460	
8.4.9 CLYDE Rail Spike	461	
8.5.0 Withdrawal Resistance of Nails	461	
8.6.0 Wood Screws—Common Types and Withdrawal Resistance	465	
8.6.1 Wood Screw Sizing	465	
8.6.2 Dimensions of Wood Screws Chart	466	
8.6.3 Basic Types of Wood Screw Drives	467	
8.6.4 Screw Head Types and Shapes	468	
8.6.5 Screw Thread and Point Types	470	
8.6.6 Type 316 Stainless Steel Deck Screws	471	
8.6.7 Type 302 Stainless Steel Bugle Head Screws with Square Drive	472	
8.6.8 Lateral and Withdrawal Resistance of Lag Screws	473	
8.7.0 Bolts in Wood	476	
8.8.0 Wood Adhesives Characterized as to Expected Performance	479	
8.8.1 Categories of Selected Wood Species According to Ease of Bonding	480	
8.8.2 Strength Properties of Various Types of Adhesives	481	
8.9.0 Fasteners Installed in Hollow Masonry Units	484	
8.9.1 Fasteners Installed in Solid Masonry Units	485	
8.9.2 Power-Driven Fasteners for Masonry Units	486	
8.9.3 Masonry Fastener Selection Chart	487	
8.10.0 Bolt Fasteners for Structural Steel—Identification Markings and Strength Requirements	488	
8.10.1 Standard Thread Pitches	492	
8.10.2 Suggested Starting Torque Values for ASTM and SAE Grade Bolts	494	
8.10.3 AASHTO to ASTM Conversion Chart	497	
8.10.4 ASTM A563 Nut Compatibility Chart	497	
8.11.0 Tension Shear—TC Bolts	498	

8.1.0 Nail Sizes—Common Wire Nails

Size	Length (inches)	Gauge	Number in a Pound	Safe Lateral Load* Required Penetration (inches)	Safe Lateral Load* Load (pounds) in Douglas Fir	Resistance to withdrawal in pounds per inch of penetration perpendicular to the grain, into the main member, in Douglas Fir
2d	1	15	876			
3d	1¼	14	568			
4d	1½	12½	316			
5d	1¾	12½	271			
6d	2	11½	181	1	70	27
7d	2¼	11½	161			
8d	2½	10¼	106	1¼	100	32
9d	2¾	10½	96			32
10d	3	9	69	1½	120	36
12d	3¼	9	64	1 5/8	130	36
16d	3½	8	49	1 5/8	160	40
20d	4	6	31	2	190	48
30d	4½	5	24	2¼	230	52
40d	5	4	18	2 1/3	270	56
50d	5½	3	14	2¾	310	61
60d	6	2	11	3	360	67

*For nails inserted perpendicular to the grain. For nails driven parallel to the grain or toe-nailed, the load should not be more than 2/3 of the value in column 6.
Source: U.S. Department of Agriculture.

8.1.1 Bright Common Nails, Box Nails, Annular Nails—Length and Diameter in United States and Metric

Nails

Nails are the most common mechanical fastenings used in wood construction. There are many types, sizes, and forms of nails (Fig. 8–1). The load equations presented in this chapter apply for bright, smooth, common steel wire nails driven into wood when there is no visible splitting. For nails other than common wire nails, the loads can be adjusted by factors given later in the chapter.

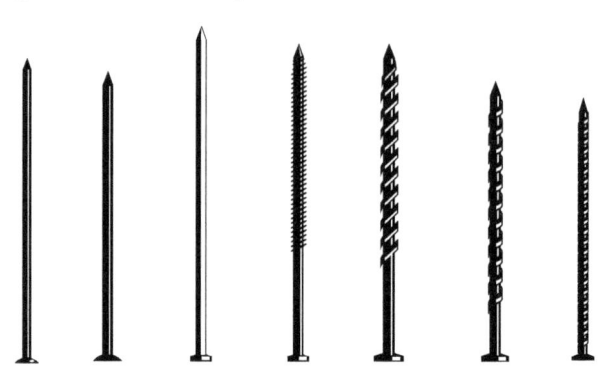

FIGURE 8–1 Various types of nails: (left to right) bright smooth wire nail, cement coated, zinc-coated, annularly threaded, helically threaded, helically threaded and barbed, and barbed.

Nails in use resist withdrawal loads, lateral loads, or a combination of the two. Both withdrawal and lateral resistance are affected by the wood, the nail, and the condition of use. In general, however, any variation in these factors has a more pronounced effect on withdrawal resistance than on lateral resistance. The serviceability of joints with nails laterally loaded does not depend greatly on withdrawal resistance unless large joint distortion is tolerable.

The diameters of various penny or gauge sizes of bright common nails are given in Table 8–1. The penny size designation should be used cautiously. International nail producers sometimes do not adhere to the dimensions of Table 8–1. Thus penny sizes, though still widely used, are obsolete. Specifying nail sizes by length and diameter dimensions is recommended. Bright box nails are generally of the same length but slightly smaller diameter (Table 8–2), while cement-coated nails such as coolers, sinkers, and coated box nails are slightly shorter (3.2 mm [1/8 in.]) and of smaller diameter than common nails of the same penny size. Helically and annularly threaded nails generally have smaller diameters than common nails for the same penny size (Table 8–3).

Withdrawal Resistance

The resistance of a nail shank to direct withdrawal from a piece of wood depends on the density of the wood, the diameter of the nail, and the depth of penetration. The surface condition of the nail at the time of driving also influences the initial withdrawal resistance.

TABLE 8–1 Sizes of Bright Common Wire Nails

Size	Gauge	Length (mm [in.])	Diameter (mm [in.])
6d	11-1/2	50.8 (2)	2.87 (0.113)
8d	10-1/4	63.5 (2-1/2)	3.33 (0.131)
10d	9	76.2 (3)	3.76 (0.148)
12d	9	82.6 (3-1/4)	3.76 (0.148)
16d	8	88.9 (3-1/2)	4.11 (0.162)
20d	6	101.6 (4)	4.88 (0.192)
30d	5	114.3 (4-1/2)	5.26 (0.207)
40d	4	127.0 (5)	5.72 (0.225)
50d	3	139.7 (5-1/2)	6.20 (0.244)
60d	2	152.4 (6)	6.65 (0.262)

TABLE 8–2 Sizes of Smooth Box Nails

Size	Gauge	Length (mm [in.)])	Diameter (mm [in.])
3d	14-1/2	31.8 (1-1/4)	1.93 (0.076)
4d	14	38.1 (1-1/2)	2.03 (0.080)
5d	14	44.5 (1-3/4)	2.03 (0.080)
6d	12-1/2	50.8 (2)	2.49 (0.098)
7d	12-1/2	57.2 (2-1/4)	2.49 (0.098)
8d	11-1/2	63.5 (2-1/2)	2.87 (0.113)
10d	10-1/2	76.2 (3)	3.25 (0.128)
16d	10	88.9 (3-1/2)	3.43 (0.135)
20d	9	101.6 (4)	3.76 (0.148)

TABLE 8-3 Sizes of Helically and Annularly Threaded Nails

Size	Length (mm [in.])	Diameter (mm [in.])
6d	50.8 (2)	3.05 (0.120)
8d	63.5 (2-1/2)	3.05 (0.120)
10d	76.2 (3)	3.43 (0.135)
12d	82.6 (3-1/4)	3.43 (0.135)
16d	88.9 (3-1/2)	3.76 (0.148)
20d	101.6 (4)	4.50 (0.177)
30d	114.3 (4-1/2)	4.50 (0.177)
40d	127.0 (5)	4.50 (0.177)
50d	139.7 (5-1/2)	4.50 (0.177)
60d	152.4 (6)	4.50 (0.177)
70d	177.8 (7)	5.26 (0.207)
80d	203.2 (8)	5.26 (0.207)
90d	228.6 (9)	5.26 (0.207)

Source: U.S. Department of Agriculture.

8.1.2 Nail Sizes—Common Wire Spikes

Size	Length (inches)	Gauge	Number per Pound
10d	3	6	41
12d	3¼	6	38
16d	3½	5	30
20d	4	4	23
30d	4½	3	17
40d	5	2	13
50d	5½	1	10
60d	6	1	8

By permission: tools@sizes.com

Formerly 7″ wire spikes were 0 gauge (0.3065″), and 8″ and 9″ spikes were 00 gauge (0.331″). Now 7″, 8″, and 9″ spikes are 5/16 inch in diameter and 10″ and 12″ spikes are 3/8 inch.

8.1.3 Nail Sizes—Casing Nails

These nails are used where the nail head must be hidden. They have small heads and smaller diameters than common nails.

Casing nails have a conical head, sometimes cupped, and are somewhat thicker than a finishing nail. They are sometimes sold already painted and are used to attach trim.

Size	Length (inches)	Gauge	Number per Pound
2d	1	15½	1,010
3d	1¼	14½	635
4d	1½	14	473
5d	1¾	14	406
6d	2	12½	236

Size	Length (inches)	Gauge	Number per Pound
7d	2¼	12½	210
8d	2½	11½	145
9d	2¾	11½	132
10d	3	10½	94
12d	3¼	10½	87
16d	3½	10	71
20d	4	9	52
30d	4½	9	46
40d	5	8	35

By permission: tools@sizes.com

8.1.4 Nail Sizes—Finishing Nails

fasteners > nails

Finishing nails are used where the nail head must be hidden. They have small heads and smaller diameters than common nails.

Finishing nails, seen in profile, have a barrel-shaped head with a small diameter and a dimple on the top. After the nail is driven almost flush with the surface, the point of a nail set is placed in the dimple and the head driven below the surface. The resulting small hole can be filled with putty. Outdoors, in time the hole will tend to close by itself when the wood fibers swell.

Size	Length (inches)	Gauge	Number per Pound
2d	1	16½	1,351
3d	1¼	15½	807
4d	1½	15	584
5d	1¾	15	500
6d	2	13	309
7d	2¼	13	238
8d	2½	12½	189
9d	2¾	12½	172
10d	3	11½	121
12d	3¼	11½	113
16d	3½	11	90
20d	4	10	62

By permission: tools@sizes.com

8.1.5 Nail Sizes—Deformed Shank Nails

Deformed shank nails are those that have helical shanks, resembling a screw, or that have shanks covered by rings and grooves. Deformed shanks are used in the hope of increasing the resistance of the nail to being pulled out, compared to plain shank nails.

Lawrence Soltis reports that the withdrawal resistance of ring shank nails is about 40% greater than that of common nails. The benefit is much greater in wood subject to repeated changes in moisture content; under those conditions deformed shank nails can be four times better than a plain shank nail of the same diameter. He also suggests a basis for choosing between a ring or a helical shank: "In general, annularly threaded nails sustain larger withdrawal loads, and helically threaded nails sustain greater impact withdrawal work values than do other nail forms."

Besides this series, many other types of nails are available with ring or spiral shanks.

Size	Length (inches)	Gauge	Diameter (inches)
3d	1¼	12½	0.099
4d	1½	12½	0.099
6d	2	11	0.120
8d	2½	11	0.120
10d	3	10	0.135
12d	3¼	10	0.135
16d	3½	9	0.148
20d	4	7	0.177
30d	4½	7	0.177
40d	5	7	0.177
50d	5½	7	0.177
60d	6	7	0.177
70d	7	5	0.207
80d	8	5	0.207

By permission: tools@sizes.com

8.1.6 Nail Sizes—Roofing Nails

Roofing nails are short, diamond-pointed steel nails with a wide flat head. The shank may be barbed, and they are often galvanized.

A peculiarity of the roofing nails standard is that it is the only type of nail with zero tolerance for undersize heads.

Length (inches)	Gauge	Head diameter (inches)
¾	12	0.375
	11	0.375
	11	0.438
	10	0.469
	9½	0.484
	9	0.500
	8	0.500
7/8	12	0.375
	11	0.375
	11	0.438
	11	0.500
	10	0.469
	9½	0.484
	9	0.500
	8	0.500
1	12	0.281
	12	0.375
	11	0.375
	11	0.438
	11	0.500
	10	0.469
	9½	0.484
	9	0.500
	8	0.500

Length (inches)	Gauge	Head diameter (inches)
1 1/8	12	0.375
	11	0.438
	10	0.469
	9½	0.484
	9	0.500
	8	0.500
1¼	12	0.375
	11	0.312
	11	0.375
	11	0.438
	11	0.500
	10	0.469
	9½	0.484
	9	0.500
	8	0.500
1½	12	0.375
	11	0.375
	11	0.438
	11	0.500
	10	0.469
	9½	0.484
	9	0.500
	8	0.500
1¾	12	0.375
	11	0.375
	11	0.438
	11	0.500
	10	0.469
	9½	0.484

By permission: tools@sizes.com

8.1.7 Nail Sizes—Joist Hanger Nails

fasteners > nails

These short, stout nails are used to install sheet metal connectors, including joist hangers. The appropriate size is specified by the connector manufacturer, and frequently the connector is packaged with the necessary number of nails.

They are galvanized, and versions in stainless steel are also available.

Smooth Shanks

	Length (inches)	Gauge	Head Diameter (inches)	No. in a Pound
8d	1½	10¼	9/32	147
10d	1½	9	5/16	123

By permission: tools@sizes.com

Joist hanger nails are also available with deformed shanks for greater holding power.

8.1.8 Cut Nails—Various Configurations

Cut nails possess great durability. They are hard to pull out because the wood fibers are pushed downward and wedge against the nails, thus greatly reducing loosening. Once your customers have used them, they will prefer them for all kinds of work.

Click on any of the products listed below for details.

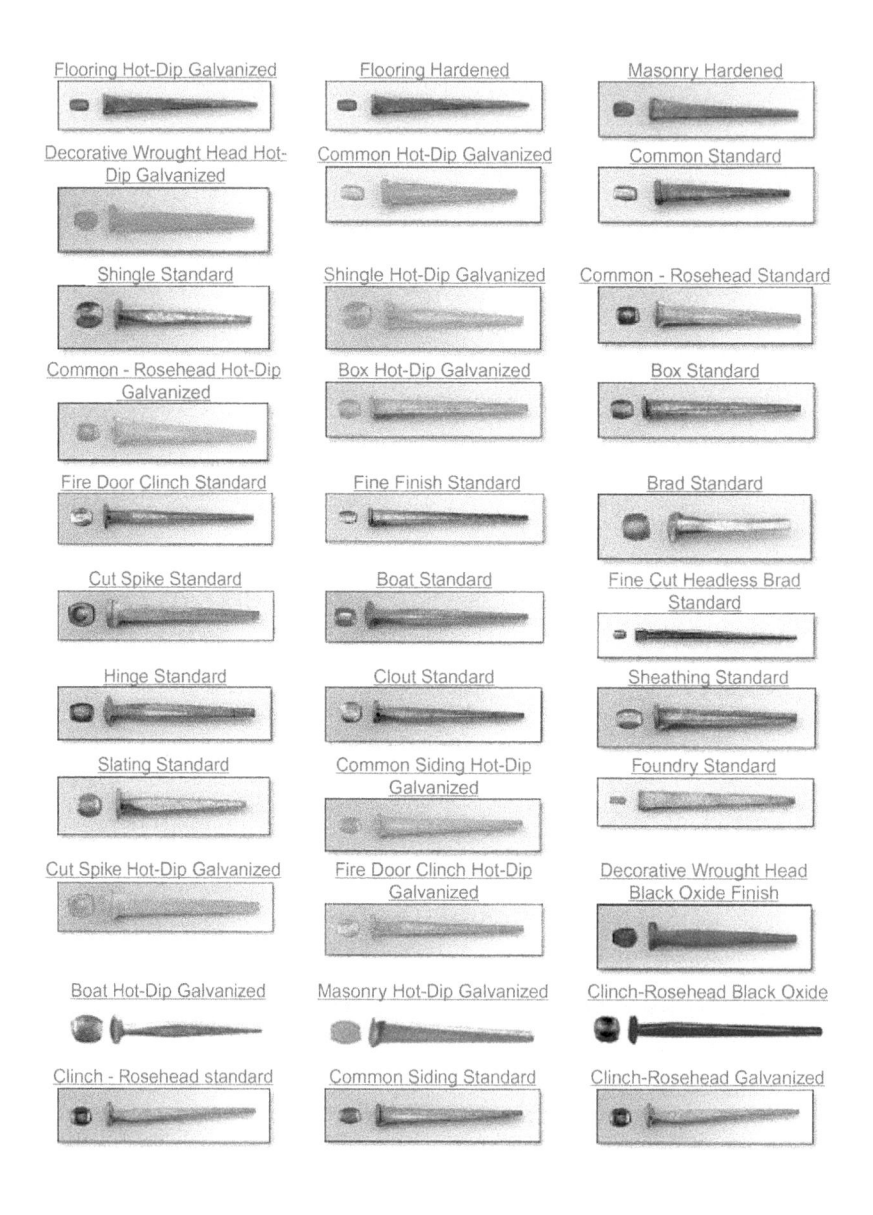

Flooring Hot-Dip Galvanized

Flooring Hardened

Masonry Hardened

Decorative Wrought Head Hot-Dip Galvanized

Common Hot-Dip Galvanized

Common Standard

Shingle Standard

Shingle Hot-Dip Galvanized

Common - Rosehead Standard

Common - Rosehead Hot-Dip Galvanized

Box Hot-Dip Galvanized

Box Standard

Fire Door Clinch Standard

Fine Finish Standard

Brad Standard

Cut Spike Standard

Boat Standard

Fine Cut Headless Brad Standard

Hinge Standard

Clout Standard

Sheathing Standard

Slating Standard

Common Siding Hot-Dip Galvanized

Foundry Standard

Cut Spike Hot-Dip Galvanized

Fire Door Clinch Hot-Dip Galvanized

Decorative Wrought Head Black Oxide Finish

Boat Hot-Dip Galvanized

Masonry Hot-Dip Galvanized

Clinch-Rosehead Black Oxide

Clinch - Rosehead standard

Common Siding Standard

Clinch-Rosehead Galvanized

Wheeling-LaBelle Cut Nails

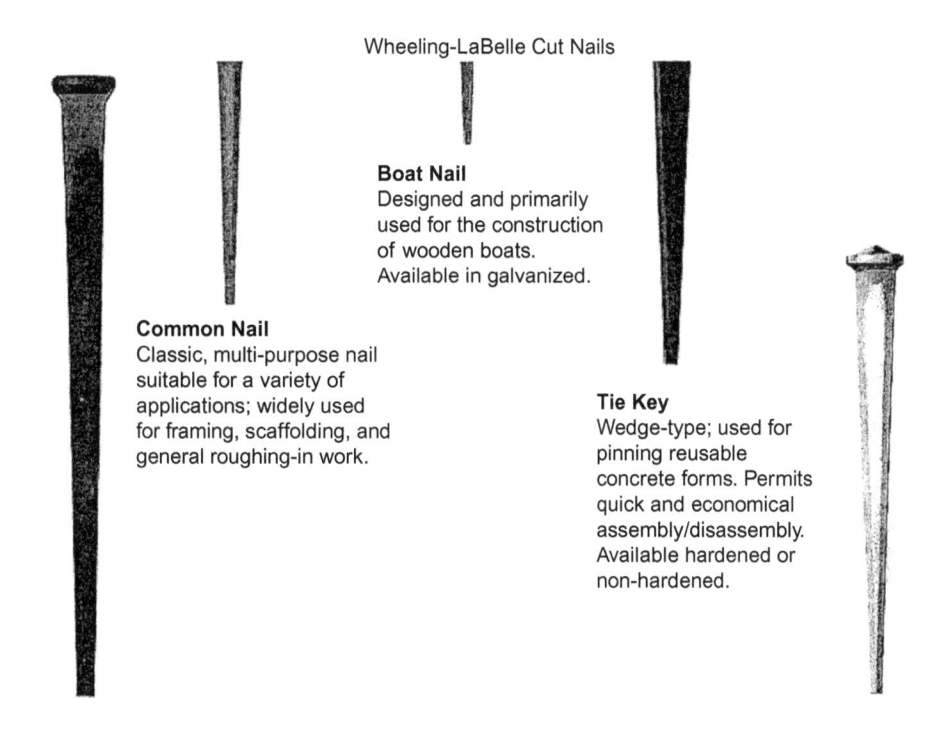

Boat Nail
Designed and primarily used for the construction of wooden boats. Available in galvanized.

Common Nail
Classic, multi-purpose nail suitable for a variety of applications; widely used for framing, scaffolding, and general roughing-in work.

Tie Key
Wedge-type; used for pinning reusable concrete forms. Permits quick and economical assembly/disassembly. Available hardened or non-hardened.

8.2.0 Instructions on Nail Selection and Usage for Exposed Wood Structures

Fasteners for Exposed Wood Structures

Robert H. Falk, P.E. and Andrew J. Baker

Abstract

This paper provides an overview of the use of fasteners that are appropriate for exposed wood structures. Several types of fasteners are reviewed, and physical and chemical explanations for fastener corrosion are provided. Recommendations for long-term performance are given.

Introduction

One of the most important considerations in building a wood outdoor structure is its performance as a structural system. Unlike the skeleton of a conventionally framed wood house, many outdoor structures are built without the sheathing, siding, and roof that provide structural stability and protection from the environment. Fully exposed to the degrading effects of the weather, structural members must be properly designed and connected to ensure long-term, safe performance. Structural support comes from not only the proper sizing and placement of the posts, beams, joists, and other members used in the construction, but from the connection of these members. This paper discusses the fasteners recommended for use in outdoor structures.

Fastener Types

The overall integrity of any wood structure depends on how its components are held together. Therefore, it makes little sense to properly size the wood members only to improperly fasten them. The most common fasteners for wood construction are nails, screws, lag screws, and bolts. Metal straps and hangers of various types are also available. Fasteners used for wood construction are typically manufactured from mild steel, although many types and sizes can be made from stainless steel, brass, and bronze. Nails and screws are the most common type of fasteners for attaching members in light-frame structures. For fastening heavy members of an outdoor structure, such as the beam to the posts, lag screws or bolts are the fasteners of choice.

Holding power and corrosion protection are probably the two most important concerns when choosing fasteners. Improperly specified fasteners can loosen when the wood shrinks and swells as a result of moisture cycling of exposed lumber. Rusting of steel fasteners not only weakens the fastener, but the chemical reactions involved in corrosion can also weaken the wood surrounding the fastener.

Nails. Smooth-shanked nails can lose some of their withdrawal resistance when exposed to wetting and drying cycles, resulting in nail pop-up and loosening of connections. Through the use of a wet-service, strength-reduction factor, the National Design Specification (NDS) (AFPA, 1991) accounts for wood shrinkage from around smooth-shanked nails as the moisture content changes from wet to dry. However, better performance in withdrawal can be expected by using deformed shank nails to resist the effects of severe wetting and drying of exposed wood structures. Two commonly available deformed shank fasteners with the capacity to retain withdrawal resistance are spirally grooved and annularly grooved (ring-shanked) nails.

Screws. Common wood screws have been used to fasten wood for decades; however, the more recently developed multipurpose screw has found common use in wood-deck construction, primarily to fasten deck boards to joists. These fasteners have a thread design that can be driven fast, and they have good holding power. Unlike common wood screws, they are straight shanked. Commonly available in 2- to 3-in. (50- to 75-mm) lengths, multipurpose screws are available with a Phillips head or square recess head and are most easily driven with a power drill.

Multipurpose screws are not intended to fasten joist hangers to beams and will not equal the design capacity of the hanger. Only manufacturer-specified hanger nails should be used to attach hangers.

Screws have an advantage over nails in that they are more easily withdrawn to remove defective or damaged members. They are also effective in drawing down cupped or twisted decking boards into a flat position, and will resist withdrawal over time.

Lag Screws and Bolts. Lag screws are commonly used to fasten one member to a thicker member where a through bolt cannot be used. Pilot holes must be drilled for lag screws, and the screw must be fully inserted to be effective. According to the NDS (AFPA, 1991), for softwood species typically used in outdoor structures, pilot holes should be about 60 to 70% of the diameter of the screw for the threaded portion, and the full diameter for the unthreaded shank. Make sure the lag screw is long enough so that at least half of its length penetrates the thicker member.

Bolts offer more rigidity and typically more load-carrying capacity than lag screws. However, their use is obviously limited to situations where a hole can be drilled completely through the members to be connected.

Holes drilled for bolts should be no more than 1/16 in. (2 mm) larger in diameter than the size of the bolt used. As with lag screws, washers should be used under both the head of the bolt and nut to distribute the bolt force over a larger area and limit crushing of the wood. Machine bolts are a better choice than carriage bolts because carriage bolts are manufactured for use without washers. Dome-head bolts, typically used in heavy timber construction, are also a good choice. After drilling holes for fasteners, it is important to immediately saturate the holes with a preservative, such as copper napthenate. After about one year, bolts should be retightened, and thereafter checked for tightness every year or so.

Joist Hangers and Metal Straps. Joist hangers, metal straps, and other hardware are often used in outdoor wood construction; however, most are intended for indoor use. Although typically electroplated with zinc, their long-term corrosion resistance in exposed environments is unknown. Some manufacturers make these products from stainless steel or apply heavy coatings of galvanizing to increase longevity.

Source: USDA Forest Service.

Wood/Metal Interaction

Wood and metal are compatible in most construction; however, if there is sufficient moisture at the wood-to-metal interface, some corrosion can be expected with susceptible metals. The corrosion of metal in contact with moist wood is an electrochemical process. The rate and amount of corrosion depend on the metal, the conductivity of the wood, and the duration and temperature of the surrounding environmental conditions. The risk of corrosion depends somewhat on the wood species, presence of external corrosive contaminants, and condition of the wood (untreated or treated with certain chemicals). Not only does moist wood in contact with metal cause some corrosion, but the chemical byproducts of corrosion can result in a slow deterioration of the wood adjacent to the metal. As a result, the fastener will lose cross section, and there will be some enlargement of the hole around the fastener. Additionally, most woods are slightly acidic, which may accelerate the corrosion of the steel (or galvanized coating).

If the moisture content of the wood is less than about 18%, the metal corrosion rate is minimal (Baker, 1988). Remember, however, that only the moisture of the wood in contact with the metal is important. This means that metal, such as a fastener that has been cooled by ambient conditions and kept cool by the surrounding wood, can corrode when it becomes wet by condensation, such as on a warm, humid day following a cool night in the early spring. The condensed moisture wets the wood at the wood-to-metal interface. At first, this results in only an iron stain on the surrounding wood surface, but in time, the iron will chemically damage the wood structure and weaken the joint. This type of corrosion is responsible for the failure of many unheated wood structures, such as barns and sheds, in areas with humid days and cool nights.

Some fasteners are corrosion resistant because of a protective coating, and some are resistant because of the properties of the metal or alloy. A fastener can be resistant to corrosion in one environment, but corrode in another. A good example is aluminum, which will perform well in untreated wood exposed to the environment, but will corrode rapidly in wood treated with preservatives that contain copper.

Coated Steel Fasteners

Most steel fasteners are uncoated because they are intended to be used in protected environments (indoors). Obviously, if these fasteners are exposed to the weather, they can rapidly corrode. In the mildest of cases, this corrosion can lead to unsightly staining of the wood. In more severe cases, it can cause complete disintegration of the fastener and a total loss of structural strength (Baker, 1988).

Several types of coatings are used to protect steel fasteners. These include chromate paint, plastic, ceramic, and metal coatings (galvanizing). Adhesive-type coatings (e.g., paint and plastic) can flake off when driving the fastener, compromising the protection it was intended to provide.

General Guidelines for Fastener Use

1. At a minimum, use hot-dipped galvanized metal in outdoor wood structures.
2. Use stainless steel fasteners for added durability in severe exposures.
3. Always fasten a thinner member to a thicker one.
   (a) A nail should be long enough to penetrate the receiving member a distance twice the thickness of the thinner member.
   (b) A lag screw should penetrate the larger member by at least half the length of the screw.
4. Reduce splitting of boards when nailing by:
   (a) Placing nails no closer to the edge than ½ the board thickness and no closer to the end than the board thickness.
   (b) Pre-drilling nail holes.
   (c) Blunting the nail point.
   (d) Using greater spacing between nails.
   (e) Staggering nails in each row to prevent splitting along the grain.
5. Avoid end-grain nailing when possible.
6. When drilling holes for lag screws or bolts, saturate the hole with a preservative such as copper napthenate to prevent the migration of decay fungi into the untreated part of the member.
7. Use washers with bolts and lag screws to reduce crushing of the wood.
8. Tighten bolts and lag screws one year after construction and thereafter check tightness periodically.

While the coating protects by providing a barrier between the steel and the environment, galvanized coatings sacrificially corrode to protect the steel. When the coating is gone, the steel will begin to corrode. The galvanizing can be applied by electroplating, mechanical plating, or single- or double-dipping the fastener in molten zinc (hot dipped). The thickness of the coating is very important; a thicker coating provides additional protection.

Most manufacturers coat fasteners to the standard ASTM A153-87 (ASTM, 1987), which specifies a minimum coverage of 0.85 oz/ft^2 (259 g/m^2) of zinc. This is probably thick enough for most outdoor structures in dry-weather areas. Although the corrosion process typically proceeds over many years, our research shows that commonly available coated fasteners simply do not have a thick enough plating for long-term protection (20 yr) in severe (underground or high humidity) environments (Baker, 1992). Thicker coatings [1.0 oz/ft^2 (305 g/m^2)] are available and should be used in wetter situations. Coating specifications should be available from the fastener manufacturer.

Unfortunately, many building contractors use only electroplated nails for outdoor construction because they are readily available for use in pneumatic nail guns. Hot-dipped galvanized fasteners are produced for pneumatic nail guns, but their availability is limited.

Stainless Steel, Copper, and Aluminum Fasteners

The chemical properties of stainless steel make it resistant to corrosion. Although more expensive than hot-dipped galvanized fasteners, stainless steel is a more durable option, particularly for outdoor structures that are located in high-humidity areas or that remain wet for much of the time. Research has shown that little long-term degradation of stainless steel fasteners occurs even in the most severe exposure conditions (Baker, 1992). Also, the use of stainless steel fasteners reduces the possibility of staining around the fastener.

Although stainless steel fasteners are available in several grades, the American Iron and Steel Institute's (AISI) 300 series (e.g., 302, 303, 304, and 316) is appropriate for use in outdoor wood structures. While the price of stainless steel fasteners and hardware can be several times higher than the price of mild steel, their use is justified. The relatively small cost increase to the overall structure adds significantly to its reliability and long-term performance in severe conditions.

Copper, usually of rather high purity, and an alloy, silicon bronze, are often used to fasten wood (wood shakes, shingles, and in boat construction), although usually not in conventional structural applications.

Aluminum is suitable for use with untreated wood and wood treated with an oil-type preservative. However, aluminum should never be used in contact with wood that is treated with a waterborne preservative that contains copper, such as chromated copper arsenate (CCA), ammoniacal copper zinc arsenate (ACZA), or ammoniacal copper arsenate (ACA).

Corrosion in Untreated Wood

For an isolated fastener in moist wood, crevice corrosion can occur (Baker, 1988). This is the type of corrosion observed in crevices along riveted and welded seams of metal tanks and pipes. The head of a steel fastener in moist wood usually acts as a cathode, and the shank in the "crevice" serves as the anode. At the anode, iron goes into solution in the form of ferrous ions. As corrosion proceeds, hydroxides of iron precipitate. This leaves an excess of hydrogen ions in the surrounding water and the pH decreases. In addition, the wood fiber is chemically degraded as the ferrous ions are oxidized to ferric ions (Baker, 1988).

Dissimilar metals that are in physical contact with each other can result in galvanic corrosion, in which the corrosion of the least corrosion-resistant metal increases and the corrosion of the most corrosion-resistant metal decreases. Because of this, the washer, nut, and bolt or lag screw should be manufactured from the same metal.

Corrosion in Preservative-Treated Wood

Oil-type preservatives. Corrosion of metals in wood treated with oil-type preservatives is usually not a problem because the presence of heavy oils tends to inhibit corrosion. This is especially true in construction situations in which the holes for the fasteners are bored prior to treating.

When the preservative has not penetrated to the center of the wood member, such as in large beams or posts, fasteners are driven into moist, untreated wood, and fastener corrosion can occur.

Waterborne preservatives. Waterborne preservatives that contain copper cause corrosion of some metals. Corrosion in moist copper-treated wood is directly related to the presence of copper ions because they will "plate out" on a fastener that is more electronegative than copper. When this happens, a galvanic corrosion cell consisting of the fastener and the deposited copper is formed and the fastener corrodes (Baker, 1988).

Geographic Location

Because a variety of climates and exposure conditions exist, local conditions should dictate proper fastener selection. In the United States, climates range from subtropical to desert to arctic. This has a large effect on the corrosion rates of metal fasteners in wood. Where the climate is moist and warm, corrosion rates are the highest; where it is cold and/or dry, corrosion rates are the lowest. The corrosion rates can differ by a factor of five to ten.

The average outdoor humidity in North America varies depending on location and season. In areas of higher average humidity and warmer temperatures (e.g., the southeastern United States, portions of the Midwest, and along the coasts), the hazard of fastener corrosion (and wood decay fungi attack) is greatest. Even in dry areas of the country, an outdoor structure that is very near or over water, or for some reason is wetted much of the time, can have a high moisture content, thus promoting corrosion and decay.

Recommended Fasteners and Hardware

For treated or untreated wood that is above grade in structures exposed to weather, we suggest hot-dipped galvanized steel fasteners with at least 0.85 oz of zinc per ft^2 (259 g/m^2). This recommendation also applies for hardware used within the structure (joist hangers, straps). For wood that is below grade and treated with a preservative that contains copper, or that is in contact with saltwater, we suggest the use of AISI stainless steel Type 304, copper, or silicon bronze. Note that engineering design values are not published for copper and silicon bronze.

References

American Forest and Paper Association (AFPA), 1991. Design values for wood construction—A supplement to the 1991 national design specification for wood construction. AFPA (formerly the National Forest Products Association), Washington, DC, 51 p.

American Society for Testing and Materials (ASTM), 1987. Standard specification for zinc coating (hot dip) on iron and steel hardware. ASTM D153-87. ASTM, Philadelphia, PA.

Baker, A.J., 1988. Corrosion of metals in preservative-treated wood. *In* Proceedings of wood protection techniques and the use of treated wood in construction. October 28–30, 1987, Memphis, TN. Forest Products Society, Madison, WI, pp. 99–101.

Baker, A.J., 1992. Corrosion of nails in CCA- and ACA-treated wood in two environments. Forest Products Journal 42 (9), 39–41.

R. H. Falk, Structural Engineel; and A. J. Baker, Chemical Engineel, USDA Forest Service, Forest Products Laboratory, Madison, WI.

The Forest Products Laboratory is maintained in cooperation with the University of Wisconsin. This article was written and prepared by U.S. government employees on official time, and it is therefore in the public domain and not subject to copyright.

8.3.0 How Did the "Penny Weight" Nail Designation Originate?

The Penny Nail

This article in *The Ironmonger* from 1915 tells us the story of the "penny" nail.

In this case, the researcher examined the records dated 1477 from the Church of St Mary-at-Hill in the City of London.

Although many different handmade nails were in use at the time which had specific names, a large proportion were named simply by the number of pence paid for a hundred nails.

For example, four penny nails were those of which a hundred were purchased for 4d. (The 'd' stands for pence in the days when sterling was denominated in pounds, shillings and pence - £ s d).

The account records of the Church of St Mary-at-Hill show

'ffor a c of v peny nayle vd'

The 'c' is the Roman numeral for 'hundred' and 'v' is the Roman numeral for 'five'.

The amount of money paid for a hundred nails—fourpence, fivepence, sixpence—is thought to depend on the size of the nail. The larger the nail, the more expensive it was. The largest nail appears to have been the tenpenny nail, also referred to as the 'fyve stroke nayle'—possibly because it took five strokes of the hammer to get it home.

This nomenclature for nails 4d, 5d, 6d, etc., is still in use today particularly in the United States but relates only to the size of the nails, not the price!

Size	Length	Size	Length	Size	Length
2d	1"	8d	2 1/2"	30d	4 1/2"
3d	1 1/4"	9d	2 3/4"	40d	5"
4d	1 1/2"	10d	3"	50d	5 1/2"
5d	1 3/4"	12d	3 1/4"	60d	6"
6d	2"	16d	3 1/2"	80d	7"
7d	2 1/4"	20d	4"	100d	8"

By permission: Glasgow Steel Nail, Glasgow, Scotland

For those who would like to work out the true cost today, the article tells us that the medieval penny would have been the equivalent of around 1s 6d in value in 1915. Government sources suggest that prices have risen over 61-fold since 1914, so a medieval penny might be worth around £4.50 today.

8.4.0 About Nails—Historic and Otherwise

About Nails

Hand-forged nails were the first manufactured nails, and they date back to biblical times. As people first used hewn beams, timbers, planks, and whole logs to build with, the early hand-made nails were spikes. With the development of the split wood shingle, nails of about 1" long came into use. When sawyers, and then sawmills, began cutting dimension lumber, the sizes and varieties of nails greatly expanded. Thus, over time, nails developed in different sizes, shapes, and used different heads to fasten lumber and wood.

Nails have always been in demand. Some blacksmiths made only nails, and they were called "Nailers." Nails were so scarce (and expensive) in pre-1850 America that people would burn dilapidated buildings just to sift the ashes for nails.[1] They did so because pulling the nails would have damaged most of them. After the nails were recovered, a blacksmith could easily straighten any nails that had been bent during construction.

We still use the term "penny" when referring to a nail's size. It is believed that this term came into use in the early 1600s in England.[2] The English monetary unit was the Pound Sterling (£), which was divided into Shillings and Pence. The cost of 100 nails in Pence in the 1600s is how we refer to nail sizes to this day. For example, 100 small nails that sold for 4 pence were called 4d nails (4d is the abbreviation of 4 pence). 100 larger nails that sold for 16 pence are 16d nails. And so on.

Source: Appalachian Blacksmiths Association.

Setting the price of nails did not standardize their size. But it is apparent that the price of nails was constant, or near constant, for a long period of time, and thus, led to standard sizes as a result. For quite some time, nails have been sold by the pound—usually 1 lb. and 5 lb. boxes for small finishing and specialty nails and 50 lb. cartons for framing nails such as 8d and 16d. Nails are also sold by keg weight.

The cut nail made its appearance in the mid-1700s. For example, Thomas Jefferson established a nail factory at his Monticello plantation as a way to increase his farm income. His nail factory made both hand-forged and cut nails. It would not be until the middle-1800s that cut nails began dominating the marketplace. Cut nails are not actually "cut"—they are sheared from steel plate that is the thickness of the nail shank. Although routinely referred to as "square nails," the cutting machine tapers the nail shank as it is sheared from the steel plate. A second machine forms the head of a cut nail. The square nails in the above photograph are made in this manner. With the hand-forged nail, all four sides are tapered. With the cut nail, two sides are parallel because they represent the thickness of the plate they were sheared from.

Cut nails could be manufactured much faster than hand-forged nails. As the process was mechanized, the cost per nail was less. However, cut nail factories employed operators and attendants for each machine so the process was still labor-intensive. The noise in those mills was deafening as well. Cut nails had their heyday from about 1820 (development of the Type B nail) to 1910, the advent of the wire nail.

Wire nails are round. Steel wire is fed into a machine that grips the wire, cuts it, makes the head, and chisels the point, all in one operation. This process is totally mechanized, requiring only someone to turn the machine on and off. Wire nail machines can make thousands of nails per minute.

Wire nails have all but replaced the cut nail. Cut nails are still used but mainly for restoration and masonry work. Though wire nails are cheaper to produce, the cut nail has a holding power of approximately four times to its modern, round cousin. Compared on that basis, cut nails win the day easily.

In modern construction, more and more nail-driving is being done with air-operated nail guns. Nails of nearly all sizes are available. However, since the air-nailing gun is large and cumbersome, it is most often used to fasten sheathing, such as plywood, to the framing. The nails are prepared to fit in the air gun's clip or nail sleeve (much like a stapler and the way staples are loaded) and are driven one-at-a-time. The air gun nail resembles the cut nail of old with the exception that the head is "T"-shaped rather than battened on all four sides.

8.4.1 Cut Floor Brads

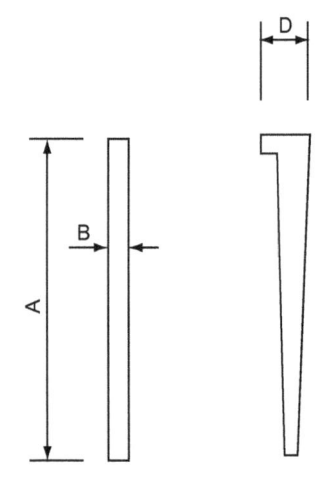

Dimensions and Tolerances (mm)					
A	+/−	B	+/−	D	+/−
40	1.0	2.5	.18	7.5	1.0
45	1.0	2.5	.18	7.5	1.0
50	1.0	2.5	.18	7.5	1.0
60	1.5	3.0	.22	9.0	1.0
65	1.5	3.2	.22	9.6	1.0
75	1.5	3.2	.22	9.6	1.0

Material Specification: Mild Steel EN 10111:1998:DD11, DD13

Galvanizing Specification: BS EN ISO 1461:1999

The above meets the requirements for Cut Floor Brads to BS1202:Part 1:1974 as amended

The tolerance on the "B" dimension refers to the thickness of the steel prior to manufacture. The process of manufacture may cause these tolerances to vary.

By permission: Glasgow Steel Nail, Glasgow, Scotland

Section | 8 Fasteners for Wood and Steel—Calculations for Selection 457

8.4.2 Palm Holdfast

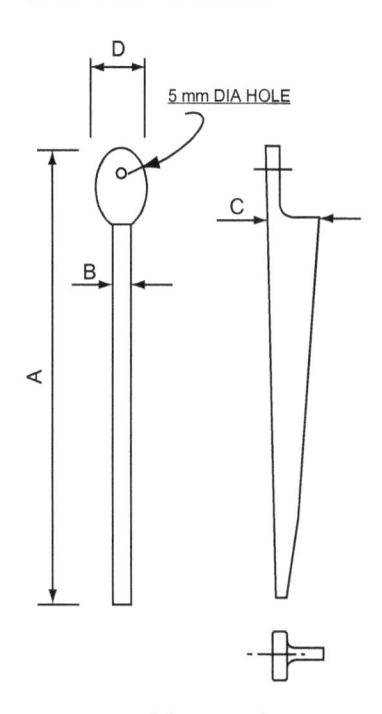

Dimensions and Tolerances (mm)							
A	+/−	B	+/−	C	+/−	D	+/−
150	2.0	6.0	.28	17.0	3.0	23.0	3.0

Material Specification: Mild Steel Up to 6mm EN 10111:1998 : DD11, DD13 Over 6mm EN 10025:2004:S275

Galvanizing Specification: BS EN ISO 1461:1999

The tolerance on the "B" dimension refers to the thickness of the steel prior to manufacture. The process of manufacture may cause these tolerances to vary.

By permission: Glasgow Steel Nail, Glasgow, Scotland

8.4.3 Moulder Brad

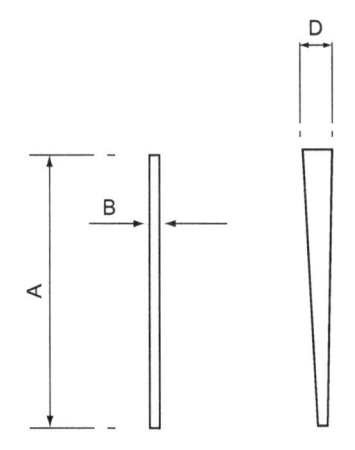

Dimensions and Tolerances (mm)					
A	+/−	B	+/−	D	+/−
50	1.0	2.0	.18	4.5	1.0
60	1.0	2.0	.18	6.5	1.0

Material Specification: Mild Steel EN 10111:1998 :DD11, DD13

Galvanizing Specification: BS EN ISO 1461:1999

The tolerance on the "B" dimension refers to the thickness of the steel prior to manufacture. The process of manufacture may cause these tolerances to vary.

By permission: Glasgow Steel Nail, Glasgow, Scotland

8.4.4 Flat Countersunk Head Spike

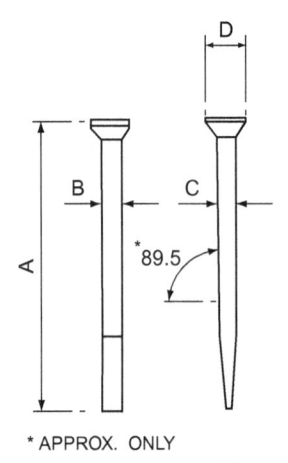

Dimensions and Tolerances (mm)							
A	+/−	B	+/−	C	+/−	D	+/−
80	1.5	5.0	.24	5.0	1.0	10.0	1.0
90	1.5	5.0	.24	5.0	1.0	10.0	1.0
100	1.5	6.0	.28	6.0	1.0	12.0	1.2
110	2.0	6.0	.28	6.0	1.0	12.0	1.2
120	2.0	6.0	.28	6.0	1.0	12.0	1.2
130	2.0	6.0	.28	6.0	1.0	12.0	1.5
130	2.0	8.0	.30	8.0	1.5	16.0	1.5
140	2.0	6.0	.28	6.0	1.5	12.0	1.5
140	2.0	8.0	.30	8.0	1.5	16.0	2.0
160	3.0	8.0	.30	8.0	1.5	16.0	2.0
180	3.0	10.0	.33	10.0	1.5	20.0	3.0
200	3.0	10.0	.33	10.0	1.5	20.0	3.0

Material Specification: Mild Steel Up to 6mm EN 10111:1998:DD11, DD13 Over 6mm EN 10025 : 2004 : S275

Galvanizing Specification: BS EN ISO 1461:1999

The tolerance on the "B" dimension refers to the thickness of the steel prior to manufacture. The process of manufacture may cause these tolerances to vary.

By permission: Glasgow Steel Nail, Glasgow, Scotland

* APPROX. ONLY

8.4.5 Rosehead Fine Shank

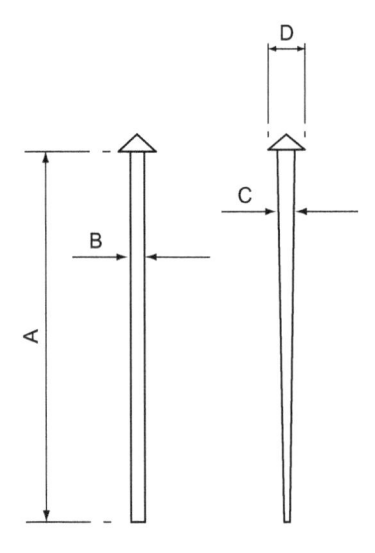

Dimensions and Tolerances (mm)							
A	+/−	B	+/−	C	+/−	D	+/−
50	1.5	3.0	.22	4.0	0.5	7.5	1.0
65	1.5	3.0	.22	4.0	0.5	7.5	1.0
75	1.5	3.2	.22	4.2	0.5	7.5	1.0
100	1.5	4.0	.22	4.7	0.5	8.5	1.0

Material Specification: Mild Steel EN 10111:1998:DD11, DD13

Galvanizing Specification: BS EN ISO 1461:1999

The tolerance on the "B" dimension refers to the thickness of the steel prior to manufacture. The process of manufacture may cause these tolerances to vary.

By permission: Glasgow Steel Nail, Glasgow, Scotland

Section | 8 Fasteners for Wood and Steel—Calculations for Selection

8.4.6 Rosehead Square Shank Spike

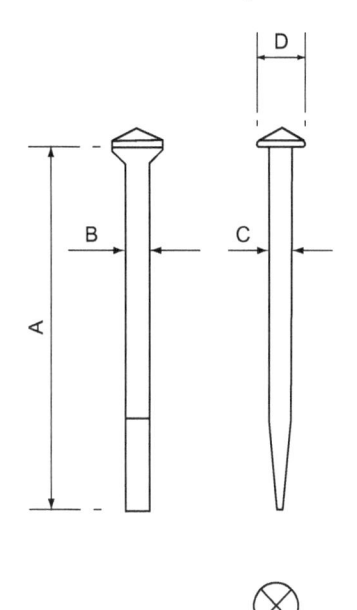

Dimensions and Tolerances (mm)							
A	+/−	B	+/−	C	+/−	D	+/−
40	1.0	3.2	.18	3.2	0.5	6.4	0.6
50	1.0	4.0	.22	4.0	0.5	8.0	0.8
65	1.5	4.0	.22	4.0	0.5	8.0	0.8
65	1.5	5.0	.24	5.0	0.5	10.0	1.0
75	1.5	5.0	.24	5.0	0.5	10.0	1.0
75	1.5	6.0	.27	6.0	0.5	12.0	1.5
90	1.5	5.0	.24	5.0	1.0	10.0	1.0
90	1.5	6.0	.27	6.0	1.0	12.0	1.5
100	1.5	6.0	.27	6.0	1.0	12.0	1.5
100	1.5	8.0	.30	8.0	1.0	16.0	2.0
115	2.0	8.0	.30	8.0	1.0	16.0	2.0
125	2.0	6.0	.27	6.0	1.0	12.0	1.5
125	2.0	8.0	.30	8.0	1.0	16.0	2.0
150	2.0	8.0	.30	8.0	1.0	16.0	2.0
150	2.0	10.0	.33	10.0	1.5	20.0	2.0
175	3.0	8.0	.30	8.0	1.0	16.0	2.5
175	3.0	10.0	.33	10.0	1.5	20.0	2.5
200	3.0	8.0	.30	8.0	1.0	16.0	2.5
200	3.0	10.0	.33	10.0	1.5	20.0	3.0
225	3.0	12.0	.36	12.0	1.5	24.0	3.5
243	3.0	12.0	.36	12.0	1.5	24.0	3.5

Material Specification: Mild Steel Up to 6mm EN10111:1998:DD11, DD13 Over 6mm EN 10025 : 2004 : S275

Galvanizing Specification: BS EN ISO 1461:1999

The tolerance on the "B" dimension refers to the thickness of the steel prior to manufacture. The process of manufacture may cause these tolerances to vary.

By permission: Glasgow Steel Nail, Glasgow, Scotland

8.4.7 Décor Nail

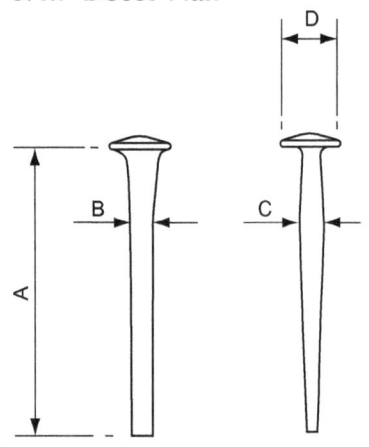

Dimensions and Tolerances (mm)							
A	+/−	B	+/−	C	+/−	D	+/−
35	1.0	3.2	.28	3.2	0.4	8.0	1.5
35	1.0	5.0	.28	5.0	0.5	12.0	1.5
40	1.0	5.0	.28	5.0	0.5	12.0	1.5
50	1.0	5.0	.28	5.0	0.5	12.0	1.5
65	1.5	5.0	.28	5.0	1.0	12.0	2.0
75	1.5	6.0	.28	6.0	1.0	15.0	2.0

Material Specification : Mild Steel Up to 6mm EN 10111:1998 DD11, DD13 Over 6mm EN 10025:2004:S275

Galvanizing Specification: BS EN ISO 1461:1999

The tolerance on the "B" dimension refers to the thickness of the steel prior to manufacture. The process of manufacture may cause these tolerances to vary.

By permission: Glasgow Steel Nail, Glasgow, Scotland

8.4.8 Boat Nail

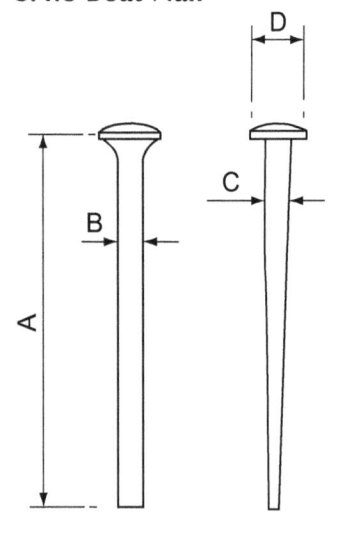

Dimensions and Tolerances (mm)							
A	+/−	B	+/−	C	+/−	D	+/−
50	1.0	5.0	.24	4.0	1.0	10.0	1.0
65	1.5	5.0	.24	4.0	1.0	10.0	1.0
75	1.5	6.0	.27	4.0	1.0	12.0	1.5
100	1.5	8.0	.27	6.0	1.0	15.0	1.5
125	2.0	10.0	.30	8.0	1.5	20.0	2.0
150	2.0	10.0	.33	8.0	1.5	20.0	2.5
175	3.0	10.0	.33	8.0	1.5	20.0	2.5
200	3.0	10.0	.33	8.0	1.5	20.0	2.5

Material Specification: Mild Steel Up to 6mm EN 10111:1998:DD11, DD13 Over 6mm EN 10025 :2004 : S275

Galvanizing Specification: BS EN ISO 1461:1999

The tolerance on the "B" dimension refers to the thickness of the steel prior to manufacture. The process of manufacture may cause these tolerances to vary.

By permission: Glasgow Steel Nail, Glasgow, Scotland

8.4.9 CLYDE Rail Spike

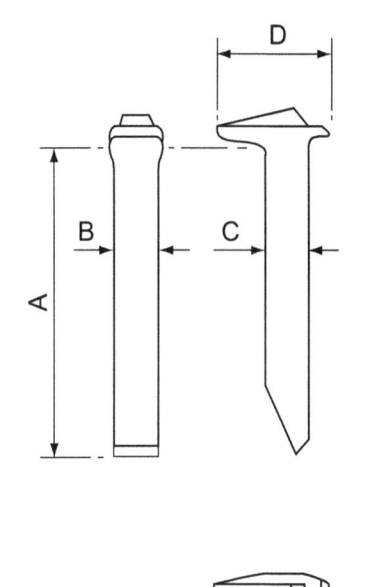

Dimensions and Tolerances (mm)							
A	+/−	B	+/−	C	+/−	D	+/−
50	1.0	8.0	.30	8.0	1.0	27.0	3.0
65	1.5	8.0	.30	8.0	1.0	27.0	3.0
65	1.5	10.0	.33	10.0	1.5	30.0	3.0
75	1.5	8.0	.30	8.0	1.0	27.0	3.0
75	1.5	10.0	.33	10.0	1.5	30.0	3.0
75	1.5	12.0	.36	12.0	1.5	35.0	3.5
90	1.5	10.0	.33	10.0	1.5	30.0	3.0
90	1.5	12.0	.36	12.0	1.5	35.0	3.5
100	1.5	10.0	.33	10.0	1.5	30.0	3.0
100	2.0	12.0	.36	12.0	1.5	35.0	3.5
100	2.0	15.0	.36	15.0	1.5	45.0	4.5
115	2.0	12.0	.36	12.0	1.5	35.0	3.5
115	2.0	15.0	.36	15.0	1.5	45.0	4.5
125	2.0	12.0	.36	12.0	1.5	35.0	3.5
125	2.0	15.0	.36	15.0	1.5	45.0	4.5
150	2.0	12.0	.36	12.0	1.5	35.0	3.5
150	2.0	15.0	.36	15.0	1.5	45.0	4.5

Material Specification: Mild Steel To EN 10025 : 2004 : S275
Galvanizing Specification: BS EN ISO 1461:1999

The tolerance on the "B" dimension refers to the thickness of the steel prior to manufacture. The process of manufacture may cause these tolerances to vary.

By permission: Glasgow Steel Nail, Glasgow, Scotland

8.5.0 Withdrawal Resistance of Nails

The general equation indicates that the dense, heavy woods offer greater nail-withdrawal resistance than the ones of lighter weight. This does not mean that the lighter species are not qualified for uses requiring high-withdrawal resistance. As a rule, the lighter species do not split as readily as the dense ones; thus lighter woods offer an opportunity for increasing the diameter, length, and number of the nails to compensate for the wood's lower nail-holding properties.

In practically all species, nails driven into green wood and pulled before any seasoning takes place will offer about the same withdrawal resistance as nails driven into seasoned wood and pulled soon after driving. However, if common smooth-shank nails are driven into green wood that is allowed to season or into seasoned wood that is subjected to cycles of wetting and drying before the nails are pulled, they lose a major part of their withdrawal resistance. In seasoned wood that is subjected only to moisture changes from normal atmospheric variations, the withdrawal resistance of smooth-shank nails also diminishes in time. On the other hand, tests indicate that, when moisture conditions cause nails to rust, withdrawal resistance is very erratic; it may be regained or even increased over the immediate withdrawal resistance, Under all conditions of use, the withdrawal resistance of nails varies so widely that it is difficult to evaluate their behavior. The withdrawal loads for plain nails driven into wood that is subjected to wide alternating changes in moisture content may be as much as 75% below the values given by the general formula.

The specific gravity of various species of wood and their relative resistance to the withdrawal of smooth-shank nails are given in Table 1. The numerical value of $6900 \, G^{5/2}$ has been calculated for each species. The load per inch of penetration immediately after driving may be obtained by multiplying this nail-withdrawal factor by the diameter, $\underline{D}$. For example, Table 1 shows a value of 790 D for an eightpenny common nail (0.131-inch diameter) in ponderosa pine. Multiplying 790 times 0.131 gives a value of 103 pounds per inch of penetration.

Source: U.S. Department of Agriculture—Forest Service.

TABLE 1 Nail-Withdrawal Resistance

Hardwoods		
	Specific Gravity[1]	Relative Nail Load[2]
Ash, black	0.53	1380 D
Ash, commercial white	.61	2900 D
Aspen, bigtooth	.41	740 D
Aspen, quaking	.40	690 D
Basswood, American	.40	690 D
Beech, American	.67	2550 D
Birch, sweet	.71	2900 D
Birch, yellow	.66	2410 D
Chestnut, American	.45	970 D
Cottonwood, black	.37	550 D
Cottonwood, eastern	.43	830 D
Elm, American	.55	1520 D
Elm, rock	.66	2410 D
Elm, slippery	.57	1720 D
Hackberry	.56	1590 D
Hickory, pecan	.65	2350 D
Hickory, true	.74	3240 D
Magnolia, southern	.53	1380 D
Maple, black	.62	2140 D
Maple, red	.55	1520 D
Maple, silver	.51	1310 D
Maple, sugar	.68	2620 D
Oak, commercial red	.66	2410 D
Oak, commercial white	.71	2990 D
Sweetgum	.53	1380 D

TABLE 1 Nail-Withdrawal Resistance—Cont'd

	Hardwoods	
	Specific Gravity	Relative Nail Load
Sycamore, American	.54	1450 D
Tupelo, black	.55	1520 D
Tupelo, water	.52	1310 D
Yellow-poplar	.43	830 D
	Softwoods	
	Specific Gravity[1]	Relative Nail Load[2]
Alaska-cedar	0.46	970 D
Baldcypress	.48	1100 D
Douglas-fir, Coast-type	.51	1310 D
Douglas-fir, Rocky Mountain-type	.45	970 D
Fir, balsam	.41	740 D
Fir, commercial white	.41	740 D
Hemlock, eastern	.43	830 D
Hemlock, western	.44	900 D
Larch, western	.59	1860 D
Pine, eastern white	.37	550 D
Pine, lodgepole	.43	830 D
Pine, Ponderosa	.42	790 D
Pine, red	.51	1310 D
Pine, southern yellow	.59	1860 D
Pine, sugar	.38	620 D
Pine, western white	.42	790 D
Port-Orford-cedar	.44	900 D
Redcedar, western	.34	470 D
Redwood (old-growth)	.42	790 D
Spruce, Engelmann	.35	500 D
Spruce, red	.41	740 D
Spruce, Sitka	.42	790 D
Spruce, white	.45	970 D
White-cedar, Atlantic	.35	500 D
White-cedar, northern	.32	410 D

[1] Based on weight and volume when ovendry.
[2] Load in pounds. D = nail diameter in inches.

Nail diameter varies for different types of nails. Here are diameters of bright common wire nails:

Penny	Gage	Diameter	Penny	Gage	Diameter
4	12-1/2	0.098	12	9	0.148
6	11-1/2	.113	16	8	.162
8	10-1/4	.131	20	6	.192
10	9	.148			

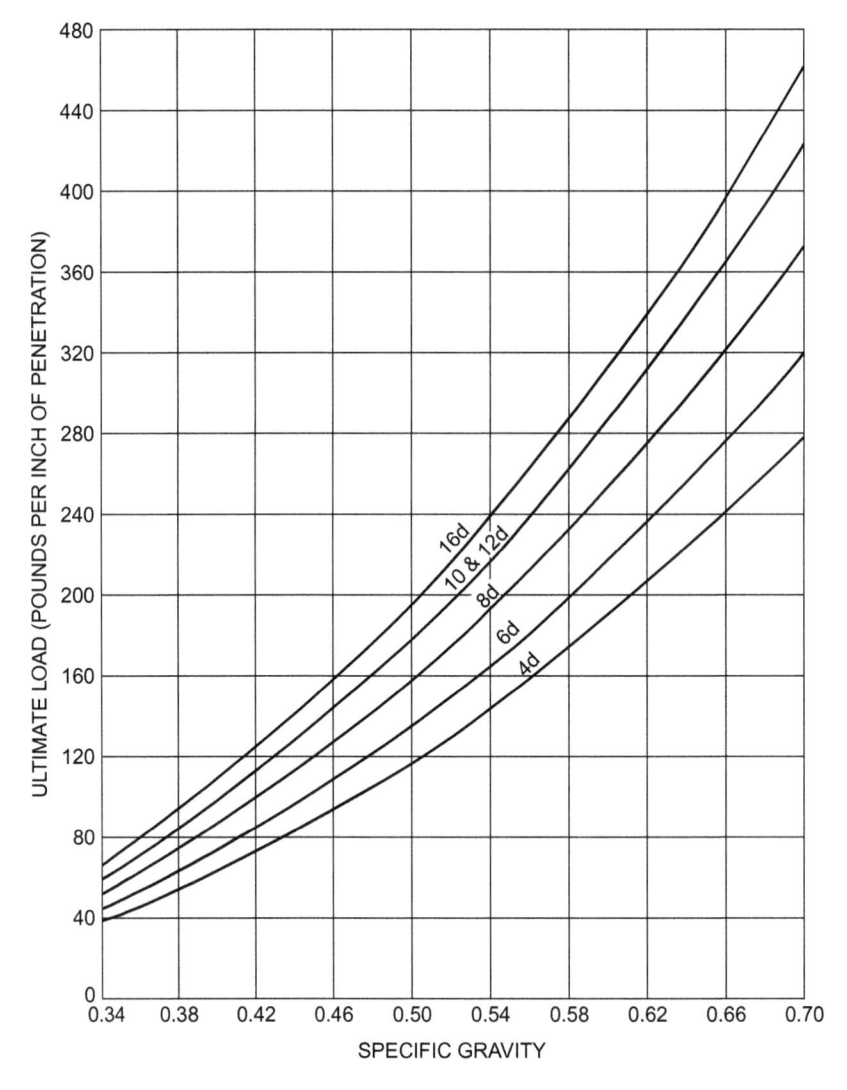

FIGURE 1 Load required to withdraw common nails from wood of different specific gravities immediately after nails were driven. Specific gravity is based on weight and volume of ovendry wood.

8.6.0 Wood Screws—Common Types and Withdrawal Resistance

Wood Screws

The common types of wood screws have flat, oval, or round heads. The flathead screw is most commonly used if a flush surface is desired. Ovalhead and roundhead screws are used for appearance, and roundhead screws are used when countersinking is objectionable. The principal parts of a screw are the head, shank, thread, and core (Fig. 8–5). The root diameter for most sizes of screws averages about two-thirds the shank diameter. Wood screws are usually made of steel, brass, other metals, or alloys, and may have specific finishes such as nickel, blued, chromium, or cadmium. They are classified according to material, type, finish, shape of head, and diameter or gauge of the shank.

Current trends in fastenings for wood also include tapping screws. Tapping screws have threads the full length of the shank and may have some advantage for certain specific uses.

Withdrawal Resistance

Experimental Loads

The resistance of wood screw shanks to withdrawal from the side grain of seasoned wood varies directly with the square of the specific gravity of the wood. Within limits, the withdrawal load varies directly with the depth of penetration of the threaded portion and the diameter of the screw, provided the screw does not fail in tension. The screw will fail in tension when its strength is exceeded by the withdrawal strength from the wood. The limiting length to cause a tension failure decreases as the density of the wood increases since the withdrawal strength of the wood increases with density. The longer lengths of standard screws are therefore superfluous in dense hardwoods.

Source: U.S. Department of Agriculture.

8.6.1 Wood Screw Sizing

Screw Sizing

The general size of a screw is given a number. As the number increases, so does the size of the entire screw, both head size and shaft size–but not length. Therefore, a #8 screw is about twice the size of a #4 screw, but may be the same length. Wake Up!—This is important.

Most wood screws have a common "pitch" to the threads, but some have a thread with a steeper incline. We simply call this a "fast" thread, but they are technically Type A screws. Think of this as a road going up a mountain; the steeper the road, the sooner you get to the top. Most of the screws for mounting hinges are self-tapping

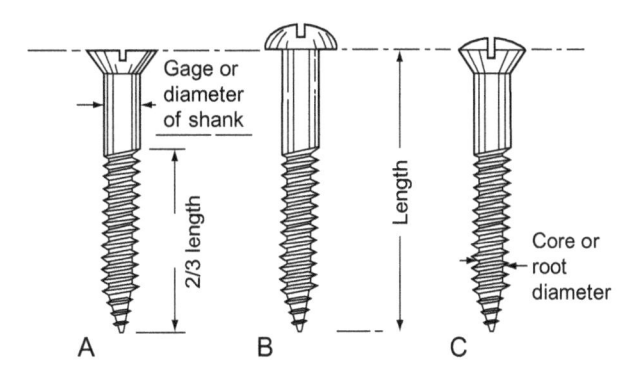

FIGURE 8–5 Common types of wood screws: A, flathead; B, roundhead; and C, ovalhead.

(they tap their own mating threads in wood) type AB (they have more threads per inch and are more effective in brittle materials like wood than Type A). For more information on this, see screw types section.

Also in reference to threads you will see "8-32." This is the common knob and pull screw thread. The "8" refers to the size (diameter) of the screw, and the "32" means it has 32 threads to the inch. The diameter is measured at the shank of the screw.

Screws are sized by gauge number and length. For example, an 8-gauge screw with 32 threads per inch and 1" in length would be written as: 8-32 × 1". However, most wood screws do not include the threads per inch measurement and would just be listed as 8 × 1". If the gauge number is not known, simply measure the diameter of the shank in inches and round to nearest listed number on chart below for screw number identification.

Source: D.Lawless Hardware- hingeddummy.info.

8.6.2 Dimensions of Wood Screws Chart

Dimensions of Wood Screws Chart

	Shank*	Diameter†			Root Diameter	
Gauge Number	Max. Head Diameter	Basic Decimal Size	Nearest Fractional Equivalent	Average Decimal Size	Nearest Fractional Equivalent	Threads per Inch
0	.119	.060	1/16	.040	3/64	32.00
1	.146	.073	5/64	.046	3/64	28.00
2	.172	.086	3/32	.054	1/16	26.00
3	.199	.099	7/64	.065	1/16	24.00
4	.225	.112	7/64	.075	5/64	22.00
5	.252	.125	1/8	.085	5/64	20.00
6	.279	.138	9/64	.094	3/32	18.00
7	.305	.151	5/32	.102	7/64	16.00
8	.332	.164	5/32	.112	7/64	15.00
9	.358	.177	11/64	.122	1/8	14.00
10	.385	.190	3/16	.130	1/8	13.00
11	.411	.203	13/64	.139	9/64	12.00
12	.438	.216	7/32	.148	9/64	11.00
14	.491	.242	1/4	.165	5/32	10.00
16	.544	.268	17/64	.184	3/16	9.00
18	.597	.294	19/64	.204	13/64	8.00
20	.650	.320	5/16	.233	7/32	8.00
24	.756	.372	3/8	.260	1/4	7.00

*Shank diameter is measured on the smooth portion of the screw above the threads.
†Root diameter is measured between the threads and does not include the thread height.
Source: D.Lawless Hardware- hingeddummy.info.

Section | 8   Fasteners for Wood and Steel—Calculations for Selection

- The length of screw is taken from the surface of the material to the point of the screw. See illustration below.

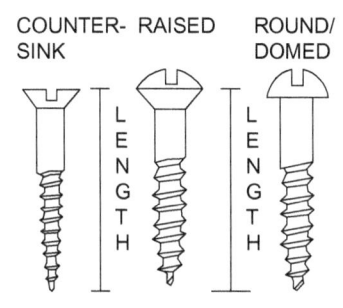

- Thus, a 6-gauge screw with 15 threads per inch and 1 1/4″ in length would be written as: 6-15 × 1 1/4″

And here is an extra tid bit: this is a formula for obtaining the diameter when you only have the screw number.

Multiply the screw Number by 13 and add ".060.

Examples -

(No.) 8 × 13 = .104 + ".060 = ".164
(No.) 2 × 13 = .026 + ".060 = ".086.

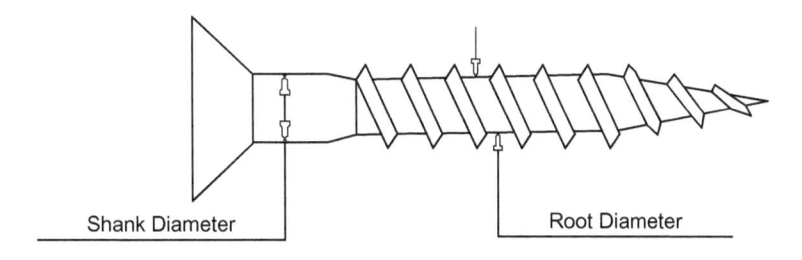

8.6.3 Basic Types of Wood Screw Drives

Basic Types of Wood Screws

Wood screws are classified by the type of drive, the shape of head, its length, and whether it is designed for wood or metal; this page refers to wood screws only.

Types of Drives

Driver refers to the indented shape on the screw head used to turn the screw. There are many different types of drives. Here we are only covering the four most used drives: slotted (flathead), Phillips (crosshead), square, and pozidriv. A brief description of each drive is presented below with a picture at the bottom to illustrate each one's unique design.

SLOTTED/FLATHEAD: This is the original screw drive. You find these everywhere, though the practice of using screws with slotted drives is on the decline because the screwdriver slips out of the slot, particularly when you are applying heavy torque to really tighten down (or loosen, for that matter) these types of screws.

PHILLIPS/CROSSHEAD: This screw drive type is very popular—and again, you find them in a very wide range of applications. Common sizes are Phillips #1, #2, and #3. The most common Phillips size is #2.

SQUARE: Square recess screw drives are being used more and more as they are very resistant to cam-out, which is a fancy way of saying the tip of the tool does not slip out and mar the screwhead very easily. It is commonly found in two sizes: Square #2 and #3.

POZIDRIV: This screwhead isn't seen very often in the United States, though it is very common in Europe. It looks a lot like a Phillips screwhead, but it includes four more contact points. Common sizes are Pozidriv #1, #2, and #3.

SLOTTED/FLATHEAD   PHILLIPS/CROSSHEAD   SQUARE   POZIDRIV

8.6.4 Screw Head Types and Shapes

Screw head types refer to the shape of the head at the top of the screw. There are many different head types as well. Again, we will adequately cover head types, but only the most generally used. The shape of the screw head can be described as flat (countersunk), oval, round, pan, truss, button (dome), and so on. Detailed descriptions for each of the ones listed are presented below. Looking at the drawings should make the designs of each self-explanatory.

FLAT/COUNTERSUNK: Supplied to standard dimensions with an 80′ to 82′ angle to be used where finished surfaces require a flush fastening unit (concealed below woods surface). The countersunk portion offers good centering possibilities.

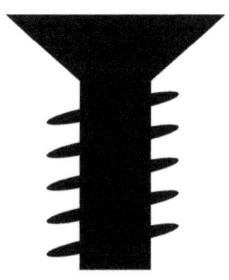

OVAL: Fully specified as "oval countersunk," this head is identical to the standard flat head, but possesses, in addition, a rounded, neat-appearing upper surface for attractiveness of design.

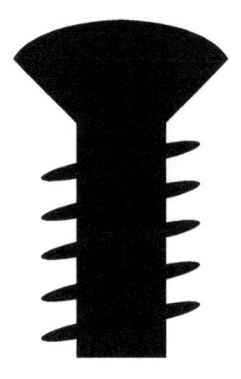

ROUND: Not recommended for new design (see pan head). The round head rests on the surface of wood. This was the most universally used in the past.

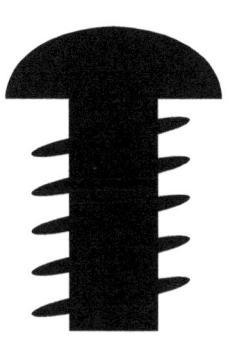

PAN: Recommended for new designs to replace round, truss, and binding heads. Provides a low large diameter head, but with characteristically high outer edge along the outer periphery of the head where driving action is most effective for high tightening torques. Slightly different head contour where supplied with recessed head.

TRUSS: Also known as oven head, stove head, and oval binding head. A low, neat-appearing, large-diameter head having excellent design qualities, and as illustrated can be used to cover larger diameter clearance holes in sheet metal when additional play in assembly tolerance is required. Suggest pan head as a substitute.

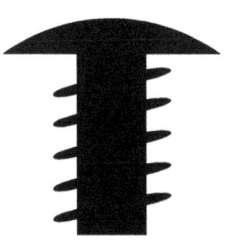

BUTTON/DOME: Cylindrical with a rounded top.

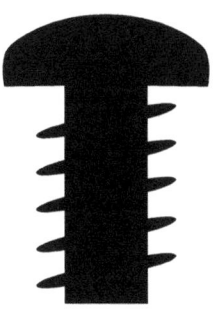

8.6.5 Screw Thread and Point Types

Most wood screws have a common "pitch" to the threads, but some have a thread with a steeper incline. We simply call this a "fast" thread, but they are technically Type A screws. Think of this as a road going up a mountain; the steeper the road, the sooner you get to the top.

Also in reference to threads you will see "8-32." This is the common knob and pull screw thread. The "8" refers to the size (diameter) of the screw, and the "32" means it has 32 threads to the inch. Most of the screws for mounting hinges are self-tapping (they tap their own mating threads in wood) type AB (they have more threads per inch and are more effective in brittle materials like wood than Type A)

Self-Tapping Screws

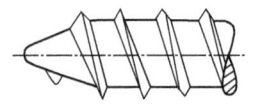

TYPE A POINT: A THREAD FORMING SCREW WITH SHARP POINT AND COARSE THREAD (FEWER THREADS PER INCH) FOR USE WITH LIGHT MATERIAL.

TYPE AB POINT: A THREAD FORMING SCREW WITH SHARP POINT AND FINE THREAD (MORE THREADS PER INCH) FOR USE ON LIGHT AND HEAVY MATERIAL.

TYPE B POINT: LARGER ROOT DIAMETER WITH FINER THREAD PITCH FOR HEAVIER USE

Threading on the shank is designed specifically for wood; wood threads have a tapped screw whereas sheet metal screws have mainly a parallel thread. Wood-type screws are also normally used for securing into wall plugs. Screws for chipboard usually have two threads the full length of the shank.

Miscellaneous

The most interesting screw here, if there is such a thing, is a variable-length break-off knob screw. If you do not know exactly what length screw will work, order these and break them off where you need them. A unique design allows this to be done without damaging the threads.

- And here is that famous break-off knob screw. Break-off screw 8-32 × 1 3/4″ zinc plated. Break-off points 1″, 1 1/4″, and 1 1/2″.

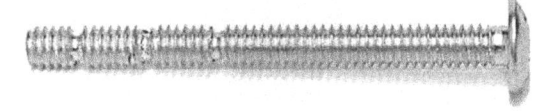

- Perfect solution when you're not quite sure which screw length you require.
- Simply grip the section below the break point you choose with a pair of pliers and break off.
- Do not grip threaded section you want to use with pliers as this would damage the threads.

Source: D. Lawless Hardware- hinged dummy.info.

8.6.6 Type 316 Stainless Steel Deck Screws

Woodpeckers Flat Head Deck Screw with Nibs

- Revolutionary 4-corner thread with raised ridge point for fast penetration, reduced drive torque, and outstanding holding power
- Self-countersinking Flat Head with Nibs
- Star Drive 6 Lobe
- Type 316 Stainless Steel for the ultimate in corrosion resistance
- Approved for use in the new ACQ lumber
- Square recess driver bits
- Click here for Nail Gun Reference Chart
- Smart-Bit™ pre-drill & counters inking bit

Type 316 Stainless Steel—*SALT WATER SAFE*

Length	Screw Size	Star Drive	Box of	Weight	Code	Price	Qty
1-5/8"	#8	T20	4000	28 lb	158FXN86A	$	
			350	2.5 lb	158FXN86B	$	
2"	#8	T20	3000	25 lb	200FXN86A	$	
			350	3 lb	200FXN86B	$	
2-1/2"	#10	T25	1750	23 lb	212FXN106A	$	
			350	6 lb	212FXN106B	$	
3"	#10	T25	1750	27 lb	300FXN106A	$	
			350	5.5 lb	300FXN106B	$	
3-1/2"	#10	T25	1000	19 lb	312FXN106A	$	
			250	6.7 lb	312FXN106B	$	
4"	#12	T27	750	28 lb	400FXN126A	$	
			100	3.8 lb	400FXN126B	$	

***AISI Grade 316 Stainless Steel** has lower carbon content than 305 or 302HQ and contains molybdenum for superior corrosion resistance in salt water and other highly corrosive environments. It is the grade of stainless steel to be specified in any seaside application.

By permission: Manasquanfasteners.com

8.6.7 Type 302 Stainless Steel Bugle Head Screws with Square Drive

Type 302HQ Stainless Steel Bugle Head Square Drive Wood Screws

- Type 17 notched point for fast penetration
- Self-countersinking bugle head
- Square drive recess reduces driver cam-out
- Threaded approximately 2/3 of shank on most sizes
- Approved for the new pressure treated lumber
- Available in Type 302HQ and 316 stainless steel
- Deep coarse threads for better hold
- Approved for use in the new ACQ lumber
- Click here for Fastener Estimator
- Smart-Bit™ pre-drill and countersinking bit
- Square recess driver bits

Type 302HQ Stainless Steel

Length	Screw Size	Square Drive	Carton Count	Code	Price pr/100	Price pr/100 min 1000	Full Carton	Enter Qty in 100s
1-1/4"	#6	#1	5000	114B6Q	$	$	$	
1-5/8"	#6	#1	4000	158B6Q	$	$	$	
2"	#6	#1	3000	200B6Q	$	$	$	
1-5/8"	#8	#2	4000	158B8Q	$	$	$	
2"	#8	#2	3000	200B8Q	$	$	$	
2-1/4"	#8	#2	3000	214B8Q	$	$	$	
2-1/2"	#8	#2	2000	212B8Q	$	$	$	
3"	#8	#2	1500	300B8Q	$	$	$	

Type 302HQ Stainless Steel—Cont'd

Length	Screw Size	Square Drive	Carton Count	Code	Price pr/100	Price pr/100 min 1000	Full Carton	Enter Qty in 100s
2"	#10	#2	2500	200B10Q	$	$	$	
2-1/2"	#10	#2	2000	212B10Q	$	$	$	
3"	#10	#2	1500	300B10Q	$	$	$	
3-1/2"	#10	#2	1000	312B10Q	$	$	$	
4"	#10	#2	1000	400B10Q	$	$	$	
2"	#12	#3	2000	200B12Q	$	$	$	
2-1/2"	#12	#3	1500	212B12Q	$	$	$	
3"	#12	#3	1500	300B12Q	$	$	$	
3-1/2"	#12	#3	1000	312B12Q	$	$	$	
2"	#14	#3	1500	200B14Q	$	$	$	
2-1/2"	#14	#3	1000	212B14Q	$	$	$	
3"	#14	#3	1000	300B14Q	$	$	$	
3-1/2"	#14	#3	500	312B14Q	$	$	$	
4"	#14	#3	500	400B14Q	$	$	$	

By permission: Manasquanfasteners.com

8.6.8 Lateral and Withdrawal Resistance of Lag Screws

Lag Screws

Lag screws are commonly used because of their convenience, particularly where it would be difficult to fasten a bolt or where a nut on the surface would be objectionable. Commonly available lag screws range from about 5.1 to 25.4 mm (0.2 to 1 in.) in diameter and from 25.4 to 406 mm (1 to 16 in.) in length. The length of the threaded part varies with the length of the screw and ranges from 19.0 mm (3/4 in.) with the 25.4- and 31.8-mm (1- and 1-1/4-in.) screws to half the length for all lengths greater than 254 mm (10 in.). Lag screws have a hexagonal-shaped head and are tightened by a wrench (as opposed to wood screws, which have a slotted head and are tightened by a screw driver). The following equations for withdrawal and lateral loads are based on lag screws having a base metal average tensile yield strength of about 310.3 MPa (45,000 lb/in²) and an average ultimate tensile strength of 530.9 MPa (77,000 lb/in²).

Withdrawal Resistance

The results of withdrawal tests have shown that the maximum direct withdrawal load of lag screws from the side grain of seasoned wood may be computed as

$$p = 125.4 G^{3/2} D^{3/4} L \quad \text{(metric)} \tag{8-14a}$$

$$p = 8,100 G^{3/2} D^{3/4} L \quad \text{(inch-pound)} \tag{8-14b}$$

where p is maximum withdrawal load (N, lb), D shank diameter (mm, in.), G specific gravity of the wood based on ovendry weight and volume at 12% moisture content, and L length (mm, in.) of penetration of the threaded part. (The NDS and LRFD use ovendry weight and volume as a basis.) Equation (8–14) was developed independently of Equation (8–10) but gives approximately the same results.
Source: U.S. Department of Agriculture.

Lag screws, like wood screws, require prebored holes of the proper size (Fig. 8–6). The lead hole for the shank should be the same diameter as the shank. The diameter of the lead hole for the threaded part varies with the density of the wood: for low-density softwoods, such as the cedars and white pines, 40 to 70% of the shank diameter; for Douglas-fir and Southern Pine, 60% to 75%; and for dense hardwoods, such as oaks, 65% to 85%. The smaller percentage in each range applies to lag screws of the smaller diameters, and the larger percentage to lag screws of larger diameters. Soap or similar lubricants should be used on the screw to facilitate turning, and lead holes slightly larger than those recommended for maximum efficiency should be used with long screws.

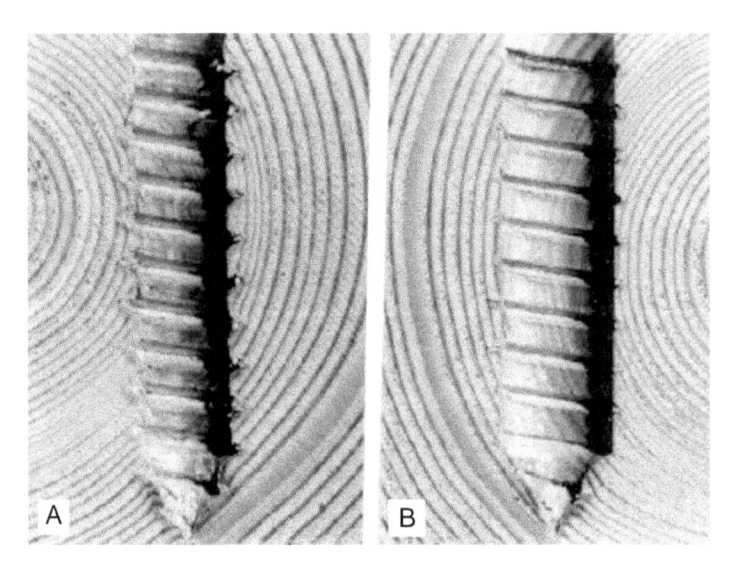

FIGURE 8–6 A, Clean-cut, deep penetration of thread made by lag screw turned into a lead hole of proper size, and B, rough, shallow penetration of thread made by lag screw turned into oversized lead hole.

In determining the withdrawal resistance, the allowable tensile strength of the lag screw at the net (root) section should not be exceeded. Penetration of the threaded part to a distance about seven times the shank diameter in the denser species (specific gravity greater than 0.61) and 10 to 12 times the shank diameter in the less dense species (specific gravity less than 0.42) will develop approximately the ultimate tensile strength of the lag screw. Penetrations at intermediate densities may be found by straight-line interpolation.

The resistance to withdrawal of a lag screw from the end-grain surface of a piece of wood is about three-fourths as great as its resistance to withdrawal from the side-grain surface of the same piece.

Lateral Resistance

Pre-1991

The experimentally determined lateral loads for lag screws inserted in the side grain and loaded parallel to the grain of a piece of seasoned wood can be computed as

$$p = KD^2 \tag{8-15}$$

where p is the proportional limit lateral load (N, lb) parallel to the grain, K a coefficient depending on the species-specific gravity, and D shank diameter of the lag screw (mm, in.). Values of K for a number of specific gravity ranges can be found in Table 8–4. These coefficients are based on average results for several ranges of specific gravity for hardwoods and softwoods. The loads given by this equation apply when the thickness of the side member is 3.5 times the shank diameter of the lag screw, and the depth of penetration in the main member is 7 times the diameter in the harder woods and 11 times the diameter in the softer woods. For other thicknesses, the computed loads should be multiplied by the factors listed in Table 8–10.
The thickness of a solid wood side member should be about one-half the depth of penetration in the main member.

When the lag screw is inserted in the side grain of wood and the load is applied perpendicular to the grain, the load given by the lateral resistance equation should be multiplied by the factors listed in Table 8–11.

TABLE 8–10 Multiplication Factors for Loads Computed from Equation (7–15)

Ratio of Thickness of Side Member to Shank Diameter of Lag Screw	Factor
2	0.62
2.5	0.77
3	0.93
3.5	1.00
4	1.07
4.5	1.13
5	1.18
5.5	1.21
6	1.22
6.5	1.22

TABLE 8–11 Multiplication Factors for Loads Applied Perpendicular to Grain Computed from Equation (7–15) with Lag Screw in Side grain of Wood

Shank diameter of Lag Screw (mm [in.])	Factor
4.8 (3/16)	1.00
6.4 (1/4)	0.97
7.9 (5/16)	0.85
9.5 (3/8)	0.76
11.1 (7/16)	0.70
12.7 (1/2)	0.65
15.9 (5/8)	0.60
19.0 (3/4)	0.55
22.2 (7/8)	0.52
25.4 (1)	0.50

8.7.0 Bolts in Wood

Bolts

Bearing Stress of Wood under Bolts

The bearing stress under a bolt is computed by dividing the load on a bolt by the product LD, where L is the length of a bolt in the main member and D is the bolt diameter. Basic parallel-to-grain and perpendicular-to-grain bearing stresses have been obtained from tests of three-member wood joints where each side member is half the thickness of the main member. The side members were loaded parallel to grain for both parallel- and perpendicular-to-grain tests. Prior to 1991, bearing stress was based on test results at the proportional limit. Since 1991, bearing stress has been based on test results at a yield limit state, which is defined as the 5% diameter offset on the load–deformation curve.

The bearing stress at proportional limit load is largest when the bolt does not bend, that is, for joints with small L/D values. The curves of Figures 8–8 and 8–9 show the reduction in proportional limit bolt-bearing stress as L/D increases. The bearing stress at maximum load does not decrease as L/D increases, but remains fairly constant, which means that the ratio of maximum load to proportional limit load increases as L/D increases.

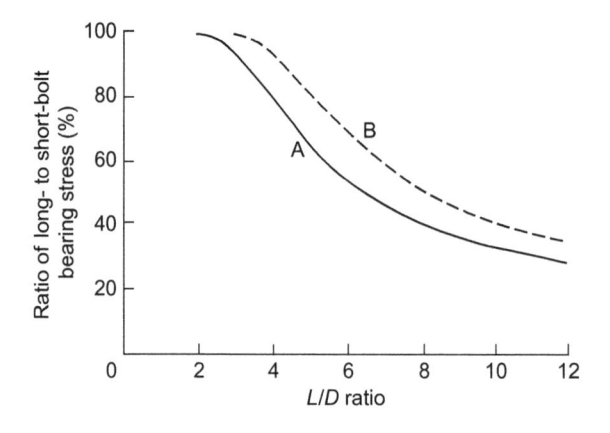

FIGURE 8–8 Variation in bolt-bearing stress at the proportional limit parallel to grain with L/D ratio. Curve A, relation obtained from experimental evaluation; curve B, modified relation used for establishing design loads.

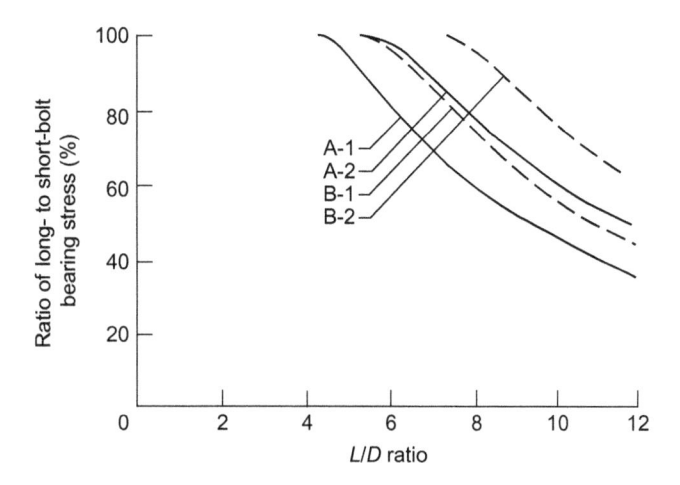

FIGURE 8–9 Variation in bolt-bearing stress at the proportional limit perpendicular to grain with L/D ratio. Relations obtained from experimental evaluation for materials with average compression perpendicular stress of 7,860 kPa (1,140 lb/in^2) (curve A–1) and 3,930 kPa (570 lb/in^2) (curve A–2). Curves B–1 and B–2, modified relations used for establishing design loads.

To maintain a fairly constant ratio between maximum load and design load for bolts, the relations between bearing stress and L/D ratio have been adjusted as indicated in Figures 8–8 and 8–9.

The proportional limit bolt-bearing stress parallel to grain for small L/D ratios is approximately 50% of the small clear crushing strength for softwoods and approximately 60% for hardwoods. For bearing stress perpendicular to the grain, the ratio between bearing stress at proportional limit load and the small clear proportional limit stress in compression perpendicular to grain depends on bolt diameter (Fig. 8–10) for small L/D ratios.

Species compressive strength also affects the L/D ratio relationship, as indicated in Figure 8–9. Relatively higher bolt proportional-limit stress perpendicular to grain is obtained with wood low in strength (proportional limit stress of 3,930 kPa (570 lb/in^2) than with material of high strength (proportional limit stress of 7,860 kPa [1,140 lb/in^2]). This effect also occurs for bolt-bearing stress parallel to grain, but not to the same extent as for perpendicular-to-grain loading.

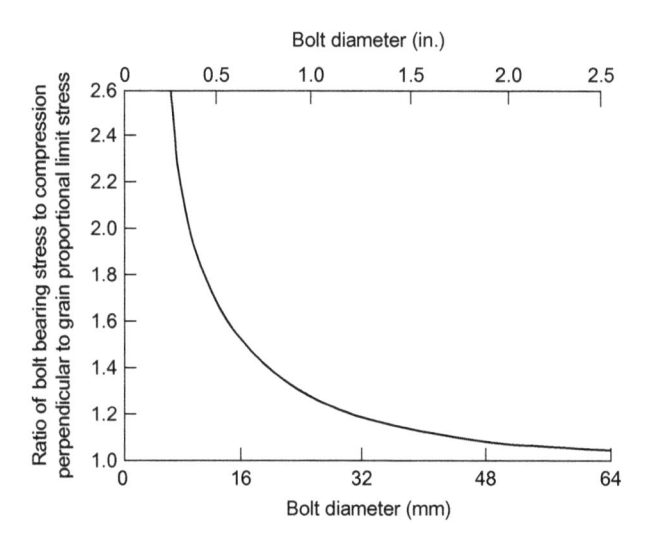

FIGURE 8–10 Bearing stress perpendicular to the grain as affected by bolt diameter.

Steel Side Plates

When steel side plates are used, the bolt-bearing stress parallel to grain at joint proportional limit is approximately 25% greater than that for wood side plates. The joint deformation at proportional limit is much smaller with steel side plates. If loads at equivalent joint deformation are compared, the load for joints with steel side plates is approximately 75% greater than that for wood side plates. Pre-1991 design criteria included increases in connection strength with steel side plates; post-1991 design criteria include steel side plate behavior in the yield model equations.

For perpendicular-to-grain loading, the same loads are obtained for wood and steel side plates.

Bolt Quality

Both the properties of the wood and the quality of the bolt are factors in determining the strength of a bolted joint. The percentages given in Figures 8–8 and 8–9 for calculating bearing stress apply to steel machine bolts with a yield stress of 310 MPa (45,000 lb/in^2). Figure 8–11 indicates the increase in bearing stress parallel to grain for bolts with a yield stress of 862 MPa (125,00 lb/in^2).

Effect of Member Thickness

The proportional limit load is affected by the ratio of the side member thickness to the main member thickness (Fig. 8–12).

Pre-1991 design values for bolts are based on joints with the side member half the thickness of the main member. The usual practice in design of bolted joints is to take no increase in design load when the side members are greater than half the thickness of the main member. When the side members are less than half the thickness of the main member, a design load for a main member that is twice the thickness of the side member is used. Post-1991 design values include member thickness directly in the yield model equations.

Two-Member, Multiple-Member Joints

In pre-1991 design, the proportional limit load was taken as half the load for a three-member joint, with a main member the same thickness as the thinnest member for two-member joints.

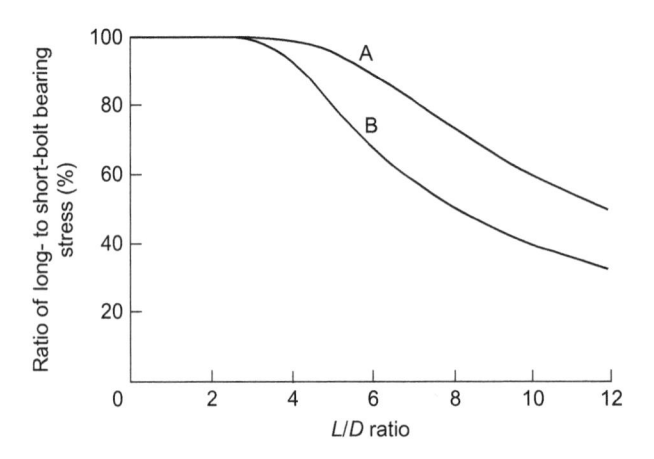

FIGURE 8–11 Variation in the proportional limit bolt-bearing stress parallel to grain with L/D ratio. Curve A, bolts with yield stress of 861.84 MPa (125,000 lb/in^2); curve B, bolts with yield stress of 310.26 MPa (45,000 lb/in^2).

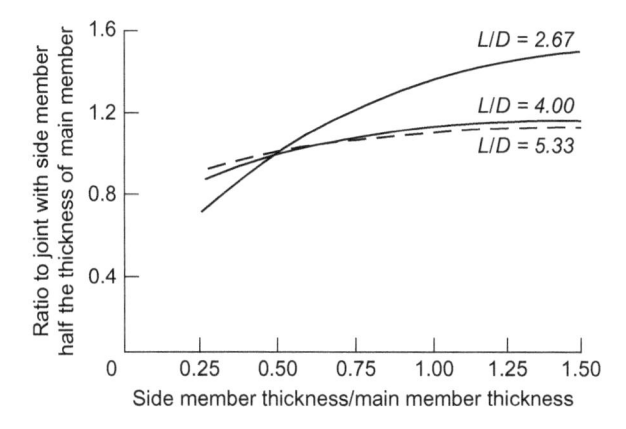

FIGURE 8–12 Proportional limit load related to side member thickness for three-member joints. Center member thickness was 50.8 mm (2 in.).

For four or more members in a joint, the proportional limit load was taken as the sum of the loads for the individual shear planes by treating each shear plane as an equivalent two-member joint.

8.8.0 Wood Adhesives Characterized as to Expected Performance

TABLE 8–2 Wood Adhesives Categorized According to Their Expected Structural Performance at Varying Levels of Environmental Exposure[a,b]

Structural Integrity	Service Environment	Adhesive Type
Structural	Fully exterior (withstands long-term water soaking and drying)	Phenol-formaldehyde Resorcinol-formaldehyde Phenol-resorcinol-formaldehyde Emulsion polymer/isocyanate Melamine-formaldehyde
	Limited exterior (withstands short-term water soaking)	Melamine-urea-formaldehyde Isocyanate Epoxy
	Interior (withstands short-term high humidity)	Urea-formaldehyde Casein
Semistructural	Limited exterior	Cross-linked polyvinyl acetate Polyurethane
Nonstructural	Interior	Polyvinyl acetate Animal Soybean Elastomeric construction Elastomeric contact Hot-melt Starch

[a] Assignment of an adhesive type to only one structural/service environment category does not exclude certain adhesive formulations from falling into the next higher or lower category.
[b] Priming wood surfaces with hydroxymethylated resorcinol coupling agent improves resistance to delamination of epoxy, isocyanate, emulsion polymer/isocyanate, melamine and urea, phenolic, and resorcinolic adhesives in exterior service environment, particularly bonds to CCA-treated lumber.

8.8.1 Categories of Selected Wood Species According to Ease of Bonding

U.S. Hardwoods	U.S. Softwoods		Imported Woods
Bond easily[a]			
Alder	Fir	Balsa	Hura
Aspen	White	Cativo	Purpleheart
Basswood	Grand	Courbaril	Roble
Cottonwood	Noble	Determa[b]	
Chestnut, American	Pacific		
Magnolia	Pine		
Willow, black	Eastern white		
	Western white		
	Redcedar, western		
	Redwood		
	Spruce, Sitka		
Bond well[c]			
Butternut	Douglas-fir	Afromosia	Meranti (lauan)
Elm	Larch, western[d]	Andiroba	Light red
American	Pine	Angelique	White
Rock	Sugar	Avodire	Yellow
Hackberry	Ponderosa	Banak	Obeche
Maple, soft	Redcedar, eastern	Iroko	Okoume
Sweetgum		Jarrah	Opepe
Sycamore		Limba	Peroba rosa
Tupelo		Mahogany	Sapele
Walnut, black		African	Spanish-cedar
Yellow-poplar		American	Sucupira
			Wallaba
Bond satisfactorily[e]			
Ash, white	Yellow-cedar	Angelin	Meranti (lauan), dark red
Beech, American	Port-Orford-cedar	Azobe	Pau marfim
Birch	Pines, southern	Benge	Parana-pine
Sweet		Bubinga	Pine
Yellow		Karri	Caribbean
Cherry			Radiata
Hickory			Ramin
Pecan			
True			
Madrone			
Maple, hard			
Oak			
Red[b]			
White[b]			
Bond with difficulty[f]			
Osage-orange		Balata	Keruing
Persimmon		Balau	Lapacho
		Greenheart	Lignumvitae
		Kaneelhart	Rosewood
		Kapur	Teak

[a] Bond very easily with adhesives of a wide range of properties and under a wide range of bonding conditions.
[b] Difficult to bond with phenol-formaldehyde adhesive.
[c] Bond well with a fairly wide range of adhesives under a moderately wide range of bonding conditions.
[d] Wood from butt logs with high extractive content is difficult to bond.
[e] Bond satisfactorily with good-quality adhesives under well-controlled bonding conditions.
[f] Satisfactory results require careful selection of adhesives and very close control of bonding conditions; may require special surface treatment.

8.8.2 Strength Properties of Various Types of Adhesives

Type	Form and Color	Preparation and Application	Strength Properties	Typical Uses
Natural origin				
Animal, protein	Solid and liquid; brown to white bondline	Solid form added to water, soaked, and melted; adhesive kept warm during application; liquid form applied directly; both pressed at room temperature; bonding process must be adjusted for small changes in temperature	High dry strength; low resistance to water and damp atmosphere	Assembly of furniture and stringed instruments; repairs of antique furniture
Blood, protein	Solid and partially dried whole blood; dark red to black bondline	Mixed with cold water, lime, caustic soda, and other chemicals; applied at room temperature; pressed either at room temperature or 120°C (250°F) and higher	High dry strength; moderate resistance to water and damp atmosphere and to microorganisms	Interior-type softwood plywood, sometimes in combination with soybean adhesive; mostly replaced by phenolic adhesive
Casein, protein	Powder with added chemicals; white to tan bondline	Mixed with water; applied and pressed at room temperature	High dry strength; moderate resistance to water, damp atmospheres, and intermediate temperatures; not suitable for exterior uses	Interior doors; discontinued use in laminated timbers
Soybean, protein	Powder with added chemicals; white to tan, similar color in bondline	Mixed with cold water, lime, caustic soda, and other chemicals; applied and pressed at room temperatures, but more frequently hot pressed when blended with blood adhesive	Moderate to low dry strength; moderate to low resistance to water and damp atmospheres; moderate resistance to intermediate temperatures	Softwood plywood for interior use, now replaced by phenolic adhesive. New fast-setting resorcinol-soybean adhesives for finger jointing of lumber being developed
Lignocellulosic residues and extracts	Powder or liquid; may be blended with phenolic adhesive; dark brown bondline	Blended with extender and filler by user; adhesive cured in hot-press 130°C to 150°C (266°F to 300°F) similar to phenolic adhesive	Good dry strength; moderate to good wet strength; durability improved by blending with phenolic adhesive	Partial replacement for phenolic adhesive in composite and plywood panel products
Synthetic origin				
Cross-linkable polyvinyl acetate emulsion	Liquid, similar to polyvinyl acetate emulsions but includes copolymers capable of cross-linking with a separate catalyst; white to tan with colorless bondline	Liquid emulsion mixed with catalyst; cure at room temperature or at elevated temperature in hot press and radio-frequency press	High dry strength; improved resistance to moisture and elevated temperatures, particularly long-term performance in moist environment	Interior and exterior doors; moulding and architectural woodwork; cellulosic overlays

(Continued)

Type	Form and Color	Preparation and Application	Strength Properties	Typical Uses
Elastomeric contact	Viscous liquid, typically neoprene or styrene-butadiene elastomers in organic solvent or water emulsion; tan to yellow	Liquid applied directly to both surfaces, partially dried after spreading and before pressing; roller-pressing at room temperature produces instant bonding	Strength develops immediately upon pressing, increases slowly over a period of weeks; dry strengths much lower than those of conventional wood adhesives; low resistance to water and damp atmospheres; adhesive film readily yields under static load	On-the-job bonding of decorative tops to kitchen counters; factory lamination of wood, paper, metal, and plastic sheet materials
Elastomeric mastic (construction adhesive)	Putty-like consistency, synthetic or natural elastomers in organic solvent or latex emulsions; tan, yellow, gray	Mastic extruded in bead to framing members by caulking gun or like pressure equipment; nailing required to hold materials in place during setting and service	Strength develops slowly over several weeks; dry strength lower than conventional wood adhesives; resistant to water and moist atmospheres; tolerant of outdoor assembly conditions; gap-filling; nailing required to ensure structural integrity	Lumber to plywood in floor and wall systems; laminating gypsum board and rigid foam insulating; assembly of panel system in manufactured homes
Emulsion polymer/ isocyanate	Liquid emulsion and separate isocyanate hardener; white with hardener; colorless bondline	Emulsion and hardener mixed by user; reactive on mixing with controllable pot-life and curing time; cured at room and elevated temperatures; radio-frequency curable; high pressure required	High dry and wet strength; very resistant to water and damp atmosphere; very resistant to prolonged and repeated wetting and drying; adheres to metals and plastics	Laminated beams for interior and exterior use; lamination of plywood to steel metals and plastics; doors and architectural materials
Epoxy	Liquid resin and hardener supplied as two parts; completely reactive leaving no free solvent; clear to amber; colorless bondline	Resin and hardener mixed by user; reactive with limited pot-life; cured at room or elevated temperatures; only low pressure required for bond development	High dry and wet strength to wood, metal, glass, and plastic; formulations for wood resist water and damp atmospheres; delaminate with repeated wetting and drying; gap-filling	Laminating veneer and lumber in cold-molded wood boat hulls; assembly of wood components in aircraft; lamination of architectural railings and posts; repair of laminated wood beams and architectural building components; laminating sports equipment; general purpose home and shop

Type	Form and Color	Preparation and Application	Strength Properties	Typical Uses
Hot melt	Solid blocks, pellets, ribbons, rods, or films; solvent-free; white to tan; near colorless bondline	Solid form melted for spreading; bond formed on solidification; requires special application equipment for controlling melt and flow	Develops strength quickly on cooling; lower strength than conventional wood adhesives; moderate resistance to moisture; gap-filling with minimal penetration	Edge-banding of panels; plastic lamination; patching; film and paper overlays; furniture assembly; general purpose home and shop
Isocyanate	Liquid containing isomers and oligomers of methylene diphenyl diisocyanate; light brown liquid and clear bondline	Adhesive applied directly by spray; reactive with water; requires high temperature and high pressure for best bond development in flake boards	High dry and wet strength; very resistant to water and damp atmosphere; adheres to metals and plastics	Flakeboards; strand-wood products
Melamine and melamine-urea	Powder with blended catalyst; may be blended up to 40% with urea; white to tan; colorless bondline	Mixed with water; cured in hot press at 120°C to 150°C (250°F to 300°F); particularly suited for fast curing in high-frequency presses	High dry and wet strength; very resistant to water and damp atmospheres	Melamine-urea primary adhesive for durable bonds in hardwood plywood; end-jointing and edge-gluing of lumber; and scarf joining softwood plywood
Phenolic	Liquid, powder, and dry film; dark red bondline	Liquid blended with extenders and fillers by user; film inserted directly between laminates; powder applied directly to flakes in composites; all formulations cured in hot press at 120°C to 150°C (250°F to 300°F) up to 200°C (392°F) in flakeboards	High dry and wet strength; very resistant to water and damp atmospheres; more resistant than wood to high temperatures and chemical aging	Primary adhesive for exterior softwood plywood, flakeboard, and hardboard
Polyvinyl acetate emulsion	Liquid ready to use; often polymerized with other polymers; white to tan to yellow; colorless bondline	Liquid applied directly; pressed at room temperatures and in high-frequency press	High dry strength; low resistance to moisture and elevated temperatures; joints yield under continued stress	Furniture; flush doors; plastic laminates; panelized floor and wall systems in manufactured housing; general purpose in home and shop
Polyurethane	Low viscosity liquid to high viscosity mastic; supplied as one part; two-part systems completely reactive; color varies from clear to brown; colorless bondline	Adhesive applied directly to one surface, preferably to water-misted surface; reactive with moisture on surface and in air; cures at room temperature; high pressure required, but mastic required only pressure from nailing	High dry and wet strength; resistant to water and damp atmosphere; limited resistance to prolonged and repeated wetting and drying; gap-filling	General-purpose home and shop; construction adhesive for panelized floor and wall systems; laminating plywood to metal and plastic sheet materials; specialty laminates; installation of gypsum board

(Continued)

Type	Form and Color	Preparation and Application	Strength Properties	Typical Uses
Resorcinol and phenol-resorcinol	Liquid resin and powdered hardener supplied as two parts; phenol may be copolymerized with resorcinol; dark red bondline	Liquid mixed with powdered or liquid hardener; resorcinol adhesives cure at room temperatures; phenol-resorcinols cure at temperatures from 21°C to 66°C (70°F to 150°F)	High dry and wet strength; very resistant to moisture and damp atmospheres; more resistant than wood to high temperature and chemical aging.	Primary adhesives for laminated timbers and assembly joints that must withstand severe service conditions.
Urea	Powder and liquid forms; may be blended with melamine or other more durable resins; white to tan resin with colorless bondline	Powder mixed with water, hardener, filler, and extender by user; some formulations cure at room temperatures, others require hot pressing at 120°C (250°F); curable with high-frequency heating	High dry and wet strength; moderately durable under damp atmospheres; moderate to low resistance to temperatures in excess of 50°C (122°F)	Hardwood plywood; furniture; fiberboard; particle-board; underlayment; flush doors; furniture cores

8.9.0 Fasteners Installed in Hollow Masonry Units

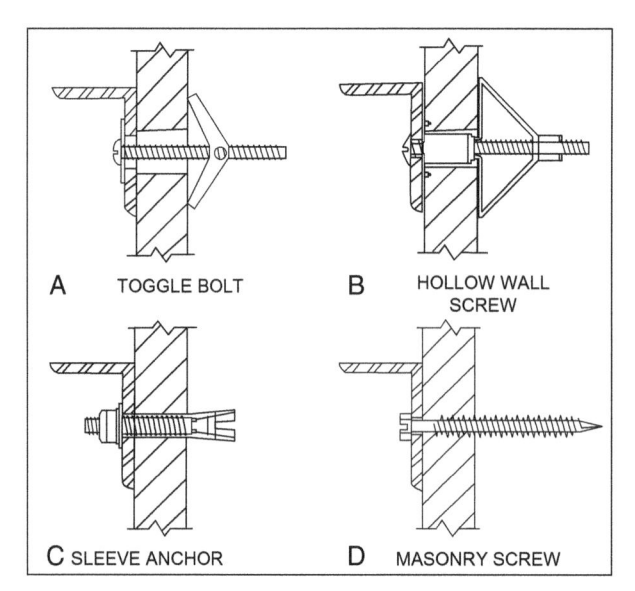

A TOGGLE BOLT
B HOLLOW WALL SCREW
C SLEEVE ANCHOR
D MASONRY SCREW

Fasteners Installed in Hollow Units

| A   TOGGLE BOLTS | B   HOLLOW WALL SCREWS |

| C   SLEEVE ANCHORS | D   MASONRY SCREWS |

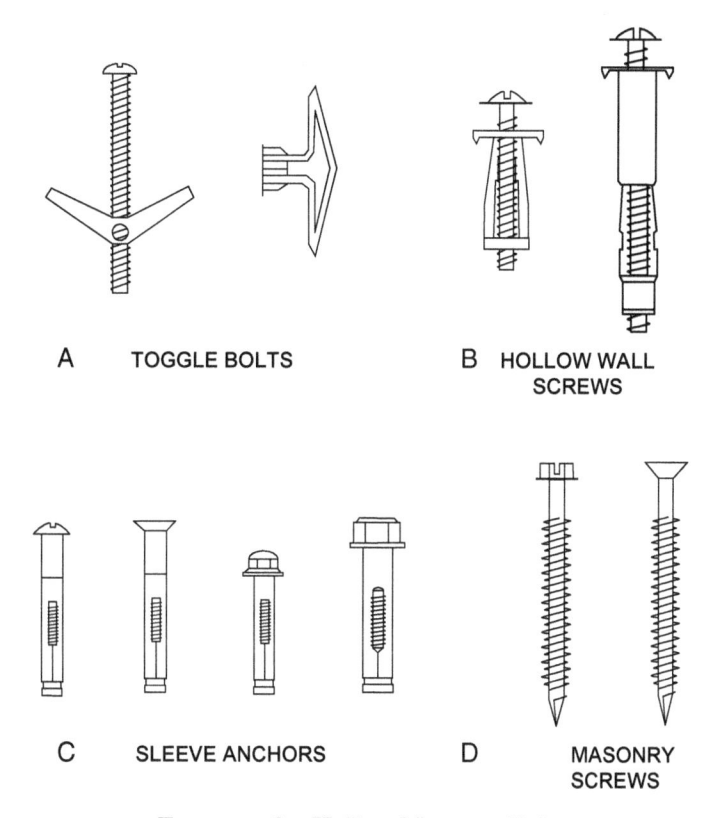

Fasteners for Hollow Masonry Units

Source: Brick Industry Association.

8.9.1 Fasteners Installed in Solid Masonry Units

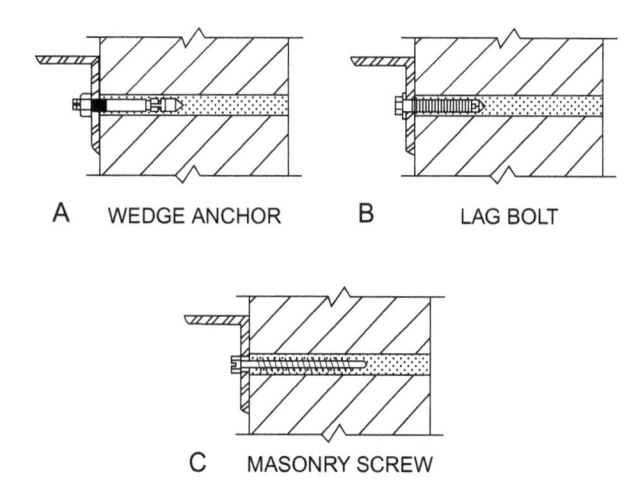

| A   WEDGE ANCHOR | B   LAG BOLT |

C   MASONRY SCREW

Fasteners Installed in Solid Masonry Units

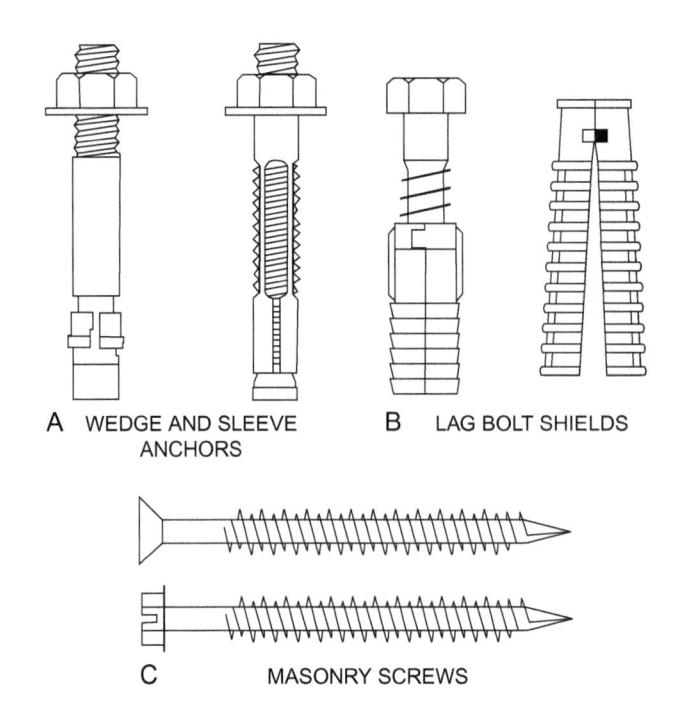

A WEDGE AND SLEEVE ANCHORS

B LAG BOLT SHIELDS

C MASONRY SCREWS

Fasteners for Solid Masonry Units

Source: Brick industry Association.

8.9.2 Power-Driven Fasteners for Masonry Units

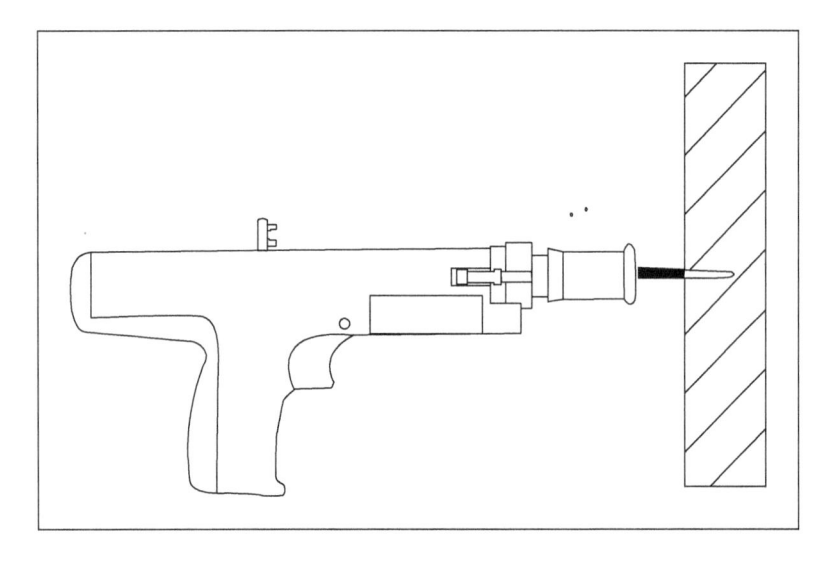

Power-Driven Fastening Tool

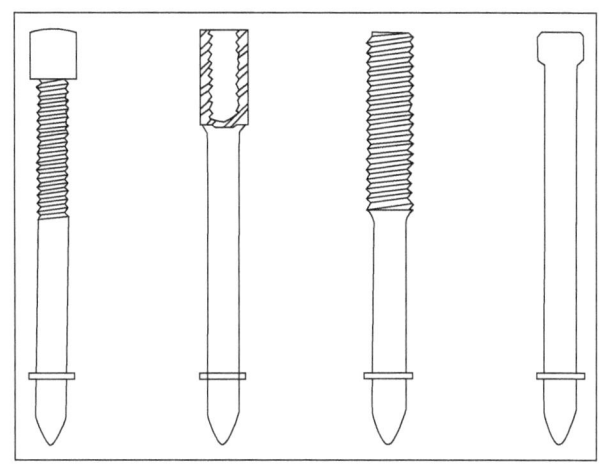

Section | 8 Fasteners for Wood and Steel—Calculations for Selection

FIG. 10 Power-Driven Pins

Power-driven fasteners require special installation equipment, safety equipment, and inspection procedures. For this reason, the manufacturer should be contacted to determine proper equipment and installation specifications

Source: Brick Industry Association.

8.9.3 Masonry Fastener Selection Chart

Fastener	Brick Type			Installation Location			Fixture Weight		
	Solid brick (cored)	Solid Brick (uncored)	Hollow Brick	Head Joint	Bed Joint	Unit Face	Light	Medium	Heavy
Wooden Blocks	X	X	X	X			X	X	
Metal Wall Plugs	X	X	X	X	X		X	X	
Screw Shields and Plugs	X	X	X	X	X		X		
Toggle Bolts			X		X	X	X	X	X
Hollow Wall Screws			X		X	X	X	X	
Screws	X	X	X	X	X	X	X	X	
Sleeve Anchors	X	X	X	X	X	X	X	X	X
Wedge Anchors	X	X		X	X		X	X	X
Lag Shields	X	X		X	X		X	X	X
Masonry Nails	X	X	X	X	X		X	X	
Powder-Driven Fasteners	X	X		X	X		X	X	
Adhesives	X	X	X	Surface Applied			X		

Environment

Environmental factors may have a definite impact on the long-term service life of fasteners and should be considered in their selection. Environmental factors do not, in general, influence the type of fastener selected, but should affect the choice of fastener based on the material from which the fastener is made. Corrosion is a major concern, especially when fasteners are exposed to the elements or when fasteners are used in areas where contact with corrosive agents is likely.

Steel fasteners used for applications under normal exposure conditions should be galvanized (zinc-coated) to resist corrosion. Lead, copper-coated or brass fasteners also provide adequate corrosion resistance for normal exposures. In applications where fasteners are subject to severe exposure conditions or exposed to chemicals, stainless steel fasteners should be used.

Aesthetics

In most applications, the fastener or fasteners installed will be hidden by the attachment (i.e., cabinets, baseboards, electrical boxes or furring), and the physical appearance of the fastener (usually the head of a screw or bolt) will not be of importance. However, when fasteners are used to attach privacy partitions, lighting fixtures or rails, the head of the fastener is usually visible and required to match or accent the finish of the fixture. In these cases, finished screws or bolts (i.e., chrome or brass-plated, solid brass or painted) can be purchased to match the fixtures. The manufacturers should be contacted to determine the availability and range of finishes available in their products.

Source: Brick Industry Association.

8.10.0 Bolt Fasteners for Structural Steel—Identification Markings and Strength Requirements

Fastener Identification Markings Bolt Strength Requirements

| Grade Marking | Specification | Material and Treatment | Nominal Size (In.) | Mechanical Properties | | | Hardness Rockwell | |
				Proof Load Min (ksi)	Yield Strength Min (ksi)	Tensile Strength Min (ksi)	Min	Max
PB	SAE J429 Grade 1	Low or Medium Carbon Steel	¼" - 1½"	33	36	60	B70	B100
307A PB	A307 Grade A		¼" - 4"	-	-	60	B69	B100
307B PB	A307 Grade B			-	-	60 min 100 max	B69	5

Fastener Identification Markings Bolt Strength Requirements—Cont'd

Grade Marking	Specification	Material and Treatment	Nominal Size (In.)	Mechanical Properties			Hardness Rockwell	
				Proof Load Min (ksi)	Yield Strength Min (ksi)	Tensile Strength Min (ksi)	Min	Max
307C PB	A307 Grade C			-	36	58 min 80 max	-	-
One End Green								
AB36 PB	F1554 Grade 36			-	36	58 min 80 max	-	-
AB55 PB	F1554 Grade 55			-	55	75 min 95 max	-	-
One End Yellow								
PB	SAE J429 Grade 2		¼" - ¾"	55	57	74	B80	B100
			⅞" - 1½"	33	36	60	B70	B100
B8 PB	A193/A320 Grade B8	AISI SS304 Stainless	No Restrictions	-	30	75	-	B96
B8M PB	A193/A320 Grade B8M	AISI SS316 Stainless Steel						

(Continued)

Fastener Identification Markings Bolt Strength Requirements—Cont'd

Grade Marking	Specification	Material and Treatment	Nominal Size (In.)	Mechanical Properties			Hardness Rockwell	
				Proof Load Min (ksi)	Yield Strength Min (ksi)	Tensile Strength Min (ksi)	Min	Max
A325 PB	A325 Type 1	Medium Carbon Steel, Q & T	½" - 1"	85	92	120	C24	C35
			1" - 1½"	74	81	105	C19	C31
PB	A325 Type 3	Atmospheric Corrosion Resistant Steel, Q & T						
A325 PB	SAE J429 Grade 5	Medium Carbon Steel, Q & T	¼" - 1"	85	92	120	C25	C34
			1" - 1½"	74	81	105	C19	C30
A449 PB	A449 Type 1		¼" - 1"	85	92	120	C25	C34
			1" - 1½"	74	81	105	C19	C30
			1" - 3"	55	58	90	B91	B100
B7 PB	A193 Grade B7	Medium Carbon Alloy Steel, Q & T	¼" - 2½"	-	105	125	-	C35
			2" - 4"		95	115		C35
			4" - 7"		75	100		C35
BC PB	A354 Grade BC		¼" - 2½"	105	109	125	C26	C36
			2" - 4"	95	94	115	C22	C33

Fastener Identification Markings Bolt Strength Requirements—Cont'd

Grade Marking	Specification	Material and Treatment	Nominal Size (In.)	Mechanical Properties			Hardness Rockwell	
				Proof Load Min (ksi)	Yield Strength Min (ksi)	Tensile Strength Min (ksi)	Min	Max
AB105 PB (One End Red)	F1554 Grade 105		¼" - 3"	-	105	125 min 150 max	-	-
L7 PB	A320 Grade L7	AISI 4140, 4142, or 4145	¼" - 2½"	-	105	125	-	-
L43 PB	A320 Grade L43	AISI 4340	¼" - 4"	-	105	125	-	-
A 490 PB	A490 Type 1	Medium Carbon Alloy Steel, Q & T	½" - 1½"	120	130	150 min 173 max	C33	C38
A 490 PB	A490 Type 3	Atmospheric Corrosion Resistant Steel, Q & T						
PB	SAE J429 Grade 8	Medium Carbon Alloy Steel, Q & T	¼" - 1½"	120	130	150	C33	C39

(Continued)

Fastener Identification Markings Bolt Strength Requirements—Cont'd

Grade Marking	Specification	Material and Treatment	Nominal Size (In.)	Mechanical Properties			Hardness Rockwell	
				Proof Load Min (ksi)	Yield Strength Min (ksi)	Tensile Strength Min (ksi)	Min	Max
BD PB	A354 Grade BD		¼" - 2½"	120	130	150	C33	C39
			2" - 4"	105	115	140	C31	C39

1. All specifications are ASTM unless otherwise noted.
2. All specifications shall be marked by the manufacturer with a unique identifier to identify the manufacturer or private label distributor, as appropriate.
3. Q & T—Quenched and Tempered.
4. Stamping of F1554 and A307 grade C bolts is a supplemental requirement while color coding is required.
5. Although markings are shown on hex heads, grade markings apply equally to products with other head configurations.
6. All Grade BD products shall be marked "BD." In addition to the "BD" marking, the product may be marked with six radial lines 60Å apart.

8.10.1 Standard Thread Pitches

Standard Thread Pitches

Thread series cover designations of diameter/pitch combinations that are measured by the number of threads per inch (TPI) applied to a single diameter.

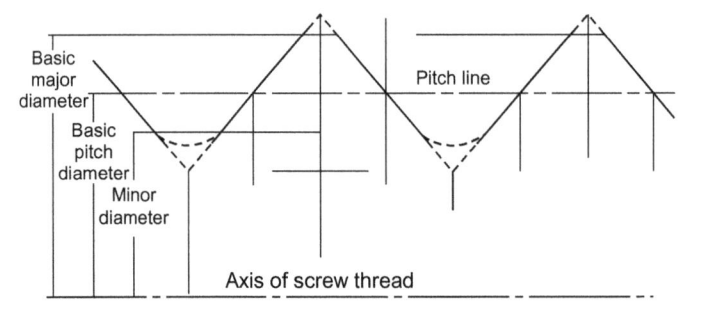

- **Coarse Thread Series (UNC/UNRC)** is the most common designation for general application bolts and nuts. Coarse thread is beneficial because it is less likely to cross thread, more tolerant in adverse conditions, and facilitates quick assembly.
- **Fine Thread Series (UNF/UNRF)** is commonly used in precision applications. Because of the larger tensile stress areas, fine threads have high tension strength. However, a longer engagement is required for fine thread applications than for coarse series threads to prevent stripping.
- **8 - Thread Series (8UN)** is the specified thread forming method for several ASTM standards, including A193 B7, A193 B8/B8M, and A320. This series is used for diameters one inch and above.

Section 8 Fasteners for Wood and Steel—Calculations for Selection   493

Coarse Thread Series—UNC				Fine Thread Series—UNF				8-Thread Series—8UN			
Nominal Size and Threads Per In.	Basic Pitch Dia.	Section at Minor Dia.	Tensile Stress Area	Nominal Size and Threads Per In.	Basic Pitch Dia.	Section at Minor Dia.	Tensile Stress Area	Nominal Size and Threads Per In.	Basic Pitch Dia.	Section at Minor Dia.	Tensile Stress Area
	In.	Sq in.	Sq in.		In.	Sq in.	Sq in.		In.	Sq in.	Sq in.
—	—	—	—	0 - 80	0.0519	0.00151	0.00180	—	—	—	—
1 - 64	0.0629	0.00218	0.00263	1 - 72	0.0640	0.00237	0.00278	—	—	—	—
2 - 56	0.0744	0.00310	0.00370	2 - 64	0.0759	0.00339	0.00394	—	—	—	—
3 - 48	0.0855	0.00406	0.00487	3 - 56	0.0874	0.00451	0.00523	—	—	—	—
4 - 40	0.0958	0.00496	0.00604	4 - 48	0.0985	0.00566	0.00661	—	—	—	—
5 - 40	0.1088	0.00672	0.00796	5 - 44	0.1102	0.00716	0.00830	—	—	—	—
6 - 32	0.1177	0.00745	0.00909	6 - 40	0.1218	0.00874	0.01015	—	—	—	—
8 - 32	0.1437	0.01196	0.0140	8 - 36	0.1460	0.01285	0.01474	—	—	—	—
10 - 24	0.1629	0.01450	0.0175	10 - 32	0.1697	0.0175	0.0200	—	—	—	—
12 - 24	0.1889	0.0206	0.0242	12 - 28	0.1928	0.0226	0.0258	—	—	—	—
1/4 - 20	0.2175	0.0269	0.0318	1/4 - 28	0.2268	0.0326	0.0364	—	—	—	—
5/16 - 18	0.2764	0.0454	0.0524	5/16 - 24	0.2854	0.0524	0.0580	—	—	—	—
3/8 - 16	0.3344	0.0678	0.0775	3/8 - 24	0.3479	0.0809	0.0878	—	—	—	—
7/16 - 14	0.3911	0.0933	0.1063	7/16 - 20	0.4050	0.1090	0.1187	—	—	—	—
1/2 - 13	0.4500	0.1257	0.1419	1/2 - 20	0.4675	0.1486	0.1599	—	—	—	—
9/16 - 12	0.5084	0.162	0.182	9/16 - 18	0.5264	0.189	0.203	—	—	—	—
5/8 - 11	0.5660	0.202	0.226	5/8 - 18	0.5889	0.240	0.256	—	—	—	—
3/4 - 10	0.6850	0.302	0.334	3/4 - 16	0.7094	0.351	0.373	—	—	—	—
7/8 - 9	0.8028	0.419	0.462	7/8 - 14	0.8286	0.480	0.509	—	—	—	—
1 - 8	0.9188	0.551	0.606	1 - 12	0.9459	0.625	0.663	1 - 8	0.9188	0.551	0.606
1-1/8 - 7	1.0322	0.693	0.763	1-1/8 - 12	1.0709	0.812	0.856	1-1/8 - 8	1.0438	0.728	0.790
1-1/4 - 7	1.1572	0.890	0.969	1-1/4 - 12	1.1959	1.024	1.073	1-1/4 - 8	1.1688	0.929	1.000
1-3/8 - 6	1.2667	1.054	1.155	1-3/8 - 12	1.3209	1.260	1.315	1-3/8 - 8	1.2938	1.155	1.233
1-1/2 - 6	1.3917	1.294	1.405	1-1/2 - 12	1.4459	1.521	1.581	1-1/2 - 8	1.4188	1.405	1.492
—	—	—	—	—	—	—	—	1-5/8 - 8	1.5438	1.68	1.78
1-3/4 - 5	1.6201	1.74	1.90	—	—	—	—	1-3/4 - 8	1.6688	1.98	2.08
—	—	—	—	—	—	—	—	1-7/8 - 8	1.7938	2.30	2.41
2 - 4-1/2	1.8557	2.30	2.50	—	—	—	—	2 - 8	1.9188	2.65	2.77
2-¼ - 4-½	2.1057	3.02	3.25	—	—	—	—	2-1/4 - 8	2.1688	3.42	3.56
2-1/2 - 4	2.3376	3.72	4.00	—	—	—	—	2-1/2 - 8	2.4188	4.29	4.44
2-3/4 - 4	2.5876	4.62	4.93	—	—	—	—	2-3/4 - 8	2.6688	5.26	5.43
3 - 4	2.8376	5.62	5.97	—	—	—	—	3 - 8	2.9188	6.32	6.51
3-1/4 - 4	3.0876	6.72	7.10	—	—	—	—	3-1/4 - 8	3.1688	7.49	7.69
3-1/2 - 4	3.3376	7.92	8.33	—	—	—	—	3-1/2 - 8	3.4188	8.75	8.96
3-3/4 - 4	3.5876	9.21	9.66	—	—	—	—	3-3/4 - 8	3.6688	10.11	10.34
4 - 4	3.8376	10.61	11.08	—	—	—	—	4 - 8	3.9188	11.57	11.81

By permission: Portland Bolt and Manufacturing, Portland, OR

8.10.2 Suggested Starting Torque Values for ASTM and SAE Grade Bolts

Suggested Starting Torque Values ASTM A307

Bolt Size	TPI	Proof Load (lbs)	Clamp Load (lbs)	Tightening Torque (ft lbs)		
				Waxed	Galv	Plain
1/4	20	1145	859	2	4	4
5/16	18	1886	1415	4	9	7
3/8	16	2790	2093	7	16	13
7/16	14	3827	2870	10	26	21
1/2	13	5108	3831	16	40	32
9/16	12	6552	4914	23	58	46
5/8	11	8136	6102	32	79	64
3/4	10	12024	9018	56	141	113
7/8	9	15200	11400	83	208	166
1	8	20000	15000	125	313	250
1 1/8	7	25200	18900	177	443	354
1 1/4	7	32000	24000	250	625	500
1 3/8	6	38100	28575	327	819	655
1 1/2	6	46400	34800	435	1088	870
1 3/4	5	68400	51300	748	1870	1496
2	4½	90000	67500	1125	2813	2250
2 1/4	4½	117000	87750	1645	4113	3291
2 1/2	4	144000	108000	2250	5625	4500
2 3/4	4	177480	133110	3050	7626	6101
3	4	214920	161190	4030	10074	8060
3 1/4	4	255600	191700	5192	12980	10384
3 1/2	4	299880	224910	6560	16400	13120
3 3/4	4	347760	260820	8151	20377	16301
4	4	398880	299160	9972	24930	19944

SAE Grade 2

Bolt Size	TPI	Proof Load (lbs)	Clamp Load (lbs)	Tightening Torque (ft lbs)		
				Waxed	Galv	Plain
1/4	20	1750	1313	3	7	5
5/16	18	2900	2175	6	14	11
3/8	16	4250	3188	10	25	20
7/16	14	5850	4388	16	40	32
1/2	13	7800	5850	24	61	49
9/16	12	10000	7500	35	88	70
5/8	11	12400	9300	48	121	97
3/4	10	18400	13800	86	216	173
7/8	9	15200	11400	83	208	166
1	8	20000	15000	125	313	250
1 1/8	7	25200	18900	177	443	354
1 1/4	7	32000	24000	250	625	500
1 3/8	6	38100	28575	327	819	655
1 1/2	6	46400	34800	435	1088	870

ASTM A325 / ASTM A449 / SAE Grade 5

Bolt Size	TPI	Proof Load (lbs)	Clamp Load (lbs)	Tightening Torque (ft lbs)		
				Waxed	Galv	Plain
1/4	20	2700	2025	4	11	8
5/16	18	4450	3338	9	22	17
3/8	16	6600	4950	15	39	31
7/16	14	9050	6788	25	62	49
1/2	13	12050	9038	38	94	75
9/16	12	15450	11588	54	136	109
5/8	11	19200	14400	75	188	150
3/4	10	28400	21300	133	333	266
7/8	9	39250	29438	215	537	429
1	8	51500	38625	322	805	644
1 1/8	7	56450	42338	397	992	794
1 1/4	7	71700	53775	560	1400	1120
1 3/8	6	85450	64088	734	1836	1469
1 1/2	6	104000	78000	975	2438	1950
1 3/4	5	104500	78375	1143	2857	2286
2	4½	137500	103125	1719	4297	3438
2 1/4	4½	178750	134063	2514	6284	5027
2 1/2	4	220000	165000	3438	8594	6875
2 3/4	4	271150	203363	4660	11651	9321
3	4	328350	246263	6157	15391	12313

ASTM A193 B7

Bolt Size	TPI	Proof Load (lbs)	Clamp Load (lbs)	Tightening Torque (ft lbs)		
				Waxed	Galv	Plain
1/4	20	3350	2513	5	13	10
5/16	18	5500	4125	11	27	21
3/8	16	8150	6113	19	48	38
7/16	14	11150	8363	30	76	61
1/2	13	14900	11175	47	116	93
9/16	12	19100	14325	67	168	134
5/8	11	23750	17813	93	232	186
3/4	10	35050	25288	164	411	329
7/8	9	48500	36375	265	663	530
1	8	63650	47738	398	995	796
1 1/8	7	80100	60075	563	1408	1126
1 1/4	7	101750	76313	795	1987	1590
1 3/8	6	121300	90975	1042	2606	2085
1 1/2	6	147550	110663	1383	3458	2767
1 3/4	5	199500	149625	2182	5455	4364
2	4½	262500	196875	3281	8203	6563
2 1/4	4½	341250	255938	4799	11997	9598
2 1/2	4	420000	315000	6563	16406	13125
2 3/4	4	468500	351263	8050	20124	16100
3	4	567150	425363	10634	26585	21268
3 1/4	4	674500	505875	13701	34252	27402

ASTM A193 B7—Cont'd

Bolt Size	TPI	Proof Load (lbs)	Clamp Load (lbs)	Tightening Torque (ft lbs)		
				Waxed	Galv	Plain
3 1/2	4	791350	593513	17311	43277	34622
3 3/4	4	917700	688275	21509	53771	43017
4	4	1052600	789450	26315	65788	52630

ASTM A354-BD / ASTM A490 / SAE Grade 8

Bolt Size	TPI	Proof Load (lbs)	Clamp Load (lbs)	Tightening Torque	
				Waxed	Plain
1/4	20	3800	2850	6	12
5/16	18	6300	4725	12	25
3/8	16	9300	6975	22	44
7/16	14	12750	9563	35	70
1/2	13	17050	12788	53	107
9/16	12	21850	16388	77	154
5/8	11	27100	20325	106	212
3/4	10	40100	30075	188	376
7/8	9	55450	41588	303	606
1	8	72700	54525	454	909
1 1/8	7	91550	68663	644	1287
1 1/4	7	120000	90000	938	1875
1 3/8	6	138600	103950	1191	2382
1 1/2	6	168600	126450	1581	3161
1 3/4	5	228000	171000	2494	4988
2	4½	300000	225000	3750	7500
2 1/4	4½	390000	292500	5484	10969
2 1/2	4	480000	360000	7500	15000
2 3/4	4	517650	388238	8897	17794
3	4	626850	470138	11753	23507
3 1/4	4	745500	559125	15143	30286
3 1/2	4	874650	655988	19133	38266
3 3/4	4	1014300	760725	23773	47545
4	4	1052600	789450	26315	52630

Notes:
1. Values calculated using industry accepted formula $T = KDP$ where T = Torque, K = torque coefficient (dimensionless), D = nominal diameter (inches), P = bolt clamp load, lb.
2. K values: **waxed** (e.g., pressure wax as supplied on high strength nuts) = .10, hot dip galvanized = .25, and **plain** non-plated bolts (as received) = 0.20.
3. Torque has been converted into ft/lbs by dividing the result of the formula by 12.
4. All calculations are for Coarse Thread Series (UNC).
5. Grade 2 calculations only cover fasteners 1/4"–3/4" in diameter up to 6" long; for longer fasteners the torque is reduced significantly.
6. Clamp loads are based on 75% of the minimum proof loads for each grade and size.
7. Proof load, stress area, yield strength, and other data are based on IFI 7th Edition (2003). Technical Data N-68, SAE J429, ASTM A307, A325, A354, A449, and A490.

The above estimated torque calculations are only offered as a guide. Use of its content by anyone is the sole responsibility of that person and they assume all risk. Due to many variables that affect the torque–tension relationship like human error, surface texture, lubrication, etc., the only way to determine the correct torque is through experimentation under actual joint and assembly conditions.

8.10.3 AASHTO to ASTM Conversion Chart

AASHTO to ASTM Conversions

AASHTO is an acronym for American Association of State Highway and Transportation Officials. It is common for bolt specifications to be called out on construction plans with an AASHTO designation for state highway projects. Many of these designations can be directly converted to an ASTM equivalent. The following table lists some of the more common fastener-related specifications and their ASTM equivalents.

AASHTO Grade	ASTM Equivalent	Description
M-111	A123	Hot-dip galvanizing of iron and steel products
M-164	A325	Structural bolt
M-183	A36	Raw material, low carbon steel
M-222	A588	Raw material, weathering steel
M-223	A572	Raw material, high strength / low alloy
M-232	A153	Hot-dip galvanizing of fasteners
M-253	A490	Structural bolt
M-291	A563	Nut specification possessing many grades
M-292	A194	High strength, heavy hex nut
M-293	F436	Hardened washer
M-314	F1554	Anchor bolt specification with 3 grades

By permission: Portland Bolt and Manufacturing, Portland, OR

8.10.4 ASTM A563 Nut Compatibility Chart

ASTM A563 Nut Compatibility Chart

Download a print version

			A563 Grade and ANSI Nut Style			
			Recommended		Suitable	
Grade of Bolt	Surface Finish	Nominal Size Inches	Hex	Heavy Hex	Hex	Heavy Hex
A307 Grade A & C	Any	¼ to 1½	A		B,D,DH	A,B,C,D,DH,DH3
		>1½ to 2		A	A	C,D,DH,DH3
		>2 to 4		A		C,D,DH,DH3
A307 Grade B	Any	¼ to 1½		A	B,D,DH	A,B,C,D,DH,DH3
		>1½ to 2		A	A	
		>2 to 4		A		
A325 Type 1	Plain	½ - 1½		C		C3,D,DH,DH3
	Galvanized	½ - 1½		DH		
A325 Type 3	Plain	½ - 1½		C3		DH3

Download a print version—Cont'd

Grade of Bolt	Surface Finish	Nominal Size Inches	A563 Grade and ANSI Nut Style			
			Recommended		Suitable	
			Hex	Heavy Hex	Hex	Heavy Hex
A354 Grade BC	Plain	¼ to 1½		C	D,DH	C3,D,DH,DH3
		>1½ to 4		C		C3,D,DH,DH3
	Galvanized	¼ to 1½		DH		
		>1½ to 4		DH		
A354 Grade BD	Plain	¼ to 1½		DH	DH	D,DH,DH3
		>1½ to 4		DH		DH3
A449	Plain	¼ to 1½	B		D,DH	B,C,C3,D,DH,DH3
		>1½ to 3		A		C,C3,D,DH,DH3
	Galvanized	¼ to 1½		DH	D,DH	D
		>1½ to 3		DH		D
A490 Type 1	Plain	½ - 1½		DH		DH3
A490 Type 3	Plain	½ - 1½		DH3		
A687	Plain	- 3		D		DH,DH3
	Galvanized	- 3		DH		
F1554 Grade 36	Any	¼ to 1½	A		B,D,DH	A,B,C,D,DH,DH3
		>1½ to 4		A		C,D,DH,DH3
F1554 Grade 55	Plain	¼ to 1½	A		B,D,DH	A,B,C,D,DH,DH3
		>1½ to 4		A		C,D,DH,DH3
	Galvanized	¼ to 4		A		C,D,DH,DH3
F1554 Grade 105	Plain	¼ to 1½	D			DH,DH3
		>1½ to 3		DH		DH3
	Galvanized	¼ to 3		DH		DH3

By permission: Portland Bolt and Manufacturing, Portland, OR

8.11.0 Tension Shear—TC Bolts

Structural Steel Fastening System

Unytite Inc., a QS 9000/ISO 9002 registered facility located in Peru, Illinois, is a manufacturer of "Structural Fastening Systems" for the Heavy Construction (High Rise, Bridge, Road, and Industrial Building applications), Petro Chemical (Refinery, Pipeline, and Chemical Industries), Heavy Equipment, Rail Car, and Tractor-Trailer O.E.M.'s (Original Equipment Manufacturer).

The unique Tension Control Fastening System is a three-piece fastening assembly comprised of a button head design bolt with a 12-point pintail, a high-strength heavy hex nut, and a hardened flat washer. When installed with a dual socket electric shear wrench, the outer socket applies the turning force to the nut, while the inner socket holds the bolt in place by gripping the 12-point spline tip. When the forces reach or exceed the designed torque-tension coefficient, the 12-point spline tip will shear off, leaving the bolt and nut securing the application at the proper tension.

Installation Procedure

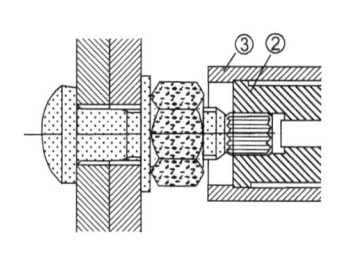

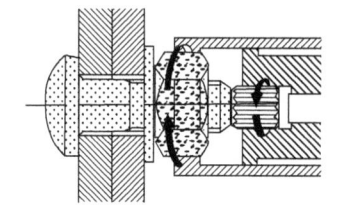

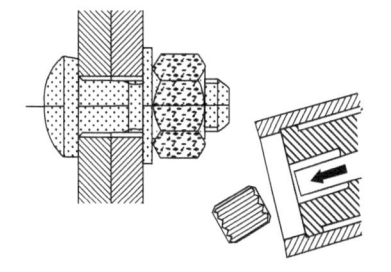

1. *Fit the inner socket of the shear wrench over the spline on the bolt and push forward until the outer socket engages completely with the nut.*
2. *Pull the larger trigger on the wrench. The inner socket will hold the bolt in place, while the outer socket tightens the nut. The spline will shear off when proper tension is reached.*
3. *Remove the wrench from the nut and pull the ejection trigger. This will eject the spline from the inner socket of the wrench. The installation is now complete and may be verified visually.*

Handling—Storage—Installation

1. *All structural fasteners should be protected from dirt and moisture at the job site. No more than the amount of bolts to be used that day should be removed from the container, or protected storage. Remaining bolts at the end of the day should be returned to the correct container. Dirty or rusted bolts should not be used.*
2. *Place all the bolts into the connection, with a washer <u>under the nut</u> in <u>standard</u> and <u>short slotted holes</u>. For <u>long slotted</u> and <u>oversize holes</u>, a <u>washer should be placed under the head of the bolt and under the nut</u>. Washer and nut identification markings should always face the opposite direction of the connection.*
3. *Bring all the fasteners in the connection to a snug tight condition, starting with the most rigid part of the connection.*

(The above recommendations by AISC apply to all A325 and A490 fasteners regardless of Installation methods)

Determining Proper Bolt Length

To determine the proper length of fastener that is needed, refer to the chart at right for the proper length to add to the grip. The bolt length should be adjusted to the next 1/4 inch for washer thickness.

(NOTE: 3-5 bolt threads should be within the structural member to prevent thread from running out.)

Nominal Bolt Size	Length Added to Grip
5/8"	7/8"
3/4"	1"
7/8"	1-1/8"
1"	1-1/4"
1-1/8"	1-1/2"

L = Bolt Length
LG = Grip Length
LA = Length Added to Grip
Source: UNYTITE, Peru, ILL.

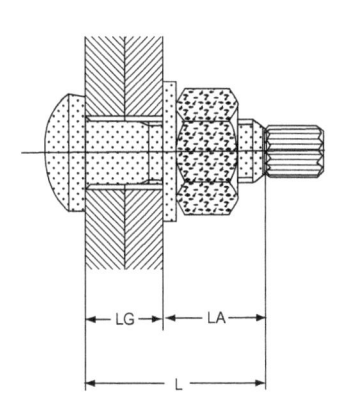

UNYTITE, INC. tension control bolts are designed, manufactured, and tested to conform to ASTM (American Society for Testing and Materials) F-1852, A-325 and A-490, AISC (American Institute of Steel Construction), FHWA* (Federal Highway Administration) and the most demanding customer specifications.

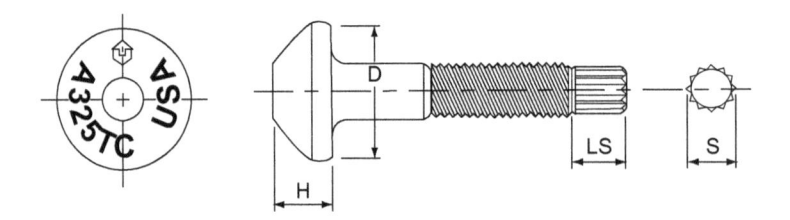

ASTM F-1852 Dimensions for Twist-Off Structural Bolt

Normal Size or Basic Product Diameter	H Height Max	H Height Min	D Bearing Surface Diameter Min	LS Length of Spline Ref.	S Width across Flats Ref
5/8" 0.625	0.403	0.378	1.102	0.60	0.43
3/4" 0.750	0.483	0.455	1.338	0.65	0.53
7/8" 0.875	0.563	0.531	1.535	0.72	0.61
1" 1.000	0.627	0.591	1.771	0.80	0.70
1-1/8" 1.125	0.718	0.658	1.991	0.90 A	0.80 A

A - The spline length (LS) and across the flat (S) dimensions are used for reference only. The grooved spline design may vary in size and shape.

ASTM A325 (F1852) Mechanical Properties

	Bolt ASTM A325 Type 1				Nut A563 DH		Washer F436
	Proof Load	Tensile Strength	Hardness Brinell	Hardness Rockwell	Proof Load	Hardness	Hardness
5/8"-11	19,200	27,100	253 ~ 319	HRC 25~34	39,550	HRC 24~38	HRC 38~45
3/4"-10	28,400	40,100			58,450		
7/8"-9	39,250	55,450			80,850		
1"-8	51,500	72,700			106,050		
1-1/8"-7	56,450	80,100	223~286	19~30	133,525		

---

* Production to meet FHWA on customer request.

A325 (F1852) Fastener Tension

Nominal Diameter	1 AISC Table 4 Design Tension Min lbf	2 AISC Installed Fastener Tension Min lbf	3 UNYTITE Fastener Tension Min lbf
5/8"-11	19,000	19,950	23,000
3/4"-10	28,000	29,400	33,000
7/8"-9	39,000	40,950	44,000
1"-8	51,000	53,550	57,000
1-1/8"-7	56,000	58,800	65,000

ASTM A490 Mechanical Properties

	Bolt				Nut		Washer	
	ASTM A490 Type 1				A563 DH		F436	
		Tensile Strength		Hardness	Proof Load	Hardness	Hardness	
	Proof Load	Max	Min	Rockwell				
3/4"-10	40,100	56,800	50,100	HRC 33~38	58,450	HRC 24~38	HRC 38~45
7/8"-9	55,450	78,550	69,300		80,850		
1"-8	72,700	103,000	90,900		106,050		
1-1/8"-7	91,550	129,700	114,450		133,525		

A490 Fastener Tension

Nominal Diameter	1 AISC Table 4 Design Tension Min lbf	2 AISC Installed Fastener Tension Min lbf	3 UNYTITE Fastener Tension Min lbf
3/4"-10	35,000	36,750	39,900
7/8"-9	49,000	51,450	55,200
1"-8	64,000	67,200	74,000
1-1/8"-7	80,000	84,000	95,400

1 - AISC minimum design specification.
2 - AISC 8 (d) (3) minimum installed tension for alternate design fastener.
3 - UNYTITE Tension Control Bolt Designed Fastener Tension. (Average Fastener Tension of 5 samples per each assembly lot)

Section 9

Calculations to Determine the Effectiveness and Control of Thermal and Sound Transmission

9.0.1	Heat Transmission Modes	504
9.0.2	Definitions and Thermal Property Symbols	504
9.0.3	R- and U-Values—Measuring the Resistance of the Flow of Heat and the Measurement of Heat Conductivity	506
9.0.4	Basic Types of Insulation—Where Applicable—Installation Methods—Advantages	506
9.1.0	Sample R-Value of Materials	508
9.1.1	Sample R-Value Calculations for Masonry Wall Assemblies	509
9.1.2	R-Values for Blanket-Batt Insulation	510
9.1.3	Calculating the R-Values for Wall Assemblies	510
9.1.4	Properties of Solid Unit Masonry and Concrete Walls	513
9.1.5	Properties of Hollow Unit Masonry Walls	514
9.2.0	Exterior Brick and Block Cavity Wall R-Values	515
9.2.1	An Exterior Masonry Wall Assembly with a Total R-Value of 20.21	516
9.2.2	An Exterior Masonry Wall Assembly with a Total R-Value of 28.21	517
9.2.3	Concrete Block Walls Utilizing Perlite Cavity Fill as an Insulator	517
9.3.0	Effective R-Values on Wood-Metal Framing Assemblies	519
9.3.1	Metal Framing Factors	520
9.3.2	Standard Air Film R-Values	520
9.4.0	Framed Wall Assemblies—U-Factors for Size/Spacing of Wood-Metal Studs	521
9.5.0	Acoustics 101—Reflection, Absorption, Isolation—the Methods by Which Sound Can Be Identified, Measured, and Controlled	525
9.5.1	Reverberation Time Creating a Buildup of Noise	526
9.5.2	Isolation—Measured by Sound Transmission Class (STC)	527
9.5.3	Impact Insulation Class—IIC—Blocking Noise from Being Transmitted Floor-to-Floor	527
9.5.4	More Sound Absorption Factors for Building Materials and Finishes	528
9.5.5	Absorption Coefficients for Various Wall and Floor Coverings	530
9.6.0	Checklist for Masking Open Space Systems	530
9.6.1	Use of Loudspeakers and Formula for Installation Spacing	531
9.7.0	Decibel Levels of Some Common Sounds	532
9.7.1	ANSI Recommended Levels for Various Types of Occupancy	532
9.7.2	Decibel Comparison Chart of Environmental Noises	533
9.7.3	OSHA Daily Permissible Noise-Level Exposure	534
9.7.4	Perceptions of Increases in Noise Levels	534
9.7.5	Sound Levels of Music	535

9.8.0	Sound Transmission Coefficient (STC) of Various Types of Insulated Partitions	538	9.8.3	How Insulation, Staggered Studs, Resilient Channels, and Added Layers of Drywall Can Affect STC Ratings	542
9.8.1	Testing for STC of Residential Carpet over Joist and Wood Subfloor	540	9.9.0	Impact Insulation Class—II—What Is This?	543
9.8.2	STC Ratings for Masonry Walls	541			

9.0.1 Heat Transmission Modes

It is important to know how heat is transferred in fish holds. Heat is transferred by conduction, convection, or radiation, or by a combination of all three. Heat always moves from warmer to colder areas; it seeks a balance. If the interior of an insulated fish hold is colder than the outside air, the fish hold draws heat from the outside. The greater the temperature difference, the faster the heat flows to the colder area.

Conduction. By this mode, heat energy is passed through a solid, liquid, or gas from molecule to molecule in a material. In order for the heat to be conducted, there should be physical contact between particles and some temperature difference. Therefore, thermal conductivity is the measure of the speed of heat flow passed from particle to particle. The rate of heat flow through a specific material will be influenced by the difference of temperature and by its thermal conductivity.

Convection. By this mode, heat is transferred when a heated air/gas or liquid moves from one place to another, carrying its heat with it. The rate of heat flow will depend on the temperature of the moving gas or liquid and on its rate of flow.

Radiation. Heat energy is transmitted in the form of light, as infrared radiation or another form of electromagnetic waves. This energy emanates from a hot body and can travel freely only through completely transparent media. The atmosphere, glass, and translucent materials pass a significant amount of radiant heat, which can be absorbed when it falls on a surface (e.g., the ship's deck surface on a sunny day absorbs radiant heat and becomes hot). It is a well-known fact that light-colored or shiny surfaces reflect more radiant heat than black or dark surfaces; therefore the former will be heated more slowly.

In practice, the entry of heat into fish holds/fish containers is the result of a mixture of the three modes mentioned above, but the most significant mode is by conduction through walls and flooring.

Source: Food and Agriculture Organization of the United Nations.

9.0.2 Definitions and Thermal Property Symbols

The thermal properties of insulating materials and other common fishing vessel construction materials are known or can be accurately measured. The amount of heat transmission (flow) through any combination of materials can be calculated. However, it is necessary to know and understand certain technical terms to be able to calculate heat losses and understand the factors that are involved.

By convention, the ending -ity means the property of a material, regardless of its thickness, and the ending -ance refers to the property of a specific body of given thickness.

Heat Energy

One kilocalorie (1 kcal or 1000 calories) is the amount of heat (energy) needed to raise the temperature of one kg of water by one degree Celsius (°C). The SI standard unit for energy is Joule (J). One kcal is approximately 4.18 kJ (this varies slightly with temperature). Another unit is the Btu (British thermal unit). One Btu corresponds roughly to 1 kJ.

Thermal Conductivity

In simple terms this is a measure of the capacity of a material to conduct heat through its mass. Different insulating materials and other types of material have specific thermal conductivity values that can be used to measure their insulating effectiveness. It can be defined as the amount of heat/energy (expressed in kcal, Btu or J) that can be conducted in unit time through unit area of unit thickness of material, when there is a unit temperature difference. Thermal conductivity can be expressed in kcal m^{-1} °C^{-1}, Btu ft^{-1} °F^{-1}, and in the SI system in watt (W) m^{-1} °C^{-1}. Thermal conductivity is also known as the k-value.

Coefficient of Thermal Conductance "l" (kcal m^{-2} h^{-1} °C^{-1})

This is designated as I (the Greek letter lambda) and defined as the amount of heat (in kcal) conducted in one hour through 1 m^2 of material, with a thickness of 1 m, when the temperature drop through the material under conditions of steady heat flow is 1 °C. The thermal conductance is established by tests and is the basic rating for any material. I can also be expressed in Btu ft^{-2} h^{-1} °F^{-1} (British thermal unit per square foot, hour, and degree Fahrenheit) or in SI units in W m^{-2} Kelvin (K)$^{-1}$.

Thermal Resistivity

The thermal resistivity is the reciprocal of the k-value (1/k).

Thermal Resistance (R-value)

The thermal resistance (R-value) is the reciprocal of l (1/l) and is used for calculating the thermal resistance of any material or composite material. The R-value can be defined in simple terms as the resistance that any specific material offers to the heat flow. A good insulation material will have a high R-value. For thicknesses other than 1 m, the R-value increases in direct proportion to the increase in thickness of the insulation material. This is x/l, where x stands for the thickness of the material in meters.

Coefficient of Heat Transmission (U) (kcal m^{-2} h^{-1} °C^{-1})

The symbol U designates the overall coefficient of heat transmission for any section of a material or a composite of materials. The SI units for U are kcal per square meter of section per hour per degree Celsius, the difference between inside air temperature and outside air temperature. It can also be expressed in other unit systems. The U coefficient includes the thermal resistances of both surfaces of walls or flooring, as well as the thermal resistance of individual layers and air spaces that may be contained within the wall or flooring itself.

Permeance to Water Vapor (pv)

This is defined as the quantity of water vapor that passes through the unit of area of a material of unit thickness, when the difference of water pressure between both faces of the material is the unit. It can be expressed as g cm mmHg^{-1} m^{-2} day^{-1} or in the SI system as g m MN^{-1} s^{-1} (grams meter per mega Newton per second).

Resistance to Water Vapor (rv)

This is the reciprocal of the permeance to water vapor and is defined as rv = 1/pv.
   Source: Food and Agriculture Organization of the United Nations.

9.0.3 R- and U-Values—Measuring the Resistance to the Flow of Heat and the Measure of Heat Conductivity

A measure of the resistance of building materials and structures to the flow of heat; the higher the R-value, the better the substance is as thermal insulation.

$$R - \text{value} = \frac{\text{temperature difference} \times \text{area} \times \text{time}}{\text{heat loss}}$$

where the temperature difference is in degrees Fahrenheit, the area is in square feet, the time is in hours, and the heat loss is in Btus. If you know the R-value of a partition, you can use this formula to find the heat loss.

Relation to U-value

The reciprocal of the R-value (1/R) is known as the U-value. The higher the U-value, the better the conduction of heat.

In Europe it is customary to use U-values instead of R-values. There, U-values are defined by the equation:

$$U - \text{value} = \frac{\text{watts}}{\text{kelvin} \times \text{meters}^2}$$

This is not the reciprocal of the American R-value (kelvin instead of degrees Fahrenheit, meters instead of feet, etc.). To convert an American R-value into a European U-value, divide 1 by the R-value, then multiply the result by 5.682. To convert a European U-value to an American R-value, multiply by 0.176, then divide 1 by the result.

9.0.4 Basic Types of Insulation—Where Applicable—Installation Methods—Advantages

Form	Insulation Materials	Where Applicable	Installation Method(s)	Advantages
Blanket: batts and rolls	Fiberglass Mineral (rock or slag) wool plastic fibers Natural fibers	Unfinished walls, including foundation walls, and floors and ceilings.	Fitted between studs, joists, and beams.	Do-it-yourself. Suited for standard stud and joist spacing, which is relatively free from obstructions.
Concrete block insulation	Foam beads or liquid foam: • Polystyrene • Polyisocyanurate or polyiso • Polyurethane Vermiculite or perlite pellets	Unfinished walls, including foundation walls, for new construction or major renovations.	Involves masonry skills.	Autoclaved aerated concrete and autoclaved cellular concrete masonry units have 10 times the insulating value of conventional concrete.

Form	Insulation Materials	Where Applicable	Installation Method(s)	Advantages
Foam board or rigid foam	Polystyrene Polyisocyanurate or polyiso Polyurethane	Unfinished walls, including foundation walls; floors and ceilings; unvented low-slope roofs.	Interior applications: must be covered with 1/2-inch gypsum board or other building-code approved material for fire safety. Exterior applications: must be covered with weatherproof facing.	High insulating value for relatively little thickness. Can block thermal short circuits when installed continously over frames or joists.
Insulating concrete forms (ICFS)	Foam boards or form blocks	Unfinished walls, including foundation walls, for new construction.	Installed as part of the building structure.	Insulation is literally built into the home's walls, creating high thermal resistance.
Loose-fill	Cellulose Fiberglass Mineral (rock or slag) wool	Enclosed existing wall or open new wall cavities; unfinished attic floors; hard-to-reach places.	Blown into place using special equipment; sometimes poured in.	Good for adding insulation to existing finished areas, irregularly shaped areas, and around obstructions.
Reflective system	Foil-faced kraft paper, plastic film, polyethylene bubbles, or cardboard	Unfinished walls, ceilings, and floors.	Foils, films, or papers: fitted between wood-frame studs, joists, and beams	Do-it-yourself. All suitable for framing at standard spacing. Bubble-form suitable if framing is irregular or if obstructions are present. Most effective at preventing downward heat flow; however, effectiveness depends on spacing.
Rigid fibrous or fiber insulation	Fiberglass Mineral (rock or slag) wool	Ducts in unconditioned spaces and other places requiring insulation that can withstand high temperatures.	HVAC contractors fabricate the insulation into ducts either at their shops or at the job sites.	Can withstand high temperatures.
Sprayed foam and foamed-in-place	Cementitious phenolic Polyisocyanurate Polyurethane	Enclosed existing wall or open new wall cavities; unfinished attic floors.	Applied using small spray containers or in larger quantities as a pressure sprayed (foamed-in-place) product.	Good for adding insulation to existing finished areas, irregularly shaped areas, and around obstructions.
Structural insulated panels (SIPs)	Foam board or liquid foam insulation core Straw core insulation	Unfinished walls, ceilings, floors, and roofs for new construction.	Builders connect them together to construct a house.	SIP-built houses provide superior and uniform insulation compared to more traditional construction methods; they also take less time to build.

Source: U.S. Department of Energy.

9.1.0 Sample R-Value of Materials

Properties of Materials

Material	R-Value Per Inch of Thickness	R-Value for Thickness Listed
4" Clay Brick		0.44
4" Block (115#/ft^3) = 72% solid		1.19
6" Block (115#/ft^3) = 59% solid		1.25
8" Block (115#/ft^3) = 54% solid		1.45
10" Block (115#/ft^2) = 52% solid		1.55
12" Block (115#/ft^2) = 48% solid		1.65
6" Block (115#/ft^2) = 59% solid/filled with perlite		3.95
8" Block (115#/ft^2) = 54% solid/filled with perlite		4.65
10" Block (115#/ft$^{2)}$ = 52% solid/filled with perlite		5.65
12" Block (115#/ft$^{2)}$ = 48% solid/filled with perlite		7.05
1" Polyisocyanurate	8.0	
1" Extruded polystyrene	5.0	
1" Expanded polystyrene	4.0	
1" of Perlite	2.70	
Exterior air film (winter)		0.17
Interior air film		0.68
Dead air space (3/4" to 4") (winter)		0.97
3/4" reflective air space		2.89
1/2" dry wall		0.45
3 1/2" Batt (R - 11)		11.00
3 5/8" (R - 13)		13.00
1 1/2" (R - 5)		5.00
6" Batt (R - 19)		19.00
6 1/2" Batt (R - 22)		22.00
9" Batt (R - 30)		30.00
12" Batt (R - 38)		38.00

Source:www.maconline.org

9.1.1 Sample R-Value Calculations for Masonry Wall Assemblies

Sample R-Value Calculations

Brick Veneer on Wood Frame (residential and single family usage)

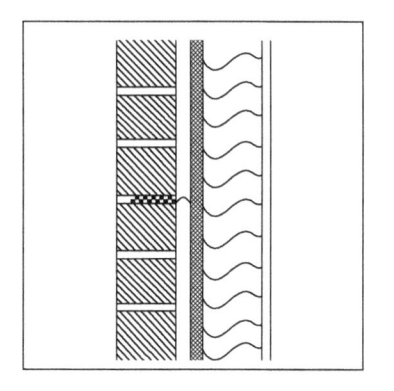

R of the outside air film	0.17
R of a 4″ brick	0.44
R of 1″ reflective air space	2.89
R of 3/4″ polyisocyanurate	5.60
R of 3 1/2″ batt insulation	11.00
R of 1/2″ drywall	0.45
R of the inside air film	0.68
R of the total wall	21.23
U of the wall	0.047

Solid Loadbearing Masonry Wall (midrise and multifamily usage)

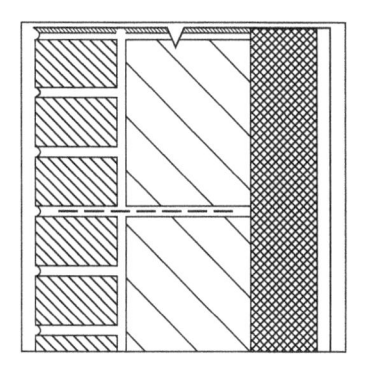

R of the outside air film	0.17
R of a 4″ brick	0.44
R of 6″ block	1.25
R of 3″ expanded polystyrene	12.00
R of 1/2″ drywall	0.45
R of the inside air film	0.68
R of the total wall	14.99
U of the wall	0.066

Brick and Block Cavity Wall (Quality construction for schools, commercial/industrial multifamily and high-rises)

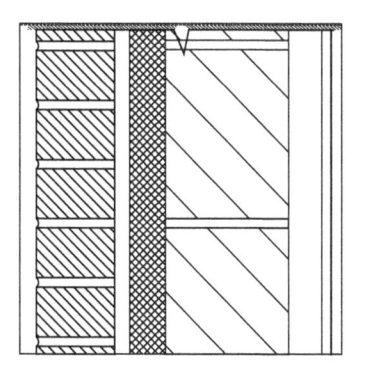

If you were designing the wall shown on the left—a wall composed of a 3 5/8″ brick, 3/4″ air space, an unknown thickness of an unspecified Type of Rigid insulation, a 5 5/8″ block, 1 1/2″ furring for 1/2″ plaster dry-wall—what insulation would you select? The insulation that gives you the best dollar value for the R-value desired.

9.1.2 R-Values for Blanket-Batt Insulation

Blanket (Batt and Roll) Insulation

Blanket insulation—the most common and widely available type of insulation—comes in the form of batts or rolls. It consists of flexible fibers, most commonly fiberglass. You also can find batts and rolls made from mineral (rock and slag) wool, plastic fibers, and natural fibers, such as cotton and sheep's wool.

Batts and rolls are available in widths suited to standard spacing of wall studs and attic or floor joists. Continuous rolls can be hand-cut and trimmed to fit. They are available with or without facings. Manufacturers often attach a facing (such as kraft paper, foil-kraft paper, or vinyl) to act as a vapor barrier and/or air barrier. Batts with a special flame-resistant facing are available in various widths for basement walls where the insulation will be left exposed. A facing also helps facilitate fastening during installation. However, it's recommended that you use unfaced batts if you're reinsulating over existing insulation.

Standard fiberglass blankets and batts have a thermal resistance or R-values between R-2.9 and R-3.8 per inch of thickness. High-performance (medium-density and high-density) fiberglass blankets and batts have R-values between R-3.7 and R-4.3 per inch of thickness. See the following table for an overview of these characteristics.

TABLE 1 Fiberglass Batt Insulation Characteristics*

Thickness (inches)	R-Value	Cost (cents/sq. ft.)
3 1/2	11	12–16
3 5/8	13	15–20
3 1/2 (high density)	15	34–40
6 to 6 1/4	19	27–34
5 1/4 (high density)	21	33–39
8 to 8 1/2	25	37–45
8 (high density)	30	45–49
9 1/2 (standard) j	30	39–43
12	30	55–60

*This table is for comparison only. Determine actual thickness, R-value, and cost from manufacturer and/or local building supplier.

9.1.3 Calculating the R-Values for Wall Assemblies

Use the following R-value table to help you determine the R-value of your wall or ceiling assemblies. To obtain a wall or ceiling assembly R-value, you must add the R-values of the individual components together. See the following example:

Calculating Assembly Wall R-Value*

Component	R-Value Studs	R-Value Cavity	Assembly R-Value
Wall—Outside Air Film	0.17	0.17	
Siding—Wood Bevel	0.80	0.80	
Plywood Sheathing—1/2"	0.63	0.63	
3 1/2" Fiberglass Batt		11.00	
3 1/2" Stud	4.38		
1/2" Drywall	0.45	0.45	

Component	R-Value Studs	R-Value Cavity	Assembly R-Value
Inside Air Film	0.68	0.68	
Percent for 16" o.c. + Additional studs	15%	85%	
Total Wall Component R-Values	7.12	13.73	
Wall Component U-Values	0.1404	0.0728	
Total Wall Assembly R-Value			12.05

Formula: Assembly R-value = 1 / (Assembly U-value) = 1 / (U-studs x % + U-cavity x %)
*This example is just for wood frame construction. Steel studs are a more complicated calculation.

R-Value Table

Material	R/Inch	R/Thickness
Insulation Materials		
Fiberglass Batt	3.14-4.30	
Fiberglass Blown (attic)	2.20-4.30	
Fiberglass Blown (wall)	3.70-4.30	
Rock Wool Batt	3.14-4.00	
Rock Wool Blown (attic)	3.10-4.00	
Rock Wool Blown (wall)	3.10-4.00	
Cellulose Blown (attic)	3.13	
Cellulose Blown (wall)	3.70	
Vermiculite	2.13	
Autoclaved Aerated Concrete	1.05	
Urea Terpolymer Foam	4.48	
Rigid Fiberglass (> 4lb/ft3)	4.00	
Expanded Polystyrene (bead board)	4.00	
Extruded Polystyrene	5.00	
Polyurethane (foamed-in-place)	6.25	
Polyisocyanurate (foil-faced)	7.20	
Construction Materials		
Concrete Block 4"		0.80
Concrete Block 8"		1.11
Concrete Block 12"		1.28
Brick 4" Common		1.80
Brick 4" Face		0.44
Poured Concrete	0.08	
Soft Wood Lumber	1.25	
2" nominal (1 1/2")		1.88
2×4 (3 1/2")		4.38
2×6 (5 1/2")		6.88
Cedar Logs and Lumber	1.33	
Sheathing Materials		
Plywood	1.25	
1/4"		0.31
3/8"		0.47
1/2"		0.63
5/8"		0.77

(Continued)

Material	R/Inch	R/Thickness
3/4"		0.94
Fiberboard	2.64	
1/2"		1.32
25/32"		2.06
Fiberglass (3/4")		3.00
(1")		4.00
(1 1/2")		6.00
Extruded Polystyrene (3/4")		3.75
(1")		5.00
(1 1/2")		7.50
Foil-faced Polyisocyanurate (3/4")		5.40
(1")		7.20
(1 1/2")		10.80
Siding Materials		
Hardboard (1/2")		0.34
Plywood (5/8")		0.77
(3/4")		0.93
Wood Bevel Lapped		0.80
Aluminum, Steel, Vinyl (hollow backed)		0.61
(w/ 1/2" Insulating board)		1.80
Brick 4"		0.44
Interior Finish Materials		
Gypsum Board (drywall 1/2")		0.45
(5/8")		0.56
Paneling (3/8")		0.47
Flooring Materials		
Plywood	1.25	
(3/4")		0.93
Particle Board (underlayment)	1.31	
(5/8")		0.82
Hardwood Flooring	0.91	
(3/4")		0.68
Tile, Linoleum		0.05
Carpet (fibrous pad)		2.08
(rubber pad)		1.23
Roofing Materials		
Asphalt Shingles		0.44
Wood Shingles		0.97
Windows		
Single Glass		0.91
w/storm		2.00
Double insulating glass (3/16") air space		1.61
(1/4" air space)		1.69
(1/2" air space)		2.04
(3/4" air space)		2.38
(1/2" w/ Low-E 0.20)		3.13
(w/ suspended film)		2.77
(w/ 2 suspended films)		3.85
(w/ suspended film and low-E)		4.05
Triple insulating glass		2.56
(1/4" air spaces)		
(1/2" air spaces)		3.23
Addition for tight fitting drapes or shades, or closed blinds		0.29

Material	R/Inch	R/Thickness
Doors		
Wood Hollow Core Flush (1 3/4")		2.17
Solid Core Flush (1 3/4")		3.03
Solid Core Flush (2 1/4")		3.70
Panel Door w/ 7/16" Panels (1 3/4")		1.85
Storm Door (wood 50% glass)		1.25
(metal)		1.00
Metal Insulating (2" w/ urethane)		15.00
Air Films		
Interior Ceiling		0.61
Interior Wall		0.68
Exterior		0.17
Air Spaces		
1/2" to 4" approximately		1.00

By permission: www.coloradoenergy.org

9.1.4 Properties of Solid Unit Masonry and Concrete Walls

Type		\multicolumn{10}{c}{Layer Thickness, inches}									
		3	4	5	6	7	8	9	10	11	12
LW CMU	U	na	0.71	0.64	Na	na	na	na	na	na	na
	Rw	na	1.4	1.6	Na	na	na	na	na	na	na
	HC	na	7.00	8.75	Na	na	na	na	na	na	na
MW CMU	U	na	0.76	0.70	Na	na	na	na	na	na	na
	Rw	na	1.3	1.4	Na	na	na	na	na	na	na
	HC	na	7.67	9.58	Na	na	na	na	na	na	na
NW CMU	U	0.89	0.82	0.76	Na	na	na	na	na	na	na
	Rw	1.1	1.2	1.3	Na	na	na	na	na	na	na
	HC	6.25	8.33	10.42	Na	na	na	na	na	na	na
Clay Brick	U	0.80	0.72	0.66	Na	na	na	na	na	na	na
	Rw	1.3	1.4	1.5	Na	na	na	na	na	na	na
	HC	6.30	8.40	10.43	Na	na	na	na	na	na	na
Concrete	U	0.96	0.91	0.86	0.82	0.78	0.74	0.71	0.68	0.65	0.63
	Rw	1.0	1.1	1.2	1.2	1.3	1.4	1.4	1.5	1.5	1.6
	HC	7.20	9.60	12.00	14.40	16.80	19.20	21.60	24.00	26.40	28.80

Notes:
LW CMU is a Light Weight Concrete Masonry Unit per ASTM C 90 or 55, Calculated at 105 PCF density
MW CMU is a Medium Weight Concrete Masonry Unit per ASTM C 90 or 55, Calculated at 115 PCF density
NW CMU is a Normal Weight Concrete Masonry Unit per ASTM C 90 or 55, Calculated at 125 PCF density
Clay Brick is a Clay Unit per ASTM C 62, Calculated at 130 PCF density
Calculations based on Energy Calculations and Data, CMACN, 1986.
Values include air films on Inner and outer surfaces.
Source: Berkeley Solar Group; Concrete Masonry Association of California and Nevada.

9.1.5 Properties of Hollow Unit Masonry Walls

	Type		Solid Grout	Core Treatment	
				Partly Grouted with Ungrouted Cells	
				Empty	Insulated
12"	LW CMU	U	0.51	0.43	0.30
		Rw	2.0	2.3	3.3
		HC	23	14.8	14.8
	MW CMU	U	0.54	0.46	0.33
		Rw	1.9	2.2	3.0
		HC	23.9	15.6	15.6
	NW CMU	U	0.57	0.49	0.36
		Rw	1.8	2.0	2.8
		HC	24.8	16.5	16.5
10"	LW CMU	U	0.55	0.46	0.34
		Rw	1.8	2.2	2.9
		HC	18.9	12.6	12.6
	MW CMU	U	0.59	0.49	0.37
		Rw	1.7	2.1	2.7
		HC	19.7	13.4	13.4
	NW CMU	U	0.62	0.52	0.14
		Rw	1.6	1.9	2.4
		HC	20.5	14.2	14.2
8"	LW CMU	U	0.62	0.50	0.37
		Rw	1.6	2.0	2.7
		HC	15.1	9.9	9.9
	MW CMU	U	0.65	0.53	0.41
		Rw	1.5	1.9	2.4
		HC	15.7	10.5	10.5
	NW CMU	U	0.69	0.56	0.44
		Rw	1.4	1.8	2.3
		HC	16.3	11.1	11.1
	Clay Unit	U	0.57	0.47	0.39
		Rw	1.8	2.1	2.6
		HC	15.1	11.4	11.4
6"	LW CMU	U	0.68	0.54	0.44
		Rw	1.5	1.9	2.3
		HC	10.9	7.9	7.9
	MW CMU	U	0.72	0.58	0.48
		Rw	1.4	1.7	2.1
		HC	11.4	8.4	8.4

Section | 9 Calculations to Determine the Effectiveness and Control of Thermal and Sound Transmission 515

		Core Treatment		
Type		Solid Grout	Partly Grouted with Ungrouted Cells	
			Empty	*Insulated*
NW CMU	U	0.76	0.61	0.52
	Rw	1.3	1.6	1.9
	HC	11.9	8.9	8.9
Clay Unit	U	0.65	0.52	0.45
	Rw	1.5	1.9	2.2
	HC	11.1	8.6	8.6

Notes:
LW CMU is a Light Weight Concrete Masonry Unit per ASTM C 90, Calculated at 105 PCF density
MW CMU is a Medium Weight Concrete Masonry Unit per ASTM C 90, Calculated at 115 PCF density
NW CMU is a Normal Weight Concrete Masonry Unit per ASTM C 90, Calculated at 125 PCF density
Clay Unit is a Hollow Clay Unit per ASTM C 652, Calculated at 130 PCF density
Values include air films on inner and outer surfaces.
Calculations based on Energy Calculations and Data, CMACN, 1986.
Grouted Cells at 32" × 48" in Partly Grouted Walls
Source: Berkeley Solar Group; Concrete Masonry Association of California and Nevada.

9.2.0 Exterior Brick and Block Cavity Wall R-Values

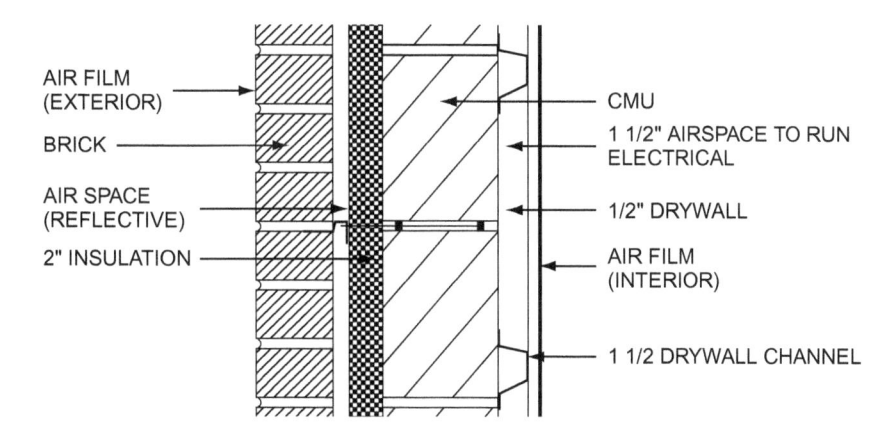

4" Clay Brick	0.44
4" Block (115#.ft^3) = 72% solid	1.19
6" Block (115#.ft^3) = 59% solid	1.34
8" Block (115#.ft^3) = 54% solid	1.51
10" Block (115#.ft^3) = 52% solid	1.61
12" Block (115#.ft^3) = 48% solid	1.72
1/2" Drywall	0.45
Exterior air film (winter)	0.17

(Continued)

Interior air film	0.68					
Dead air space (3/4″ to 4″) (winter)	0.97					
* Reflective air space	2.8					
Insulation type thickness (inches)	1/2″	1″	1 1/2″	2″	2 1/2″	3″
Polyisocyanurate (foil face) Dow Tuff R™ /Thermax™	4.0	8	12	16	18	21.6
Extruded Polystyrene Dow, Owens Corning	-	5	7.5	10	12.5	15

*Use this value when insulation has a foil backing directly adjacent to air space.

Exterior air film	0.17
4″ Brick	0.44
R of the reflective air space	2.80
R of 2″ Dow, Tuff R-C, polyisocyanurate insulation	16.0
6″ CMU	1.34
1 1/2″ air space	0.97
1 /2″ Drywall	0.45
Interior Air Film	0.68

Source: Berkeley Solar Group, Concrete Masonsry Association of California and Nevada.

9.2.1 An Exterior Masonry Wall Assembly with a Total R-Value of 20.21

BRICK VENEER OVER 4″ WOOD FRAME CONSTRUCTION

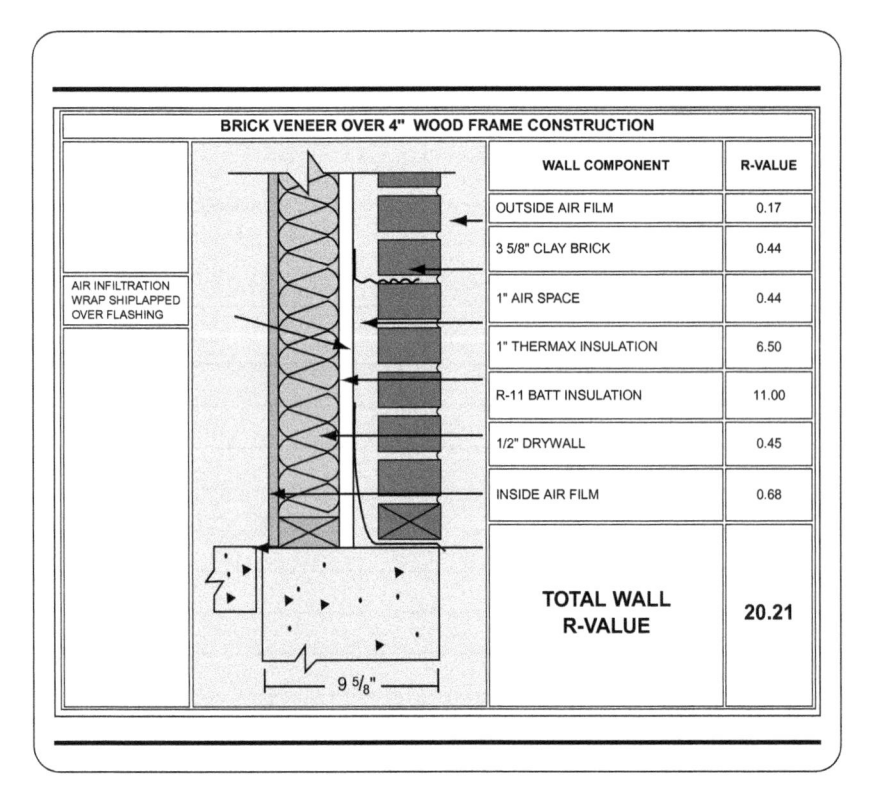

AIR INFILTRATION WRAP SHIPLAPPED OVER FLASHING

WALL COMPONENT	R-VALUE
OUTSIDE AIR FILM	0.17
3 5/8″ CLAY BRICK	0.44
1″ AIR SPACE	0.44
1″ THERMAX INSULATION	6.50
R-11 BATT INSULATION	11.00
1/2″ DRYWALL	0.45
INSIDE AIR FILM	0.68
TOTAL WALL R-VALUE	**20.21**

9 5/8″

9.2.2 An Exterior Masonry Wall Assembly with a Total R-Value of 28.21

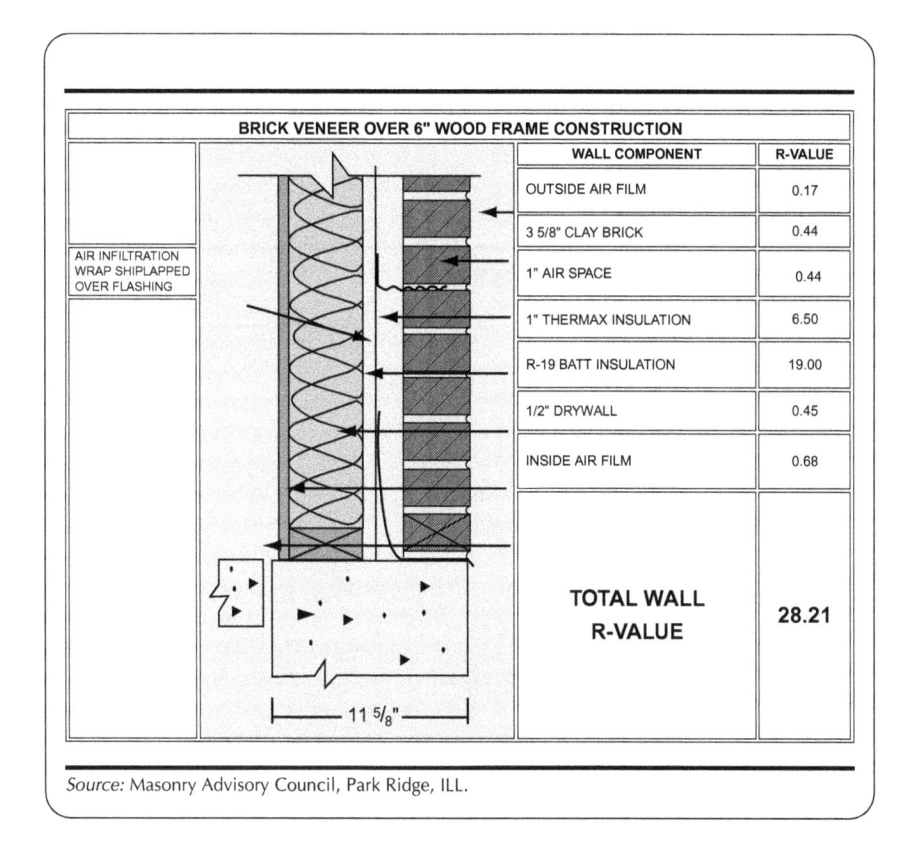

BRICK VENEER OVER 6" WOOD FRAME CONSTRUCTION	
WALL COMPONENT	R-VALUE
OUTSIDE AIR FILM	0.17
3 5/8" CLAY BRICK	0.44
1" AIR SPACE	0.44
1" THERMAX INSULATION	6.50
R-19 BATT INSULATION	19.00
1/2" DRYWALL	0.45
INSIDE AIR FILM	0.68
TOTAL WALL R-VALUE	28.21

AIR INFILTRATION WRAP SHIPLAPPED OVER FLASHING

11 5/8"

Source: Masonry Advisory Council, Park Ridge, ILL.

9.2.3 Concrete Block Walls Utilizing Perlite Cavity Fill as an Insulator

Insulation is essential in all construction for energy conservation. The original cost of installing perlite loose fill insulation can be recovered quickly due to substantial reductions in heating and air condition energy consumption. In addition, perlite loose fill insulation cuts installation costs since it is lightweight and pours easily and quickly in place without need for special installation equipment or skills. The insulation may be poured directly into walls or emptied into a simply wood or metal hopper, which can be slid along the wall to direct the free-flowing perlite into cores of cavities, thus insulating all voids and air pockets.

Concrete Block Walls—U-Values and R-Values

		Block Only			
		Uninsulated		Insulated	
Block Thickness in Inches	Density of Block	R	U	R	U
6	80 pcf Lightweight	2.64	.38	6.75	.15
	125 pcf Sand & Gravel	2.05	.49	3.86	.26
8	80 pcf Lightweight	2.86	.35	9.07	.11
	125 pcf Sand & Gravel	2.21	.45	5.06	.20

(Continued)

		Block Only			
Block Thickness in Inches	Density of Block	Uninsulated		Insulated	
		R	U	R	U
10	80 pcf Lightweight	3.00	.33	11.02	.09
	125 pcf Sand & Gravel	2.31	.45	5.95	.17
12	80 pcf Lightweight	3.12	.32	13.44	.08
	125 pcf Sand & Gravel	2.38	.42	7.17	.15

Values given are approximate and include the effect of inside and outside air film resistances.
Source: The Schundler Company, Metuchen, NJ.

Installation

a. The insulation must be installed in the following locations:
   - In the cores of all exterior (and interior) hollow masonry units.
   - In the cavity between all exterior (and interior) masonry walls.
   - Between exterior masonry walls and interior furring.
b. The insulation shall be poured directly into the wall at any convenient internal. Wall sections under doors and windows shall be filled before sills are placed.
c. All holes and openings in the wall through which insulation can escape shall be permanently sealed or caulked prior to installation of the insulation. Copper, galvanized steel, or fiberglass screening shall be used in all weep holes.

(The inclusion of weep holes is considered good construction design practice to allow passage of any water that might penetrate the cavities or core spaces of wall construction.)

Approximate Coverage[a]

Sq. Ft. of Wall Area[b]	1" Cavity	2" Cavity	3" Cavity	6" Block	8" Block	12" Block
100	2	4	6	5	7	12
500	10	21	31	23	33	58
1,000	21	42	62	46	65	118
2,000	42	84	124	96	130	236
3,000	63	126	186	138	195	354
5,000	105	210	310	230	325	590
7,000	147	294	434	322	455	826
10,000	210	420	620	460	650	1,180

[a] 4-cubic foot bags required to fill.
[b] A standard 8" × 16" block equals 0.89 sq. ft.
Multiply the number of blocks times 0.89 to calculate the total square footage needed.

9.3.0 Effective R-Values on Wood-Metal Framing Assemblies

Type Actual Thick	Frame	\multicolumn{22}{c}{Furring Space R-Value without Framing Effects}
		0	1	2	3	4	5	6	7	8	9	10	11	12	13	14	15	16	17	18	19	20	21
Any	None	0.5	1.5	2.5	3.5	4.5	5.5	6.5	7.5	8.5	9.5	10	11.5	12.5	13.5	14.5	15.5	16.5	17.5	18.5	19.5	20.5	21.5
0.5"	Wood	1.3	1.3	1.9	2.4	2.7	na	na	na	na	na	na	na	na	na	na	na	na	na	na	na	na	na
	Metal	0.9	0.9	1.1	1.1	1.2	na	na	na	na	na	na	na	na	na	na	na	na	na	na	na	na	na
0.75"	Wood	1.4	1.4	2.1	2.7	3.1	3.5	3.8	na	na	na	na	na	na	na	na	na	na	na	na	na	na	na
	Metal	1.0	1.0	1.3	1.4	1.5	1.5	1.6	na	na	na	na	na	na	na	na	na	na	na	na	na	na	na
1.0"	Wood	1.3	1.5	2.2	2.9	3.4	3.9	4.3	4.6	4.9	na	na	na	na	na	na	na	na	na	na	na	na	na
	Metal	1.0	1.1	1.4	1.6	1.7	1.8	1.8	1.9	1.9	na	na	na	na	na	na	na	na	na	na	na	na	na
1.5"	Wood	1.3	1.5	2.4	3.1	3.8	4.4	4.9	5.4	5.8	6.2	6.5	6.8	7.1	na	na	na	na	na	na	na	na	na
	Metal	1.1	1.2	1.6	1.9	2.1	2.2	2.3	2.4	2.5	2.5	2.6	2.6	2.7	na	na	na	na	na	na	na	na	na
2"	Wood	1.4	1.5	2.5	3.3	4.0	4.7	5.3	5.9	6.4	6.9	7.3	7.7	8.1	8.4	8.7	9.0	9.3	na	na	na	na	na
	Metal	1.1	1.2	1.7	2.1	2.3	2.5	2.7	2.8	2.9	3.0	3.1	3.2	3.2	3.3	3.3	3.4	3.4	na	na	na	na	na
2.5"	Wood	1.4	1.5	2.5	3.4	4.2	4.9	5.6	6.3	6.8	7.4	7.9	8.4	8.8	9.2	9.6	10.0	10.3	10.6	10.9	11.2	11.5	na
	Metal	1.2	1.3	1.8	2.3	2.6	2.8	3.0	3.2	3.3	3.5	3.6	3.6	3.7	3.8	3.9	3.9	4.0	4.0	4.1	4.1	4.1	na
3"	Wood	1.4	1.5	2.5	3.5	4.3	5.1	5.8	6.5	7.2	7.8	8.3	8.9	9.4	9.9	10.3	10.7	11.1	11.5	11.9	12.2	12.5	12.9
	Metal	1.2	1.3	1.9	2.4	2.8	3.1	3.3	3.5	3.7	3.8	4.0	4.1	4.2	4.3	4.4	4.4	4.5	4.6	4.6	4.7	4.7	4.8
3.5"	Wood	1.4	1.5	2.6	3.5	4.4	5.2	6.0	6.7	7.4	8.1	8.7	9.3	9.8	10.4	10.9	11.3	11.8	12.2	12.6	13.0	13.4	13.8
	Metal	1.2	1.3	2.0	2.5	2.9	3.2	3.5	3.8	4.0	4.2	4.3	4.5	4.6	4.7	4.8	4.9	5.0	5.1	5.1	5.2	5.2	5.3
4"	Wood	1.4	1.6	2.6	3.6	4.5	5.3	6.1	6.9	7.6	8.3	9.0	9.6	10.2	10.8	11.3	11.9	12.4	12.8	13.3	13.7	14.2	14.6
	Metal	1.2	1.3	2.0	2.6	3.0	3.4	3.7	4.0	4.2	4.5	4.6	4.8	5.0	5.1	5.2	5.3	5.4	5.5	5.6	5.7	5.8	5.8
4.5"	Wood	1.4	1.6	2.6	3.6	4.5	5.4	6.2	7.1	7.8	8.5	9.2	9.9	10.5	11.2	11.7	12.3	12.8	13.3	13.8	14.3	14.8	15.2
	Metal	1.2	1.3	2.1	2.6	3.1	3.5	3.9	4.2	4.5	4.7	4.9	5.1	5.3	5.4	5.6	5.7	5.8	5.9	6.0	6.1	6.2	6.3
5"	Wood	1.4	1.6	2.6	3.6	4.6	5.5	6.3	7.2	8	8.7	9.4	10.1	10.8	11.5	12.1	12.7	13.2	13.8	14.3	14.8	15.3	15.8
	Metal	1.2	1.4	2.1	2.7	3.2	3.7	4.1	4.4	4.7	5.0	5.2	5.4	5.6	5.8	5.9	6.1	6.2	6.3	6.5	6.6	6.7	6.8
5.5"	Wood	1.4	1.6	2.6	3.6	4.6	5.5	6.4	7.3	8.1	8.9	9.6	10.3	11.0	11.7	12.4	13.0	13.6	14.2	14.7	15.3	15.8	16.3
	Metal	1.3	1.4	2.1	2.8	3.3	3.8	4.2	4.6	4.9	5.2	5.4	5.7	5.9	6.1	6.3	6.4	6.6	6.7	6.8	7.0	7.1	7.2

All furring thickness values given are actual dimensions.
All values include 0.5" gypbd on the inner surface; interior surface resistances not included.

24" OC Furring   24 Gage, Z-type Metal Furring   Douglas-Fir Larch Wood Furring, density = 34.9 lb/cu. ft.
Insulation assumed to fill the furring space.
Source: Berkeley Solar Group; Concrete Masonry Association of California and Nevada.

9.3.1 Metal Framing Factors

Metal Framing Factors

Stud Spacing	Stud Depth	Insulation R-Value	Framing Factor
16" O.C.	4"	R-7	0.522
		R-11	0.403
		R-13	0.362
		R-15	0.328
	6"	R-19	0.325
		R-21	0.300
		R-22	0.287
		R-25	0.263
24" O.C.	4"	R-7	0.577
		R-11	0.458
		R-13	0.415
		R-15	0.379
	6"	R-19	0.375
		R-21	0.348
		R-22	0.335
		R-25	0.308

9.3.2 Standard Air Film R-Values

Air Films[1]

	Wall	Roof Flat[2]	Roof 45° Angle[3]	Floor
Inside	0.68	0.61	0.62	0.92
Outside	0.17	0.17	0.17	0.17
Air Spaces[4]				
0.5 inch	0.77	0.73	0.86	0.77
0.75 inch	0.84	0.75	0.81	0.85
1.5 inch	0.87	0.77	0.80	0.94
3.5 inch[5]	0.85	0.80	0.82	1.00

NOTE: Values from ASHRAE Handbook of Fundamentals, 1993 edition, Chapter 22, Tables 1 & 2.
[1] Assumes a nonreflective surface emittance of 0.90 and winter heat flow direction.
[2] Use the "Flat" roof R-values for roof angles between horizontal and 22 degrees.
[3] Use the 45 degree roof R-values for roof angles between 23 and 60 degrees.
[4] Assumes mean temperature of 90 degrees Fahrenheit, temperature difference of 10 degrees Fahrenheit, surface emittance of 0.82, and winter heat flow direction.
[5] Use these R-values for air spaces greater than or equal to 3.5 inches, such as attics.
Source: Berkeley Solar Group, Concrete Masonry Association of California and Nevada.

9.4.0 Framed Wall Assemblies—U-factors for Size/Spacing of Wood-Metal Studs

Framing Type and Spacing	Framing Cavity R-Value	Insulated Sheathing R-Value	Wood Wall U-Factor	Metal Wall U-Factor
2×4 @ 16" O.C	11 (compressed)	0	0.098	0.202
		4	0.068	0.112
		5	0.064	0.101
		7	0.056	0.084
		8.7	0.051	0.073
	13	0	0.088	0.195
		4	0.063	0.109
		5	0.059	0.099
		7	0.052	0.082
		8.7	0.048	0.072
	15	0	0.081	0.189
		4	0.059	0.108
		5	0.055	0.097
		7	0.049	0.077
		8.7	0.045	0.071
2×4 @ 24" O.C	11	0	0.094	0.173
		4	0.066	0.102
		5	0.062	0.093
		7	0.055	0.078
		8.7	0.050	0.069
	13	0	0.085	0.165
		4	0.061	0.099
		5	0.057	0.090
		7	0.051	0.077
		8.7	0.047	0.068
	15	0	0.077	0.158
		4	0.056	0.097
		5	0.053	0.088
		7	0.047	0.071
		8.7	0.044	0.067
2×6 @ 16" O.C.	19 (compressed)	0	0.065	0.158
		4	0.058	0.098
		5	0.048	0.089
		7	0.043	0.075
		8.7	0.040	0.067
	21	0	0.059	0.157
		4	0.046	0.096
		5	0.044	0.088
		7	0.041	0.075
		8.7	0.037	0.066
	22 (compressed)	0	0.062	0.158
		4	0.048	0.097
		5	0.045	0.088
		7	0.041	0.075
		8.7	0.038	0.067

(Continued)

Framing Type and Spacing	Framing Cavity R-Value	Insulated Sheathing R-Value	Wood Wall U-Factor	Metal Wall U-Factor
2×6 @ 24" O.C.	19 (compressed)	0	0.062	0.135
		4	0.048	0.088
		5	0.045	0.081
		7	0.042	0.070
		8.7	0.039	0.062
	21	0	0.056	0.130
		4	0.044	0.086
		5	0.042	0.079
		7	0.039	0.068
		8.7	0.036	0.061
	22 (compressed)	0	0.058	0.132
		4	0.046	0.086
		5	0.043	0.079
		7	0.040	0.068
		8.7	0.037	0.061
2×8 @ 16" O.C.	19	0	0.059	0.145
		4	0.047	0.092
		5	0.044	0.084
		7	0.041	0.072
		8.7	0.038	0.064
	22	0	0.054	0.140
		4	0.043	0.090
		5	0.041	0.082
		7	0.038	0.071
		8.7	0.035	0.063
	25	0	0.050	0.136
		4	0.040	0.088
		5	0.038	0.081
		7	0.035	0.070
		8.7	0.033	0.062
	30 (compressed)	0	0.048	0.135
		4	0.039	0.088
		5	0.037	0.081
		7	0.035	0.070
		8.7	0.032	0.062
2×8 @ 24" O.C.	19	0	0.056	0.122
		4	0.045	0.082
		5	0.043	0.076
		7	0.040	0.066
		8.7	0.037	0.059
	22	0	0.051	0.117
		4	0.041	0.080
		5	0.040	0.074
		7	0.036	0.064
		8.7	0.034	0.058

Framing Type and Spacing	Framing Cavity R-Value	Insulated Sheathing R-Value	Wood Wall U-Factor	Metal Wall U-Factor
	25	0	0.047	0.113
		4	0.038	0.078
		5	0.037	0.072
		7	0.034	0.063
		8.7	0.032	0.057
	30 (compressed)	0	0.046	0.112
		4	0.037	0.077
		5	0.036	0.072
		7	0.034	0.063
		8.7	0.031	0.057
2×10 @ 16" O.C.	30	0	0.041	0.120
		4	0.035	0.081
		5	0.033	0.075
		7	0.031	0.065
		8.7	0.029	0.059
	38 (compressed)	0	0.040	0.199
		4	0.033	0.080
		5	0.032	0.074
		7	0.030	0.065
		8.7	0.028	0.058
2×10 @ 24" O.C	30 (compressed)	0	0.039	0.099
		4	0.033	0.071
		5	0.032	0.066
		7	0.038	0.058
		8.7	0.028	0.053
	38	0	0.038	0.097
		4	0.032	0.070
		5	0.031	0.066
		7	0.029	0.058
		8.7	0.027	0.053

Source: State of California.

Framing Type and Spacing	Framing Cavity R-Value	Insulated Sheathing R-Value	Wood Wall U-Factor	Metal Wall U-Factor
2×4 @ 16" O.C.	11 (compressed)	0	0.098	0.202
		4	0.068	0.122
		5	0.064	0.101
		7	0.056	0.084
		8.7	0.051	0.073
	13	0	0.088	0.195
		4	0.063	0.109
		5	0.059	0.099
		7	0.052	0.082
		8.7	0.048	0.072

(Continued)

Framing Type and Spacing	Framing Cavity R-Value	Insulated Sheathing R-Value	Wood Wall U-Factor	Metal Wall U-Factor
	15	0	0.081	0.189
		4	0.059	0.108
		5	0.055	0.097
		7	0.049	0.077
		8.7	0.045	0.071
2×4 @ 24" O.C.	11	0	0.094	0.173
		4	0.066	0.102
		5	0.062	0.081
		7	0.055	0.078
		8.7	0.050	0.069
	13	0	0.085	0.165
		4	0.061	0.099
		5	0.057	0.090
		7	0.051	0.077
		8.7	0.047	0.068
	15	0	0.077	0.158
		4	0.056	0.097
		5	0.053	0.088
		7	0.047	0.071
		8.7	0.044	0.067
2×6 @ 16" O.C.	19 (compressed)	0	0.065	0.158
		4	0.058	0.098
		5	0.048	0.089
		7	0.043	0.075
		8.7	0.040	0.067
	21	0	0.059	0.157
		4	0.046	0.096
		5	0.044	0.088
		7	0.041	0.075
		8.7	0.037	0.066
	22 (compressed)	0	0.062	0.158
		4	0.048	0.097
		5	0.045	0.088
		7	0.041	0.075
		8.7	0.038	0.067
2×6 @ 24" O.C.	19 (compressed)	0	0.062	0.135
		4	0.048	0.088
		5	0.045	0.081
		7	0.042	0.070
		8.7	0.039	0.062
	21	0	0.056	0.130
		4	0.044	0.086
		5	0.042	0.079
		7	0.039	0.068
		8.7	0.036	0.061
	22 (compressed)	0	0.058	0.132
		4	0.046	0.086
		5	0.043	0.079
		7	0.040	0.068
		8.7	0.037	0.061

Source: State of California.

9.5.0 Acoustics 101—Reflection, Absorption, Isolation—the Methods by Which Sound Can be Identified, Measured, and Controlled

Reflections

Reflected sound strikes a surface or several surfaces before reaching the receiver. These reflections can have unwanted or even disastrous consequences. Although reverberation is due to continued multiple reflections, controlling the reverberation time in a space does not ensure that the space will be free from problems from reflections.

Reflective corners or peaked ceilings can create a "megaphone" effect, potentially causing annoying reflections and loud spaces. Reflective parallel surfaces lend themselves to a unique acoustical problem called standing waves, creating a "fluttering" of sound between the two surfaces.

Reflections can be attributed to the shape of the space as well as the material on the surfaces. Domes and concave surfaces cause reflections to be focused rather than dispersed, which can cause annoying sound reflections. Absorptive surface treatments can help to eliminate both reverberation and reflection problems.

Noise Reduction Coefficient (NRC)

The Noise Reduction Coefficient (NRC) is a single-number index for rating how absorptive a particular material is. Although the standard is often abused, it is simply the average of the midfrequency sound absorption coefficients (250, 500, 1000, and 2000 Hertz rounded to the nearest 5%). The NRC gives no information as to how absorptive a material is in the low and high frequencies, nor does it have anything to do with the material's barrier effect (STC).

Sound Transmission Class (STC):

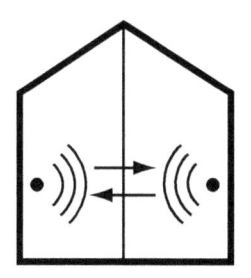

The Sound Transmission Class (STC) is a single-number rating of a material's or assembly's barrier effect. Higher STC values are more efficient in reducing sound transmission. For example, loud speech can be understood fairly well through an STC 30 wall but should not be audible through an STC 60 wall. The rating assesses the airborne sound transmission performance at a surface before reaching the receiver. These reflections can have unwanted or even disastrous consequences. Although reverberation is due to continued multiple reflections, controlling the reverberation time in a space does not ensure the space will be free from problems from reflections.

Reflective corners or peaked ceilings can create a "megaphone" effect, potentially causing annoying reflections and loud spaces. Reflective parallel surfaces lend themselves to a unique acoustical problem called standing waves, creating a "fluttering" of sound between the two surfaces.

Reflections can be attributed to the shape of the space as well as the material on the surfaces. Domes and concave surfaces cause reflections to be focused rather than dispersed, which can cause annoying sound reflections. Absorptive surface treatments can help to eliminate both reverberation and reflection problems.

9.5.1 Reverberation Time Creating a Buildup of Noise

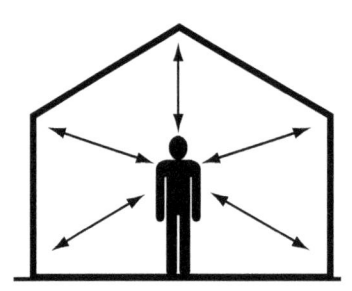

In an enclosed space, sound source stops emitting energy, and it takes some time for the sound to become inaudible. This prolongation of the sound in the room caused by continued multiple reflections is called reverberation.

Reverberation time plays a crucial role in the quality of music and the ability to understand speech in a given space. When room surfaces are highly reflective, sound continues to reflect or reverberate. The effect of this condition is described as a live space with a long reverberation time. A high reverberation time will cause a buildup of the noise level in a space. The effects of reverberation time on a given space are crucial to musical conditions and understanding speech. It is difficult to choose an optimum reverberation time in a multifunction space, as different uses require different reverberation times. A reverberation time that is optimum for a music program could be disastrous to the intelligibility of the spoken word. Conversely, a reverberation time that is excellent for speech can cause music to sound dry; flat assemblies should provide specific sound transmission class (STC) ratings when separating a core learning space from an adjacent space:

- **STC-45** if the adjacent space is a corridor, staircase, office or conference room.
- **STC-50** if the adjacent space is another core learning space, speech clinic, health care room or outdoors.
- **STC-53** if the adjacent space is a restroom.
- **STC-60** if the adjacent space is a music room, mechanical equipment room, cafeteria, gymnasium, or indoor swimming pool.
- Classroom doors should be rated as **STC-30** or more, and music room doors as **STC-40** or more. Entry doors across a corridor should be staggered to minimize noise transmission.
- STC ratings ranging from **45-60** are outlined for assemblies separating ancillary spaces from adjacent spaces.
- (*Note:* Open-plan classroom designs will not meet the requirements of this standard.)

By permission: acoustics.com

Achieving a specific STC rating depends highly on the materials and the installation methods used. Wall and ceiling assemblies can be specified and detailed to meet a required STC rating. This is the architect or designer's responsibility. However, specifying an STC level is not all that will be required. It is important to note that sound transmission can be strongly affected by sound leakage through penetrations, joints, and over or around the structure.

The number and location of penetrations through the wall, as well as the number and location of electrical outlet, should be considered in the design. In order to meet a specified STC, installation methods become crucial. Placement and installation instructions for the electrical system are given within Annex B in order to limit sound transfer between rooms. For single stud walls, electrical boxes should not be located within the same stud space. For staggered or dual-stud walls, boxes should be separated by at least 24″. If back-to-back electrical boxes cannot be avoided, they should be enclosed in full gypsum board enclosures that do not contact the framing of the other row of studs. In addition, all joints and air gaps should be sealed air tight with caulking or acoustical sealant.

As mentioned previously, background noise is a major concern in learning facilities. STC ratings will help to limit the background

By permission: Acoustics.com

9.5.2 Isolation—Measured by Sound Transmission Class (STC)

The amount of airborne sound blocked from transmitting through a partition is measured in a sound transmission class (STC) rating. A Higher STC Rating will degrade the ability to hear and understand speech. Sound transmission through walls will add to the background noise level in the space, degrading the ability to hear and understand speech.

Single or composite walls, floor-ceiling and roof-ceiling noise levels within a space (depending on the effect of sound transmission on the background noise level). It may be necessary to increase a required STC rating in order to meet a specified background noise level requirement.

Sound transmission problems can be avoided or lessened by good *I* site selection and good space planning.

Typical, single-stud construction will not meet the required STC ratings. The walls will most likely require staggered or dual-stud construction and/or multiple layers of drywall. (There are also specialty products that can help ensure compliance.) It is also important to note that acoustical ceiling tiles will not prevent sound transmission over the wall. Walls surrounding core learning spaces should extend to the deck of the building structure in order to adequately control sound transmission.

Carefully consider the placement of electrical outlets. Do not place them back-to-back. Again it will be important to work with your electrical engineer in order to specify installation instructions that will limit sound transmission. Specify on your drawings for contractors to seal all joints and penetrations with an acoustical sealant.

Most importantly, do not locate mechanical equipment rooms, restrooms, music rooms, gymnasiums, cafeterias, or any other noisy space adjacent to a class room or core learning space.

9.5.3 Impact Insulation Class—IIC—Blocking Noise from Being Transmitted Floor-to-Floor

Impact Insulation Class (IIC) is a rating for the ability of a floor-ceiling assembly to block impact/structure-borne noise from transmitting to the space below. A floor-ceiling assembly with a low IIC rating will potentially cause distracting noise in the room below, leading to possible annoyance and problems with communication.

- IIC ratings for floor-ceiling assemblies above core learning spaces should be at least **IIC-45** and preferably **IIC-50** (measured without carpeting on the floor).

- In new construction, gymnasia, dance studios, or other high floor impact activities shall not be located above core learning spaces.
- In existing facilities **IIC-65-70** (depending on the volume of 9.5.3-p.2 the space below) is recommended if gymnasia, dance studios, or other high floor impact activities are located above core learning spaces.

IIC is a major concern for multistory educational facilities. The floor-ceiling system should be specified and constructed in order to meet the specified IIC rating. Installing carpet on the floor above will help reduce the impact of sounds. It may be necessary to isolate the finished floor from the structural floor or to isolate the ceiling from the floor above. For any vibrating machinery located on the floor above or on the roof structure, rubber pads or spring systems should be installed. As with all requirements in the standard, it is the architect's or designer's responsibility to take the necessary steps in specification and design, but careful construction and installation will be necessary to ensure compliance.

This is only a concern for multistory schools. In most cases, installing carpet on the floor above will dramatically improve the IIC rating. In order to achieve the specified levels, a separate hard lid ceiling assembly could be required. Ideally, this would be completely isolated from the floor structure above. The classroom below may still need an acoustically absorptive ceiling treatment in order to meet the required reverberation time. Working with your mechanical engineer, be sure to specify appropriate vibration dampening measures for mechanical equipment.

By permission: acoustics.com

9.5.4 More Sound Absorption Factors for Building Materials and Finishes

Floor Materials	125 Hz	250 Hz	500 Hz	1000 Hz	2000 Hz	4000 Hz
Concrete or tile	0.01	0.01	0.15	0.02	0.02	0.02
Linoleum/vinyl tile on concrete	0.02	0.03	0.03	0.03	0.03	0.03
Wood on joists	0.05	0.11	0.10	0.07	0.06	0.07
Parquet on concrete	0.04	0.04	0.07	0.06	0.06	0.07
Carpet on concrete	0.02	0.06	0.14	0.37	0.60	0.65
Carpet on foam	0.08	0.24	0.57	0.69	0.71	0.73

Seating Materials	125 Hz	250 Hz	500 Hz	1000 Hz	2000 Hz	4000 Hz
Fully occupied— fabric upholstered	0.60	0.74	0.88	0.96	0.93	0.85
Occupied wooden pews	0.57	0.61	0.75	0.86	0.91	0.86
Empty—fabric upholstered	0.49	0.66	0.80	0.88	0.82	0.70
Empty metal/wood seats	0.15	0.19	0.22	0.39	0.38	0.30

Wall Materials	125 Hz	250 Hz	500 Hz	1000 Hz	2000 Hz	4000 Hz
Brick: unglazed	0.03	0.03	0.03	0.04	0.05	0.07
Brick: unglazed & painted	0.01	0.01	0.02	0.02	0.02	0.03
Concrete block— coarse	0.36	0.44	0.31	0.29	0.39	0.25
Concrete block— painted	0.10	0.05	0.06	0.07	0.09	0.08
Curtain: 10 oz/sq yd fabric molleton	0.03	0.04	0.11	0.17	0.24	0.35
Curtain: 14 oz/sq yd fabric molleton	0.07	0.31	0.49	0.75	0.70	0.60

Wall Materials	125 Hz	250 Hz	500 Hz	1000 Hz	2000 Hz	4000 Hz
Curtain: 18 oz/sq yd fabric molleton	0.14	0.35	0.55	0.72	0.70	0.65
Fiberglass: 2" 703 no airspace	0.22	0.82	0.99	0.99	0.99	0.99
Fiberglass: spray 5"	0.05	0.15	0.45	0.70	0.80	0.80
Fiberglass: spray 1"	0.16	0.45	0.70	0.90	0.90	0.85
Fiberglass: 2" rolls	0.17	0.55	0.80	0.90	0.85	0.80
Foam: Sonex 2"	0.06	0.25	0.56	0.81	0.90	0.91
Foam: SDG 3"	0.24	0.58	0.67	0.91	0.96	0.99
Foam: SDG 4"	0.33	0.90	0.84	0.99	0.98	0.99
Foam: polyur. 1"	0.13	0.22	0.68	1.00	0.92	0.97
Foam: polyur. 1/2"	0.09	0.11	0.22	0.60	0.88	0.94
Glass: 1/4" plate large	0.18	0.06	0.04	0.03	0.02	0.02
Glass: window	0.35	0.25	0.18	0.12	0.07	0.04
Plaster: smooth on tile/brick	0.013	0.015	0.02	0.03	0.04	0.05
Plaster: rough on lath	0.02	0.03	0.04	0.05	0.04	0.03
Marble/Tile	0.01	0.01	0.01	0.01	0.02	0.02
Sheetrock 1/2" 16" on center	0.29	0.10	0.05	0.04	0.07	0.09
Wood: 3/8" plywood panel	0.28	0.22	0.17	0.09	0.10	0.11

Ceiling Materials	125 Hz	250 Hz	500 Hz	1000 Hz	2000 Hz	4000 Hz
Acoustic Tiles	0.05	0.22	0.52	0.56	0.45	0.32
Acoustic Ceiling Tiles	0.70	0.66	0.72	0.92	0.88	0.75
Fiberglass: 2" 703 no airspace	0.22	0.82	0.99	0.99	0.99	0.99
Fiberglass: spray 5"	0.05	0.15	0.45	0.70	0.80	0.80
Fiberglass: spray 1"	0.16	0.45	0.70	0.90	0.90	0.85
Fiberglass: 2" rolls	0.17	0.55	0.80	0.90	0.85	0.80
Wood	0.15	0.11	0.10	0.07	0.06	0.07
Foam: Sonex 2"	0.06	0.25	0.56	0.81	0.90	0.91
Foam: SDG 3"	0.24	0.58	0.67	0.91	0.96	0.99
Foam: SDG 4"	0.33	0.90	0.84	0.99	0.98	0.99
Foam: polyur. 1"	0.13	0.22	0.68	1.00	0.92	0.97
Foam: polyur. 1/2"	0.09	0.11	0.22	0.60	0.88	0.94
Plaster: smooth on tile/brick	0.013	0.015	0.02	0.03	0.04	0.05
Plaster: rough on lath	0.02	0.03	0.04	0.05	0.04	0.03
Sheetrock 1/2" 16" on center	0.29	0.10	0.05	0.04	0.07	0.09
Wood: 3/8" plywood panel	0.28	0.22	0.17	0.09	0.10	0.11

Miscellaneous Materials	125 Hz	250 Hz	500 Hz	1000 Hz	2000 Hz	4000 Hz
Water	0.008	0.008	0.013	0.015	0.020	0.025
People (adults)	0.25	0.35	0.42	0.46	0.5	0.5

Absorption Coefficients (a) of Bldg.Matls and Finishes- p.2
Source: Sengpielaudio Sound Engineering Studio, Germany.

9.5.5 Absorption Coefficients for Various Wall and Floor Coverings

Material	128 Hz	256 Hz	512 Hz	1,024 Hz	2,048 Hz	4,096 Hz
Draperies hung straight, in contact with wall, cotton fabric, 10 oz. per square yard	0.04	0.05	0.11	0.18	0.30	0.44
The same, velour, 18 oz. per square yard	0.05	0.12	0.35	0.45	0.40	0.44
Same as above, hung 4 inches from wall	0.09	0.33	0.45	0.52	0.50	0.44
Felt, all hair, contact with wall	0.13	0.41	0.56	0.69	0.65	0.49
Rock wool (1 inch)	0.35	0.49	0.63	0.80	0.83	–
Carpet on concrete (0.4 inch)	0.09	0.08	0.21	0.26	0.27	0.37
Carpet, on 1/8 inch felt, on concrete (0.4 inch)	0.11	0.14	0.37	0.43	0.27	0.27
Concrete, unpainted	0.010	0.012	0.016	0.019	0.023	0.035
Wood sheeting, pine (0.8 inch)	0.10	0.11	0.10	0.08	0.08	0.11
Brick wall, painted	0.012	0.013	0.017	0.020	0.023	0.025
Plaster, lime on wood studs, rough finish (1/2 inch)	0.039	0.056	0.061	0.089	0.054	0.070

Source: Simon Fraser University, Vancouver, British Columbia, Canada.

9.6.0 Checklist for Masking Open Space Systems

How Masking Systems Work

Masking systems provide ambient background sound that reduces exposure to distracting office noises by emitting a discreet, electronically generated sound through specially installed, unobtrusive speakers. When installed properly, employees won't be aware of the pink noise being generated around them, but they will be able to focus on their work without unwanted sound distractions. Of course, carefully choosing office furniture, wall treatments, and flooring systems will also contribute to a productive work area.

Checklist of Masking Systems for Open Plans

- Ideally, speakers should be in enclosures located just above suspended ceilings, aimed upward toward hard plenum surfaces. If sound-absorbing insulation is applied to the underside of the structural deck, speakers should be aimed downward (or possibly sideways), or speaker enclosures that reflect sound downward should be used.
- Plenums should have uncomplicated air duct layouts and smooth sound-reflecting structural surfaces to allow wider spacings between loudspeakers.

- Coverage should include adjacent areas (or zones) so that occupants moving about the building will not notice the masking system.
- Masking should not exceed a sound level of 45 to 50 dBA because occupants tend to raise voices to compensate, thus defeating the intended masking effects. Occupants may also begin to complain about the sound level when it exceeds 50 dBA.
- To reduce the likelihood that occupants will notice background masking, consider installation procedures that initially operate the system at low sound levels. Then gradually increase the level by about 1 dB each day until the desired masking level is achieved in a week to 10 days or longer.
- A well-designed masking system deliberately garbles the sound it produces and therefore should not be used for paging and routine office functions.
- Provisions can be made to reduce masking noise levels during off-hours to enhance the ability of security personnel to hear unusual sounds.
- Be sure to consider the consequences of background masking on the usability of open plans by hearing-impaired persons. For example, when background noise levels exceed 30 dBA, hearing-impaired persons (even when using hearing aids) have far more difficulty understanding speech than do normal-hearing persons.

In addition to avoiding excessive noise levels, background noise from electronic sound masking systems in open-office plans should have a neutral tonal quality. This may be facilitated by designing the system to simulate familiar building sounds, such as the air flow at diffusers and registers of HVAC systems. The electronically produced sound levels in the finished room should be no higher than necessary to mask unwanted intruding speech and so that pronounced hisses are avoided. The sound level of the masking system should be neither too high nor too low, and the snectrum should roll off at the high end of the frequency range.

9.6.1 Use of Loudspeakers and Formula for Installation Spacing

In open plans, loudspeakers can usually be hidden in plenums above suspended ceilings. This strategy can achieve uniform masking sound throughout the room. Be careful when designing this kind of installation because openings for return or supply air in ceilings and luminaries can be noticeable sound leaks, which make it difficult to achieve uniform masking sound.

For preliminary planning, loudspeaker spacing, S, can be found by:

$$S = 1.4\,(2D + H - 4)$$

where S = spacing between loudspeakers (ft)
D = plenum depths (ft)
H = floor-to-ceiling height (ft)

Closer spacings may be required when spray-on, sound-absorbing fire protection, or insulation is applied to the underside of structural decks, or when complicated air duct layouts or deep structural members obstruct plenums.

By permission: Acoustics.com

9.7.0 Decibel Levels of Some Common Sounds

Threshold of hearing	0 dB	Motorcycle (30 feet)	88 dB
Rustling leaves	20 dB	Food blender (3 feet)	90 dB
Quiet whisper (3 feet)	30 dB	Subway (inside)	94 dB
Quiet home	40 dB	Diesel truck (30 feet)	100 dB
Quiet street	50 dB	Power mower (3 feet)	107 dB
Normal conversation	60 dB	Pneumatic riveter (3 feet)	115 dB
Inside car	70 dB	Chainsaw (3 feet)	117 dB
Loud singing (3 feet)	75 dB	Amplified Rock and Roll (6 feet)	120 dB
Automobile (25 feet)	80 dB	Jet plane (100 feet)	130 dB

Typical Average Decibel Levels (dBA) of Some Common Sounds

STC	What Can Be Heard
25	Normal speech can be understood quite easily and distinctly through wall
30	Loud speech can be understood fairly well, normal speech heard but not understood
35	Loud speech audible but not intelligible
40	Onset of "privacy"
42	Loud speech audible as a murmur
45	Loud speech not audible; 90% of statistical population not annoyed
50	Very loud sounds such as musical instruments or a stereo can be faintly heard; 99% of population not annoyed.
60+	Superior soundproofing; most sounds inaudible

Source: U.S. Department of Energy.

9.7.1 ANSI Recommended Levels for Various Types of Occupancy

Occupancy	NCB Curve
Broadcast Studios (distant microphone pickup used)	10
Concert halls, opera houses, and recital halls (listening to faint musical sounds)	10–15
Large auditoriums, large drama theatres, and large churches (for very good speech articulation)	15–20
TV and broadcast studios (close microphone pickup only)	15–25
Private Residences:	
Bedrooms	25–30
Apartments	28–38
Family rooms and living rooms	28–38
Schools:	
Lecture and classrooms	25–30
Open-plan classrooms	33–37

Occupancy	NCB Curve
Hotels/Motels:	
Individual rooms or suites	28–33
Meeting/banquet rooms	25–35
Halls, corridors, lobbies	38–43
Service support areas	38–48
Office Buildings:	
Executive offices	25–30
Conference rooms (large)	25–30
Conference rooms (small) and private offices	30–35
General secretarial areas	38–43
Open-plan areas	35–40
Business machines/computers	38–43
Public circulation	38–48
Hospitals and clinics:	
Private rooms	25–30
Wards	30–35
Operating rooms	25–30
Laboratories	33–43
Corridors	33–43
Public areas	38–43
Small auditoriums	25–30
Movie theatres	27–37
Churches	30–35
Courtrooms	33–37
Libraries	33–37
Restaurants	38–43
Light maintenance shops, industrial plant control rooms, kitchens, and laundries	43–53
Shops, garages	50–60

Source- ANSI/Public documents.

9.7.2 Decibel Comparison Chart of Environmental Noises

Here are some interesting numbers, collected from a variety of sources, that help one to understand the volume levels of various sources and how they can affect our hearing.

Environmental Noise	
Weakest sound heard	0 dB
Whisper quiet library	30 dB
Normal conversation (3-5′)	60–70 dB
Telephone dial tone	80 dB
City traffic (inside car)	85 dB

(Continued)

Train whistle at 500', Truck Traffic	90 dB
Subway train at 200'	95 dB
Level at which sustained exposure may result in hearing loss	*90–95 dB*
Power mower at 3'	107 dB
Snowmobile, motorcycle	100 dB
Power saw at 3'	110 dB
Sandblasting, loud rock concert	115 dB
Pain begins	*125 dB*
Pneumatic riveter at 4'	125 dB
Even short-term exposure can cause permanent damage—Loudest recommended exposure WITH hearing protection	*140 dB*
Jet engine at 100', Gun Blast	140 dB
Death of hearing tissue	180 dB
Loudest sound possible	194 dB

9.7.3 OSHA Daily Permissible Noise-Level Exposure

OSHA Daily Permissible Noise Level Exposure	
Hours per day	Sound level
8	90 dB
6	92 dB
4	95 dB
3	97 dB
2	100 dB
1.5	102 dB
1	105 dB
.5	110 dB
.25 or less	115 dB

By permission: Galen Carol Audio, San Antonio, TX

9.7.4 Perceptions of Increases in Noise Levels

Perceptions of Increases in Decibel Level	
Imperceptible change	1 dB
Barely perceptible change	3 dB
Clearly noticeable change	5 dB
About twice as loud	10 dB
About four times as loud	20 dB

9.7.5 Sound Levels of Music

Sound Levels of Music

Normal piano practice	60–70 dB
Fortissimo Singer, 3'	70 dB
Chamber music, small auditorium	75–85 dB
Piano Fortissimo	84–103 dB
Violin	82–92 dB
Cello	85–111 dB
Oboe	95–112 dB
Flute	92–103 dB
Piccolo	90–106 dB
Clarinet	85–114 dB
French horn	90–106 dB
Trombone	85–114 dB
Tympani & bass drum	106 dB
Walkman on 5/10	94 dB
Symphonic music peak	120–137 dB
Amplifier rock, 4-6'	120 dB
Rock music peak	150 dB

NOTES:
- One-third of the total power of a 75-piece orchestra comes from the bass drum.
- High-frequency sounds of 2-4000 Hz are the most damaging. The uppermost octave of the piccolo is 2048-4096 Hz.
- Aging causes gradual hearing loss, mostly in the high frequencies.
- Speech reception is not seriously impaired until there is about 30 dB loss; by that time damage is severe.

Sound	Noise Level (dBA)	Effect
Jet Engines (Near)	140	
Shotgun Firing	130	
Jet Takeoff (100-200 Ft.)	130	
Rock Concert (Varies)	110–140	Threshold of pain (125 dB)
Oxygen Torch	121	
Discotheque/Boom Box	120	Threshold of sensation (120 dB)
Thunderclap (Near)	120	
Stereo (Over 100 Watts)	110–125	
Symphony Orchestra	110	Regular exposure of more than 1 minute risks permanent hearing loss (over 100 dB)
Power Saw (Chain Saw)	110	
Jackhammer	110	
Snowmobile	105	
Jet Fly-over (1000 Ft.)	103	

(Continued)

Sound	Noise Level (dBA)	Effect
Electric Furnace Area	100	No more than 15 minutes of unprotected exposure recommended (90-100 dB)
Garbage Truck/Cement Mixer	100	
Farm Tractor	98	
Newspaper Press	97	
Subway, Motorcycle (25 Ft)	88	Very annoying
Lawnmower, Food Blender	85–90	Level at which hearing damage (8 hrs.) begins (85dB)
Recreational Vehicles, TV	70–90	
Diesel Truck (40 Mph, 50 Ft.)	84	
Average City Traffic Noise	80	Annoying; interferes with conversation; constant exposure may cause damage
Garbage Disposal	80	
Washing Machine	78	
Dishwasher	75	
Vacuum Cleaner	70	Intrusive; interferes with telephone conversation
Hair Dryer	70	
Normal Conversation	50–65	
Quiet Office	50–60	Comfortable (under 60 dB)
Refrigerator Humming	40	
Whisper	30	Very quiet
Broadcasting Studio	30	
Rustling Leaves	20	Just audible
Normal Breathing	10	
	0	Threshold of normal hearing (1000-4000 Hz) 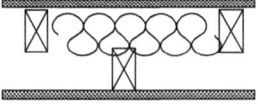
2 × 4 studs, 5/8″ gyp (2 layers total), Batt insulation	56–59	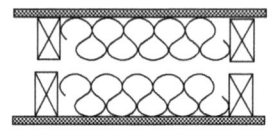

By permission: Galen Carol Audio, San Antonio, TX

Section | 9 Calculations to Determine the Effectiveness and Control of Thermal and Sound Transmission

3. Metal stud walls perform better than wood stud walls.

(*NOTE:* This only applies to single-stud assemblies. For double-stud assemblies, there is virtually no difference.)

Description	Estimated STC Rating	Wall Assembly
2 × 4 stud, 5/8" gyp (2 layers total), Batt insulation	34–39	
3 5/8" metal studs, 5/8" gyp (2 layers total), Batt insulation	43–44	

4. Resilient channel can improve the STC rating of an assembly.

(*NOTE*: These ratings are based on laboratory tests. Because of the special care required when installing resilient channels, actual results could be substantially lower.)

Description	Estimated STC Rating	Wall Assembly
2 × 4 stud, 5/8" gyp (2 layers total), Batt insulation	34–39	
2 × 4 stud, 5/8" gyp (2 layers total), Resilient Channel, Batt insulation	45–52	

5. Adding additional layers of drywall can improve the STC rating of an assembly.

Description	Estimated STC Rating	Wall Assembly
2 × 4 stud, 5/8" gyp (2 layers total), Batt insulation	34–39	

9.8.0. Sound Transmission Coefficient (STC) of Various Types of Insulated Partitions

1. Insulation will noticeably improve the STC rating of an assembly.

Description	Estimated STC Rating	Wall Assembly
3 5/8" metal studs, 5/8" gyp (2 layers total), No insulation	38–40	
3 5/8" metal studs, 5/8" gyp (2 layers total), Batt insulation	43–44	

2. Staggered or double-stud walls are higher rated than single-stud walls.

Description	Estimated STC Rating	Wall Assembly
2 × 4 stud, 5/8" gyp (2 layers total), Batt insulation	34–39	
Staggered studs, 5/8" gyp (2 layers total), Batt insulation	46–47	
2 × 4 studs, 5/8" gyp (2 layers total), Batt insulation	56–59	

3. Metal stud walls perform better than wood stud walls.

(*NOTE*: This only applies to single-stud assemblies. For double-stud assemblies, there is virtually no difference.)

Description	Estimated STC Rating	Wall Assembly
2 × 4 stud, 5/8" gyp (2 layers total), Batt insulation	34–39	
3 5/8" metal studs, 5/8" gyp (2 layers total), Batt insulation	43–44	

By permission: Acoustics.com

Section | 9 Calculations to Determine the Effectiveness and Control of Thermal and Sound Transmission

4. Resilient channel can improve the STC rating of an assembly.

(*NOTE:* These ratings are based on laboratory tests. Because of the special care required when installing resilient channels, actual results could be substantially lower.)

Description	Estimated STC Rating	Wall Assembly
2 × 4 stud, 5/8" gyp (2 layers total), Batt insulation	34–39	
2 × 4 stud, 5/8" gyp (2 layers total), Resilient Channel, Batt insulation	45–52	

5. Adding additional layers of drywall can improve the STC rating of an assembly.

Description	Estimated STC Rating	Wall Assembly
2 × 4 stud, 5/8" gyp (2 layers total), Batt insulation	34–39	
3 5/8" metal studs, 5/8" gyp (3 layers total), Batt insulation	39–40	
2 × 4 stud, 5/8" gyp (4 layers total), Batt insulation	43–45	

6. Drywall between double studs can dramatically reduce the STC rating of an assembly.

Description	Estimated STC Rating	Wall Assembly
2 × 4 studs, 5/8" gyp (4 layers total), Batt insulation	44–45	

Description	Estimated STC Rating	Wall Assembly
2 × 4 studs, 5/8" gyp (2 layers total), Batt insulation	56–59	
2 × 4 studs, 5/8" gyp (3 layers total), Batt insulation	59–60	
2 × 4 studs, 5/8" gyp (4 layers total), Batt insulation	58–63	

STC RATINGS FOR MASONRY WALLS

STC ratings for masonry/CMU walls are based on weight of the block, on whether or not the cells are filled, and on what material it is filled with.

Estimated STC Ratings for CMU Walls

Wall Thickness, in.	Hollow Unite		Grout Filled		Sand Filled	
	Weight	STC	Weight	STC	Weight	STC
4	20	44	38	47	32	46
6	32	46	63	51	50	49
8	42	48	86	55	68	52
10	53	50	109	60	86	55

The STC rating of a CMU wall can be estimated based on its weight using the following formula:

$STC = 0.18W + 40$

where W = pounds per square foot (psf)

9.8.1 Testing for STC of Residential Carpet over Joist and Wood Subfloor

Observations: B-2

1. Impact Noise Ratings were all lower than those found in Test Series B-1. Various cushion/carpet combinations yielded substantially lower INR (Impact Noise Ratings) values for wood joist floors than for concrete floors.
2. Test Series B-1 has already shown that as pile weight increases, the INR increases. The assumption is also true with wood joist construction, but probably with lower relative ratings.

Testing for Sound Transmission Class (STC) of a Residential Carpet Installed with Cushion over a Joist and Plywood Subfloor

Tested Materials: Carpet and cushion—25 ounces per square yard 100% nylon cut pile residential carpet installed over bonded polyurethane bonded cushion ½ inch thickness with 6.0 lb/ft³ density.

Test Floor: Open Joist 2000 system, 13 inches deep installed, 24 inches on center. Subfloor 5/8 inch thick T&G plywood. Bridging of continuous 2 × 4 inch wood nailed to bottom chord and the sides of the diagonals with 2-inch long nails. Cellulose insulation with density of 1.6 pcf, 51/2 inches thick was used. Resilient channels of 24-gauge galvanized steel placed 16 inches on center and attached to bottom chords with screws. Ceiling of gypsum board of 5/8 inches thick. Sheets fastened to resilient channels by means of 1 ½ inch screws, spaced 6 inches on center. Joints taped and finished with two layers of compound.

Procedure: Sound transmission loss was determined per ASTM E90-87, Standard Test Method for Laboratory Measurement of Sound Transmission Loss in Building Partitions, by mounting and perimeter sealing the test specimen (carpet over cushion) as a partition between two reverberation rooms. Sound is introduced in one of the rooms (the source room), and measurements are made of noise reduction between the source room and the receiving room. The rooms are so arranged and constructed that the only significant sound transmission between them is through the test specimen.

Results of Test: Sound Transmission Classification was found to be 49 per ASTM E413-94, Classification for Rating Sound Insulation.

Overall Conclusions about Carpet and Sound

Carpet is highly effective in controlling noise in buildings by absorbing airborne sound, reducing surface noise generation, and reducing the impact of sound transmission to rooms below. Properly specified carpet/cushion combinations have proven to handle the vast majority of sound absorption requirements in architectural spaces. Specifying for critical areas such as theaters, broadcast studios, and open-plan office areas may require full details of impact insulation properties and noise absorption characteristics.

Source: Carpet & Rug Institute, Dalton, GA.

9.8.2 STC Ratings for Masonry Walls

STC ratings for masonry/CMU walls are based on weight of the block and on whether or not the cells are filled and what material it is filled with.

Estimated STC Ratings for CMU Walls

Wall Thickness, in.	Hollow Units		Grout Filled		Sand Filled	
	Weight	STC	Weight	STC	Weight	STC
4	20	44	38	47	32	46
6	32	46	63	51	50	49
8	42	48	86	55	68	52
10	53	50	109	60	86	55

The STC rating of a CMU wall can be estimated based on its weight using the following formula:

$$STC = 0.18W + 40$$

where W = pounds per square foot (psf)

By permission: racoustics.com

9.8.3 How Insulation, Staggered Studs, Resilient Channels, and Added Layers of Drywall Can Affect STC Ratings

Below are the STC ratings of various wall assemblies, each presented to help illustrate concepts, improvements, and rules of thumb. The estimated ratings are based on laboratory test results from various compendiums of STC ratings. It is recommended that you consult a professional acoustician for more detailed information or to analyze the specifics of your project/assembly.

To view different wall assemblies, click on each point below that may apply to your project.

1. Insulation will noticeably improve the STC rating of an assembly.
2. Staggered or double-stud walls are higher rated than single-stud walls.
3. Metal stud walls perform better than wood stud walls.
4. Resilient channels can improve the STC rating of an assembly.
5. Adding additional layers of drywall can improve the STC rating of an assembly.
6. Drywall between double studs can dramatically reduce the STC rating of an assembly.

1. Insulation will noticeably improve the STC rating of an assembly.

Description	Estimated STC Rating	Wall Assembly
3 5/8" metal studs, 5/8" gyp (2 layers total), No insulation	38–40	
3 5/8" metal studs, 5/8" gyp (2 layers total), Batt insulation	43–44	

2. Staggered or double-stud walls are higher rated than single-stud walls.

Description	Estimated STC	Wall Assembly
2 × 4 stud, 5/8" gyp (2 layers total), Batt insulation	34–39	
Staggered studs, 5/8" gup (2 layers total), Batt insulation	46–47	
3 5/8" metal studs, 5/8" gyp (3 layers total), Batt insulation	39–40	
2 × 4 stud, 5/8" gyp (4 layers total), Batt insulation	43–45	

By permission:acoustics.com

6. Drywall between double studs can dramatically reduce the STC rating of an assembly.

Description	Estimated STC	Wall Assembly
2×4 studs, 5/8" gyp (4 layers total), Batt insulation	44–45	
2×4 studs, 5/8" gyp (2 layers total), Batt insulation	56–59	
2×4 studs, 5/8" gyp (3 layers total), Batt insulation	59–60	
2×4 studs, 5/8" gyp (4 layers total), Batt insulation	59–63	

9.9.0 Impact Insulation Class —IIC—What Is This?

Impact Insulation Class (IIC)

IIC–What Is It?

In a multilevel home or business, when a floor covering in one of the upper rooms is impacted, by dropping an object or moving furniture, for example, the impact creates a vibration that travels through the floor, subfloor, and through the ceiling to the room below. These vibrations result in unwanted and annoying sounds in those rooms. This is called impact sound transmission. Floor coverings with a high IIC rating help to reduce impact sound transmissions to lower levels, thus reducing or eliminating those bothersome noises. The lowest IIC rated floors/ceiling assemblies come in at around 25, and the highest rated systems can come in at 85 and up.

Common Guidelines to Use When Selecting the Proper IIC Rating for Your Space

IIC 50—The least amount of impact sound transmission reduction considered effective. Some occupants would be dissatisfied with this level of sound transmission.

IIC 60—Considered a medium level of impact sound transmission reduction.

IIC 65—Considered a high level of impact sound transmission reduction that would satisfy most occupants.

By permission findanyfloor.com/sound/SoundControl.xhtml

Section 10

Interior Finishes

10.1.0	Gypsum Drywall Panels—Types, Thickness, Width, Length	546	
10.1.1	Wall Framing and Drywall Panel Measurements—U.S. and Metric	547	
10.1.2	Gypsum Wall Panel Coverage Calculator	547	
10.1.3	Fastener/Compound/Tape Calculator	549	
10.1.4	Drywall Finishing Guide	550	
10.2.0	Calculations to Determine Sealer and Filler Yield	552	
10.3.0	Understanding Wall Coverings—Types and Usage	554	
10.3.1	Basic Types of Fabric-Backed Vinyl Wall Coverings	556	
10.3.2	Formulas for Estimating Wall Covering Quantities	556	
10.3.3	Calculating How Much Wallpaper Is Required	561	
10.3.4	Wall Covering Adhesives	561	
10.4.0	General Information to Calculate Various Types of Floor and Wall Tiles	562	
10.4.1	Calculating Requirements for Ceiling Tile	564	
10.4.2	Painting Ceilings and Walls	565	
10.5.0	Types of Carpeting	565	
10.5.1	Calculating the Amount of Carpet Required—Rooms 8–35 Feet in Length and 13–20 Feet in Width	567	
10.6.0	Solid Hardwood Flooring	568	
10.6.1	Janka Wood Hardness Scale for Wood Flooring Species	568	
10.6.2	Laminate Flooring	570	
10.7.0	Finishing of Interior Wood	572	
10.7.1	Opaque, Transparent Finishes, Stains	573	
10.7.2	Fillers and Sealers	574	
10.7.3	Finishes for Floors	575	
10.7.4	Finishes for Items Used for Food	576	
10.7.5	Finishes for Butcher Blocks and Cutting Boards	577	
10.7.6	Wood Cleaners and Brighteners	578	
10.7.7	Paint Strippers	578	
10.8.0	Characteristics of Selected Woods for Painting	580	
10.9.0	When Calculating and Measuring for Interior Trim and Millwork—Learn Tool Basics	581	

10.1.0 Gypsum Drywall Panels—Types, Thickness, Width, Length

Panel Sizes

Drywall Type	Thickness	Width	Length
Standard gypsum drywall panels	1/4"	4'	8', 10', 12'
	5/16"	4'	8', 10', 12'
	3/8"	4'	8', 10', 12'
↓	1/2"	4'	8' to 16'
Fire-resistant drywall Type X for commercial use where multiple layers are required for extended fire wall rating durations	1/2"	4'	8' to 16'
Fire-resistant drywall Type X	5/8"	4'	8' to 16'
Fire-resistant backer board Type X	1/2"	4'	8'
↓	5/8"	4'	8'
Fire-resistant shaft liner drywall	1"	23-7/8"	8' to 12'
Water-resistant drywall	1/2"	4'	8', 10', 12'
Fire- and water-resistant drywall	1/2"	4'	8', 10', 12'
↓	5/8"	4'	8', 10', 12'
Foil-back panels	3/8"	4'	6' to 16"
↓	1/2"	4'	6' to 16"
Foil-back Type X	5/8"	4'	6' to 16"
54" wide panels	1/2"	54"	8', 10', 12'
↓	5/8"	54"	8', 10', 12'
Exterior ceiling panels (soffit board)	1/2"	4'	8', 9', 10'
↓	5/8"	4'	8', 9', 10'
High-strength ceiling panels	1/2"	4'	8', 9', 10'
Drywall panels for factory decoration	5/16"	4'	cut to specified size
Sound deadening panels	1/4"	4'	8'
Tile-backing panels	1/4"	32", 4'	4'
↓	1/2"	4'	5', 8'
Type X fire-resistant tile-backing panels	5/8"	4'	8'
Exterior sheathing	1/2"	4'	8', 9', 10'
↓	5/8"	4'	8', 9', 10'
Veneer-base drywall (blue board)	3/8"	4'	8' to 16"
↓	1/2"	4'	8' to 16"

Section | 10 Interior Finishes

10.1.1 Wall Framing and Drywall Panel Measurements—U.S. and Metric

Wall Framing Spacing—U.S. and Metric
- 16 inches = 406 mm
- 24 inches = 510 mm

Framing Fasteners
- 2 inch centers = 51 mm
- 2 1/2 inch centers = 64 mm
- 6 inch centers = 153 mm
- 7 inch center = 178 mm
- 8 inch center- 203 mm
- 12 inch centers = 305 mm
- 16 inch centers = 406 mm
- 24 inch centers = 610 mm

Gypsum Wallboard Panel Size and Thickness—U.S. and Metric

Panel width
- 2 feet = 610 mm (61 cm)
- 4 feet = 1219 mm (121.9 cm)

Panel length
- 4 feet = 1219 mm or 1.219 meters
- 5 feet = 1524 mm or 1.524 meters
- 6 feet = 1828 mm or 1.828 meters
- 8 feet = 2428 mm or 2.428 meters
- 10 feet = 3048 mm or 3.048 meters
- 12 feet = 3658 mm or 3.658 meters

Panel Thickness
- ¼ inch = 6.4 mm
- 3/8 inch = 9.54 mm
- ½ inch = 12.7 mm
- 5/8 inch = 15.9 mm
- One inch = 25.4 mm

10.1.2 Gypsum Wall Panel Coverage Calculator

Gypsum Panel Coverage Calculator

No. of Panels	Size of Panels		
	4' × 8'	4' × 10'	4' × 12'
10	320 sq. ft.	400 sq. ft.	480 sq. ft.
11	352	440	528
12	384	480	576
13	416	520	624
14	448	560	672
15	480	600	720
16	512	640	768
17	544	680	816
18	576	720	864
19	608	760	912

Gypsum Panel Coverage Calculator—Cont'd

No. of Panels	Size of Panels		
	4' × 8'	4' × 10'	4' × 12'
20	640	800	960
21	672	840	1008
22	704	880	1056
23	736	920	1104
24	768	960	1152
25	800	1000	1200
26	832	1040	1248
27	864	1080	1296
28	896	1120	1344
29	928	1160	1392
30	960	1200	1440
31	992	1240	1488

Use the following table to determine the maximum frame spacing for direct application of gypsum panels to wood framing.

Frame Spacing for Single-Layer Application

Board Thickness	Location	Application Method[1]	Max. Frame Spacing on center in.
3/8"	ceiling[2][3]	perpendicular[1]	16
	sidewall	parallel or perpendicular	16
1/2"	ceiling[4]	parallel[3]	16
		perpendicular	24[5]
	sidewall	parallel or perpendicular	24
5/8"	ceiling[4]	parallel[3]	16
		perpendicular	24
	sidewall	parallel or perpendicular	24

For Sheetrock Brand Interior Ceiling Panels—Sag-Resistant

Board Thickness	Location	Application Method[1]	Max. Frame Spacing on center in.
1/2"	ceilings	parallel or perpendicular	24

[1] Long edge position relative to framing.
[2] Not recommended below unheated spaces.
[3] Not recommended if water-based texturing material is to be applied.
[4] Sheetrock Brand Gypsum Panels—Water Resistant are not recommended for ceiling where framing is greater than 12" on center for single layer resilient application where file is to be supplied.
[5] Max. spacing 16" on center if water-based texturing material to be applied.

This information is a copyrighted work of USG Corporation.

Section | 10 Interior Finishes

10.1.3 Fastener/Compound/Tape Calculator

Planning the Job

To estimate the quantity of fasteners, compound, and tape you will need, use the following table.

Fastener/Compound/Tape Calculator

With this amount of SHEETROCK Brand Gypsum Panels sq. ft	Use this amount of wallboard nails[1] lb.	Or this amount of Type W Screws[2] lb.	Use this amount of SHEETROCK Brand Joint Tape ft.
100	0.6	0.3	37
200	1.1	0.6	74
300	1.6	0.9	111
400	2.1	1.2	148
500	2.7	1.4	185
600	3.2	1.6	222
700	3.7	1.9	259
800	4.2	2.2	296
900	4.8	2.4	333
1000	5.3	2.7	370

With this amount of SHEETROCK Brand Gypsum Panels sq. ft.	Use this amount of SHEETROCK Brand All-Purpose Ready Mixed Joint Compound[3] lb.	Use this amount of SHEETROCK Brand Lightweight All-Purpose Ready Mixed Joint Compound (PLUS 3)[3] gal.	Use this amount of SHEETROCK Brand First Coat gal.
100	14	0.9	0.3
200	28	1.9	0.6
300	41	2.8	0.9
400	55	3.8	1.1
500	69	4.7	1.4
600	83	5.6	1.7
700	97	6.6	2.0
800	110	7.5	2.3
900	124	8.5	2.6
1000	138	9.4	2.9

[1] Spaced 7" on ceiling; 8' on wall. Reduce by 50% for adhesive/nail-on application.
[2] Spaced 12" on ceiling: 16" on wall.
[3] Coverage figures shown here approximate the amount of joint compound needed to treat the flat joints, inside corners and outside corners using metal corner bead, in a typical room. Coverage can vary widely depending on factors such as condition of substrate, tools used, application methods, and other job factors.

This information is a copyrighted work of USG Corporation.

10.1.4 Drywall Finishing Guide

Scope

The following industry specifications describe various levels of gypsum board finishes prior to the application of decoration. The recommended level of finish for gypsum board wall and ceiling surfaces varies depending on their location in the structure, type of paint to be applied, and final decoration to be applied and can also be dependent on the type and strength of illumination striking the surface. Each recommended level of finish is described with typical applications.

Terminology

Accessories: Metal or plastic beads, trim or molding used to protect, conceal, or decorate corners, edges, or the abutments of the gypsum board construction.

Back-Roll: Rolling a spray painted surface (while still wet) with a paint roller immediately following spray application.

Critical or Severe Lighting: Strong side lighting from windows or surface-mounted light fixtures. Wall and ceiling areas abutting window mullions or skylights, long hallways, or atriums with large surface areas flooded with artificial and/or natural lighting are a few examples of critical lighting areas. Strong side lighting from windows or surface-mounted light fixtures may reveal even minor surface imperfections. Light striking the surface obliquely, at a very slight angle, greatly exaggerates surface irregularities. If critical lighting cannot be avoided, the effects can be minimized by skim coating the gypsum board surfaces, by decorating the surface with medium to heavy textures, or by the use of draperies and blinds that soften shadows. In general: gloss, semigloss, and enamel paint finishes highlight surface defects; textures hide minor imperfections.

Joint Photographing or Telegraphing: The shadowing of the finished joint areas through the surface decoration.

Source: K. E. McNurney, Inc., Dimi, CA.

Paint: Any pigmented liquid, liquefiable, or mastic composition designed for application to a substrate as a thin layer that is converted to an opaque solid film after application. Used for protection, decoration, or identification or to serve some functional purpose, such as filing or concealing surface irregularities.

Primer/Sealer Drywall: A paint material formulated to fill the pores and equalize the suction difference between gypsum board surface paper and the compound used on finished joints, angles, fastener heads, accessories, and over skim coatings. A good primer / sealer (*note:* I always recommend a good quality paint as the primer coat because the "PVA" drywall sealers lack enough solids) formulated with higher binder solids, applied undiluted, is typically specified for new gypsum board surfaces prior to the application of texture materials and gloss, semigloss, and flat latex wall paints. An alkali and moisture-resistant primer and a tinted enamel undercoat may be required under enamel paints. Always consult with the finish paint manufacturer for specific recommendations.

Primer/Sealer Wall Coverings: White, self-sizing water base, "universal" (all-purpose) wall covering primers have recently been introduced into the marketplace for use on new gypsum surfaces. It is claimed that these products are drywall strippable, bind poor latex paint, allow hanging over glossary surfaces and existing vinyls, hide wall colors, and are water washable.

Properly Painted Surface: A surface that is uniform in appearance, color, and sheen. It is one that is free of foreign material, lumps, skins, runs, sags, holidays, misses, strike-through, or insufficient coverage. It is a surface that is free of chips, splatters, spills, or over spray that was caused by the contractors' workforce. Compliance with the criteria of a "properly painted surface" should be determined when viewed without magnification at a distance of 5 feet or more, under normal lighting conditions, and from a normal viewing position. (*Note:* A surface uniform in appearance, color, and sheen may not be sufficiently achieved with a coat of primer or a single coat of topcoat.)

Skim Coat: A thin coat of joint compound over the entire surface to fill imperfections in the joint work, smooth the paper texture, and provide a uniform surface for decorating.

Spotting: Method used to cover fastener heads (nails, screws, staples) with joint compound.

Texture: A decorative treatment of gypsum board surfaces.

Texturing: Regular or irregular patterns typically produced by applying a mixture of joint compound and water, or proprietary texture materials, including latex base texture paint, to a gypsum board surface previously coated with primer/sealer. Texture material is applied by brush, roller, spray, trowel, or a combination of these tools, depending on the desired result. Textured wall surfaces are normally overpainted with the desired finish; overpainting of textured ceiling may not be deemed necessary where an adequate amount of material is applied to provide sufficient hiding properties. A primer/sealer may not be required with certain proprietary texture materials. Always consult with the manufacturer of the texture material for specific recommendations.

Tool Marks and Ridges: A smooth surface may be achieved by lightly sanding or wiping the joint compound down with a dampened sponge. Great care should be exercised to ensure that the nap of the gypsum board facing paper is not raised during sanding operations.

Topcoat: The finish coat(s) of a coating system, formulated for appearance and or environmental resistance.

Wall Covering: Any type of paper, vinyl, fabric, or specialty material that is pasted onto a wall or ceiling in a wide array of colors, patterns, textures, and performance characteristics, such as washability and abrasion resistance.

Levels of Gypsum Board Finish

LEVEL 0: No taping, finishing, or accessories required. This level of finish may be useful in temporary construction or whenever the final decoration has not been determined.

LEVEL 1: All joints and interior angles shall have tape embedded in joint compound. Surface shall be free of excess joint compound. Tool marks and ridges are acceptable. This level of finish, often referred to as "fire taping," is frequently specified in plenum areas above ceilings, in attics, in areas where the assembly would generally be concealed, or in building service corridors and other areas not normally open to public view. Accessories (cornerbead, base shoe, other trims) are optional at specifier discretion in *corridors* and other areas with pedestrian traffic.

LEVEL 2: All joints and interior angles shall have tape embedded in joint compound and one separate coat of joint compound applied over all joints, angles, fastener heads, and accessories. Surface shall be free of excess joint compound; tool marks and ridges are acceptable. This level of finish is specified where water-resistant drywall is used as a substrate for tile; it may be specified in garages, warehouse storage, or similar areas where surface appearance is not of primary concern.

LEVEL 3: All joints and interior angles shall be tape embedded in joint compound, and two coats of joint compound applied over all joints, angles, fastener heads, and accessories. All joint compound shall be smooth and free of tool marks and ridges. *Note:* It is recommended that the prepared surface be coated with a primer/sealer prior to the application of final finishes. This level of finish is typically specified in appearance areas that are to receive heavy or medium texture (spray or hand applied) finishes before final painting, or where heavy-grade wall coverings are to be applied as the final decoration.

LEVEL 4: All joints and interior angles shall be tape embedded in joint compound and three coats of joint compound applied over all joints, angles, fastener heads, and accessories. All joint compound shall be smooth and free of tool marks and ridges. *Note:* It is recommended that the prepared surface be coated with a primer/sealer prior to the application of final finishes. This level of finish is typically specified where light textures or wall coverings are to be applied, or economy is of concern.

LEVEL 5: All joints and interior angles shall be tape embedded in joint compound and three separate coats of joint compound applied over all joints, angles, fastener heads and accessories. A thin skim coat of joint compound, or a material manufactured especially for this purpose, shall be applied to the surface. The surface shall be smooth and free of tool marks and ridges. *Note:* It is recommended that the prepared surface be coated with a primer/sealer prior to the application of final finishes. This level of finish is recommended where gloss, semigloss, enamel, or nontextured flat paints are specified or where severe side-lighting conditions occur. This highest quality finish is the most effective method to provide a uniform surface and minimize the possibility of joint photographing and of fasteners showing through the final decoration. (*Note:* Application of primer/paint products over Level 4 and Level 5 smooth finish) Industry experience demonstrates that an effective method for achieving a visually uniform surface for both the primer and topcoat is spray application immediately followed by back rolling or roller application using good roller techniques such as finishing in one direction and using roller types and naps recommended by the paint manufacturer.

Resources

ASTM C-840 - Standard Specifications for Application and Finishing of Gypsum Board. American Society for Testing and Materials.
GA-505 - Gypsum Board Terminology. Gypsum Association.
GA-214-96 - Recommended Levels of Gypsum Board Finish. Gypsum Association.
FSCT - Coatings Encyclopedia Dictionary. Federation of Societies for Coatings Technology.
DWFC - Recommended Specification for Preparation of Gypsum Board Surfaces Prior to Texture Application. Drywall Finishing Council Inc.
DWFC - Interior Job Condition Specifications for the Application of Drywall Joint Compounds, Drywall Textures, and Paint/Coatings. Drywall Finishing Council Inc.

10.2.0 Calculations to Determine Sealer and Filler Yield

Crack Sealant and Filler Yield Calculator

Crack/Joint Width × Depth (inches)	Feet per Gallon	Crack/Joint Width × Depth (inches)	Feet per Gallon	Crack/Joint Width × Depth (inches)	Feet per Gallon	Crack/Joint Width × Depth (inches)	Feet per Gallon
1/8 × 3/8	410.7 / 308.0	3/8 × 3/8	136.9 / 102.7	5/8 × 3/8	82.1 / 61.6	7/8 × 3/8	58.7 / 44.0
1/8 × 1/2	246.4 / 205.3	3/8 × 1/2	82.1 / 68.4	5/8 × 1/2	49.3 / 41.1	7/8 × 1/2	35.2 / 29.3
1/8 × 5/8	176.0 / 154.0	3/8 × 5/8	58.7 / 51.3	5/8 × 5/8	35.2 / 30.8	7/8 × 5/8	25.1 / 22.0
1/8 × 3/4	308.0	3/8 × 3/4	154.0	5/8 × 3/4	102.7	7/8 × 3/4	77.0
1/8 × 7/8	205.0 / 154.0	3/8 × 7/8	102.7 / 77.0	5/8 × 7/8	68.4 / 51.3	7/8 × 7/8	51.3 / 38.5
1/8 × 1	123.2 / 102.7	3/8 × 1	61.6 / 51.3	5/8 × 1	41.1 / 34.2	7/8 × 1	30.8 / 25.7
1/4 × 1/4	88.0 / 77.0	1/2 × 1/4	44.0 / 38.5	3/4 × 1/4	29.3 / 25.7	1 × 1/4	22.0 / 19.3
1/4 × 3/8		1/2 × 3/8		3/4 × 3/8		1 × 3/8 / 1 × 1/2	
1/4 × 1/2		1/2 × 1/2		3/4 × 1/2		1 × 5/8 / 1 × 3/4	

Crack Sealant and Filler Yield Calculator—Cont'd

Crack/Joint Width × Depth (inches)	Feet per Gallon	Crack/Joint Width × Depth (inches)	Feet per Gallon	Crack/Joint Width × Depth (inches)	Feet per Gallon	Crack/Joint Width × Depth (inches)	Feet per Gallon
1/4 × 5/8		1/2 × 5/8		3/4 × 5/8		1 × 7/8	
						1 × 1	
1/4 × 3/4		1/2 × 3/4		3/4 × 3/4			
1/4 × 7/8 1/4 × 1		1/2 × 7/8		3/4 × 7/8			
		1/2 × 1		3/4 × 1			

Band-Aid* Coverage:		Band-Aid Configuration	Feet per Gallon	Band-Aid Configuration	Feet per Gallon
*i.e., material squeegeed on surface. Does not include crack volume.		1/16" × 2"	154.0	3/32" × 2"	102.4
		1/16" × 3"	102.7	3/32" × 3"	68.4
		1/16" × 4"	77.0	3/32" × 4"	51.4

Wallcovering Basics
Coversion Table

For the convenience of our clients, OMNOVA Solutions provides the following dimensional conversion table — which converts square feet to lineal yards—for our 54" Commercial Wallcovering. The table should prove helpful when estimating and ordering the amount of material needed for a specific project.

Helpful Guidelines

- Standard OMNOVA Commercial Wall Covering is 54" or 4.5 feet wide.
- As a result, there are 13.5 square feet in every lineal yard of our wall coverings: 3 feet (1 yard) in height × 4.5 feet in width equals 13.5 square feet.
- To convert lineal yards to square feet, multiply the number of lineal yards by 13.5. (Example: 5 lineal yards multiplied by 13.5 yields 67.5 square feet.)
- To convert square feet to lineal yards, divide the number of square feet by 13.5. (Example: 108 square yards divided by 13.5 yields 8 lineal yards.)
- Our wall covering is stocked and shipped in standard roll sizes of 30 and 60 lineal yards. (A cutting charge applies for yardage orders that are less than standard roll size.)

For 54" widths

Conversion Table

Square Feet	Lineal Yards
13.5	1
67.5	5
135	10
270	20
405	30
607.5	45
810	60
1620	120
2700	200
3240	240
6750	500

10.3.0 Understanding Wall Coverings—Types and Usage

Overview

Wall coverings can be used in virtually any residential or contract environment. Since there are many different types of wall coverings on the market—some for very specific uses—it is important to understand the qualities of each type and in what type of environment the wall coverings will be used.

Key Points

- When selecting wall coverings, the first variable to consider is the amount of traffic the area will receive.
- Paper and natural wall coverings are most appropriate where traffic is minimal. They are more delicate than their vinyl counterparts, yet offer ample durability and a special style to a variety of placements.
- Vinyl and synthetic textiles, with their maximum durability and ease of cleaning, are especially appropriate for hospitals, sporting arenas, schools, and other high-traffic situations.
- While wall coverings are categorized by residential and contract segments, it is not uncommon to use residential wall coverings in contract settings, like assisted living facilities for a homey feel, for instance, or to use contract wall coverings in some of today's more avant-garde homes.

Contract Wall Coverings

Contract wall coverings are produced specifically for use in hotels, apartment buildings, office buildings, retail outlets, schools, and hospitals. They are manufactured to meet or surpass minimum physical and performance characteristics set forth in Federal Specifications CCC-W-408.

The most popular types of wall coverings for contract installations are as follows:

- **Vinyl Coated Paper**—wall coverings that have a paper substrate on which the decorative surface has been sprayed or coated with an acrylic type vinyl or polyvinyl chloride (PVC).
- **Paper-Backed Vinyl/Solid Sheet Vinyl**—wall coverings that have a paper (pulp) substrate laminated to a solid decorative surface. These types of wall coverings are very durable since the decorative surface is a solid sheet of vinyl. They are classified as scrubbable and peelable.
- **Fabric Backed Vinyl**—wall coverings that have a woven substrate of fabric or a nonwoven synthetic substrate. In either case, the substrate is laminated to a solid vinyl decorative surface.

General categories of these types of wall coverings include the following:

- **Type I (Light Duty)**—for use in office areas, hospital patient rooms, and hotel rooms. Also intended for ceilings and areas of light abrasion.
- **Type II (Medium/Heavy Duty)**—for use in foyers, lounges, corridors, and classrooms, or areas of average to heavy scuffing. Can also be used as wainscot.

Specialty Wall Coverings

A special category of wall coverings is used in highly specialized circumstances or for areas of light traffic. Many of these types of wall coverings have been replaced by vinyl wall coverings that simulate the same look with greater durability. Nonetheless, many of the types of wall coverings outlined below have historical importance and can be produced by specialty manufacturers or custom firms. They are highly decorative and appropriate for use in any contract area where a dramatic look is desired.

- **String Effects**—wall coverings that have very fine vertical threads laminated to a paper-type substrate. Most suitable for offices, boardrooms, and areas of light traffic.
- **Natural Textile Wall Coverings**—natural textiles usually laminated to a backing to enhance dimensional stability and to prevent the adhesive from coming through to the surface. These backings are usually acrylic or paper. Textiles are manufactured in a variety of widths and are constructed of natural fibers. Natural textiles can be finely designed or coarse in texture depending on the desired look.
- **Polyolefin/Synthetic Textile Wall Coverings**—woven and nonwoven-looking wall coverings developed to give the aesthetic appearance of a natural textile while adding an increased value in stain and abrasion resistance. These products are generally put up with an acrylic or paper backing. Many of these products are comprised of polyolefin yarns, which are olefin fibers made from polymers or copolymers of propylene. These types of wall coverings are appropriate for higher traffic areas.
- **Acoustical Wall Coverings**—designed for use on vertical surfaces, panels, operable walls, and anyplace sound reduction is a primary factor, such as meeting rooms, offices, theaters, auditoriums, restaurants as well as corridors and elevator lobbies. These products are predominantly made of man-made polyester and olefin fibers, and are tested for a special sound-absorption rating known as a Noise Reduction Coefficient (NCR) rating. This rating indicates the amount of sound absorbed into the wall. The higher the number, the more noise absorption.

Types and Usage

- **Cork and Cork Veneer**—variegated texture with no definite pattern or design. Cork veneer is shaved from cork planks or blocks and laminated to a substrate that may be colored or plain. Offers some degree of sound resistance; can be used as bulletin boards.
- **Digital Wall Covering**—borders, murals, and wall covering. Unlimited supplies of designs, ideas, and colors. Digital wall coverings allow the person the freedom to express any theme, style, or design on a ground of his or her choice.
- **Wood Veneer**—wood wall coverings mostly laminated to fabric backing. They are usually made in sheets 18 to 24 inches wide and provided in any length up to 144 inches long. Due to characteristics relative to environmental and grain matching, wood veneers are used mostly in the office or conference room environment along with some other specialty areas, such as large columns.
- **Foils**—a thin sheet of metallic material with a paper or fabric substrate. Popular in the 1960s and 1970s. Require a very smooth surface and extreme care when installing. Usage is limited; highly decorative.
- **Mylar** (by DuPont) — wall coverings ground made of vacuum-metallized polyester film laminated to a substrate. Offers a highly reflective surface with an appearance similar to foil with less stiffness.

- **Flocks**—resemble cut velvet and very popular in the sixteenth and seventeenth centuries. Also popular in the 1970s. Produced by laminated shredded fibers to paper, vinyl, Mylar, or foil. Highly decorative, period wall coverings; limited abrasion resistance. For use in low-traffic areas.
- **Underliner**—blank stock-type wall coverings. Comes in different weights such as light, medium, and heavy. Can be plain paper stock or a nonwoven-type material. Liner can be used on almost any wall surface, such as plaster, sheetrock (drywall), paneling, and cinder block. Its purpose is to provide a smooth surface for the installation of wall coverings.

Source: Wallcovering Association, Chicago, ILL.

10.3.1 Basic Types of Fabric-Backed Vinyl Wall Coverings

Wall Covering Basics: Types of Wall Covering

There are three general categories of fabric-backed vinyl wall covering—Type I, Type II, and Type III, which refer to the weight and performance associated with the wall coverings in these categories.

Type I

Type I is a light-duty wall covering. The exact weight of the Type I vinyl wall covering to specify depends on the application or the decorative effects to be achieved.

- Type I lightweight vinyl wall covering is considered to have a weight less than 15 oz. per linear yard (based on a 54″ width) and is produced with a scrim backing or nonwoven fabric. It is intended for use on interior commercial walls where a combination of wall protection and design effect is desired.
- A Type I heavyweight vinyl wall covering is considered to have a weight between 15 oz. and 19 oz. per linear yard (based on a 54″ width) and is produced with a scrim backing or nonwoven fabric. It is recommended for use in corridors or offices where moderate traffic is expected.

The 15 oz. per linear yard (based on a 54″ width) wall covering is the most widely specified weight of Type I vinyl wall coverings.

Type II

Type II is a medium-duty wall covering. It weighs between 20 and 32 oz. per linear yard (based on a 54″ width) and is produced on Osnaburg (poly-cotton or polyester) or nonwoven fabric backing. Type II is considered the "work-horse" among vinyl wall coverings and is typically specified for areas where greater than normal traffic and surface abrasion is evident or expected. It is ideal for offices, hospital wards, public areas and rooms in hotels, lounges, dining rooms, public corridors, and classrooms.

The 20 oz. per linear yard (based on a 54″ width) is the most widely specified weight of Type II vinyl wall coverings.

Type III

Type III is a heavy-duty wall covering. It generally weighs in excess of 33 oz. per linear yard (based on a 54″ width) and is made with a drill fabric backing. Type III wall covering is recommended as a wall protection for areas exposed to extraordinarily hard use, vehicular traffic, and abrasive conditions.

Source: omnova.com

10.3.2 Formulas for Estimating Wall Covering Quantities

Formulas and Estimating

- Measuring Accurately
- Square Foot Area Method

Section | 10 Interior Finishes

- Stairways or Cathedral Walls
- Estimating Commercial Square Footage

Measuring before Estimating Wall Covering Needs

The most important step in estimating wall covering is accurate measurements. Use a yardstick or cloth tape measure. Take measurements in feet rounding off to the next highest half foot or for doors, windows and ceiling height. If a wall is unusually broken up with a fireplace or built-in book shelves, detailed measurements will be beneficial in figuring the square footage of wall covering needed.

Measure wall height from floor to ceiling. Exclude baseboards and moldings. Measure length of windows. Find the total square feet of the wall(s) by multiplying ceiling height by total wall length covered. (Standard doors are about 3 × 7 feet or 21 square feet; standard windows about 3 × 4

These calculations give the total number of square feet to be covered. Using this the number of wall covering can be determined.

For example:

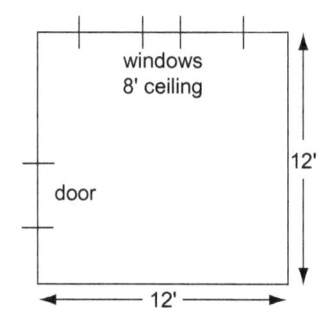

In the above figure, each wall is 12′ long with an 8′ ceiling. Multiply 12 × 8 = 96 square feet for e (since there are four walls with 96 square feet each) = 384 total square feet for the room.

Metric Single Roll

Repeat Length	Usable Yield
0″ to 6″	25 sq. ft.
7″ to 12″	22 sq. ft.
13″ to 18″	20 sq. ft.
19″ to 23″	18 sq. ft.

Estimate Square Footage:

Ceiling height [0] × Total Wall Length [0]

- [Calculate Square Footage]

= [   ] TOTAL SQUARE FEET

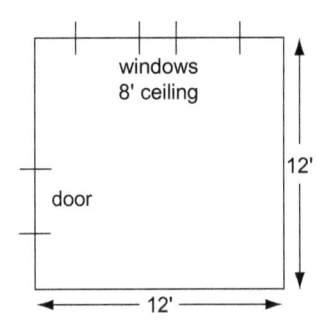

Using the above diagram as an example, figure the amount of wall covering that will be needed for 384 square feet has not taken into account the square footage of the doors and windows. Subtract each opening such as 21 square feet for the door and 12 square feet for each of the windows. 384 sq. 12 + 12 = 45) = 339 square feet of wall space that will be covered with wall covering. If you are using a repeat of 8 inches, figure that each metric single roll will contain 22 square feet of usable walk amount of wall space from above that will be hung) divided by 22 square feet (from Usable Yield round up to 16 metric single rolls (msr) that will be needed to hang the example room (8 metric rolls)

The equation would look like this:

384 sq. ft. (room size)
−21 sq. ft. (one standard door)
−12 sq. ft. (one standard window)
−12 sq. ft. (one standard window)

= 339 sq. ft. of wall space that will be hung

339 sq. ft. / 22 sq. ft. = 15.4 rolls to hang the room, rounded up to 16 rolls

Calculate Square Foot Area Method:

Room Size: [0]

Number of Standard Doors: [0]

Number of Standard Windows: [0]

[Calculate Square Footage]

= [   ] TOTAL SQUARE FEET

Back to top

Stairways or Cathedral Walls

When estimating a wall that has a diagonal, remember that there will be extra waste to allow for this pitch. There are two different types of stairways to figure: one with a horizontal ceiling line, and a ceiling line that parallels the fall of the steps.

Section | 10   Interior Finishes

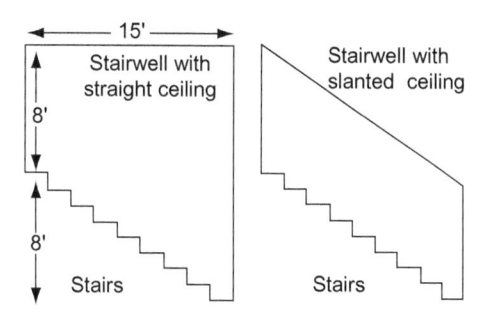

In both cases, the first step is to divide the wall into either squares or rectangles to determine the above, the upstairs ceiling height is 8″, and the downstairs ceiling height is 8″. Next, measure wall width horizontally from the top of the stairs to an imaginary vertical line which in the example is 15″. Taking the top rectangle, figure 8″ × 15″ = 120 sq. ft. Next, figure this sq. ft., but since a portion of this wall area is under the stairs, multiply the bottom rectangle sq. ft standard. Add the two figures together to arrive at the square feet that needs to be hung with will look as follows:

8″ × 15″ = 120 sq. ft. (top rectangle)
8″ × 15″ = 120 sq. ft. × 65% = 78 sq. ft. (bottom rectangle)
120 sq. ft. + 78 sq. ft. = 198 sq. ft.

Once you have the square footage figured, estimate the amount of wall covering just as you would the usable square feet for the particular pattern from the Usable Yield Chart and then dividing square footage of each roll to determine number of rolls required.

Formulas and Estimating

For example, if one were using a wall covering with a repeat of 14″, each msr would contain 20 square feet. The equation would look as follows: 198 sq. ft. / 20 sq. ft. = 9.9 msr rounded to 10 msr. If the stair the first example in finding the width and length of the imaginary rectangle or square. The next rectangle/square figures multiplied by 65% to find the square feet of wall area. The equation will look like this:

8″ × 15″ = 120 sq. ft. (top rectangle)
8″ × 15″ = 120 sq. ft. (bottom rectangle)
120 sq. ft. + 120 sq. ft. = 240 sq. ft.
240 sq. ft. × 65% = 156 sq. ft. of wall area to be covered

Using the same pattern with a repeat of 14″, each msr would contain 20 square feet of usable would look as follows:

156 sq. ft. /20 sq. ft. = 7.8 msr rounded to 8 msr

A cathedral ceiling would be estimated the same way, squaring the top rectangle, multiplying then adding that figure to the square feet of the bottom rectangle.

Calculate Square Footage for Stairways with Straight Ceiling

Square Footage of Top Rectangle: [0]

Square Footage of Bottom Rectangle: [0]

[Calculate Square Footage]

= [0]   TOTAL SQUARE FEET

Calculate Square Footage for Cathedral Ceilings

Square Footage of Top Rectangle: [0]

Square Footage of Bottom Rectangle: [0]

[Calculate Square Footage]

= [0] TOTAL SQUARE FEET

Back to top

Estimating Commercial Square Footage

After the wall covering has been determined from the specification, now figure the square footage of the job. Once the width is known, the number of square feet in a lineal yard for that particular is determined. An important formula to remember is as follows:

Width divided by 12 = number of feet
Number of feet multiplied by 3 (1 yard) = square feet/width (square feet per lineal yard)
Divide square feet of wall space to cover by square feet/width

For example:
54-inch-wide material used to cover 1500 square feet is figured
54 divided by 12 = 4.5
4.5 multiplied by 3 = 13.5 square feet per lineal yard
1500 divided by 13.5 = 111.11

1500 square feet of wall space would require 112 yards without waste.

Once the width and the square footage for the width are known, any amount can be determined. If the yardage for a particular width is known, and the width of the material is changed, to convert from one width to another, yardage.

For example:
150 yards of 54-inch-wide material
54 inches is 13.5 square feet per yard (54/12 = 4.5 × 3 = 13.5)
150 yards multiplied by 13.5 = 2,025 square feet of wall space to cover

Formulas and Estimating

36/12 = 3
3 × 3 = 9 square feet
2,025 divided by 9 = 225 yards of 36-inch wide wall covering instead of original 1 material.

These are exact yardage amounts and do not allow for waste caused by pattern repeat. However, a matching pattern with a large repeat would require additional material pattern without a match. All contractors should be aware of the pattern, width, and match before ordering.

Estimate Commercial Square Footage:

Width of Material (in inches): [0]

Square Footage to be covered: [0]

[Calculate Material Needed]

= [ ] YARDS NEEDED (Without Waste)

Section | 10 Interior Finishes

10.3.3 Calculating How Much Wallpaper Is Required

Wallpaper is usually sold in double rolls, each roll of which contains 30 sf.

The following calculations are based on determining wall square footage and assuming various ceiling heights for each size room.

Room Size	W/8 Foot Ceiling	w/9' Ceiling	w/10' Ceiling	w/12' Ceiling	Single Rolls for Ceiling
8 × 10	10	11	12	14	3
9 × 12	11	13	14	17	4
10 × 10	11	13	15	17	4
10 × 12	12	13	15	18	4
10 × 14	13	14	16	19	5
12 × 14	14	16	18	21	6
12 × 16	15	17	19	22	7
12 × 18	16	18	20	24	8
12 × 20	17	19	21	26	8
14 × 16	16	19	20	24	8
14 × 18	17	19	21	26	8
14 × 20	18	20	22	27	10
14 × 22	19	22	24	29	11
16 × 18	19	20	23	28	10
16 × 20	19	22	24	29	11
16 × 22	21	23	25	30	12
16 × 24	21	24	27	32	13
18 × 20	21	23	25	30	12

As a contingency add 10% for waste.

10.3.4 Wall Covering Adhesives

There are many types of wall covering adhesives, each formulated for various performance characteristics. These characteristics fall into two general categories:

- How they bond the wall covering to the wall surface.
- How they apply to the wall covering (this is a major consideration for the installer).

Wall covering adhesives are formulated for specific applications. Some adhesives are formulated for lightweight and delicate fabrics, whereas others are designed to adhere to heavyweight vinyl and acoustical coverings.

Adhesives vary in level of wet-tack, solids, open-time, strippability, and ease of application. All wall covering adhesives contain a biocide system. These systems are designed to prevent bacterial contamination and mildew/fungal infestation both "in-the-can" and in the dried adhesive. Wall covering adhesives are generally applied on the back of the wall covering either by roller or pasting machine. However, nonwoven acoustical wall coverings require the adhesive be applied to the wall using a brush or roller. Consult the manufacturer's installation instructions.

There are four main categories of adhesives:

- **Prepasted Activators**—were created to assist in the hanging of prepasted wall coverings by activating the existing adhesive, increasing slip, and extending the open-time available (helpful when matching patterns). Using activators is generally considered less messy, minimizes seam splitting, and reduces seam lifting.
- **Clear Adhesives**—are either corn or wheat based. They are generally considered to have more open-time and are easier to clean up than clay-based adhesives. Clear adhesives are designed for both the retail customer and commercial installer.
- **Clay Adhesives**—like clear adhesives are starched based, but clay is added as a filler to increase the wet-tack and level of solids. Clay adhesives are generally not recommended for retail customers. They are considered to have a higher level of wet-tack and are more difficult to clean up versus clear adhesives.
- **Vinyl-Over-Vinyl**—and border adhesives contain synthetic polymers and are specifically formulated to bond to vinyl. In addition to the synthetic polymers, these adhesives may contain some starch and other ingredients to assist in their application. These adhesives require extra attention during application because once they are dried they are permanent. Vinyl-over-vinyl adhesives are specifically designed to hang new wall coverings to existing wall coverings or borders to wall coverings.

Source: Wallcovering Association, Chicago, ILL.

10.4.0 General Information to Calculate Various Types of Floor and Wall Tiles

Floor and wall tile

The following paragraphs include information pertaining to the various types of tile and their installation procedures.

The number of tiles needed is calculated by performing the following procedures:

First, calculate the square feet of the area to be tiled. If you are using 12-inch-square tiles, the total floor area (in square feet) equals the total number of tiles needed, plus an additional 10 percent waste factor. If another size of tiles is being used, multiply the area by 144 to convert to square inches. Then divide that number by the area (square inches) of the tiles to find the required amount (include a 10 percent waste factor).

Example: You are using tiles 9 × 9 inches. To tile a floor 12 feet long and 9 feet wide—

Multiply the room dimensions to find the area: *12 feet × 9 feet = 108 square feet*

Multiply the area by 144: *108 × 144 = 15,552 square inches*

Calculate the area of the tile: *9 inches × 9 inches = 81 square inches*

Divide the room area (square inches) by the tile area (square inches): *15,552 divided by 81 = 192 tiles*

Add 10 percent waste factor: *192 + 19 = 211 tiles required*

Resilient floor tile

Resilient floor tile is durable, easily maintained, comfortable and attractive, and low cost. It is made of rubber, vinyl, linoleum, and asphalt. Common sizes of this tile are either 9 × 9 inches or 12 × 12 inches.

A notched trowel (used for spreading adhesive) and a tile cutter are required for installation. To lay out and install resilient floor tile, perform the following procedures:

Locate the center of the end walls of the room. Establish a main centerline by snapping a chalk line between these two points. Lay out another centerline at right angles to the main centerline. This line may be established using a framing square or the triangulation method. With the centerline established, make a trial layout of the tiles along the centerlines. Measure the distance between the wall and the last tile. If this measurement is less than 1/2 tile, move the centerline half the width of the tile, closer to the wall. This adjustment will eliminate the need to install border tiles that are too narrow. Since the original centerline is moved exactly half the tile size, the border tile will remain uniform on opposite sides of the room. Check the layout along the other centerline in the same way.

Spread adhesive over one quarter of the total area, starting with the quarter farthest from the door and working toward the door. Ensure that the floor surface is clean before you spread the adhesive. Spread up to the chalk lines but do not cover them. Be sure to use a notched trowel with the notch depth recommended by the manufacturer of the adhesive. Allow the adhesive to take an initial set before setting the first tile. The time required will vary, depending on the type of adhesive used.

Source: Constructionknowledge.org

Start laying the tiles at the center of the room. Make sure the edges of the tiles are aligned with the chalk line. Lay rows by width, stair-stepping additional rows and ensuring that the tiles are tight against one another in a cross-grained pattern unless otherwise specified. After all of the full tiles have been laid, install the border or edge tiles around the room. To lay out a border tile, place a loose tile over the last tile in the outside row with the grains running in opposite directions (if using a cross-grained pattern). Then, take another tile and place it in position against the wall and mark a pencil line on the first tile. Cut the tile along the marked line.

After all the tiles have been installed, remove any excess adhesive using a cleaner or solvent and procedures approved by the manufacturer.

Ceramic and other specialty tiles

This tile is used extensively where sanitation, stain resistance, easy cleaning, and low maintenance are desired. Types of tile include ceramic, mosaic, paver, quarry, brick-veneer, cement-bodied, marble, and other stone tiles. These can be used for both interior and exterior flooring. Tile is used on both walls and floors. Field tile is regular tile placed on all courses in the main field of an installation. Trim tile is a specially shaped tile used to border and complete the main field of tile; it is available in a wide variety of shapes, sizes, and colors to match field tile.

Tiles come with two types of finishes—glazed and unglazed. Glazed tiles are coated with a glaze before firing to give the tile color and to preserve its surface. They may be fired to a smooth or textured finish. Glazed tiles are most commonly used for walls but may also be applied to floors and countertops. They are used mainly for interiors. Unglazed tiles are fired without a glaze coating. They derive their color from the clay from which they are made. Adhesives used are Thinset or Organic Mastic. Thinset is a powdered cement-based product that is mixed with either water, a latex or acrylic additive, or epoxy. It is very versatile. Organic Mastic is premixed in a solvent or latex base. It may deteriorate if exposed to heat or water.

Grout is a powder made from sand and cement and is used to seal the cracks between the tiles. It is mixed with either water or, to increase durability, an additive. It is available in a variety of colors.

The following tools and equipment are required for installing ceramic and specialty tiles:

- A *striking tool is* used to compact the grout into the joints.
- A *beating block* is a board used to even the tile surface after it has been set.
- A *square-notched trowel* is used to spread adhesive.
- A *pointing trowel is* used to spread adhesive in tight spots.
- A *tile cutter is* used to score the tile surface so that it can be snapped by applying pressure to the score.
- A *fine file or tile stone is* used to smooth rough edges after cutting tile.
- A *time nipper is* used to clip tile and cut irregular openings.
- A *squeegee or sponge is* used to remove excess grout from the tile surface.
- A *sponge float or rubber-faced trowel* is used to spread grout over the surface.
- An *electric tile saw is* similar to a mason saw. It is used to make clean, accurate cuts.
- To lay out and install ceramic and specialty tiles, perform the following procedures:
- Check the area to be tiled to determine if it is square. If the area is slightly out of square, minor changes in the layout can accommodate these conditions. If the area is seriously out of square, the process stops for any required structural repairs or surface preparation. If the framing problems are serious, it may not be possible to tile the area.

Draw the layout on paper. Layout depends greatly on the pattern desired and the type, size, and shape of the tile being used. Use as many full tiles and as few cut tiles as possible. Place cut tile away from visual focal points (doorways, thresholds, and so forth); tiles should be set symmetrically for a more attractive finish and appearance.

Place reference lines on the floor or wall. Once the layout has been established on paper, transfer it to the floor or wall. A reference line should be snapped to mark the rows of cut tiles around the perimeter. A grid of reference lines should be snapped to enclose all full tiles in sections no larger than 3 square feet.

To install tiles, first spread adhesive over a small area or section (3 × 3 feet), making sure to spread it just up to the lines so that the lines will still be visible. Align the first tile against a 90° intersection in the grid and press it gently into the adhesive. After each course of tile is applied, use a beating block to level the surface. After all the tiles are set, allow the adhesive to set the required time, according to manufacturer's instructions. Prepare the grout and spread it over the tile surface, ensuring that the joints are filled. When the grout begins to dry, clean the tile with a damp sponge. After the grout has dried, wipe off the haze with a clean rag or towel. After the grout has completely dried and hardened (approximately 72 hours), a grout sealer may be applied.

10.4.1 Calculating Requirements for Ceiling Tile

Suspended ceilings are primarily designed for acoustical control; however, ceilings are also lowered to save on heating and air conditioning expenses; finish off exposed joints; and cover damaged plaster.

Acoustical tile

Acoustical tile absorbs sound, reduces noise, reflects light, and resists flame. Its thickness ranges from 3/16 to 3/4 inch; its width from 12 to 30 inches; and its length from 12 to 60 inches. The most common size panels used are 2 × 2 feet and 2 × 4 feet.

Grid-system components

The grid-system components used in suspended ceilings include the following: the *main tee* (12-foot lengths), the *cross tee* (2- and 4-foot lengths), the *wall angle* (10-foot lengths), the *splice plate* (available in aluminum only), *suspending devices,* and *suspending wire.*

Suspending devices include screw eyelets; suspending hooks and nails; 8d common nails or larger, driven into wood joists and bent into a U-shape; and an approved Hilti fastener for concrete or steel.

Suspending wire includes 16-gauge anneal wire placed at 4-foot intervals and attached to suspending devices at the ceiling and to the main tees in the grid system.

Installation

First, lay out the grid pattern. This is based on the ceiling's length and width at the new ceiling height. If the ceiling's length or width is not divisible by 2 feet, increase to the next higher dimension divisible by 2 feet. For example, if a ceiling measures 12 feet 7 inches × 10 feet 4 inches, the dimensions should be increased to 14 feet × 12 feet for layout purposes. Draw the layout on paper. Make sure that the main tees run perpendicular to the joists. Position the main tees so that the border panels at the room's edges are equal and as large as possible. Draw in cross tees so that the border panels at the room's ends are equal and as large as possible. Determine the number of pieces of wall angle by dividing the perimeter by 10 and adding 1 additional piece for any fraction. Determine the number of main tees and cross tees by counting them on the grid pattern layout.

Next, establish the ceiling height. Mark a line around the entire room at the desired height to serve as a reference line. There must be a minimum of 2 inches between the new ceiling and the existing ceiling. Ensure that this line is level and marked continuously so that it meets at intersecting corners. Next, install the wall angle. Secure the wall angle along the reference line, ensuring that it is level.

Install the suspension wire. Suspension wires are required every 4 feet along the main tees and on each side of all splices. Attach the wires to the suspending devices. The wires should be cut at least 2 feet longer than the distance between the old and new ceiling. Now, install the main tees. Main tees need to be laid out from the center to ensure that the slots line up with the cross-tee locations. Cut them where appropriate. Tees 12 feet long or less are installed by resting the ends on opposite wall angles and inserting the suspension wire. Tees over 12 feet long must be cut to ensure that the cross tees will not intersect the main tee at a splice joint. Rest the cut end on the wall angle and attach suspension wires along the tee. Make necessary splices and continue attaching suspension wires along the tee until the tee rests on the opposite wall angle. Ensure that the main tees are level and secured before continuing.

Install the cross tees. Cut and install border tees on one side of the room. Install the remaining cross tees according to the grid-pattern layout. At opposite wall angles, install the remaining border tiles. Finally, install the acoustical panels. Install the full-size panels first. Handle panels with care and ensure that the surfaces are kept clean from hand prints and smudges. If you are working on a large project, work from several cartons to avoid a noticeable change of uniformity. Cut and install the border panels.

Source: Constructionknowledge.org

10.4.2 Painting Ceilings and Walls

The following tools and equipment are required for painting:

- Paint brushes, wall, 2 to 4 inches wide
- Paint roller with cover
- Paint pan
- Stepladder
- Paddle (stir stick)
- Rags
- Paint, latex, flat
- Bucket of water

Prepare the paint for application. Remove the cover from the paint container. Remove any film layer from the top of the paint. Using the paddle (stir stick), mix the paint thoroughly, in a figure-8 motion.

Scrape off and break up any unsettled matter on the bottom or lower sides of the container. Pour the paint into the paint pan until it is 2/3 full.

Ceiling

Brush a narrow strip of paint around the perimeter of the ceiling along the inside edges where the wall and the ceiling meet. Using a roller, paint the remaining portion of the ceiling. Cross roll to ensure complete paint coverage without voids.

Walls

Brush a narrow strip of paint along the inside corners of the wall and corner post. Cut in around all trim and baseboards with a trim brush. Using a roller, paint the remaining portion of the wall and corner post. The corner post may be painted with a brush. When the first coat has completely dried, apply a second coat in the same way. Ensure that the entire surface is covered and without voids.

10.5.0 Types of Carpeting

Having a basic understanding of the various types of carpet construction will help you find the best carpet for your specific situation. For example, some types of carpeting do better in high-traffic areas than others. Some carpets have a more casual appearance, while others look more formal.

Carpeting is basically constructed in two types of piles: **loop pile** or **cut pile**. **Patterned carpet** is a combination of both loop and cut yarns or variations of loops set to different pile heights.

Today's carpet manufacturers and fiber producers are combining softer fibers and better built-in stain resistance with special backings that block odors, spills, and more! These types of features are built-in to many of the better quality carpets offered by the leading carpet manufacturers. For example: see our page on Mohawk SmartStrand Carpet or Shaw Carpets with SoftBac Platinum for some major advancements in carpet construction technology.

Most carpeting today is tufted rather than woven. The tufting process is similar to using a large sewing machine that is about 13 feet wide with 100 of needles that sew the yarn into a synthetic backing. The majority of carpet sold today are made from nylon fibers (such as Solutia's Wear-Dated ® carpet fibers) and twisted into yarns and then tufted into carpet.

How long a carpet will retain its like-new texture and appearance is based on the type of carpet fiber, how tight the yarns are twisted and heatset, and the pile density. Also, how tightly the tufts are packed together will also affect the long-term durability of a carpet. Obviously, not cleaning your carpets regularly, or using improper cleaning methods will affect the appearance and life of your carpet as well.

See **Berber carpets** for information about berber carpet styles.

Below are listed some of the common types of carpet constructions and their main features:

By permission: floorfacts.com

Types of carpeting | carpet construction

Cut Pile Saxony Carpets

- Generally made in solid colors
- Surface has a smooth appearance
- Generally made with nylon, wool, or polyester fibers
- Good performance and appearance
- Works well with traditional or formal room settings

Textured Cut Pile Saxony Carpet

- Surface appearance is textured
- Stylish, casual appearance
- Won't show vacuum cleaner marks or footprints
- Very popular carpet style today
- Good choice for active areas of the home

Frieze Carpet

- Very textured, knobby surface appearance
- Extremely durable and excellent wearing
- Yarns very tightly twisted
- Will cost more than textured cut pile carpets
- Great for active areas of the home

Cut and Uncut Patterned Carpet

- Intermixed loops and cut pile, creates a patterned design
- Loops are shorter than the cut pile, creating a carved appearance
- Usually constructed in multicolor designs
- Helps hide footprints and traffic patterns
- Great choice for a variety of room settings

Multilevel Loop Carpets

- Have several different heights of loops
- Generally multicolored
- Very durable, casual appearance
- Offered in many unique-looking designs and patterns
- Helps hides traffic patterns
- Great for family rooms, basements, etc.

Level Loop Carpets

- Loops are same height and generally multicolored
- Usually made from polypropylene (olefin) carpet fibers
- Often called **Indoor-Outdoor** or **Commercial Carpet**
- Casual appearance but extremely durable
- Great for family rooms or basements

10.5.1 Calculating the Amount of Carpet Required—Rooms 8–35 Feet in Length and 13–20 Feet in Width

Your Room Size Length	Width	Carpets Are 12 Feet Wide	13 ft	14 ft	15 ft	16 ft	17 ft	18 ft	19 ft	20 ft
8 ft		10.66 yd	11.55	12.44	13.33	14.22	15.11	16	16.88	17.77
9 ft		12 yd	13yd	14yd	15yd	16yd	17	18	19	20
10 ft		13.33	14.44	15.55	16.66	17.77	18.88	20	21.11	22.22
11 ft		14.66	15.88	17.11	18.33	19.55	20.77	22	23.22	24.44
12 ft		16	17.33	18.66	20	21.33	22.66	24	25.33	26.66
13 ft		17.33	18.77	20.22	21.66	23.11	24.55	26	27.44	28.88
14 ft		18.66	20.22	21.77	23.33	24.88	26.44	28	29.55	31.11
15 ft		20	21.66	23.33	25	26.66	28.33	30	31.66	33.33
16 ft		21.33	23.11	24.88	26.66	28.44	30.22	32	33.77	35.55
17 ft		22.66	24.55	26.44	28.33	30.22	32.11	34	35.88	37.77
18 ft		24	26	28	30	32	34	36	38	40
19 ft		25.33	27.44	29.55	31.66	33.77	35.88	38	40.11	42.22
20 ft		26.66	28.88	31.11	33.33	35.35	37.77	40	42.22	44.44
21 ft		28	30.33	32.66	35	37.33	39.66	42	44.33	46.66
22 ft		29.33	31.77	34.22	36.66	39.11	41.55	44	46.44	48.88
23 ft		30.66	33.22	35.77	38.33	40.88	43.44	46	48.55	51.11
24 ft		32	34.66	37.33	40	42.66	45.33	48	50.66	53.33
25 ft		33.33	36.11	38.88	41.66	44.44	47.22	50	52.77	55.55
26 ft		34.66	37.55	40.44	43.33	46.22	49.11	52	54.88	57.77
27 ft		36	39	42	45	48	51	54	57	60
28 ft		37.33	40.44	43.55	45.66	49.77	52.88	56	59.11	62.22
29 ft		38.66	41.88	45.11	48.33	51.55	54.77	58	61.22	64.44
30 ft		40	43.33	46.66	50	53.33	56.66	60	63.33	66.66
31 ft		41.33	44.77	48.22	51.66	55.11	58.55	62	65.44	68.88
32 ft		42.66	46.22	49.77	53.33	56.88	60.44	64	67.55	71.11
33 ft		44	47.66	51.33	55	58.66	62.33	66	69.66	73.33
34 ft		45.33	49.11	52.88	56.33	60.44	64.22	68	71.77	75.55
35 ft		46.66	50.55	54.44	58.66	62.22	66.11	70	73.88	77.77

By permission: www.abccarpets.com

10.6.0 Solid Hardwood Flooring

Solid wood floors have been used for centuries and never seem to lose their charm and warmth. We generally think of solid hardwood floors as a 3/4" thick plank that comes in a narrow 2 1/4" strip and has to be finished on the job site. This is the classic hardwood strip floor.

Today, manufacturers offer solid hardwood floors in a variety of widths, thicknesses, finishes, and wood species. The most common North American hardwood species used for solid wood flooring are red oak, white oak, ash, and maple, but you can also get solid hardwood flooring in many exotic wood species, such as Brazilian cherry, tiger wood, Australian cypress, and many others from around the world. Red oak is still the most popular and commonly used hardwood floor.

When we talk about unfinished wood flooring, we generally think of solid wood floors. Unfinished solid oak floors come in several different qualities. These qualities are clear, select and better, #1 common, and #2 common. The clear has no visual blemishes or knots and is extremely expensive. The select and better quality has some small knots and very little dark graining, while the #1 common and #2 common have more knots and more dark graining. When buying an unfinished solid wood floor, make sure you know which quality you are buying.

Solid wood planks are cut out as a solid block right from the tree. The wood blocks are then sawn into solid flooring planks with tongue and grooves edges. The planks are than either prefinished at the factory or placed into unfinished bundles of varying lengths.

All solid wood floors will react to the presence of moisture. In the winter heating months the lack of humidity can cause solid wood floors to contract, which leaves unsightly gaps between each plank. In the summer months when the humidity is higher the wood planks will expand and the gapping will disappear. If there is too much moisture present, the wood planks may cup or buckle. This is why it is so important to leave the proper expansion gap along all vertical walls and to acclimate the solid wood planks prior to installation. (Engineered wood planks are not nearly as affected by humidity as solid wood floors.)

By permission: floorfacts.com

10.6.1 Janka Wood Hardness Scale for Wood Flooring Species

Thus a common use of the Janka Hardness Scale is to determine a woods suitability as a wood for hardwood flooring. The higher the number, the greater its resistance to denting as it lives life.

For the woodworker it will be a good indicator of the difficulty of sawing and nailing—or maybe just moving the wood around your shop...

California Redwood	420
Douglas Fir	660
Southern Yellow Pine (loblolly & short leaf)	690
Honduran Mahogany	800
African Mahogany	830
South American Lacewood	840
Southern Yellow Pine (longleaf)	870
Black Cherry	950
American Black Walnut	1010
American Black Walnut Hardwood Flooring	
Peruvian Walnut	1080
Brazilian Eucalyptus	1125

Species	
Teak	1155
Bamboo (carbonized)	1180
Larch	1200
Heart Pine	1225
Caribbean Heart pine	1240
Yellow Birch	1260
Red Oak (Northern)	1290
American Beech	1300
Ash	1320
White Oak	1360
Australian Cypress	1375
Bamboo (natural)	1380
Royal Mahogany	1400
Hard Maple	1450
African Walnut/Sappelle	1500
Brazilian Maple	1500
Zebrawood	1575
Wenge	1630
Brazilian Oak	1650
Bamboo	1650
Patens	1691
Peruvian Maple	1700
Kempas	1710
African Pedauk (Padeuk/African Cherry)	1725
Bolivian Rosewood /Morado	1780
Hickory/Pecan	1820
Kempas	1854
Purpleheart	1860
Jarrah	1910
Amendoim	1912
Merbau	1925
African Rosewood (Bubinga)	1980
Grapia	2053
Jarrah	2082
Purple Heart	2090
Tigerwood	2160
Burma Mahogany	2170
Amberwood	2200
Cabreuva (Santos Mahogany)	2200
Caribbean Rosewood	2300
Mesquite	2345
Brazilian Cherry (Jatoba)	2350
Peruvian Cherry	2350
Red Walnut	2450
African Cedar / Bosse	2600
Patagonian Rosewood	2800
Bloodwood	2900
Brazilian Rosewood (Tamarindo)	3000
Brazilian Redwood	3190
Tiete Rosewood	3280
Cumaru (Brazilian Teak)	3540
Southern Chestnut, Tiete Chestnut	3540
Lapacho (usually grouped with IPE Wood) Ipe Lumber	3640
Bolivian Cherry	3650

(Continued)

African Pearwood/Moabi	3680
Brazilian Walnut / Ipe Ipe Decking	3680
Brazilian Ebony	3692
Patagonian Rosewood	3840
Brazilian Tiger Mahogany	3840
Curupy	3880

Source: woodsthe best.com

10.6.2 Laminate Flooring

Laminate Flooring Construction Review

Laminate flooring comes in both planks and square tiles. It is constructed with several different layers of various materials that are *thermofused* together to form the laminated flooring planks and tiles.

The four basic laminate flooring construction layers are as follows:

- **Wear Layer**—This is the transparent top surface that protects the floor from scratching, staining, and scuffing and also protects the printed design layer below. The wear layer is a combination of **melamine with aluminum oxide particles** which makes it extremely durable.
  - **In-Register Embossing**—Many manufacturers have developed specialized methods of texturizing the top layer (called *in-register embossing*) to add more authentic realism to the flooring. Many also offer beveled plank edges to give the floors even more of a realistic appearance.
  - **AC Ratings**—Laminate flooring manufacturers have also adopted a method of scoring the durability of the top layer to help consumers with choosing the right laminate floor for their situation. This is called the *AC Ratings*. **AC** stands for **Abrasion Coefficient.** The **AC Ratings go from AC1 to AC5**, with AC5 being the best. Both the in-store samples and laminate flooring cartons should have their AC Rating marked for consumers to see. For very active areas and kids' playrooms, it's best to choose a laminate floor with an AC Rating of AC3 or greater.
- **Photographic Image Layer**—This is the photographic image layer of either a real hardwood plank, ceramic tile, stone, or some other material. The photographic images are extremely clear, vibrant, and realistic. Combined with texturizing the top layer, this creates a true, authentic looking, natural floor appearance. For example, some laminate designs are actually photographic images of old historical floors.
- **Inner Core Layer**—The inner core is generally made from high-density fiberboard and is also used to form the tongue and groove edges for locking laminated planks together. The core is also the base for the photographic image and wear layer. Most manufacturers also saturate the inner core with melamine resins or a water-resistant sealer to help protect the inner core from moisture.
- **Backing Layer**—This layer is fused to the inner core to add stability and create a barrier that helps protect the planks from moisture and warping. Like the inner core the backing is also treated with some sort of water-resistant sealer.

Note: The inner core combined with the backing layer is what really makes up the overall thickness of each plank. Planks generally range from 8 mm to 12 mm in thickness. The thicker planks are more rigid and help overcome minor irregularities in the subflooring.

By permission: floorfacts.com

Types of Laminate Floors

Today, there are several different types of laminate floors to choose from, as well as different shapes, thicknesses, and installation locking systems. What they all have in common is they must be floated over the subfloor and have a print layer to give a realistic appearance of a real natural floor material. Laminate floors can be installed on all grade levels and over fully cured concrete slabs, wood subfloors, and some types of existing hard surfaces floors.

Basic Types of Laminate Floor Construction

High-Pressure Laminate (HPL)— planks are usually fused together in either a one- or two-step process. Several layers are first glued together, and then these layers are combined with the remaining materials and than glued and fused into a plank. This gives a harder finish, more durable plank than DPL.

Direct Pressure Laminate (DPL)—all materials are fused together in one step, which reduces the costs of manufacturing.

Laminate Floor Thickness—laminate floors come in various thicknesses from around 7–8 mm to 12 mm thick. The thicker planks will be slightly more costly but should be sturdier and more durable, especially if the subfloor is not perfectly level.

Laminate Floor Locking Systems

Depending on the manufacturer, you will find several different types of locking systems for installing the planks together. There are mechanical locking systems (reinforced from underneath by an aluminum, mechanical locking system), specially designed tongue and groove fiber core locking systems, and a few tongue and groove pre-glued systems as well. By far the most popular are the specially designed tongue and groove locking systems that are part of the middle fiber core of each plank. These floors are often referred to as: **glueless laminate floors**. Most of these glueless floors are snapped together by holding the plank at a 45 degree angle and pressing the tongue into the groove of the other plank.

Laminate Floor Styles

Laminate floor styles have improved dramatically over the past few years. The print layers have become much better and more realistic looking. Also, some manufacturers have added what is called "embossed in register." This means the plank's surface has the realistic graining and textures found in natural flooring products. Some laminate wood planks now have micro-beveled edges giving the look of many hardwood floors. Obviously, the more realism and more rich design styles will cost more than the lesser grades.

Laminate floor designs are offered in wood plank designs, ceramic tile designs, and natural stone and slate patterns. The tile patterns are usually in squares, although some are offered in long rectangular planks. The patterns will repeat every 3–4 planks in the box. So be sure to lay the planks out and do some dry laying to view the design before actual installation.

Laminate Construction

The North American Laminate Flooring Association (NALFA) has a Certification Seal for laminate flooring manufacturers. The Seal certifies that the laminate floor has passed a rigorous and demanding series of ANSI tests designed to evaluate the performance, durability, strength, and overall quality of the laminate flooring. Look for it on the manufacturer's sample boards.

The laminated planks are usually fused together in either a one- or two-step process. In the two-step process several layers are first glued together, and then these layers are combined with the remaining materials and then glued and fused into a plank. This method is called High Pressure Laminate (HPL). The other method is that in which all materials are fused together in one step, and this is called Direct Pressure Laminate (DPL).

AC Ratings Overview

- **AC1**—floors with this rating are suitable for low-traffic areas, such as bedrooms
- **AC2**—floors suitable to low to medium traffic, such as living rooms or dining rooms
- **AC3**—floors suitable for most areas in homes, including hallways and light commercial
- **AC4**—anywhere in the home as well as commercial buildings. For example: an office or store
- **AC5**—can be used in heavy-traffic commercial areas.

Laminate Flooring Installation Systems

The planks have tongue and grooved edges on all four sides to secure the planks together. Today, most laminate floors use some sort of **glueless locking system**, often referred to as "clic" floors. Glueless laminate floors can go almost anywhere in the home and are ideal for do-it-yourself projects.

The two main glueless locking systems either involve a tongue and groove that is reinforced from underneath by an aluminum, mechanical locking system or a tongue-and-groove glueless locking system built right into the middle core that allows the planks to snap or click together during installation.

Some other laminate floors have a tongue that was pre-glued at the factory with a specially formulated, water-resistant glue. Once the tongue is moistened with a wet sponge, it activates the glue and locks the planks together. Laminate floors are also offered that require specially formulated glue to be applied to the tongue and groove at the time of the installation to secure the planks to one another.

Laminate Flooring Definitions for Some Commonly Used Terms

- **Backing**—is usually a melamine plastic layer used to give additional structural stability and more moisture protection to the planks.
- **Core**—generally made from high-density fiber board (HDF), particleboard, or plastic, the core adds impact resistance, and forms the tongue and groove locking system. Melamine plastic resins are also impregnated in the core by some of the manufacturers to improve the moisture resistance of the core.
- **Melamine**—is a plastic-type resin used throughout the construction process to add durability and stability to the laminated planks.
- **Print Film**—is also called the decorative layer and gives the floor the appearance of a real hardwood or tile. Some manufacturers have been able to replicate the old wood floors found only in some old historical buildings.
- **Wear layer**—is a tough clear melamine layer with aluminum oxide particles. Using heat and pressure, the wear layer becomes an incredibly hard and durable finish. The resin-filled wear layer is so dense it becomes extremely difficult to stain, scratch, or burn.
- **Underlayment**—is a clear thin plastic sheet that is installed over the substrate before the laminate floor is floated. The plastic sheet helps the laminate floor to float freely above the substrate.

10.7.0 Finishing of Interior Wood

Interior finishing differs from exterior finishing primarily in that interior woodwork usually requires much less protection against moisture but more exacting standards of appearance and cleanability. A much wider range of finishes and finish methods are possible indoors because weathering does not occur. Good finishes used indoors should last much longer than paint or other coatings on exterior surfaces. The finishing of veneered panels and plywood may still require extra care because these wood composites tend to surface check.

Much of the variation in finishing methods for wood used indoors is caused by the wide latitude in the uses of wood— from wood floors to cutting boards. There is a large range of finishing methods for just furniture. Factory finishing of furniture is often proprietary and may involve more than a dozen steps. Methods for furniture finishing will not be included in this publication; however, most public libraries contain books on furniture finishing. In addition, product literature often contains recommendations for application methods. This section will include general information on wood properties, some products for use in interior finishing, and brief subsections on finishing of wood floors and kitchen utensils.

Color change of wood can sometimes cause concern when using wood in interiors, particularly if the wood is finished to enhance its natural appearance. This color change is a natural aging of the newly cut wood, and nothing can be done to prevent it, except, of course, to keep the wood in the dark. The color change is caused by visible light, not the UV radiation associated with weathering. It is best to keep all paintings and other wall coverings off paneling until most of the color change has occurred. Most of this change occurs within two to three months, depending on the light intensity. If a picture is removed from paneling and there is a color difference caused by shadowing by the picture, it can be corrected by leaving the wood exposed to light. The color will even out within several months.

To avoid knots, the use of finger-jointed lumber has become common for interior trim. As with exterior wood, the quality of the lumber is determined by the poorest board. Pieces of wood for finger-jointed lumber often come from many different trees that have different amounts of extractives and resins. These extractives and resins can discolor the finish, particularly in humid environments such as bathrooms and kitchens. When finishing finger-jointed lumber, it is prudent to use a high-quality stain-blocking primer to minimize discoloration.

10.7.1 Opaque, Transparent Finishes, Stains

Opaque Finishes

The procedures used to paint interior wood surfaces are similar to those used for exterior surfaces. However, interior woodwork, especially wood trim, requires smoother surfaces, better color, and a more lasting sheen. Therefore, enamels or semigloss enamels are preferable to flat paints. Imperfections such as planer marks, hammer marks, and raised grain are accentuated by high-gloss finishes. Raised grain is especially troublesome on flat-grain surfaces of the denser softwoods because the hard bands of latewood are sometimes crushed into the soft earlywood in planning, and later expand when the wood moisture content changes. To obtain the smoothest wood surface, it is helpful to sponge it with water, allow it to dry thoroughly, and sand before finishing. Remove surface dust with a tack cloth. In new buildings, allow woodwork adequate time to come to equilibrium moisture content in the completed building before finishing the woodwork.

To effectively paint hardwoods with large pores, such as oak and ash, the pores must be filled with wood filler (see subsection on wood fillers). The pores are first filled and sanded; then interior primer/sealer, undercoat, and top coat are applied. Knots, particularly in the pines, should be sealed with shellac or a special knot-sealer before priming to retard discoloration of light-colored finishes by colored resins in the heartwood of these species. One or two coats of undercoat are next applied, which should completely hide the wood and also provide a surface that can be easily sanded smooth. For best results, the surface should be sanded just before applying the coats of finish. After the final coat has been applied, the finish may be left as is, with its natural gloss, or rubbed to a soft sheen.

Transparent Finishes

Transparent finishes are often used on hardwoods and some softwood trim and paneling. Most finish processes consist of some combination of the fundamental operations of sanding, staining, filling, sealing, surface coating, and sometimes waxing. Before finishing, planer marks and other blemishes on the wood surface that would be accentuated by the finish must be removed.

Stains

Some softwoods and hardwoods are often finished without staining, especially if the wood has an attractive color. When stain is used, however, it often accentuates color differences in the wood surface because of unequal absorption into different parts of the grain pattern. With hardwoods, such emphasis of the grain is usually desirable; the best stains for this purpose are dyes dissolved in either water or solvent. The water-soluble stains give the most pleasing results, but they raise the grain of the wood and require extra sanding after they dry.

The most commonly used stains are those that do not raise grain and are dissolved in solvents that dry quickly. These stains often approach the water-soluble stains in clearness and uniformity of color. Stains on softwoods color the earlywood more strongly than the latewood, reversing the natural gradation in color unless the wood has been initially sealed. To give more nearly uniform color, softwoods may be coated with penetrating clear sealer before applying any type of stain. This sealer is often called a "wash coat."

If stain absorbs into wood unevenly causing a blotchy appearance, the tree was probably infected with bacteria and/or blue-stain fungi prior to being cut for lumber. Once the log is cut into lumber, the infection occurs across grain boundaries and makes infected areas more porous than normal wood. When such areas are stained, they absorb excessive amounts of stain very quickly, giving the wood an uneven blotchy appearance. Although this problem is not very common, should it occur it can be difficult to fix. Blue stain on lumber can easily be seen; the infected pieces can either be discarded or sealed before staining. However, bacteria-infected areas cannot be detected prior to staining. If the wood is to be used for furniture or fine woodwork, it might be a good idea to check the lumber, before planing, by applying a stain. Pieces on which the stain appears blotchy should not be used. Sealing the lumber with varnish diluted 50/50 with mineral spirits prior to staining may help; commercial sealers are also available. Bacteria or blue-stain infection may occur in the sapwood of any species, but it seems to be more problematic with the hardwoods because these species tend to be used for furniture, cabinets, and fine woodwork.

Source: U.S. Department of Agriculture.

10.7.2 Fillers and Sealers

In hardwoods with large pores, the pores must be filled, usually after staining and before varnish or lacquer is applied, if a smooth coating is desired. The filler may be transparent and not affect the color of the finish, or it may be colored to either match or contrast with the surrounding wood. For finishing purposes, hardwoods may be classified as shown in Table 15–6. Hardwoods with small pores may be finished with paints, enamels, and varnishes in exactly the same manner as softwoods. A filler may be a paste or liquid, natural or colored. Apply the filler by brushing it first across and then with the grain. Remove surplus filler immediately after the glossy wet appearance disappears. First, wipe across the grain of the wood to pack the filler into the pores; then, wipe with a few light strokes along the grain. Allow the filler to dry thoroughly and lightly sand it before finishing the wood.

Sealers

Sealers are thinned varnish, shellac, or lacquer that are used to prevent absorption of surface coatings and to prevent the bleeding of some stains and fillers into surface coatings, especially lacquer coatings. Lacquer and shellac sealers have the advantage of drying very quickly.

Surface Coats

Transparent surface coatings over the sealer may be gloss varnish, semigloss varnish, shellac, nitrocellulose lacquer, or wax. Wax provides protection without forming a thick coating and without greatly enhancing the natural luster of the wood. Other coatings are more resinous, especially lacquer and varnish; they accentuate

TABLE 15–6 Classification of Hardwoods by Size of Pores[a]

Large Pores	Small Pores
Ash	Aspen
Butternut	Basswood
Chestnut	Beech
Elm	Cherry
Hackberry	Cottonwood
Hickory	Gum
Lauan	Magnolia
Mahogany	Maple
Mahogany, African	Red alder
Oak	Sycamore
Sugarberry	Yellow-poplar
Walnut	

[a]Birch has pores large enough to take wood filler effectively, but small enough to be finished satisfactorily without filling.

the natural luster of some hardwoods and seem to give the surface more "depth." Shellac applied by the laborious process of French polishing probably achieves this impression of depth most fully, but the coating is expensive and easily marred by water. Rubbing varnishes made with resins of high refractive index for light (ability to bend light rays) are nearly as effective as shellac. Lacquers have the advantages of drying rapidly and forming a hard surface, but more applications of lacquer than varnish are required to build up a lustrous coating. If sufficient film buildup is not obtained and the surface is cleaned often, such as the surface of kitchen cabinets, these thin films can fail.

Varnish and lacquer usually dry to a high gloss. To decrease the gloss, surfaces may be rubbed with pumice stone and water or polishing oil. Waterproof sandpaper and water may be used instead of pumice stone. The final sheen varies with the fineness of the powdered pumice stone; coarse powders make a dull surface and fine powders produce a bright sheen. For very smooth surfaces with high polish, the final rubbing is done with rottenstone and oil. Varnish and lacquer made to produce a semigloss or satin finish are also available.

Flat oil finishes commonly called Danish oils are also very popular. This type of finish penetrates the wood and does not form a noticeable film on the surface. Two or more coats of oil are usually applied; the oil may be followed by a paste wax. Such finishes are easily applied and maintained but they are more subject to soiling than is a film-forming type of finish. Simple boiled linseed oil or tung oil are also used extensively as wood finishes.

10.7.3 Finishes for Floors

Wood possesses a variety of properties that make it a highly desirable flooring material for homes, factories, and public buildings. A variety of wood flooring products are available, both unfinished and prefinished, in many wood species, grain characteristics, flooring types, and flooring patterns.

The natural color and grain of wood floors accentuate many architectural styles. Floor finishes enhance the natural beauty of wood, protect it from excessive wear and abrasion, and make the floor easier to clean. The finishing process consists of four steps: sanding the surface, applying a filler (for open-grain woods), staining to achieve a desired color effect, and finishing. Detailed procedures and specified materials depend to a great extent on the species of wood used and finish preference.

Careful sanding to provide a smooth surface is essential for a good finish because any irregularities or roughness in the surface will be accentuated by the finish. Development of a top-quality surface requires sanding in several steps with progressively finer sandpaper, usually with a machine unless the area is small. When sanding is complete, all dust must be removed with a vacuum cleaner and then a tack cloth. Steel wool should not be used on floors unprotected by finish because minute steel particles left in the wood later cause iron stains. A filler is required for wood with large pores, such as oak and walnut, if a smooth, glossy varnish finish is desired (Table 15–6).

Stains are sometimes used to obtain a more nearly uniform color when individual boards vary too much in their natural color. However, stains may also be used to accent the grain pattern. The stain should be an oil-based or non–grain-raising type. Stains penetrate wood only slightly; therefore, the finish should be carefully maintained to prevent wearing through to the wood surface; the clear top-coats must be replaced as they wear. It is difficult to renew the stain at worn spots in a way that will match the color of the surrounding area.

Finishes commonly used for wood floors are classified as sealers or varnishes. Sealers, which are usually thinned varnishes, are widely used for residential flooring. They penetrate the wood just enough to avoid formation of a surface coating of appreciable thickness. Wax is usually applied over the sealer; however, if greater gloss is desired, the sealed floor makes an excellent base for varnish. The thin surface coat of sealer and wax needs more frequent attention than do varnished surfaces. However, re-waxing or resealing and waxing of high-traffic areas is a relatively simple maintenance procedure, as long as the stained surface of the wood hasn't been worn.

Varnish may be based on phenolic, alkyd, epoxy, or polyurethane resins. Varnish forms a distinct coating over the wood and gives a lustrous finish. The kind of service expected usually determines the type of varnish. Varnishes especially designed for homes, schools, gymnasiums, or other public buildings are available. Information on types of floor finishes can be obtained from flooring associations or individual flooring manufacturers.

The durability of floor finishes can be improved by keeping them waxed. Paste waxes generally provide the best appearance and durability. Two coats are recommended, and if a liquid wax is used, additional coats may be necessary to get an adequate film for good performance.

Source: U.S. Department of Agriculture.

10.7.4 Finishes for Items Used for Food

The durability and beauty of wood make it an attractive material for bowls, butcher blocks, and other items used to serve or prepare food. A finish also helps keep the wood dry, which makes it less prone to harbor bacteria and less likely to crack. When wood soaks up water, it swells; when it dries out, it shrinks. If the wood dries out rapidly, its surface dries faster than the inside, resulting in cracks and checks. Finishes that repel water will decrease the effects of brief periods of moisture (washing), making the wood easier to clean.

Finishes that form a film on wood, such as varnish or lacquer, may be used but they may eventually chip, crack, and peel. Penetrating finishes, either drying or nondrying, are often a better choice for some products.

Types of Finish

Sealers and Drying Oils

Sealers and drying oils penetrate the wood surface, then solidify to form a barrier to liquid water. Many commercial sealers are similar to thinned varnish. These finishes can include a wide range of formulations including polyurethane, alkyds, and modified oils. Unmodified oils such as rung, linseed, and walnut oil can also be used as sealers if they are thinned to penetrate the wood.

Nondrying Oils

Nondrying oils simply penetrate the wood. They include both vegetable and mineral oils. Vegetable oils (such as olive, com, peanut, and safflower) are edible and are sometimes used to finish wood utensils. Mineral (or paraffin) oil is a nondrying oil from petroleum. Since it is not a natural product, it is not prone to mildew or to harboring bacteria.

Paraffin Wax

Paraffin wax is similar to paraffin oil but is solid at room temperature. Paraffin wax is one of the simplest ways to finish wood utensils, especially countertops, butcher blocks, and cutting boards.

Eating Utensils

Wood salad bowls, spoons, and forks used for food service need a finish that is resistant to abrasion, water, acids, and stains and a surface that is easy to clean when soiled.

Appropriate finishes are varnishes and lacquers, penetrating wood sealers and drying oils, and nondrying vegetable oils.

Many varnishes and lacquers are available, and some of these are specifically formulated for use on wood utensils, bowls, and/or cutting boards. These film-forming finishes resist staining and provide a surface that is easy to keep clean; however, they may eventually chip, peel, alligator, or crack. These film-forming finishes should perform well if care is taken to minimize their exposure to water. Utensils finished with such finishes should never be placed in a dishwasher.

Penetrating wood sealers and drying oils may also be used for eating utensils. Some of these may be formulated for use on utensils. Wood sealers and oils absorb into the pores of the wood and fill the cavities of the wood cells. This decreases the absorption of water and makes the surface easy to clean and more resistant to scratching compared with unfinished wood. Penetrating wood sealers are easy to apply and dry quickly. Worn places in the finish may be easily refinished. Some of these finishes, particularly drying oils, should be allowed to dry thoroughly for several weeks before use.

Nondrying vegetable oils are edible and are sometimes used to finish wood utensils. They penetrate the wood surface, improve its resistance to water, and can be refurbished easily. However, such finishes can become rancid and can sometimes impart undesirable odors and/or flavors to food.

Of these finish types, the impermeable varnishes and lacquers may be the best option for bowls and eating utensils; this kind of finish is easiest to keep clean and most resistant to absorption of stains.

Note: Whatever finish is chosen for wood utensils used to store, handle, or eat food, it is important to be sure that the finish is safe and not toxic (poisonous). Also be sure that the finish you select is recommended for use with food or is described as food grade. For information on the safety and toxicity of any finish, check the label, contact the manufacturer and/or the Food and Drug Administration, or consult your local extension home economics expert or county agent.

10.7.5 Finishes for Butcher Blocks and Cutting Boards

One of the simplest treatments for wood butcher blocks and cutting boards is the application of melted paraffin wax (the type used for home canning). The wax is melted in a double-boiler over hot water and liberally brushed on the wood surface. Excess wax, which has solidified on the surface, can be melted with an iron to absorb it into the wood, or it may be scraped off. Refinishing is simple and easy. Other penetrating finishes (sealers, drying and nondrying oils) may also be used for butcher blocks and cutting boards. As mentioned in the subsection on eating utensils, vegetable oils may become rancid. If a nondrying oil is desired, mineral oil may be used. Film-forming finishes are not recommended for butcher blocks or cutting boards.

10.7.6 Wood Cleaners and Brighteners

The popularity of wood decks and the desire to keep them looking bright and new has led to a proliferation of commercial cleaners and brighteners. Mildew growth on unpainted and painted wood continues to be the primary cause of discoloration. Although it can be removed with a dilute solution of household bleach and detergent, many commercial products are available that can both remove mildew and brighten the wood surface.

The active ingredient in many of these products is sodium percarbonate (disodium peroxypercarbonate). This chemical is an oxidizing agent, as is bleach, and it is an effective mildew cleaner. It also helps brighten the wood surface. Some cleaners and brighteners are reported to restore color to wood. It is not possible to add color to wood by cleaning it. Removing the discoloration reveals the original color. Brightening the wood may make it appear as if it has more color. Once all the colored components of the wood surface have been removed through the weathering process, the surface will be a silvery gray. If color is desired after weathering occurs, it must be added to the wood by staining.

In addition to sodium percarbonate, other oxidizing products may contain hydrogen peroxide by itself or in combination with sodium hydroxide. If sodium hydroxide is used without a brightener, it will darken the wood. Commercial products are also formulated with sodium hypochlorite and/or calcium hypochlorite (household bleach is a solution of sodium hypochlorite). These products usually contain a surfactant or detergent to enhance the cleansing action of the oxidizing agent. Other types of brighteners contain oxalic acid. This chemical removes stains caused by extractives bleed and iron stains and also brightens the wood, but it is not very effective for removing mildew.

10.7.7 Paint Strippers

Removing paint and other film-forming finishes from wood is a time-consuming and often difficult process. It is generally not done unless absolutely necessary to refinish the wood. Removing the finish is necessary if the old finish has extensive cross-grain cracking caused by buildup of many layers of paint, particularly oil-based paint. If cracking and peeling are extensive, it is usually best to remove all the paint from the affected area. Total removal of paint is also necessary if the paint has failed by intercoat peeling. It may be necessary to remove paint containing lead; however, if the paint is still sound and it is not illegal to leave it on the structure, it is best to repaint the surface without removing the old paint.

This discussion of paint strippers is limited to film-forming finishes on wood used in structures. Removing paint from furniture can be done using the same methods as described here. Companies that specialize in stripping furniture usually immerse the furniture in a vat of paint stripper and then clean and brighten the wood. This procedure removes the paint very efficiently.

Some of the same methods can be used for the removal of interior and exterior paint. Because of the dust caused by mechanical methods or the fumes given off by chemical strippers, it is extremely important to use effective safety equipment, particularly when working indoors. A good respirator is essential, even if the paint does not contain lead (see Lead-Based Paint).

Note: The dust masks sold in hardware stores do not block chemical fumes and are not very effective against dust.

Two general types of stripping methods are discussed here: mechanical and chemical. The processes are discussed in general terms primarily in regard to their effect on wood; some attention is given to their ease of use and safety requirements. Consult product literature for additional information on appropriate uses and safety precautions.

Source: U.S. Department of Agriculture.

Mechanical Methods

Finishes can be removed by scraping, sanding, wet or dry sandblasting, spraying with pressurized water (power washing), and using electrically heated pads, hot air guns, and blow torches. Scraping is effective only in removing loosely bonded paint or paint that has already partially peeled from the wood. This method is generally used when paint needs to be removed only from small areas of the structure, and it is usually combined with sanding to feather the edge of the paint still bonded to the wood (see Lead-Based Paint).

When the paint is peeling and partially debonded on large areas of a structure, the finish is usually removed by power washing or wet sandblasting. These methods work well for paint that is loosely bonded to the wood. If the paint is well bonded, complete removal can be difficult without severely damaging the wood surface. The pressure necessary to debond paint from the wood can easily cause deep erosion of the wood. The less dense earlywood erodes more than the dense latewood, leaving behind a surface consisting of latewood, which is more difficult to repaint. Power washing is less damaging to the wood than is wet or dry sandblasting, particularly if low pressure is used. If high pressure is necessary to remove the paint, it is probably bonded well enough that it does not need to be removed for normal refinishing. If more aggressive mechanical methods are required, wet sandblasting can remove even well-bonded paint, but it causes more damage to the wood than does water blasting. Dry sandblasting is not very suitable for removing paint from wood because it can quickly erode the wood surface along with the paint, and it tends to glaze the surface.

A number of power sanders and similar devices are available for complete paint removal. Many of these devices are suitable for removing paint that contains lead; they have attachments for containing the dust. Equipment that has a series of blades similar to a power hand-planer is less likely to "gum up" with paint than equipment that merely sands the surface. Some of this equipment is advertised in the *Old House Journal* and the *Journal of Light Construction*. Please consult the manufacturers' technical data sheets for detailed information to determine the suitability of their equipment for your needs and to meet government regulations on lead-containing paint.

Paint can be removed by heating and then scraping it from the wood, but this method must not be used for paint that contains lead. Paint can be softened by using electrically heated pads, hot air guns, or blow torches. Heated pads and hot air guns are slow methods, but they cause little damage to the wood. Sanding is still necessary, but the wood should be sound after the paint is removed. Blow torches have been used to remove paint and, if carefully used, do not damage the wood. Blow torches are extremely hazardous; the flames can easily ignite flammable materials beneath the siding through gaps in the siding. These materials may smolder, undetected, for hours before bursting into flame and causing loss of the structure.

Note: Removing paint with a blow torch is not recommended.
Source: U.S. Department of Agriculture.

Chemical Methods

If all the paint needs to be removed, then mechanical methods should be used in concert with other methods, such as chemical paint strippers. For all chemical paint strippers, the process involves applying paint stripper, waiting, scraping off the softened paint, washing the wood (and possibly neutralizing the stripper), and sanding the surface to remove the wood damaged by the stripper and/or the raised grain caused by washing. Chemical paint strippers, though tedious to use, are sometimes the most reasonable choice. A range of paint strippers are available. Some are extremely strong chemicals that quickly remove paint but are dangerous to use. Others remove the paint slowly but are safer. With the exception of alkali paint stripper (discussed below), there appears to be an inverse correlation between how safe a product is and how fast it removes the paint.

Solvent-Based Strippers

Fast-working paint strippers usually contain methylene chloride, a possible carcinogen that can burn eyes and skin. Eye and skin protection and a supplied-air respirator are essential when using this paint stripper. Paint strippers that have methylene chloride can remove paint in as little as 10 minutes. Because of concerns with methylene chloride, some paint strippers are being formulated using other strong solvents; the same safety precautions should be used with these formulations as with those containing methylene chloride. To remain effective in removing paint, a paint stripper must remain liquid or semiliquid; slow-acting paint stripers are often covered to keep them active. Solvent-type strippers contain a wax that floats to the surface to slow the evaporation of the solvent. Covering the paint stripper with plastic wrap also helps to contain the solvent.

10.8.0 Characteristics of Selected Woods for Painting

Wood Species	Specific Gravity[a] Green/Dry	Shrinkage (%)[b]		Paint-holding Characteristic (I, best; V, worst)[c]		Weathering		Color of Heartwood
		Flat Grain	Vertical Grain	Oil-based Paint	Latex Paint	Resistance to Cupping (1, most; 4, least)	Conspicuousness of Checking (1, least; 2, most)	
Softwoods								
Baldcypress	0.42/0.46	6.2	3.8	I	I	1	1	Light brown
Cedars								
Incense	0.35/0.37	5.2	3.3	I	I	–	–	Brown
Northern white	0.29/0.31	4.9	2.2	I	I	–	–	Light Brown
Port-Orford	0.39/0.43	6.9	4.6	I	I	–	1	Cream
Western red	0.31/0.32	5	2.4	I	I	1	1	Brown
Yellow	0.42/0.44	6	2.8	I	I	1	1	Yellow
Douglas-fir[d]	0.45/0.48[e]	7.6	4.8	IV	II	2	2	Pale red
Larch, western	0.48/0.52	9.1	4.5	IV	II	2	2	Brown
Pine								
Eastern white	0.34/0.35	6.1	2.1	II	II	2	2	Cream
Ponderosa	0.38/0.42	6.2	3.9	III	II	2	2	Cream
Southern[d]	0.47/0.51[f]	8	5	IV	III	2	2	Light brown
Sugar	0.34/0.36	5.6	2.9	II	II	2	2	Cream
Western white	0.36/0.38	7.4	4.1	II	II	2	2	Cream
Redwood, old growth	0.38/0.40	4.4	2.6	I	I	1	1	Dark brown
Spruce, Engelmann	0.33/0.35	7.1	3.8	III	II	2	2	White
Tamarack	0.49/0.53	7.4	3.7	IV	–	2	2	Brown
White fir	0.37/0.39	7.0	3.3	III	–	2	2	White
Western hemlock	0.42/0.45	7.8	4.2	III	II	2	2	Pale brown
Hardwoods								
Alder	0.37/0.41	7.3	4.4	III	–	–	–	Pale brown
Ash, white	0.55/0.60	8	5	V or III	–	4	2	Light brown
Aspen, bigtooth	0.36/0.39	7	3.5	III	II	2	1	Pale brown
Basswood	0.32/0.37	9.3	6.6	III	–	2	2	Cream
Beech	0.56/0.64	11.9	5.5	IV	–	4	2	Pale brown
Birch, yellow	0.55/0.62	9.5	7.3	IV	–	4	2	Light brown
Butternut	0.36/0.38	6.4	3.4	V or III	–	–	–	Light brown
Cherry	0.47/0.50	7.1	3.7	IV	–	–	–	Brown
Chestnut	0.40/0.43	6.7	3.4	V or III	–	3	2	Light brown
Cottonwood, eastern	0.37/0.40	9.2	3.9	III	II	4	2	White

Wood Species	Specific Gravity Green/Dry	Shrinkage (%)		Paint-holding Characteristic (I, best; V, worst)		Weathering		Color of Heartwood
		Flat Grain	Vertical Grain	Oil-based Paint	Latex Paint	Resistance to Cupping (1, most; 4, least)	Conspicuousness of Checking (1, least; 2, most)	
Elm, American	0.46/0.50	9.5	4.2	V or III	–	4	2	Brown
Hickory, shagbark	0.64/0.72	11	7	V or IV	–	4	2	Light brown
Lauan plywood	–^{g}	8	4	IV	–	2	2	Brown
Magnolia, southern	0.46/0.50	6.6	5.4	III	–	2	–	Pale brown
Maple, sugar	0.56/0.63	9.9	4.8	IV	–	4	2	Light brown
Oak								
White	0.60/0.68	8.8	4.4	V or IV	–	4	2	Brown
Northern red	0.56/0.63	8.6	4.0	V or IV	–	4	2	Brown
Sweetgum	0.46/0.52	10.2	5.3	IV	III	4	2	Brown
Sycamore	0.46/0.49	8.4	5	IV	–	–	–	Pale brown
Walnut	0.51/0.55	7.8	5.5	V or III	–	3	2	Dark brown
Yellow-poplar	0.40/0.42	8.2	4.6	III	II	2	1	Pale brown

^{a}Specific gravity based on weight oven dry and volume at green or 12% moisture content.
^{b}Value obtained by drying from green to oven dry.
^{c}Woods ranked in Group V have large pores that require wood filler for durable painting. When pores are properly filled before painting, Group II applies. Vertical-grain lumber was used for cedars and redwood. Other species were primarily flat-grain. Decrease in paintability is caused by a combination of species characteristics, grain orientation, and greater dimensional change of flat-grain lumber. Flat-grain lumber causes at least a 1-unit decrease in paintability.
^{d}Lumber and plywood.
^{e}Coastal Douglas-fir.
^{f}Loblolly, shortleaf, specific gravity of 0.54/0.59 for longleaf and slash.
^{g}Specific gravity of different species varies from 0.33 to 0.55.

10.9.0 When Calculating and Measuring for Interior Trim and Millwork—Learn Tool Basics

Marking and Measuring Tool Basics

If you're like most woodworkers, you've spent a lot of time picking out the best woodworking machinery, hand planes, chisels, scrapers, rasps, clamps, and all manner of specialized jigs, tools, and accessories that help make your work more accurate and go more smoothly. In the process, have you given much thought to the most fundamental tools in your shop—your marking and measuring tools?

It's worth taking stock of your marking and measuring tool kit. Most of the common problems in woodworking—joints that fit badly, out of square frames and casework, and so on—can be traced back to marking and measuring errors. And the majority of marking and measuring errors are rooted in a simple matter of using the wrong marking and measuring tools. In short, a tape measure just wasn't designed for making the close-tolerance measurements that many woodworking projects require. In this article we'll take a look at a few of the most common marking and measuring tasks in woodworking and at some of the tools that make the process easy, intuitive, and accurate. Then, to help you get set up with a basic kit, we'll pick out a few essentials from Rockler's broad selection of marking and measuring tools.

Measuring and Marking Linear Dimensions

In most woodworking projects, measuring and marking linear dimensions is the first crucial step, and depending on the project, it can make for some exacting work. In projects that involve intricate joinery and small, close-fitting parts, measuring and marking errors as small as a few 100 this of an inch can turn up later as gaps in joints, misaligned parts, and a host of other less-than-appealing results.

Measuring from point A to point B is a simple process, but your results still depend on how accurately you are able to translate a measurement into a physical mark on a piece of wood. If you've ever tried to hold a tape measure flat on a board while you accurately mark off a measurement, you know that just getting a clearly defined mark in exactly the right spot can be a surprising challenge. For precise measuring and marking, the tool you use needs to be readable and, of course, accurately calibrated. Going beyond that, the best distance measuring tools offer a little help in getting the mark in the right place.

Incra Precision Marking Rules are famous for their accuracy, lay flat, and have an easy-to-read scale. But what makes them the tool of choice for linear measuring is that they make it virtually impossible to put a mark in the wrong place. Incra rules are made with micro fine guide holes positioned at 1/32" increments so that, used in conjunction with a mechanical pencil or a metal scribe, you'd really have to try hard to put a measurement mark anywhere other than exactly where it's supposed to be.

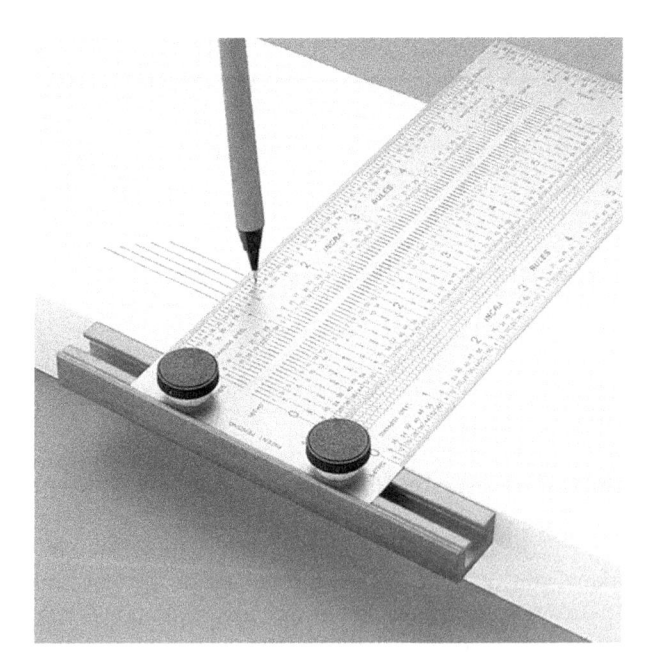

Special features of some tools further simplify one of the most common measuring tasks in woodworking—measuring and marking a distance from an edge. Tools like the Incra Precision T-Rule and Precision Bend Rule take care of positioning "point A" in the "point A to B" measuring formula, while still offering the famous Incra accuracy and easy-to-use design. All that's left for the woodworker to make a perfectly positioned mark at a precise distance from ah edge is to get the scribe or mechanical pencil in the right guide hole and make a mark.

By permission: rockler.com

Section | 10 Interior Finishes

Measuring Squareness

When you put a square on the end of a board to check a cut, you're trusting the "known" 90 degree angle of the square to tell you something about the piece of wood. But how square is a square? Some manufacturers tell you. Crown Hand Tools' rosewood and brass Try-Miter Square is manufactured in accordance with British Standard 3322, which means that it is accurate within a tolerance range of .01 mm/cm, or a little over 2 thousandths of an inch over its 6" blade. Part of the benefit of knowing the tight tolerance range of the square you are using is knowing that it is manufactured in accordance with a standard. A good many squares on the market don't boast a manufacturing standard at all.

Do fractions of a degree really matter? Often, inaccuracies in angle measurements that you are not even aware of multiply in accordance with the number of slightly off-square marks and cuts you make over the course of a project. When you are joining a large number of parts a tenth of a degree here and a tenth of a degree there, it really starts to add up. Remember, too, that you'll use a square to check the angle of your table saw blade and miter saw fence alignment. Slightly off-square angle settings on these tools are multiplied by two every time you make a joint or join two boards and can noticeably affect the flatness of edge glue-ups and miter joints.

The Crown try-miter square is especially handy because it also allows you to check and mark the second most common angle in woodworking—45 degrees. A combination square takes that versatility and adds to it with a graduated scale and a blade that can be adjusted and locked into position to gauge depth or distance relative to the edge of a material. The combination square was borrowed from the machinist's tool chest years ago, and because of its all-around usefulness, it remains the "workhorse" square in most woodworking shops.

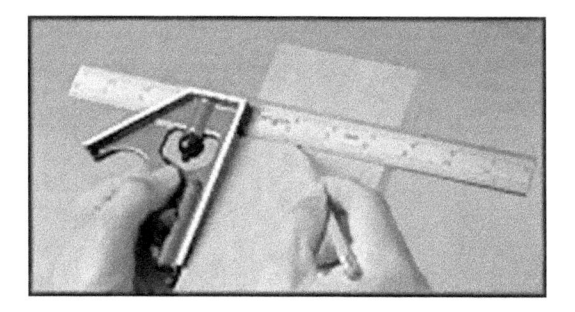

Angles Other Than 90 and 45 Degrees

The 90 and 45 degree angles may be the most common in woodworking, but they're not the only ones that come up. For marking angles other than 90 and 45 degrees, most woodworkers use a <u>sliding bevel</u>, "T-bevel." The T-bevel's sliding blade is infinitely positionable and has the added benefit of giving you four possible handle-to-blade angle orientations when the tool is set up in the "T" shape (with some of the blade on either side of the handle).

If you're shopping for a T-bevel, it's important to look for one that has a good lock-down mechanism so that you don't run the risk of accidentally moving the angle setting while you're using the tool. This classic example by Crown with a rosewood body and steel blade cinches down more than well enough to hold a setting for as long as you need it to.

Measuring Angles

A T-bevel is strictly an angle marking tool; it doesn't tell you anything about the measurement of the angle. For that you need an angle measuring tool, and there you have a few choices. But to simplify matters, we can divide angle measurement into two basic tasks. In general, you'll either want to set a tool or make a mark at a known angle, or you'll want to know the angle measurement of an existing angle, like the angle of a corner where two walls meet.

For cases where you need to set up a marking tool—like a T-bevel—with a known angle, the Mastergage Universal Angle Guide is about as good as it gets. The guide is laid out with a computer-guided laser etched angle scale in 1/2 degree increments on heavy gauge aluminum and makes it easy to transfer angle measurements to a T-bevel with dead-on accuracy. A tool that gives you accurate angle settings quickly and easily, like the Universal Angle Guide, is indispensable in working with the odd angles that turn up in complicated joinery projects or in any project where corners meet at other than the "usual" angles.

On the other hand, if you need an accurate measurement of an existing angle, you won't do better than the Starrett Protractor/Angle Finder. The tool is calibrated to read both inside and outside corners and quickly not only gives you the angle of the corner, but also offers the correct miter setting for your saw. Starrett is one of the most trusted names in calibration and measurement tools, so you can be confident that the Protractor/Angle Finder's precision matches its speed. A tool like the Starrett angle finder is an essential angle measurement tool for fitting your work into the real world, which, as anyone who's ever installed cabinets, crown molding, or any kind of trim will tell you, isn't always laid out in perfect 90 degree angles.

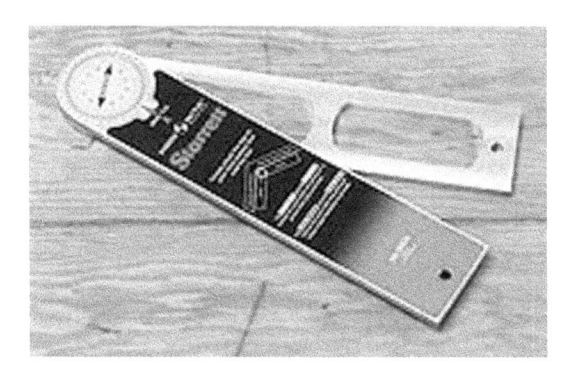

Measuring Depth, Gap, and Thickness

Often, woodworking projects require that you measure a short distance with extreme accuracy. Fine tuning the depth of a rabbet, checking the width of a dado, and measuring the thickness of veneer or stock all call for a tool that will give you extremely precise short distance readings. For these necessarily finicky measuring tasks, you really can't go wrong with a digital caliper.

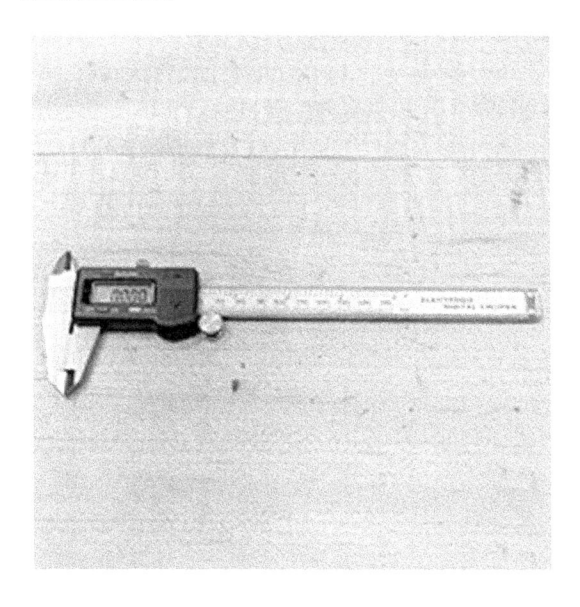

The digital caliper is the latest advancement in a precision measuring tool that migrated from the machinist's tool kit into the wood shop years ago. They're equipped with sets of jaws that measure inside and outside dimensions with accuracy in the 1000ths of an inch range, and a probe that slides down from the bottom of the tool to gauge depth with equal precision. Calipers are also available in models that have a dial readout and a standard calibrated scale, but the modern digital variety is so easy to use and read that most woodworkers find the slight upcharge for the feature well worth the price.

Putting Together a Basic Measuring and Marking Kit

At this point, we've just scratched the surface of marking and measuring. There are many other truly useful marking and measuring tools, many of which are extremely helpful in specialized measuring tasks. Wood turners will want to take a look at the J-Square Center Finder and a Wood Turner's Caliper Set. If your projects have you marking off a lot of curved shapes, then you might want to add a set of French Curves or a Flexible Curve to your marking tool collection. And we'd also like to point out that tools that help you get the most accurate results from your woodworking machinery, like calibration tools and precision fences and miter gauges, are in essence measuring tools.

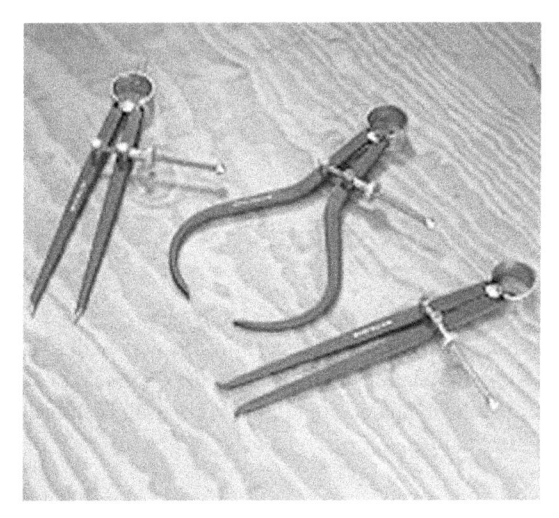

But our purpose is to help you get set up with the marking and measuring tools that we think are "standard equipment" for any woodworking operation. Here are the ten tools that will cover the most common and important marking and measuring tasks in woodworking:

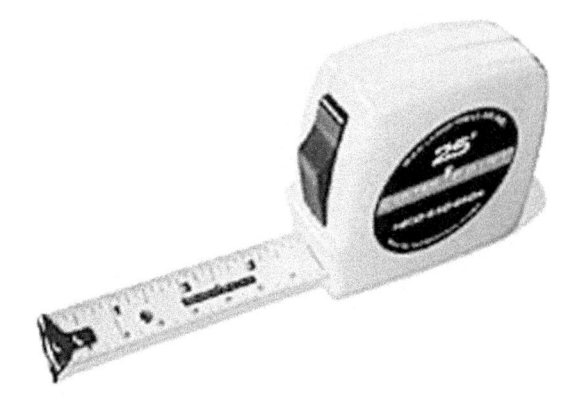

1. Tape Measure. Did we give you the impression that we don't like tape measures? A tape measure will always be an irreplaceable tool.
2. Incra Precision Marking Rule. As we've pointed out, you can't beat this tool for precision measuring and marking.
3. Incra Precision T-Rule. Measuring and marking a distance from an edge is one of the most common tasks in woodworking. The Incra T-Rule is *the* tool for measurements up to twelve inches.
4. Cabinetmaker's Pencil Set. A quality graphite pencil that sharpens to a micro-fine point for accurate marking.
5. Try -Miter Square or Engineer Square. Judging squareness is so central to woodworking that we think every shop should have a tool that does it accurately.
6. Combination Square. One of the most versatile marking and measuring tools ever introduced into woodworking.
7. T-Bevel. For years and years the T-bevel has been the tool for marking angles other than 90 degrees.
8. Mastergage Universal Angle Guide or Incra Precision Protractor. Either tool will give you the angle measurement precision you need for complex projects.
9. Starrett Protractor/Angle Finder. The best tool for dead-on accurate angle readings.
10. Digital Caliper. There really isn't any other way to get precise measurements of the depth of your rabbets, the width of your dadoes, or the thickness of your stock.

# Section 11

Plumbing and HVAC Calculations

11.0.0	Water Supply Force Units (WSFUs) Established by the Uniform Plumbing Code Determines the Water Supply Required for Proper Functioning of Plumbing Fixtures. Developing Plumbing Fixtures that Conserve Water, at an Economic Cost, is the Biggest Challenge Facing the Construction Industry Today	590	11.2.3 Current and Proposed Commercial Clothes Washers Water/Energy Usage Rates	599
11.0.1	Mean Daily Residential Water Use as Determined at 12 Study Sites	591	11.2.4 Commercial Dishwashers—Only Current Energy Star, Water Sense Specifications Prevail	600
11.0.2	Calculate Usage Based upon Maximum Allowable Water Efficiency Standards for Toilets, Kitchen and Lavatory Faucets, and Shower Heads	591	11.2.5 Automatic Commercial Ice Makers— No Current Standards—Proposed Only for 2010	601
11.0.3	Calculate Water Usage of Various Types of Low- and High-Volume Toilets	592	11.3.0 U.S. Green Building Council LEED (R)—Plumbing Fixture Water Efficiency Goals	602
11.0.4	Reported Savings Due to Use of Low-Flow Toilets in Four Studies	592	11.3.1 Preexisting State and Local Standards for Water-Efficient Plumbing Fixtures	605
11.1.0	Evolution of Low-Flow Toilet Testing Procedures	593	11.3.2 Projected Reduction in Walter Consumption, 2010–2020—With and Without Daily Savings	606
11.1.1	Three Common Types of Toilet Construction and Related Efficiencies	594	11.4.0 Where Does Our Water Come From? Volume of Earth's Oceans	607
11.2.0	The National Efficiency Standards and Specifications for Residential and Commercial Water Using Fixtures Enacted in 1992 and Updated in 2005—Relating to Residential Fixtures	596	11.5.0 How Much Water Do We Use on Average?	607
			11.5.1 Create a Personal Water Usage Chart	608
			11.6.0 Calculating the Size of Storage and Heat Pump Water Heaters	609
			11.6.1 Calculating the Cost of a Demand, Storage, or Heat Pump Water Heater	610
11.2.1	Current and Proposed Residential Dishwasher Standards	597	11.7.0 HVAC—Understanding and Calculating Relative Humidity	611
11.2.2	Current and Proposed Commercial Plumbing Fixture Water Usage Rates	598	11.7.1 HVAC—Understanding and Calculating Dewpoint	612
			11.7.2 Methods of Calculating Heating Efficiency—Combined Heat and Power (CHP)	612
			11.7.3 How Much Moisture Can the Air "Hold"?	616

11.7.4	General Heating Formulas—Energy Required to Heat, Offset Losses	617	
11.7.5	General Heating Formulas—Energy Required to Heat Air Flow	617	
11.7.6	Formula to Convert Actual CFM (ACFM) to Standard Cubic Feet per Minute (SCFM)	617	
11.8.0	Estimated Average Fuel Conversion Efficiency of Common Heating Appliances	618	
11.8.1	Reading Those Gas Meters	619	
11.9.0	Calculating Home Heating Energy—Gas versus Electric Resistance Heating	619	
11.10.0	Comparing Fuel Costs of Heating and Cooling Systems—Gas, Electric, Kerosene, Wood, Pellets	621	
11.11.0	Residential Ground Source Heat Pump (GSHP) Savings versus Electric, Gas, and Fuel Oil	630	
11.11.1	Paybacks for Residential GSHP Economics	631	
11.11.2	Commercial Ground Source Heat Pump (GSHP) Savings versus Electric, Gas, Fuel Oil	631	
11.11.3	Paybacks for Commercial GSHP Economics	632	

11.0.0 Water Supply Force Units (WSFUs) Established by the Uniform Plumbing Code Determines the Water Supply Required for Proper Functioning of Plumbing Fixtures. Developing Plumbing Fixtures that Conserve Water, at an Economic Cost, is the Biggest Challenge Facing the Construction Industry Today

The Water Supply Fixture Units (WSFUs) are defined by the Uniform Plumbing Code (UPC) and can be used to determine the water supply to fixtures and their service systems.

Individual Fixtures	Minimum Fixture Branch Pipe Size (inch)	Water Supply Fixture Units	
		Private Installations	Public Installations
Bathtub	1/2	4	4
Bathtub with 3/4" fill valve	3/4	10	10
Bidet	1/2	1	
Dishwasher, domestic	1/2	1.5	1.5
Drinking fountain	1/2	0.5	0.5
Hose Bibb	1/2	2.5	2.5
Lavatory	1/2	1	1
Bar sink	1/2	1	2
Clinic faucet sink	1/2	3	
Kitchen sink, domestic	1/2	1.5	1.5
Laundry sink	1/2	1.5	1.5
Service or mop basin	1/2	1.5	3
Washup basin	1/2	2	
Shower head	1/2	2	2
Urinal with flush tank	1/2	2	2
Wash fountain	3/4	4	
Water closet with gravity tank	1/2	2.5	2.5
Water closet with flushometer tank	1/2	2.5	2.5
Water cooler	1/2	0.5	0.5

1 WFSU = 1 GPM = 3.79 liter/min
Source: Uniform Plumbing Code.

Section | 11 Plumbing and HVAC Calculations

11.0.1 Mean Daily Residential Water Use as Determined at 12 Study Sites

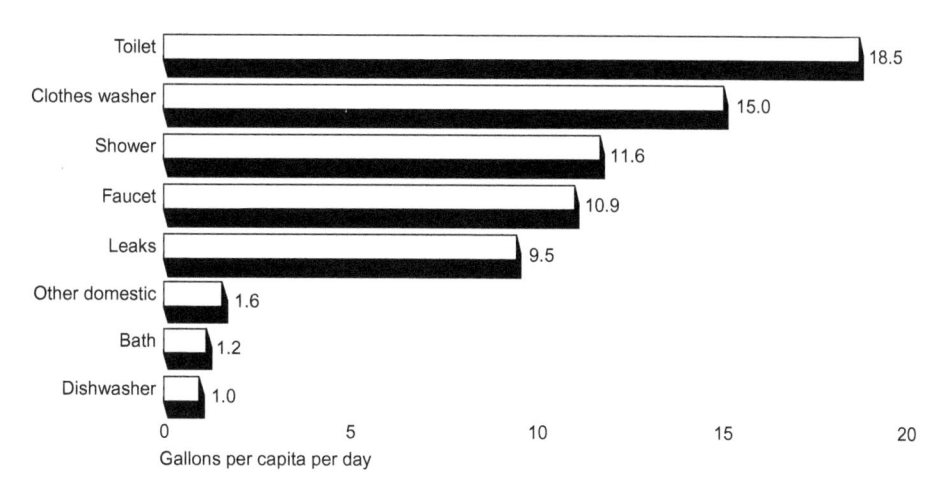

Fixture	Gallons per capita per day
Toilet	18.5
Clothes washer	15.0
Shower	11.6
Faucet	10.9
Leaks	9.5
Other domestic	1.6
Bath	1.2
Dishwasher	1.0

FIGURE 1 Mean Daily Residential Water Use at 12 Study Sites. *Source: Residential End Uses of Water*, American Water Works Association Research Foundation (1999), p. xxv.

11.0.2 Calculate Usage Based upon Maximum Allowable Water Efficiency Standards for Toilets, Kitchen and Lavatory Faucets, and Shower Heads

National Water Efficiency Standards	
Fixture Type	Maximum Allowable Water Use
Toilets, including gravity tank-type toilets,[a] flushometer tank toilets,[b] and electromechanical hydraulic toilets[c]	1.6 gallons per flush
Kitchen and lavatory faucets (or replacement aerators[d])	2.5 gallons per minute, when measured at a flowing water pressure of 80 pounds per square inch
Showerheads	2.5 gallons per minute, when measured at a flowing water pressure of 80 pounds per square inch
Urinals	1.0 gallon per flush

[a] A gravity tank-type toilet is designed to flush by gravity only with water supplied to the bowl.
[b] A flushometer tank toilet is designed to flush using a flushometer valve, which is attached to a pressurized water supply pipe and, when actuated, opens the line for direct water flow into the bowl at a rate and predetermined quantity needed to properly operate the toilet.
[c] An electromechanical hydraulic toilet is designed to flush using electronically controlled devices, such as air compressors, pumps, motors, or macerators in place of or as an aid to gravity in flushing the toilet bowl.
[d] An aerator is an apparatus for controlling water flow (e.g., from faucets).

Under the Department of Energy's regulations, water-efficient plumbing fixtures must meet the standards for maximum water consumption. For each model of a regulated plumbing fixture, manufacturers and private labelers must submit a compliance statement to the Department to certify that the model complies with the applicable water conservation standard and that all required testing has been conducted according to the test requirements prescribed in the regulations. In addition, the Department's regulations prohibit manufacturers and private labelers from distributing in commerce any fixture that does not meet the water conservation standard prescribed under the Energy Policy Act of 1992, and provide for the assessment of a civil penalty of not more than $110 per violation.

Source: U.S General Accounting Office.

11.0.3 Calculate Water Usage of Various Types of Low- and High-Volume Toilets

Water Use by Type of Toilet

Household Toilet Types	Average Gallons per Flush	Average water use		
		Number of Households	Gallons per Toilet per Day	Gallons per Capita per Day
Low-flow only	<2.0	101	24.2	9.6
Mix of low-flow and higher-volume	2.0 – 3.5	311	45.4	17.6
Higher-volume only	>4.0	776	47.9	20.1
All households		1188	45.2	18.5

Legend
< means less than
> means greater than
Source: Residential End Uses of Water, American Water Works Association Research Foundation, pp. 131–132.

In addition to the comprehensive study by the American Water Works Association's Research Foundation, a number of studies have used similarly sophisticated equipment to measure water flow to individual appliances at a small number of households. The purpose of these studies was to estimate whether water-efficient fixtures reduce water consumption in residences and if so, by how much. Toilets consume the most water in residences, and, as such, they have been the focus of the greatest attention, but showerheads, faucets, and clothes washers have also been considered in these studies, although clothes washers will not be subject to national standards until 2004. The studies all agree that compared with older toilets, ultra-low-flow toilets save significant amounts of water, easily overwhelming any changes in user practices (such as the frequency of flushing).

Source: U.S. General Accounting Office.

11.0.4 Reported Savings Due to Use of Low-Flow Toilets in Four Studies

While it is widely believed that the installation of water-efficient showerheads and sink faucets also results in significant savings, the studies we reviewed are not in complete agreement on this point. The East Bay, California, study reported savings of 1.7 gallons per capita per day with low-flow showerheads—about one-third of the savings resulting from toilet replacement. The Tampa, Florida, study reported savings of 3.6 gallons per capita per day—more than half of the savings from toilet replacement.

Reported Savings Attributable to Low-Flow Toilets in Studies Using Precise Measurements

Location	Date Published	Number of Households	Water Use in Toilets (gal. per capita per day)			
			Before Retrofit	After Retrofit	Amount Saved	Percentage Saved
Boulder, CO[a]	May 1996	14	15.9	7.6	8.3	52
East Bay Municipal Utility District, CA	Oct. 1991	25	12.8[c]	6.7[c]	5.3[c]	41[c]
Seattle, WA[b]	July 2000 (draft)	37	18.8	8.1	10.6	57
Tampa, FL	Feb. 1993	25	13.3	7.2	6.1	46

[a]In this study, half of the toilets were replaced with low-flow toilets and half were not; the reported savings were obtained by averaging the results for all toilets—higher-volume and low-flow. For the purpose of this table, we computed the water use and the amount of savings on the basis of the results for the replaced toilets.

[b]We obtained a copy of the draft report on this study. Because the authors are still finalizing the report, we did not have all of the information that would be useful in evaluating the results of this study.

[c]Because the study did not explicitly report the average water use before and after retrofit, we estimated these values by multiplying the average volume per flush by the number of flushes per person. The difference between these values does not equal the amount of savings reported in the study, which was measured separately for each toilet before averaging and, thus, is more accurate.

Sources

Boulder: *Project Report: Measuring Actual Retrofit Savings and Conservation Effectiveness Using Flow Trace Analysis.* Prepared for: City of Boulder, Colorado, Utilities Division, Office of Water Conservation, by Aquacraft Water Engineering & Management (May 16, 1996).

East Bay: *East Bay Municipal Utility District Water Conservation Study.* Prepared for: East Bay MUD, Oakland, California, A. Aher et al., Stevens Institute of Technology, Building Technology Laboratory, T. P. Konen, Director, Report No. R 219 (Oct. 1991).

Seattle: Draft report prepared for EPA and Seattle Public Utilities, by P. Mayer and W. DeOreo, Aquacraft, Inc., private communication from P. Mayer (July 2000).

Tampa: *The Impact of Water Conserving Plumbing Fixtures on Residential Water Use Characteristics: A Case Study in Tampa, Florida.* Prepared for: City of Tampa Water Department, Water Conservation Section, by Stevens Institute of Technology and Ayres Associates, T. P. Konen and D. L. Anderson, Principal Investigators (Feb. 1993).

11.1.0 Evolution of Low-Flow Toilet Testing Procedures

Evolution of Industry Testing Requirements for Low-Flow Toilets

Test Name	Test Description	1990 Edition	1995 Edition	Pending Revision, 2000[a]
Water consumption per flush	To determine average water consumption: average consumption shall not exceed 1.6 gallons.	New	Same	Same
Maximum water consumption per flush	To determine maximum water consumption after adjusting trim components for maximum water use: average water consumption shall not exceed 2.4 gallons.	N/A	N/A	New
Ball test	To determine solids removal: 100 polypropylene balls are placed in toilet bowl; 75 must be removed in initial flush.	Same	Same	Deleted; combined with granule test
Granule test	To determine solids removal: 2500 polyethylene disc-shaped pellets are placed in toilet bowl; not more than 125 may remain after initial flush.	Same	Same	Adds 100 nylon balls; not more than 3 are allowed to remain after initial flush

(Continued)

Evolution of Industry Testing Requirements for Low-Flow Toilets—Cont'd				
Test Name	Test Description	1990 Edition	1995 Edition	Pending Revision, 2000[a]
Ink line test	To determine rim washing: a water soluble ink is marked on a bowl's surface; after initial flush, no line segment can exceed 1/2 inch, and aggregate of all segments may not exceed 2 inches.	Same	Same	A second line is added 2 inches below rim jets; this line is completely washed away
Dye test	To determine water exchange: a dye is added to bowl; 100% dilution must occur after initial flush.	Same	Same	Deleted
Trap seal test	To determine if trap seal works properly: fixture must return to full trap seal after each flush.	Same	Same	Same
Mixed media test	To determine solids removal: 12 sponges and 10 paper balls are used; not more than 4 sponges or balls may remain after initial flush.	N/A	N/A	New
Drain line carry test	To determine length of transport of solid wastes: fixture must carry waste a minimum of 40 feet in the drain line.	New	Same	Same
Overflow test	To determine leakage of gravity tank-type toilets: tank fill valve is opened to maximum flow for 5 minutes; fixture shall not leak.	N/A	N/A	New
Water rise test	To determine wetting of person sitting on seat during flush: a vertically positioned rod is placed 3 inches under the bowl rim; during flush, water should not touch rod.	New	Same	Deleted
Rim top and seat fouling test	To determine soiling of rim top and seat: a plate is placed over toilet bowl; no water shall splash on plate during flushing.	N/A	New	Deleted

Source: U.S. General Accounting Office.

11.1.1 Three Common Types of Toilet Construction and Related Efficiencies

There are three common varieties of toilets: gravity flow, (siphon-jet) flush valve, and pressurized tank systems. Similarly, there are four common varieties of urinals: the siphonic jet urinal, washout/wash-down urinals, blowout urinals, and waterless urinals. All of these must meet federal water efficiency standards, though waterless urinals go far beyond the conservation minimums. Composting toilets also use no water, but potential applications are generally limited to national park facilities and small highway rest stops.

Opportunities

The vast majority of toilets and urinals in federal facilities were installed at a time when there was little or no regard for using water efficiently. Consequently, there are ample opportunities to make significant savings in water usage. Complete replacement is the desired option. Retrofit of existing toilets and urinals is a second choice that may be more attractive if there are budget constraints. While retrofits reduce the amount of water used per flush, most fixtures were not designed to use reduced amounts of water and their performance may suffer. Only complete replacement of porcelain fixtures ensures that, even with less water, they can still perform efficiently and effectively.

Technical Information

Toilets account for almost half of a typical building's water consumption. Americans flush about 4.8 billion gallons (18.2 billion liters) of water down toilets each day, according to the U.S. Environmental Protection Agency. According to the Plumbing Foundation, replacing all existing toilets with 1.6 gallons (6 liters) per flush, ultra-low-flow (ULF) models would save almost 5500 gallons (25,000 liters) of water per person each year. A widespread toilet replacement program in New York City apartment buildings found an average 29% reduction in total water use for the buildings studied. The entire program, in which 1.3 million toilets were replaced, is estimated to be saving 60–80 million gallons (230–300 million liters) per day.

There is a common perception that ULF toilets do not perform adequately. A number of early 1.6-gallons-per-flush (gpf) (6-liter) gravity-flush toilets that were simply adapted from 3.5- gpf (16-liter) models—rather than being designed from the ground up to operate effectively with the ULF volume—performed very poorly, and some low-cost toilets today still suffer from that problem. But studies show that most 1.6-gpf (6-liter) toilets work very well. Where flush performance is a particular concern, or water conservation beyond that of a 1.6-gpf (6-liter) model is required, pressurized-tank toilets, vacuum toilets, and dual-flush toilets should be considered. Carefully choose toilet models based on recommendations from industry surveys or experienced plumbers and facility managers. You may also want to contact some managers of facilities that have already installed the toilets under consideration.

While some retrofit options for toilets reduce water use (see next page), none of these modifications will perform as effectively or use as little water as quality toilets manufactured after January 1, 1994. These retrofits will merely allow the fixture to operate using less water until it is replaced.

Even greater water conservation can be achieved in certain (limited) applications with composting toilets. Because of the size of composting tanks, lack of knowledge about performance, local regulatory restrictions, and higher first-costs, composting toilets are rarely an option except in certain unique applications, such as national park facilities. Composting toilets are being used very successfully, for example, at Grand Canyon National Park.

With urinals, water conservation well beyond the standard 1.0-gpf (4.5-liter) performance for new products can be obtained using waterless urinals. These products, available from the Waterless Company, use a special trap with a lightweight biodegradable oil that lets urine and water pass through but prevents odors from escaping into the restroom; there are no valves to fail, and clogging does not cause flooding.

Projected Water Savings from Installing Waterless Urinals

Building Type	No. Males	No. Urinals	Uses/ Day	Gal/ Flush	Days/ Year	Ann. Water Gallons	Savings/Urinal Liters
Small Office	25	1	3	3.0	260	58,500	220,000
New const.	25	1	3	1.0	260	19,500	73,800
Restaurant	150	3	1	3.0	360	54,000	204,000
New const.	150	3	1	1.0	360	18,000	68,100
School	300	10	2	3.0	185	33,300	126,000
New const.	300	10	2	1.0	185	11,100	42,000

Source: Environmental Building News, February 1998.

11.2.0 The National Efficiency Standards and Specifications for Residential and Commercial Water Using Fixtures Enacted in 1992 and updated in 2005—Relating to Residential Fixtures

Fixtures and Appliances	EPAct 1992, EPAct 2005 (or backlog NAECA updates)		WaterSense® or ENERGY STAR®		Consortium for Energy Efficiency	
	Current Standard	Proposed/Future Standard	Current Specification	Proposed/Future Specification	Current Specification	Proposed/Future Specification
Residential Toilets	1.6 gpf[1]		WaterSense 1.28 gpf with at least 350 gram waste removal[2]		No specification	
Residential Bathroom Faucets	2.2 gpm at 60 psi[3]		WaterSense 1.5 gpm at 60 psi (no less than 0.8 gpm at 20 psi)[4]		No specification	
Residential Showerheads	2.5 gpm at 80 psi		No specification		No specification	
Residential Clothes Washers	MEF $\geq$ 1.26 ft^3/kWh/cycle *No specified water use factor	Proposed to DOE Asst. Sec. jointly by AHAM and efficiency advocates to be effective in 2011 MEF $\geq$ 1.26 ft^3/kWh/cycle WF $\leq$ 9.5 gal/cycle/ft^3	ENERGY STAR (DOE) MEF $\geq$ 1.72 ft^3/kWh/cycle; WF $\leq$ 8.0 gal/cycle/ft^3	ENERGY STAR (DOE) Effective July 1, 2009: MEF $\geq$ 1.8 ft^3/kWh/cycle WF $\leq$ 7.5 gal/cycle/ft^3 Effective January 1, 2011: MEF $\geq$ 2.0 ft^3/kWh/cycle WF $\leq$ 6.0 gal/cycle/ft^3	**Tier 1:** MEF $\geq$ 1.80 ft^3/kWh/cycle; WF $\leq$ 7.5 gal/cycle/ft^3 **Tier 2:** MEF $\geq$ 2.00 ft^3/kWh/cycle; WF $\leq$ 6.0 gal/cycle/ft^3 **Tier 3:** MEF $\geq$ 2.20 ft^3/kWh/cycle; WF $\leq$ 4.5 gal/cycle/ft^3	

[1] EPAct 1992 standard for toilets applies to both commercial and residential models.
[2] See WaterSense HET specification at http://www.epa.gov/watersense/docs/spechet508.pdf.
[3] EPAct 1992 standard for faucets applies to both commercial and residential models.
[4] See WaterSense specification for lavatory faucets at http://www.epa.gov/watersense/docs/faucet spec508.pdf.

DOE: Department of Energy
EPA: Environmental Protection Agency
EPAct 1992: Energy Policy Act of 1992
EPAct 2005: Energy Policy Act of 2005

EF: energy factor
ft^3: cubic feet
gal: gallons
gpm: gallons per minute

gpf: gallons per flush
kWh: kilowatt hour
MEF: modified energy factor
MaP: maximum performance

NAECA: National Appliance Energy Conservation Act
psi: pounds per square inch
WF: water factor

Section | 11 Plumbing and HVAC Calculations

11.2.1 Current and Proposed Residential Dishwasher Standards

Fixtures and Appliances	EPAct 1992, EPAct 2005 (or backlog NAECA updates)		WaterSense® or ENERGY STAR®		Consortium for Energy Efficiency	
	Current Standard	Proposed/Future Standard	Current Specification	Proposed/Future Specification	Current Specification	Proposed/Future Specification
Residential Dishwashers[5]	*Standard models:* EF ≥ 0.46 cycles/kWh *Compact models:* EF ≥ 0.62 cycles/kWh *No specified water use factor (Energy Independence & Security Act of 2007: As of January 1, 2010 *Standard models:* 355 kWh/year WF ≤ 6.5 gallons/cycle *Compact models:* 260 kWh/year WF ≤ 4.5 gallons/cycle)	New standards under development: DOE scheduled final action: March 2009; Stakeholder meeting held 4/27/2006 Proposed to DOE Asst. Sec. jointly by AHAM and efficiency advocates to be effective in 2010 *Standard models:* 355 kWh/year (.62 EF + 1 watt standby) WF ≤ 6.5 gallons/cycle *Compact models:* 260 kWh/year WF ≤ 4.5 gallons/cycle	ENERGY STAR (DOE) *Standard models.* EF ≥ 0.65 cycles/kWh *Compact models:* EF ≥ 0.88 cycles/kWh *No specified water use factor	Proposed to DOE Asst. Sec. jointly by AHAM & efficiency advocates to be effective in 2009 *Standard models:* 324 kWh/year (0.68 EF + 1 watt standby) WF ≤ 5.8 gallons/cycle *Compact models:* 234 kWh/year WF ≤ 4.0 gallons/cycle Phase Two (Proposed by DOE): July 1, 2011 *Standard models:* 307 kWh/yr 5.0 gallons/cycle *Compact models:* 222 kWh/yr 3.5 gallons/cycle	*Standard models:* **Tier 1:** EF ≥ 0.65 cycles/kWh; maximum 339 kWh/year **Tier 2:** EF ≥ 0.68 cycles/kWh; maximum 325 kWh/year *Compact models:* **Tier 1:** EF ≥ 0.88 cycles/kWh; maximum 252 kWh/year *No specified water use factor	In December 2006, CEE announced they will consider adding a water factor in future dishwasher specifications

[5]Standard models: capacity is greater than or equal to eight place settings and six serving pieces; Compact models: capacity is less than eight place settings and six serving pieces

DOE: Department of Energy
EPA: Environmental Protection Agency
EPAct 1992: Energy Policy Act of 1992
EPAct 2005: Energy Policy Act of 2005

EF: energy factor
ft³: cubic feet
gal: gallons
gpm: gallons per minute

gpf: gallons per flush
kWh: kilowatt hour
MEF: modified energy factor
MaP: maximum performance

NAECA: National Appliance Energy Conservation Act
psi: pounds per square inch
WF: water factor

11.2.2 Current and Proposed Commercial Plumbing Fixture Water Usage Rates

Fixtures and Applainces	EPAct 1992, EPAct 2005 (or backlog NAECA updates)		WaterSense® or ENERGY STAR®		Consortium for Energy Efficiency	
	Current Standard	Proposed/Future Standard	Current Specification	Proposed/Future Specification	Current Specification	Proposed/Future Specification
Commercial Toilets	1.6 gpf[6]		No Specification[7]		No specification	
Urinals	1.0 gpf		No specification		No specification	
Commercial Faucets	2.2 gpm at 60 psi NOTE: Superseded by national plumbing codes (UPC, IPC, and NSPC) for all "public" lavatories: 0.5 gpm maximum.[8] 0.25 gallon per cycle for metering faucets		WaterSense specification applicable to private lavatories (e.g. hotel room bathrooms)[9] 1.5 gpm at 60 psi (no less than 0.8 gpm at 20 psi)			

[6] EPAct 1992 standard for toilets applies to both commercial and residential models.
[7] No specification for flushometer valve toilets. WaterSense Specification (http://www.epa.gov/watersense/docs/spechet508.pdf) applicable to tank type toilets found in some commercial applications.
[8] In addition to EPAct requirements, the American Society of Mechanical Engineers standard for public lavatory faucets is 0.5 gpm at 60 psi (ASME A112.18.1-2005). Public lavatory faucets are those intended for the unrestricted use of more than one individual (including employees) in assembly occupancies, business occupancies, public buildings, transportation facilities, schools and other educational facilities, office buildings, restaurants, bars, other food service facilities, mercantile facilities, manufacturing facilities, military facilities, and other facilities that are not intended for private use.
[9] Specification for bathroom sink faucets and faucet accessories (e.g., aerators, flow regulators, laminar devices), applicable to some commercial situations (e.g., hotel room bathrooms). Not applicable to public lavatory faucets (see footnote 8).

DOE: Department of Energy
EPA: Environmental Protection Agency
EPAct 1992: Energy Policy Act of 1992
EPAct 2005: Energy Policy Act of 2005

EF: energy factor
ft^3: cubic feet
gal: gallons
gpm: gallons per minute

gpf: gallons per flush
kWh: kilowatt hour
MEF: modified energy factor
MaP: maximum performance

NAECA: National Appliance Energy Conservation Act
psi: pounds per square inch
WF: water factor

11.2.3 Current and Proposed Commercial Clothes Washers Water/Energy Usage Rates

Fixtures and Appliances	EPAct 1992, EPAct 2005 (or backlog NAECA updates)		WaterSense® or ENERGY STAR®		Consortium for Energy Efficiency	
	Current Standard	Proposed/Future Standard	Current Specification	Proposed/Future Specification	Current Specification	Proposed/Future Specification
Commercial Clothes Washers (Family-sized)	MEF $\geq$ 1.26 ft^3/kWh/cycle; WF $\leq$ 9.5 gal/cycle/ft^3	New standards under development: DOE scheduled final action: January 2010; Stakeholder meeting held 4/27/2006	ENERGY STAR (DOE) MEF $\geq$ 1.72 ft^3/kWh/cycle; WF $\leq$ 8.0 gal/cycle/ft^3	Effective July 1, 2009 MEF $\geq$ 1.8 ft^3/kWh/cycle Effective January 1, 2011 MEF $\geq$ 2.0 ft^3/kWh/cycle WF $\leq$ 6.0 gal/cycle/ft^3	**Tier 1:** MEF $\geq$ 1.80 ft^3/kWh/cycle; WF $\leq$ 7.5 gal/cycle/ft^3 **Tier 2:** MEF $\geq$ 2.00 ft^3/kWh/cycle; WF $\leq$ 6.0 gal/cycle/ft^3 **Tier 3:** MEF $\geq$ 2.20 ft^3/kWh/cycle; WF $\leq$ 4.5 gal/cycle/ft^3	

DOE: Department of Energy
EPA: Environmental Protection Agency
EPAct 1992: Energy Policy Act of 1992
EPAct 2005: Energy Policy Act of 2005

EF: energy factor
ft^3: cubic feet
gal: gallons
gpm: gallons per minute

gpf: gallons per flush
kWh: kilowatt hour
MEF: modified energy factor
MaP: maximum performance

NAECA: National Appliance Energy Conservation Act
psi: pounds per square inch
WF: water factor

Source: U.S. Environmental Protection Agency.

11.2.4 Commercial Dishwashers—Only Current Energy Star, Water Sense Specifications Prevail

Fixtures and Appliances	EPAct 1992, EPAct 2005 (or backlog NAECA updates)		WaterSense® or ENERGY STAR®		Consortium for Energy Efficiency	
	Current Standard	Proposed/Future Standard	Current Specification	Proposed/Future Specification	Current Specification	Proposed/Future Specification
Commercial Dishwashers	No standard		ENERGY STAR (EPA) Water Consumption; Idle Energy: *Under counter.* Hi Temp: ≤ 1.0 gal/rack; ≤ 0.9 kW Lo Temp: ≤ 1.70 gal/rack; ≤ 0.5 kW *Stationary Single Tank Door.* Hi Temp: ≤ 0.95 gal/rack; ≤ 1.0 kW Lo Temp: ≤ 1.18 gal/rack; ≤ 0.6 kW *Single Tank Conveyor.* Hi Temp: ≤ 0.70 gal/rack; ≤ 2.0 kW Lo Temp: ≤ 0.79 gal/rack; ≤ 1.6 kW *Multiple Tank Conveyor.* Hi Temp: ≤ 0.54 gal/rack; ≤ 2.6 kW Lo Temp: ≤ 0.54 gal/rack; ≤ 2.0 kW		No specification	

DOE: Department of Energy
EPA: Environmental Protection Agency
EPAct 1992: Energy Policy Act of 1992
EPAct 2005: Energy Policy Act of 2005

EF: energy factor
ft³: cubic feet
gal: gallons
gpm: gallons per minute

gpf: gallons per flush
kWh: kilowatt hour
MEF: modified energy factor
MaP: maximum performance

NAECA: National Appliance Energy Conservation Act
psi: pounds per square inch
WF: water factor

Source: U.S. Environmental Protection Agency.

11.2.5 Automatic Commercial Ice Makers—No Current Standards—Proposed Only for 2010

Fixtures and Appliances	EPAct 1992, EPAct 2005 (or backlog NAECA updates)		WaterSense® or ENERGY STAR®		Consortium for Energy Efficiency	
	Current Standard	Proposed/Future Standard	Current Specification	Proposed/Future Specification	Current Specification	Proposed/Future Specification
Automatic Commercial Ice Makers[10]	No standard	Effective **1/1/2010**: Energy and condenser water efficiency standards vary by equipment type on a sliding scale depending on harvest rate and type of cooling (see link to additional information at end of this table)	No specification	ENERGY STAR (EPA) Effective **1/1/08**: Energy and water efficiency standards vary by equipment type on a sliding scale depending on harvest rate. Water cooled machines excluded from Energy Star (see link to additional information at end of this table)	Energy and water (potable and condenser) standards are tiered and vary by equipment type on a sliding scale depending on harvest rate and type of cooling (see link to additional information at end of this table)	
Pre-rinse Spray Valves	Flow rate ≤ 1.6 gpm (no pressure specified; no performance requirement)		No specification	Proposed ENERGY STAR specification abandoned after standard established in EPAct 2005 WaterSense specification under consideration	No specification (program guidance recommends 1.6 gpm at 60 psi and a cleanability requirement)	

[10]Optional standards for other types of automatic ice makers are also authorized under EPAct 2005.

DOE: Department of Energy
EPA: Environmental Protection Agency
EPAct 1992: Energy Policy Act of 1992
EPAct 2005: Energy Policy Act of 2005

EF: energy factor
ft³: cubic feet
gal: gallons
gpm: gallons per minute

gpf: gallons per flush
kWh: kilowatt hour
MEF: modified energy factor
MaP: maximum performance

NAECA: National Appliance Energy Conservation Act
psi: pounds per square inch
WF: water factor

Source: U.S. Environmental Protection Agency.

11.3.0 U.S. Green Building Council LEED (R)—Plumbing Fixture Water Efficiency Goals

The Water Efficiency category of the LEED® rating system is the least emphasized, with a potential of three LEED® points obtainable through Innovative Waste Water Technologies and Water Use Reduction. The LEED® rating system for Water Use Reduction is based on the U.S. Energy Policy Act of 1992. This Act set maximum plumbing fixture flow rates.

Energy Policy Act of 1992	
Fixture:	**Maximum Flow Rate:**
Water Closet	1.6 Gallons Per Flush (GPF)
Urinals	1.0 Gallons Per Flush (GPF)
Faucets	2.5 Gallons Per Minute (GPM)
Shower Heads	2.5 Gallons Per Minute (GPM)

We now have a basis to evaluate sustainable features in a nonresidential building. The purpose of LEED® is to make buildings more efficient and sustainable than the maximum required levels. I will not differentiate between new or existing construction, it should be clear which features will be easiest to implement in the type of building you are evaluating or designing.

Innovative Waste Water Technologies: WE Credit 2, 1-Point

(Based on LEED®-NC Version 2.2 Reference Guide)

Intent:

To reduce generation of wastewater and potable water demand, while increasing the local aquifer recharge.

Requirements:

Option 1

Reduce potable water use for building sewage conveyance by 50% through the use of water-conserving fixtures (water closets and urinals) or nonpotable water (captured rainwater, recycled graywater, and on-site or municipally treated wastewater).

OR

Option 2

Treat 50% of wastewater on-site to tertiary standards. Treated water must be filtered or used on-site.

    My experience is that the collection of rainwater is a fairly straightforward method of obtaining this credit point. I recommend the collection of rainwater from the roof rather than parking lots, which contain oils and other hazardous waste contaminants that are more difficult to filter and handle for disposal to be used in the plumbing system. The collection of rainwater requires that the rainwater drainage system be collected and piped to a collection tank(s) (underground or above ground). The rainwater is in most cases pumped from the collection tank through a series of filters (5 micron to collect the large particles and 50 micron to collect any other solids) and then passed through a UV sterilizer to kill any bacteria. The collected rainwater is now ready to be used to flush the water closets and urinals. Keep in mind that this is nonpotable and should be treated as such. In my designs I provide a backup connection to the potable water system in the building in the event no rainwater is available. The backup connection of potable water is protected by a reduced-pressure backflow device to protect the building potable water system from the cross-connected reclaimed rainwater. This system is most likely to be designed in new construction, since an existing building retrofit would be cost prohibitive.

Example Case:

In this example, we will show potable water calculations for sewage flows for a nonresidential building with an occupant capacity of 100 (50 males and 50 females). The calculation is based on a typical 8-hour workday. Male occupants are assumed to use water closets once and urinals twice a day. Female occupants are assumed to use water closets three times.

Baseline Case

Fixture Type	Daily Uses	Flow Rate (GPF)	Occupants	Sewage Generation (GAL)
Water Closet (Male)	1	1.6	50	80
Water Closet (Female)	3	1.6	50	240
Urinal (Male)	2	1.0	50	100

Total Daily Volume (GAL) 420
Annual Work Days 260
TOTAL ANNUAL VOLUME (GAL) 109,200

Design Case

Fixture Type	Daily Uses	Flow Rate (GPF)	Occupants	Sewage Generation (GAL)
Water Closet (Male)	1	1.2	50	60
Water Closet (Female)	3	1.2	50	180
Urinal (Male)	2	0.5	50	50

Total Daily Volume (GAL) 290
Annual Work Days 260
Annual Volume (GAL) 75,400
Rainwater Volume (GAL) (25,000)
TOTAL ANNUAL VOLUME (GAL) 50,400

The baseline case flow rates use the maximum flow rates based on the U.S. Energy Policy Act of 1992. Using a combination of water-conserving fixtures and rainwater collection, the design case building indicates a 54% reduction in potable water volume used for sewage conveyance; this therefore qualifies for the one point credit.

Baseline Case

Fixture Type	Daily Uses	Flow Rate (GPF)	Duration (flush)	Auto Controls (N/A)	Occupants	Water Use (GAL)
Water Closet (Male)	1	1.6	1		50	80
Water Closet (Female)	3	1.6	1		50	240
Urinal (Male)	2	1.0	1		50	100
Fixture Type	Daily Uses	Flow Rate (GPM)	Duration (Sec)	Auto Controls	Occupants	Water Use (GAL)
Lavatory	3	2.5	15	No	100	188

Total Daily Volume (GAL) 608
Annual Work Days 260
TOTAL ANNUAL VOLUME (GAL) 158,080

Design Case

Fixture Type	Daily Uses	Flow Rate (GPF)	Duration (flush)	Auto Controls (N/A)	Occupants	Water Use (GAL)
Water Closet (Male)	1	1.2	1		50	60
Water Closet (Female)	3	1.2	1		50	180
Urinal (Male)	2	0.5	1		50	50
Fixture Type	Daily Uses	Flow Rate (GPM)	Duration (Sec)	Auto Controls	Occupants	Water Use (GAL)
Lavatory	3	0.5	10	Yes	100	25

Total Daily Volume (GAL) 315
Annual Work Days 260
TOTAL ANNUAL VOLUME (GAL) 81,900

The baseline case flow rates use the maximum flow rates based on the U.S. Energy Policy Act of 1992. Using a combination of water conserving fixtures the design case building indicates a 48% reduction in potable water volume; this therefore qualifies for the one-point credit for the 20% reduction plus an additional one point for exceeding the 30% water-use reduction for a total of two LEED® points. What makes this so simple is the fact that standard fixtures are available from all the major fixture manufacturers to meet these criteria. If we only used a 0.5 GPF urinal and water-saving metering faucets on the lavatories, we would still realize a 35% water reduction without changing the water closets.

The LEED® point system also allows for **Innovative & Design Process** points (maximum of 4). These allow the designer to submit to the USGBC an innovative design concept that might not be covered within the existing point structure, such as Press-Fit copper piping or CSST gas piping. These systems may qualify for an innovative credit point because they use recyclable materials and are solder-less and oil-less, which is environmental friendly.

11.3.1 Preexisting State and Local Standards for Water-Efficient Plumbing Fixtures

Preexisting State and Local Standards for Water-Efficient Plumbing Fixtures and Their Status If National Standards Were Repealed

Sixteen states and six localities had water efficiency standards for at least two of the plumbing fixtures regulated under the Energy Policy Act before the national standards took effect in 1994. Table shown below compares the standards adopted by each jurisdiction with the national standards.

State and Local Standards for Water-Efficient Plumbing Fixtures

State/Locality	Effective Date[a]	Water Efficiency Standard				
		Ultra-Low-Flow Toilets (gal. per flush)	Low-Flow Showerhead (gal. per minute)	Kitchen Faucets (gal. per min.)	Lavatory Faucets (gal. per min.)	Urinals (gal. per flush)
National Standard	Jan. 1, 1994	1.6	2.50	2.5	2.5	1.0
States						
Arizona	Jan. 1, 1993	1.6	2.50	2.5	2.0	None
California	Jan. 1, 1992	1.6	2.50	2.5	None	1.0
Connecticut[b]	Jan. 1, 1990	1.6	2.50	2.5	2.5	1.0
Delaware	Apr. 1, 1992	1.6	2.50	2.5	2.0	1.0
Georgia	Apr. 1, 1992	1.6	2.50	2.5	2.0	1.0
Maryland	Apr. 1, 1992	1.6	2.50	2.5	2.0	1.0
Massachusetts	Mar. 2, 1989	1.6	3.00	None	None	1.0
Nevada	Mar. 1, 1993	1.6	2.50	2.5	2.5	1.0
New Jersey[b]	July 1, 1991	1.6	3.00	3.0	3.0	1.5
New York[b]	Jan. 1, 1992	1.6	3.00	None	2.0	1.0
North Carolina[b]	Jan. 1, 1993	1.6	3.00	3.0	3.0	1.5
Oregon	July 1, 1993	1.6	2.50	2.5	2.5	1.0
Rhode Island[b]	Mar. 1, 1991	1.6	2.50	2.0	2.0	1.0
Texas	Jan. 1, 1992	1.6	2.75	2.2	2.2	1.0
Utah	July 1, 1992	1.6	2.50	None	None	None
Washington	July 1, 1993	1.6	2.50	2.5	2.5	1.0
Localities						
Dade County, FL	Jan. 1, 1992	1.6	2.50	2.5	2.0	1.0
Denver, CO	Mar. 1, 1992	1.6	2.50	2.2	2.2	1.0
District of Columbia	Jan. 1, 1992	1.6	2.50	2.5	2.0	1.0
Hillsborough County, FL	Mar. 26, 1992	1.6	2.50	2.2	2.2	1.0
Palm Beach, FL[b]	Apr. 1, 1991	1.6	3.00	None	None	1.5
Tampa, FL[b]	June 1, 1990	2.0	2.50	2.0	2.0	1.0

11.3.2 Projected Reduction in Walter Consumption 2010–2020—With and Without Daily Savings

Projected Reduction in Water Consumption by 2010 and 2020, by Location

Projected Water Use and Savings in Millions of Gallons per Day

Location	Population	Year	Average Daily Water Use		Projected Daily Water Savings	
			Without Water Efficiency Standards	With Water Efficiency Standards	Amount	Percentage
Austin, TX, City of Austin Water & Wastewater Utility	650,000	2010	167.5	160.8	6.7	4.0
		2020	230.9	215.2	15.7	6.8
Boston, MA, Boston Water & Sewer Commission	650,000	2010	84.2	81.0	3.2	3.8
		2020	85.1	79.4	5.7	6.7
Cary, NC, Town of Cary	84,779	2010	16.1	15.1	1.0	6.2
		2020	23.1	21.0	2.1	9.1
Clarksburg, WVa, Clarksburg Water Board	19,000	2010	2.9	2.8	0.1	3.4
		2020	3.6	3.4	0.2	5.5
Fort Worth, TX, Fort Worth Water Department	753,116	2010	170.0	166.6	3.4	2.0
		2020	178.8	172.9	5.9	3.3
Laurel, MD, Washington Suburban Sanitary Commission	1,700,000	2010	206.3	199.7	6.6	3.2
		2020	224.1	212.0	12.1	5.4
Los Angeles, CA, Los Angeles Department of Water & Power	3,800,000	2010	560.6	542.1	18.5	3.3
		2020	560.3	527.8	32.5	5.8
Michigan City, IN, Michigan City Department of Water	41,000	2010	12.1	11.7	0.4	3.3
		2020	12.7	12.0	0.7	5.5
Oceanside, CA, City of Oceanside Water	157,869	2010	37.5	36.6	0.9	2.4
		2020	46.2	44.4	1.8	3.9
Phoenix, AZ, Phoenix Water Services	1,252,425	2010	341.4	331.5	9.9	2.9
		2020	393.6	375.1	18.5	4.7
Pinellas County, FL, Pinellas County Utilities	643,191	2010	84.4	80.6	3.8	4.5

11.4.0 Where Does Our Water Come From? Volume of Earth's Oceans

Ice	29,492	2.2	Much of this ice is in the Antarctic
Groundwater	6,733	0.5	Underground aquifers, deep wells
Lakes	242	0.02	Provide drinking water, irrigation water, fish and recreation
Soil Moisture	74	0.005	This is being used by our crops, trees, and surface vegetation
Water Vapor in the Atmosphere	14	0.001	Clouds, fog, and dew
Rivers	1.3	0.0001	Provide water for drinking, irrigation, and recreation

Adapted from: Environment Canada.

One Estimate of Global Water Distribution

Water Source	Water Volume, in cubic miles	Water Volume, in cubic kilometers	Percent of Fresh Water	Percent of Total Water
Oceans, Seas, & Bays	321,000,000	1,338,000,000	–	96.5
Ice caps, Glaciers, & Permanent Snow	5,773,000	24,064,000	68.7	1.74
Groundwater	5,614,000	23,400,000	–	1.7
Fresh	2,526,000	10,530,000	30.1	0.76
Saline	3,088,000	12,870,000	–	0.94
Soil Moisture	3,959	16,500	0.05	0.001
Ground Ice & Permafrost	71,970	300,000	0.86	0.022
Lakes	42,320	176,400	–	0.013
Fresh	21,830	91,000	0.26	0.007
Saline	20,490	85,400	–	0.006

11.5.0 How Much Water Do We Use on Average?

Managing Water Resources

Having a reliable and safe supply of fresh water is very important for us to stay alive. The way that many of us live our lives means that we use a lot more fresh water than we need just to stay alive. If the number of people in the world increases and they all use more fresh water, we will not be able to supply all the fresh water that is wanted. One way of helping to solve this problem is for each of us to use fresh water more carefully.

Activity 1: Making an estimate

Here's how to estimate how much water you use each day:

Fill in the first column of the table below for the volume of water that you use, on average, each day. Then calculate how much water you use in a year.

In some cases you will need to work out your **share** of the water used in your home. These cases are marked with a * in the table. For example, if your family uses a washing machine to wash clothes three times a week this uses 3 × 120 = 360 liters per week. On average, this is 360 ÷ 7 = 51 liters per day. If there are five members of your family, this is just over 10 liters per day for each member of the family.

You will probably find the following average values helpful:

- Brushing teeth: 0.01–1 liter
- Cooking a meal: 1–5 liters
- Flushing toilet: 5–10 liters
- Washing hands: 1–3 liters
- Dish washer: 30–50 liters

- Washing machine: 30–100 liters
- Shower: 1–40 liters
- Cleaning car: 5–200 liters
- Drinking: 1–2 liters
- Watering garden: 1–17 liters per m^2
- Bath: 50–150 liters

(Washing clothes or dishes by hand only uses a quarter as much water.)
To convert these UNESCO litres to gallons

1 liter = 0.264 gallons
3 liters = 0.79 gallons
5 liters = 1.3 galloons
10 liters = 2.64 gallons
17 liters = 4.49 gallons
30 liters = 7.925 gallons
40 liters = 10.57 gallons
50 liters = 13.21 gallons
150 liters = 39.6 galloons
200 liters = 52.8 gallons

11.5.1 Create a Personal Water Usage Chart

Activity Handout How Much Water do You Use?
Survey: How Much Water Do You Use?

▶ DIRECTIONS This is a survey to find how much water you use in your home during one full week. Place a tally mark in the Times per Day column every time someone in your family does the activity.

Activity	Times per Day							Weekly Total	Water per Activity*	Total Water Used
	Sun	Mon	Tues	Wed	Thurs	Fri	Sat			
Toilet Flushing								____	× 5 gallons =	____
Short Shower (5-10 minutes)								____	× 25 gallons =	____
Long Shower (>10 minutes)								____	× 35 gallons =	____
Tub Bath								____	× 35 gallons =	____
Teeth Brushing								____	× 2 gallons =	____
Washing Dishes with Running Water								____	× 30 gallons =	____
Washing Dishes Filling a Basin								____	× 10 gallons =	____
Using Dishwasher								____	× 20 gallons =	____
Washing Clothes								____	× **40 gallons** =	____
									GRAND TOTAL =	____

NOTE: Another significant seasonal water use is lawn and garden watering. This survey deals with daily water use in the home, but most of us use additional amounts of water at school, at work, and other places throughout the day.
*These are estimated values.
Source: U.S. Environmental Protection Agency.

11.6.0 Calculating the Size of Storage and Heat Pump Water Heaters

U.S. Department of Energy—Energy Efficiency and Renewable Energy Energy Savers

Sizing Storage and Heat Pump (with Tank) Water Heaters

To properly size a storage water heater—including a heat pump water heater with a tank— for your home, use the water heater's first hour rating (FHR). The FHR is the amount of hot water in gallons the heater can supply per hour (starting with a tank full of hot water). It depends on the tank capacity, source of heat (burner or element), and the size of the burner or element.

The EnergyGuide Label lists the FHR in the top left corner as "Capacity (first hour rating)." The Federal Trade Commission requires an EnergyGuide Label on all new conventional storage water heaters but not on heat pump water heaters. Product literature from a manufacturer may also provide the FHR. Look for water heater models with a FHR that matches within 1 or 2 gallons of your peak hour demand—the daily peak 1-hour hot water demand for your home.

Do the following to estimate your peak hour demand:

- Determine what time of day (morning, noon, evening) you use the most hot water in your home. Keep in mind the number of people living in your home.
- Use the worksheet below to estimate your maximum usage of hot water during this one hour of the day—this is your peak hour demand. *Note:* The worksheet does not estimate total daily hot water usage.

The worksheet example shows a total peak hour demand of 46 gallons. Therefore, this household would need a water heater model with a first hour rating of 44 to 48 gallons.

Worksheet for Estimating Peak Hour Demand/First Hour Rating

Use	Average Gallons of Hot Water per Usage		Times Used during 1 Hour		Gallons Used in 1 Hour
Shower	12	×		=	
Bath	9	×		=	
Shaving	2	×		=	
Hands and face washing	4	×		=	
Hair shampoo	4	×		=	
Hand dishwashing	4	×		=	
Automatic dishwasher	14	×		=	
Food preparation	5	×		=	
Wringer clothes washer	26	×		=	
Automatic clothes washer	32	×		=	
			Total Peak Hour Demand	=	

Example

3 showers	12	×	3	=	36
1 shave	2	×	1	=	2
1 shampoo	4	×	1	=	4
1 hand dishwashing	4	×	1	=	4
Peak Hour Demand				=	46

Adapted from information from the Air Conditioning, Heating, and Refrigeration Institute

Before selecting a storage water heater, you also want to consider the following:

- Fuel type and availability
- Energy efficiency
- Cost.

If you haven't yet considered what type of water heater might be best for your home, here are your options:

- Conventional storage water heater
- Demand (tankless or instantaneous) water heater
- Heat pump water heaters
- Solar water heater
- Tankless coil and indirect water heaters

11.6.1 Calculating the Cost of a Demand, Storage, or Heat Pump Water Heater

U.S. Department of Energy—Energy Efficiency and Renewable Energy Energy Savers

Estimating a Storage, Demand, or Heat Pump Water Heater's Costs

When considering a water heater model for your home, estimate its annual operating cost. Then compare costs with other more and/or less energy-efficient models. This will help you determine the energy savings and payback period of investing in a more energy-efficient model, which will probably have a higher purchase price.

Before you can choose and compare the costs of various models, you need to determine the correct size water heater for your home. If you haven't done this already, see the following:

- Sizing a Demand (Tankless or Instantaneous) Water Heater
- Sizing Storage and Heat Pump (with Tank) Water Heaters

Calculating Annual Operating Cost

To estimate the annual operating cost of a storage, demand (tankless or instaneous), or heat pump (not geothermal heat pump) water heater, you need to know the following about the model:

- Energy factor (EF)
- Fuel type and cost (your local utility can provide current rates)

Then, use the following calculations:

For gas and oil water heaters

You need to know the unit cost of fuel by Btu (British thermal unit) or therm. (1 therm = 100,000 Btu)

$365 \times 41045/EF \times$ Fuel Cost (Btu) = estimated annual cost of operation

OR

$365 \times 0.4105/EF \times$ Fuel Cost (therm) = estimated annual cost of operation

Example: A natural gas water heater with an EF of .57 and a fuel cost of $0.00000619/Btu

$365 \times 41045/.57 \times \$0.00000619 = \$163$

For electric water heaters, including heat pump units

You need to know or convert the unit cost of electricity by kilowatt-hour (kWh).

$365 \times 12.03/EF \times$ Electricity Cost by kWh = estimated annual cost of operation

Example: A heat pump water heater with an EF of 2.0 and a electricity cost of $0.0842/kWh

$365 \times 12.03/2.0 \times \$0.0842 = \$185$

Comparing Costs and Determining Payback

Once you know the purchase and annual operating costs of the water heater models you want to compare, you can use the following table to determine the cost savings and payback of the more energy-efficient model(s).

Models	Price of Water Heater	EF	Estimated Annual Operating Cost
Model A			
Model B (higher EF)			
Additional cost of more efficient model (Model B)			Price of Model B - Price of Model A = $Additional Cost of Model B
Estimated annual operating cost savings (Model B)			Model B Annual Operating Cost - Model A Annual Operating Cost = $Model B's Cost Savings per Year
Payback period for Model B			$Additional Cost of Model B/$Model B's Cost Savings per Year = Payback period/years

Example

Comparison of two gas water heaters with a local fuel cost of $0.60 per therm.

Models	Price of Water Heater	EF	Estimated Annual Operating Cost
Model A	$165	0.54	$166
Model B	$210	0.58	$155
Additional cost of more efficient model (Model B)			$210 − $165 = $45
Estimated annual operating cost savings (Model B)			$166 − $155 = $11 per year
Payback period for Model B			$45/$11 per year = 4.1 years

Other Costs

If you want to include installation and maintenance costs, consult the manufacturer(s) and a qualified contractor to help estimate these costs. These costs will vary among system types and sometimes even from water heater model to model.

11.7.0 HVAC—Understanding and Calculating Relative Humidity

Relative Humidity

The amount of water vapor in the air at any given time is usually less than that required to saturate the air. The relative humidity is the percentage of <u>saturation humidity</u>, generally calculated in relation to saturated vapor density.

$$\text{Relative Humidity} = \frac{\text{actual vapor density}}{\text{saturation vapor density}} \times 100\%$$

The most common units for vapor density are gm/m³. For example, if the actual vapor density is 10 g/m³ at 20°C compared to the saturation vapor density at that temperature of 17.3 g/m³, then the relative humidity is

$$R.H. = \frac{10 g/m^3}{17.3 g/m^3} \times 100\% = 57.8\% \qquad \textbf{Calculation}$$

Careful! There are dangers and possible misconceptions in these common statements about relative humidity.

Relative humidity is the amount of moisture in the air compared to what the air can "hold" at that temperature. When the air can't "hold" all the moisture, then it condenses as dew.

What's the Problem?

Saturation vapor pressure   Dewpoint   Relative humidity calculation

11.7.1 HVAC—Understanding and Calculating Dewpoint

Dewpoint

If the air is gradually cooled while maintaining the moisture content constant, the relative humidity will rise until it reaches 100%. This temperature, at which the moisture content in the air will saturate the air, is called the dew point. If the air is cooled further, some of the moisture will condense.

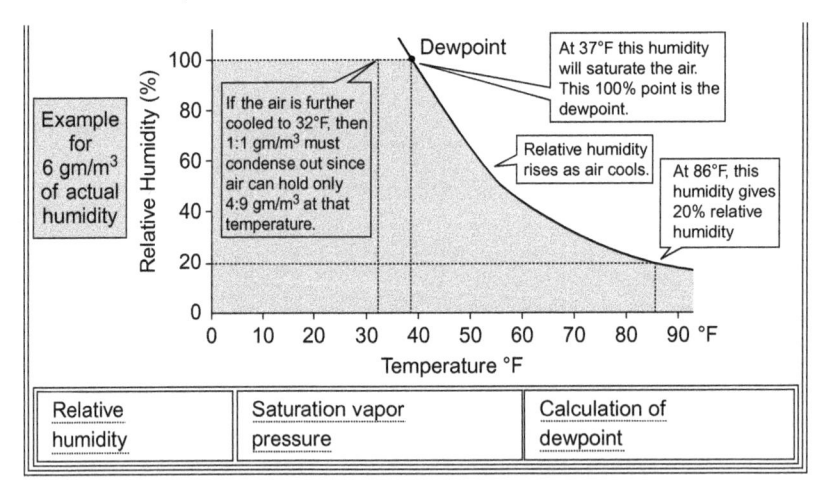

Relative humidity	Saturation vapor pressure	Calculation of dewpoint

11.7.2 Methods of Calculating Heating Efficiency—Combined Heat and Power (CHP)

Methods for Calculating Efficiency

CHP is an efficient and clean approach to generating power and thermal energy from a single fuel source. It is used either to replace or supplement conventional separate heat and power (SHP) (i.e., central station electricity available via the grid and an onsite boiler or heater).

- CHP System Efficiency Defined
- Key Terms Used in Calculating CHP Efficiency
- Calculating Total System Efficiency
- Calculating Effective Electric Efficiency
- Which CHP Efficiency Metric Should You Select?

Basic Information

- Catalog of CHP Technologies
- Biomass CHP
- Efficiency Benefits
- Reliability Benefits
- Environmental Benefits
- Economic Benefits

CHP System Efficiency Defined

Every CHP application involves the recovery of otherwise wasted thermal energy to produce additional power or useful thermal energy. Because CHP is highly efficient, it reduces emissions of traditional air pollutants and carbon dioxide, the leading greenhouse gas associated with global climate change.

Efficiency is a prominent metric used to evaluate CHP performance and compare it to SHP.

The illustration below illustrates the potential efficiency gains of CHP when compared to SHP.

Conventional Generation versus CHP: Overall Efficiency

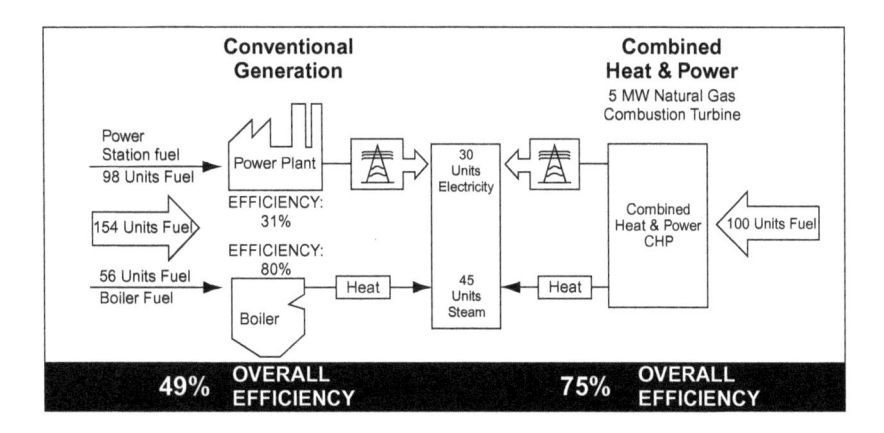

In this example of a typical CHP system, to produce 75 units of useful energy, the conventional generation or separate heat and power systems use 154 units of energy–98 for electricity production and 56 to produce heat—resulting in an overall efficiency of 49%. However, the CHP system needs only 100 units of energy to produce the 75 units of useful energy from a single fuel source, resulting in a total system efficiency of 75%.

Source: U.S. Environmental Protection Agency.

Key Terms Used in Calculating CHP Efficiency

Calculating a CHP system's efficiency requires an understanding of several key terms, described as follows:

- **CHP system.** The CHP system includes the unit in which fuel is consumed (e.g., turbine, boiler, engine), the electric generator, and the heat recovery unit that transforms otherwise wasted heat to usable thermal energy.

- **Total fuel energy input (Q_{FUEL}).** The thermal energy associated with the total fuel input. Total fuel input is the sum of all the fuel used by the CHP system. The total fuel energy input is often determined by multiplying the quantity of fuel consumed by the heating value of the fuel.

    Commonly accepted heating values for natural gas, coal, and diesel fuel are:

- 1020 Btu per cubic foot of natural gas
- 10,157 Btu per pound of coal
- 138,000 Btu per gallon of diesel fuel
- **Net useful power output (W_E).** Net useful power output is the gross power produced by the electric generator minus any parasitic electric losses—in other words, the electrical power used to support the CHP system. (An example of a parasitic electric loss is the electricity that may be used to compress the natural gas before the gas can be fired in a turbine.)
- **Net useful thermal output (ΣQ_{TH}).** Net useful thermal output is equal to the gross useful thermal output of the CHP system minus the thermal input. An example of thermal input is the energy of the condensate return and makeup water fed to a heat recovery steam generator (HRSG). Net useful thermal output represents the otherwise wasted thermal energy that was recovered by the CHP system.

Gross useful thermal output is the thermal output of a CHP system *utilized* by the host facility. The term utilized is important here. Any thermal output that is not used should not be considered. Consider, for example, a CHP system that produces 10,000 pounds of steam per hour, with 90% of the steam used for space heating and the remaining 10% exhausted in a cooling tower. The energy content of 9,000 pounds of steam per hour is the gross useful thermal output.

Calculating Total System Efficiency

The most commonly used approach to determining a CHP system's efficiency is to calculate *total system efficiency*. Also known as *thermal efficiency*, the total system efficiency (η_o) of a CHP system is the sum of the net useful power output (W_E) and net useful thermal outputs (ΣQ_{TH}) divided by the total fuel input (Q_{FUEL}), as shown below:

$$\eta_o = \frac{W_E + \Sigma Q_{TH}}{Q_{FUEL}}$$

The calculation of total system efficiency is a simple and useful method that evaluates what is produced (i.e., power and thermal output) compared to what is consumed (i.e., fuel). CHP systems with a relatively high net useful thermal output typically correspond to total system efficiencies in the range of 60 to 85%.

Note that this metric does not differentiate between the value of the power output and the thermal output; instead, it treats power output and thermal output as additive properties with the same relative value. In reality and in practice, thermal output and power output are not interchangeable because they cannot be converted easily from one to another. However, typical CHP applications have coincident power and thermal demands that must be met. It is reasonable, therefore, to consider the values of power and thermal output from a CHP system to be equal in many situations.

Calculating Effective Electric Efficiency

Effective electric efficiency calculations allow for a direct comparison of CHP to conventional power generation system performance (e.g., electricity produced from central stations, which is how the majority of electricity is produced in the United States). Effective electric efficiency (ξEE) can be calculated using the equation below, where (W_E) is the net useful power output, (ΣQ_{TH}) is the sum of the net useful thermal outputs, (Q_{FUEL}) is the total fuel input, and η equals the efficiency of the conventional technology that otherwise would be used to produce the useful thermal energy output if the CHP system did not exist:

$$\varepsilon_{EE} = \frac{W_E}{Q_{FUEL} - \sum(Q_{TH}/\alpha)}$$

For example, if a CHP system is fired by natural gas and produces steam, then α represents the efficiency of a conventional natural gas-fired boiler. Typical α values for boilers are 0.8 for natural gas-fired boiler, 0.75 for a biomass-fired boiler, and 0.83 for a coal-fired boiler.

The calculation of effective electric efficiency is essentially the CHP net electric output divided by the additional fuel the CHP system consumes over and above what conventional systems would have used to produce the thermal output for the site. In other words, this metric measures how effectively the CHP system generates power once the thermal demand of a site has been met.

Typical effective electrical efficiencies for combustion turbine-based CHP systems are in the range of 51 to 69%. Typical effective electrical efficiencies for reciprocating engine-based CHP systems are in the range of 69 to 84%.

Which CHP Efficiency Metric Should You Select?

The selection of an efficiency metric depends on the purpose of calculating CHP efficiency.

- If the objective is to compare CHP system energy efficiency to the efficiency of a site's SHP options, then the **total system efficiency metric** may be the right choice. Calculation of SHP efficiency is a weighted average (based on a CHP system's net useful power output and net useful thermal output) of the efficiencies of the SHP production components. The separate power production component is typically 33% efficient grid power. The separate heat production component is typically a 75 to 85% efficient boiler.
- If CHP electrical efficiency is needed for a comparison of CHP to conventional electricity production (i.e., the grid), then the **effective electric efficiency metric** may be the right choice. Effective electric efficiency accounts for the multiple outputs of CHP and allows for a direct comparison of CHP and conventional electricity production by crediting that portion of the CHP system's fuel input allocated to thermal output.

Both the total system and effective electric efficiencies are valid metrics for evaluating CHP system efficiency. They both consider all the outputs of CHP systems and, when used properly, reflect the inherent advantages of CHP. However, since each metric measures a different performance characteristic, use of the two different metrics for a given CHP system produces different values.

For example, consider a gas turbine CHP system that produces steam for space heating with the following characteristics:

Fuel Input (MMBtu/hr)	41
Electric Output (MW)	3.0
Thermal Output (MMBtu/hr)	17.7

Using the total system efficiency metric, the CHP system efficiency is 68% (3.0*3.413 + 17.7)/41).
Using the effective electric efficiency metric, the CHP system efficiency is 54% (3.0*3.413)/(41 − (17.7/0.8).
This is not a unique example; a CHP system's total system efficiency and effective electric efficiency often differ by 5 to 15%.

NOTE: Many CHP systems are designed to meet a host site's unique power and thermal demand characteristics. As a result, a truly accurate measure of a CHP system's efficiency may require additional information and broader examination beyond what is described in this document.

11.7.3 How Much Moisture Can the Air "Hold"?

Careful! There are dangers and possible misconceptions in these common statements about relative humidity.

Relative humidity is the amount of moisture in the air compared to what the air can "hold" at that temperature. When the air can't "hold" all the moisture, then it condenses as dew.

Of all the statements about relative humidity that I have heard in everyday conversation, the above is probably the most common. It may represent understanding of the phenomenon and has some common-sense utility, but it may represent a complete misunderstanding of what is going on physically. The air doesn't "hold" water vapor in the sense of having some attractive force or capturing influence. Water molecules are actually lighter and higher speed than the nitrogen and oxygen molecules that make up the bulk of the air, and they certainly don't stick to them and are not in any sense held by them. If you examine the thermal energy of molecules in the air at a room temperature of 20°C, you find that the average speed of a water molecule in the air is over 600 m/s or over 1400 miles/hr! You are not going to "hold" that molecule!

Another possibly helpful perspective would be to consider the space between air molecules under normal atmospheric conditions. From knowledge of atomic masses and gas densities and the modeling of the mean free path of gas molecules, we can conclude that the separation between air molecules at atmospheric pressure and 20°C is about 10 times their diameter. They will typically travel on the order of 30 times that separation between collisions. So water molecules in the air have a lot of room to move about and are not "held" by the air molecules.

By permission:hyperphysics.phy-astr.gsu.edu

When one says that the air can "hold" a certain amount of water vapor, the fact that is being addressed is that a certain amount of water vapor can be resident in the air as a constituent of the air. The high-speed water molecules act, to a good approximation, as particles of an ideal gas. At an atmospheric pressure of 760 mm Hg, you can express the amount of water in the air in terms of a partial pressure in mm Hg that represents the vapor pressure contributed by the water molecules. For example, at 20°C, the saturation vapor pressure for water vapor is 17.54 mm Hg, so if the air is saturated with water vapor, the dominant atmospheric constituents nitrogen and oxygen are contributing most of the other 742 mm Hg of the atmospheric pressure.

But water vapor is a very different type of air constituent than oxygen and nitrogen. Oxygen and nitrogen are always gases at Earth temperatures, having boiling points of 90K and 77K, respectively. Practically, they always act as ideal gases. But extraordinary water has a boiling point of 100°C = 373.15K and can exist in solid, liquid, and gaseous phases on the Earth. It is essentially always in a process of dynamic exchange of molecules between these phases. In air at 20°C, if the vapor pressure has reached 17.54 mm Hg, then as many water molecules are entering the liquid phase as are escaping to the gas phase, so we say that the vapor is "saturated." It has nothing to do with the air "holding" the molecules, but common usage often suggests that. As the air approaches saturation, we say that we are approaching the "dewpoint." The water molecules are polar and will exhibit some net attractive force on each other and therefore begin to depart from ideal gas behavior. By collecting together and entering the liquid state, they can form droplets in the atmosphere to make clouds, or near the surface to form fog, or on surfaces to form dew.

Another approach that might help clarify the point that air does not actually "hold" water is to note that the relative humidity really has nothing to do with the air molecules (i.e., N_2 and O_2). If a closed flask at 20°C had liquid water in it but no air at all, it would reach equilibrium at the saturated vapor pressure 17.54 mm Hg. At that point it would have a vapor density of 17.3 gm/m^3 of pure water vapor in the gas phase above the water surface. But if you had just removed the air and sealed the container with liquid water in it, you might have a situation where there was only 8.65 gm/m^3 resident in the gas phase at that particular moment. We would say that the relative humidity in the flask is 50% at that point because the resident water vapor density is half its saturation density. That is exactly the same thing we would say if the air were present—8.65 gm/m^3 of water vapor in the air at 20°C represents 50% relative humidity. Under these conditions, water molecules would be evaporating from the surface into the gas phase faster than they would be entering the water surface, so the vapor pressure of the water vapor above the surface would be rising toward the saturation vapor pressure.

11.7.4 General Heating Formulas—Energy Required to Heat, Offset Losses

General heating formulas

Energy required for heat up:

$$\frac{M \times c \times Delta\ T}{3214} = \text{kW hours/hours for heat up} = \text{kW}$$

M = Weight of Material in lbs.
c = Specific Heat of Material
Delta T = Desired temperature - starting temperature
hrs = heat up time expressed in hours

Energy required to offset losses:

$$\frac{\text{Losses(Watts)/square ft} \times \text{area (square ft)}}{1000} = \text{kW required to offset losses}$$

Energy required to change state:

$$\frac{\text{Weight(lbs.)} \times \text{latent heat of fusion}}{3412 \times \text{heat up time(in hrs.)}} = \text{kW}$$

11.7.5 General Heating Formulas—Energy Required to Heat Air Flow

Air heating formulas

Energy required to heat air flow: (approx.)

$$\text{kW} = \frac{\text{SCFM} \times (T2 - T1)°F}{2500}$$

kW = Energy (Kilowatts) required to heat flow
SCFM = Air flow rate in standard cubic feet per minute
T_2 = Temperature at heater exit
T_1 = Temperature at heater inlet
1.2 = Efficiency factor (assumes 20% energy is lost)

11.7.6 Formula to Convert Actual CFM (ACFM) to Standard Cubic Feet per Minute (SCFM)

Converting actual CFM (ACFM) to standard cubic feet per minute (SCFM):

$$\text{SCFM} = \text{ACFM} \times ((Pg + Patm)/Patm) \times ((Tref + 460)/(Tact + 460))$$

SCFM = Air flow in standard cubic feet per minute (corrected for temp. and pressure)
ACFM = Air flow in actual cubic feet per minute (uncorrected)
Pg = Gage pressure (psig)
Patm = Atmospheric pressure (14.7 psia)
Tact = Actual air temperature (°F)
Tref = Reference air temperature (70°F)

Constants and conversion formulas:

Density of Air at 32°F and 14.7 psia = 1.293 kg/m^3 (.081 lbs/ft^3)
Kilowatts × 3412 = BTU/hr

Meters³/hour × 35.3 = ft³/hr
Liters/min × 2.12 = ft³/hr
CFH/60 = CFM
°C = 5/9 (°F - 32)
°F = 9/5 °C + 32

11.8.0 Estimated Average Fuel Conversion Efficiency of Common Heating Appliances

Estimated Average Fuel Conversion Efficiency of Common Heating Appliances

Fuel Type—Heating Equipment	Efficiency (%)
Coal (bituminous)	
Central heating, hand-fired	45.0
Central heating, stoker-fired	60.0
Water heating, pot stove (50 gal.)	14.5
Oil	
High-efficiency central heating	89.0
Typical central heating	80.0
Water heater (50 gal.)	59.5
Gas	
High-efficiency central furnace	97.0
Typical central boiler	85.0
Minimum efficiency central furnace	78.0
Room heater, unvented	99.0
Room heater, vented	65.0
Water heater (50 gal.)	62.0
Electricity	
Baseboard, resistance	99.0
Central heating, forced air	97.0
Central heating, heat pump	200+
Ground source heat pump	300+
Water heaters (50 gal.)	97.0
Wood and Pellets	
Franklin stoves	30.0 - 40.0
Stoves with circulating fans	40.0 - 70.0
Catalytic stoves	65.0 - 75.0
Pellet stoves	85.0 - 90.0

Armed with these numbers, several "Evaluation Tools" above will allow you to calculate the cost of energy delivered to your home for different fuels and systems. The actual cost of fuels is best determined by examining your utility bills or contacting your utility and by contacting local suppliers of other fuels. The DOE's Energy Information Administration also tracks some fuel prices.

Section | 11 Plumbing and HVAC Calculations

11.8.1 Reading Those Gas Meters

Calculating Fuel Consumption and Carbon Footprint of the Glass Studio

Calculating approximate fuel consumption and carbon output involves several things. With gas, it is necessary to isolate each piece of equipment, and, using a gas meter, clock the actual consumption of the unit, both in various phases of firing and as a total for the day or week. Those using natural gas are likely to have a gas meter for the studio. The older meters are of the multiple spinning dial variety where the consumption can be measured quickly on the dial clocking half a cubic foot, or perhaps 2 cubic feet.

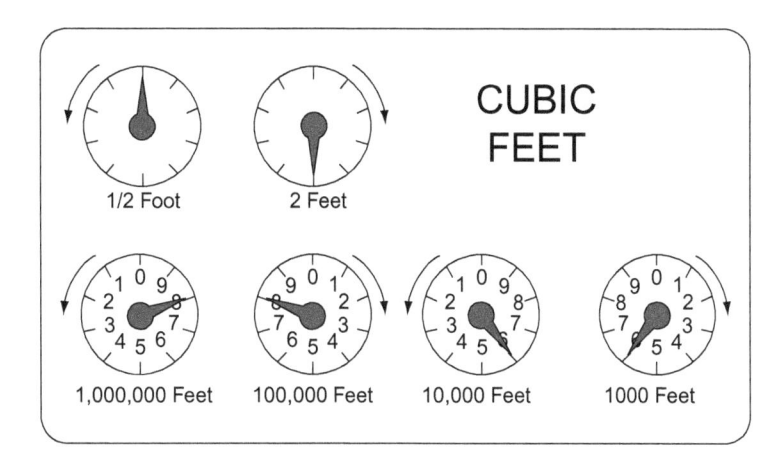

Newer models look more like an odometer, and the numbers turn only after 100 cubic feet have passed through. This is cheaper for the gas company and easy for the meter reader, but it makes clocking the consumption of individual units tedious.

1	2	3	4	5	6	7	8	9	1	2	3

11.9.0 Calculating Home Heating Energy—Gas versus Electric Resistance Heating

Calculating Home Heating Energy

Heat transfer from your home can occur by conduction, convection, and radiation. It is typically modeled in terms of conduction, although infiltration through walls and around windows can contribute a significant additional loss if they are not well sealed. Radiation loss can be minimized by using foil-backed insulation as a radiation barrier.

The U.S. heating and air conditioning industry uses almost entirely the old British and U.S. common units for their calculations. For compatibility with the commonly encountered quantities, this example will be expressed in those units.

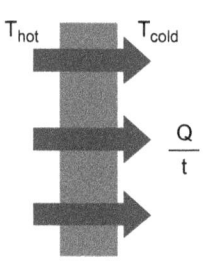

$$\text{Heat loss rate} = \frac{Q}{t} = \frac{(\text{Area}) \times (T_{inside} - T_{outside})}{\text{Thermal resistance of wall}}$$

If Q/t is in BTU/hr
Area in ft^2
$T_{in} - T_{out}$ in °F
then the thermal resistance is the "R-factor" quoted by insulation manufacturers. The units of the R-factor are

$$\frac{\text{ft}^2 \times °F}{\text{BTU/hr}}$$

For standard R11 wall insulation, you lose 1/11 BTU/hr per square foot of wall space, per degree Fahrenheit temperature difference.

I. Calculate wall-loss rate in BTUs per hour.

For a 10 ft by 10 ft room with an 8 ft ceiling, with all surfaces insulated to R19 as recommended by the U.S. Department of Energy, with inside temperature 68°F and outside temperature 28°F:

$$\text{Heat loss rate} = \frac{(320\,\text{ft}^2) \times (68°F - 28°F)}{19 \frac{\text{ft}^2 \times °F}{\text{BTU/hr}}} = 674\,\text{BTU/hr}$$

II. Calculate loss per day at these temperatures.

By permission: hyperphysics.phy-astr.gsu.ed

$$\text{Heat loss per day} = (674\,\text{BTU/hr})(24\,\text{hr}) = 16168\,\text{BTU}$$

Note that this is just the loss through the walls. The loss through the floor and ceiling is a separate calculation, and usually involves different R-values.

III. Calculate loss per "degree day."

This is the loss per day with a one degree difference between inside and outside temperature.

$$\text{Loss per degree day} = Q = \frac{(320\,\text{ft}^2) \times (1°F)}{19 \frac{\text{ft}^2 \times °F}{\text{BTU/hr}}} \times 24\,\text{hr/day} = 404 \frac{\text{BTU}}{\text{degree day}}$$

If the conditions of case II prevailed all day, you would require 40 degree-days of heating, and therefore require 40 degree-days × 404 BTU/degree-day = 16168 BTU to keep the inside temperature constant.

IV. Calculate heat loss for entire heating season.

The typical heating requirement for an Atlanta heating season, September to May, is 2980 degree-days (a long-term average).

$$\text{Heat loss} = Q = 404 \frac{\text{BTU}}{\text{degree day}} \times 2980\,\text{degree days} = 1.20\,\text{million BTU}$$

The typical number of degree-days of heating or cooling for a given geographical location can usually be obtained from the weather service.

V. Calculate heat loss per heating season for a typical uninsulated southern house in Atlanta.

The range of loss rates given by DOE for uninsulated typical dwellings is 15,000 to 30,000 BTU/degree-day. Choosing 25,000 BTU/degree-day:

$$\text{Heat loss} = Q = 25{,}000 \frac{\text{BTU}}{\text{degree day}} \times 2980 \text{ degree days} = 74.5 \text{ million BTU}$$

VI. Calculate annual heating cost.

Assume natural gas cost of $12 per million BTU in a furnace operating at 70% efficiency.

$$\left[\frac{74.5 \text{ million BTU}}{0.70}\right]\left[\frac{\$12}{\text{million BTU}}\right] = \$1277$$

Assume electric resistance heating at 100% efficiency*, 9¢/kWh.

$$(7.45 \times 10^7 \text{ BTU})(2.93 \times 10^{-4} \text{ kWh/BTU})(\$.09/\text{kWh}) = \$1965$$

Assume an electric heat pump with coefficient of performance = 3

$$\frac{\$1965}{3} = \$655$$

When you are heating with natural gas, you are using the primary fuel at your house, and this is clearly preferable to using electric resistance heating, which is wasteful of the high-quality delivered electric energy. By using an electric heat pump, at least in the southern United States, you can get a coefficient of performance of about 3 and just about balance off that 3:1 loss in the generation process. In the above example, the electric heat pump is considerably cheaper, but that may be an artifact of the currently high natural gas prices. Over the last 25 years or so, natural gas and electric heat pump heating have stayed comparable in cost.

11.10.0 Comparing Fuel Costs of Heating and Cooling Systems—Gas, Electric, Kerosene, Wood, Pellets

---

*100% efficiency for using electricity in your house to produce heat is a common marketing ploy by electric utility companies. It is misleading because you have to burn about 3 units of primary fuel to deliver 1 unit of electric energy to the house because of the thermal bottleneck in electricity generation. So 100% efficient use at your house is about 33% efficient in the use of the primary fuel.

Comparing Fuel Costs of Heating and Cooling Systems

June 2003

Introduction

One of the most common questions posed to energy specialists at Engineering Extension asks for a comparison between costs to operate different heating and cooling systems. It might be a comparison of a furnace to a heat pump, a regular furnace to a high-efficiency furnace, or a wood burning stove to a pellet stove.

There are two components to cost: the initial cost to purchase and install the system, and the ongoing fuel cost. In general, higher efficiency equipment costs more initially but saves operating costs. To determine the purchase price, get bids from one or more contractors. Be certain bids include all costs to make the system fully functional, including duct work, thermostats, and chimneys. The accompanying fact sheets will help you compare the cost of fuel for several types of heating and cooling systems.

Annual cost of delivering heating and cooling to a home depends on cost of the fuel, the efficiency with which the system converts the fuel source into heating or cooling energy, and the quantity of heating and cooling required. The following section, "Estimating the Cost of Heating or Cooling," allows you to estimate the cost of one million Btus for several fuels and system types. However, if you want to compare annual estimated costs for two or more fuels, then you will also need to estimate the heating load of your home.

What Is the Price of Fuel?

Fuel prices vary between suppliers, may change seasonally, and are affected by world events. To estimate fuel costs, you can either contact your local utility or supplier or you can use past billings.

To estimate natural gas costs from your utility bill, divide the monthly charge by the consumption, usually measured in MCF (MCF = 1000 Cubic feet). The cost should be between $3 and $12 per MCF. Use a winter bill so that meter charges are spread out over several units of gas. If your bill shows gas consumption in CCF (CCF = 100 Cubic feet), you will need to multiply the gas cost by 10 to get it in dollars per MCF.

To obtain an average $ per kilowatt hour (kWh), divide the total monthly cost by the consumption in kWh. Use a midwinter bill if you want to estimate heating costs and a midsummer bill if you want to estimate cooling costs. The cost for electricity in Kansas varies from $0.04 to $0.15 per kWh.

Propane, fuel oil, wood, and pellets are sold in simple units and should be easy to determine.

Estimating the Cost of Heating or Cooling

If you just want to compare operating costs of different systems, you can use Tables 1 through 7 to directly determine the cost of delivering one million Btus (MBTUs) of heating and Table 8 for one MBTU of cooling. For example, you could compare the cost of delivering one MBTU to your home from a high-efficiency natural gas furnace to the cost of delivering one MBTU from wood in a modern wood stove. There are several measures of system efficiency. A brief explanation is provided in the description of the tables.

Table 1 is for natural gas furnaces and boilers. There are three efficiency levels, and gas prices range from $5 to $15 per thousand cubic feet (MCF). If your furnace was installed before about 1985, use the "older equipment" column. If you have a modern but normal-efficiency unit, use the 78% column. The last column is for high-efficiency (condensing) equipment.

Tables 2 and 3 are similar to Table 1, but are for propane and fuel oil, respectively.

Modern natural gas, propane, and fuel oil furnaces and boilers receive an annual fuel-utilization efficiency (AFUE) rating. Older units were not rated but an assumed performance of 65% is reasonable.

TABLE 1 Natural Gas Heating Costs—$ per MBTU Delivered for Three Appliance Efficiencies

	Furnace or Boiler efficiency		
Gas Price $/MCF.	65% (low) Older Equipment	AFUE = 78% (average) Current Minimum	AFUE = 95% High Efficiency
$5.00	$7.69	$6.41	$5.26
$5.50	$8.46	$7.05	$5.79
$6.00	$9.23	$7.69	$6.32
$6.50	$10.00	$8.33	$6.84
$7.00	$10.77	$8.97	$7.37
$7.50	$11.54	$9.62	$7.89
$8.00	$12.31	$10.26	$8.42
$8.50	$13.08	$10.90	$8.95
$9.00	$13.85	$11.54	$9.47
$9.50	$14.62	$12.18	$10.00
$10.00	$15.38	$12.82	$10.53
$10.50	$16.15	$13.46	$11.05
$11.00	$16.92	$14.10	$11.58
$11.50	$17.69	$14.74	$12.11
$12.00	$18.46	$15.38	$12.63
$12.50	$19.23	$16.03	$13.16
$13.00	$20.00	$16.67	$13.68
$13.50	$20.77	$17.31	$14.21
$14.00	$21.54	$17.95	$14.74
$14.50	$22.31	$18.59	$15.26
$15.00	$23.08	$19.23	$15.79

TABLE 2 Propane Heating Costs—$ per MBTU Delivered for Three Appliance Efficiencies

	Furnace or Boiler Efficiency		
Propane Price $/gal.	65% (low) Older Equipment	AFUE = 78% (average) Current Minimum	AFUE = 95% High Efficiency
$0.60	$10.14	$8.24	$6.94
$0.65	$10.99	$8.93	$7.52
$0.70	$11.83	$9.62	$8.10
$0.75	$12.68	$10.30	$8.68
$0.80	$13.52	$10.99	$9.25
$0.85	$14.37	$11.68	$9.83
$0.90	$15.22	$12.36	$10.41
$0.95	$16.06	$13.05	$10.99
$1.00	$16.91	$13.74	$11.57
$1.05	$17.75	$14.42	$12.15
$1.10	$18.60	$15.11	$12.72
$1.15	$19.44	$15.80	$13.30
$1.20	$20.29	$16.48	$13.88
$1.25	$21.13	$17.17	$14.46
$1.30	$21.98	$17.86	$15.04

(Continued)

TABLE 2 Propane Heating Costs—$ per MBTU Delivered for Three Appliance Efficiencies—Cont'd

Propane Price $/gal.	Furnace or Boiler Efficiency		
	65% (low) Older Equipment	AFUE = 78% (average) Current Minimum	AFUE = 95% High Efficiency
$1.35	$22.82	$18.54	$15.62
$1.40	$23.67	$19.23	$16.19
$1.45	$24.51	$19.92	$16.77
$1.50	$25.36	$20.60	$17.35
$1.55	$26.20	$21.29	$17.93
$1.60	$27.05	$21.98	$18.51

TABLE 3 Fuel Oil Heating Costs — $ per MBTU Delivered for Three Appliance efficiencies

Oil Price $/gallon	Furnace or Boiler Efficiency		
	65% (low) Older Equipment	AFUE = 78% (average) Current Minimum	AFUE = 86% High Efficiency
$0.70	$7.76	$6.47	$5.87
$0.75	$8.32	$6.93	$6.29
$0.80	$8.87	$7.39	$6.71
$0.85	$9.43	$7.86	$7.13
$0.90	$9.98	$8.32	$7.55
$0.95	$10.54	$8.78	$7.96
$1.00	$11.09	$9.24	$8.38
$1.05	$11.65	$9.71	$8.80
$1.10	$12.20	$10.17	$9.22
$1.15	$12.76	$10.63	$9.64
$1.20	$13.31	$11.09	$10.06
$1.25	$13.87	$11.55	$10.48
$1.30	$14.42	$12.02	$10.90
$1.35	$14.97	$12.48	$11.32
$1.40	$15.53	$12.94	$11.74
$1.45	$16.08	$13.40	$12.16
$1.50	$16.64	$13.87	$12.58
$1.55	$17.19	$14.33	$12.99
$1.60	$17.75	$14.79	$13.41
$1.65	$18.30	$15.25	$13.83
$1.70	$18.86	$15.71	$14.25

Table 4 is for electric heat. The price per MBTU for electric resistance heat includes both baseboard and central resistance heating systems. Sections for air-source heat pumps, groundwater heat pumps, and ground-loop heat pumps are provided, and each contains three performance levels.

Air-source heat pumps are the most common heat pump. They have an inside blower and coil with an outside compressor and coil, and look like a conventional air conditioner. Use an air-source heat pump heating seasonal performance factor (HSPF) of 5 for older heat pumps, 6.8 for an average-performance unit, and 9.4 if you have or plan to buy a superior-performance unit.

Ground-loop and groundwater are both geothermal heat pump systems. A ground-loop heat pump, Figure 1, circulates water through buried piping loop. Coefficient of performance (COP) is the measure of performance for geothermal heat pumps. A COP of 3.1 would be appropriate for an older or low-performance system; a COP of 3.5 is representative of average equipment sold today; and a system with a COP of 4.2 would represent superior performance.

Unlike a ground-loop system that circulates water in a piping system, a groundwater heat pump, Figure 2, draws water from a well, extracts heat from the water in the winter or rejects heat to it in the summer, and then discharges the water, typically to another well. The heat pump is normally located inside, but there will be one or two wells associated with its operation. Groundwater heat pumps also use COP as a measure of performance, with a COP of 3.2 for an older or low-performance system, 4.1 for average performance, and 4.7 for superior performance. In many cases, the same equipment is used for both ground-loop and groundwater systems. They are rated with different COPs because of the differences between ground-loop and groundwater temperatures.

TABLE 4 Electric Heating Costs — $ per MBTU Delivered for Several Appliances and Performance Levels

Electricity $/kWh	Electric Resistance	Air-source Heat Pump Performance			Ground-loop Heat Pump Performance			Groundwater Heat Pump Performance		
		HSPF = 5.0 (low) Older Equipment	HSPF = 6.8 (average) Current Minimum	HSPF = 9.4 (superior)	COP = 3.1 (low)	COP = 3.5 (average)	COP = 4.2 (superior)	COP = 3.6 (low)	COP = 4.1 (average)	COP = 4.7 (superior)
$0.040	$11.73	$8.00	$5.88	$4.26	$4.21	$3.74	$3.11	$3.54	$3.10	$2.70
$0.045	$13.20	$9.00	$6.62	$4.79	$4.74	$4.21	$3.49	$3.98	$3.49	$3.04
$0.050	$14.66	$10.00	$7.35	$5.32	$5.26	$4.67	$3.88	$4.42	$3.88	$3.38
$0.055	$16.13	$11.00	$8.09	$5.85	$5.79	$5.14	$4.27	$4.87	$4.26	$3.72
$0.060	$17.60	$12.00	$8.82	$6.38	$6.32	$5.61	$4.66	$5.31	$4.65	$4.05
$0.065	$19.06	$13.00	$9.56	$6.91	$6.84	$6.07	$5.05	$5.75	$5.04	$4.39
$0.070	$20.53	$14.00	$10.29	$7.45	$7.37	$6.54	$5.43	$6.19	$5.43	$4.73
$0.075	$21.99	$15.00	$11.03	$7.98	$7.89	$7.01	$5.82	$6.64	$5.81	$5.07
$0.080	$23.46	$16.00	$11.76	$8.51	$8.42	$7.48	$6.21	$7.08	$6.20	$5.41
$0.085	$24.93	$17.00	$12.50	$9.04	$8.95	$7.94	$6.60	$7.52	$6.59	$5.74
$0.090	$26.39	$18.00	$13.24	$9.57	$9.47	$8.41	$6.99	$7.96	$6.98	$6.08
$0.095	$27.86	$19.00	$13.97	$10.11	$10.00	$8.88	$7.38	$8.41	$7.36	$6.42
$0.100	$29.33	$20.00	$14.71	$10.64	$10.53	$9.35	$7.76	$8.85	$7.75	$6.76
$0.105	$30.79	$21.00	$15.44	$11.17	$11.05	$9.81	$8.15	$9.29	$8.14	$7.09
$0.110	$32.26	$22.00	$16.18	$11.70	$11.58	$10.28	$8.54	$9.73	$8.53	$7.43
$0.115	$33.72	$23.00	$16.91	$12.23	$12.11	$10.75	$8.93	$10.18	$8.91	$7.77
$0.120	$35.19	$24.00	$17.65	$12.77	$12.63	$11.21	$9.32	$10.62	$9.30	$8.11
$0.125	$36.66	$25.00	$18.38	$13.30	$13.16	$11.68	$9.70	$11.06	$9.69	$8.45
$0.130	$38.12	$26.00	$19.12	$13.83	$13.68	$12.15	$10.09	$11.50	$10.08	$8.78
$0.135	$39.59	$27.00	$19.85	$14.36	$14.21	$12.62	$10.48	$11.95	$10.47	$9.12
$0.140	$41.06	$28.00	$20.59	$14.89	$14.74	$13.08	$10.87	$12.39	$10.85	$9.46

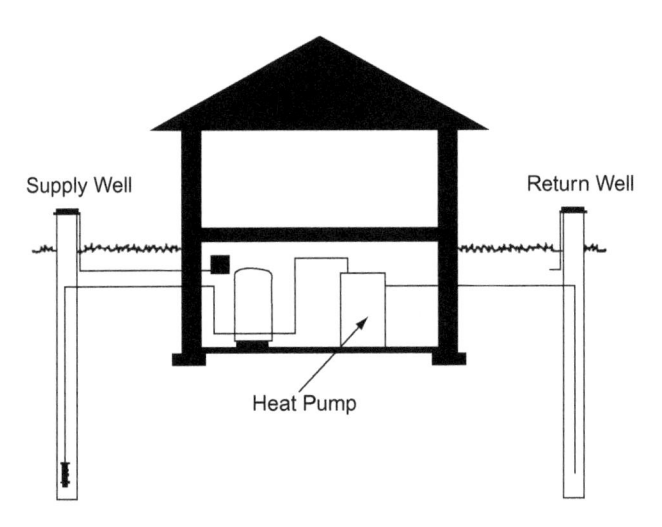

FIGURE 1 Groundwater heat pump.

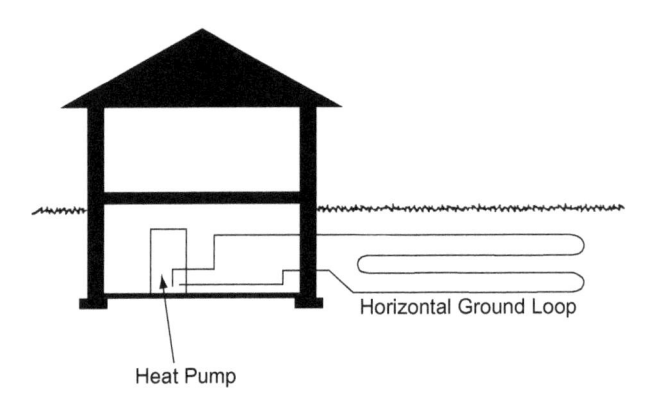

FIGURE 2 Groundwater heat pump.

Table 5 is used to estimate the cost per MBTU for unvented kerosene heaters. They are 100% efficient because all of the heat is delivered to the home. If you are using a vented kerosene appliance, use Table 3.

TABLE 5 Kerosene Heating Costs—$ per Million MBTU Delivered

	Unvented Kerosene Heater
Kerosene Price $/Gallon	Unvented Equipment
$1.00	$7.56
$1.10	$8.31
$1.20	$9.07
$1.30	$9.83
$1.40	$10.58
$1.50	$11.34

Unvented Kerosene Heater	
Kerosene Price $/Gallon	Unvented Equipment
$1.60	$12.09
$1.70	$12.85
$1.80	$13.61
$1.90	$14.36
$2.00	$15.12
$2.10	$15.87
$2.20	$16.63
$2.30	$17.38
$2.40	$18.14
$2.50	$18.90
$2.60	$19.65
$2.70	$20.41
$2.80	$21.16
$2.90	$21.92
$3.00	$22.68

Estimate the cost of delivered heating energy

Example: Compare the cost of heat from a propane furnace to the cost of heat from an air-source heat pump.
First, you will need to know the cost of both fuels and efficiencies of the systems. Follow this example to learn how to use Tables 1 through 8.
Table 2 is for propane appliances. Assuming you have an old propane furnace, the efficiency will be about 65%. If you pay $0.90 per gallon for propane, the cost per million Btus (MBTUs) will be $15.22.
Table 4 is for electric appliances. Compare this to the cost of heating with an average-efficiency, air-source heat pump with electricity costing $0.07 per kilowatt hour (kWh). The cost per MBTU will be about $10.29.
Delivered heat from the heat pump costs about two-thirds that of propane.

Table 6 will allow you to estimate the cost per MBTU for several wood heating appliances. The species of wood, cost per cord, and appliance efficiency are all important to getting an accurate estimate. The efficiency ratings provided are typical but may vary between manufacturers. Several common wood species are listed, with cord costs ranging from $80 to $140. There are sections of the table for open fireplaces; pre-1980 wood stoves; masonry heaters; and post-1980, EPA-certified wood stoves. For more details on solid-fuel heating appliances, obtain a copy of *Solid-Fuel Heating Appliances* online at *www.engext.ksu.edu/*. Look under publications.
Table 7 provides heating cost estimates for pellet- and corn-burning appliances.
Table 8 will estimate the cost of providing one MBTU of cooling for air conditioners and heat pumps. A seasonal energy efficiency rating (SEER) is the performance measure for modern air conditioners and air-source heat pumps. Older units may not be rated, and a SEER of 7 is reasonable for estimating operating costs.

Estimating Annual Costs

Once you have determined the cost per MBTU for any fuel, you can estimate annual heating or cooling costs. It is important to remember these are estimates only; lifestyle, actual housing conditions, house configuration, and other factors can greatly influence heating and cooling costs.

TABLE 6 Wood Heating Costs—$ per Million BTU for Several

	Wood Heating Appliance Efficiency							
	10%—typical open fireplace				50%—typical central boiler, furnace, or pre-1980 wood stove			
Price per cord	$80	$100	$120	$140	$80	$100	$120	$140
Species								
Cottonwood	$50.63	$63.29	$75.95	$88.61	$10.13	$12.66	$15.19	$17.72
Elm, American	$40.00	$50.00	$60.00	$70.00	$8.00	$10.00	$12.00	$14.00
Hackberry	$37.74	$47.17	$56.60	$66.04	$7.55	$9.43	$11.32	$13.21
Honeylocust	$29.96	$37.45	$44.94	$52.43	$5.99	$7.49	$8.99	$10.49
Maple, Silver	$42.11	$52.63	$63.16	$73.68	$8.42	$10.53	$12.63	$14.74
Oak, Red	$32.52	$40.65	$48.78	$56.91	$6.50	$8.13	$9.76	$11.38
Osage Orange	$24.32	$30.40	$36.47	$42.55	$4.86	$6.08	$7.29	$8.51
	60% — typical masonry heater				70% — typical EPA-certified wood stoves and inserts			
Price per cord	$80	$100	$120	$140	$80	$100	$120	$140
Species								
Cottonwood	$8.44	$10.55	$12.66	$14.77	$7.23	$9.04	$10.85	$12.66
Elm, American	$6.67	$8.33	$10.00	$11.67	$5.71	$7.14	$8.57	$10.00
Hackberry	$6.29	$7.86	$9.43	$11.01	$5.39	$6.74	$8.09	$9.43
Honeylocust	$4.99	$6.24	$7.49	$8.74	$4.28	$5.35	$6.42	$7.49
Maple, Silver	$7.02	$8.77	$10.53	$12.28	$6.02	$7.52	$9.02	$10.53
Oak, Red	$5.42	$6.78	$8.13	$9.49	$4.65	$5.81	$6.97	$8.13
Osage Orange	$4.05	$5.07	$6.08	$7.09	$3.47	$4.34	$5.21	$6.08

Wood Species, Heating Appliance Efficiencies, and Cord Wood Costs

TABLE 7 Pellet and Corn Hating Costs — $ per MBTU

Pellet price		Typical Pellet Stove	Corn Price	Typical Corn Stove
Price per 40-pound Bag	Price per Ton		Price per Bushel	
$2.50	$125	$9.77	$1.50	$5.05
$3.00	$150	$11.73	$2.00	$8.42
$3.50	$175	$13.68	$2.50	$11.78
$4.00	$200	$15.63	$3.00	$15.15

Table 9[1] provides estimates of heating and cooling requirements of homes in Kansas. Three levels of home efficiency are listed. Standard practice represents homes as they have generally been constructed in Kansas, energy code-compliant applies to a home that would meet modern energy codes, and energy efficient represents homes where high performance was a major design goal. There are also three climate areas listed.

Based on the type of home and location, choose the appropriate index. Multiply it by the size of your home (square feet of living space) and the cost of your fuel in $ per MBTU, then divide by 1000 to estimate annual costs. If you live in an older, poorly insulated and weatherized home, your heating costs will be higher than those estimated by this method. To estimate savings for using higher performance equipment or other fuels, calculate the costs for each and compare.

---

1. *Ground-Source Heat Pumps, An Efficient Choice for Residential and Commercial Use*, J. Mark Hannifan, Joe E King, AIA, 1995.

Section | 11 Plumbing and HVAC Calculations

TABLE 8 Electric Cooling Costs—$ per MBTU Cooling for Several Appliances and Performance Levels

Electricity $/kWh	Air Conditioner or Air-source Heat Pump Performance			Groundwater Heat Pump Performance			Ground-loop Heat Pump Performance		
	SEER = 7 (low) Older Equipment	SEER = 12 (average)	SEER = 15 (superior)	EER = 16 (low)	EER = 19 (average)	EER = 24 (superior)	EER = 13 (low)	EER = 16 (average)	EER = 20 (superior)
$0.040	$5.71	$4.00	$2.67	$2.61	$2.22	$1.78	$3.18	$2.62	$2.13
$0.045	$6.43	$4.50	$3.00	$2.93	$2.50	$2.00	$3.58	$2.95	$2.39
$0.050	$7.14	$5.00	$3.33	$3.26	$2.78	$2.22	$3.98	$3.28	$2.66
$0.055	$7.86	$5.50	$3.67	$3.58	$3.06	$2.44	$4.38	$3.61	$2.92
$0.060	$8.57	$6.00	$4.00	$3.91	$3.33	$2.67	$4.78	$3.94	$3.19
$0.065	$9.29	$6.50	$4.33	$4.23	$3.61	$2.89	$5.18	$4.27	$3.46
$0.070	$10.00	$7.00	$4.67	$4.56	$3.89	$3.11	$5.57	$4.59	$3.72
$0.075	$10.71	$7.50	$5.00	$4.89	$4.17	$3.33	$5.97	$4.92	$3.99
$0.080	$11.43	$8.00	$5.33	$5.21	$4.44	$3.56	$6.37	$5.25	$4.25
$0.085	$12.14	$8.50	$5.67	$5.54	$4.72	$3.78	$6.77	$5.58	$4.52
$0.090	$12.86	$9.00	$6.00	$5.86	$5.00	$4.00	$7.17	$5.91	$4.78
$0.095	$13.57	$9.50	$6.33	$6.19	$5.28	$4.22	$7.56	$6.23	$5.05
$0.100	$14.29	$10.00	$6.67	$6.51	$5.56	$4.44	$7.96	$6.56	$5.32
$0.105	$15.00	$10.50	$7.00	$6.84	$5.83	$4.67	$8.36	$6.89	$5.58
$0.110	$15.71	$11.00	$7.33	$7.17	$6.11	$4.89	$8.76	$7.22	$5.85
$0.115	$16.43	$11.50	$7.67	$7.49	$6.39	$5.11	$9.16	$7.55	$6.11
$0.120	$17.14	$12.00	$8.00	$7.82	$6.67	$5.33	$9.55	$7.87	$6.38
$0.125	$17.86	$12.50	$8.33	$8.14	$6.94	$5.56	$9.95	$8.20	$6.65
$0.130	$18.57	$13.00	$8.67	$8.47	$7.22	$5.78	$10.35	$8.53	$6.91
$0.135	$19.29	$13.50	$9.00	$8.79	$7.50	$6.00	$10.75	$8.86	$7.18
$0.140	$20.00	$14.00	$9.33	$9.12	$7.78	$6.22	$11.15	$9.19	$7.44

TABLE 9 Annual Heat and Cooling Indices—1000 Btus/square foot

	Heating			Cooling		
	Northwest	Central	Southeast	Northwest	Central	Southeast
Current practice	50	45	40	11	13	14
Energy code compliant	36	32	29	10	11	12
Energy efficient	28	25	23	9	10	11

This material was prepared with the support of the U.S. Department of Energy (DOE) Grant No. DE-FG48-97R802102. However, any opinions, findings, conclusions, or recommendations expressed herein are those of the authors and do not necessarily reflect the views of DOE.

Estimating Annual Costs

Example: Estimate the annual cost of heating a 2000-square-foot home in rural Sedgwick County. The home was built in the 1960s. The home owner is considering both propane and an air-source heat pump. Fuel costs were determined in the previous example to be $15.22/MBTU for propane and $10.29/MBTU for a heat pump.

The heating index for the home would be 45. Annual heating costs would be

$$\frac{2{,}000 \times 45 \times 15.22}{1{,}000} = \$1{,}370 \text{ for propane, and}$$

$$\frac{2{,}000 \times 45 \times 10.29}{1{,}000} = \$925 \text{ for the heat pump.}$$

11.11.0 Residential Ground Source Heat Pump (GSHP) Savings versus Electric, Gas, and Fuel Oil

Residential GSHP Annual Savings

Conventional System	Mean Annual Savings (%)			
	Number	Energy	Number	Dollars
Electric Resistance Heat/AC[a]	21	57%	18	54%
Air-Source Heat Pump	33	31%	21	31%
Natural Gas Furnace/AC[b]	17	67%	21	18%
Oil Furnace/AC[b]	6	71%	9	33%
Other (propane, unspecified)	7	46%	7	39%

[a.] AC means with electric air conditioning.
[b.] Natural gas or oil furnaces with electric air conditioning had annual operating costs less than GSHP systems for 23% of the case studies.
The mean annual dollar savings of GSHPs shown above may appear attractive; however, due to the relatively low-annual operating costs of conventional energy systems, it is difficult in many cases to recover the additional incremental cost (ground loop) of GSHP systems.
Residential GSHP system peak demand reduction compared to single-zone electric resistance heating for 13 case studies ranged from 5.3 kW to 10.4 kW with a mean of 7.2 kW.

School Ground-Source Heat Pumps

The potential for savings of GSHP systems in schools is documented in 26 case studies, which include 54% vertical ground-coupled, 19% groundwater, 12% horizontal ground-coupled, and 15% other types of systems.

School GSHP Annual Savings

Conventional System	Mean Annual Savings (%)			
	Number	Energy	Number	Dollars
Electric Resistance Heat	2	51%	3	45%
Natural Gas	3	61%	1	13%
Fuel Oil	1	76%	1	58%

Source: Study conducted by U.S. Department of Energy via Grant C92-12025-001.

11.11.1 Paybacks for Residential GSHP Economics

Residential GSHP Economics

Conventional System	Number	Simple Payback (yrs) Range	Simple Payback (yrs) Mean
Electric Resistance Heat/AC	4	2.7 to 6.8	4.4
Air-Source Heat Pump	3	2.0 to 9.5	5.9
Natural Gas/AC	9	4.2 to 24.1	11.6
Fuel Oil/AC	6	1.4 to 7.1	4.4
Other	5	2.0 to 6.8	4.3

The biggest barrier to faster paybacks of GSHP systems is the incremental cost of the ground loop.

Since residential GSHP systems are usually included in the mortgage, a break-even value of electric rates is a more meaningful value than simple payback.

In January 1995, data became available on the cost of purchasing and installing residential GSHP systems. Based on this data, an earlier analysis of a new well-insulated home, and a 30-year fixed rate mortgage at 8%, the electrical break-even rates were calculated for two different climate zones. In the colder zone, the break-even rates were \$0.061/kWh for vertical and \$0.058/kWh for horizontal ground-coupled systems. In the warmer zones, they were \$0.097 and \$0.084, respectively. Electric rates in excess of these break-even values would result in the GSHP system having a positive cash flow to the homeowner.

Simple paybacks for school systems were reported in only 5 out of 23 case studies. These simple paybacks ranged from 5 to 14 years for electric resistance heating, 3.5 years for natural gas systems, and 5 to 7 years for others.

Case studies for commercial buildings reported simple paybacks for 17 out of 46 GSHP systems. The range of simple paybacks was 1.3 to 4.7 years, with a mean of 2.8 years.

For commercial buildings, all but four of the simple paybacks represent buildings located in northern climates.

Caution should be used in arriving at economic conclusions for any of the three groups presented in this summary. This is due to variables of climate, ground characteristics, GSHP system type, equipment efficiency, sizing, complex utility rate structures, and a variety of economic analysis methods used in the case studies.

Benefits reported for using GSHP systems in schools are: addition of mechanical cooling, improved control, and simplicity of maintenance and repair. In southern climates, the benefits include: elimination of cooling towers, outdoor equipment, mechanical rooms, and ductwork.

11.11.2 Commercial Ground Source Heat Pump (GSHP) Savings versus Electric, Gas, Fuel Oil

Case studies (46) documented for commercial GSHP systems ranged in capacity from 30 to 4700 tons. These systems employed vertical ground-coupled (43%), groundwater (35%), horizontal ground-coupled (11%), and other (11%) types of ground systems. Commercial GSHP systems were monitored in 84% of the case studies, and conventional energy systems represent only 20% of the comparisons.

The average annual energy savings of GSHP systems ranged from 40 to 72%, and dollar savings ranged from 31 to 56%.

11.11.3 Paybacks for Commercial GSHP Economics

	Mean Annual Savings (%)			
Conventional System	Number	Energy	Number	Dollars
Electric Resistance Heat/AC	6	59%	5	56%
Air-Source Heat Pump	3	40%	3	37%
Natural Gas	4	69%	4	49%
Fuel Oil	6	72%	7	31%

The savings attributable to the use of GSHP systems in commercial buildings vary over a wide range. In addition to parameters common to all GSHP applications, unique to commercial buildings are building use, internal heat gains, and more complex rate structures.

Predictions of savings to be achieved with a GSHP system are a very site-specific endeavor for commercial buildings.

Economics

The economics of residential GSHP were reported as simple paybacks in only 15% of the 184 case studies. A favorable simple payback is considered to be less than 5 years.

Residential GSHP system simple paybacks ranged from 1.4 to 24.1 years, and the mean was 7.0 years.

Commercial GSHP Economics			
Conventional System	Number	Range Years	Mean Years
Electric Resistance Heat	5	1.3 - 3.0	2.3
Natural Gas	3	1.9 - 4.7	3.4
Fuel Oil	7	2.2 - 4.5	
Other	2	2.5 - 2.7	2.5

When considering a GSHP system for either new or retrofit situations, it is imperative that a deliberate economic analysis be performed.

Utility Programs

Ground-source heat pumps (GSHPs) are one of many technologies utilities are considering or implementing for demand-side management (DSM), especially aimed at improving the efficiency with which customers use electricity. Information was developed on the status of DSM programs for GSHPs including: utility/contacts, marketing, barriers to market entry, incentives, number of units installed in service areas, and benefits. A total of 57 utilities and rural electric cooperatives out of 178 investigated were reported to have DSM programs involving GSHPs.

Marketing techniques employed by utilities included utility publications and seminars (36%), newspaper and radio/TV advertising (16%), test/demonstrations (10%), education (6%), home shows (6%), and other (26%).

The primary market penetration barrier cited by utilities was the first cost of installation of GSHP systems, especially the incremental cost (median cost of $700 to $900/ton of the ground loop). Other barriers include low-annual cost of natural gas, lack of manufacturers, dealers and loop installers, customer resistance to heat pump technology, and regulatory problems.

Utilities have designed a number of incentive packages to encourage the installation of GSHPs. The most common are cash rebates to customers (mean values of $208/ton and $382/unit) and trade ally ($200/ton). In many cases, the utility specifies a minimum Seasonal Energy Efficiency Ratio (SEER), usually 10 or greater, energy audits, or minimum insulation standards. Other types of incentives include special financing, discounted energy rates, and free ground loop installation.

More than 18,800 GSHP systems (3-ton equivalent) were reported by 35 utilities out of the 57 contacted. The types of systems installed include: horizontal ground-coupled (46%), undefined systems (40%), groundwater systems (7%), and vertical ground-coupled (7%).

Section 12

Electrical Formulas and Calculations

12.0.0	Converting Watts to Volts	636	12.3.4	Recommended Product Specifications Proposed by the State of Florida	648
12.0.1	Converting Watts to Amps	636	12.3.5	Appliance and Equipment Efficiency Ratings Explained—EER, SEER, COP, HSPF, AFUE	650
12.0.2	Converting Amps to Watts	636			
12.0.3	Converting Horsepower to Amps	636			
12.0.4	Converting KVA to Amps	637	12.4.0	Electric Generators—Understanding Your Power Needs	652
12.0.5	Converting Kw to Amps	637			
12.0.6	Symbols for Electrical Terms	637	12.4.1	Electric Generator and Power Generator Safety	653
12.1.0	Ohm's Law	638			
12.1.1	Resistors Networks—Terminology and How Identified	638	12.4.2	Typical Specifications for a Residential Emergency Generator	653
12.1.2	Inductor Networks—Terminology and How Identified	638	12.4.3	The Left-Hand Generator Rule	654
			12.4.4	The Right-Hand Generator Rule	654
12.1.3	Capacitor Networks—Terminology and How Identified	638	12.5.0	Dielectrics and Dielectric Constants of Various Materials	655
12.1.4	Watt's Law	638			
12.1.5	Calculating Reactance	639	12.5.1	Wire Gauges Table—AWG Gauge—Ft/Ohm Calculations	656
12.1.6	Resonance in RLC Series Circuit— Explanation and Formula	639			
			12.5.2	Wire Gauge Comparison Chart—AWG— Strandings/Wire Diameter, Overall Diameter	658
12.1.7	Bandwidth and Quality Factor	639			
12.1.8	Wavelength–Explained	640			
12.1.9	Frequency and Time–Explained	640	12.5.3	Resistance in Ohms per 1000 Feet of Conductor—Aluminium and Copper	660
12.1.10	Impedance of a Circuit—Formula and Rules for Circuits with Both Capacitive and Inductive Reactances	640			
			12.5.3.1	Solid and Concentric Stranding of Class B and Class C Strandings	661
12.2.0	Rules for Amp Draw per Horsepower at Voltage Ranging from 115 V to 575 V	640	12.5.4	Copper to Aluminium Conversion Tables	663
			12.5.5	Conduit inside Diameters and Electrical Conductor Areas—U.S. to Metric Conversion	665
12.2.1	How to Figure Out the Phase for a Certain Circuit Number	641			
12.3.0	Typical Wattage of Various Appliances	641	12.5.6	Conduit Weight Comparisons—Rigid, EMT, PVC	667
12.3.1	Typical Start-up and Running Wattage for Tools and Equipment	642	12.5.7	Recommended Power and Ground Cable Sizes—By Power and Distance	667
12.3.2	Estimating Appliance and Home Electronic Energy Use	644	12.6.0	Types of Transformers	668
			12.6.1	Dry-Type Transformers—KVA Ratings—Single- and Three-Phase	669
12.3.3	Appliance Energy Use Chart Based on an Operating Cost of $0.095 per KWh	646	12.7.0	Enclosure Types for All Locations	670

12.0.0 Converting Watts to Volts

Source: www.bugclub.org

1. Convert Watts to Volts:

   Voltage = Watts/AMPS

   $E = P \div I$

12.0.1 Converting Watts to Amps

Source: www.bugclub.org

2. Convert Watts to AMPS:

   AMPS = Watts/Voltage

   $I = P \div E$

   Example:
   2300 WATTS = 2300w divided by 120v = 19.1 AMPS
   (for 3 Phase divide by 1.73)

12.0.2 Converting Amps to Watts

Source: www.bugclub.org

3. Convert AMPS to Watts:

   Watts = Voltage × Amps

   $P = E \times I$

   Example: 19.1 AMPS multiplied by 120v = 2300 Watts
   (for 3 phase multiply by 1.73)

12.0.3 Converting Horsepower to Amps

Source: www.bugclub.org

4. Convert Horse Power to AMPS:

   HORSEPOWER = (V × A × EFF) ÷ 746
   EFFICIENCY = (746 × HP) ÷ (V × A)

   Multiply Horse Power by 746 w (1 HP = 746 Watts)
   Find Circuit Voltage and Phase

   Example:
   30 HP at 480 (3 Phase) - 746 multiplied by 30 = 22380
   22380 divided by 480 (3 Phase) = 46.5
   46.5 divided by 1.73 = 29.5 AMPS
   Multiply all the motor loads by 1.50% and go to the next circuit size.

Section | 12   Electrical Formulas and Calculations

12.0.4 Converting KVA to Amps

Source: www.bugclub.org

5. Convert KVA to AMPS:

   Multiply KVA by 1000/voltage

   Example:
   30 KVA multiplied by 1000v = 30,000 Watts
   30,000 Watts divided by 480 = 62.5 AMPS
   (for 3 phase divide by 1.73)

12.0.5 Converting Kw to Amps

Source: www.bugclub.org

6. Convert KW to AMPS:

   Multiply KW by 1000/voltage and then by power factor

   Example:
   30KW multiplied by 1000v = 30,000
   30,000 divided by 480 = 62.5 × .90 = 56.25 AMPS
   (for 3 phase divide by 1.73)

12.0.6 Symbols for Electrical Terms

Source: www.bugclub.org

E = VOLTS or (V = VOLTS)
P = WATTS or (W = WATTS)
R = OHMS or (R = RESISTANCE)
I = AMPERES or (A = AMPERES)
HP = HORSEPOWER
PF = POWER FACTOR
kW = KILOWATTS
kWh = KILOWATT HOUR
VA = VOLT-AMPERES
kVA = KILOVOLT-AMPERES
C = CAPACITANCE
EFF = EFFICIENCY (expressed as a decimal)

12.1.0 Ohm's Law

Source: www.ziplink.net

- E = Voltage - measured in Volts
- I = Current - measured in Amperes
- R = Resistance - measured in Ohms
   - $E = I * R$, answer in Volts
   - $I = E/R$, answer in Amperes
   - $R = E/I$, answer in Ohms

12.1.1 Resistors Networks—Terminology and How Identified

Resistor Networks

- Resistors are labeled by a # following the R : R1...R2...Rn
- Rn = Continues for all the resistors you have
- Rt = Resistance total
- Resistors in SERIES add : Rt = R1 + R2 + R3...+ Rn
- Two (2) Resistors in PARALLEL : Rt = (R1 * R2)/(R1 + R2) This the Product <u>divided</u> by the Sum
- More than Two Resistors in PARALLEL: Rt = 1/(1/R1 + 1/R2...+1/Rn). This is called the reciprocal formula.
- Another form of this formula is : 1/Rt = 1/R1 + 1/R2...+1/Rn

  Source: www.ziplink.net

12.1.2 Inductor Networks—Terminology and How Identified

Inductor Networks

- The rules for Inductors are exactly like those for Resistors.
- Inductors are labeled by a # following the L : L1...L2...Ln.
- Ln = Continues for all the Inductors you have.
- Lt = Inductance total.
- Inductors in SERIES add : Lt = L1 + L2 + L3...+ Ln.
- Two (2) Inductors in PARALLEL : Lt = (L1 * L2)/(L1 + L2) This the Product <u>divided</u> by the Sum.
- More than Two Inductors in PARALLEL: Lt = 1/(1/L1 + 1/L2...+1/Ln) This is called the reciprocal formula.
- Another form of this formula is : 1/Lt = 1/L1 + 1/L2...+1/Ln.

  Source: www.ziplink.net

12.1.3 Capacitor Networks—Terminology and How Identified

Capacitor Networks

- The rules for Capacitor Networks are exactly opposite those of Resistors and Inductors.
- Capacitors are labeled by a # following the C : C1...C2...Cn.
- Cn = Continues for all the Capacitors you have.
- Ct = Capacitance total.
- Capacitors in PARALLEL add : Ct = C1 + C2 +C3....+ Cn.
- Two (2) Capacitors in SERIES : Ct = (C1 * C2)/(C1 + C2) This is the Product <u>divided</u> by the Sum.
- More than Two Capacitors in SERIES : Ct = 1/(1/C1 + 1/C2...+1/Cn). This is called the reciprocal formula.
- Another form of this formula is : 1/Ct = 1/C1 + 1/C2.....+1/Cn.

  Source: www.ziplink.net

12.1.4 Watt's Law

- P = Power - measured in Watts
- I = Current - measured in Amperes
- E = Voltage - measured in Volts
- R = Resistance - measured in Ohms

- P = I * E, answer in Watts (easy as "pie")
- P = I*I*R, answer in Watts (read as I squared R)
- P = E*E/R, answer in Watts (read as E squared, divided by R)
- Note: Use Ohm's Law to derive other formulas for I, R, and E, using the formulas above.

Source: ziplink.net

12.1.5 Calculating Reactance

- **Inductive Reactance**
  - Pi = π = 3.14
  - f = Frequency, in Hertz
  - L = Inductance of coil, in Henries
  - Xl = Inductive Reactance, in Ohms
    - Xl = 2πfL, answer in Ohms
    - **Note: Xl is a linear function; it increases as frequency or inductor value increases.**
- **Capacitive Reactance**
  - Pi = π = 3.14
  - f = Frequency, in Hertz
  - C = Capacitance in Farads
  - Xc = Capacitive Reactance, in Ohms
    - **Xc** = $\frac{1}{2\pi fC}$, answer in Ohms
    - Note: Xc is inversely proportional to frequency and capacitance. If either frequency or capacitance increases, Xc decreases.

Source: ziplink.net

12.1.6 Resonance in RLC Series Circuit—Explanation and Formula

Resonance in RLC Series Circuit

- Resonance occurs when Xc and XL are equal to each other.
  - Fo = Frequency of resonance, in Hertz
  - Pi = π = 3.14
  - L = Inductance in Henries
  - C = Capacitance in Farads
  - **Fo** = $\frac{1}{2\pi\sqrt{LC}}$

12.1.7 Bandwidth and Quality Factor

- Q = Quality Factor of the circuit
- BW = Bandwidth in Hertz
- XL = Inductive Reactance of the Inductor
- R = Resistance of Inductor plus any external Resistance
- Fo = Frequency of Resonance
  - **Q = Xl/R**, Quality Factor
  - **BW = Fo/Q**, measured in Hertz
  - Note: As Quality Factor *increases* the Bandwidth *decreases*, but has **more gain**. And the opposite is true as well. When the Quality Factor decreases, the Bandwidth increases, but has **less gain**.

Source: ziplink.net

12.1.8 Wavelength–Explained

Wavelength

- Wavelength is how long a wave is for a given Frequency.
- **The speed of light is 186,000 miles per second or about 300,000,000 meters per second.**
- f is the Frequency in Hertz/Second (or Cycles/Second).
- Wavelength = (300,000,000 meters/second)/(f in Cycles/Second).
- The answer is in meters/cycle.

Source: ziplink.net

12.1.9 Frequency and Time–Explained

Frequency and Time

- f is the Frequency in Hertz/Second or Cycles/Second
- t is time in Seconds
- f = 1/t, answer in Hertz/Second or Cycles/Second.
- t = 1/f, answer in Seconds

Source: ziplink.net

12.1.10 Impedance of a Circuit—Formula and Rules for Circuits with Both Capacitive and Inductive Reactances

Impedance of a Circuit

- Xc = Capacitive Reactance, in Ohms
- Xl = Inductive Reactance, in Ohms
- R = Resistance, in Ohms
- Z = Impedance, in Ohms
  - For a Capacitive circuit
  - $z = \sqrt{R^2 + Xc^2}$
    - For an Inductive Circuit
    - $z = \sqrt{R^2 + Xl^2}$
- Rules for when you have a circuit with both Capacitive and Inductive Reactances
- When Xl is larger than Xc, use the formula in figure (a). If Xc is Larger than Xl, use the formula in Figure (b).
- a. $z = \sqrt{R^2 + (Xl - Xc^2)}$ b. $z = \sqrt{R^2 + (Xc - Xl^2)}$

12.2.0 Rules for Amp Draw per Horsepower at Voltage Ranging from 115 V to 575 V

- At 575 volts, a 3-phase motor draws 1 amp per horsepower.
- At 460 volts, a 3-phase motor draws 1.27 amps per horsepower.
- At 230 volts, a 3-phase motor draws 2.5 amps per horsepower.
- At 230 volts, a single-phase motor draws 5 amps per horsepower.
- At 115 volts, a single phase motor draws 10 amps per horsepower.

*These above are approximations.

- 746 watts = 1 HP

12.2.1 How to Figure Out the Phase for a Certain Circuit Number

An easy way to figure the phase of a circuit number is to divide it by 6. *If it divides evenly, it is always "C" phase*.

Let's say you have circuit number 27. Divide it by 6. Six will go into 27 four times with a reaminder of 3. Normally, panels are labeled with the odds on the left and the evens on the right (*see Panel layout example below*). So 3 is the second one down from the top on the odd side; therefore it will be "B" phase. Let's try another one. Say your circuit number is 50. Divide it by 6. Six will go into 50 eight times with a remainder of 2. So the correct phase for circuit 50 will be "A" phase. This is taken in consideration if the three-phase system is phased A,B,C left to right, top to bottom. This is the normal phasing of a system.

Example of Panel Layout:

A - ckt 1	A - ckt 2
B - ckt 3	B - ckt 4
C - ckt 5	C - ckt 6
A - ckt 7	A - ckt 8
B - ckt 9	B - ckt 10
C - ckt 11	C - ckt 12

By permission: www.elec-toolbox.com

12.3.0 Typical Wattage of Various Appliances

Here are some examples of the range of nameplate wattages for various household appliances:

- Aquarium = 50–1210 Watts
- Clock radio = 10
- Coffee maker = 900–1200
- Clothes washer = 350–500
- Clothes dryer = 1800–5000
- Dishwasher = 1200–2400 (using the drying feature greatly increases energy consumption)
- Dehumidifier = 785
- Electric blanket- *Single/Double* = 60 / 100
- Fans
  - Ceiling = 65–175
  - Window = 55–250
  - Furnace = 750
  - Whole house = 240–750
- Hair dryer = 1200–1875
- Heater *(portable)* = 750–1500
- Clothes iron = 1000–1800
- Microwave oven = 750–1100
- Personal computer
  - CPU - awake/asleep = 120/30 or less
  - Monitor - awake/asleep = 150/30 or less
  - Laptop = 50
- Radio *(stereo)* = 70–400
- Refrigerator *(frost-free, 16 cubic feet)* = 725
- Televisions (color)
  - 19" = 65–110
  - 27" = 113

- 36" = 133
- 53"-61" Projection = 170
- Flat screen = 120
- Toaster = 800–1400
- Toaster oven = 1225
- VCR/DVD = 17–21/20–25
- Vacuum cleaner = 1000–1440
- Water heater *(40 gallon)* = 4500–5500
- Water pump *(deep well)* = 250–1100
- Water bed *(with heater, no cover)* = 120–380

12.3.1 Typical Start-up and Running Wattage for Tools and Equipment

Tool	Start-up Wattage	Running Wattage
Air compressor-1/2 HP	2000	1000
Air compressor-1 HP	4500	1500
Air compressor-2 HP	7700	2800
Battery charger	0	3000
Chainsaw, electric-1/2 HP	0	900
Drill, 1/4" -3 amps	400	300
Drill, 3/8" - 4 amps	600	440
Drill, 1/2 "-5.4 amps	900	600
Floor polisher-16", 3/4 HP	3100	1400
Grinder, Bench-6"	1000	720
Grinder, Bench-10"	3600	1600
Hammer, demolition	3300	1900
Hedge trimmer-18"	0	400
Impact wrench-1/2"	1200	750
Impact wrench-1"	2400	1400
Mixer, 55 gallons-1/4 HP	1200	700
Pump, submersible-400 GPH	400	200
Pump, submersible-900 GPH	400	500
Saw, band-14"	1400	1100
Saw, circular-8 1/4" heavy duty	3000	1800
Saw, mitre-10 amp	2000	1100
Saw, mitre-15 amp	3000	1650
Saw, Table-9" radial	3000	1500
Saw, Table-10" radial	4500	1800

Tool	Start-up Wattage	Running Wattage
Saw, Skill type- 7 ¼"	2600	1800
Trimmer, Heavy duty- 9'	0	350
Trimmer, Heavy duty-12"	0	500
Vacuum—wet-dry- 1.7 HP	0	900
Vacuum—wet-dry-2 ½ HP	0	1300
Washer—Hi-pressure- 5/8 HP	2700	900
Washer—Hi-pressure- 1 HP	3600	1200
Washer—Hi-Pressure- 1 ½ HP	4300	1450
Washer—Hi-Pressure- 3 HP	9300	3125
Fan Duty motors- 1/8 HP	600	400
Fan Duty motors- 1/6 HP	850	550

The above start-up and running wattage will vary according to model, make, and efficiency of the appliance and tool, and these calculations are based on an average and are to be viewed as "approximations" and not as a way to determine wiring or circuitry sizes.

Wattage Requirement—Typical Both for Start-up and for Running Appliances and Residential Equipment

Equipment	Start-Up Wattage	Running Wattage
Blanket, electric	0	400
Blender	0	200
Bread maker	2300	600
Broiler, electric	0	1350
Broom, electric	0	500
CD Player/speaker	0	100
Clothes dryer-gas	720	650
Clothes dryer-electric	1800	750
Coffee maker	1000	550
Computer	0	720–900
Computer printer	0	350–720
Copy machine	0	1600
Dehumidifier	800	650
Dishwasher, cool dry	1400	700
Dishwasher, hot dry	1400	1450
Freezer	2200	700

(Continued)

Equipment	Start-Up Wattage	Running Wattage
Fry pan, electric	0	1300
Furnace	500–2350	300–875
Garage door opener	1100–1400	550–725
Hair dryer	0	900
Hot tub heater	1700	1700
Hot tub pump	950	800
Iron	0	1000
Light, incandescent	0	100
Light, fluorescent	125	90
Light, flood	0	500
Microwave oven	800	625
Range, electric-6" elements	0	1500
Range, electric-8" elements	0	1200
Refrigerator	2200	700
Toaster-2 slice	0	1050
Toaster- 4 slice	0	1650
Pump, sump- ½ hp	2150	1050
Pump, sump-1/3 hp	1300	800
Radio	0	200
Security alarm panel	0	200
Space heater (average)	0	800
Television	0	300
Vacuum cleaner	0	1100
Washer-Dryer	3000	2000

12.3.2 Estimating Appliance and Home Electronic Energy Use

Source: U.S. Department of Energy

Energy Savers

Estimating Appliance and Home Electronic Energy Use

If you're trying to decide whether to invest in a more energy-efficient appliance or you'd like to determine your electricity loads, you may want to estimate appliance energy consumption.

Formula for Estimating Energy Consumption

You can use this formula to estimate an appliance's energy use:

(Wattage × Hours Used Per Day) ÷ 1000 = Daily Kilowatt-hour (kWh) consumption
(1 kilowatt (kW) = 1,000 Watts)

Multiply this by the number of days you use the appliance during the year for the annual consumption. You can then calculate the annual cost to run an appliance by multiplying the kWh per year by your local utility's rate per kWh consumed.

Note: To estimate the number of hours that a refrigerator actually operates at its maximum wattage, divide the total time the refrigerator is plugged in by three. Refrigerators, though turned "on" all the time, actually cycle on and off as needed to maintain interior temperatures.

Examples

Window fan:

(200 Watts × 4 hours/day × 120 days/year) ÷ 1000
= 96 kWh × 8.5 cents/kWh
= $8.16/year

Personal Computer and Monitor:

(120 + 150 Watts × 4 hours/day × 365 days/year) ÷ 1000
= 394 kWh × 8.5 cents/kWh
= $33.51/year

Wattage

You can usually find the wattage of most appliances stamped on the bottom or back of the appliance, or on its nameplate. The wattage listed is the maximum power drawn by the appliance. Since many appliances have a range of settings (for example, the volume on a radio), the actual amount of power consumed depends on the setting used at any one time.

If the wattage is not listed on the appliance, you can still estimate it by finding the current draw (in amperes) and multiplying that by the voltage used by the appliance. Most appliances in the United States use 120 volts. Larger appliances, such as clothes dryers and electric cooktops, use 240 volts. The amperes might be stamped on the unit in place of the wattage. If not, find a clamp-on ammeter—an electrician's tool that clamps around one of the two wires on the appliance—to measure the current flowing through it. You can obtain this type of ammeter in stores that sell electrical and electronic equipment. Take a reading while the device is running; this is the actual amount of current being used at that instant.

When measuring the current drawn by a *motor*, note that the meter will show about three times more current in the first second that the motor starts than when it is running smoothly.

12.3.3 Appliance Energy Use Chart Based on an Operating Cost of $0.095 per KWh

Appliance Energy Use Chart

The Appliance Energy Use Chart presented below is designed to give you an idea of how much electricity is consumed by many of the most common household appliances. Except where noted, the figures used in the chart have been based on the typical efficiency levels of appliances found in Springfield homes audited by the CWLP Energy Experts and on the price per <u>kilowatt-hour</u> paid by the "average" CWLP residential customer. Appliances with efficiency levels much lower or higher than the norm might consume significantly more or less energy than indicated on this table.

To translate the usages given in this chart into energy dollars, simply multiply the appliance's kilowatt-hour (kWh) usage by your average price per kWh (see the following Note for more about this) and the amount or number of times you use the appliances over a specific period.

NOTE: Based on current electric rates and the State Utility Tax, plus the average fuel adjustment charge for the previous year, the average annual cost per kWh of electricity paid by CWLP's regular (not all-electric) residential electric customers is approximately 9.5¢. For all-electric residential customers, the average annual cost is about 8.9¢ per kWh. (*Cost-per-kWh estimates were last updated September 30, 2008.*)

For instance, using the average cost-per-kWh provided in the Note above and the energy consumption information provided in the Appliance Energy Use Chart, we can calculate that it will cost a regular (Rate 30) CWLP residential electric customer about $2.57 a month to watch a 21-inch color television for an average of three hours a day (approximately 90 hours each month).

$$0.3 \text{ kwh/hr} \times \$0.095 \text{ per kwh} \times 90 \text{ hrs/mo.} = \$2.57 \text{ per mo.}$$

Source: City water Light and Power, Springfield, Illinois

An addition to helping you determine the approximate cost of operating your various appliances over time, the Appliance Energy Use Chart can help you realize how changes in your energy use habits—such as using appropriately sized stove burners, substituting a microwave oven for a conventional oven, or turning off lights, TVs, and other appliances when they aren't needed—can help you control your monthly energy costs.

Appliance Energy Use Chart

Appliance	kWh Usage	Operating Cost (@ 9.5¢/kWh)
Kitchen		
Toaster	0.04 kWh / serving	less than 1¢ / serving
Microwave oven	0.75 kWh / hr	7¢ / hr
Electric frying pan	1.2 kWh / hr	11¢ / hr
Coffee maker	0.2 kWh / pot	2¢ / pot
Range burner (large)	2.4 kWh / hr	23¢ / hr
Range burner (small)	1.2 kWh / hr	11¢ / hr
Oven (baking or roasting)	3.2 kWh / hr	30¢ / hr

Appliance Energy Use Chart—Cont'd

Appliance	kWh Usage	Operating Cost (@ 9.5¢/kWh)
Oven (broiling)	3.6 kWh / hr	34¢ / hr
Oven (self-cleaning cycle)	10 kWh / clean	95¢ / clean
Refrigerator (pre-2002, manual defrost)	63 kWh / month	$5.99 / month
Refrigerator (pre-2002, frost-free)	168 kWh / month	$15.96 / month
Refrigerator (2002 or newer)	82 kWh / month	$7,79 / month
Deep freezer (frost free)	1835 kWh / month	$17.39 / month
Deep freezer (manual defrost)	135 kWh / month	$12.83 / month
Dishwasher	1 kWh / load	9.5¢ / load
Living Room/Office/Family Room		
Television (21-inch color)	0.3 kWh / hr	3¢ / hr
Stereo	0.15 kWh / hr	1¢ / hr
Computer with monitor (average)	0.09 kWh / hr	1¢ / hr
Computer with monitor (sleep mode)	0.02 kWh / hr	less than 1¢ / hr
Fan	0.2 kWh / hr	2¢ / hr
Room space heater (1500 watt)	1.5 kWh / hr	14¢ / hr
Bedroom		
Waterbed heater	120 kWh / month	$11.40 / month
Electric blanket	1 kWh / night	9.5¢ / night
Basement/Utility Room		
Washing machine (excluding water)	0.25 kWh / load	2¢ / load
Clothes dryer (electric)	2.7 kWh / load	35¢ / load
Water heater (for average family of 4)	400 kWh / month	$38.00 / month
Dehumidifier	0.76 kWh / hr	7¢ / hr
Air conditioner (central, 10 SEER)	1.2 kWh / hr / ton	11¢ / hr / ton
Air conditioner (central, 14 SEER)	0.85 kWh / hr / ton	8¢ / hr / ton
	MISCELLANEOUS	
Light bulb (100-watt incandescent)	0.1 kWh / hr	4¢ / 4 hrs
Light bulb (25-watt CFL, 100-watt equiv.)	0.25 kWh / hr	1¢ / 4 hrs

12.3.4 Recommended Product Specifications Proposed by the State of Florida

Appendix A

TABLE A-1 Summary of Recommended Product Standards for Florida

Product	Annual Average Baseline Energy per Product	Baseline Energy Units	Annual Operating Cost	Recommended Standard	Basis for Standard	Annual Average Energy Savings per Product	Savings Energy Units
Bottle-type water dispensers	854	kWh/yr	$97.36	Max. 1.2 kWh/day standby energy	Energy Star and CEC Title 20	266	kWh/yr
Commercial boilers	9246	therms/yr	$19,877.83	Min 0.81 thermal efficiency	Proposal to ASHRAE	481	therms/yr
Commercial hot food holding cabinets	2402	kWh/yr	$273.83	Max. idle energy rate 40 W/ft³	Energy Star and CEC Title 20	1815	kWh/yr
Compact audio products	64	kWh/yr	$7.34	Max. 2.0 W standby energy	Energy Star and CEC Title 20	53	kWh/yr
DVD players and recorders	26	kWh/yr	$3.02	Max 3.0 W standby energy	Energy Star and CEC Title 20	11	kWh/yr
Televisions (added by FSEC)* **	442	kWh/yr	$50.39	Max 3.0 W standby, Calif. prop. Title 20	Energy Star and CEC Title 20	215	kWh/yr
Metal halide lamp fixtures	2015	kWh/yr	$229.69	Pulse-start ballast	Pulse-start ballast	307	kWh/yr
Residential pool heaters*	1125	therms/yr	$2,418.75	Min. 80% thermal effic. & electric ignition	DOE 2004	213.75	therms/yr
Portable electric spas (hot tubs)	2500	kWh/yr	$285.00	Max. 5 V$^{(2/3)}$ standby energy	CEC Title 20	250	kWh/yr
	110	fan kWh	$12.54	2% electricity ratio	GAMA/CEE specification	28	fan kWh
Residential furnaces and residential boilers* (baseline AFUE=80)	165	therms/yr	$354.75	Boilers min. 84 AFUE; Furnaces/nat. gas. Tier 1 min. 80 AFUE, Tier 2 min. 90 AFUE/oil min. 83 AFUE	For boilers, significant current sales; for Tier 1 nat. gas. furnaces, non-condensing max; for Tier 2 nat. gas. furnaces, condensing; for oil furnaces, significant current sales.	1.98	therms/yr

Section 12 — Electrical Formulas and Calculations

Item	Value	Unit	Cost	Description	Notes	Savings	Unit
Residential pool pumping*	4200	kWh/yr	$478.80	No split-phase or capacitor start–induction run types; 2-speeds; 2" pipe and motor sizing	2-speed pump; friction losses	2688	kWh/yr
Single-voltage external AC to DC power supplies	38	kWh/yr	$4.36	Varies with size	CEC Title 20 (Tier 1) and other states' standards	4.1	kWh/yr
State-regulated incandescent reflector lamps	209	kWh/yr	$23.88	Varies with size	EPAct 1992 standard with MA exemptions	61	kWh/yr
Walk-in refrigerators and freezers	18,859	kWh/yr	$2,149.94	Typical installation from CEC case study	CEC Title 20 with a few modifications	8,220	kWh/yr
Ceiling fans (added by FSEC)	317	kWh/yr	$36.14	Airflow and lighting efficiency minimums	Energy Star	142	kWh/yr
Incandescent lighting (added by FSEC)	55	kWh/yr	$6.27	Phase-out of incandescent lighting	Three states have similar plan/time line	37	kWh/yr
Res. water heaters (added by FSEC)	2300	kWh/yr	$262.20	Electric EF > 2 (phased in over time)	Technology is available	1,150	kWh/yr
	133	therms/yr	$285.95	Gas EF minimum EF 0.80 (2012)		46	therms/yr

Note: All data supplied by ACEEE (American Council for an Energy Efficient Economy) except televisions, part of residential pool pumps, ceiling fans, incandescent lighting and electric water heaters supplied by FSEC (Florida Solar Energy Center). **For televisions, average energy savings/unit includes 15% energy savings from interaction with *Original ACEEE data modified based on FSEC experience and judgment.
Source: State of Florida.

12.3.5 Appliance and Equipment Efficiency Ratings Explained—EER, SEER, COP, HSPF, AFUE

Appliance and Equipment Efficiency Ratings

When referring to the efficiency of an appliance or energy system, we are actually talking about how much energy that system must use to perform a certain amount of work. The higher its energy consumption per unit of output, the less efficient the system is. For example, an air conditioner that requires 750 watts of electricity to provide 6000 Btu of cooling will be less efficient than one that can provide the same amount of cooling for only 500 watts. The most common ratings applied to energy systems are EER and SEER for most central cooling systems; COP for some heat pumps and chillers; HSPF for all-electric heat pumps in their heating modes; and AFUE for gas furnaces and boilers.

Efficiency Rating	Definition	Explanation
EER	... (**E**nergy **E**fficiency **R**atio) is the measure of how efficiently a cooling system will operate when the outdoor temperature is at a specific level (usually 95°F). A higher EER means the system is more efficient. The term EER is most commonly used when referring to window and unitary air conditioners and heat pumps, as well as water-source and geothermal heat pumps.	The formula for calculating EER is: $$EER = \frac{Btu/hr \text{ of cooling at } 95°}{watts \text{ used at } 95°}$$ For instance, if you have a window air conditioner that draws 1500 watts of electricity to produce 12,000 Btu per hour of cooling when the outdoor temperature is 95°, it would have an EER of 8 (12,000 divided by 1500). A unit drawing 1200 watts to produce the same amount of cooling would have an EER of 10 and would be more energy efficient. Using this same example, you can see how efficiency can affect a system's operating economy. First, you'll need to determine the total amount of electricity—measured in kilowatt-hours (kWh)—the unit will consume over a period of time. (A kilowatt-hour is defined as 1000 watts used for one hour. This is the measure by which your monthly utility bills are calculated.) To do this, let's assume you operate your 8 EER window air conditioner—drawing 1500 watts at any given moment—for an average of 12 hours every day during the summer (*1500 watts × 12 hours*). At this rate, it will use 18,000 watt-hours or 18 kWh each day, leading to a total consumption of 540 kWh over the course of a 30-day month (*18 kWh × 30 days*). At a summer electric rate of 8.51¢ per kWh, it would cost you about $46 to operate that window air conditioner each month (*540 kWh × $0.0851*)—not including fuel adjustment and state utility tax. At the same time, the 1200-watt, 10 EER system, consuming 14.4 kWh per day and 432 kWh per month, would cost you about $37, a 20% savings over the less efficient model.
SEER	... (**S**easonal **E**nergy **E**fficiency **R**atio) measures how efficiently a residential central cooling system (air conditioner or heat pump) will operate over an entire cooling season, as opposed to at a single outdoor temperature. SEER is calculated based on the total amount of cooling (in Btu) the system will provide over the entire season divided by the total number of watt-hours it will consume.	The formula for calculating SEER is: $$SEER = \frac{hr \text{ of cooling at}}{seasonal \text{ watt} - hours \text{ used}}$$ As with EER, a higher SEER reflects a more efficient cooling system. By federal law, every split cooling system manufactured in or imported into the United States today must have a seasonal energy efficiency ratio of at least 13.0.
COP	... (Coefficient of Performance) is the measurement of how efficiently a heating or	COP can be calculated by two different methods. In the first, you divide the Btu of heat produced by the heat pump by the

Efficiency Rating	Definition	Explanation
	cooling system (particularly a heat pump in its heating mode and a chiller for cooling) will operate at a single outdoor temperature condition. When applied to the heating modes of heat pumps, the temperature condition is usually 47°F.	Btu equivalent of electricity that is required to produce the heat. This formula is stated: $$COP = \frac{\text{Btu of heat produced at } 47°F}{\text{Btu worth of electricity used at } 47°F}$$ For instance, let's assume a heat pump uses 4000 watts of electricity to produce 42,000 Btu per hour (Btu/hr) of heat when it is 47° outside. To determine its COP, you would first convert the 4000 watts of electrical consumption into its Btu/hr equivalent by multiplying 4000 times 3.413 (the number of Btu in one watt-hour of electricity). Then, you would divide your answer—13,648 Btu/hr—into the 42,000 Btu/hr heat output. This would show your heat pump to have a 47°F COP of 3.08. This means that for every Btu of electricity the system uses, it will produce a little more than three Btu of heat when the outdoor temperature is 47°F. The second formula is most frequently used to determine chiller efficiency. Using this calculation method, you would divide 3.516 by the number of kilowatts (kW) per ton used by the system. This formula is stated: $$COP = \frac{3.516}{kW/ton}$$ For example, a chiller that consumes 0.8 kW per ton of capacity would have a COP of 4.4 (3.516 divided by 0.8). On the other hand, a chiller that uses 0.5 kW per ton, would have a COP of 7 (3.516 divided by 0.5). The higher the COP, the more efficient the system.
HSPF	... (**H**eating **S**easonal **P**erformance **F**actor) is the measurement of how efficiently all residential and some commercial all-electric heat pumps will operate in their heating mode over an entire normal heating season. HSPF is determined by dividing the total number of Btu of heat produced over the heating season by the total number of watt-hours of electricity that is required to produce that heat.	The formula for calculating HSPF is: $$HSPF = \frac{\text{Btu of heat produced over the heating season}}{\text{watts} - \text{hours of electricity used over the season}}$$ The higher the HSPF, the more efficient the system. Most all-electric heat pumps installed in Springfield today probably have HSPFs in the 7.0 to 8.0 range, meaning they operate with seasonal efficiencies of anywhere from 205 to 234%. (To convert the HSPF number into a percentage, divide the HSPF by 3.413, the number of Btu in one watt-hour of electricity.) That means that, for every Btu worth of energy they use over the entire heating season, these systems will put out anywhere from 2.05 to 2.34 Btu of heat. Compare this to electric furnaces, which have nominal efficiencies of 100% (for each Btu worth of electricity used, they put out 1.0 Btu of heat), or new gas furnaces, which have efficiency ratings of about 80 to 97% (for each Btu worth of gas used, they put out 0.8 to 0.97 Btu of heat). (*NOTE*: When comparing energy systems that use different primary fuel sources with different costs per Btu, It is important that you understand that higher operating efficiency is not necessarily equivalent to better operating economy. Although an electric furnace might work with greater efficiency than a gas furnace, it might or might not be more economical to operate. That will depend on the prices of electricity and natural gas.)

(Continued)

Efficiency Rating	Definition	Explanation
AFUE	... (**A**nnual **F**uel **U**tilization **E**fficiency) is the measurement of how efficiently a gas furnace or boiler will operate over an entire heating season. The AFUE is expressed as a percentage of the amount of energy consumed by the system that is actually converted to useful heat. For instance, a 90% AFUE means that for every Btu worth of gas used over the heating season, the system will provide 0.9 Btu of heat. The higher the AFUE, the more efficient the system.	When comparing efficiencies of various gas furnaces, it is important to consider the AFUE, not the steady-state efficiency. Steady state refers to the efficiency of the unit when the system is running continuously, without cycling on and off. Since cycling is natural for the system over the course of the heating season, steady state doesn't give a true measurement of the system's seasonal efficiency. For instance, gas furnaces with pilot lights have steady-state efficiencies of 78 to 80%, but seasonal efficiencies—AFUEs—closer to 65%. Virtually all gas forced-air furnaces installed in this area from the 1950s through the early 1980s had AFUEs of around 65%. Today, federal law requires most gas furnaces manufactured and sold in the United States to have minimum AFUEs of 78%. (Mobile home furnaces and units with capacities under 45,000 Btu are permitted somewhat lower AFUEs.) Gas furnaces and boilers now on the market have AFUEs as high as 97%.

Source: City Water Light and Power, Springfield, ILL.

12.4.0 Electric Generators—Understanding Your Power Needs

In order to choose the right emergency power source and to size it properly, you need to understand something about the power requirements of the devices you plan to operate.

The basic unit of power measurement is the watt, and with an emergency power source there are two wattage ratings that are important: steady-state wattage and surge wattage. A normal 60-watt incandescent light bulb requires, as you would expect, 60 watts, and it requires that wattage both when you turn it on and while it is running. A ceiling fan motor, on the other hand, might require 150 watts to get it started and 75 watts while it is running. That extra wattage to start the motor is called the surge wattage and is typical of anything that contains an electric motor. Here are the usual wattages of some of the devices found in a typical household:

Device	Typical Wattage	Surge Wattage
Light bulb	60 watts	60 watts surge
Fan	75 watts	150 watts surge
Small black/white television	100 watts	150 watts surge
Color television	300 watts	400 watts surge
Home computer and monitor	400 watts	600 watts surge
Electric blanket	400 watts	400 watts surge
Microwave oven	750 watts	1,000 watts surge
Furnace fan	750 watts	1,500 watts surge
Refrigerator	1,200 watts	2,400 watts surge
Well pump	2,400 watts	3,600 watts surge
Electric water heater	4,500 watts	4,500 watts surge
Whole-house A/C or heat pump	15,000 watts	30,000 watts surge

Source: xgpower, Whitehouse, NJ.

12.4.1 Electric Generator and Power Generator Safety

When buying a diesel generator, make a list of the lights and equipment that will be running off the generator. Total the wattage requirements to determine the capacity of the generator. Compare wattage requirement and the price of the generator. If the generator is to be connected directly to the electrical system, then it is advisable to hire a qualified technician to install the transfer switch. Ensure that the generator has adequate storage capacity, longer usage time at a stretch, overload protection, and auto shut-off facility.

Portable generators are critical to have in emergencies or for use in areas where there is no traditional electricity. Always keep in mind that there are safety issues related to the proper use of the generators. Portable generators can cause electrocution if they are left in the rain or sitting in water. Keep the generator under a canopy where it is protected, but not totally enclosed. It must still have adequate ventilation. Never touch the generator when you are standing in water or your hands are wet. Never run extension cords through water of any kind. Keep all unauthorized people away from the unit while it is operating.

Make sure all extension cords are of high enough rating for whatever load they are to carry. Also check for fraying, exposed wires, or areas where the cord may be underneath something else and is hard to see. Many people will put cords under rugs or furniture, but this can harm the cord and hide any defects in the cord.

Check cords when the portable generator is operating to make sure they are not overheating. Overheating is an indication of either too much load for the cord or damaged wires inside the cord. Even though the cords may be rated for the load you have, if compressors are turning on and off, the load may be temporarily increasing beyond the rating.

Even with the best alternative system such as solar energy or wind energy, sometimes the weather may not cooperate and will land us in darkness when the main power source fails. To keep our power system running, a backup engine generator may be just the thing. For businesses with sensitive computer networks, or homes with critical medical equipment, a backup power system such as a diesel generator may be necessary even when grid power is available.

When buying a diesel generator, make a list of the lights and equipment that will be running off the generator. Total the wattage requirements to determine the capacity of the generator. Compare wattage requirement and the price of the generator. If the generator is to be connected directly to the electrical system, then it is advisable to hire a qualified technician to install the transfer switch. Ensure that the generator has adequate storage capacity, longer usage time at a stretch, overload protection, and auto shut-off facility.

Source: xgpower, Whitehouse, NJ.

12.4.2 Typical Specifications for a Residential Emergency Generator

Our generator packs plenty of power to run heavy-load equipment such as heaters or air compressors. Auto decompression ensures easy starting; dependable, maintenance-free ignition and brushless alternator; large, super-quiet muffler; black, tubular steel frame and easy-to-read fuel gauge. This super-fuel-efficient generator gets up to 11 hours of continuous operation on a tank of gas.

Generator Specifications:

- Type: Brushless AVR—Single Phase
- AC Frequency : 60 HZ
- Maximum AC Output : 6700 Watts
- Rated AC Output : 6000 Watts
- Rated / Maximum AC Current : 50.0/55.0 Amps @ 120 Volts
- Driving Method : Direct
- Two 110-volt and two 220-volt outlets

Engine Specifications:

- Type : 4-Stroke, OHV, Air Cooled
- Rated Horsepower : 12HP @ 3600 rpm
- Bore × Stroke : 66 mm × 50 mm
- Displacement : 357cc/12HP
- Cooling : Forced Air
- Lubrication : Wet Sump
- Oil Capacity : 0.63 qt
- Starting Method : Recoil
- Ignition System : Transistor Controlled
- Fuel Tank Capacity : 5.5 gal (20.8 Liters)
- Recommended Fuel : Unleaded Regular
- Continuous Operating Hours : 11.3
- Dry Weight : 185 lb (84 kg)
- Noise Level @ 25 ft : 73.5 dBA
- Package size: 27 in.L × 21 in.W × 22.5 in.H

Source: xgpower, Whitehouse, NJ.

12.4.3 The Left-Hand Generator Rule

The left-hand generator rule can be used to determine the relationship of the motion of the conductor in a magnetic field to the direction of the induced current. To use the left-hand rule, place the thumb, forefinger, and center finger at right angles to each other. The **forefinger** *points in the direction of the field flux*, assuming that magnetic lines of force are in a direction of north and south. The **thumb** *points in the direction of thrust*, or movement of the conductor, and the **center finger** *shows the direction of the current induced into the armature.*

Here is an easier way to remember this:

Thumb = Thrust
Forefinger = Flux
Center finger = Current

By permission: www.elec-toolbox.com

12.4.4 The Right-Hand Generator Rule

Right-Hand Motor Rule

The right-hand generator rule is used to determine the rotation of the armature when the magnetic field polarity of the pole pieces and the direction of current flow through the armature are known. The thumb indicates the direction of thrust or movement of the armature. The forefinger indicates the direction of the field flux assuming that flux lines are in a direction of north to south, and the center finger indicates the direction of current flow through the armature.

Here is an easier way to remember this:

Thumb = Thrust (direction of armature rotation)
Forefinger = Field (direction of magnetic field)
Center finger = Current (direction of armature current)

By permission:www.elec-tool-box.com

12.5.0 Dielectrics and Dielectric Constants of Various Materials

Dielectrics are materials that do not conduct electricity well, if at all. Materials have different dielectric constants at room temperatures. A low dielectric constant of a material means that the material has a low ability to polarize and hold a charge. A high dielectric material is good at holding a charge and is therefore the preferred dielectric for electronic capacitors.

This chart lists the minimum and maximum dielectric constant of various materials, and these values change as temperatures change.

Material	Min.	Max.
Air	1	1
Amber	2.6	2.7
Asbestos fiber	3.1	4.8
Bakelite	5	22
Barium Titanate	100	1250
Beeswax	2.4	2.8
Cambric	4	4
Carbon Tetrachloride	2.17	2.17
Celluloid	4	4
Cellulose Acetate	2.9	4.5
Durite	4.7	5.1
Ebonite	2.7	2.7
Epoxy Resin	3.4	3.7
Ethyl Alcohol	6.5	25
Fiber	5	5
Formica	3.6	6
Glass	3.8	14.5
Glass Pyrex	4.6	5
Gutta Percha	2.4	2.6
Isolantite	6.1	6.1
Kevlar	3.5	4.5
Lucite	2.5	2.5
Mica	4	9
Micarta	3.2	5.5
Mycalex	7.3	9.3
Neoprene	4	6.7
Nylon	3.4	22.4
Paper	1.5	3
Paraffin	2	3

(Continued)

Material	Min.	Max.
Plexiglass	2.6	3.5
Polycarbonate	2.9	3.2
Polyethylene	2.5	2.5
Polyimide	3.4	3.5
Polystyrene	2.4	3
Porcelain	5	6.5
Quartz	5	5
Rubber	2	4
Ruby Mica	5.4	5.4
Selenium	6	6
Shellac	2.9	3.9
Silicone	3.2	4.7
Slate	7	7
Soil dry	2.4	2.9
Steatite	5.2	6.3
Styrofoam	1.03	1.03
Teflon	2.1	2.1
Titanium Dioxide	100	100
Vaseline	2.16	2.16
Vinylite	2.7	7.5
Water distilled	34	78
Waxes, Mineral	2.2	2.3
Wood dry	1.4	2.9

Version 1.1.2

12.5.1 Wire Gauges Table—AWG Gauge—Ft/Ohm Calculations

Wire Gauges Information

Gauge AWG	Ft. / Ohm @ 77 °F	Ohms/1000 ft @ 77 °F	Ft/Ohm @ 149°F	Ohms/1000 ft @ 149°F	AMP @ 140°F	Dia. in mils (1000th in)	Dia. in mm	Wt. in Lbs. per 1000 ft
0000	20000	0.050	17544	0.057	195	460.0	11.684	641.0
000	15873	0.063	13699	0.073	165	410.0	10.414	508.0
00	12658	0.079	10870	0.092	145	365.0	9.271	403.0
0	10000	0.100	8621	0.116	125	325.0	8.255	319.0
1	7936	0.126	6849	0.146	110	289.0	7.348	253.0

Wire Gauges Information—Cont'd

Gauge AWG	Ft. / Ohm @ 77 °F	Ohms/1000 ft @ 77 °F	Ft/Ohm @ 149°F	Ohms/1000 ft @ 149°F	AMP @ 140°F	Dia. in mils (1000th in)	Dia. in mm	Wt. in Lbs. per 1000 ft
2	6289	0.159	5435	0.184	95	258.0	6.544	201.0
3	4975	0.201	4310	0.232	85	229.0	5.827	159.0
4	3953	0.253	3425	0.292	70	204.0	5.189	126.0
5	3135	0.319	2710	0.369		182.0	4.621	100.0
6	24811	0.403	2151	0.465	55	162.0	4.115	79.5
7	968	0.508	1706	0.586		144.0	3.665	63.0
8	1560	0.641	1353	0.739		128.0	3.264	50.0
9	1238	0.808	1073	0.932	40	114.0	2.906	39.6
10	980.4	1.02	847.5	1.18	30	102.0	2.588	31.4
11	781.3	1.28	675.7	1.48		91.0	2.305	24.9
12	617.3	1.62	534.8	1.87	25	81.0	2.053	19.8
13	490.2	2.04	423.7	2.36		72.0	1.828	15.7
14	387.6	2.58	336.7	2.97	20	64.0	1.628	12.4
15	307.7	3.25	266.7	3.75		57.0	1.450	9.86
16	244.5	4.09	211.4	4.73		51.0	1.291	7.82
17	193.8	5.16	167.8	5.96		45.0	1.150	6.2
18	153.6	6.51	133.2	7.51	-	40.0	1.024	4.92
19	121.8	8.21	105.5	9.48		36.0	0.912	3.90
20	96.2	10.4	84.0	11.9		32.0	0.812	3.09
21	76.3	13.1	66.2	15.1		28.5	0.723	2.45
22	60.6	16.5	52.6	19.0		25.3	0.644	1.94
23	48.1	20.8	41.7	24.0	-	22.6	0.573	1.54
24	38.2	26.2	33.1	30.2		20.1	0.511	1.22
25	30.3	33.0	26.2	38.1		17.9	0.455	0.970
26	24.0	41.6	20.8	48.0		15.9	0.405	0.769
27	19.0	52.5	16.5	60.6		14.2	0.361	0.610
28	15.1	66.2	13.1	76.4	-	12.6	0.321	0.484
29	12.0	83.4	10.4	96.3		11.3	0.286	0.384
30	9.5	105	8.3	121		10.0	0.255	0.304
31	7.5	133	6.5	153		8.9	0.227	0.241
32	6.0	167	5.2	193		8.0	0.202	0.191
33	4.7	211	4.1	243	-	7.1	0.180	0.152
34	3.8	266	3.3	307		6.3	0.160	0.120
35	3.0	335	2.6	387		5.6	0.143	0.095
36	2.4	423	2.0	488		5.0	0.127	0.0757
37	1.9	533	1.6	616		4.5	0.113	0.0600
38	1.5	673	1.3	776		4.0	0.101	0.0476
39	1.2	848	1.0	979	-	3.5	0.090	0.0377
40	0.93	1070	0.81	1230		3.1	0.080	0.0200

By permission: cgnetwork.com

Version 3.0.0

12.5.2 Wire Gauge Comparison Chart—AWG—Strandings/Wire Diameter, Overall Diameter

This table is the comparison chart for both **AWG** and metric wire sizes, including the makeup and construction of electrical wire. You can compare AWG to SWG and metric square millimeter cable sizes with the equivalent circular **MIL** conductor sizes.

Electrical Wire Gauge Comparison Table

Circ. Mils	Equivalent Circ. Mils	Awg. Size	Metric Wire Size mm²	Stranding/Wire Dia. per Strand in	Stranding/Wire Dia. per Strand mm	Approximate Overall Diameter in	Approximate Overall Diameter mm
-	987	-	0.50	1/.032	1/.813	.032	0.81
1020	-	20	-	7/.0121	7/.307	.036	0.91
-	1480	-	0.75	1/.039	1/.991	.039	0.99
1620	-	18	-	1/.0403	1/1.02	.040	1.02
1620	-	18	-	7/.0152	7/.386	.046	1.16
-	1974	-	1.0	1/.045	1/1.14	.045	1.14
-	1974	-	1.0	7/.017	7/.432	.051	1.30
2580	-	16	-	1/.0508	1/1.29	.051	1.29
2580	-	16	-	7/.0192	7/.488	.058	1.46
-	2960	-	1.5	1/.055	1/1.40	.055	1.40
-	2960	-	1.5	7/.021	7/5.33	.063	1.60
4110	-	14	-	1/.0641	1/1.63	.064	1.63
4110	-	14	-	7/.0242	7/.615	.073	1.84
-	4934	-	2.5	1/.071	1/1.80	.071	1.80
-	4934	-	2.5	7/.027	7/.686	.081	2.06
6530	-	12	-	1/.0808	1/2.05	.081	2.05
6530	-	12	-	7/.0305	7/.775	.092	2.32
-	7894	-	4	1/.089	1/2.26	.089	2.26
-	7894	-	4	7/.034	7/.864	.102	2.59
10380	-	10	-	1/.1019	1/2.59	.102	2.59
10380	-	10	-	7/.0385	7/.978	.116	2.93
-	11840	-	6	1/.109	1/2.77	.109	2.77
-	11840	-	6	7/.042	7/1.07	.126	3.21
13090	-	9	-	1/.1144	1/2.91	.1144	2.91
13090	-	9	-	7/.0432	7/1.10	.130	3.30
16510	-	8	-	1/.1285	1/3.26	.128	3.26
16510	-	8	-	7/.0486	7/1.23	.146	3.70
-	19740	-	10	1/.141	1/3.58	.141	3.58
-	19740	-	10	7/.054	7/1.37	.162	4.12
20820	-	7	-	1/.1443	1/3.67	.144	3.67
20820	-	7	-	7/.0545	7/1.38	.164	4.15
26240	-	6	-	1/.162	1/4.11	.162	4.11
26240	-	6	-	7/.0612	7/1.55	.184	4.66
-	31580	-	16	7/.068	7/1.73	.204	5.18
33092	-	5	-	7/.0688	7/1.75	.206	5.24
41740	-	4	-	7/.0772	7/1.96	.232	5.88
-	49340	-	25	7/.085	7/2.16	.255	6.48
-	49340	-	25	19/.052	19/1.32	.260	6.60
52620	-	3	-	7/.0867	7/2.20	.260	6.61

Electrical Wire Gauge Comparison Table—Cont'd

Circ. Mils	Equivalent Circ. Mils	Awg. Size	Metric Wire Size mm²	Stranding/Wire Dia. per Strand		Approximate Overall Diameter	
				in	mm	in	mm
66360	-	2	-	7/.0974	7/2.47	.292	7.42
-	69070	-	35	7/.100	7/2.54	.300	7.62
-	69070	-	35	19/.061	19/1.55	.305	7.75
83690	-	1	-	19/.0664	19/1.69	.332	9.43
-	98680	-	50	19/.073	19/1.85	.365	9.27
105600	-	1/0	-	19/.0745	19/1.89	.373	9.46
133100	-	2/0	-	19/.0837	19/2.13	.419	10.6
-	138100	-	70	19/.086	19/2.18	.430	10.9
167800	-	3/0	-	19/.094	19/2.39	.470	11.9
167800	-	3/0	-	37/.0673	37/1.71	.471	12.0
-	187500	-	95	19/.101	19/2.57	.505	12.8
-	187500	-	95	37/.072	37/1.83	.504	12.8
211600	-	4/0	-	19/.1055	19/2.68	.528	13.4
-	237.8mcm	-	120	37/.081	37/2.06	.567	14.4
250mcm	-	-	-	37/.0822	37/2.09	.575	14.6
300mcm	-	-	150	37/.090	37/2.29	.630	16.0
350mcm	-	-	-	37/.0973	37/2.47	.681	17.3
-	365.1mcm	-	185	37/.100	37/2.54	.700	17.8
400mcm	-	-	-	37/.104	37/2.64	.728	18.5
-	473.6mcm	-	240	37/.114	37/2.90	.798	20.3
-	473.6mcm	-	240	61/.089	61/2.26	.801	20.3
500mcm	-	-	-	37/.1162	37/2.95	.813	20.7
500mcm	-	-	-	61/.0905	61/2.30	.814	20.7
-	592.1mcm	-	300	61/.099	61/2.51	.891	22.6
600mcm	-	-	-	61/.0992	61/2.52	.893	22.7
700mcm	-	-	-	61/.1071	61/2.72	.964	24.5
750mcm	-	-	-	61/.1109	61/2.82	.998	25.4
750mcm	-	-	-	91/.0908	91/2.31	.999	25.4
-	789.4mcm	-	400	61/.114	61/2.90	1.026	26.1
800mcm	-	-	-	61/.1145	61/2.91	1.031	26.2
800mcm	-	-	-	61/.0938	91/2.38	1.032	26.2
1000mcm	986.8mcm	-	500	61/.1280	61/3.25	1.152	29.3
1000mcm	-	-	-	91/.1048	91/2.66	1.153	29.3
-	1233.7mcm	-	625	91/.117	91/2.97	1.287	32.7
1250mcm	-	-	-	91/.1172	91/2.98	1.289	32.7
1250mcm	-	-	-	127/.0992	127/2.52	1.290	32.8
1500mcm	-	-	-	91/.1284	91/1.284	1.412	35.9
1500mcm	-	-	-	127/.1087	127/.1087	1.413	35.9
-	1578.8mcm	-	800	91/.132	91/.132	1.452	36.9
-	1973.5mcm	-	1000	91/.147	91/3.73	1.617	41.1
2000mcm	-	-	-	127/.1255	127/3.19	1.632	41.5
2000mcm	-	-	-	169/.1088	169/2.76	1.632	41.5

Version 1.0.1

12.5.3 Resistance in Ohms per 1000 Feet of Conductor—Aluminium and Copper

Resistance in Ohms per 1000 feet per conductor at 20°C and 25°C of solid wire and class B concentric strands copper and aluminum conductor

Conductor Size, AWG or kcmil	Annealed Uncoated Copper Annealed Aluminum								Annealed Coated Copper							
	Solid				Stranded Class B				Solid				Stranded Class B			
	20°C			25°C*		20°C		25°C*		20°C		25°C*		20°C		25°C*	
	CU	AL	CU	AL	CU	AL	CU	AL	CU	CU	CU	CU
24	25.7000	----	26.2000	----	----	----	----	----	26.8000	27.3000	----	----
22	16.2000	----	16.5000	----	----	----	----	----	16.9000	17.2000	----	----
20	10.1000	----	10.3000	----	10.30000	----	10.50000	----	10.5000	10.7000	11.00000	11.20000
19	8.0500	----	8.2100	----	----	----	----	----	8.3700	8.5300	----	----
18	6.3900	----	6.5100	----	6.51000	----	6.64000	----	6.6400	6.7700	6.92000	7.05000
16	4.0200	----	4.1000	----	4.10000	----	4.18000	----	4.1800	4.2600	4.35000	4.44000
14	2.5200	4.1400	2.5700	4.220	2.57000	----	2.62000	----	2.6200	2.6800	2.68000	2.73000
12	1.5900	2.6000	1.6200	2.660	1.62000	2.65000	1.65000	2.70000	1.6200	1.6800	1.68000	1.72000
10	0.9990	1.6400	1.0200	1.670	1.02000	1.67000	1.04000	1.70000	1.0400	1.0600	1.06000	1.08000
9	0.7920	1.3000	0.8080	1.320	0.80800	1.33000	0.82400	1.35000	0.8160	0.8310	0.84000	0.85700
8	0.6280	1.0300	0.6410	1.050	0.64100	1.05000	0.65400	1.07000	0.6460	0.6590	0.66600	0.67900
7	0.4980	.8170	0.5080	.833	0.51800	.83300	0.51800	0.85000	0.5130	0.5230	0.52800	0.53900
6	0.3950	.6480	0.4030	.661	0.40300	.66100	0.41000	0.67400	0.4070	0.4150	0.41900	0.42700
5	0.3130	.5140	0.3190	.524	0.32000	.52400	0.32600	0.53500	0.3230	0.3290	0.33300	0.33900
4	0.2480	.4070	0.2530	.415	0.25300	.41600	0.25900	0.42400	0.2560	0.2610	0.26400	0.26900
3	0.1970	.3230	0.2010	.330	0.20500	.33000	0.20500	0.33600	0.2030	0.2070	0.20900	0.21300
2	0.1560	.2560	0.1590	.261	0.15900	.26200	0.16200	0.26700	0.1610	0.1640	0.16600	0.16900
1	0.1240	.2030	0.1260	.207	0.12600	.20600	0.12900	0.21100	0.1280	0.1300	0.13100	0.13400
1/0	0.0982	.1610	0.1000	.164	0.10000	.16500	0.10200	0.16800	0.1010	0.1030	0.10400	0.10600
2/0	0.0779	.1280	0.0795	.130	0.07950	.13100	0.08110	0.13300	0.0798	0.0814	0.08270	0.08430
3/0	0.0618	.1010	0.0630	.103	0.06300	.10300	0.06420	0.10500	0.0633	0.0645	0.06560	0.06680
4/0	0.0490	.0803	0.0500	.082	0.05000	.08210	0.05090	0.08360	0.0502	0.0512	0.05150	0.05250
250	----	----	----	----	0.04230	.06950	0.04310	0.07080	----	----	0.04400	0.04490
300	----	----	----	----	0.03530	.05790	0.03600	0.05900	----	----	0.03670	0.03740
350	----	----	----	----	0.03020	.04960	0.03080	0.05050	----	----	0.03140	0.03200
400	----	----	----	----	0.02640	.04340	0.02700	0.04420	----	----	0.02720	0.02780
500	----	----	----	----	0.02120	.03480	0.02160	0.03540	----	----	0.02180	0.02220
600	----	----	----	----	0.01760	.02900	0.01800	0.02950	----	----	0.01840	0.01870
750	----	----	----	----	0.01410	.02320	0.01440	0.02360	----	----	0.01450	0.01480
1000	----	----	----	----	0.01060	.01740	0.01080	0.01770	----	----	0.01090	0.01110
1250	----	----	----	----	0.00846	.01390	0.00863	0.01420	----	----	0.00871	0.00888
1500	----	----	----	----	0.00705	.01160	0.00719	0.01180	----	----	0.00726	0.00740
1750	----	----	----	----	0.00604	.00992	0.00616	0.01010	----	----	0.00622	0.00634
2000	----	----	----	----	0.00529	.00869	0.00539	0.00885	----	----	0.00544	0.00555
2500	----	----	----	----	0.00427	.00702	0.00436	0.00715	----	----	0.00440	0.00448

By permission: The Okonite Company

12.5.3.1 Solid and Concentric Stranding of Class B and Class C Strandings

Compact and Compressed Diameters

Conductor Size AWG or kcmil	Number of Wires	Compact Diameter (inches)	Compress Diameter (inches)
8	7	0.134	0.141
6	7	0.169	0.178
4	7	0.213	0.225
2	7	0.268	0.283
1	19	0.299	0.322
1/0	19	0.336	0.361
2/0	19	0.376	0.406
3/0	19	0.423	0.456
4/0	19	0.475	0.512
250	37	0.520	0.558
300	37	0.570	0.611
350	37	0.616	0.661
400	37	0.659	0.706
500	37	0.736	0.789
600	61	0.813	0.866
750	61	0.908	0.968
800	61	0.938	1.000
900	61	0.999	1.060
1000	61	1.060	1.117

Solid and Concentric Stranding

Conductor Size AWG or kcmil	Circular Mil Cross-Sectional Area	Sq. MM	Conductor Diameter Mils	Solid Conductor Weight (lb./M ft.) Aluminum	Solid Conductor Weight (lb./M ft.) Copper	Class "B" Stranding Number of Wires	Class "B" Stranding Diameter of Each Wire Mils	Class "B" Stranding Conductor Diameter (in.)	Class "C" Stranding Number of Wires	Class "C" Stranding Diameter of Each Wire Mils	Class "C" Stranding Conductor Diameter (in.)	Conductor wgt. Class "B" & "C" Strandings (lb./M ft.) Aluminum	Conductor wgt. Class "B" & "C" Strandings (lb./M ft.) Copper
24	404	0.205	20.1	—	1.22	7	7.6	0.023	—	—	—	—	1.24
22	640	0.324	25.3	—	1.94	7	9.6	0.029	—	—	—	—	1.98
20	1,020	0.519	32.0	0.942	3.10	7	12.1	0.036	—	—	—	—	3.15
19	1,290	0.653	35.0	1.19	3.90	7	13.6	0.041	—	—	—	—	3.98
18	1,620	0.823	40.3	1.49	4.92	7	15.2	0.046	—	—	—	—	5.01
16	2,580	1.310	50.8	2.38	7.81	7	19.2	0.058	—	—	—	—	7.97
14	4,110	2.080	64.1	3.78	12.44	7	24.2	0.073	19	14.7	0.074	—	12.7
12	6,530	3.310	80.8	6.01	19.77	7	30.5	0.092	19	18.5	0.093	6.13	20.2
10	10,380	5.260	101.9	9.56	31.43	7	38.5	0.116	19	23.4	0.117	9.75	32.0
9	13,090	6.630	114.4	12.04	39.63	7	43.2	0.130	19	26.2	0.131	12.30	40.4
8	16,510	8.370	128.5	15.20	50.00	7	48.6	0.146	19	29.5	0.146	15.15	51.0
7	20,820	10.550	144.3	19.16	63.03	7	54.5	0.164	19	33.1	0.166	19.16	64.2
6	26,240	13.300	162.0	24.15	79.40	7	61.2	0.184	19	37.2	0.186	24.60	81.0
5	33,090	16.670	181.9	30.45	100.20	7	68.8	0.206	19	41.7	0.209	31.10	102.0
4	41,740	21.150	204.3	38.44	126.40	7	77.2	0.232	19	46.9	0.235	39.20	129.0
3	52,620	26.670	229.4	48.43	159.30	7	86.7	0.260	19	52.6	0.263	49.40	162.0
2	66,360	33.620	257.6	61.07	200.90	7	97.4	0.292	19	59.1	0.296	62.30	205.0
1	83,690	44.210	289.3	77.03	253.30	19	66.4	0.332	37	47.6	0.333	78.60	258.0
1/0	105,600	53.490	324.9	97.15	319.60	19	74.5	0.373	37	53.4	0.374	99.10	326.0
2/0	133,100	67.430	364.8	122.50	402.90	19	83.7	0.419	37	60.0	0.420	125.00	411.0
3/0	167,800	85.010	409.6	154.40	507.90	19	94.0	0.470	37	67.3	0.471	157.00	518.0
4/0	211,600	107.200	460.0	194.70	640.50	19	105.5	0.528	37	75.6	0.529	199.00	653.0
250	—	127.000	—	—	—	37	82.2	0.575	61	64.0	0.576	235.00	772.0
300	—	152.000	—	—	—	37	90.0	0.630	61	70.1	0.631	282.00	926.0
350	—	177.000	—	—	—	37	97.3	0.681	61	75.7	0.681	329.00	1081.0
400	—	203.000	—	—	—	37	104.0	0.728	61	81.0	0.729	376.00	1235.0
500	—	253.000	—	—	—	37	116.2	0.813	61	90.5	0.815	469.00	1544.0
600	—	304.000	—	—	—	61	99.2	0.893	91	81.2	0.893	563.00	1853.0
750	—	380.000	—	—	—	61	110.9	0.998	91	90.8	0.999	704.00	2316.0
1000	—	507.000	—	—	—	61	128.0	1.152	91	104.8	1.153	939.00	3088.0
1250	—	633.000	—	—	—	91	117.2	1.289	127	99.2	1.290	1173.00	3859.0
1500	—	760.000	—	—	—	91	128.4	1.412	127	108.7	1.413	1408.00	4631.0
1750	—	887.000	—	—	—	127	117.4	1.526	169	101.8	1.527	1643.00	5403.0
2000	—	1010.000	—	—	—	127	125.5	1.632	169	108.8	1.632	1877.00	6175.0
2500	—	1263.000	—	—	—	127	140.3	1.824	169	121.6	1.824	2370.00	7794.0

12.5.4 Copper to Aluminium Conversion Tables

Copper to Aluminium Conversion Tables

3-Conductor 75°C Copper or Aluminum Circuitry (Without Ground Wire)

Amps	Copper		Aluminum	
	Wire Size	Conduit Size	Wire Size	Conduit Size
20	#14	1/2"	#12	1/2"
25	#12	1/2"	#10	3/4"
35	#10	3/4"	#8	3/4"
50	#8	3/4"	#6	1"
65	#6	1"	#4	1 1/4"
85	#4	1"	#2	1 1/4"
100	#3	1 1/4"	#1	1 1/4"
115	#2	1 1/4"	1/0	1 1/2"
130	#1	1 1/4"	2/0	2"
150	1/0	1 1/2"	3/0	2"
175	2/0	1 1/2"	4/0	2 1/2"
200	3/0	2"	250	2 1/2"
230	4/0	2"	300	2 1/2"
255	250	2 1/2"	400	3"
285	300	2 1/2"	500	3"
310	350	2 1/2"	500	3"
335	400	3"	600	3 1/2"
380	500	3"	750	4"
420	600	3"	(2) 300	4'
460	700	3 1/2"	(2) 350	5"
475	750	4"	(2) 400	5"

4-Conductor 75°C Copper or Aluminum Circuitry (Without Ground Wire)

Amps	Copper		Aluminum	
	Wire Size	Conduit Size	Wire Size	Conduit Size
20	#14	1/2"	#12	1/2"
25	#12	1/2"	#10	3/4"
35	#10	3/4"	#8	3/4"
50	#8	3/4"	#6	1"
65	#6	1"	#4	1 1/4"
85	#4	1 1/4"	#2	1 1/4"
100	#3	1 1/4"	#1	1 1/2"
115	#2	1 1/4"	1/0	2"
130	#1	1 1/2"	2/0	2"
150	1/0	1 1/2"	3/0	2"
175	2/0	2"	4/0	2 1/2"
200	3/0	2"	250	2 1/2"
230	4/0	2 1/2"	300	3"
255	250	2 1/2"	400	3"
285	300	3"	500	3 1/2"

(Continued)

4-Conductor 75°C Copper or Aluminum Circuitry (Without Ground Wire)—Cont'd

Amps	Copper		Aluminum	
	Wire Size	Conduit Size	Wire Size	Conduit Size
310	350	3"	500	3 1/2"
335	400	3"	600	4"
380	500	3 1/2"	750	4"
420	600	4"	(2) 300	4"
460	700	4"	(2) 350	5"
475	750	4"	(2) 400	5"

By permission: www.mc2-ice.com

3-Conductor 75°C Copper or Aluminum Circuitry (With Ground Wire)

Amps	Copper			Aluminum		
	Wire Size	Conduit Size	Ground	Wire Size	Conduit Size	Ground
20	#14	1/2"	#16	#12	1/2"	#12
25	#12	1/2"	#14	#10	3/4"	#12
35	#10	3/4"	#12	#8	3/4"	#10
50	#8	3/4"	#10	#6	1"	#8
65	#6	1"	#8	#4	1 1/4"	#6
85	#4	1"	#8	#2	1 1/4"	#6
100	#3	1 1/4"	#8	#1	1 1/4"	#6
115	#2	1 1/4"	#6	1/0	1 1/2"	#4
130	#1	1 1/4"	#6	2/0	2"	#4
150	1/0	1 1/2"	#6	3/0	2"	#4
175	2/0	1 1/2"	#6	4/0	2 1/2"	#4
200	3/0	2"	#6	250	2 1/2"	#4
230	4/0	2"	#4	300	2 1/2"	#2
255	250	2 1/2"	#4	400	3"	#2
285	300	2 1/2"	#4	500	3"	#1
310	350	2 1/2"	#3	500	3"	#1
335	400	3"	#3	600	3 1/2"	#1
380	500	3"	#3	750	4"	#1
420	600	3"	#2	(2) 300	4"	1/0
460	700	3 1/2"	#2	(2) 350	5"	1/0
475	750	4"	#2	(2) 400	5"	1/0

4-Conductor 75°C Copper or Aluminum Circuitry (With Ground Wire)

Amps	Copper			Aluminum		
	Wire Size	Conduit Size	Ground	Wire Size	Conduit Size	Ground
20	#14	1/2"	#16	#12	1/2"	#12
25	#12	1/2"	#14	#10	3/4"	#12
35	#10	3/4"	#12	#8	3/4"	#10

4-Conductor 75°C Copper or Aluminum Circuitry (With Ground Wire) – Cont'd

		Copper			Aluminum		
Amps	Wire Size	Conduit Size	Ground	Wire Size	Conduit Size	Ground
50	#8	3/4"	#10	#6	1"	#8
65	#6	1"	#8	#4	1 1/4"	#6
85	#4	1 1/4"	#8	#2	1 1/4"	#6
100	#3	1 1/4"	#8	#1	1 1/2"	#6
115	#2	1 1/4"	#6	1/0	2"	#4
130	#1	1 1/2"	#6	2/0	2"	#4
150	1/0	1 1/2"	#6	3/0	2"	#4
175	2/0	2"	#6	4/0	2 1/2"	#4
200	3/0	2"	#6	250	2 1/2"	#4
230	4/0	2 1/2"	#4	300	3"	#2
255	250	2 1/2"	#4	400	3"	#2
285	300	3"	#4	500	3 1/2"	#1
310	350	3"	#3	500	3 1/2"	#1
335	400	3"	#3	600	4"	#1
380	500	3 1/2"	#3	750	4"	#1
420	600	4"	#2	(2) 300	4"	1/0
460	700	4"	#2	(2) 350	5"	1/0
475	750	4"	#2	(2) 400	5"	1/0

12.5.5 Conduit inside Diameters and Electrical Conductor Areas—U.S. to Metric Conversion

Metric Converstions

1 meter = 39.37 inches
1 centimeter = 0.39 inch
1 millimeter = 0.039 inch
1 inch = 0.025 meter
1 inch = 2.564 centimeters
1 inch = 25.641 millimeters

Conduit Inside Diameters

Trade Size	Inches	Millimeters
1/2	0.622	15.8
3/4	0.824	20.9
1	1.049	26.6
1 1/4	1.380	35.0
1 1/2	1.610	40.9
2	2.067	52.5
2 1/2	2.469	62.7
3	3.068	77.9
3 1/2	3.548	90.1
4	4.026	102.3
5	5.047	128.2
6	6.065	154.1

Electrical Conductor Areas United States

Size (AWG)	Cir Mills (area)	Sq MM (area)
18	1,620	0.82
16	2,580	1.30
14	4,110	2.08
12	6,530	3.30
10	10,380	5.25
8	16,510	8.36
6	26,240	13.29
4	41,740	21.14
3	52,620	26.65
2	66,360	33.61
1	83,690	42.39
1/0	105,600	53.49
2/0	133,100	67.42
3/0	167,800	85.00
4/0	211,600	107.19
250	250,000	126.64
300	300,000	151.97
350	350,000	177.3
400	400,000	202.63
500	500,000	253.29

Closest European Sizes

Size—Sq MM	Area—Cir Mils
0.75	1,480
1.00	1,974
1.50	2,961
2.50	4,935
4.00	7,896
6.00	11,844
10.0	19,740
16.0	31,584
25.0	49,350
—	—
35.0	69,090
50.0	98,700
—	—
70.0	138,180
95.0	187,530
120.0	236,880
150.0	296,100
—	—
185.0	365,190
240.0	473,760

12.5.6 Conduit Weight Comparisons—Rigid, EMT, PVC

(All weights are per 100 feet of conduit)

Size	Rigid Steel	IMC	Rigid Aluminum	EMT	PVC Sch 40	PVC Sch 80	PVC Coated* Rigid Steel
1/2"	80	60	28	29	16	21	87
3/4"	108	82	37	45	22	29	115
1"	160	116	55	65	33	42	166
1 1/4"	208	150	72	96	46	61	217
1 1/2"	254	182	89	111	56	71	262
2"	344	242	119	141	74	98	367
2 1/2"	550	401	188	215	117	149	557
3"	710	493	246	260	153	200	724
3 1/2"	855	573	296	365	185	246	917
4"	1000	683	350	390	219	292	1056
5"	1335	—	479	—	298	400	1535
6"	1845	—	630	—	385	510	2025

*20 mil thickness.
Note: The above weights were taken from several manufacturers' catalogues, include the threaded coupling when applicable, and have been rounded off to the closest whole number.

12.5.7 Recommended Power and Ground Cable Sizes—By Power and Distance

To calculate the proper power and ground cable sizes, find the distance of the power cable along the top row. If your measurement is between two measurements, use the higher one. Next, find the total power the cable must support on the left. If your measurement is between two measurements, use the higher one. The size listed where your two measurements meet is the recommended cable size.

Recommended Cable Size by Power and Distance

Total RMS Power (watts)	Distance				
	4 feet	8 feet	12 feet	16 feet	20 feet
100	10 gauge	10 gauge	8 gauge	8 gauge	4 gauge
200	10 gauge	8 gauge	8 gauge	4 gauge	4 gauge
400	8 gauge	8 gauge	4 gauge	4 gauge	4 gauge
600	8 gauge	4 gauge	4 gauge	4 gauge	4 gauge
800	4 gauge	4 gauge	4 gauge	2 gauge	2 gauge
1000	4 gauge	4 gauge	2 gauge	2 gauge	2 gauge
1400	4 gauge	2 gauge	2 gauge	2 gauge	2 gauge

Current Draw by Power		Power & Ground Cable Specs	
Total RMS Power (watts)	Current Amps	Cable Size Wire Gauge	Current Capacity Amperage (amps)
100	16	1/0	350
200	32	2	225
400	64	4	150
600	96	8	100
800	128	10	60
1000	160	12	40
1200	172	14	25
1400	188	16	15

| | | Distance | | | |
Total RMS Power (watts)	4 feet	8 feet	12 feet	16 feet	20 feet
600	96	8		100	
800	128	10		60	
1000	160	12		40	
1200	172	14		25	
1400	188	16		15	

Note: *These figures have been rounded off for easy reference. Ground cables should be the same size (or larger) as the power cable.*

12.6.0 Types of Transformers

Transformer types

Transformers can be constructed so that they are designed to perform a specific function. A basic understanding of the various types of transformers is necessary to understand the role transformers play in today's nuclear facilities.

EO 1.4   STATE the applications of each of the following types of transformers:

   a. Distribution
   b. Power
   c. Control
   d. Auto
   e. Isolation
   f. Instrument potential
   g. Instrument current

Types of Transformers

Transformers are constructed so that their characteristics match the application for which they are intended. The differences in construction may involve the size of the windings or the relationship between the primary and secondary windings. Transformer types are also designated by the function the transformer serves in a circuit, such as an isolation transformer.

Distribution Transformer

Distribution transformers are generally used in electrical power distribution and transmission systems. This class of transformer has the highest power, or volt-ampere ratings, and the highest continuous voltage rating. The power rating is normally determined by the type of cooling methods the transformer may use. Some commonly used methods of cooling are by using oil or some other heat-conducting material. Ampere rating is increased in a distribution transformer by increasing the size of the primary and secondary windings; voltage ratings are increased by increasing the voltage rating of the insulation used in making the transformer.

Power Transformer

Power transformers are used in electronic circuits and come in many different types and applications. Electronics or power transformers are sometimes considered to be those with ratings of 300 volt-amperes and below. These transformers normally provide power to the power supply of an electronic device, such as in power amplifiers in audio receivers.

Source: Integrated Publishing.

12.6.1 Dry-Type Transformers—KVA Ratings—Single- and Three-Phase

Single Phase

KVA Rating	Amperes		
	120V	240V	480V
1	8.33	4.16	2.08
1.5	12.5	6.24	3.12
2	16.66	8.33	4.16
3	25	12.5	6.1
5	41	21	10.4
7.5	62	31	15.6
10	83	42	21
15	124	62	31
25	208	104	52
37.5	312	156	78
50	416	208	104
75	624	312	156
100	830	415	207
167	1390	695	348
200	1660	833	416

Three Phase

KVA Rating	Amperes			
	208V	240V	480V	600V
3	8.3	7.2	3.6	2.9
6	16.6	14.4	7.2	5.8
9	25.0	21.6	10.8	8.7
15	41.6	36	18	14.4
30	83.0	72	36	28.8
45	125	108	54	43
75	208	180	90	72
112.5	312	270	135	108
150	415	360	180	144
200	554	480	240	192
225	625	540	270	216
300	830	720	360	288
400	1110	960	480	384
500	1380	1200	600	487
750	2080	1800	900	720

12.7.0 Enclosure Types for All Locations

Industry Standards

Enclosure Types for All Locations

National Electrical Manufacturers Association (NEMA Standard 250)
NEMA/EEC to IEC

TYPE	Intended Use and Description
Type 1	General-purpose enclosures are suitable for general-purpose application indoors, where atmospheric conditions are normal. These enclosures serve as protection against falling dust, but are **not** dust tight.
Type 2	Driptight (indoor) enclosures are similar to NEMA 1 enclosures, with the addition for drip shields, and are suitable for application where condensation may be severe, such as that encountered in cooling rooms or laundries.
Type 3	Dust, rain-proof, and sleet-resistant enclosures provide proper protection against **windblown** dust and weather hazards such as rain, sleet, or snow. They are suitable for applications outdoors on ship docks, canal locks, construction work, and for applications in subways and tunnels; use indoors where dripping water is a problem.
Type 3R	Dust, rain-proof, and sleet-resistant enclosures provide proper protection against **falling** dirt and weather hazards such as rain, sleet, or snow. They are suitable for applications outdoors on ship docks, canal locks, construction work, and for applications in subways and tunnels; use indoors where dripping water is a problem.
Type 4	Water-tight enclosures are suitable for dairies, breweries, etc., where the enclosure may be subjected to large amounts of water from any angle. (**They are not submersible**.)
Type 4X	Corrosion-resistant enclosures satisfy the same requirements as NEMA 4; in addition, they are suitable for food-processing plants, dairies, refineries, and other industries where corrosion is prominent.
Type 6	Submersible enclosures are suitable for application where the equipment may be subject to submersion, such as quarries, mines, and manholes. The enclosure design will depend on the specified conditions of pressure and time. Hazardous location enclosures—Class II, Group E, F, or G. These enclosures are designed to meet the requirements of the "Canadian Electrical Code" Part I for Class II hazardous locations, and CSA codes section 18 Class II Group E,f, and G.
Type 9	Class II Group E—atmosphere containing metal dust Class II Group F—atmosphere containing carbon black, coal, or coke dust. Class II Group G—atmosphere containing flour, starch, or grain dust.
Type 12	Industrial use enclosures are **oil tight**. Hammond type 12 enclosures meet JIC standard and also satisfy requirements of NEMA. The cover is held in place with screws, bolts, or other suitable fasteners, with a continuous gasket construction. The fastener parts are held captive when the door is opened. There are no holes through the enclosures for mounting or attaching controls inside the enclosure, and no conduit knock-outs or openings.
Type 13	Mounting feet, brackets, or other mounting means are provided. These enclosures are suitable for application to machine tools and other industrial processing machines where oil, coolants, water, filings, dust, or lint may enter, seep into, or infiltrate the enclosure through mounting holes, unused conduit knockouts, or holes used for mounting equipment with the enclosure.

The preceding descriptions are not intended to be complete representations of the National Electrical Manufacturers Association standards for enclosures.

Underwriter Laboratories Inc. (UL 50 and UL 508)

TYPE	Intended Use and Description
Type 1	Indoor use primarily to provide protection against contact with the enclosed equipment and against limited amount of falling dirt.
Type 2	Indoor use to provide a degree of protection against limited amounts of falling water and dirt.
Type 3	Outdoor use to provide a degree of protection against **windblown** dust and **windblown** rain; undamaged by the formation of ice on the enclosure.
Type 3R	Outdoor use to provide a degree of protection against **falling** rain; undamaged by the formation of ice on the enclosure.
Type 4	Either indoor or outdoor use to provide a degree of protection against falling rain, splashing water, and hose-directed water; undamaged by the formation of ice on the enclosure.
Type 4X	Same as type 4 except this one is **corrosion resistant**.
Type 6	Indoor or outdoor use to provide a degree of protection against entry of water during temporary submersion at a limited depth; undamaged by formation of ice on the enclosure.
Type 12	Indoor use to provide a degree of protection against dust, dirt, fiber flyings, dripping water, and external condensation of noncorrosive liquids.
Type 13	Indoor use to provide a degree of protection against lint, dust seepage, external condensation and spraying of water, oil, and noncorrosive liquids.

Index

Note: Page numbers followed by *f* indicate figures and *t* indicate tables.

A

AASHTO. *See* American Association for State Highway Transportation Officials
Above-grade, non-load-bearing, 262
Abrasions resisting, 336
   steel sheets, 340*t*
Absolute value, 108
AC ratings, 570
   overview, 572
ACA. *See* Ammoniacal copper arsenate
Acceleration, 130
   average, 117*t*
   constant-acceleration
      circular motion, 117*t*
      linear motion, 117*t*
   gravity and, 132
Accelerators, 212
ACI. *See* Standard American Concrete Institution
Acid pickling, 336
Acoustical tile, 564
Acoustical wall coverings, 555
Acoustics, 525–526
Acres
   hectares, conversion to, 39*t*
   square feet
      conversion from, 43*t*
      conversion to, 44*t*
Act of Congress, 3, 8
Actual dimensions, 253
ACZA. *See* Ammoniacal copper zinc arsenate
Adhesives
   clay, 562
   clear, 562
   Organic Mastic, 563
   strength properties of, 481–484
   Thinset, 563
   for wall coverings, 561–562
   wood, 479, 479*t*
Admixtures
   air-entrainers, 223–224
   chemical, 218
      concrete, 212–213
   concrete, 212–213
   mineral, 224

retarding, 237
water-reducing, 221
Aesthetics, 488
African hardwoods, 397
African mahogany, 392
AFUE. *See* Annual fuel utilization efficiency
Aggregate moisture content, 215
Air
   Blaine air permeability test, 218
   buoyancy, 12
   constituent of, 616
   content control, 224
   content estimation, 214, 214*t*
   heating formulas, 617
   standard air film, 520
Air-entrainers, 212, 218, 223–224
   admixtures, 223–224
   factors affecting, 224
   recommendations for, 224
Air-nailing gun, 456
Air-source heat pumps, 624
AISI. *See* American Iron and Steel Institute
Alam, Mahbub, 81
Algebra, 81–82
Alloying materials, 266
Aluminum, 147, 451
   channels, 331–332
   copper, conversion from, 663–665
   expanded metal grating, 342–343
   fasteners, 452–453
   plank sections, 348–349
   rectangular bar grating, 344–345, 346–347
   structural angles, 330–332
   structural beams, 333–335
American Association for State Highway Transportation Officials (AASHTO), 218
   aggregate and soil terminology, 165–166
   ASTM, conversion to, 497
American Iron and Steel Institute (AISI), 453
American Society for Testing Materials (ASTM), 218, 260
   A325, fastener tension, 501

A490
   fastener tension, 501
   mechanical properties, 501
   AASHTO, conversion to, 497
   aggregate and soil terminology, 165–166
   nut compatibility chart, 497*t*
   PCC types, 216–217
   steel designations, 268–270, 271–272
American Society of Mechanical Engineers (ASME), 275
American Softwood Lumber Standard, 375
American Water Works Association, 592
American wire gauge (AWG), 67
Ammoniacal copper arsenate (ACA), 453
Ammoniacal copper zinc arsenate (ACZA), 453
Amorphous peat (muck), 161, 163
Amps
   draws of, 640
   horsepower, converting from, 636
   KVA, converting from, 637
   KW, converting from, 637
   watts
      converting from, 636
      converting to, 636
Anchor bolts
   embedded
      galvanized, 247–248
      plain finish, 245–247
   in stock plain finish, 224, 245*t*, 246*t*
Angles, 105, 198
   base, 94
   double, 106
   formulas
      relationship, 198
      small angle, 96–97
   marking, 584
   measuring, 584–585
   of N-gon, 99–100
   right, 100*f*
      hypotenuse of, 86
   of slope, 199
   steel, 283
   structural
      aluminum, 330–332
      steel, 326–327

673

Angles (*Continued*)
  units of, 6
  vertex, 94
Angular impulse, 117*t*
Angular momentum, 117*t*
Angular sizes, 97–98
Angular speed, 117*t*
Annual fuel utilization efficiency (AFUE), 650–652
Annual Meeting of the National Conference on Weights and Measures, 1
Annular nails, 442–443
  threaded, 444*t*
Apitong, 396–397
Apothecary system
  units of liquid volume, 17*t*
  units of mass, 18*t*
Apparatus, calibration of, 195*t*
Appearance lumber, 363–365
  grades of, 363*t*
  signs of, 364*t*
Appliances
  efficiency
    fuel conversion of, 618
    ratings of, 650–652
  energy use of, 644
    chart, 646–647, 646*t*
    wattage of, 641–642
Arabic dates, 71
Archimedes, 83–84
Architects, 10
Arcseconds, 98
Area
  of circle, 83–84, 89–90
    givens, 83–84
    goals, 83
    knowns and assumptions, 84
    method, 84
  of cone, theorem, 93
  of cylinder, 87, 92, 139
  formulas of, 104
  of parallelogram, 85
  of prisms, 91–92
  of sphere, theorem, 93–95
  surface area
    of cone, 99
    of cube, 99
    of cuboid, 99
    formulas of, 99
    of frustum, 88–89
    of sphere, 99
  of trapezoid, 86
  of triangle, 86
  units of, 15*t*, 16*t*
    international measure, 20*t*
    survey measure, 21*t*
The Area Engineer, 239

Arithmetical systems of numbers, 10
Aromatic red cedar, 397
ASME. *See* American Society of Mechanical Engineers
ASTM. *See* American Society for Testing Materials
Atmospheric pressure, 616
Attrition milling, 437
Avoirdupois system, 11, 34
AWG. *See* American wire gauge

B

Babylonians, 4, 10
Backing, 572
  layer, 570
Back-roll, 550–551
Baker, Andrew J, 449
Baldrige National Quality Program, 1
Bandwidth, 639
Banked circular tracks, 117*t*
Bar grating, 348–349
  rectangular, 344–345, 346–347
Bar size tees, 328
Barcol hardness test, 150
Base angles, 94
Base units, 7
Base-loaded plants, 227
Batt insulation, 510
  fiberglass, 510*t*
Beams
  of framing lumber, 367*t*
  I beams, 293–294, 293*t*
  junior, 293–294, 294*t*
  section properties of, 365–367
  steel wide flange, 282
    weight and size of, 284–289
  structural, aluminum, 333–335
  types of, 279
  typical characteristics of, 280
  wide flange, 276–277
    size and weight of, 284–293
    steel, 282
Bearing
  above-grade, non-load-bearing, 262
  California bearing ratio, 178
  load-bearing
    solid, 509
    test, 178
  soil, capacity of, 181–182
  stress, 476*f*, 477*f*, 478*f*
    under bolts, 476, 478*f*
Beating block, 563
Berber carpets, 566
Bernoulli's equation, 117*t*
BF. *See* Board foot
BHP. *See* Brake horsepower
Binary subdivisions, 9

Binomial, 81
Biological factors, of soil, 173
BIPM. *See* International Bureau of Weights and Measures
Black gum, quartered, 389
Blackbody radiation, 117*t*
Blaine air permeability test, 218
Blanket insulation, 510
Blasting, 156
Bleeding, 228
Blended cement, 219–220, 219*t*, 229
Block cavity wall, 509
  R-values of, 515–516
Blocky soil, 163
Board foot (BF), 381
  conversion factors of, 382*t*
  in lumber bundle, 382
Boat nail, 460
BOCA. *See* National building code
Boiling point, of water, 616
Bolts, 450–451
  anchor
    embedded, 245–248
    in stock plain finish, 224, 245*t*, 246*t*
  bearing stress under, 476, 478*f*
  determining proper length of, 499–500
  fasteners, for structural steel, 488–492
  quality of, 478
  twist-off structural, 500
  in wood, 476–479
Bonding agents, 213
Bonds, 251–252
  patterns of, 251–252
Boring, 373–375
  guidelines, 373
Borrow excavation, 157
Boulder soil, 163
Box nails, 442–443, 443*t*
Brads
  cut floor, 456
  moulder, 457
Brahe, Tycho, 132
Brake horsepower (BHP), 79
Break-off knob screw, 470
Bricks
  dimensions of, 253
  face, 248–249
  modular, 249, 252–253
  nonmodular, 249
  number of, 250*t*
    calculating, 250–251
  positions of, 250
  sizes, 248–249, 254
    nominal modular, 254
  types of, by material, 252

Index

Brick masonry, 264
Brick wall, 509
  R-values of, 515–516
Bridge plank, 390–391
Bright common nails, 442–443
Brighteners, of wood, 578
Brinell hardness test, 148
Britain, measures of
  apothecaries, 19
  capacity, 18–19
  systems, 9
  units, 18–19
British thermal unit (Btu), 504
Btu. *See* British thermal unit
Building codes, 373
Buoyancy, 117t
Bureau of Reclamation, U.S, 226
Bush, George H. W, 34
Butcher blocks, finishes for, 577
Button/dome screw head, 469

C

C channels, 280
  dimensions, 323–324
  weight and size of, 319–320, 321
    metric, 323
Cabinet grade lumber, 380
Cabinetmaker's pencil set, 587
CABO. *See* Council of American Building Officials
Calcareous soil, 163
Calibration tools, 586
California bearing ratio (CBR), 178
Canada, lumber standards of, 375
Capacitance, 117t
Capacitive reactance, 639
Capacitors, 117t
  combinations, 117t
  networks, 638
Capacity
  bearing, of soil, 181–182
  British measures of, 18–19
  of cylindrical tanks, 138–139
  of rectangular tanks, 146–147
  of round tanks, 146
  standards of, 12–13
    calibrations of, 13
  units of, 22t, 28t, 30t
    equivalents, 28t
Carat, 4, 5
Carbon footprint, of glass studio, 619
Carbon steel, 267
Cargo tonnage, 13
Carnot cycle, 117t
Carpet
  amount required, calculating, 567
  berber, 566

cut pile, 566
  saxony, 566
  textured saxony, 566
  frieze, 566
  loop pile, 566
    level, 567
  multilevel loop, 567
  patterned, 566
    cut and uncut, 566
  sound and, 541
  STC and, 540–541
  types of, 565–567
Cash-flow series calculations, 138
Casing nails, 444–445
Cathedral ceilings, square footage for, 560
Cathedral walls, 558–559
Catwalk grating, 342–343
CBR. *See* California bearing ratio
CCA. *See* Chromated copper arsenate
Ceilings
  cathedral, 560
  materials, absorption factors of, 529t
  painting, 565
  square footage of, 559, 560
  tile, 564–565
    grid-system components, 564
    installation, 564–565
Celsius, Fahrenheit
  conversion from, 53t
  conversion to, 54t
Cement
  blended, 219–220, 219t, 229
  composition, 217–218
  concrete and, 219t
  content, 215
  expansive, 220–221
  portland cement concrete, 213
    ASTM types, 216–217
    chemical admixtures and, 212–213
    modified, 220–221
    physical properties of, 218–219
    types of, 216–217, 217t
    waterproof, 263–264
  types of, 221
  water-cement ratio, 215, 215t
Cemented soil, 163
Centimeter, inch
  conversion from, 35t
  conversion to, 36t
Centripetal force, 117t
Ceramic tile, 563–564
CGPM. *See* General Conference on Weights and Measures
Channels
  aluminum, 331–332
  C, 280

dimensions of, 323–324
  weight and size of, 319–320, 321, 323
hat, 280
hemmed, 280
J, 280
lipped, 280
plain, 281
resilient, 537
  STC ratings and, 542–543
ship and car, 322
stainless steel, 280
steel, 280
  applications of, 280–281
  identifying, 282–283
  weight and size of, 319–320
U, 280
Charging, 117t
Chemicals
  additives, 212–213
  admixtures, 218
  concrete, 212–213
  paint removing, 579
  surfactants, 223
Chipped grain, 386
CHP. *See* Combined heat and power
Chromated copper arsenate (CCA), 453
Church of St Mary-at-Hill, 454
CIPM. *See* International Committee of Weights and Measures
Circle
  area and circumference of, 83–84, 89–90
    givens, 83–84
    goals, 83
    knowns and assumptions, 84
    method, 84
  radius of
    circumscribed, 104
    inscribed, 104
Circuit
  impedance of, 640
  phases of, 641
  RC, 117t
Circular motion, 131
Circular unbanked tracks, 117t
Circumference, of circle, 83–84, 89–90
Clay, 160
  adhesives, 562
  fat soil, 163
  silt-clay materials, 171–172
Cleaners, of wood, 578
Clear adhesives, 562
Climate, 173
Clinkers, 216
Clothes washers, water usage by, 599
CLYDE rail spike, 461
CMUs. *See* Concrete masonry units
Coarse thread series (UNC/UNRC), 492

Coarse-grained soils, 159
Coarseness factor chart, 223
Coated steel fasteners, 451–452
Coating, 337t
   galvanized, 336–337, 452
   types of, 452
Cobble soil, 163
Coefficient of performance (COP), 625, 650–652
Cofunctions, 113–115
Cohesionless soil, 162t
Cohesive soil, 162t
Coil steel, 336
Coiled Rebar, 243–244
Coke, 266
Cold joints, 232
Cold rolled steel, 336–337, 341t
Cold rounds, 329
Color
   change, of wood, 573
   of soil, 161
   sorting, of lumber, 383
Columns, 374
Combination square, 583–584, 587
Combined heat and power (CHP), 612–615
   conventional generation *vs.*, 613
   efficiency of, 613–614
   system, 613
      efficiency, 613
Common excavation, 156
Common Roof-Framing Errors, 374
Common wire nails, 442
Common wire spikes, sizes, 444
Compaction
   equipment types, 186–189
   measuring devices, 189
   roller-type, 185
   of soil, calculating, 177–179
   tests, 177–179
Compound
   calculator, 549
   interest, 109
Compression
   parallel to grain, 355
   perpendicular to grain, 354–355
Compressive strength, 215t
Comsel grade, 384
Concentric strandings, 661–662, 662t
Concrete
   admixtures, 212–213
      chemical, 212–213
   block walls, 517–518, 517t
   cement and, 219t
   Class B, 233
   Class BZ, 233
   Class D, 233

Class DT, 233
Class E, 234
Class H, 234
Class HT, 234
Class P, 234
Class S35, 234
Class S40, 234
Class S50, 235
compressive strength, 221–222
fly ash, 229
hardened, 228
inspection of, 239
lightweight, mix design, 235–236
perlite, 235–236
   general considerations of, 236
   mix designs of, 235t
   mix instructions of, 236
portland cement concrete, 213
   ASTM types, 216–217
   blended, 219–220, 219t, 229
   chemical admixtures and, 218
   modified, 220–221
   physical properties of, 218–219
   waterproof, 263–264
production of, 212
ready-mix, 238
reinforcing bars
   designations, 244
   size and weight of, 243
retarders and, 237
self-stressing, 220
small batches of, 238
   calculations for, 239
strength, 222
structural, 233–235
superplasticizers and, 225
types of, 216–217, 217t
walls, 513
workability of, 228
Concrete masonry units (CMUs), profiles and dimensions of, 258–259
Conduction, 504
Conductivity, 430t
   heat, 506
   thermal, 505
Conduit
   inside diameters, 665–666
   weight comparisons, 667
Cone
   area of, theorem, 93
   surface area of, 99
      total, 99
   volume of, 87
      frustum, 88
      theorem, 93
Constant rise, 200

Constant-acceleration
   circular motion, 117t
   linear motion, 117t
Continuity, of fluid flow, 117t
Contract wall covering, 554–555
Convection, 504
Conversion factors, 57t
   of BF, 382t
   cooking and, 73–76
Cooling
   cost of, 622–627
   electric, 629t
   indices, 629t
   systems, fuel costs of, 621, 622–630
COP. *See* Coefficient of performance
Copper, 147
   ACA, 453
   ACZA, 453
   aluminum, conversion to, 663–665
   CCA, 453
   fasteners, 452–453
Core, 572
Cork, 555
   veneer, 555
Corn heating, costs of, 628t
Corrosion
   crevice, 453
   inhibitors, 213
   of metal, 451
   in preservative-treated wood, 453
   protection, 450
   resistance to, 267, 451
   in untreated wood, 453
Cosine, 100, 101
   law of, 102, 111, 112–113, 125
Coulomb's law, 117t
Council of American Building Officials (CABO), 435
Coursing
   horizontal, 256–258
   vertical, 255–256
Critical lighting, 550
Cross tee, 564
Crossing plank, 390–391
Crown Hand Tools, 583
Cube
   surface area of, 99
   volume of, 85
Cubit, 5
Cuboid, surface area of, 99
Customary units, 9
Cuts
   configurations of, 202f
   of lumber
      plain sawn, 353
      quarter sawn, 352
      rift sawn, 352–353

Index

Cut floor brads, 456
Cut nails, 448–449, 455–456
Cut pile carpet, 566
  saxony, 566
    textured, 566
Cutting boards, finishes for, 577
Cylinder
  area of, 87, 139
    theorem, 92
  volume of, 87, 139
    theorem, 92
Cylindrical tanks
  capacity of, 138–139
  volume of, 138–141
  work sheet, 139–141

D

Danish oils, 575
Davis, R. E, 226
de Broglie matter waves, $117t$
Dead weight tonnage, 14
Decibels
  comparison chart, 533–534
  levels, of common sounds, 532
  scale, $117t$
Decimals
  converting to, 340–341
  subdivisions, 9
Décor nail, 460
Deformed shank nails, 445–446
Demand-side management (DSM), 632
Density
  calculating, 205–210
  dry, 190–193, $192f$
    in situ, 193–196
    wet *vs.*, 208–210
  formula, $117t$
  of material, $204t$
  MDF, 437, 438
  moisture and, 190
  saturation vapor, 612
  of soil
    measurement of, 194, $196t$
    moisture content *vs.*, $179f$
    tests, 179–180
  vapor, 616
  wet, $209f$
    dry *vs.*, 208–210
  of wood, $371t$, $372t$
Department of Agriculture, U.S, 1, 174, 462
Department of Energy, U.S, 591–592
Depth, measuring, 585–586
Derived units, 7
Deutsches Institute für Normung (DIN), 275
Dewpoint, 612, 616
Dicalcium silicate, 217
Dielectrics, 655–656

Difference formulas, 106
Digital caliper, 585–586, 587
Digital wall covering, 555
Dimensions
  actual, 253
  of bricks, 253
  of C channels, 323–324
  change, coefficients of, 405–407
  of CMUs, 258–259
  green, 407
  linear, 582
  nominal, 253
  of round high-strength steel, 317–318
  soft and hard metric, 257–258
  specified, 253
  of stress-graded lumber, 369
  of wood screws, $466t$
DIN. *See* Deutsches Institute für Normung
Direct pressure laminate (DPL), 571
Discharging, $117t$
Dishwashers
  specifications of, 600
  standards of, 597
Displacement tonnage, 14
Distance formula, 109
Distribution transformers, 668
Distributive function, 81
Dolomitic limestone, 263–264
Doppler effect, $117t$
Double angles, 106
Double-stud walls, 538
DPL. *See* Direct pressure laminate
Dram, 9
Dressed lumber, 378–379, $379t$
Drives
  Phillips/crosshead, 467
  Pozidriv, 468
  slotted/flathead, 467
  square, 468
  types of, 467–468
Dry density, 190–193, $192f$
  in situ, 193–196
  wet density *vs.*, 208–210
Dry soil, 163
Dry type transformers, 669
Dry volume, units of, $17t$
Drying oils, 576
Drywall
  additional layers of, 537
    STC ratings and, 542–543
  finishing guide, 550–552
  gypsum panels, 546
  scope, 550
  sealer, 550
  terminology, 550–552

DSM. *See* Demand-side management
Duodecimal subdivisions, 9
Durometer hardness test, 150

E

Earth
  moving
    equipment production, 183–185
    formulas for, 184–185
    quantities hauled, 184
Eating utensils, 576–577
Economy grade lumber, 380
EER. *See* Energy efficiency ratio
Effective electric efficiency, 614–615
  metric, 615
Efficiency, $117t$
  AFUE, 650–652
  of appliances, 650–652
  calculating, 612
  of CHP, 613–614
  EER, 650–652
  effective electric, 614–615
  of equipment, 650–652
  of fuel conversion, 618
  of heat engine, maximum, $117t$
  overall, 613
  of plumbing, 605
  ratios, 650–652
  SEER, 650–652
  thermal, 614
  total system, 614, 615
  of water, 602
Egypt, 4
8-Thread series (8UN), 492
Elasticity, modulus of, 355
Electric cooling, $629t$
Electric field, $117t$
Electric generators, 652
  safety and, 653
Electric heating, $625t$
Electric power, $117t$
Electric tile saw, 563
Electric water heaters, 610
Electrical conductor areas, 665–666
Electrical terms, symbols for, 637
Electricity, 135–136
Elizabeth (queen), 5
Emergency generators, 653–654
EN. *See* Euronorom
Enclosures, 670–671
Energy, 132–133
  consumption of, 645
  heat, 504
  home heating, 619–621
  kinetic, 133
    rotational, $117t$

Energy (*Continued*)
  mass-energy equivalence, 117*t*
  mechanical, 117*t*
  of particle, 117*t*
  of photon, 117*t*
  potential, 117*t*
  released, by fusion, 117*t*
  savers, 644–645
  SEER, 650–652
  thermal, 616
  total fuel input, 614
  usage of, 644
    chart, 646–647, 646*t*
Energy efficiency ratio (EER), 650–652
Energy Policy Act, 591–592, 602*t*, 603–604
EnergyGuide Label, 609
Engineers, 10
Engineer square, 587
English water ton, 13
Entropy, 117*t*
Environmental factors, 488
Environmental noise, 533*t*
Environmental Protection Agency, 595
Equilateral triangle, 94–95
Equilibrium moisture content, of wood, 404
Equipment, efficiency ratings of, 650–652
Equivalents, 25–31
Euronorom (EN), 275
Evaporation, 616
Excavated materials, use of, 157
Excavations
  borrow, 157
  common, 156
  limits, 157
  measurement and payment, 158
  over-excavation, 157
  rock, 156
  rudiments of, 156–159
    classification, 156
    scope, 156
  unclassified, 156
Executive Order 12770, 34
Expanded metal grating, 341
  aluminum, 342–343
Expansion, thermal, 117*t*
Expansive cement, 220–221
Exponential functions, 84–85
Exterior plywood, grade names of, 433, 434*t*

F

F1F. *See* FAS one face
Fabric-backed vinyl, 554
  wall coverings, 556
Face brick, 248–249

Fahrenheit
  Celsius
    conversion from, 54*t*
    conversion to, 53*t*
  Rankine
    conversion from, 55*t*
    conversion to, 54*t*
Falk, Robert H, 449
FAS grade, 384
FAS one face (F1F), 384
Fast thread, 465–466
Fasteners
  aesthetics and, 488
  aluminum, 452–453
  of bolts, for structural steel, 488–492
  calculator, 549
  coated steel, 451–452
  copper, 452–453
  corrosion resistant, 451
  environmental factors of, 488
  in hollow masonry units, 484–485
  identification markings, 488*t*
  powder-driven, 486–487
  recommended, 454
  selection chart, 487–488
  in solid masonry units, 485–486
  stainless steel, 452–453
  tension, 501
  types of, 450–451
  usage of, guidelines, 452
  in wood structures, 449–454
Fat clay soil, 163
Feet, 5, 8
  board feet, 381, 382, 382*t*
  inch, conversion from, 34*t*
  meter
    conversion from, 38*t*
    conversion to, 37*t*
  square feet
    acres and, 43*t*, 44*t*
    square inch and, 40*t*, 41*t*
    square mile and, 42*t*, 43*t*
Fiber stress
  in bending, 353
  in tension, 354
Fiberglass batt insulation, 510*t*
Fibrous peat, 160
Field Guide to Common Framing Errors, 374
Field tile, 563
Figured wood, 390
Fill
  configurations of, 201*f*
  soil, 163
Fillers, 574–575

Filler yield, 552–553
Financial formulas, 136–138
  symbols and variables in, 137*t*
Fine thread series (UNF/UNRF), 492
Fine-grained soils, 160
Finish lumber, 361–362
Finishes
  for butcher blocks, 577
  for cutting boards, 577
  for eating utensils, 576–577
  flat oil, 575
  of floors, 575–576
  interior, 545–587
  opaque, 573
  removing, mechanical methods of, 579
  transparent, 573
  types of, 576–577
Finishing nails, 445
First Law of Thermodynamics, 117*t*
Fissured soil, 163
Flat countersunk head spike, 458
Flat oil finishes, 575
Flat/countersunk screw heads, 468
Flexible curves, 586
Flexural strength, 232
Flitches, 396
Flocks, 556
Floor joists, 373
  cuts, maximum sizes for, 374*t*
  softwood, 373–375
Flooring
  cut brads, 456
  finishes of, 575–576
    durability, 576
  laminate, 570–572
    commonly used terms, 570–572
    construction, 571–572
    installation systems, 572
    locking systems, 571
    styles, 571
    thickness, 571
    types of, 571
  materials, absorption factors of, 528*t*
  particleboard, 437
  solid hardwood, 568
  tile, 562–564
    resilient, 562–563
Fluid flow, continuity of, 117*t*
Flux, 266
Fly ash, 226–229
  classifications of, 227
  highway construction and, 229
  mix design of, 227–229
  specifications of, 227

Index

Foils, 555
Force, 130
   centripetal, $117t$
   friction, $117t$
   Newton's equation of, 130
   WSFUs, 590
Formulas
   absolute value, 108
   of air heating, 617
   of angles, 96–97, 198
   of area, 104
   completing the square, 108
   of density, $117t$
   of difference, 106
   distance, 109
   double angle, 106
   financial, 136–138
   future value, 138
   geometric, 85–91
   of grade, ratio, and angle relationships, 198
   of heating, 617
   Heron's, 104
   midpoint, 109
   Mollweide's, 103
   for moving earth, 184–185
   Newton's, 103–104
   of physics, 117–130
   power-reducing, 106
   present value, 136–137
   of pressure, 79
   product-to-sum, 107
   quadratic, 107, 124
   sum, 106
   sum-to-product, 107
   of surface area, 99
   triangle inequality, 109
   of trigonometry, 102–109
   for wall coverings, 556–558, 559
   of weights, $117t$
   of work, $117t$
Foundations, settlement under, 181–182
Frame grade, 389
Frame spacing, $548t$
Framed wall assemblies, 521–524
Framing
   common errors of, 374
   metal, 520
   wood-metal, 519
Framing lumber, 367–440
   design values of
      beams and stringers, $367t$
      posts and timbers, $368t$
   sizes of, 360
Freeze-thaw durability, 232
French curves, 586

Frequency, 640
Frequent soil, 163
Friable soil, 163
Friction force, $117t$
Frieze carpet, 566
Frustum
   surface area of, 88–89
   volume of
      cone, 88
      pyramid, 87
Fuel
   AFUE, 650–652
   consumption of, 619
   conversion, efficiency of, 618
   costs of, 621, 622–630, $624t$
   oil heating, $624t$
   price of, 622
   total energy input, 614
Furlong, 5
Future value formulas, 138

G

Gabion retaining walls, 203
Galvanized coatings, 336–337, 452
Galvanized steel, $340t$
Gap, measuring of, 585–586
Gas
   heating, $117t$
   ideal gas law, $117t$
   meters, 619
   natural, $622t$
   water heaters, 610
   work and, $117t$
Gauge inches, converting to decimals, 340–341
General Conference on Weights and Measures (CGPM), 6, 7
Generators
   electric, 652
      safety and, 653
   emergency, 653–654
   left hand rule, 654
   right hand rule, 654
Geographic location, 453–454
Geometric formulas, 85–91
Geothermal heat pump systems, 625
Glass bead method, 207–208
Glass block mortar, mixture calculations for, 263
Glass studio, carbon footprint of, 619
Glazed tile, 563
Glueless locking system, 572
Grade
   cabinet, 380
   calculating, 197–198
   comsel, 384
   economy, 380

FAS, 384
   frame, 389
   maps, 197
   MSR, $431t$
   No. 1C, 380
   No. 2AC, 380
   of particleboard, $436t$, 437, $437f$
   of plywood, 433, $434t$, 435, $435f$
   prime, 384
   relationship formulas, 198
   rules of, 388–389
   stress-graded lumber, 358–360, 369
Grain, 5
Granular materials, 170–171
Graphing, 81
Grating
   bar, 348–349
      rectangular, 344–345, 346–347
   catwalk, 342–343
   expanded metal, 341
      aluminum, 342–343
   plank, 348
   rectangular bar, aluminum, 344–345, 346–347
Gravel, 159, 170
   group names for, 168
Gravitation, 132
   universal, $117t$
Gravity, 462
   acceleration and, 132
   of wood, specific, 462
Greek Parthenon, 4
Green Building Council, U.S, 602
Green wood
   dimensional change of, 407
   hardwoods, 386–388
   moisture content of, 370–372
Grid-system components, 564
Gross tonnage, 13
Ground cable, 667–668
Ground granulated blast-furnace slag, 231–233
   batching, 233
   handling, 233
   mix design of, 231–232
   specifications of, 231
   storage, 233
Ground source heat pump (GSHP), 630
   economics of, 632
   paybacks for, 631, 632–633
   savings of, 631
   in schools, 630
   utility programs, 632–633
Ground-loop heat pump, 625
Groundwater heat pump, 625, $626f$
Grout, 563

GSHP. *See* Ground source heat pump
Gum
   black, quartered, 389
   red
      plain, 389–390
      quartered, 390
      sap, quartered, 389
Gunter's units of measurement, 16t
Gypsum
   board finish
      levels of, 551–552
      resources, 552
   drywall panels, 546
   wall panel, coverage calculator, 547–548
   wallboard, panel sizes and thickness, 547

H

Halstead, 229
Hand Test, 180
Hardboard
   property requirements of, 438, 438t
   siding, properties of, 439–440
Hardness, 148–150
   Janka Scale, 568–570
   scales, comparison of, 151
   testing for, 147–151
      Barcol, 150
      Brinell, 148
      comparison chart, 152t
      Durometer, 150
      Knoop, 149
      Mohs, 148
      Rebound, 150
      Rockwell, 148–149
      Rockwell (superficial), 149
      Scleroscope, 150
      Vickers, 149
Hardware, recommended, 454
Hardwood, 381–384
   African, 397
   classification of, 575t
   construction, 391–392
   drying of, 408
   hearts, 391
   mixed, 391
   North American, 397
   Philippine, 396–397
   solid flooring, 568
   strengths of, 398–403
   tropical American, 397
   weights of, 386–388
Harmonic motion, 117t
Hat channels, 280
Heat
   CHP, 612–615
   conductivity, measure of, 506
   conventional generation of, 613
   energy, 504
   flow, resistance to, 506
   loss of, 620, 621
   SHP, 612–613
   transmission
      coefficient of, 505
      modes, 504
Heat pumps
   air-source, 624
   geothermal, 625
   ground source, 630
      economics of, 632
      paybacks for, 631, 632–633
      savings of, 631
      in schools, 630
      utility programs, 632–633
   ground-loop, 625
   groundwater, 625, 626f
   water heaters, 609–610
Heating
   of air, 617
   annual cost of, 621, 627–628, 629–630
   appliances, fuel conversion efficiency of, 618
   corn, 628t
   cost of, 622–627
   delivered, 627
   electric, 625t
   formulas of, 617
   fuel costs, 621, 622–630, 624t
   gas, 117t
   home, 619–621
   indices, 629t
   kerosene, 626t
   liquid, 117t
   natural gas, 622t
   pellet, 628t
   propane, 623t
   solids, 117t
   wood, 628t
Heating seasonal performance factor (HSPF), 624, 650–652
Heavy Duty plank grating, 348
Heavy ripping equipment, 156
Hectares
   acres, conversion from, 39t
   square mile
      conversion from, 48t
      conversion to, 48t
Helically nails, 444t
Hemmed channels, 280
Heron's formula, 104
Hexagon, 99–100
High calcium limestone, 263–264
High-pressure laminate (HPL), 571

Highways. *See also* American Association for State Highway Transportation Officials
   construction of
      fly ash concrete in, 229
      soil classification for, 169–172
   projects, 497
   VHTRC, 229
Hindus, 10
Holding power, 450
Hollings Manufacturing Extension Partnership, 1
Hollow masonry units, fasteners in, 484–485
Hollow unit masonry walls, 514–515
Home electronic energy use, 644
Home heating energy, 619–621
Hooke's Law, 117t
Horizontal coursing, 257–258
   calculating, 256–257
Horizontal shear, 354
Horsepower
   amps, converting to, 636
   brake, 79
   power unit, 79–80
   water, 79
Hot rolled rounds, 329
Hot rolled steel, 340t
   shape designations, 64t
Hot-pressed plywood, 407
HPL. *See* High-pressure laminate
HSPF. *See* Heating seasonal performance factor
Humidity, 454
   relative, 611–612, 616
   saturation, 611–612
Hundredweight, 5–6, 34
Hungry Horse Dam, 226
Hydrated lime, 260

I

I beams, 293t
   weight and size of, 293–294
Ice makers, 601
Ideal gas law, 117t
Identities
   basic, 105
   Pythagorean, 105, 110
   quotient, 110
IIC. *See* Impact insulation class
Impact insulation class (IIC), 527–528, 543–544
Impedance, of circuit, 640
Imperial bushel, 18–19
Imperial gallon, 18–19
Imperial system, 34
Impulse, 117t, 134
   angular, 117t

Index

Inch, 5, 9
  centimeter
    conversion from, 36t
    conversion to, 35t
  feet, conversion to, 34t
  gauge, 340–341
  psi, 222
  square, 40t, 41t
  TPI, 492–494
Inch-pound, 420t
Incra Precision Marking Rules, 582, 587
Incra Precision Protractor, 587
Incra Precision T-Rule, 587
Index of refraction, 117t
Induced voltage, 117t
Inductive reactance, 639
Inductors, 117t
  networks, 638
Industrial Clears, 385
Inertia, 117t
Infrastructure, 226
Inner core layer, 570
Innovative & Design Process points, 605
Innovative Waste Water Technologies, 602
  baseline case, 603t
  design case, 603t, 604t
  intent of, 602
  requirements of, 602–603
In-register embossing, 570
Insulation
  approximate coverage, 518t
  basic types of, 506–507
  batt, 510
    fiberglass, 510t
  blanket, 510
  IIC, 527–528, 543–544
  installation of, 518
  partitions, STC of, 538–540
  roll, 510
  STC ratings and, 542–543
Interest rates, converting, 138
Interior
  finishes, 545–587
  plywood, grade names of, 433
  trim, 581–587
  wood, finishing of, 572–573
Internal combustion units, 80
International Bureau of Weights and Measures (BIPM), 6, 7–8
International Committee of Weights and Measures (CIPM), 6
International Metric Convention, 6
International Prototype Kilogram, 7
International Prototype Meter, 7
International System of Units (SI), 6–7, 34
  derived units, 125t
  fundamental units of, 125t

Inverse functions, 109
Iron, 147
  molten, 266
  ore, 266
The Ironmonger, 454
Isolation, 527

J

J channels, 280
Janka Hardness Scale, 568–570
Japanese Industrial Standards (JIS), 275–277
Jefferson, Thomas, 455–456
JIS. *See* Japanese Industrial Standards
Joints, 478–479
Joint photographing, 550
Joint telegraphing, 550
Joist hangers, 451
  nails, 447
Joists
  floor, 373, 374t
  section properties of, 365–367
  softwood floor, 373–375
Joule, 504
Journal of Light Construction, 579
J-Square Center Finder, 586
Junior beams, 293–294, 294t

K

KE. *See* Kinetic energy
Kepler, Johannes, 132
Kerosene heating, 626t
KEY to METALS database, 276
Kiln-dried hardwood, 386–388
Kiln-dried lumber, 382
Kilocalorie, 504
Kilogram, 7
  pound
    conversion from, 51t
    conversion to, 52t
Kilometer, mile
  conversion from, 49t
  conversion to, 50t
Kilovolt-amperes (KVA), amps, converting to, 637
Kilowatts (KW), amps, converting to, 637
Kinetic energy (KE), 133
  rotational, 117t
Kirchhoff's Law, 117t
Kitchen equivalents, 73t
Knoop hardness test, 149
Knots, avoiding, 573
KVA. *See* Kilovolt-amperes
KW. *See* Kilowatts

L

Lacquer, 575
Lag screws, 450
  lateral resistance of, 473–475
  penetration by, 474f
  withdrawal resistance of, 473–475
Laminate, high-pressure, 571
Laminate flooring, 570–572
  commonly used terms, 572
  construction, 571–572
  installation systems, 572
  locking systems, 571
  styles of, 571
  thickness, 571
  types of, 571
Laminated soil, 163
Lateral resistance, 473–475
Law
  Coulomb's, 117t
  Hooke's, 117t
  ideal gas, 117t
  Kirchhoff's, 117t
  Lenz's, 117t
  of metric conversion, 34
  Ohm's, 117t, 135, 637
  Poiseuille's, 117t
  radioactive decay rate, 117t
  Snell's, 117t
  Watt's, 638–639
Law of Cosines, 102, 111, 112–113, 125
Law of Motion, 129, 133–134
Law of Physics, 131
Law of Sines, 102, 111, 112, 125
Law of Tangents, 103
Law of Universal Gravitation, 132
Layback, 200
Layer soil, 163
LEED rating system, 602
Left hand generator rule, 654
LEM. *See* Linear equivalent mass
Length
  of bolts, 499–500
  of lumber, 56t
  relativistic contraction, 117t
  of screws, 467
  standards of, 10–11
    calibration of, 11
  units of, 15t, 16t
    equivalents, 26t
    international measure, 19t
    survey measure, 20t
    wavelength, 640
Lens soil, 163
Lenz's law, 117t
Light Series plank grating, 348
Lighting, critical and severe, 550

Lightweight concrete, mix design, 235–236
Lime
  hydrated, 260
  mortar
    mixture calculations for, 262–263
    types, 263–264
  putty, 260
Limestone, 263–264
Linear dimensions, measuring and marking, 582
Linear equations, 82
Linear equivalent mass (LEM), 128–129
Linear momentum, 117t
Linear sizes, 97–98
Linear speed, 117t
Lines, 109
Lipped channels, 280
Liquid
  conversion of, 58t
  heating, 117t
  measure, 9
    equivalents, 74t
Liquid volume, units of, 15t
  apothecaries, 17t
Liter, 7
Load-bearing test, 178
Loads, multiplication factors for, 475t, 476t
Logarithms, 108
  common, 114t, 116
Long ton, 13, 34
Loop pile, 566
  level, 567
Lorentz transformation factor, 117t
Loudspeakers, 531
Low-carbon sheet steel, 336
Lumber
  appearance, 363–365
    grades of, 363t
    standard signs of, 364t
  board feet, in bundle, 382
  cabinet grade, 380
  Canadian standards of, 375
  color sorting of, 383
  cuts of
    plain sawn, 353
    quarter sawn, 352
    rift sawn, 352–353
  dressed, 378–379, 379t
  economy grade, 380
  FAS grade, 384
  finish
    minimum dressed sizes of, 361–362
    nominal sizes of, 361–362
  framing, 367–440
    beams, 367t
    design values of, 367t, 368t
    sizes of, 360
  grade of
    determining, 383–384
    rules, 388–389
  industry abbreviations, 356–358
  kiln-dried, 382
  length, conversion of, 56t
  mine, 391
  MSR, 431–433
    design values of, 431t, 432–433, 432t
    grade designation of, 431t
  No. 1C, 380
  No. 2AC, 380
  patterns of, 375
  planed, 382
  producers council, 432
  purchase of, 377
  red one face, 383
  ripping of, 373
  rough sawn, thickness for, 381
  sap, 383
  sizes for, 385–386
  softwood, 356, 375
    drying of, 408
    grading, 375–376
    species of, 358–360, 375
  stress-graded, 358–360
    dimensions of, 369
  surfaced, 385–386
    thickness for, 382
  worked
    dressed sizes of, 362t, 377
    typical, 375–377
Lumber Standard, 375

M

Machine stress-rated lumber (MSR lumber), 431–433
  design values of, 431t, 432–433
    psi, 432t
  grade designation of, 431t
Magnetic field, 117t
Magnification, 117t
Mahogany
  African, 392
  Philippine, 394–395
    shorts, 395–396
  pin wormy, 392–393
  shorts, 393, 395–396
  strips, 394
Main tee, 564
Manufacturing imperfections, 386
Maple, Pacific coast, 397
Maps
  grade, calculating, 197
  of soil, 175–177

Marking
  angles, 584
  kit, 586–587
  of linear dimensions, 582
  tools, 581
Marl, 161
Masking systems, 530–531
Masonry
  brick, 264
  concrete units, 258–259
  hollow units, 484–485
  solid units, 485–486, 513
Masonry wall
  assemblies, 516
    R-value calculations for, 509
  solid loadbearing, 509
  STC ratings for, 540–544
Mass
  center of, 117t
  linear equivalent, 128–129
  relativistic increase, 117t
  standards of, 11–12
    calibrations of, 12
    maintenance and preservation of, 13
  true, 12
  units of, 5–6, 15t, 24t
    apothecaries, 18t
    equivalents, 31t
    Troy, 18t
  weight and, 11–12
Mass-energy equivalence, 117t
Masteroaqe Universal Angle Guide, 585, 587
Mathematics, basic, 81–82
Maximum aggregate size, 214
MDF. *See* Medium-density fiberboard
Mean free path, 616
Measurement, 4
  of angles, 584–585
  British units of, 18–19
  of depth, 585–586
  of excavation, 158
  of gap, 585–586
  kit, 586–587
  knowledge of, 4
  of linear dimensions, 582
  liquid, 9
  of power, 652–653
  of soil, 194, 196t
  of squareness, 583
  standards of, 3, 4
  systems of, 3–14
    early history of, 4
  of thickness, 585–586
  ton, 13

Index

tools of, 581
units of, 4–10
   Gunter's, 16*t*
   U.S, 16–18
for wall coverings, 557
of water, 79
Mechanical advantage, 117*t*
Mechanical energy, 117*t*
Medium-density fiberboard (MDF), 437
   property requirements of, 438
Melamine, 572
Member thickness, 478, 479*f*
Mesh
   size, 65–67
   welded wire, 244–245
Metal
   common materials from, 147
   corrosion rate of, 451
   expanded grating, 341
      aluminum, 342–343
   framing, 520
      wood-metal, 519
   hardness, testing for, 147–151
   KEYS to METAL database, 276
   scraps, 451
   stud walls, 537, 538
   wood, interaction with, 451
Meter, 7, 10
   feet
      conversion from, 37*t*
      conversion to, 38*t*
   square meter
      square mile and, 46*t*, 47*t*
      square yard and, 45*t*
   tri-meter square, 583, 587
Metric Conversion Act, 8
   effective electric efficiency, 615
Metric Conversion Law, 34
Metric single roll wall coverings, 557
Metric system, 6–8
   C channels, weight and size of, 323
   conversions, U.S, 274–275
   dimensions, soft and hard, 257–258
   equivalents, 75*t*
   International Metric Convention, 6
   rectangular high-strength steel
      sections, weight and size of, 306–310
   round high-strength steel
      dimensions and section properties of, 317–318
      weight and size of, 315–316
   square high-strength steel sections, weight and size of, 299–302
   standards of, 7
   total system efficiency, 615

units of, 7, 14–15
   in U.S, 8
   conversions, 274–275
wide flange beams, weight and size of, 289–293
wood
   mechanical properties of, 416–420, 426–429
   strength properties of, 416*t*
Metric ton, 13
Micropowder grit, 65*t*
Microsilica, 229
Midpoint formula, 109
Mile, 10
   kilometer
      conversion from, 50*t*
      conversion to, 49*t*
   square mile
      hectares and, 34*t*, 48*t*
      square feet and, 42*t*, 43*t*
      square meter and, 46*t*, 47*t*
Millwork, 581–587
Mine lumber, 391
Mineral admixtures, 224
Miter gauges, 586
Mixing water, 214, 214*t*, 215
Modular bricks, 249, 252–253
Modulus, of elasticity, 355
Mohs hardness test, 148
Moist soil, 163
Moisture content, 451
   aggregate, 215
   equilibrium, of wood, 404
   of green wood, 370–372
   recommended, 407–409, 408*t*
   of soil, 161, 179, 190–193
      density *vs.*, 179*f*
Moisture density relation, 190
Mollweide's formulas, 103
Molten iron, 266
Momentum, 133
   angular, 117*t*
   change in, 117*t*
   linear, 117*t*
Monomials, 81
Mortar
   glass block, 263
   lime
      mixture calculations for, 262–263
      types, 263–264
   minimum compressive strengths, 264
   mixes, 260
      types of, 260–262
   pointing, 262
   Type K, 262
   Type M, 261
   Type N, 261

Type O, 262
Type S, 261
Motion, 117*t*
   circular, 131
   laws of, 129, 133–134
   two-dimensional, 134–135
   three-dimensional, 134–135
Mottled soil, 163
Moulder brad, 457
MPa (megapascals), 222
MSR lumber. *See* Machine stress-rated lumber
Muck. *See* Amorphous peat
Multilevel loop carpet, 567
Multiple-member joints, 478–479
Music, sound levels of, 535–537
Mylar, 555

N

Nails, 442–443, 450
   air-nailing gun, 456
   annular, 442–443
      threaded, 444*t*
   boat, 460
   box, 442–443, 443*t*
   bright common, 442–443
   casing, 444–445
   common wire, 442
      spikes, 444
   cost of, 455
   cut, 448–449, 455–456
   décor, 460
   deformed shank, 445–446
   diameters of, 464–465
   finishing, 445
   helically, 444*t*
   history of, 455–456
   joist hanger nails, 447
   nomenclature for, 455
   penny, 454–455
   roofing, 446–447
   selection and usage of, 449
   smooth box nails, 443*t*
   square, 455–456
   types of, 442*f*
   wire, 456
   withdrawal resistance of, 443, 461–464, 462*t*
Nailers, 455
NALFA. *See* North American Laminate Flooring Association
National building code (BOCA), 373
National Cooperative Soil Survey, 173
National Design Specification (NDS), 373, 450

National Evaluation Service (NES), 435
National Institute of Standards and Testing (NIST), 3, 6
 founding of, 1
 Handbook 44, 1, 2f
Natural gas heating, 622t
Natural textile wall coverings, 555
NDS. *See* National Design Specification
NES. *See* National Evaluation Service
Net tonnage, 14
Net useful power output, 614
Net useful thermal output, 614
New York State Department of Public Works, 223
Newton, Isaac
 force equation of, 130
 formulas of, 103–104
 Law of Universal Gravitation, 132
 Laws of Motion, 133–134
 Laws of Physics, 131
 Second Law of Motion, 117t, 129
Newtons, 130, 222
N-gon, angles of, 99–100
Nickel, 147
NIST. *See* National Institute of Standards and Testing
No. 1C. *See* Number 1 Common grade
No. 2AC. *See* Number 2A Common grade
Noise
 environmental, 533t
 levels, 535t
  perception of increases, 534
  permissible exposure, 534
Noise reduction coefficient (NRC), 525
Nominal dimensions, 253
Nondrying oils, 577
Nonmodular bricks, 249
North American hardwoods, 397
North American Laminate Flooring Association (NALFA), 571
Notching, 373–375
 guidelines, 373
 ripping of, 373
NRC. *See* Noise reduction coefficient
Number 1 Common grade (No. 1C), 380
Number 1 white, 383
Number 2 white, 383
Number 2A Common grade (No. 2AC), 380

O

Occasional soil, 163
Ohm's Law, 117t, 135, 637
Oil-type preservatives, 453
Old House Journal, 579

Omnibus Trade and Competitiveness Act, 8
One-Family Dwelling Code, 373
Opaque finishes, 573
Organic Mastic, 563
Organic soil, 163
OSB. *See* Oriented stand board
Oval screw heads, 468
Oven temperatures, 74t
Over-excavation, 157

P

Pacific coast maple, 397
Pacific coast red alder, 397
Padded rollers, 188
Paint, 550
 removing
  chemical methods, 579
  mechanical methods of, 579
 strippers, 578–580
  solvent-based, 579–580
Painting
 of ceilings, 565
 of walls, 565
 wood for, 580–581
Palm holdfast, 457
Pan screw heads, 469
Paper, 76t
Paper-backed vinyl, 554
Paraffin wax, 577
Parallelogram
 area of, 85
 theorems of, 95–96
 vector addition, 134–135
Parent material, 173
Particles
 energy of, 117t
 size, conversion chart, 66t
Particleboard
 flooring, 437
 grade
  requirements, 436t
  stamps, 437f
Parting soil, 163
Pascal's triangle, 107
Patterned carpet, 566
 cut and uncut, 566
Pavers, figuring, 251
Payment
 calculations, 137
 for excavation, 158
 number of, 138
PCA. *See* Portland Cement Association
PCC. *See* Portland cement concrete

Peat, 160
 amorphous, 161, 163
 fibrous, 160
Pellet heating, 628t
Pendulum, 117t
Penny nail, 454–455
Penny size designation, 443
Percentage, 198
Periodic waves, 117t
Perlite cavity fill, 517–518
Perlite concrete, 235–236
 general considerations of, 236
 mix designs, 235t
 mix instructions, 236
Permeance, to water vapor, 505
Phase change, heat of, 117t
Philippine mahogany, 394–395
Phillips/crosshead drive, 467
Photoelectric effect, 117t
Photographic image layer, 570
Photographing, joint, 550
Photon, energy of, 117t
Physics
 circular motion, 131
 concepts of, 130–131
 electricity, 135–136
 energy, 132–133
 formulas of
  basic, 117–130
  sheet, 117–124
 gravitation, 132
  universal, 117t
 laws of, 131
 power, 133
 quantum, 117t
 reference guide, 117–124
 work, 132–133
Pi, 86–89
Pickling, acid, 336
Picture Perfect Framing, 374–375
Pigments, 213
Pilot holes, 450
Pin wormy mahogany, 392–393
Pin wormy Philippine, 395
Pipes
 economical size selection, 80t
 flow capacities, maximized, 80–81
 plastic, friction loss, 80t
 steel, 319
Pith, 388, 390
Plain channels, 281
Plain red gum, 389–390
Plain sawn lumber, 353
Planed lumber, 382
Planks
 bridge, 390–391
 crossing, 390–391

Index

grating, 348
punch/pattern, 348–349
sections
    aluminum, 348–349
    pattern availability, 348
Plastic pipes, friction loss, 80t
Plasticizers, 212
Plate steel, 334–335
Plates
    splice, 334–335
    steel side, 478
    universal mill, 327–328
    vibratory, 186
        reversible, 186–187
Plating out, 453
Plenums, 530
Plumbing fixtures
    Uniform Plumbing Code, 590–591
    water
        efficiency of, 605
        usage rates of, 598
Plutarch, 4
Plywood
    exterior, grade names of, 433, 434t
    hot-pressed, 407
    interior, grade names for, 433
    softwood, 434–435
    typical grade stamps for, 435, 435f
Point, 5
Pointing mortar, 262
Pointing trowel, 563
Poiseuille's law, 117t
Polivka, 218
Polygons, 100t
Polynomials, 81
    derivatives of, 128
    integrals of, 128–130
    prefixes of, 128t
Polyolefin wall coverings, 555
Pomeroy, 217–218
Portland Cement Association (PCA), 212
Portland cement concrete (PCC), 213
    ASTM types, 216–217
    blended, 219–220, 219t, 229
    chemical admixtures and, 212–213
    mix calculation for, 260
    modified, 220–221
    physical properties of, 218–219
    types of, 216–217
        uses and, 217t
    waterproof, 263–264
Postulates, of special relativity, 117t
Potential energy, 117t
Pound, 5
    inch-pound, 420t
    kilogram
        conversion from, 52t
        conversion to, 51t
Pounds per square inch (psi), 222
Powder-driven fasteners, 486–487
Power, 133
    cable sizes, 667–668
    CHP, 612–615
    electric, 117t
    formulas, 117t
        power-reducing, 106
    holding, 450
    horsepower
        amps and, 636
        brake, 79
        power unit, 79–80
        water, 79
    measurement of, 652–653
    net useful output, 614
    SHP, 612–613
    transformers, 669
Power washing, 579
Pozidriv, 468
Precision fences, 586
Prefixes, 14–15
    of polynomials, 128t
Prepasted activators, 562
Present value formulas, 136–137
Preservatives
    oil-type, 453
    treated wood, 453
    waterborne, 453
Pressure, 131
    atmospheric, 616
    calculations and formulas of, 79
    direct laminate, 571
    formulas of, 79
    high-pressure laminate, 571
    lateral, 193
    saturation vapor, 616
    units of, 222
    vapor, 616
    water, 192
        under, 117t
Prestressed slabs, 222
Prime grade, 384
Primer, 550
Print film, 572
Prisms
    area of, 91–92
    right area, 91–92
    volume of, 90–91
Proctor Test, 179
    alternative methods, 190t
Product standards, 648
Production, earth-moving equipment, 183–185
Product-to-sum formulas, 107
Projectile motion, 117t
Propane heating, 623t
Properly painted surface, 550
Proportions, 81
Psi. *See* Pounds per square inch
Pumping aids, 213
Pusher tractor, 156
Pyramid, volume of, 90, 91
    frustum, 87
    theorem, 91
Pythagorean
    identities, 105, 110
    Theorem, 101, 109–110

Q

Quadratic equations, 82
Quadratic formula, 107, 124
Quality factor, 639
Quantum physics, 117t
Quarter, 5–6
Quarter sawn lumber, 352
Quenched steel, 267
Quotient identities, 110

R

Radiation, 504
Radioactive decay rate law, 117t
Radius
    of circumscribed circle, 104
    of inscribed circle, 104
Rammers, 186
Rankine, Fahrenheit
    conversion from, 54t
    conversion to, 55t
Ratios, 198
    California bearing, 178
    EER, 650–652
    reciprocal, 113
    relationship formulas, 198
    SEER, 650–652
    of slope, 200t
    trigonometric, 101–102
    water-cement, 215, 215t
RC circuits, 117t
Reactance
    calculating, 639
    capacitive, 639
    inductive, 639
Ready-mix concrete, 238
Rebound hardness test, 150
Reciprocal ratios, 113
Rectangular bar grating, aluminum, 344–345, 346–347
Rectangular high-strength steel sections, weight and size of, 302–305
    metric, 306–310
Rectangular tanks, capacity of, 146–147

Red alder, 382–383
  Pacific coast, 397
Red gum
  plain, 389–390
    sawn, 390
    quartered, 390
Red one face, 383
Reefer plank grating, 348
References, 239–243
Refining, 437
Reflections, 525
Refraction, index of, 117t
Register ton, 13
Register under-deck tonnage, 14
Relative humidity, 611–612, 616
Relativistic length contraction, 117t
Relativistic mass increase, 117t
Repetitive member design values, 354
Resilient channel, 537
  STC ratings and, 542–543
Resistance, 136. *See also* Thermal resistance
  of brick walls, 515–516
  of conductors, 660
  to corrosion, 267, 451
  to flow of heat, 506
  lateral, 473–475
  to water vapor, 505
  withdrawal, 443, 461–464, 462t, 464f
    of experimental loads, 465
    of lag screws, 473–475
Resistivity, 430t
  thermal, 505
Resistors networks, 638
Resonance, 639
Retaining structures, common, 204
Retaining walls, 203
Retarders, 212, 237
  composition and mechanism of, 237
  concrete and, 237
Reverberation time, 526–527
Reversible vibratory plates, 186–187
Ribbon stripe, 389, 396
Ride-on rollers, 188
Ridges, 551
Rift sawn lumber, 352–353
Right angle, 100f
  definitions, 105
  hypotenuse of, 86
Right area prisms, 91–92
Right hand generator rule, 654
Ripping, of lumber, 373
Rock excavation, 156
Rockwell hardness test, 148–149
  superficial, 149
Rogers, Danny H, 81

Roll insulation, 510
Rollers, 187
  roller-type compaction, 185
  safety and general guidelines, 188–189
  walk-behind, 188
Romans, 10
Roman numerals, 71
Roof deck, weight of, 338–340
Roofing
  galvanized corrugated, 339t
  nails, 446–447
Roots, 83
Rosehead fine shank, 458
Rosehead square shank spike, 459
Rotational inertia, 117t
Rotational kinetic energy, 117t
Rough sawn lumber, thickness for, 381
Round high-strength steel
  metric dimensions and section properties of, 317–318
  weight and size of, 311–314
    metric, 315–316
Round screw heads, 469
Round tank
  capacity of, 146
  volume of, 142–145
Rounds, cold and hot rolled, 329
Rubber-faced trowel, 563
Rubber-tire rollers, 188
Running wattage, 642–643
Rutherford-Bohr hydrogen-like atoms, 117t
R-value. *See* Thermal resistance

S

Sand, 160
  group names for, 168
Sandblasting, 579
Sand-cone method, 177
Sanding, 575–576
Sap
  gum, quartered, 389
  lumber, 383
Saturated soil, 163
Saturation
  humidity, 611–612
  vapor
    density, 612
    pressure, 616
SBC. *See* Standard building code
Scleroscope hardness test, 150
Scope, 550
Scrap steel, 266
Scraping, 579
Screws, 450
  advantage of, 450
  lag, 450
    lateral resistance of, 473–475

    penetration by, 474f
    withdrawal resistance of, 473–475
  length of, 467
  multipurpose, 450
  self-tapping, 470
  sizing of, 466
  stainless steel deck, 471–472, 471t
  threads, 470
  types of, 467
  wood, 465
    dimensions of, 466t
    sizing of, 465–466
    types of, 465f
  woodpeckers flat head deck, 471–472
Screw heads
  button/dome, 469
  flat/countersunk, 468
  oval, 468
  pan, 469
  point types, 470
  truss, 469
  types and shapes of, 468–469
Sealers, 574–575, 576
  determining, calculations for, 552–553
  drywall, 550
  wall coverings, 550
Seasonal energy efficiency ratio (SEER), 650–652
Seating materials, absorption factors of, 528t
SEER. *See* Seasonal energy efficiency ratio
Self-tapping screws, 470
Separate heat and power (SHP), 612–613
Set-retarding, 237–238
  recommendations for, 238
Settlement, under foundations, 181–182
Severe lighting, 550
Sewer sheathing, 391
Sexagesimal system, 10
Shake, 390
Sheet piling, 391
Sheet steel, 336
  abrasion resisting, 340t
  types of, 337–338
Sheetrock, 548t
Shoring, 189
SHP. *See* Separate heat and power
Shrinkage, 405–407
SI. *See* International System of Units
Sieve analysis, 170, 222–223
  chart, 65t
Silica fume, 229–231
  availability of, 231
  handling of, 231
  microsilica, 229
  mix design, 230
  specifications for, 230

Index

Silt, 160
Silt-clay materials, 171–172
Sine, 100, 101
   law of, 102, 111, 112, 125
Single member design values,
   354
Sinusoidal motion, 117t
Skim coat, 551
Sliding bevel, 584
Slope
   angles, 199
   calculating percent of, 196–197
   layback, 200
   ratios, 200t
Slotted/flathead drive, 467
Slump, 213
   ranges of, 213t
   specifications, 213t
SM. *See* Surface measure
Small angle formula, 96–97
Smooth shanks, 447t
Smooth-drum rollers, 188
Snell's Law, 117t
Sodium percarbonate, 578
Softwood
   floor joints, 373–375
   lumber, 356, 375
      drying of, 408
      grading, 375–376
      species, 358–360, 375
   plywood, 434–435
   strengths of, 398–403
Soil. *See also* Earth
   additional descriptors, 161
   bearing capacity of, 181–182
   biological factors of, 173
   blocky, 163
   boulder, 163
   calcareous, 163
   cemented, 163
   classification, 173
      determination of, 169
      field method, 166
      flowline, 167
      groups, 170
      for highway construction, 169–172
      of sediments, 167
   coarse-grained, 198
   cobble, 163
   cohesionless, 162t
      lateral pressures in, 193
   cohesive, 162t
      lateral pressures in, 193
   color of, 161
   compaction of
      calculating, 177–179
      tests, 177–179
   density of, 179
      measurement of, 194, 196t
      moisture content *vs.*, 179f
      tests, 179–180
   desirability of, 180
   dry, 163
   dry density of, 190–193, 192f
      in situ, 193–196
   fat clay, 163
   fill, 163
      configurations of, 201f
   fine-grained, 160
   fissured, 163
   formation of, 173
      factors, 173
   fractions, 170
   frequent, 163
   friable, 163
   grain size, 164–165
   gravel, 159, 170
      group names for, 168
   great groups, names of, 176t
   laminated, 163
   lateral pressures in, 193
   layer, 163
   lens, 163
   maps of, 175–177
   moist, 163
   moisture content of, 161, 179, 190–193
      density *vs.*, 179f
   mottled, 163
   occasional, 163
   orders, 174
   organic, 160, 163
   parting, 163
   peat, 160
      amorphous, 161, 163
      fibrous, 160
   permeability of, 178, 181
   plasticity chart, 165f
   primary constituents of, 159–161
   saturated, 163
   secondary constituents, 161
   series, 173
   settlement of, 181–182
   silt, 160
   stable slope ratios for, 200t
   strength assessment, 162–163
   suborders, 175t
   taxonomy, 173–177
   terminology, 165–166
   tertiary constituents, 161
   testing, 179–180
   textural classification chart, 172f
   time and, 174
   trace, 163
   unified classification, 159–163
      symbol chart, 164f
   varved, 163
Solids, heating, 117t
Solid hardwood flooring, 568
Solid loadbearing masonry wall, 509
Solid masonry units, fasteners in, 485–486
Solid sheet vinyl, 554
Solid strandings, 661–662, 662t
Solid unit masonry, properties of, 513
Soltis, Lawrence, 445
Sound
   absorption of
      coefficients, 530
      factors, 528–529
   carpet and, 541
   cutting, 388
   decibels of, 532
   masking systems of, 530–531
   music, levels of, 535–537
   noise levels of, 535t
Sound transmission class (STC), 525–526
   of insulated partitions, 538–540
   isolation and, 527
   for masonry walls, 540–544
   testing, for residential carpet, 540–541
Soundness, 218–219
Spanish cedar, 392
Special factors, 108
Special relativity, postulates of, 117t
Special rules, 111
Specific gravity, of wood, 462
Specified dimensions, 253
Speed
   angular *vs.* linear, 117t
   average, 117t
   converting, 72–73
   minimum, 117t
   of wave, on string, 117t
Sphere
   area of, theorem, 93–95
   surface area of, 99
   volume of, 89, 93–95
Spikes
   CLYDE rail, 461
   common wire, 444
   flat countersunk head, 458
   rosehead square shank, 459
Splice plate, 564
Splits, 388
Sponge, 563
   float, 563
Spotting, 551
SPT. *See* Standard Penetration Test
Square
   combination, 583–584, 587
   completing, 108

Square (*Continued*)
 drives, 468
 engineer, 587
 tri-meter, 583, 587
Square feet
 acres
  conversion from, 44*t*
  conversion to, 43*t*
 square inch
  conversion from, 40*t*
  conversion to, 41*t*
 square mile
  conversion from, 43*t*
  conversion to, 42*t*
Square footage
 for cathedral ceilings, 560
 commercial, estimating, 560
 for stairways, with straight ceiling, 559
 of wall coverings, 557–558
  formulas and estimating, 559
Square high-strength steel sections, weight and size of, 295–298
 metric, 299–302
Square inch, square feet
 conversion from, 41*t*
 conversion to, 40*t*
Square meter
 square mile
  conversion from, 46*t*
  conversion to, 47*t*
 square yard
  conversion from, 45*t*
  conversion to, 45*t*
Square mile
 hectares
  conversion from, 48*t*
  conversion to, 48*t*
 square feet
  conversion from, 42*t*
  conversion to, 43*t*
 square meter
  conversion from, 47*t*
  conversion to, 46*t*
Square nails, 455–456
Square yard, square meter
 conversion from, 45*t*
 conversion to, 45*t*
Squareness, 583
Square-notched trowel, 563
Squeegee, 563
Staggered studs, 542–543
Staggered walls, 538
Stainless steel, 472
 bugle head screws, 472–473
 channels, 280
 deck screws, 471–472, 471*t*

fasteners, 452–453
type 302HQ, 472*t*
Stains, 574, 576
Stairways, 558–559
 with straight ceiling, square footage for, 559
Standards. *See also* American Softwood Lumber Standard; Japanese Industrial Standards; National Institute of Standards and Testing
 Canadian, 375
 of capacity, 12–13
  calibrations of, 13
 of dishwashers, 597
 early history of, 4–6
 of length, 10–11
  calibrations of, 11
 of mass, 11–12
  calibrations of, 12
  maintenance and preservation of, 13
 of measurement, 3, 4
 of metric system, 7
 origin of, 4–6
 product, 648
Standard air film, R-values of, 520
Standard American Concrete Institution (ACI), 212
Standard building code (SBC), 373
Standard Penetration Test (SPT), 162
Standard wire gauge (SWG), 67
Starrett protractor/angle finder, 585, 587
Starting torque values, suggested, 494–497
Start-up wattage, 642–643
STC. *See* Sound transmission class
Steel, 147
 angles, 283
 ASTM designations for, 268–270, 271–272
 carbon, 267
 channels, 280
  application of, 280–281
  identifying, 282–283
  weight and size of, 319–320
 coil, 336
 cold rolled, 336–337, 341*t*
 common structural shapes, 279–281
 corrosion-resistant high-strength low-alloy, 267
 galvanized, 340*t*
 high-strength low-alloy, 267
 hot-rolled, 340*t*
  shape designs of, 64*t*
 ingredients of, 266
 materials, preparation, and testing, 271–274
 mechanical properties of, 277–279, 278*t*
 pipes, 319

 plate, 334–335
 round high-strength, 311–314, 315–316
 scrap, 266
 shape designations, 267–268
 sheet, 336
  abrasion resisting, 340*t*
  types of, 337–338
 side plates, 478
 specifications, 275
 stainless
  bugle head screws, 472–473
  channels, 280
  deck screws, 471–472, 471*t*
  fasteners, 452–453
  type 302HQ, 472*t*
 standards of, 274, 275–276
 structural, 267–268
  angles, 326–327
  bolt fasteners for, 488–492
  fastening system, 498, 499
 structural members, 283
 tempered alloy, 267
 wide flange beams, 282
Stone, 5–6
Storage water heaters, 609–610
Strain, 117*t*
Strandings, 661–662, 662*t*
Stress, 117*t*
Stress-graded lumber, 358–360
 dimensions of, 369
Striking tools, 563
String effects, 555
Strips, 396
Structural angles, aluminum, 326–327, 330–332
Structural shapes, new designs and, 64–65
Structural steel
 angles, weight and size of, 326–327
 bolt fasteners for, 488–492
 fastening system, 498
  handling/storage/installation, 499
  installation procedure, 499
Structure, of soil, 161
Stud walls, 373–375
Subdivision, of units, 9–10
Sum, formulas of, 106
Sum-to-product formulas, 107
Superplasticizers, 213, 225–226
 recommendations for, 226
Surface area
 of cone, 99
 of cube, 99
 of cuboid, 99
 formulas, 99
 of frustum, 88–89
 of sphere, 99

Index

Surface coats, 574–575
Surface measure (SM), 381
Surfaced lumber, thickness for, 382
Surfacing, 385–386
Suspending devices, 564
Suspending wire, 564
SWG. *See* Standard wire gauge
Symmetry properties, 106
Synthetic textile wall coverings, 555

T

Tangents, law of, 103
Tape
　calculator, 549
　measure, 587
T-bevel, 584, 587
Technology Innovations Program, 1
Telegraphing, joint, 550
Temperature
　of oven, 74t
　of water, 79
Tempered alloy steel, 267
Tension Control Fastening System, 498
Tension shears, 498–501
Tetracalcium aluminoferrite, 217
Texture, 551
Texturing, 551
Thermal conductance, coefficient of, 505
Thermal conductivity, 504
Thermal efficiency, 614
Thermal energy, 616
Thermal expansion, 117t
Thermal property symbols, 504–505
Thermal qualities, of wood, 430
Thermal resistance (R-value), 505
　of block cavity walls, 515–516
　calculations, 509
　　for masonry wall assemblies, 509
　materials, properties of, 508
　of standard air film, 520
　table, 511t
　for wall assemblies, 510–513
Thermal resistivity, 505
Thermodynamics
　first law of, 117t
　second law of, 117t
Thickness, measuring of, 585–586
Thinset, 563
Thread pitches, standard, 492–493
Threaded annular nails, 444t
Threads per inch (TPI), 492–494
Three-dimensional motion, 134–135
Tile
　acoustical, 564
　ceiling, 564–565
　　grid-system components, 564
　　installation, 564–565

　ceramic, 563–564
　cutter, 563
　electric saw, 563
　field, 563
　floor, 562–564
　　resilient, 562–563
　glazed, 563
　specialty, 563–564
　trim, 563
　unglazed, 563
　wall, 562–564
Timbers, 368t
　products, 391
　section properties of, 365–367
Time, 640
　nipper, 563
　reverberation, 526–527
　of setting, 228
　soil and, 174
　units of, 6
　zones, U.S, 69f, 70t
Titanium, 147
Toilets
　composting, 595
　construction types of, 594–595
　efficiencies of, 594–595
　low-flow
　　savings due to, 592
　　testing procedures, 593
　technical information, 595
　ultra-low-flow, 595
　water usage by, 592
Ton, 5–6
　English water, 13
　long, 13, 34
　measurement, 13
　metric, 13
　register, 13
　specialized term usage of, 13–14
Tonnage, specialized term usage of, 13–14
Tools
　calibration, 586
　marking and measuring, 581
　marks, 551
　striking, 563
Topcoat, 551
Topography, 173
Torn grain, 386
Torque, 117t
　starting value of, 494–497
Total fuel energy input, 614
Total system efficiency, 614
　metric, 615
TPI. *See* Threads per inch
T-pots, 125–127
Trace soil, 163
Tracks

　banked circular, 117t
　circular unbanked, 117t
Transformers, 117t
　distribution, 668
　dry type, 669
　power, 669
　types of, 668–669
Transparent finishes, 573
Trapezoid, area of, 86
Treaty of the Meter, 6
Triangle
　area of, 86
　equilateral, 94–95
　inequality, 109
　isosceles, 94
　Pascal's, 107
　properties of, 111
　right, definitions, 105
Tricalcium aluminate, 217
Tricalcium silicate, 217
Trigonometry
　definitions, 124
　　inverse, 124
　formulas of, 102–109
　functions
　　basic, 100–101
　　values of, 127t
　ratios, 101–102
Trim
　interior, 581–587
　tile, 563
Tri-meter square, 583, 587
Tropical American, 392
　hardwoods, 397
Troy system, 34
　units of mass, 18t
True mass, 12
Truss screw heads, 469
Tubing, 324–325
Tupelo, quartered, 389
Twist-off structural bolts, 500
Two-dimensional motion, 134–135
Two-Family Dwelling Code, 373
Two-member joints, 478–479
Type I/II portland cements, 216

U

U channels, 280
UBC. *See* Uniform building code
Ultra-low-flow (ULF) toilets, 595
Unclassified excavation, 156
Unconsolidated sediments, 206
UNC/UNRC. *See* Coarse thread series
Uncut patterned carpet, 566
Underlayerment, 572
Underliner, 556
Underwriter Laboratories Inc, 671

Index

UNF/UNRF. *See* Fine thread series
Unglazed tile, 563
Unified soil classification, 159–163
   symbol chart, 164f
Uniform building code (UBC), 373
Uniform Plumbing Code (UPC), 590–591
United States (U.S.)
   Bureau of Reclamation, 226
   Department of Agriculture, 1, 174, 462
   Department of Energy, 591–592
   Green Building Council, 602
   metric system in, 8
      conversions, 274–275
   systems of measurement in, 9
   time zones
      converting, 69t, 70t
      map of, 69f
   units of measurement, 16–18
Units. *See also* International System of Units
   of angle, 6
   of area, 15t, 16t
      international measure, 20t
      survey measure, 21t
   base, 7
   of capacity, 22t, 28t, 30t
      equivalents, 28t
   classes of, 7
   conversion of, 98
   customary, 9
   derived, 7
   development of, 5–6
   of dry volume, 17t
   early history of, 4–6
   of internal combustion, 80
   of length, 15t, 16t
      equivalents, 26t
      international measure, 19t
      survey measure, 20t
   of liquid volume, 15t, 17t
   of mass, 5–6, 15t, 24t
      apothecaries, 18t
      equivalents, 31t
      Troy, 18t
   of measurement, 4–10, 19–25
      Gunter's, 16t
      U.S, 16–18
   of metric system, 7, 14–15
   origin of, 4–6
   of pressure, 222
   subdivision of, 9–10
   of time, 6
   of volume, 15t, 16t, 21t
      dry volume measure, 22t
   WSFUs, 590
Universal Atlas Cement
      Company, 223
Universal gravitation, 117t
Universal mill plates, weight and size of, 327–328
Unytite, Inc, 498
UPC. *See* Uniform Plumbing Code
Urinals, 594
   water conservation of, 595–596
   waterless, 595t
Utility boards, 391–392
U-value, 506

V

Value
   absolute, 108
   design values
      of framing lumber, 367t, 368t
      of MSR lumber, 431t, 432–433, 432t
      repetitive member, 354
      single member, 354
   future, 138
   present, 136–137
   starting torque, 494–497
   of trigonometry functions, 127t
   U-value, 506
Vapor density, 616
Vapor pressure, 616
Variable rise, 200
Varnish, 575, 576
Varved soil, 163
Vectors
   addition, 134–135
   components of, 117t
   rectangular components of, 135
   resolution of, 135
Velocity, 130
   average, 117t
Veneer
   cork, 555
   wood, 555
Vertex angle, 94
Vertical coursing height, calculating, 255–256
Vessel tonnage, 13
VHTRC. *See* Virginia Highway and Transportation Research Council
Vibration control, 182–183
Vibratory plates, 186
   reversible, 186–187
Vickers hardness test, 149
Vinsol resin, 223
Vinyl
   coated paper, 554
   fabric-backed, 554
      wall coverings, 556
   paper-backed, 554
   solid sheet, 554
Vinyl-over-vinyl, 562

Virginia Highway and Transportation Research Council (VHTRC), 229
Voltage, induced, 117t
Volts, watts, converting from, 636
Volume
   of cone, 87, 88
      theorem, 93
   of cube, 85
   of cylinder, 87, 139
      theorem, 92
   of cylindrical tanks, 138–141
   dry, 17t
   of prisms, 90–91
   of pyramid, 90, 91
      frustum, 87
      theorem, 91
   of round tank, 142–145
   of sphere, 89, 93–95
   units of, 15t, 16t, 21t
      dry volume measure, 22t
      equivalents, 28t, 30t
      liquid volume measure, 15t, 22t
   weight, conversion to, 62–64

W

Wagner Turbidimeter, 218
Wail materials, absorption factors of, 528t
Walk-behind rollers, 188
Walls
   angle, 564
   assemblies, R-values for, 510–513
   cathedral, 558–559
   masonry
      assemblies, 504, 509
      solid loadbearing, 509
      STC ratings, 541
   painting, 565
   tile, 562–564
Wall coverings, 551
   acoustical, 555
   adhesives, 561–562
   basics, 553–554
   contract, 554–555
   conversion table, 554
   digital, 555
   fabric-backed vinyl, 556
   helpful guidelines, 553
   key points, 554
   measurements for, 557
   metric single roll, 557
   natural textile, 555
   overview, 554–556
   polyolefin, 555
   quantities, formulas for estimating, 556–558
   sealers, 550
   specialty, 555

Index

square footage of, 557–558
  formulas and estimating, 559
synthetic textile, 555
Type I (light duty), 555, 556
Type II (medium/heavy duty), 555, 556
Type III (heavy duty), 556
types and usage, 554, 555–556
Wallpaper, 561
Walnut, 383
Wane, 388
Waste materials, disposal of, 157
Water
  boiling point of, 616
  efficiency of, 602
    plumbing fixtures, 605
  fixtures, 596
  form conversion, 56t
  global distribution of, 607t
  heaters
    annual operating cost of, 610
    cost of, 610–611
    electric, 610
    gas, 610
    heat pump, 609–610
    oil, 610
    payback of, 611
    storage, 609–610
  measurement, 79
  mixing, 214, 214t, 215
  pressure, 192
    under, 117t
  reduced consumption of, 606
  saving opportunities, 594–595
  sources of, 607
  temperature of, 79
  usage of
    average, 607–608
    calculating, 591–592
    by clothes washers, 599
    estimating, 607–608, 609t
    mean daily residential, 591
    personal chart, 608
    by plumbing fixtures, 598
    by toilet, 592
  vapor
    permeance to, 505
    resistance to, 505
  weighing in, 207
  weight of, 76t
Water horsepower (WHP), 79
Water supply force units (WSFUs), 590
Water Use Reduction, 602
Waterborne preservatives, 453
Water-cement ratio, 215
  compressive strength and, 215t
Waterproof portland cement, 263–264
Water-reducing admixtures, 221

Wattage, 645
  of appliances, 641–642
  requirements, 643
  running, 642–643
  start-up, 642–643
Watts, 652–653
  amps
    converting from, 636
    converting to, 636
  volts, converting to, 636
Watt's Law, 638–639
Wavelength, 640
Waves, periodic, 117t
Wear layer, 570, 572
Weights, 4
  aggregate, 215
  of conduit, 667
  dead weight tonnage, 14
  equivalents, 75t
  formula, 117t
  of hardwood, 386–388
  hundredweight, 5–6, 34
  knowledge of, 4
  mass and, 11–12
  of rectangular high-strength steel
    sections, 302–305, 306–310
  of roof deck, 338–340
  of round high-strength steel, 311–314,
    315–316
  of square high-strength steel sections,
    295–298, 299–302
  of steel channels, 319–320
  of structural steel angles, 326–327
  of universal mill plates, 327–328
  volume, conversion from, 62–64
  of water, 76t
  of wide flange beams, 284–293
Welded wire mesh, designations of, 244–245
Wet density, 209f
  dry density vs., 208–210
Wheel tractor-scraper, 156
WHP. See Water horsepower
Wide flange beams, 276–277
  steel, 282
  weight and size of, 284–289
    metric, 289–293
Wire gauge
  comparison chart, 658–659
  converting, 67–68
  table, 656–657
Wire nails, 456
Withdrawal resistance, 443, 461–464, 462t,
    464f
  experimental loads, 465
  of lag screws, 473–475
Wood. See also Specific types and species
  adhesives, 479, 479t

bolts in, 476–479
bonding of, 480
brighteners of, 578
cleaners, 578
color change of, 573
composite boards, classification of, 436
density of, 371t, 372t
dimensional change of, 405–407
drying of, 407–409
  defects, 409
  successful, 408
equilibrium moisture content of, 404
figured, 390
green
  dimensional change of, 407
  moisture content of, 370–372
hardwood, 381–384
  African, 397
  classification of, 575t
  construction, 391–392
  drying of, 408
  hearts, 391
  mixed, 391
  North American, 397
  Philippine, 396–397
  solid flooring, 568
  strengths of, 398–403
  tropical American, 397
  weights of, 386–388
heating, costs of, 628t
interior, finishing of, 572–573
mechanical properties of, 410–415
  inch-pound, 420t
  metric, 416–420, 426–429
metal, interaction with, 451
moisture content of, 451
for painting, 580–581
physical properties of, 353–355
preservative-treated, corrosion in, 453
screws, 465
  dimensions of, 466t
  sizing of, 465–466
  types of, 465f, 467
shrinkage of, 353, 409–410
softwood
  drying of, 408
  floor joints, 373–375
  grading of, 375–376
  lumber, 356, 375
  plywood, 434–435
  species, 358–360, 375
  strengths of, 398–403
specific gravity of, 462
strength properties of, 410t
  metric, 416t
structures
  fasteners for, 449–454

Wood (*Continued*)
    integrity of, 450
    thermal qualities of, 430
    untreated, corrosion in, 453
    veneer, 555
Wood Turner's Caliper Set, 586
Wood-metal framing, 519
Woodpeckers flat head deck screw, 471–472
Word picture, 159, 164–165
Work, 132–133
    formula, $117t$
    gas and, $117t$
Workability, 213
    of fresh concrete, 228
Worked lumber
    dressed sizes of, $362t$, 377
    typical, 375–377
World War II, 6–7
WSFUs. *See* Water supply force units

Y

Yard, 5, 8, 11
    square yard
        square meter and, $45t$

Lightning Source UK Ltd.
Milton Keynes UK
UKHW052252031120
372559UK00026B/309